Sixth Edition

AGRISCIENCE

FUNDAMENTALS AND APPLICATIONS

AGRICULTURAL
FFA
EDUCATION

Sixth Edition

AGRISCIENCE

FUNDAMENTALS AND APPLICATIONS

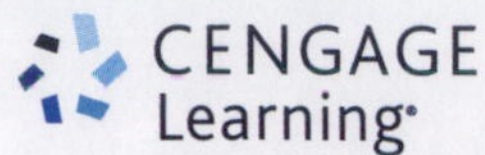

Agriscience: Fundamentals and Applications, 6th Ed.
L. DeVere Burton, PhD

General Manager: Dawn Gerrain

Associate Product Manager: Nicole Sgueglia

Sr. Director, Development: Marah Bellegarde

Sr. Product Development Manager: Larry Main

Sr. Content Developer: Laura J. Stewart

Product Assistant: Scott Royael

Executive Director of Marketing: Carolyn Pestritto

Sr. Product Marketing Manager: Leah Klein

Sr. Production Director: Wendy Troeger

Production Manager: Mark Bernard

Sr. Content Project Manager: Elizabeth C. Hough

Sr. Art Director: David Arsenault

Media Developer: Deborah Bordeaux

Cover image credits: Field: © tlorna/Shutterstock. comCover and Section Opener image credits: (top image) Windmill: © Michal Bednarek/Shutterstock. com; (second image) Sheep: © iStockphoto/Joan Wynn; (third image) Water with Boat: © Vladislav Gajic/Shutterstock.com; (bottom image) Plants: © David Kay/Shutterstock.com

Unit Opener image credits: Unit 1: © Garsya/ shutterstock.com.; Unit 2: © Don Farrall/Photodisc/ Getty Images.; Unit 3: Courtesy of USDA/ARS #K-1968-13.; Unit 4: © iStockphoto/Leah-Anne Thompson.; Unit 5: © Goodluz/ Shutterstock.com.; Unit 6: © iStockphoto/Lise Gagne.; Unit 7: © SNEHIT/ Shutterstock.com.; Unit 8: © biletskiy/ Shutterstock.com.; Unit 9: © luigi nifosi'/ Shutterstock.com.; Unit 10: © Christopher Barrett/ Shutterstock.com.; Unit 11: © Wildnerdpix/ Shutterstock.com.; Unit 12: © Vladislav Gajic/ Shutterstock.com.; Unit 13: Courtesy of Boise National Forest.; Unit 14: Courtesy of USDA/ARS #K-3663-15.; Unit 15: Courtesy of USDA/ARS.; Unit 16: Courtesy of USDA/ARS #K3212-1; Unit 17: © iStockphoto/Sandralise; Unit 18: © Alexander Raths/ Shutterstock.com; Unit 19: © tristan tan/ Shutterstock.com; Unit 20: © Tomas Pavelka/ Shutterstock.com; Unit 21: © Zeljko Radojko/ Shutterstock.com; Unit 22: © Tyler Olson/ Shutterstock.com; Unit 23: © Nadia Zagainova/ Shutterstock.com; Unit 24: © Le Do/Shutterstock. com; Unit 25: © Nessli Orpmas/Shutterstock.com; Unit 26: © iStockphoto/Alexandru Nika; Unit 27: © photomak/Shutterstock.com; Unit 28: Courtesy of USDA/ARS #K-2681-13; Unit 29: © Serdar Tibet/ Shutterstock.com; Unit 30: Courtesy of USDA/ARS #K-4166-5; Unit 31: © Zuzule/Shutterstock.com; Unit 32: © iStockphoto/Serhiy Zavalnyuk; Unit 33: Courtesy of Price Chopper Supermarkets; Unit 34: Courtesy of DeVere Burton; Unit 35: © Elena Elisseeva/Shutterstock.com; Unit 36: © mangostock/Shutterstock.com; Appendix A: © Michal Bednarek/Shutterstock.com.; Appendix B: © Michal Bednarek/Shutterstock.com.

WCN: 01-100-101

Library of Congress Control Number: 2013943109

ISBN-13: 978-1-133-68688-0

Cengage Learning
200 First Stamford Place, 4th Floor
Stamford, CT 06902
USA

Cengage Learning is a leading provider of customized learning solutions with office locations around the globe, including Singapore, the United Kingdom, Australia, Mexico, Brazil, and Japan. Locate your local office at: **www.cengage.com/global**

Cengage Learning products are represented in Canada by Nelson Education, Ltd.

To learn more about Cengage Learning, visit **www.cengage.com**

Purchase any of our products at your local college store or at our preferred online store **www.cengagebrain.com**

Printed in the United States of America
2 3 4 5 6 7 19 18 17 16 15

CONTENTS AT A GLANCE

SECTION 10 Putting It All Together / 735

CONTENTS

SECTION 3 Natural Resources Management / 131

SECTION 4 Integrated Pest Management / 271

SECTION 7 Ornamental Use of Plants / 491

SECTION 8 Animal Sciences / 559

PREFACE

Welcome to the agriscience world of the twenty-first century! *Agriscience: Fundamentals and Applications, Sixth Edition* is about a new century of agricultural and agriscience developments.

This textbook will be used by a generation of students whose lives may span two different centuries and two different millennia. It is interesting to consider that in all of the ages since humans first engaged in agricultural pursuits, nearly all of the agricultural innovations and technologies ever known to humankind have evolved in fewer than 100 years.

It is to the agriscience students of the new millennium that this textbook is dedicated, for the agriculturists, scientists, and innovators of tomorrow are today's high school students. The "millennium generation" will be called on to feed the world as the human population nearly doubles to 10 billion people. To do this, they must learn more than any other generation has ever learned, and they need to discover more ways to increase food production than any other generation has ever discovered. They must accomplish this using marginal land because many of our fertile farms have been swallowed up to build cities and towns. *Agriscience: Fundamentals and Applications, Sixth Edition* is the modern agriscience textbook that will introduce the "millennium generation" to agricultural careers. This generation will also lead the industry that the people of the United States depend on to feed and clothe them and to export surplus agricultural products to other regions of the world.

This edition of the book expands on the original text and the ideas of earlier editions. The science component has been strengthened with some new lab exercises. Statistics and text have been modified to reflect changes that have occurred since the last edition was published, and new examples of agricultural applications of science and technology have been added. The book is intended for introductory-level agriscience classes in the ninth and tenth grades.

NEED FOR AN INTRODUCTORY TEXTBOOK

This book is an introductory textbook in a series of modern secondary agricultural textbooks published by Cengage Learning. It addresses the most basic levels of agriscience using language and examples that are matched to the needs of beginning students in the natural science career pathway.

Revisions in this new edition are the work of current Cengage agriscience author L. DeVere Burton. He is also the author of three other textbooks in the agriscience series: *Agriscience & Technology, Second Edition; Fish and Wildlife: Principles of Zoology and Ecology, Third Edition*; and *Introduction to Forestry Science, Third Edition.* He also edited a new textbook titled *Environmental Science Fundamentals and Applications.* Each of these works, including this edition of *Agriscience: Fundamentals and Applications*, reflects the premise on which agricultural education was founded—that most students learn best as they apply the principles of science and agriculture to real-life problems.

ORGANIZATION

This edition of *Agriscience: Fundamentals and Applications* is organized into 10 sections and 36 units. Each section introduces the subjects that will be covered in the individual units. The text and illustrations for each section have been revised. Each unit begins with a stated objective and a list of competencies to be developed. Important terms are listed at the beginning of each unit and highlighted in the text. They are also included in the glossary at the end of the book. Each unit contains profiles on science, careers, and agriculture, and concludes with student activities and a section on self-evaluation. The book includes a complete and thorough index.

NEW FEATURES AND ENHANCED CONTENT

The science content of this edition has been strengthened by adding new examples of science applications to agriculture, and new science lab exercises have been added to the laboratory manual. Each unit includes a feature called "Hot Topics in Agriscience." Each of these features describes a scientific principle or discovery for which an agricultural application has been currently identified. "Suggested Class Activities" is another feature found at the beginning of each unit. New photographs and illustrations have been added throughout the book to bring a sharper focus to the agriscience emphasis of the text. Internet icons are featured throughout the textbook. They include key words for Internet searches on the topics of discussion. This feature will help students explore agriscience topics beyond the boundaries of this textbook.

- "Suggested Class Activities" in each unit give both the student and the instructor an innovative way to become actively involved with the content of each unit.
- "Hot Topics in Agriscience" is a standard unit feature that describes recent scientific discoveries for which an agricultural application has been identified.
- Internet icons are placed throughout each unit. These icons include key search terms that will help students and instructors explore agriscience topics beyond the scope of the textbook.

- Broad applications to science, math, agriculture, natural resources, and the environment provide the appropriate balance for the evolving agriscience curriculum.
- Hundreds of updated full-color photos and illustrations help stimulate interest and enhance learning. Photos now reflect today's digital student, and dozens of illustrations have been redrawn in full color to improve quality for the visual learner.

EXTENSIVE TEACHING/ LEARNING MATERIALS

A complete supplemental package is provided with this textbook. It is intended to assist teachers as they plan their teaching strategies by providing materials that are up to date and efficiently organized. These materials are also intended to assist students who want to explore beyond the confines of the textbook. They include the following resources:

Lab Manual

ISBN: 978-1-13368-689-7

The lab manual has been updated to correlate to the content updates made in the textbook. This comprehensive lab manual reinforces the text content. It is recommended that students complete each lab to confirm understanding of essential science content. Great care has been taken to provide instructors with low-cost, strongly science-focused labs to help meet the science-based curriculum needs of the Introductory Agriscience course in secondary schools.

New to this edition, optional Internet supplements offer additional research opportunities and educational resources to learn more about topics covered in the lab exercises. Each lab exercise has been enhanced with new photos and illustrations to help stimulate visual learning.

Classmaster CD-ROM

ISBN: 978-1-13368-734-4

This technology supplement provides the instructor with valuable resources to simplify the planning and implementation of the instructional program. It has been expanded for this edition to include the following support materials:

- Performance Objectives, Competencies to Be Developed, and Terms to Know lists with definitions for each unit
- A PDF version of the **Instructor's Manual**. The *Instructor's Manual* has been expanded for the sixth edition to provide the following materials for instructors:
  - Teaching Aids and Suggested Resources, including Suggested Class Activities and ideas for Supervised Agricultural Experiences
  - Lesson Plans for each unit
  - Answers to the Self-Evaluation questions at the end of each unit
  - Suggested essay/discussion questions

- New! Correlation guides map textbook content to the National Agriculture, Food and Natural Resources (AFNR) Career Cluster Content Standards and identify Science, Technology, Engineering, and Mathematics (STEM) focused content
- An **Image Library** with all the illustrations from the textbook; use in slide presentations or as part of classroom discussion
- A PDF of the **Lab Manual Instructor's Guide**, which provides answers to lab manual exercises and additional guidance for the instructor.
- A **computerized test bank** created in ExamView® makes generating tests and quizzes a snap. With 1500+ questions and different question formats from which to choose, you can create customized assessments for your students with the click of a button. Add your own unique questions and print rationales for easy class preparation.
- **Instructor support slide presentations** that can be customized in PowerPoint® format focusing on key points for each chapter. Approximately 700 slides (about 20–25 slides per unit) are available to accompany the textbook.

Instructor Companion Web Site

New! The instructor companion Web site provides online access to many of the instructor support materials provided on the ClassMaster CD-ROM, including the *Instructor's Manual, Lab Manual Instructor's Guide,* computerized test bank files, correlation guides, and support slides. To access the available materials, sign up for a faculty account at login.cengage.com. Add the core textbook to your bookshelf using the 13-digit ISBN that appears on the back cover of the textbook.

Agriscience CourseMate

Printed Access Card ISBN-13: 978-1-13360-496-9
Instant Access Code ISBN-13: 978-1-13360-595-9

CourseMate brings course concepts to life with interactive learning, study, and exam preparation tools that support the printed textbook. Watch student comprehension soar as your class works with the printed textbook and the textbook-specific Web site.

Agriscience CourseMate includes:

- An interactive eBook, with highlighting, note taking, and search capabilities;
- Gale's **AgricultureCollection Infotrac® database** offers access to more than 400 titles focused on agriculture and its related fields: from practical aspects of farming to cutting-edge scientific research in horticulture. The collection is updated on a continual basis in order to provide access to the latest research and news articles. The instructor and student can easily search for and access current events news articles, research papers, and videos/images related to specific topics. Use for research assignments and in-class activities.

- Interactive learning tools, including:
  - **Pre- and post-assessment quizzes** in multiple choice and true/false formats help students review unit content;
  - **Flashcards and comprehensive Study Guides** help students prepare for quizzes and exams;
  - **Web links** provide resources for additional information on major topics related to agriscience;
  - **Assignments and Worksheets** offer additional homework ideas and study content. Exercises include image labeling, terminology worksheets, and Internet activities;
  - And more!
- Engagement Tracker, a first-of-its-kind tool that monitors student engagement in the course.

Students should visit www.cengagebrain.com and look for this icon to access these resources. Instructors can access these resources via login.cengage.com.

ACKNOWLEDGMENTS

The author and publisher wish to express their appreciation to the many individuals, FFA associations, and organizations that have supplied photographs and information necessary for the creation of this text. A very special thank you goes to all the folks at the National FFA organization and the USDA photo libraries, who provided many of the excellent photographs found in this textbook. Because of their efforts, this is a better book.

The author and publisher also gratefully acknowledge the unique expertise provided by the contributing authors to the text. Their work provided the core material upon which successive editions have expanded. The contributing authors are the following:

Robert S. DeLauder,
Agriscience Instructor, Damascus, Maryland;

Thomas S. Handwerker, PhD,
Department of Agriculture, University of Maryland at Princess Anne;

Curtis F. Henry,
Business Manager, College of Agriculture, University of Maryland at College Park;

Dr. David R. Hershey,
Assistant Professor, Department of Horticulture, University of Maryland at College Park;

Robert G. Keenan,
Agriscience Instructor, Landsdowne High School, Baltimore, Maryland;

J. Kevin Mathias, PhD,
Institute of Applied Agriculture, University of Maryland at College Park;

Renee Peugh,
Biological Science Consultant, Boise, Idaho;

Regina A. Smick, EdD,
Academic Advisor and Instructor, College of Agriculture, Virginia Tech, Blacksburg; and

Gail P. Yeiser,
Instructor, Institute of Applied Agriculture, University of Maryland at College Park.

It is most appropriate to remember the work of the late Elmer L. Cooper, who authored the early editions of this textbook and whose imprint will always remain on its contents. He will be remembered as a forward-looking agriscience educator who left his indelible mark on his profession and on the lives of innumerable agriscience students.

Appreciation is expressed to Renee Peugh, who consulted with the author on various sections of the text. She also provided information on science lab materials and student activities for recent editions of *Agriscience: Fundamentals and Applications*.

A special thank you is also extended to the reviewers of this sixth edition. Their content expertise and suggestions for updates and improvements greatly enhanced the overall text.

Christina Griffith
Agriculture Instructor
Milan High School
Milan, TN

Mary Handrich
Grades 8–12 Teacher/Agriculture
Fall Creek School District
Fall Creek, WI

Elizabeth Harper
Agriscience Teacher
NAAE member
New Smyrna Beach, FL

Lowell E. Hurst
Agricultural Educator Emeritus
Watsonville High School, Ret.
Watsonville, CA

Jim Satterfield
Agriculture Teacher
Jefferson County High School
Dandridge, TN

Tom Willingham
Instructor
Blythewood High School
Blythewood, SC

HOW TO USE THIS TEXTBOOK

Welcome to the world of agriscience! This section highlights important features of this textbook. In this textbook, content is broken into units, each of which explores an important aspect of agriscience. Each unit begins with the following tools:

- A core *objective* explains the purpose of the unit.
- *Competencies to be developed* lists specific goals to meet as you read and review the unit.
- A *materials list* identifies items you will need to complete the unit.
- *Suggested class activities* provide you and your instructor with hands-on ways to explore critical concepts discussed in each unit.
- A list of *Terms to Know* identifies key vocabulary to master.

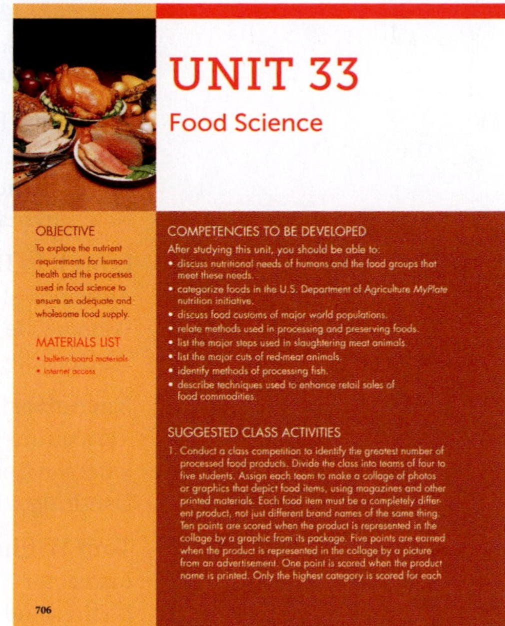

UNIT 33

Food Science

OBJECTIVE

To explore the nutrient requirements for human health and the processes used in food science to ensure an adequate and wholesome food supply.

MATERIALS LIST

- bulletin board materials
- Internet access

COMPETENCIES TO BE DEVELOPED

After studying this unit, you should be able to:

- discuss nutritional needs of humans and the food groups that meet these needs.
- categorize foods in the U.S. Department of Agriculture *MyPlate* nutrition initiative.
- discuss food customs of major world populations.
- relate methods used in processing and preserving foods.
- list the major steps used in slaughtering meat animals.
- list the major cuts of red-meat animals.
- identify methods of processing fish.
- describe techniques used to enhance retail sales of food commodities.

SUGGESTED CLASS ACTIVITIES

1. Conduct a class competition to identify the greatest number of processed food products. Divide the class into teams of four to five students. Assign each team to make a collage of photos or graphics that depict food items, using magazines and other printed materials. Each food item must be a completely different product, not just different brand names of the same thing. Ten points are scored when the product is represented in the collage by a graphic from its package. Five points are earned when the product is represented in the collage by a picture from an advertisement. One point is scored when the product name is printed. Only the highest category is scored for each

706

Throughout each unit:

- *Hot Topics in Agriscience* features describe recent scientific discoveries for which an agricultural application has been identified.

HOT TOPICS IN AGRISCIENCE SAE—INSIDE THE DAIRY BUSINESS

The Dairy Heifer Replacement Project is a program that is in place in many areas of the United States. It is designed to increase the knowledge and interest of young people in the dairy industry. The goal is to enhance life skills of its youth participants. The project begins when the participant purchases a heifer calf from a program-approved seller. For the next few months, the participant cares for the animal. Specific care must be provided, vaccinations must be given, and a magnet needs to be administered. Quality feed must be provided to ensure proper weight and good health. The participant is required to keep detailed records on the animal. In some instances, the heifer is bred to an approved sire, and the project culminates in the sale of the pregnant heifer when she is presented for show and sale. The Dairy Replacement Project is a challenging SAE that allows each participant to gain an understanding of a crucial part of the dairy industry. Contact the agriculture teacher in your high school or your local extension office for more information about such a program in your area.

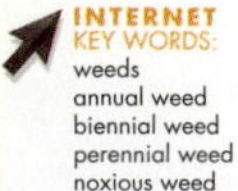

INTERNET KEY WORDS:
weeds
annual weed
biennial weed
perennial weed
noxious weed

Damage by pests to agricultural crops in the United States has been estimated to be one-third of the total crop-production potential. Therefore, an understanding of the major pest groups and their biology is required to ensure success in reducing crop losses caused by pests.

Weeds

Weeds are plants that are considered to be growing out of place (Figure 13-3). Such plants are undesirable because they interfere with plants grown for crops. The word *weed* is therefore a relative term. Corn plants growing in a soybean field or white clover growing in a field of turfgrass are examples of weeds, just as crab-

- *Internet Key Words* icons are placed throughout each unit. These icons include key search terms that will help students and instructors explore agriscience topics beyond the scope of the textbook.

SCIENCE CONNECTION THE SEARCH FOR PERFECT PLANTS

The plants we use today for food, clothing, fiber, and ornamental use are quite different from those found in the wild. Domestic plants or plants grown for a specific use have generally been selected or bred to survive better, grow faster, look different, or in some way perform differently from their ancestors in the wild. However, it is becoming increasingly apparent to plant breeders that we must have wild plants that are not closely related to our favorite domestic species to inject new characteristics into our favored domestic plants.

Pest resistance is an area that requires a continuous reserve of foreign genetic sources. This is to be expected because the very pests that we breed plants to resist are constantly adapting to our plants through survival of the fittest among their kind. Insects and disease-causing pathogens have an amazing capacity to adapt to and eventually break crop resistance. Resistant varieties usually become obsolete in 3 to 10 years. It generally takes 8 to 11 years to breed a new variety to resist the changing individuals of a given pest. Therefore, plant

Plant geneticist Keith Schertz examines grain sorghum bred for tropical climates. Bags prevent the sorghum flowers from cross-pollinating, so the plant breeder can control which plants provide the male pollen to fertilize the female part of any given plant.

- *Science Connection* features profile important science-related aspects of agriculture.
- *Key terms* are indicated in color to alert you when an important vocabulary word is used.
- Hundreds of *full-color photographs and illustrations* help you visualize the topics discussed.

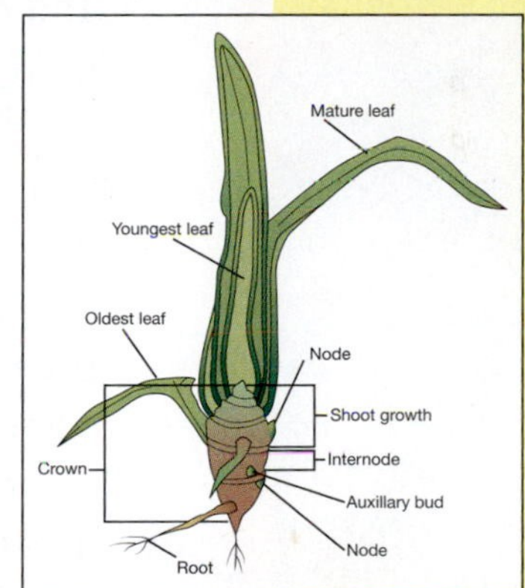

At the end of each unit, *Student Activities* provide additional hands-on activities to help you master the content. Some activities provide ideas for Supervised Agricultural Experiences for you to explore with your instructor.

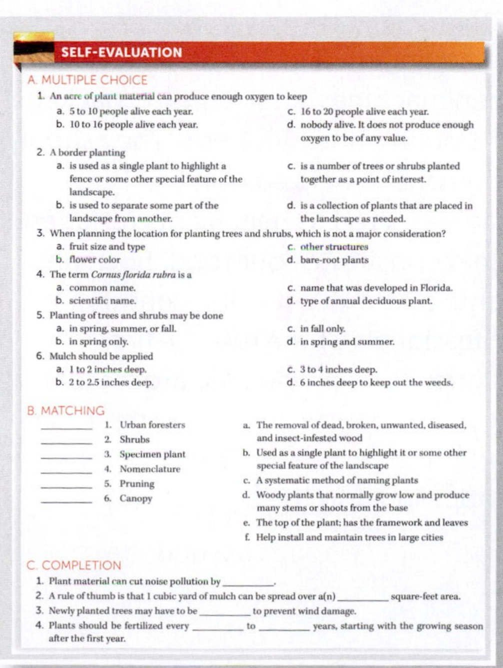

SELF-EVALUATION

A. MULTIPLE CHOICE

1. An acre of plant material can produce enough oxygen to keep
   a. 5 to 10 people alive each year.
   b. 10 to 16 people alive each year.
   c. 16 to 20 people alive each year.
   d. nobody alive. It does not produce enough oxygen to be of any value.
2. A border planting
   a. is used as a single plant to highlight a fence or some other special feature of the landscape.
   b. is used to separate some part of the landscape from another.
   c. is a number of trees or shrubs planted together as a point of interest.
   d. is a collection of plants that are placed in the landscape as needed.
3. When planning the location for planting trees and shrubs, which is not a major consideration?
   a. fruit size and type
   b. flower color
   c. other structures
   d. bare-root plants
4. The term *Cornus florida rubra* is a
   a. common name.
   b. scientific name.
   c. name that was developed in Florida.
   d. type of annual deciduous plant.
5. Planting of trees and shrubs may be done
   a. in spring, summer, or fall.
   b. in spring only.
   c. in fall only.
   d. in spring and summer.
6. Mulch should be applied
   a. 1 to 2 inches deep.
   b. 2 to 2.5 inches deep.
   c. 3 to 4 inches deep.
   d. 6 inches deep to keep out the weeds.

B. MATCHING

_________ 1. Urban foresters
_________ 2. Shrubs
_________ 3. Specimen plant
_________ 4. Nomenclature
_________ 5. Pruning
_________ 6. Canopy

a. The removal of dead, broken, unwanted, diseased, and insect-infested wood
b. Used as a single plant to highlight it or some other special feature of the landscape
c. A systematic method of naming plants
d. Woody plants that normally grow low and produce many stems or shoots from the base
e. The top of the plant; has the framework and leaves
f. Help install and maintain trees in large cities

C. COMPLETION

1. Plant material can cut noise pollution by _________.
2. A rule of thumb is that 1 cubic yard of mulch can be spread over a(n) _________ square-feet area.
3. Newly planted trees may have to be _________ to prevent wind damage.
4. Plants should be fertilized every _________ to _________ years, starting with the growing season after the first year.

Self-Evaluations allow you to review the unit content using multiple choice, matching, and completion question.

SECTION 1

BETTER LIVING THROUGH RESEARCH

Science and technology are modern miracles that have opened the doors to new areas of research, turning the dreams of humankind into realities. Space station research, new frontiers to investigate, and our never-ending quest for knowledge have exploded into many new and exciting careers.

You could become one of the people growing plants or animals in a space station high above the Earth. Or, you might become an engineer who designs the animal- or plant-growing module of the space station, or a molecular geneticist or plant breeder designing new plants to grow well in low gravity, or a food scientist developing packaging for space-grown produce. Closer to home, you might discover ways to prevent plant or animal diseases. Perhaps you will become a researcher who discovers a better way to preserve food or a safe way to sanitize fresh fruits and vegetables. You may have personal attributes and skills that will propel you to become a teacher of agriscience, giving you an opportunity to have a positive influence in the lives of many students.

One career area in ever-expanding demand is plant science. As you will learn, plants are "green machines" that capture, package, and store energy from the sun through photosynthesis. They supply food and fiber for animals and humans to help sustain life. But, human knowledge and energy are required to help plants function in the overall "green machine" that constitutes our food, fiber, and natural resources system. Students of the twenty-first century will become the agricultural professionals of the twenty-first century. They will become the agricultural producers, processors, marketers, and scientists who discover new ways to feed the citizens of the United States and the world. This will be accomplished by conducting basic research and applying it to the agricultural food system.

Agriscience in the Information Age

Whether you choose a career in plant or animal science, sales and marketing, mechanics, or processing, it is certain to be rewarding. By studying agriscience, you are opening the door to exciting educational programs and careers that contribute to better living conditions for people everywhere. What role will you play in the challenging task of producing the food and fiber that will be required by future generations?

Biosphere 2

UNIT 1

The Science of Living Things

OBJECTIVE

To recognize the major sciences contributing to the development, existence, and improvement of living things.

MATERIALS LIST

- writing materials
- newspapers and magazines
- online encyclopedias
- Internet connection

COMPETENCIES TO BE DEVELOPED

After studying this unit, you should be able to:

- define agriscience.
- discover agriscience in the world around us.
- relate agriscience to agriculture, agribusiness, and renewable natural resources.
- name the major sciences that support agriscience.
- describe basic and applied sciences that relate to agriscience.

SUGGESTED CLASS ACTIVITIES

1. Invite a retired farmer to be a guest speaker on the topic of improvements or advances in the science and technology of agricultural production that he or she has experienced during his or her career. Have the students make a list of the agricultural technologies that are discussed. Speculate on new agricultural technologies that the students may experience during their careers.
2. Obtain a copy of the application process for the National FFA Agriscience Student Award. This award offers excellent scholarship opportunities to students who plan and carry out agriscience research projects. Discuss some local agricultural problems that might be addressed by students who express interest in planning a research project in agriscience.

TERMS TO KNOW

agriscience
agriculture
agribusiness
renewable natural resources
technology
high technology
aquaculture
agricultural engineering
animal science technology
crop science
soil science
biotechnology
integrated pest management
water resources
environment
biology
chemistry
biochemistry
entomology
agronomy
horticulture
ornamentals
animal sciences
agricultural economics
agricultural education

3. In groups of four or five students, research "organic farming" on the Internet. Compare and contrast organic farming and traditional farming. In your search, you may include factors such as cost versus yield, consumer demand, or any other factor that drives production.

Life in the United States and throughout the world is changing every moment of our lives. The space we occupy, as well as the people we work and play with, may be constant for a brief time. However, these are quick to change with time and circumstances. The things we need to know and the resources we have available to use are constantly shifting as the world turns.

Humans have the gift of intelligence—the ability to learn and to know (Figure 1-1). This permits us to compete successfully with the millions of other creatures that share the Earth with us (Figure 1-2). In ages past, humans have not always fared well in this competition. Wild animals had the advantages of speed, strength, numbers, hunting skills, and superior senses over humans. These superior senses of sight, smell, hearing, heat sensing, and reproduction all helped certain animals, plants, and microbes to exercise control over humans to meet their own needs.

The cave of the cave dweller, lake of the lake dweller, and cliff of the cliff dweller indicate early human reliance on natural surroundings for basic needs (food, clothing, and shelter) (Figure 1-3). Those early homes gave humans some protection from animals and unfavorable weather. However, they were still exposed to disease, the pangs of hunger, the sting of cold, and the oppression of heat.

The world of agriscience has changed the comfort, convenience, and safety of people today. According to the USDA/Economic Research Service, Americans spent only 9.4 percent of our wages to feed ourselves in 2010 (Figure 1-4). Despite fluctuations in the percentage of income that is spent for food, the percentage of annual income spent for food in the United States has tended to decrease. People in many nations spend more than half of their incomes on food. We are fortunate

FIGURE 1-1 Humans have the gift of intelligence—the ability to learn and to know.

FIGURE 1-2 The gift of intelligence has permitted humans to compete with and benefit from animals, even though most animals are superior to humans in other ways.

FIGURE 1-3 Early humans had to rely on features in their natural environment to shield them from danger and the elements.

FIGURE 1-4 Americans spend only 9.4 percent of their disposable income on food.

that our scientists have discovered new ways to produce greater amounts of food and fiber (such as cotton) from each acre of agricultural land. They have done this by finding ways to stimulate growth and production of animals and plants and to reduce losses from diseases, insects, parasites, and storage. We have learned to preserve our food from one production cycle until the next without excessive waste; however, spoilage of stored food remains high among agriscience research priorities. The agriscience, agribusiness, and renewable natural resources of the nation provide materials for clothing, housing, and industry at an equally attractive price.

AGRISCIENCE DEFINED

Agriscience is a relatively new term that you may not find in your dictionary. **Agriscience** is the application of scientific principles and new technologies to agriculture. **Agriculture** is defined as the activities involved with the production of plants and animals and related supplies, services, mechanics, products, processing, and marketing (Figure 1-5). Actually, modern agriculture covers so

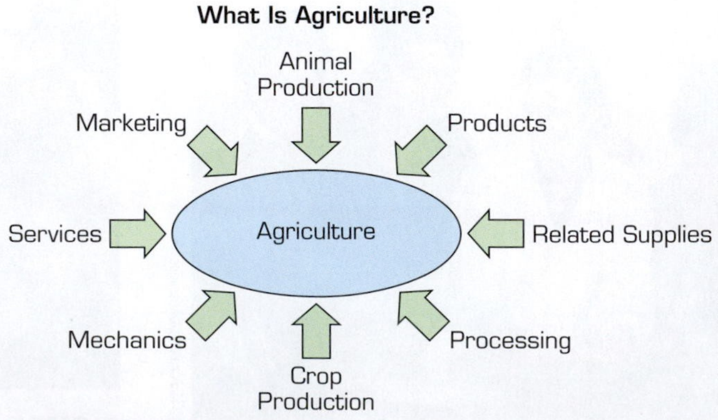

What Is Agriculture?

Animal Production
Marketing
Products
Services
Agriculture
Related Supplies
Mechanics
Processing
Crop Production

FIGURE 1-5 Agriculture consists of all the steps involved in producing a plant or animal and getting the plant or animal products to the people who consume them.

© Nate A./Shutterstock.com.

FIGURE 1-6 Farming and ranching account for approximately 12 percent of the agricultural jobs in the United States.

Courtesy of DeVere Burton.

FIGURE 1-7 Agricultural education teachers and agricultural extension educators are among those whose careers are related to agriculture.

many activities that a simple definition is not possible. Therefore, the U.S. Department of Education has used the phrase *agriculture/agribusiness and renewable natural resources* to refer to the broad range of activities in agriculture.

Agriculture generally has some tie-in or tieback to animals or plants. However, production agriculture, or farming and ranching, accounts for only 12 percent of the total jobs in agriculture (Figure 1-6). The other 88 percent of the jobs in agriculture are nonfarm and nonranch jobs, such as sales of farm equipment and supplies, plant and animal research, processing of agricultural products (Figures 1-7 and 1-8), agricultural education, and maintaining the health of plants and animals. **Agribusiness** refers to commercial firms that have developed in support of agriculture (Figure 1-9).

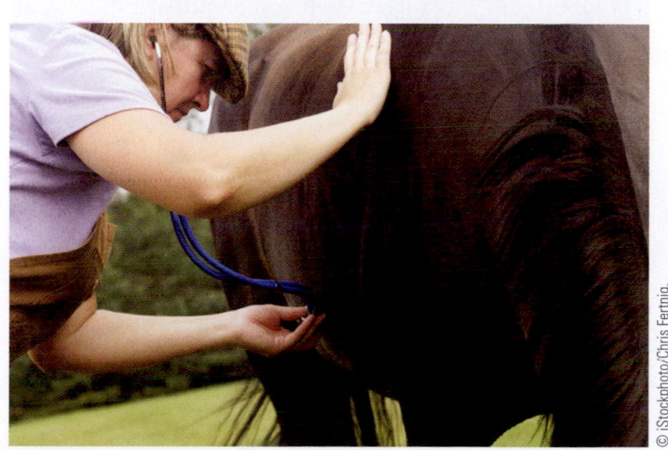
© iStockphoto/Chris Fertnig.

FIGURE 1-8 Veterinarians and veterinary technicians are people whose careers relate to agriculture in the field of animal health.

© iStockphoto/David Sucsy.

FIGURE 1-9 Agribusinesses are important to the people and the stability of most communities.

HOT TOPICS IN AGRISCIENCE WORLD FOOD CRISIS

Unemployment and high food prices drive those who are most affected to seek food from charitable organizations.

A serious food issue surfaced in late 2007 as the world supply of rice, wheat, and corn dropped to dangerously low levels. The result was a substantial worldwide increase in the purchase price for all grains. This crisis continues to this day, and prices have increased far beyond expectations. In the United States and other nations, the cost of bread and other grain products increased as food processors adjusted the price of their products to compensate for the high cost of raw materials and transportation. The cost of grain and energy has also affected the price of meats, eggs, milk, and other foods, driving the price upward.

Political turmoil across the world as a result of economic recessions and in the form of revolutions has become a serious deterrent to affordable food prices in other ways. The price of oil has been driven up by reduction in the production of crude oil, which is often associated with military conflicts in many oil-producing nations. This has raised production costs for most food items because the cost of fuel has increased and remained high.

Among the poor nations of the world and among those living on fixed incomes or in poverty here at home, obtaining enough food to meet the needs of individuals and families has become difficult. What should be done to overcome and resolve a world food crisis?

INTERNET TIPS: Forming your search into a question will narrow the results. Example: What is agriscience?

INTERNET KEY WORDS: renewable, natural resources

Renewable natural resources are the resources provided by nature that can replace or renew themselves. Examples of such resources are wildlife, trees, and fish (Figure 1-10). Some occupations in renewable natural resources are game trapper, forester, and fisher (someone who harvests fish, oysters, and other seafood).

Technology is defined as the application of science to solve a problem. The application of science to an industrial use is called *industrial technology*. *Agriscience* was coined to describe the application of high technology to agriculture. **High technology** refers to the use of electronics and state-of-the-art equipment to perform tasks and control machinery and processes (Figure 1-11). It plays an important role in the industry of agriculture.

FIGURE 1-10 As mature trees are harvested, sunlight on the forest floor stimulates the growth of seeds and seedlings, providing a renewable source of wood for the future.

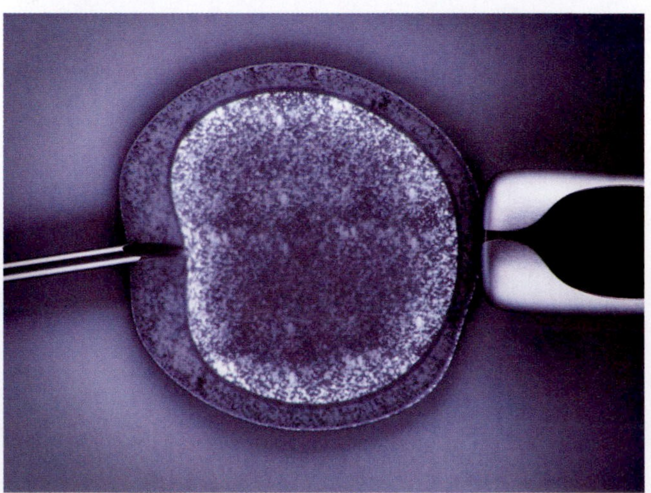

FIGURE 1-11 Advances in technology make it possible to create identical cloned animals by dividing the cell mass of a growing embryo.

HOT TOPICS IN AGRISCIENCE

AGRICULTURAL RESEARCH: FEEDING A HUNGRY WORLD

© Garsya/Shutterstock.com.

Environment refers to all the conditions, circumstances, and influences surrounding and affecting an organism or group of organisms.

The world's population reached 7 billion people in 2011 and it is projected that it will reach 8 billion in 2025. During the same period, the amount of land and fresh water per person will decrease. Food production must become much more efficient if the people of the world are to have enough food to eat. During the past 50 years, food production has increased at a rate that is greater than the increases in the domestic population; however, food shortages and famine still exist in the world. Agricultural production is driven by a worldwide market. Agricultural research has resulted in greater productivity of food, plants, and animals, and new technologies have made it possible for farmers to perform their work with greater efficiency. The key to an adequate food supply for the growing human population in the new millennium is agricultural research. New agricultural technologies that lead to the development of more efficient plants and animals and more efficient agricultural machinery will be needed. In addition, we will need to discover new food sources and maintain a healthy environment as the population approaches 10 billion people.

Agriscience includes many endeavors. Some of these are aquaculture, agricultural engineering, animal science technology, crop science, soil science, biotechnology, integrated pest management, organic foods, water resources, and environment. **Aquaculture** means the growing and management of living things in water, such as fish or oysters. **Agricultural engineering** consists of the application of mechanical and other engineering principles in agricultural uses. **Animal science technology** refers to the use of modern principles and practices for animal growth, production, and management (Figure 1-12). **Crop science**

Photo by Scott Bauer. USDA/ARS K5441-1.

FIGURE 1-12 Veterinarians use animal sciences to help keep our pets and production animals healthy.

AGRI-PROFILE CAREER AREA: AGRISCIENTIST

Courtesy of USDA/ARS K5304-17.

Students experience the wonder of living things.

Science plays an increasing role in the lives of plants and animals and the people around them. These living bodies include plants ranging in size from microscopic bacteria to the huge redwood and giant sequoia trees. They include animals from the one-celled amoeba to elephants and whales.

Only recently has science identified the nature of viruses and permitted humans to observe the submicroscopic world in which they exist. The electron microscope, radioactive tracers, computers, electronics, robotics, nanotechnology, and biotechnology are just a few of the developments that have revolutionized the world of living things. We call this the world of science. Agriscience is a part of this world. Through agriscience, humans can control their destinies better than at any time in known history.

Agriscience spans many of the major industries of the world today. Some examples are food production, processing, transportation, sales, distribution, recreation, environmental management, and professional services. Study of and experiences in a wide array of basic and applied sciences are appropriate preparations for careers in agriscience.

refers to the use of science principles in growing and managing crops. **Soil science** refers to the study of the properties and management of soil to grow plants. **Biotechnology** refers to the management of the genetic characteristics transmitted from one generation to another and its application to our needs. It may be defined as the use of cells or components of cells to produce products and processes (Figure 1-13).

The phrase **integrated pest management** refers to combining two or more different control methods to control insects, diseases, rodents, and other pests. *Organic food* is a term used for foods that have been grown without the use of chemical pesticides. **Water resources** cover all aspects of water conservation and management. Finally, **environment** refers to all the conditions, circumstances, and influences surrounding and affecting an organism or group of organisms (Figure 1-14). This generally means air, water, and soil, but it

Courtesy of USDA/ARS K-5011-19.

FIGURE 1-13 Genetic engineering and other forms of biotechnology have become some of the most important priorities in research today.

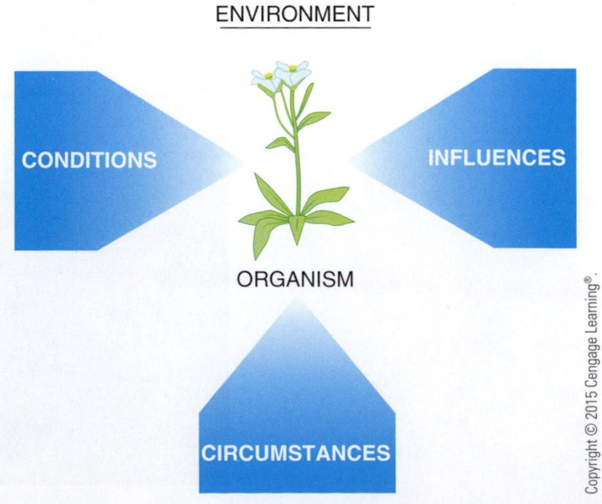

FIGURE 1-14 The term environment refers to all the conditions, circumstances, and influences surrounding and affecting an organism or group of organisms.

may also include such things as temperature, presence of pollutants, intensity of light, and other influences.

AGRISCIENCE AROUND US

Agriscience and technologies have helped humans change their living conditions from dependence on hand labor to a highly mechanized society. In the process, food and fiber production has become much more efficient. Many members of U.S. society have become free to pursue new careers in business, industry, or the arts because they are no longer required to spend all of their time finding or producing food for themselves and their families. Fewer than 2 percent of the people in the United States are farmers. On average, each farmer produces enough food for approximately 167 people. The large surplus of food that is produced in the United States is shipped to many other countries in the world.

Whether you live in the city, town, or country, you are surrounded by the world of agriscience. Plants use water and nutrients from the soil and release water and oxygen into the air. Animals provide companionship as pets and assistance with work. Both plants and animals are sources of food. Many microscopic plants and animals are silent garbage disposals (Figure 1-15). They assist in the process of decay of the unused plant and animal residue around us. This process returns nutrients to the soil and has many other benefits to our environment and our well-being.

Agriscience encompasses the wildlife of our cities and rural areas, and the fish and other life in streams, ponds, lakes, and oceans. Plants are used extensively to decorate homes, businesses, shopping malls, buildings, and grounds. When the use of one crop is lessened, another takes its place. This occurs even where the land resource changes from farm use to suburban and urban uses.

Corn has long been referred to as king among crops in the United States. Yet, in some states, including Texas and Virginia, turf grass is the number-one agricultural crop. Turf is grass that is used for decorative as well as soil-holding purposes. This change has occurred as more land is used for roads, housing, businesses, institutions, recreation, and other nonfarm uses (Figure 1-16).

Agriculture and the agriscience activities that support it extend far beyond the borders of the United States. Many nations throughout the world depend on agriscience to improve the production of their crop and livestock industries. Agriculture is a global industry, and although the United States exports many of

FIGURE 1-15A The process of composting uses bacteria in moist and aerated conditions to break down plant residue.

FIGURE 1-15B The material that remains when the composting process is complete is used to provide nutrients to crops and gardens.

Courtesy of USDA/ARS CS-0311.

FIGURE 1-16 Turf has become an important crop, especially in areas near large cities where it is used to establish new lawns.

its agricultural products, it also imports many agricultural products from other parts of the world. Many of the flowers used by florists in the United States come from South America, Colombia, and other foreign countries (Figure 1-17). Bulbs come from Holland, and meat products are imported from Argentina. Lumber is shipped from the United States to Japan, only to return in the form of plywood and other processed lumber products. A decrease in the price of sow bellies or an unexpected change in the price of grain futures in Chicago can affect business and investment around the world (Figure 1-18).

The great water-control projects on the Colorado River have permitted the transformation of the American Southwest from a desert to irrigated lands

© Igor Karon/Shutterstock.com.

FIGURE 1-17 Flowers are often imported to the United States during seasons when local florists are unable to produce them at competitive prices.

Courtesy of National FFA; FFA #18.

FIGURE 1-18 Marketing in agriscience has become big business.

FIGURE 1-19 The great Southwestern desert has been transformed into highly productive land, using irrigation water from the huge dams on the Colorado River and its tributaries.

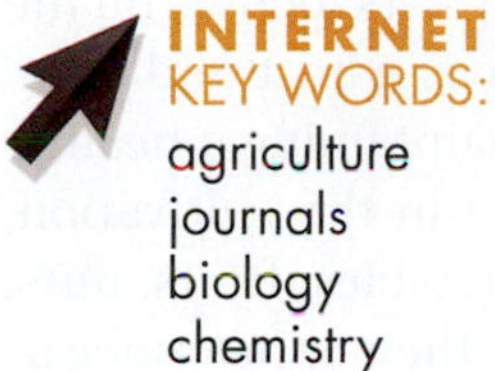

(Figure 1-19). This is now an area of intensive crop production that has stimulated national population shifts. Water management has transformed the great dust bowl of the American West into the "breadbasket" of the world.

Agriscience enterprises extend beyond farming to such fields as journalism and communications. Agricultural publications such as magazines, journals, and newspapers provide information to farmers, helping to make farm production more efficient. Radio and television programs provide similar services to agriculture. They provide a communications link among such people as agricultural specialists, agricultural extension educators, wildlife biologists, and others to communicate the latest information to farmers and other managers of natural resources. Such subjects as plants, animals, wildlife, market reports, gardening, and lawn care are popular "Saturday morning" topics.

AGRISCIENCE AND OTHER SCIENCES

Agriscience is really the application of many sciences. Colleges of agriculture and life sciences perform dual roles of conducting research and teaching students in these sciences. **Biology** is one of the three basic sciences. It derives from two Greek words: *bios*, meaning "life," and *logy* meaning "to study." It is the science that studies all living things (organisms) and the environment in which these organisms live. Biology emphasizes the structures, functions, and behaviors of all living organisms. It focuses on the traits that organisms have in common as well as their differences. Having an understanding of biology is important to you, the agriscience student. New biological discoveries can affect many areas of your life. Examples include the choice to plant a new variety of flower in the front yard, changes in the way food is processed that result in a fresher product with a long shelf life, and new medical treatments that can help keep you healthy.

Chemistry is another basic science. It is the branch of science that studies the nature and characteristics of elements or simple substances. Chemists study the changes substances undergo when they react with other substances. These changes are responsible for compounds that have been used by humans for thousands of years, whereas other compounds, like artificial sweeteners, are relatively new discoveries. Chemistry currently benefits us in many ways; for example, it is responsible for the creation of new medicines, textiles, fuels, and fertilizers.

Biochemistry is the last of the three basic sciences. It is a combination of biology and chemistry. Recall the definition of *bio*, which means life; when added to the word *chemistry*, it is easy to see that the science of biochemistry is the study of the chemical activities or processes of living organisms. These chemical activities take place in the cells and molecules of living organisms. This science is responsible for explaining things such as brain function, how genetic traits are passed from one generation to another, and how cells communicate and work together inside an organism. Applied science uses the basic sciences in practical ways. For instance, **entomology** is the science of insects, the most abundant species on the planet. Insects account for more than 3 million human deaths per year, they transmit diseases, and they are our principal competitors for food. Insects, however, are required for pollination by half the plants on Earth to produce seeds. It is important to find ways to help control problem insects safely without causing secondary problems such as halting pollination of plants.

There are many other applied sciences. Knowledge of biology, chemistry, and biochemistry is important in entomology and to the other applied sciences listed below.

Agronomy is the science of soil management and crops. Its focus is on the growth, management, and improvement of field crops such as wheat and corn. The goal is to increase food production and quality while maintaining a healthy environment. **Horticulture** is the science and art involved in the cultivation, propagation, processing, and marketing of flowers, turf, vegetables, fruits, nuts, and ornamental plants. **Ornamentals** are plants grown for their appearance or

SCIENCE CONNECTION THE AGRISCIENCE PROJECT

Photo by Scott Bauer, USDA/ARS K8329-2.

The written report completes the research project and enables others to benefit from new knowledge.

The scientific method is an excellent and widely used method for systematic inquiry and documentation of new findings. The agriscience student is encouraged to learn and use the scientific method for classroom, laboratory, and field studies. The following procedures will guide you in your quest for new knowledge in agriscience.

STEP 1. IDENTIFY THE PROBLEM

Decide precisely and specifically what it is that you wish to find out, for example, "How much nitrogen fertilizer is needed to grow healthy corn plants?" Be careful to limit your topic to a single researchable objective. Your teacher can suggest other topics that could be researched.

STEP 2. REVIEW THE LITERATURE

Reviewing the literature simply means reading up on and becoming well informed about the topic. See what is already known about it. Magazines, newspapers, reference books, encyclopedias, science journals, computer information systems, television, cooperative extension meetings, teleconferences, and personal interviews may be sources of appropriate information. Be sure to seek information on appropriate ways to conduct research on the type of problem you have chosen.

STEP 3. FORM A HYPOTHESIS

After learning as much as you can about the topic, develop a hypothesis or statement to be proven or disproved, which will solve the problem. For example, "Trout grown in 60° F water will grow faster than trout grown in 75° F water."

beauty. Examples are flowers, shrubs, trees, and grasses. Horticulture is a unique science because it also incorporates the art of plant design.

The **animal sciences** are applied sciences that involve growth, care, and management of domestic livestock. They include veterinary medicine, animal nutrition, animal reproduction, and animal production and care. Animal scientists work to discover scientific principles related to animals. Scientific principles are then applied to animal management plans to improve animal health and production.

Economics is the study of how societies use available resources to meet the needs of people. **Agricultural economics** relates to factors that affect the management of agricultural resources, including farms and agribusinesses, to meet the needs of the human population. Farm policy and international trade are important components of agricultural economics. **Agricultural education** is one of the most unique programs available to students. It offers organized instruction, supervised agricultural experience (SAE), FFA, and extension education activities. Agricultural communications, journalism, and community development are also components of agricultural education. These and other disciplines are part of the dynamic study known as agriscience.

A PLACE FOR YOU IN AGRISCIENCE

What about career opportunities in agriscience? The nation's agricultural colleges report steady demand for graduates. A U.S. Department of Agriculture (USDA) study group forecasted a national shortage of 4,000 agricultural and life sciences

STEP 4. PREPARE A PROJECT PROPOSAL

Prepare a proposal outlining how you think the project should be done. Include the timelines, facilities, and equipment required as well as anticipated costs and a description of how you will do the project.

STEP 5. DESIGN THE EXPERIMENT

Considering the information gathered in Step 2, develop a plan for carrying out the project so you can test the hypothesis. This is the most critical step in your research project. If this is not done correctly, you may invest considerable time, work, and expense and end up with incorrect or invalid conclusions. The method or procedure should be carefully thought out and discussed with your teacher or other research authorities. This is to ensure that your design will actually measure what you are testing.

STEP 6. COLLECT THE DATA

In this phase, you conduct the experiment. Here you test and/or observe what takes place and record what you measure or observe.

STEP 7. DRAW CONCLUSIONS

Summarize the results. Make all appropriate calculations. Determine if the information allows you to accept or reject the hypothesis or if the information is inconclusive.

STEP 8. PREPARE A WRITTEN REPORT

The written report provides you, your teacher, and other interested parties with a permanent record of your research. From this, you can report to your peers, get course credit, and possibly apply for awards. Perhaps you can use a computer and hone your word-processing skills. For scientists, the written report becomes a permanent document that is kept by the research institution and becomes the basis for articles in research publications for the world to see. The results become part of the "literature" on the topic.

© Alexander Raths/Shutterstock.com.

FIGURE 1-20 Plenty of good jobs are available for agricultural graduates, whether at the technical degree, bachelor degree, or graduate degree levels of education.

graduates per year toward the end of the twentieth century. The shortage became reality in the early 1990s and continues even today. Employers are offering higher salaries and more job variety to agriscience college graduates than ever before. Career opportunities continue to be strong in these fields, and they also extend into technology, as it relates to agricultural systems. Consequently, high school agricultural, horticultural, or other agriscience program participants have opportunities to obtain good jobs and have rewarding careers (Figure 1-20). These opportunities are described in later units in this text. By studying agriscience, you open the door to exciting educational programs and careers that contribute to professional satisfaction and prosperity.

STUDENT ACTIVITIES

1. Write the Terms to Know and their meanings in your notebook.
2. List examples of animals that have superior senses than do humans. Indicate the sense(s) along with the animals.
3. Write a paragraph or two on (1) cave, (2) lake, and (3) cliff dwellers. Explain how the types and locations of their homes provided protection from (1) animals and (2) unfavorable weather. Access to an online encyclopedia would be a good resource for this activity.
4. Ask your teacher to assign you to a small discussion group to talk about the responses to Activity 3.
5. Place a map of your school community on a bulletin board. Insert a colored map pin in every location of a farm, ranch, or agribusiness in your school community.
6. Talk to your County Extension Agent or another agricultural leader regarding the importance and role of agriscience, agribusiness, and renewable natural resources in your county.
7. Select one of the sciences mentioned in this unit. Prepare a written report on the meaning and nature of that science. Report to the class.

SELF-EVALUATION

A. MULTIPLE CHOICE

1. Humans have the ability to learn and know. This is known as
   a. achievement.
   b. intelligence.
   c. intuition.
   d. spontaneity.
2. The percentage of an average U.S. worker's pay that is used for food is
   a. 9.4 percent.
   b. 14.5 percent.
   c. 50 percent.
   d. 74 percent.
3. The best term to describe the application of scientific principles and new technologies to agriculture is
   a. agribusiness.
   b. renewable natural resources.
   c. farming.
   d. agriscience.
4. Harmful insects, rodents, and diseases are all referred to as
   a. animals.
   b. plants.
   c. pests.
   d. parasites.
5. Agriscience encompasses
   a. wildlife and fish.
   b. ornamental plants and trees.
   c. farms and agribusinesses.
   d. all of the above and more.
6. Irrigated lands are generally used for
   a. intensive crop production.
   b. wildlife refuges.
   c. forests.
   d. boating and fishing.
7. An example of a basic science is
   a. agronomy.
   b. aquaculture.
   c. horticulture.
   d. chemistry.
8. An example of an applied science is
   a. animal science.
   b. biochemistry.
   c. biology.
   d. chemistry.
9. One relationship of agriscience with many other sciences is that
   a. agriscience is the application of many other sciences.
   b. agriscience is entirely different from all other sciences.
   c. agriscience is an old term and is unlike other sciences.
   d. agriscience is a narrow science and is easily defined.
10. The career and job outlook in agriscience is
    a. a strong demand for college graduates.
    b. a shortage of 4,000 trained workers per year.
    c. higher salaries are being offered.
    d. all of the above.

B. MATCHING

______	1. Aquaculture	a. Commercial firms in agriculture
______	2. Renewable resource	b. Electronics and ultramodern equipment
______	3. Agribusiness	c. Growing in water
______	4. Chemistry	d. Basic science of plants and animals
______	5. High technology	e. Can replace itself
______	6. Biology	f. Characteristics of elements
______	7. Organic food	g. Space and mass around us
______	8. Environment	h. Grown without chemical pesticides

C. COMPLETION

1. Integrated pest management refers to the application of many different methods used together to ________.
2. The transformation of the American Southwest from desert to irrigated lands was made possible, in part, by water-control projects on the ________ River.
3. By studying agriscience, you open the door to exciting educational programs that may lead to ________.

UNIT 2

Better Living through Agriscience

OBJECTIVE

To determine important elements of a desirable environment and explore efforts made to improve the environment.

MATERIALS LIST

- paper
- pen or pencil
- current newspaper
- online encyclopedias
- agriscience magazines
- Internet connection

COMPETENCIES TO BE DEVELOPED

After studying this unit, you should be able to:

- describe the conditions of desirable living spaces.
- discuss the influence of climate on the environment.
- compare the influences of humans, animals, and plants on the environment.
- examine the problems of an inadequate environment.
- identify significant world population trends.
- identify significant historical developments in agriscience.
- state practices used to increase productivity in agriscience.
- identify important research achievements in agriscience.
- describe future research priorities in agriscience.

SUGGESTED CLASS ACTIVITIES

1. Discuss ways that farmwork has changed during the last 100 years. Identify several important tasks that must be done by farmers. Describe how each of those tasks was done 100 years ago. Describe how farmers perform each of those tasks today. What scientific discoveries have contributed to greater efficiency in doing farmwork today?
2. Investigate ways that new and modern farming methods have contributed to opportunities in career fields other than agriculture. How have efficient farming methods benefited all the citizens of the United States?

3. Using library materials, the Internet, or other scientific sources, learn about climates in the latitudes of 90, 60, and 30 degrees. Using the information you have obtained about these regions, determine what kinds of crops might be raised at each of these latitudes.

TERMS TO KNOW

sewage system
polluted
condominium
townhouse
famine
contaminate
parasite
reaper
combine
moldboard plow
cotton gin
corn picker
milking machine
tractor
legume
tofu
Katahdin
BelRus
Russet
Green Revolution
feedstuff
selective breeding
genetic engineering
monoclonal antibody
hybrid
laser

Living conditions in the world vary extensively. In all countries, there are some very desirable places to live and work (Figure 2-1), yet even in highly developed countries, there are pockets of poverty. How do you explain the differences in living conditions from one place to another? Why do living conditions vary from one community to another, from one neighborhood to another, or from one house to another? The wealth and preferences of individuals explain some of the differences. Yet, the environment or the area around us has much to do with the quality

FIGURE 2-1 People of all cultures appreciate pleasant surroundings for their homes and work.

of life. It also has much to do with the way we feel about ourselves and others. In addition, the way we feel about ourselves is often expressed in the way we treat and care for our environment.

VARIETY IN LIVING CONDITIONS

The population of the world reached 7 billion people in October, 2011(Figure 2-2). It is predicted that it will take 14 years for the population to increase to 8 billion in 2025. Across the world, several babies were given the distinction of being the 7 billionth human being living on the planet Earth (Figure 2-3). What will the home and community be like where "Baby 7 Billion" grows up? Will there be adequate food? Will that child be warm, but not too warm or too cold? Will the child be kept free from serious illness? Will his or her family have a house or good living space they call home? Will they have clothing to permit them to live and work outside the home in relative comfort? What will be the quality of life of others who live near the 7 billionth human being? Will that child survive, and

© Dmitry Nikolaev/Shutterstock.com.

FIGURE 2-2 In 2011, the population of the world reached and exceeded 7 billion people.

Courtesy of NASA.

FIGURE 2-3 What will life be like on planet Earth when Baby 8 Billion is born?

will he or she go on to live a happy and productive life? Positive answers to these questions would indicate a good environment for a person. These same questions should be asked for the rest of humankind.

The Homes We Live In

Homes of the Very Poor

Homes of the poorest people range in size and quality from nothing to a piece of cardboard or a scrap of wood on the ground. Many survive freezing winters with only a tattered blanket on the warm sidewalk grates of our modern cities. For others, housing may take the form of a grass hut or a shack made of wood, cardboard, plastic, or scraps of sheet metal. These people often depend on the outdoors to provide water and washing areas and as a receptacle for human waste. Large families often share such homes with pets, poultry, or other livestock (Figure 2-4).

In cities, the poor frequently live in old buildings that are in bad condition and with plumbing that does not work. Drugs, crime, poisonous lead paint, and disease are typical hazards for these people. The steamy streets and sidewalks provide little relief from oppressive summer heat.

Homes of the Less Fortunate

People with modest sources of income may have homes that are simple but provide basic protection from the elements. Such homes may be of wood, stone, masonry blocks, concrete, sheet metal, brick, or other fairly permanent materials. The presence of windows and doors may provide protection from the elements and some privacy.

These people frequently have access to water that is safe to drink, but bathrooms may be nonexistent or toilets may not have safe sewage disposal systems. A **sewage system** receives and treats human waste (Figure 2-5). To be regarded as safe, a sewage system must decompose human waste and release by-products

INTERNET KEY WORDS: sewage treatment

Courtesy Elmer Cooper/© Cengage Learning 2015®.

FIGURE 2-4 The poor of the world live in substandard housing, and it is not uncommon for homes to be shared with pets, livestock, or poultry.

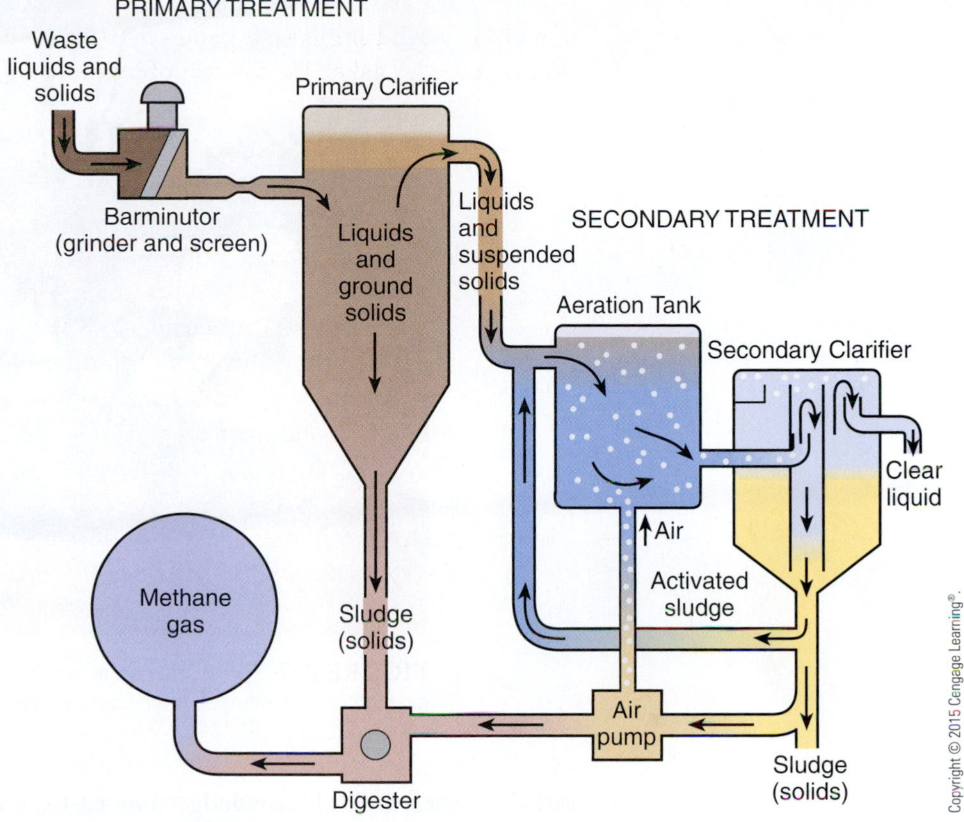

FIGURE 2-5 Safe sewage disposal is essential to good health.

that are free from harmful chemicals and disease-causing organisms (Figure 2-6). In most countries of the world, people rely on creeks or rivers to supply their drinking water, bathe the family, wash the clothes, and carry away the human waste.

People with low or modest incomes also may live in housing with bathrooms and running water. However, maintenance of the systems may be poor,

FIGURE 2-6 This modern sewage treatment plant captures energy from sewage sludge by using the sludge to produce methane gas. The gas is used as fuel for engines that drive generators, producing electricity.

FIGURE 2-7 For much of the world's population, even a single community source of safe water is a luxury.

and the users' lack of knowledge may cause conditions that are hazardous to health.

In the developing countries, the lower classes are fortunate if there is a source of safe water (water that is free of harmful chemicals and disease-causing organisms) at the village center (Figure 2-7). Modest and simple running-water facilities are generally the first evidence of community development in many such areas. Even a single faucet with unpolluted water is a major step forward for many communities. Communities that do not have a source of safe water must use whatever water they have available to them. Too often, water is **polluted** or unsafe to drink because it contains waste materials, chemicals, or unhealthful organisms.

Homes of the Middle and Upper Classes

The middle- and upper-class people of the world can afford and enjoy housing that is clean, safe, and convenient. Such living spaces are often found as single houses, in both rural and urban areas. In towns, villages, and urban areas, homes may also be in the form of townhouses, condominiums, or apartment buildings. A **condominium** is a building with many individually owned living areas or units. All living space of a single unit is generally on one floor. A **townhouse** is one of a row of houses connected by common side walls. Each unit is usually two or three stories high, giving the occupants more variety of living space.

Food

Until the 1970s, much of the world went to bed hungry. Only a few countries had sufficient food for their people. Most countries had problems with distribution. Not everyone had food of sufficient amount and quality for proper nutrition. Today, major **famines** (widespread starvation) are still a fact of life and death. Even in the modern world, serious famines have occurred in various countries.

INTERNET KEY WORDS: world food supply

United Nations scientist John Tanner concluded that, in theory, the world could feed itself; but in practice it could not. It is estimated that nearly a billion people are not getting enough food for an active working life. Although some countries enjoy an adequate food supply from their own production and imports, most nations have many individuals who do not receive proper nutrition.

Family

Family life may well be the dominant force that shapes the environment for most individuals. The family has control of the household activities and sets the priorities of its members. The family has considerable influence over maintaining attractive surroundings and promoting warm relations among individuals.

For some, the family chooses the neighborhood and community where they live. A wise choice, however, is based on having the knowledge of better opportunities and the necessary resources to move to a better living environment. For most of the world's population, the communities where individuals are born are the communities where they are raised and spend their lives.

Neighborhood and Community

The neighborhood and community have substantial influence on the environment in which we live. Some communities have tree-lined country roads with attractive fields, pastures, or woodlands to provide variety in the landscape. Other communities may have the advantage of attractive homes, businesses, or community centers (Figure 2-8). Urban areas may boast high-rise buildings for work and residence. These provide beautiful vistas of city lights or harbor scenes of commerce and recreation.

FIGURE 2-8 Good communities provide pleasant, attractive environments for living and working.

Neighborhoods and villages are parts of larger communities. These communities are influenced greatly by the families who live in the immediate area. If families work together toward common goals, they can shape the character, education, religious activities, social outlets, employment opportunities, and other broad aspects of their environment.

HOT TOPICS IN AGRISCIENCE

PRESERVING THE ENVIRONMENT USING INTENSIVE FARMING PRACTICES

Courtesy of DeVere Burton.

Responsible use of intensive farming practices may be our best option to preserve forests and other natural environments.

When the first colonists arrived in America from Europe, abundant land resources were available for food production. Trees were removed to make way for the farms. Midway through the twenty-first century, as the world's population approaches 10 billion people, the forests of the world will again become endangered unless we can continue to increase the food production of current farmland. We will need to increase the efficiencies of our existing farmlands and production methods to produce an adequate food supply for a growing world population. If we fail to increase the food production of our land, forests will probably be converted to farms because more land will be required to produce the additional food that will be needed.

Responsible use of intensive farming practices is likely to play a big role in preserving our forest lands. Applications of agricultural chemicals to crops contribute to high production of food by controlling weeds and insects. The application of fertilizers to farmland is also a proven method for sustaining high levels of production. However, good judgment must always be exercised in the application of fertilizers and chemicals to ensure that they are used safely and that they do not pollute the environment. It is an interesting paradox to consider that farm fertilizers and pesticides may be our best hope for preserving the forests and other natural environments in the world. As many foreign governments can attest, preservation of the environment becomes a low priority to people who are starving.

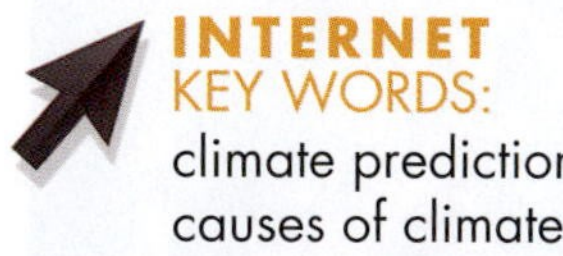
INTERNET KEY WORDS:
climate prediction
causes of climate

Climate and Topography

Climate and topography are also important factors affecting our environment. Climate is the average yearly temperature and precipitation for a region. Unlike topography, which is the physical shape of the Earth, the climate of a given area is shaped by many factors. The movement of heat by wind and ocean currents, the amount of heat absorbed from the sun, latitude, and the amount of precipitation received are all factors that influence climate. The climate affects what kinds of crops can be raised. The tropical areas of the world produce crops such as pineapples and bananas, whereas the more mild climates found in the U.S. mainland are better adapted for crops such as corn and wheat (Figure 2-9). Average annual temperatures are very high near the equator. Yet people living near ocean waters, even in tropical areas, enjoy a moderate climate with cool breezes most of the time. Inland, the inhabitants are likely to experience hot, humid weather with high rates of rainfall. The high rainfall, in turn, stimulates heavy plant growth, resulting in jungle conditions. Similarly, sea-level elevations may create balmy 80° F temperatures, whereas a short trip to the top of a nearby mountain may reveal snow on its peak (Figure 2-10).

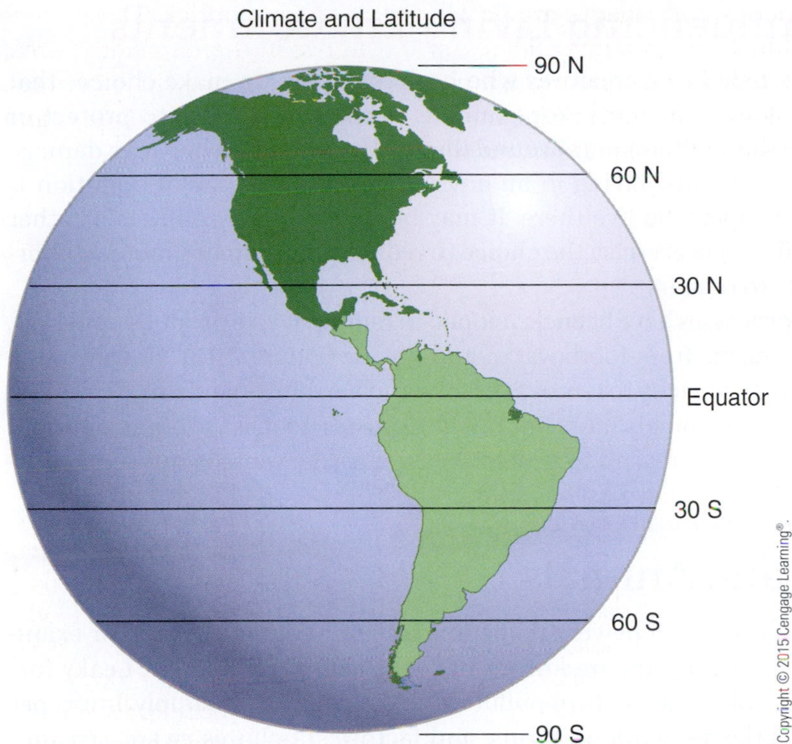

FIGURE 2-9 The distance from the equator is measured by degrees, with zero being the equator and 90 degrees North being the north pole. Distance from the equator affects the climate of a region.

FIGURE 2-10 Topography is an important factor that affects the temperature of the environment.

INTERNET KEY WORDS:
insect contamination
insects
forests
pollinating insects
pesticides
pollution

Northern areas, such as Alaska, may border on the Arctic Circle and have long, frigid winters. Yet those same latitudes enjoy summers suitable for short-season crops (Figure 2-11). People inhabit most areas of the Earth, so the climate and topography where they find themselves create environmental conditions that influence their quality of life.

FIGURE 2-11 The climate near Anchorage, Alaska, is influenced by a short growing season; however, during the growing season, the number of hours of daylight per day is high, contributing to excellent growth conditions for some food crops.

© Don Farrall/Photodisc/Getty Images.

FIGURE 2-12 Processes have been developed that will remove contaminants such as heavy metals and harmful organisms from water.

Factors Influencing Living Environments

Humans are the only living creatures who have the ability to make choices that affect their living environments. For example, people can choose to protect or even to clean up the environments around them. They may also choose to damage their living environments. Living in an environment that is free of pollution is a choice of the people who live there. It may be influenced by politics or other causes, but until people exercise the choice to reduce pollution in their neighborhoods, it is likely to remain.

Scientific processes have been developed to remove waste products, poisons, and disease organisms from food, water, and the air (Figure 2-12). However, the body is limited in its capacity to remove poisons and harmful organisms. To remain healthy, humans and animals must limit their exposure to disease organisms and poisons. One important reason to protect living environments is to keep pollutants from entering food and water supplies.

Humans and Animals

Some human activities can be very damaging to living environments. For example, exhaust fumes from cars are known to cause acid precipitation. Leaky fuel tanks pollute the soil, which in turn pollutes the drinking water supply. Improper use of chemicals (lawns, gardens, farms, and factories) pollutes rivers, streams, and lakes. Poor soil management results in erosion of the soil and contamination of streams, lakes, and reservoirs with silt.

A major problem for humans and animals is to avoid **contaminating** (adding material that will change the purity or usefulness of a substance) food and water with secretions from their own bodies. Urine and feces are liquid and solid body wastes. They are serious contaminants of food and water. Diseases are often spread by body contact, by eating impure food and water, or by breathing contaminated air.

There are serious animal diseases that spread from animal to animal by contact with body wastes. If animals have plenty of living space, this generally does not cause serious problems. But, as with humans, when animals are concentrated, health hazards increase. Fortunately, most diseases are spread among only a given species of animal and not from one species to another. For instance, most diseases of dogs do not spread to cats. Similarly, most diseases of animals do not infect humans. Brucellosis is an example of an animal disease that may be transferred from animals to humans, creating serious health problems. It kills unborn animals and humans by spreading the disease through the milk and birth fluids of infected individuals.

However, some animal disorders can cause human sickness. **Parasites** are organisms that live on or inside other organisms called hosts with no benefit to those hosts. The parasite is an unwelcome guest because it always causes some kind of harm to its host by feeding on it.

Insects

Insects impact heavily on our environment. Some cause damage to our living environment and others help to improve it. The cockroach is an unwelcome guest in many households of the world (Figure 2-13). Some cockroaches feed on human waste and then on the food of humans. They transmit disease from waste material to food and water. In poor housing conditions, they can move from household to household. In doing so, they often leave disease organisms and illnesses in their wake.

HOT TOPICS IN AGRISCIENCE PESTICIDE REDUCTION AND BIOTECHNOLOGY

© iStockphoto/BasieB.

The Colorado potato beetle can be controlled in genetically modified potatoes without the use of pesticides.

A significant and well-documented outcome derived from the adoption of genetically engineered plants is the declining use of pesticides. The USDA reported that pesticide use declined by 14 percent over an eight-year period. These findings should be no surprise considering that many genetic modifications have introduced pest resistance into some of our most widely used food plants. One such plant is the potato. This crop is highly vulnerable to the Colorado potato beetle, and an entire field can be destroyed by this pest in a very short period of time. The adult and immature forms of this beetle eat the stems and foliage and destroy the ability of the plant to engage in photosynthesis. The traditional treatment for these pests has been to apply pesticides to the crop. Potatoes that have been genetically modified to resist the beetles are able to produce a chemical that does not affect humans, but which poisons the beetles as they eat the foliage. The end result is that pesticides are no longer needed to control this pest in potato varieties that carry the resistant gene. According to the Food and Agriculture Organization of the United Nations (2004):

"Thus far, in those countries where transgenic crops have been grown, there have been no verifiable reports of them causing any significant health or environmental harm." (p. 76)

Insects cause damage to the environment in other ways. They damage and kill trees and other plants. A population of harmful insects can expand quickly, and damage to trees can be extensive unless steps are taken to control them. For example, pine beetles are capable of killing entire populations of pine trees (Figure 2-14). Imagine the damage to the environment of a community when many of the trees in parks, yards, and streets are of a single variety that is susceptible to a highly destructive insect pest. When a tree or other plant is not affected by the presence of a harmful insect, it is described as being immune. Many plants in the environment are immune to particular insects, but they are often vulnerable to other insect species.

Photo by Scott Bauer. USDA/ARS K7233-6.

FIGURE 2-13 Cockroaches are household pests that pollute living environments and contaminate foods and beverages. The cockroach feeding station, pictured here, lures these insects inside and poisons them.

Courtesy of DeVere Burton.

FIGURE 2-14 Pine beetles have destroyed a forest of pine trees that were first weakened by drought and then infested with beetles.

Not all insects are destructive to the environment. Some insects, such as bees and other pollinators, are useful to living environments. Without these insects to carry pollen from one flower to another, many plants could not reproduce. Some insects, such as the lady bird beetle, prey on harmful insects. These insects play important roles in keeping populations of harmful insects in check. Insects do have profound effects on the living environments that surround us.

Chemicals

Chemicals can have both helpful and harmful effects on living environments (Figure 2-15). Oil spills and industrial chemical discharges have caused serious problems in oceans, lakes, rivers, and streams (Figure 2-16). Improperly used chemical pesticides continue to threaten wildlife, fish, shellfish, beneficial insects, microscopic organisms, plants, animals, and humans. In the mid-1960s, American biologist Rachel Carson shocked the world with her book *Silent Spring*. This was one of the first books to provide convincing evidence of environmental damage due to pesticides.

In 1972, DDT (dichloro-diphenyl-trichloroethane) was banned in the United States because of its damaging effects on the environment. This insecticide had been used to control mosquitoes, which carried the dreaded malaria organism. DDT was also a very effective chemical used against flies, and it enjoyed widespread use in homes, on farms and ranches, and wherever flies were a problem. Yet, because it was determined that DDT was responsible for interfering with the reproduction of birds by weakening the egg shells, it had to be discontinued; safer substitutes have been found.

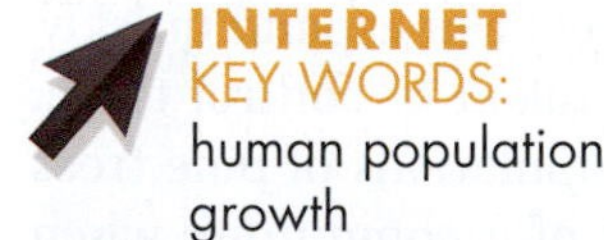

INTERNET KEY WORDS:
human population growth

Careful management and control of the different chemicals we use is absolutely essential. It requires the utmost care to avoid unacceptable damage to our environment.

Photo by Jeff Vanuga, USDA Natural Resources Conservation Service.

FIGURE 2-15 Chemicals are needed in our modern society, but they may threaten the health of animals and people if misused or abused.

© Gl0ck/Shutterstock.com.

FIGURE 2-16 Spills of petroleum or chemicals in water environments cause serious water pollution and damage populations of wild animals and plants.

OUR SHARED LIVING ENVIRONMENT

All of the members of the plant and animal kingdoms, including humans, must share the living environments that are available on Earth. Some living environments are more friendly to the inhabitants than others, and plant and animal life tends to be concentrated in the warm, temperate regions of the world. Other environments, such as the Arctic and Antarctic regions, do not support the variety of plants and animals that are found in other places. The same is true of the high-altitude, mountainous areas. One thing is certain: The living environments of the Earth will never grow any larger. In fact, the habitable regions may actually decline for many species of plants and animals as humans pave the Earth and create cities. Only a few species are able to adapt to the less favorable environments as they are crowded from their natural ranges.

As we ponder the life of "Baby 7 Billion" (see earlier), we must wonder if we are doing our part to preserve and enhance the environment. Plants, animals, insects, soil, water, and air must be kept in reasonable balance, or all will suffer. Plants are generally considered to improve living environments, but excessive plant growth can infringe on the space for humans and animals. Humans and animals can damage a plant species until it is unable to adequately reproduce itself. Too many animals in a shared environment can compete excessively with humans for food, water, and space. Some species of insects are regarded as harmful by people because they feed on desirable crops or afflict humans or livestock. However, many species of insects are beneficial to plants, animals, or humans.

Humans and animals tend to consume or remove plants, which hold soil in place and prevent erosion from wind and water (Figure 2-17). For instance, during the 1960s, most of the forests of China were cut and not replanted. Rapid and alarming soil erosion followed. The government then placed a high priority on

Courtesy of DeVere Burton.

FIGURE 2-17 Soils that are located on steep slopes are vulnerable to severe erosion problems. Permanent plant cover binds the soil particles with roots, reducing the tendency to erode.

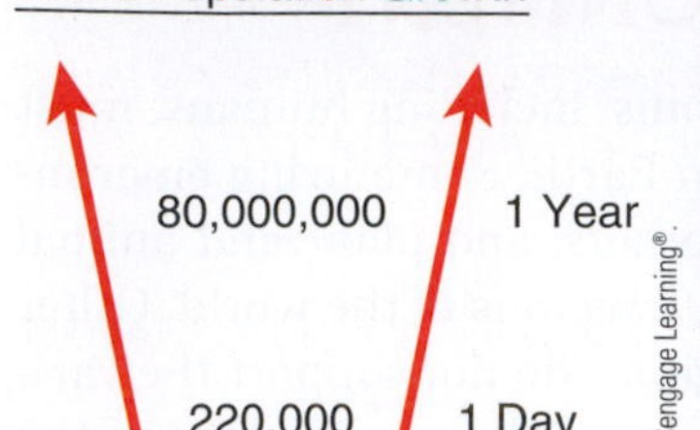

FIGURE 2-18 Earth's population reached 7 billion in April 2012 and is expected to reach 8 billion by 2025.

FIGURE 2-19 National parks and areas that are set aside as wilderness are intended to preserve the environment and the plants and animals that live there.

reforestation and reversed the trend. Soil is needed to hold nutrients until plants require them. Similarly, we need the soil to filter and store clean water for plant growth and human and animal consumption. Plants take water from the soil and release water and oxygen to the air, which benefits humans and animals.

The number of human beings in the world is growing at the rate of 150 every minute; 220,000 a day; 80 million a year (Figure 2-18). At the current rate of growth, the Earth's population will reach 8 billion by 2025. Can Earth sustain such population growth? Will humans find enough to eat? Will we learn to protect our environment, or will we destroy the system that supports life itself? Will we survive the competition of such population growth but sacrifice our quality of life? Might we, in fact, improve our quality of life by using our intelligence to stabilize and improve our environment?

Humans are the only living creatures who can choose to improve the living environments for themselves and other living organisms. The motivation to improve the environment is seldom present when people are hungry, however. As food production becomes more efficient, less of the total land area is required for the production of food. This allows some land areas and the living environments in the region to be preserved in their natural condition. For example, the national system of parks and monuments includes large areas where living environments support native plants and animals (Figure 2-19). Vast regions have been designated as "wilderness," and the type of human activity in these areas is strictly controlled to favor wild creatures. Humans also intervene in damaged environments to clean and restore them. Favorable economic conditions, in general, increase our ability and motivation to restore natural living environments and to preserve others.

AGRISCIENCE IN OUR GROWING WORLD

The keys to a prosperous future, indeed the bottom line for survival of the world's population, can be found in agriscience. Agriscience is the science of food production, processing, and distribution. It is the system that supplies fiber for building materials, rope, silk, wool, cotton, and medicines. It provides the grasses and ornamental trees and shrubs that beautify our landscapes, protect the soil, filter out dust and sound, and supply oxygen to the air.

AGRI-PROFILE CAREER AREA: ENVIRONMENTAL MANAGEMENT

© William Perugini/Shutterstock.com.

Environmental management requires skills in observation, analysis, and interpretation.

Management of the environment requires the attention of consumers as well as professionals. However, specialists in air and water quality, soils, wildlife, fire control, automotive emissions, and factory emissions all help maintain a clean environment against tremendous population pressures in many localities. Helicopter, airplane, and satellite crews gather important data for scientific analysis to help monitor the quality of our environment.

Individuals in environmental careers may work indoors or outdoors; in urban or rural settings; or in boats, planes, factories, laboratories, or parks. Careers range from laborer to professional. Environmental concerns are high on global agendas today as nations attempt to reduce global hunger and pollution.

According to the Bureau of Labor Statistics (2012–13), agriscience accounts for 14 percent of jobs in the United States. It is the mechanism that permits the United States and other developed countries of the world to enjoy high standards of living. It is the system that developing countries are using in their efforts to feed and clothe their bulging populations. People look to agriscience for the necessary technology to compete on a par with other nations in the twenty-first century. We must look to agriscience to maintain and improve our quality of life.

The United States is a major world supplier of food. It is also a major supplier of fiber for clothing and of trees for lumber, posts, pilings, paper, and wood products. The use of ornamental plants and acreage devoted to recreation has never been greater in the history of our country.

CHANGING POPULATION PATTERNS

The United Nations Organization has reported that more children than ever before are surviving to adulthood. It also indicated that adults are living longer (Figure 2-20). Together, these trends mean more population growth and more pressure on the environment. Advancements in medical science and services have made good health and longer lives a reality, but only for those who can afford good nutrition and modern health services. Similarly, through agriscience, we have made substantial gains in providing food, fiber, and shelter for the world (Figure 2-21). At the same time, the environment has stayed reasonably clean, considering the impact of bulging populations.

In the past, individuals younger than 25 years constituted the world's largest population group. This occurred because children were valued for the help they provided in making the family living. Children and young adults were engaged in a nation's labor force and provided the manpower for armies. In most countries, the young respected their elders and provided for the needs of the elderly within the family.

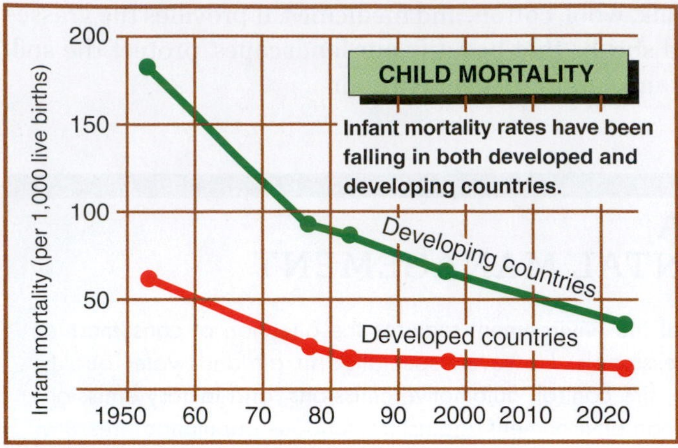

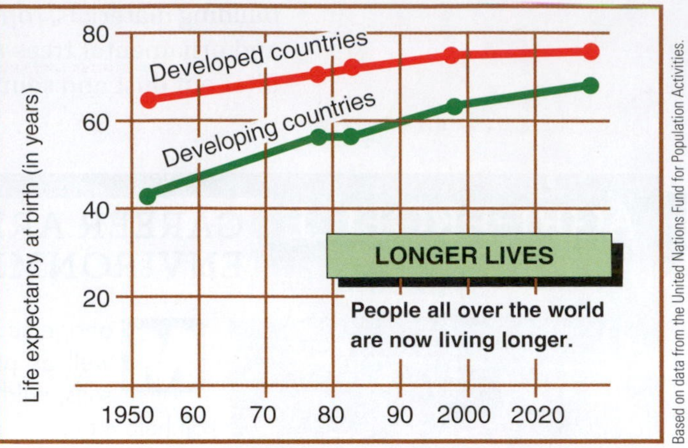

FIGURE 2-20 Worldwide child mortality rates and life expectancy estimates.

The age profiles of people in developed countries are quite different from those of developing countries. Honduras has the traditional population pattern, with its largest number of citizens younger than 5 years. The number per age group then decreases to the smallest number, which occurs in the age group older than 80 years. When the Honduras population groups are displayed by sex in a bar graph, the graph takes the shape of a pyramid (Figure 2-22). Canada's pattern is slightly different. Its greatest population group is around the 20-year mark. Its graph reminds you of a Christmas tree, with its narrow bottom and cone appearance. The population of Sierra Leone illustrates the reduction of the 20- to 30-year-old population due to civil war. Sweden, a country known for its excellent health services and high survival rate, has age brackets that are about equal. The population graph for that country resembles a column.

China has about one-fifth of the world's population, yet it has been reasonably successful at feeding its population by keeping about 70 percent of its work force on farms. In contrast, less than 2 percent of the work force in the United States is necessary to operate the nation's farms.

In the mid-1970s, China implemented a policy whereby each couple was limited to one child. They called it the 4–2–1 policy. This means that extended families consisted of four grandparents, two parents, and one child. What would be the outcome if such a policy were strictly enforced for several generations? You would expect the pyramidal shape of China's population graph to change to the shape of a Christmas tree, and, in time, to an

**#1.
LABOR REQUIRED TO PRODUCE WHEAT, CORN, AND COTTON (in hours)**

	1800	1935–39	1955–59	1980–84	1990 or Later
Wheat (100 bu.)	373	67	17	7	7
Corn (100 bu.)	344	108	20	3	2.88
Cotton (1 bale)	601	209	74	5	5

**#2.
YIELDS PER ACRE OF WHEAT, CORN, AND COTTON**

	1800	1940	1960	1985–86	1990 (Prelim)
Wheat (bu.)	15	15	20	34	39.5
Corn (bu.)	25	29	55	118	118.5
Cotton (lb.)	154	253	446	630	640.0

FIGURE 2-21 Production of most agricultural crops has increased due to better varieties and improved cultural practices. Mechanization and improved yield have greatly reduced the amount of labor per unit of production.

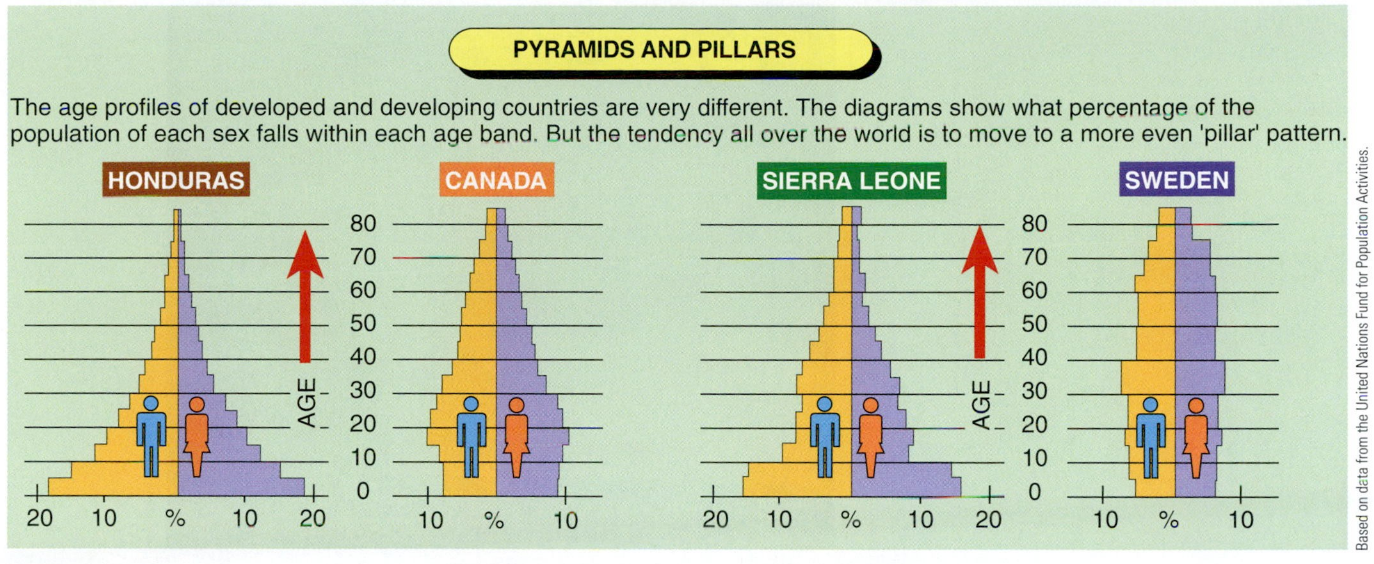

FIGURE 2-22 Age profiles and population patterns for developing and developed countries.

upside-down pyramid! What would be the implications of feeding a nation with a population of mostly elderly people?

IMPACT OF AGRISCIENCE

FIGURE 2-23 The inventions of the 1800s brought revolutionary changes in agriculture in the United States and Europe.

History records little progress in agriculture for thousands of years. Then, starting in the early 1800s, the use of iron spurred inventions that revolutionized agriculture in the United States, British Isles, and northern Europe (Figure 2-23). However, for most of the world, progress has been much slower. In some nations, government leaders are slow to implement agriscience because the nation initially would experience massive unemployment as machines displaced human labor.

Progress through Agricultural Engineering

Mechanization through inventive engineering was an important factor in the United States' agricultural development. The change from 90 percent to less than 2 percent of the workers being farmers evolved over a 200-year period. Machines helped make this possible. The old saying that "necessity is the mother of invention" suggests the relationship between an inventor's problem and the use of previously acquired skills to solve that problem. The solution is frequently a new device, machine, or process.

INTERNET KEY WORDS:
rural electricity

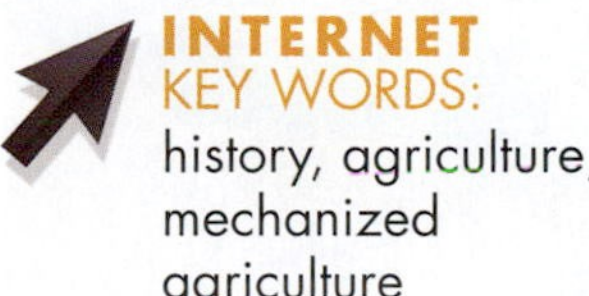

INTERNET KEY WORDS:
history, agriculture, mechanized agriculture

One of the most significant technologies to increase the efficiency of farm production was the generation and distribution of electricity to rural farming areas. Many of the labor-intensive jobs that were performed by hand 75 years ago are now performed by machines that are powered by electricity. Some examples include grain augers, milking machines, water pumps, fans, conveyor belts, power tools, and many other machines (Figure 2-24). The electric motor has revolutionized the world.

American Inventors

The United States is home to the inventors of many of the world's most important agricultural machines. In 1834, Cyrus McCormick invented the **reaper**, a machine to cut small grain (Figure 2-25A). Later, a threshing device was added

Courtesy of DeVere Burton.

FIGURE 2-24 Electric power has replaced human labor for many agricultural tasks.

to the reaper, and the new machine was called a combine. The reaper cut and bundled the grain in the field. Today, grain is harvested with a machine called a **combine**, which cuts and threshes in a single operation (Figure 2-25B). One modern combine operator can cut and thresh as much grain in one day as 100 individuals could cut and bundle in the 1830s.

Thomas Jefferson's invention of an iron plow to replace the wooden plow of the time was of great significance. Later, in 1837, a blacksmith named John Deere experienced the frustration of prairie soil sticking to the cast-iron plows of the time. It became apparent that Jefferson's invention would not work in the rich prairie soils of the Midwest. Through numerous attempts at shaping and polishing a piece of steel cut from a saw blade, the steel **moldboard plow** evolved. That plow permitted plowing of the rich, deep prairie soils for agricultural production and launched the beginning of the John Deere Company.

In 1793, Eli Whitney invented the cotton gin. The **cotton gin** separated the cotton seeds from cotton fiber. This paved the way for an expanded cotton

Courtesy Elmer Cooper/© Cengage Learning 2015®.

A

© Jose Ignacio Soto/Shutterstock.com.

B

FIGURE 2-25 Cyrus McCormick's reaper (A) led to the development of the modern grain combine (B).

Photo by Keith Weller. USDA ARS K8082-1.

FIGURE 2-26 The invention of the mechanical milking machine has greatly reduced the amount of human labor required to care for dairy animals.

and textile industry. In 1850, Edmund W. Quincy invented the mechanical **corn picker**, which removed ears of corn from the stalks. During the same era, Joseph Glidden developed barbed wire, with sharp points to discourage livestock from touching fences. This effective fencing permitted establishment of ranches with definite boundaries. In 1878, Anna Baldwin invented a **milking machine** to replace hand milking (Figure 2-26). In 1904, Benjamin Holt invented the **tractor**, which became the source of power for belt-driven machines as well as for pulling.

Formation of Machinery Companies

Many of the early inventors worked alone or with one or two partners. They were all workers in the area of agricultural mechanics and, as such, in agriscience. By the early 1900s, the inventors or other enterprising people had formed companies to produce agricultural machinery or process agricultural products. This made invention a continuing process. Successive inventions were used to improve earlier inventions and to develop new equipment and supplies to meet the needs of a changing agricultural industry.

The development of mechanical cotton pickers and corn harvesters greatly expanded the output per farm. Significant expansion of U.S. agriculture also resulted from the development of irrigation technology. Since the end of World War II, the mechanization of U.S. agriculture has moved at a breathtaking pace (Figure 2-27).

Mechanizing Undeveloped Countries

In the developing countries of the world, many engineers, teachers, and technicians have sought simple, tough, reliable machines to improve agriculture. In such countries, the United States' highly developed, complex, computerized, and expensive machinery does not work for long. Most countries do not have people

FIGURE 2-27 Since World War II, U.S. agriscience has progressed at a breathtaking pace.

trained for the variety of agriculture mechanics jobs that are needed to support U.S. agriculture.

A machine with rubber tires is useless if a tire is damaged and repair services are not available. Similarly, failure of an electronic device may cause a $200,000 machine to become junk in the hands of an unskilled person in a country without appropriate repair facilities. This is the case in most developing countries in Central and South America, Eastern Europe, Asia, and Africa. For the developing nations of the world, other aspects of agriscience must become the vehicles for advancing agricultural productivity.

Improving Plant and Animal Performance

Humans have improved on nature's support of plant and animal growth since they discovered that the loosening of soil and planting of seeds could result in new and better plants. Even before that discovery, they aided plant growth by keeping animals away from them until fruit or other plant parts edible to humans were harvested.

The human touch has permitted plants and animals to increase production and performance to the point where fewer people are needed to produce the food supply for the United States and other developed countries. Surplus food is exported to many other nations.

One of the remarkable occurrences of the twentieth century was the mechanization of agriculture. The many technologies that were developed for the agricultural industry have contributed to larger farms. Many people were displaced from their family farms because they were slow to adopt the new technologies and farming practices that were needed to make their farms more efficient. Many of these people have learned trades other than farming and have become productive citizens in other industries. Without the farming revolution of the last 60 years, our citizens would not be free to pursue other occupations. The U.S. space program is possible because our scientists do not have to produce their own food. The efficiency of U.S. farms has contributed to the freedom of our citizens to engage in many new and exciting occupations. These include the development of computers and other technologies that have resulted in the current "information age."

INTERNET KEY WORDS: soybean science

INTERNET KEY WORDS: biotechnology, agriculture

BIO-TECH CONNECTION AGRISCIENCE AND BIOTECHNOLOGY

Photo by Scott Bauer. USDA/ARS K 5011-19.

Biotechnology provides valuable mechanisms to improve life.

Agriscience is heavily impacted by biotechnology, which addresses the continuation of life. In ornamental horticulture, one uses a myriad of plants to beautify the interiors of homes, businesses, and institutions. Such plants also consume carbon dioxide gas and supply oxygen for humans, animals, insects, and other living organisms. In outdoor settings, ornamental trees, shrubs, and turfgrasses beautify our yards, streets, parks, and other public areas. Our highways rely on plants to screen off oncoming traffic, provide living hedges, absorb sound, prevent soil erosion, and create a stimulating environment to keep motorists alert. Fruits, vegetables, grains, and forage crops of gardens, ranches, and farms provide the backbone of the world's food supply. Trees provide wood, paper, and other fiber products. The productive capability of plants has been greatly improved through the efforts of scientists, technicians, and growers.

Similarly, many animal species have been modified over the centuries by humans through domestication, selection, breeding, and care. Currently, biotechnology is improving the productivity of plants and animals and providing new foods and medicines to enrich our lives. Genetic engineering enables humans to modify and utilize microorganisms in our fight against harmful insects and other pests. Today, the Earth is providing food, shelter, habitat, health care, and other essentials to more people than at any time in history. However, there is much malnutrition and starvation in the world. The causes tend to be rooted in deficiencies in government, national infrastructure, poverty, and lack of education. The technology of food production is believed to be adequate to meet the world food needs if modern technology could be applied to agricultural production opportunities throughout the world.

Improving Life through Agriscience Research

Unlocking the Secrets of the Soybean

Americans have long appreciated the extensive research on the peanut done by the American scientist George Washington Carver. Carver is credited with finding more than 300 uses for the peanut. These include food for humans, feed for livestock, cooking fats and oils, cosmetics, wallboard, plastics, paints, and explosives.

Less known are the secrets of the soybean. The Chinese have known for centuries that the soybean is a versatile plant with many uses. Calling it the "yellow jewel," the Chinese are said to have grown the soybean 3,000 years ago. The strong flavor of the soybean itself is not appealing, but the bean is a legume and is nutritious. A **legume** is a plant that hosts nitrogen-fixing bacteria. These bacteria convert nitrogen from the air to a form that can be used by plants. They also convert atmospheric nitrogen to excellent sources of protein for humans and animals.

Courtesy of the American Soybean Association.

FIGURE 2-28 The soybean is the world's most important source of vegetable oil, and it provides the basic materials for hundreds of products.

A Chinese scholar is believed to have first made tofu from soybeans in 164 BC. **Tofu** is a popular Chinese food made by boiling and crushing soybeans, coagulating the resulting soy milk, and pressing the curds into desired shapes. Today, tofu is a major food in the diet of China's huge population. Tofu contributes to a reasonably healthful diet. It can be fermented; marinated; smoked; steamed; deep-fried; sliced; shredded; made into candy; or shaped into loaves, cakes, or noodles.

Soy oil is the world's most plentiful vegetable oil. It is first extracted from the soybean, and the material that is left is processed into a protein-rich livestock feed known as soybean meal. The components of the soybean are used for hundreds of items. These range from food products to lubricants, paper, chalk, paint, printing ink, and plastics (Figure 2-28).

Baked Potatoes

Many improvements in our way of life can be traced to agriscience research. For instance, the U.S. Department of Agriculture developed many pest-resistant varieties of potatoes. A case in point is the work with the **Katahdin**, a popular potato variety of the 1930s. From the Katahdin, scientists developed the **BelRus**, a superior baking variety bred to grow well in the Northeast. In a similar manner, the **Russet** potato grown in the volcanic soils of Idaho and the Northwest has been improved through research. Selection of parent stock has increased its resistance to diseases and insects, resulting in greater yields.

Turkey for the Small Family

In your grandparents' time, Thanksgiving was probably observed by having all the relatives visit to consume the typical 30-lb. turkey. As families became smaller and more scattered, the need for such large birds decreased; but even people with small families liked turkey. The 30-lb. bird was too much, so the problem was to develop a breed of turkey that weighed 8 to 12 lb. at maturity (Figure 2-29). A solution was the Beltsville Small White turkey, named after the Beltsville Agricultural Research Center in Maryland, where the breed was developed. Further research and development have yielded meat animals with high yields of lean meat and less fat.

The Green Revolution

During the 1950s, starvation was rampant in many countries of the world. A major question was: "Could the world's agriculture sustain the new population growth?" The solution was partly in the development of new, higher-yielding, disease- and insect-resistant varieties of small grains for developing countries. The result was the **Green Revolution**, a process whereby many countries became self-sufficient in food production in the 1960s by using improved plant varieties and proven management practices.

Cultivated Blueberries

Wild blueberries were enjoyed in early times when people had time to pick the tiny berries growing in the wild. But labor costs became too high to harvest such berries for sale. The solution was development of high-quality, large-fruited blueberry varieties from the wild. This started today's new and valuable cultivated-blueberry industry.

Courtesy of USDA/ARS.

FIGURE 2-29 The Beltsville Small White turkey was developed to meet the needs of small families.

Nutritional Values

Until recently, animal and human nutrition were based on poor methods of feed and food analysis. The problem was the following: How can one recommend what to feed or what to eat if the content of food for humans and crops for livestock cannot be accurately determined? The solution was to develop detergent chemical methods for determining the nutritional value of **feedstuff** (any edible material used in the diets of animals). The procedures are now used widely throughout the world in both human and animal nutrition.

Biological Attractants

The use of chemical pesticides usually provides short-term solutions to many insect-control problems. However, it has become apparent that chemicals have some disadvantages and that additional means of control must be found. A partial solution was to discover chemicals that insects produce and give off to attract their mates (Figure 2-30). These chemicals are now produced in the laboratory. Laboratory production of these chemicals has permitted mass trapping of insects to survey insect populations for integrated pest-management programs.

Breakthroughs in Agriscience

Genetically Engineered Tomato

Calgene, an agricultural biotechnology firm in Davis, California, developed a bioengineered tomato that resists rotting. The new tomato was developed by turning off the gene that causes the tomato to soften and rot. The new tomato lasts longer on the shelf at the grocery store, retains its flavor longer, and has proven superior in taste tests.

Natural Rubber Production

Scientists at the USDA Western Regional Research Center at Albany, California, have modified a scrubby bush called "guayule" by genetically engineering methods to produce up to 1,000 kg of rubber per hectare from the plant (an increase from 200 kg/hectare from native plants). The plant looks like sagebrush. This new technology makes it possible to produce a domestic supply of natural rubber in the United States.

Bioengineered Designer Foods

By altering the genetic structure of food products, scientists have created new foods such as crispy vegetables, sweeter carrots, leaner meats, high-protein milk, longer-lasting melons, and healthier cooking oils.

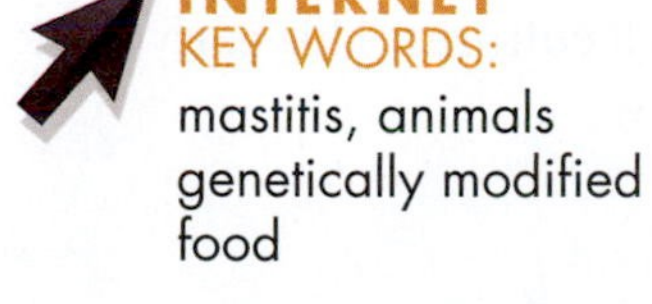

INTERNET KEY WORDS: mastitis, animals genetically modified food

Monoclonal Antibodies in Goats' Milk

Monoclonal antibodies are natural substances in blood that fight diseases and infections. Transgenic goats have been developed by inserting a gene into

© iStockphoto/GaryAlvis.

FIGURE 2-30 Attractants are valuable nonpolluting chemicals that lure insects to traps, bait, or a system of control through sterility.

HOT TOPICS IN AGRISCIENCE GENETICALLY MODIFIED FOODS

Humans have been modifying the genes of plants and animals for centuries using a technique called **selective breeding**. Plants or animals that exhibit desired genetic characteristics are selected as the parents of the next generation. For example, if a scientist wanted a wheat plant that was resistant to a disease, he or she would plant a field of wheat and expose it to the disease. Then the researchers would find the plants that survived the disease and use them as parent stock for the next generation. This process is then repeated until a variety that has a high resistance to the targeted disease is found. This process is difficult and can take years.

Genetic engineering is a technique that allows scientists to physically put specific genes into the cells of a plant or animal without the rigorous process of selective breeding. As a result, genetic engineering is much faster than selective breeding. This technique has revolutionized agriculture. One common genetically modified crop is herbicide-resistant wheat. A gene that is resistant to the deadly effects of plant-killing herbicides is positioned into the DNA of the wheat. As a result, a farmer can spray an entire field with an herbicide, killing only the weeds and not the valuable crops. Genetic modification is a powerful tool that has resulted in more efficient production of a number of agricultural products.

This new advancement does not come without controversy. The United Kingdom, as well as other countries, placed bans on genetically modified crops. Activist groups, together with media, lobbied the governments of these countries to halt the sale of genetically modified products because of perceptions that the products are not safe for people to use. The controversy over genetically modified food is not isolated to Europe. In the United States, special interest groups continue to lobby the Food and Drug Administration (FDA) to require food labeling for all foods containing bioengineered ingredients. Legislation relating to labeling of genetically engineered food products has been introduced in some of the states. Oregon voters turned down a bill that would require all genetically modified food to be reported on the food labels. In 2002, Maryland placed a ban on all genetically engineered fish.

the goats' DNA, causing them to produce and secrete up to 4 g. of the monoclonal antibody in each liter of milk. This level of antibody production is 10 to 100 times greater than traditional methods of production from cell cultures. This early work with transgenic goats produced an anticancer antibody.

Bio-diesel from Animal Fat

Excess animal fat (tallow) that is trimmed from the carcasses of meat animals is a low-value by-product of the meat-processing industry. A process has been developed that converts tallow to bio-diesel, a product much like the diesel fuel extracted from crude oil (Figure 2-31). The fat is heated to a liquid, followed by a purification process. The purified fat product is then mixed with methyl alcohol and a chemical catalyst. The bio-diesel that is produced from this process has approximately the same heating value and power potential as traditional diesel fuel, and it will burn in an ordinary diesel engine.

INTERNET KEY WORDS:
fire ant control

Mastitis Reduced

The mastitis organism has always been a serious problem for dairy farmers. Mastitis is an infection of the milk-secreting glands of cattle, goats, and other milk-producing animals. The resulting loss of milk production adds millions of dollars yearly to the cost of milk in the United States. Recent research efforts resulted in the development of abraded plastic loops for insertion into cow udders. The procedure resulted in a reduction in clinical mastitis of 75 percent.

Courtesy of DeVere Burton.

FIGURE 2-31 Forms of bio-diesel made from animal fat or plant oils show promise as a fuel for diesel engines.

The reduction in infections resulted in increased milk production, averaging nearly 4 lb. of milk per cow per day.

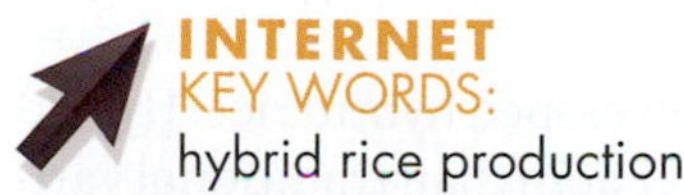

INTERNET KEY WORDS: hybrid rice production

© injun/Shutterstock.com.

FIGURE 2-32 A new synthetic material has been identified for control of fire ants by increasing the ratio of drones to worker ants.

© Michael Shake/Shutterstock.com.

FIGURE 2-33 New varieties of exotic flowers known as impatiens have been developed.

Human Nutrition

Studies in human nutrition have demonstrated the benefits of decreasing the amount of animal fat in the diet and increasing the proportion of fat from vegetable sources. This practice reduces high blood pressure and risk for heart attack. Much of the progress in human nutrition has grown out of research on animals and plants by agriscientists. Research on human nutrition has yielded new recommendations for healthful eating.

Fire Ant Control

Fire ants infest more than 320 million acres in 13 states in the United States (Wilcox and Giuliano, 2006; see Figure 2-32). Their presence in the warmer climates of the world is a constant threat to the well-being of humans and livestock. A new synthetic control for fire ants increases the ratio of nonproductive drone ants to worker ants. This ratio change gradually weakens the colony and causes it to die.

Exotic Flowers

Horticulturists, gardeners, and hobbyists will be delighted with the new varieties of impatiens (a popular, easy-to-grow, summer-flowering plant) (Figure 2-33). Plant explorers introduced exotic new germ plasm, and plant breeders developed a new technique called ovuleculture to develop hybrids and new kinds of impatiens. A **hybrid** is the offspring of a plant or animal derived from the crossing of two different species or varieties.

Courtesy of USDA/ARS K3957-11.

FIGURE 2-34 Infrared photos taken from satellites help diagnose problems such as ant infestations in fields and pastures.

Satellites and Nitrogen Gas Lasers

Nutrient deficiencies in growing corn and soybean crops are not easy to detect from the ground. A deficiency occurs when a nutrient is not available in the amounts that are needed for optimum growth. New technology now permits the monitoring from satellites of deficiencies of iron, nitrogen, potassium, and other nutrients using nitrogen gas **lasers** (devices used to determine wavelengths given off by the plants). These wavelengths indicate the levels of various nutrients in plants (Figure 2-34).

Sugar Beet

The development of new varieties is a technique that has been used in agriscience for many decades to improve plant performance (Figure 2-35). A recent breakthrough has provided a sugar beet hybrid with a high ratio of taproot weight to leaf weight. The hybrid yields about 15 percent more sugar per acre than previous varieties.

Another breakthrough in sugar beets is the release of a genetically engineered variety that causes the plant to be tolerant to the herbicide, Roundup. This makes it possible to control all of the weeds in a field without hand labor. Resistance to the new variety continues in the legal system, however, as advocacy groups attempt to prevent the variety from being produced in the United States.

© Luis Carlos Jimenez del rio/Shutterstock.com.

FIGURE 2-35 Sugar beet hybrids have improved sugar yield up to 15 percent.

Rice Hybrids

On the other side of the world, Chinese agronomists developed hybrid rice. Hybrid rice is capable of yielding up to 40 percent more rice per acre than traditional varieties. The combination of hybrid semidwarf rice varieties, improved irrigation, and chemical fertilizers has increased world rice production from 240 million metric tons in the 1960s to more than 650 million metric tons in 2007.

A study of USDA publications reveals great numbers of improved varieties, new products, and superior processes that have been discovered or developed through agriscience research.

INTERNET KEY WORDS:
agricultural discoveries

AGRISCIENCE AND THE FUTURE

In 2011, the average American farmer was capable of producing enough food and fiber for approximately 167 people. Agriscience will become even more important in the next 100 years. As the world's population increases, it will require a highly sophisticated agriscience industry to provide the food, clothing, building materials, ornamental plants, recreation areas, and open-space needs for the world's billions. Americans will have to work more in the international arena as more countries become highly competitive in agriscience and as trade barriers are removed. Research and development will continue to play a dominant role as they lead the way in agriscience expansion in the future.

The USDA has developed the following mission statement to guide the agency: We provide leadership on food, agriculture, natural resources, and related issues based on sound public policy, the best available science, and efficient management.

The new century and millennium bring a new set of challenges to United States and world agriculture. The international business economy is a dominant factor in marketing agricultural products. A major share of U.S. agricultural production is now consumed in foreign countries, and this trend is expected to increase the volume of agricultural commodities that are sold outside U.S. borders. Canada and Mexico have become two of our biggest markets. They have also become successful competitors with the United States for a world market share in some agricultural commodity markets.

The future of U.S. agriculture will require that farmers become even more efficient in the production of food and fiber crops. Animal agriculture will depend on scientific improvements in production methods and in the genetic superiority of food animals to improve the quality of animal products. The efficiency with which they are produced must also improve for animal products to continue to demand a strong market share of the food supply in a world crowded with humans.

STUDENT ACTIVITIES

1. Write the Terms to Know and their meanings in your notebook.
2. Develop a bulletin board that illustrates the components of our environment.
3. Collect newspaper articles that describe environmental problems in your community.
4. Prepare a two- or three-page paper describing a good environment in which to live. Include factors such as home, community, air, water, cleanliness, wildlife, plants, and animals.
5. Ask your teacher to invite a public health official to your class to discuss health problems in the community and how they could be reduced by improving the environment.
6. Draw a chart that illustrates some relationships among plants, animals, trees, soil, water, air, and people.
7. Look up three prominent U.S. inventors and describe the events that led to the inventions that made them famous.
8. Make a model of one of the machines that strongly influenced agriscience development.
9. Make a collage depicting some important discoveries, inventions, and developments in agriscience.
10. Assume that a bar graph depicting China's population for 1975 was pyramid-shaped, similar to that of Honduras. Make a bar graph representing what the population pattern will look like after two generations of 4–2–1 families.
11. Ask your teacher to arrange a field trip to study the variety of living environments in your community.
12. In groups of three or four students, develop an idea for a new piece of farm equipment that will improve the harvest or production of a locally grown crop, farming technique, or ranching practice.

SELF-EVALUATION

A. MULTIPLE CHOICE

1. To be regarded as safe, a sewage system must
   a. be connected to a city system.
   b. be constructed from concrete block.
   c. decompose human waste.
   d. discharge into a stream or river.

2. Safe water is
   a. any water pumped from wells.
   b. water collected from a roof.
   c. free of harmful chemicals and organisms.
   d. water taken from free-flowing rivers.
3. Apartments on one level in large buildings and owned by the residents are called
   a. condominiums.
   b. single houses.
   c. townhouses.
   d. villas.
4. The world's food production capability indicates that
   a. it could feed itself in theory, but not in practice.
   b. it probably could never keep up with population growth.
   c. food supplies outpace demand, and reduced production is recommended.
   d. widespread famine could not be helped by better distribution.
5. Contaminants of food and water include
   a. registered pesticides.
   b. contact by cockroaches.
   c. feces and urine.
   d. all of the above.
6. Although plants cannot move to food and water, they survive because of
   a. their capacity to reproduce.
   b. their ability to survive without food and water.
   c. their roots, which extract water from any material.
   d. parasites that convert water to nutrients.
7. The world's population is projected to increase to 8 billion people by
   a. 2050.
   b. 2025.
   c. 2022.
   d. 2034.
8. Agriscience in the United States
   a. accounts for 14 percent of the jobs.
   b. is likely to diminish in importance.
   c. reduces our standard of living.
   d. is being replaced by biotechnology.
9. Agricultural development would be described as
   a. rapid until about 1850.
   b. slow in most countries.
   c. very rapid in the United States in recent years.
   d. rapid in the past, but decreasing today.
10. The inventor of the iron plow was
    a. Cyrus McCormick.
    b. John Deere.
    c. Joseph Glidden.
    d. Thomas Jefferson.
11. The combine is a combination of the reaper and a
    a. corn picker.
    b. cotton gin.
    c. cotton picker.
    d. threshing device.
12. "Yellow jewel" is the name given by the Chinese to
    a. a special type of horse.
    b. a very young emperor.
    c. garden peas.
    d. soybeans.
13. Beltsville Small White is a
    a. breed of rabbit.
    b. breed of turkey.
    c. type of building.
    d. variety of soybean.
14. The great advance in world food production in the 1960s was called the
    a. biological attractants.
    b. Green Revolution.
    c. Greening of America.
    d. Great Leap Forward.

B. MATCHING

GROUP 1

_____	1. Neighborhood	a. Release oxygen into the air
_____	2. Starvation	b. A priority in China
_____	3. Parasite	c. Registered pesticides in one bay area
_____	4. Immune	d. Part of a community
_____	5. DDT	e. Not harmed by
_____	6. 10,000	f. Famine
_____	7. Cockroach	g. Lives on another organism
_____	8. Reforestation	h. Banned insecticide
_____	9. Plants	i. Feeds on human waste and food

GROUP 2

_____	1. Aerosol	a. Sense nutrient deficiencies
_____	2. Barbed wire	b. "Bug bomb"
_____	3. Coccidiosis	c. Nitrogen fixation
_____	4. Cotton gin	d. Disease of poultry
_____	5. Impatiens	e. Sharp points to discourage livestock
_____	6. Laser	f. Colorful flower
_____	7. Legume	g. Eli Whitney
_____	8. Low-input agriculture	h. Curd from soybeans
_____	9. Mastitis	i. John Deere
_____	10. Milking machine	j. New research objective
_____	11. Steel moldboard plow	k. Cyrus McCormick
_____	12. Reaper	l. Infection of milk-secreting glands
_____	13. Tofu	m. Anna Baldwin

UNIT 3
Biotechnology

OBJECTIVE

To examine elements of biotechnology.

MATERIALS LIST

- paper
- pencil or pen
- encyclopedias, agriscience magazines
- Internet access

COMPETENCIES TO BE DEVELOPED

After studying this unit, you should be able to:

- define biotechnology, DNA, and other related terms.
- compare methods of plant and animal improvement.
- discuss historic applications of biotechnology.
- explain the concept of genetic engineering.
- describe applications of biotechnology in agriscience.
- state some safety concerns and safeguards in biotechnology.

SUGGESTED CLASS ACTIVITIES

1. Introduce this unit by serving one or more small portions of bread, yogurt, or cheese to class members. Explain to the class that each of these foods is a product of biotechnology because living organisms act on the food to preserve it. Name as many food products as you can that are products of biotechnology.
2. Describe the controversy concerning the use of biotechnology methods to improve the productivity of food plants and animals. Identify some Web sites where students may learn more about the production of food by plants and animals that have been improved using genetic engineering techniques.

TERMS TO KNOW

3. Scientists have a working knowledge of the cloning process. Recently, scientists have cloned equine. Now that it is possible to clone a mammal by replicating its chromosomes, conduct a class discussion of ethics in biotechnology.

bio
improvement by selection
selective breeding
genetics
heredity
gene
generation
progeny
deoxyribonucleic acid (DNA)
nucleic acid
base
adenine (A)
guanine (G)
cytosine (C)
thymine (T)
gene splicing
recombinant DNA technology
gene mapping
clone
genetic engineering
bovine somatotropin (BST)
porcine somatotropin (PST)

INTERNET KEY WORDS: Mendel, genetics

INTERNET KEY WORDS: genetic engineering

Biotechnology has become an important tool in agriscience. It promises unprecedented advancements in plant and animal improvement, pest control, environmental preservation, and life enhancement. However, there are real dangers that this new power over life processes can lead to unmanageable consequences in careless, uninformed, or criminal hands. Therefore, governments, scientists, agencies, corporations, and individuals have moved cautiously in the pursuit of new benefits through biotechnology.

Bio means life or living; therefore, biotechnology is the application of living processes to technology. Although many definitions abound for biotechnology, one of the more popular definitions is the use of microorganisms, animal cells, plant cells, or components of cells to produce products or carry out processes with living organisms.

HISTORIC APPLICATIONS OF BIOTECHNOLOGY

Living organisms have been used for centuries to alter and improve the quality and types of food for humans and animals. Examples include the use of yeast to make bread rise, bacteria to ferment sauerkraut, bacteria to produce dozens of types of cheeses and other dairy products, and microorganisms to transform fruit and grains into alcoholic beverages (Figure 3-1). Similarly, green grasses and grains have been stored in airtight spaces and containers, such as silos, where bacteria convert sugars and starches into acids. The acids provide a desirable taste and protect the feed from spoilage by other microorganisms. The converted feed is called *silage* (Figure 3-2).

Courtesy of USDA/ARS #K36072-2.

FIGURE 3-1 Many foods owe their texture and taste to microorganisms such as yeast or bacteria.

IMPROVING PLANT AND ANIMAL PERFORMANCE

Improvement by Selection

History documents the domestication of the dog, horse, sheep, goat, ox, and other animals thousands of years ago. Improvement by selection soon followed. **Improvement by selection** means picking the best plants or animals for

© Wallentine/Shutterstock.com.

A

© Frontpage/Shutterstock.com.

B

FIGURE 3-2 Silage often consists of grains and/or green plant material preserved by the action of bacteria in an airtight environment.

producing the next generation (Figure 3-3). As people bought, sold, bartered, and traded, they were able to get animals that had desirable characteristics, such as speed, gentleness, strength, color, size, and milk production. By mating animals with characteristics that humans preferred, the offspring of those animals would tend to exhibit the characteristics of the parents and further intensify the desired characteristics. Whether by accident or through a very basic understanding of heredity, the owner was practicing **selective breeding**, or the selection of parents to get desirable characteristics in the offspring.

The chariot armies of the Egyptians and Romans, the might of the Chinese emperors, the speed of the invading barbarians into northern Europe, the strength of mounts carrying armored knights into battle, and the evasive Arabians of the desert all provide convincing testimony to early successes at breeding horses for specific purposes.

Improvement by Genetics

An Austrian monk named Gregor Johann Mendel is credited with discovering the effect of genetics on plant characteristics. **Genetics** is the science of heredity.

Courtesy of National FFA; FFA #162.

FIGURE 3-3 Improvement by selection means picking the best specimens for breeding purposes.

Heredity is the transmission of characteristics from an organism to its offspring through genes in reproductive cells. **Genes** are components of cells that determine the individual characteristics of living things. Mendel experimented with garden peas. He observed that there was definitely a pattern in the way different characteristics were passed down from one generation to another. **Generation** refers to the offspring, or **progeny**, of common parents.

In 1866, Mendel published a scientific article reporting the results of his experiments. He had discovered that certain characteristics occurred in pairs, for example, short and tall in pea plants. Furthermore, he observed that one of those characteristics seemed to be dominant over the other. If tall was the dominant characteristic, then tall plants crossed with tall or short plants produced mostly tall plants, but some plants would still be short. It was observed that the short characteristic could be hidden in tall plants in the form of a recessive gene. Such recessive genes could not express themselves in the form of a short plant unless both genes in the plant cells were the recessive genes for shortness. He also observed that short plants crossed with short plants always had short plants as offspring. This happened because there were no tall characteristics in either parent to dominate the characteristic of the offspring. Mendel's work provides an excellent example of the power of the written word. His discoveries and conclusions would have been lost if they had not been recorded. The usefulness of his discoveries was not recognized until long after his death. In 1900, other scientists reviewed his writings and built upon the observations and conclusions he had reported. Today, biologists credit his work as being the foundation for the scientific study of heredity (Figure 3-4). Principles of heredity apply to animals as well as plants. More information on the principles of heredity may be found in subsequent units of this text.

Courtesy of National FFA; FFA #199.

FIGURE 3-4 Mendel's extensive experimentation, observation, and recordkeeping provided the foundation for the modern science of genetics.

DNA—Genetic Code of Life

Of the estimated 300,000 kinds of plants and more than 1 million kinds of animals in the world, all are different in some ways. Conversely, plants and animals have certain similar characteristics that lend themselves to classification and permit prediction

AGRI-PROFILE

CAREER AREA: GENETIC ENGINEERING

Courtesy of USDA/ARS #K-1968-13.

Using a DNA probe, an animal physiologist examines film showing gene patterns of various animals.

Genetic engineering cuts across many fields of endeavor. Procedures for genetic modification of organisms have been developing for more than a decade. Biologists, microbiologists, plant breeders, and animal physiologists are some examples of specialists who might use genetic engineering in their work. The work settings include the field, laboratory, classroom, and commercial operations.

People involved in genetic engineering usually have advanced degrees, and they are highly specialized in a narrow area of research, such as cellular biology. Others may work in applied research in areas such as weevil control in small grains, nutrient requirements of small grains, or reproductive problems in dairy cattle. Others may be crop, animal, or pest-control technicians who help manage the plants or animals that are the subjects of research. Still others may work in laboratories and devote most of their lives to analysis and observation.

Because genetic engineering is a relatively new field and the applications are so numerous, opportunities are expanding as the field develops.

of characteristics of offspring by viewing the parents—that is, the individual fertilized cell, called the embryo, contains coded information that determines what that cell and its successive cells will become. The coded material in a cell is called DNA. DNA is the acronym, or abbreviation, for **deoxyribonucleic acid**.

SCIENCE CONNECTION TRAIT PREDICTABILITY

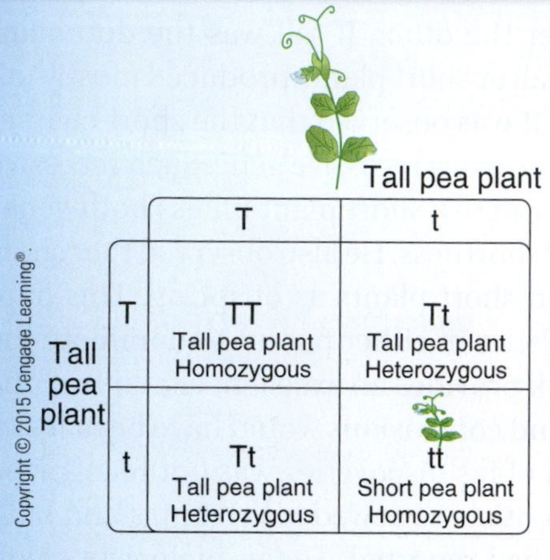

A *trait* is another word used to describe a characteristic of an organism. Mendel discovered that parents pass their own traits to their offspring. In this type of reproduction, each parent contributes one half of the genes. The result of parental combinations can be shown and predicted in a Punnett square. The Punnett square represents the results of some of Mendel's experiments. An allele is one of the forms in which gene pairs can occur. In this example, there are three possible alleles. (1) "TT" represents a pea plant that is tall. Homozygous means that an organism has the same alleles for a given trait. (2) "tt" represents a homozygous pea plant that appears short. (3) "Tt" represents a heterozygous pea plant that is tall. The gene for "tall" is expressed because "T" is dominant over "t." Because of this dominance, both "TT" and "Tt" will result in a tall plant, and "tt" will result in a short plant. Heterozygous means that an organism has different alleles for a given trait. Because the alleles for this trait are heterozygous, or different, this organism is called a hybrid. The Punnett square cross is done one box at a time; by taking the parental alleles to the left and top of each box, the offspring allele combinations are found inside each of the four boxes.

It is believed that a universal chemical language or code unites all living things. It was observed in the early 1800s that all living organisms are composed of cells, and that cells of microscopic organisms, as well as larger plants and animals, are basically the same. In 1867, Friedrich Meischer observed that the nuclei of all cells contain a slightly acidic substance. He named the substance **nucleic acid**. Later, the name was expanded to deoxyribonucleic acid, or DNA. DNA in all living cells is similar in structure, function, and composition and is the transmitter of hereditary information. A gene is a small section of DNA that is responsible for a specific trait. Chromosomes are rod-like structures made of DNA and other substances that hold genes.

DNA occurs in pairs of strands intertwined with each other and connected by chemicals called **bases**. The pairs of DNA strands may be likened to the two sides of a wire ladder. The bases may be likened to the rungs of that wire ladder. The different bases are composed of only four chemicals: (1) **adenine**, (2) **guanine**, (3) **cytosine**, and (4) **thymine**. The first letters of each name of the bases—A, G, C, and T—have become known as the genetic alphabet of the language of life (Figure 3-5).

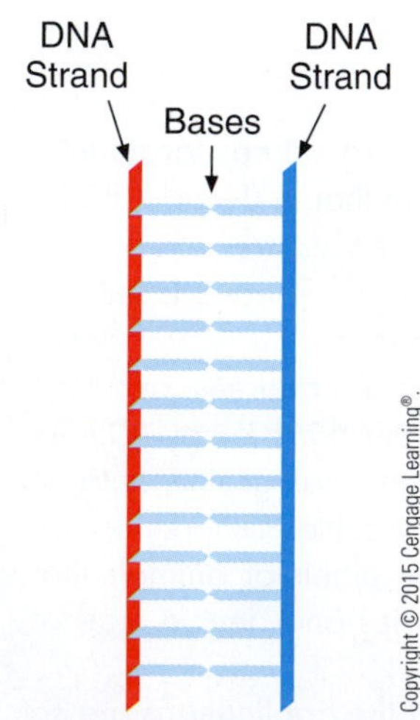

FIGURE 3-5 The components of DNA may be likened to a ladder with very close rungs.

If one end of the wire ladder is held while the other end is twisted, the resulting shape would be called a double helix. This is the shape of DNA strands in a cell. Two strands of DNA and the bases between the strands compose a specific gene (Figure 3-6). The order or sequence of the bases between the DNA strands is the code by which a gene controls a specific trait. Therefore, each rung with its accompanying side pieces of DNA constitutes a gene containing the genetic code to a single trait (Figure 3-7). The genetic material in the cells of a given microbe, plant, animal, or human can be isolated and observed. The trait or traits that a

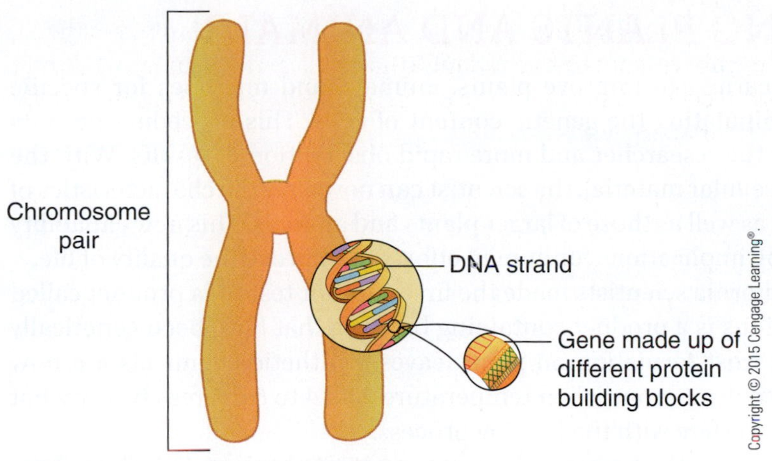

FIGURE 3-6 A gene is a small segment of DNA located at a specific site on a chromosome that influences or controls a hereditary trait.

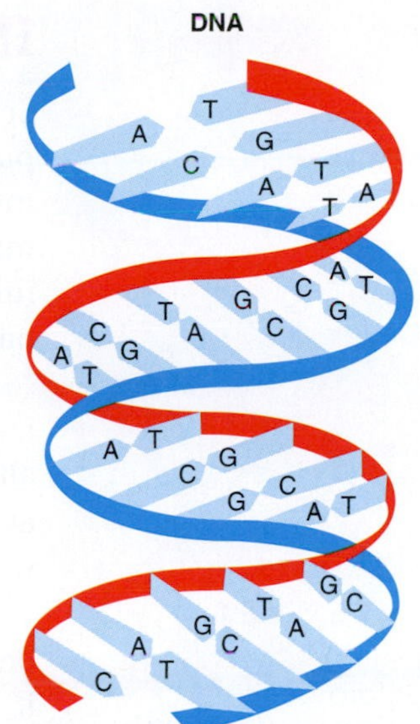

FIGURE 3-7 A chromosome is a structure containing the genetic information of a cell. The DNA is wound tightly with other substances to form the chromosome. A gene is made of DNA. There are thousands of genes on each strand of DNA.

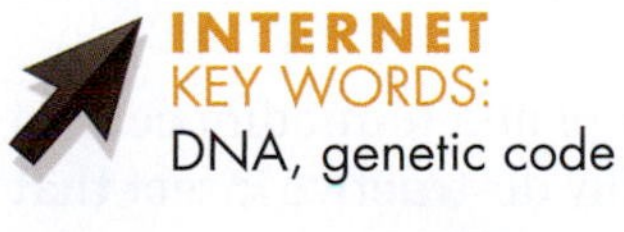

INTERNET KEY WORDS: DNA, genetic code

INTERNET KEY WORDS: transgenic animals

given gene controls can be identified, and the combination of genes that influence a single trait can be determined.

Some examples of individual traits are hair color, tendency for baldness in humans, height of plants at maturity, and tendency of females to have multiple births in contrast with single births.

As a cell divides, the DNA strands separate from each other and create duplicate strands to go to each of two new cells. Therefore, the genetic codes are duplicated and passed on from old cells to new cells as growth occurs and individuals reproduce (Figure 3-8). The process of identifying the location of a specific gene on a chromosome is called mapping.

Scientists can now identify an individual gene carrying specific genetic information and replace it with a gene containing a different genetic code. By doing so, a given characteristic or performance can be altered in the organism. For instance, plants that are susceptible to being eaten by certain insects may be altered so they will have resistance to that insect. The process of removing particular DNA segments and inserting new genes into a DNA sequence is called **gene splicing**/, or **recombinant DNA technology**. The process of finding and recording the location of genes is called **gene mapping**.

Scientists have a working knowledge of how genetic information is stored in a cell, duplicated, and passed on from cell to cell as cells divide and new organisms are formed. Furthermore, the process of transmitting genetic codes from parents to offspring and from parent to clone is common scientific knowledge. A **clone** is an exact duplicate of something. A major breakthrough was made in the early 1980s as scientists developed the process of genetic engineering. **Genetic engineering** is the movement of genetic information in the form of genes from one cell to another.

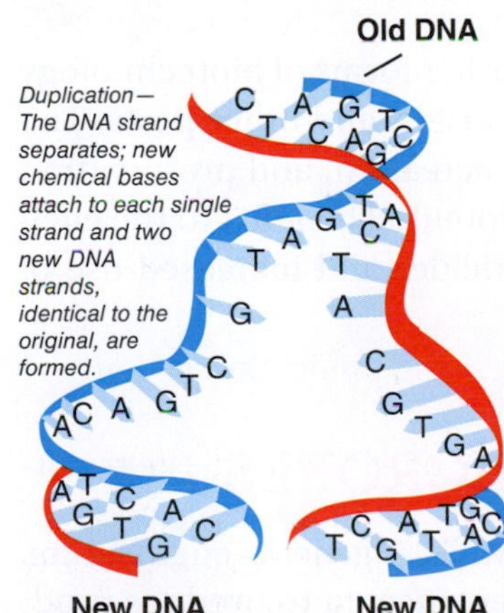

FIGURE 3-8 DNA strands divide, and bases attach themselves to the new strands to form identical genes for new cells.

IMPROVING PLANTS AND ANIMALS

Scientists have learned to improve plants, animals, and microbes for specific purposes by manipulating the genetic content of cells. This procedure permits more choices for the researcher and more rapid observation of results. With the manipulation of cellular material, the scientist can now alter the characteristics of microorganisms, as well as those of larger plants and animals. This new capability has some amazing implications for human efforts to improve the quality of life.

In 1988, California scientists made the first outdoor tests of a product called ice-minus. Ice-minus is a product containing bacteria that have been genetically altered to retard frost formation on plant leaves. Synthetic chemicals are now available to protect fruit crops when temperatures fall 4 to 6 degrees below what would normally interfere with the fruiting process.

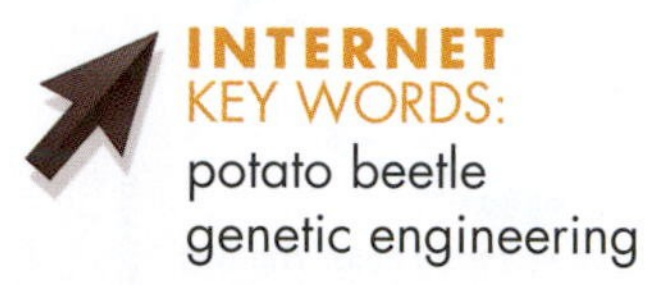

In animal science, the hormone **bovine somatotropin (BST)** has long been known for its stimulation of increased milk production in cows. However, it was not available for commercial use until bacteria were altered to produce the hormone at a reasonable cost. Another example of hormone production by genetically altered bacteria is an animal hormone called **porcine somatotropin (PST)**, which increases meat production in swine.

Each time that humans or animals are exposed to a disease, some individuals do not become infected. Sometimes an entire population is found to be resistant to a disease that is highly contagious to other populations of the same species. In some instances, the disease resistance is due to a single gene that has mutated or changed. It is now possible to identify the location of a resistant gene on a chromosome and to isolate it. This new genetic material can be transferred successfully to the chromosomes of an organism that is susceptible to the disease. Such a genetically altered individual is capable of passing the disease resistance to its offspring.

It is also possible to alter the genetic material in plants to modify certain traits. For example, the Colorado potato beetle is a highly destructive insect that can destroy all the plants in a potato field. A gene has been identified that causes potato plants to produce a substance in the leaves that is toxic to the potato beetle but which does not affect humans who consume the tuberous root. New potato varieties were created by inserting this gene into the DNA of commercial potato varieties. The new potato varieties are toxic to the beetles that eat the leaves of the modified plants.

It is now evident that genetic engineering and other forms of biotechnology hold great promise in controlling diseases, insects, weeds, and other pests. The plants and animals that we nurture for food, fiber, recreation, and preservation can benefit from biotechnology as well as humans. Potential benefits to the environment include less frequent use of chemical pesticides and increased use of biological controls (Figure 3-9).

SOLVING PROBLEMS WITH MICROBES

Microscopic plants and animals lend themselves well to genetic engineering. Microbes reproduce quickly and can be genetically engineered to produce products needed by other plants, animals, and humans. One of the first commercial products made by genetic engineering was insulin. Insulin is a chemical used by people with diabetes to control their blood sugar levels. Previously, insulin was

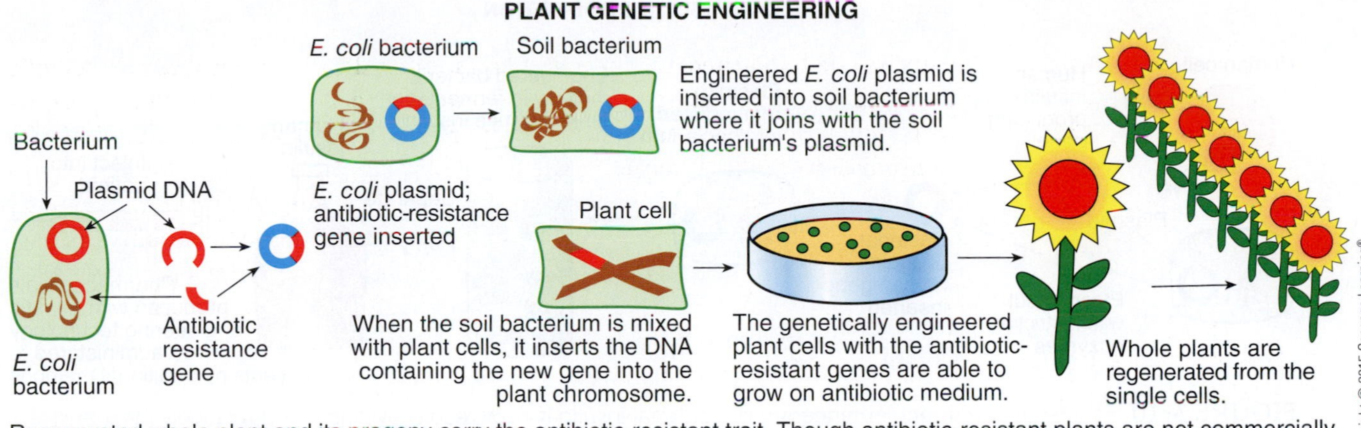

FIGURE 3-9 Plant genetic engineering is widely used in the battle against diseases, insects, weeds, and other pests.

HOT TOPICS IN AGRISCIENCE

TRANSGENIC ANIMALS—A NEW KIND OF FARMING

Transgenic animals have been modified using genetic engineering methods to express genes that are not naturally found in the animal. Because the new gene is inserted into the chromosomes of the transgenic animal, it may be passed on to its offspring. Insulin and growth hormone are some of the first products for human use to be produced in this way. Most of the medical products produced by transgenic animals consist of proteins that are separated from milk or blood. These products include Human Protein C, an anticlotting protein that dissolves blood clots in humans. Other products include hemoglobin, a blood substitute produced in transgenic pigs, and Factors VIII and XI, which cause clotting in human blood to stop severe bleeding in people who have disorders such as hemophilia. All of these drugs are tested to ensure they are safe, and they offer hope for medical breakthroughs of even greater magnitude. The transgenic animal becomes a "living drug factory" that produces human proteins in milk or blood.

INTERNET KEY WORDS: cloned animals

available only from animal pancreas tissue. It was in short supply and was very expensive. However, a bacterium called *Escherichia coli* was genetically engineered to produce insulin for human use (Figure 3-10). This important diabetes medication is isolated from a solution containing the engineered bacteria and it is purified for human use.

WASTE MANAGEMENT

Environmental pollution and the elimination of waste products from home, business, industry, utility, government, military, and other sources has become a major problem throughout the world. Landfills are becoming full, and pollutants from garbage are causing problems such as leakage into the groundwater. Old dump sites are creating new problems, waste is piling up, and sewage and chemical disposal is a constant problem. Large floating masses of garbage pollute the world's oceans. Current environmental laws mandate reductions in solid waste

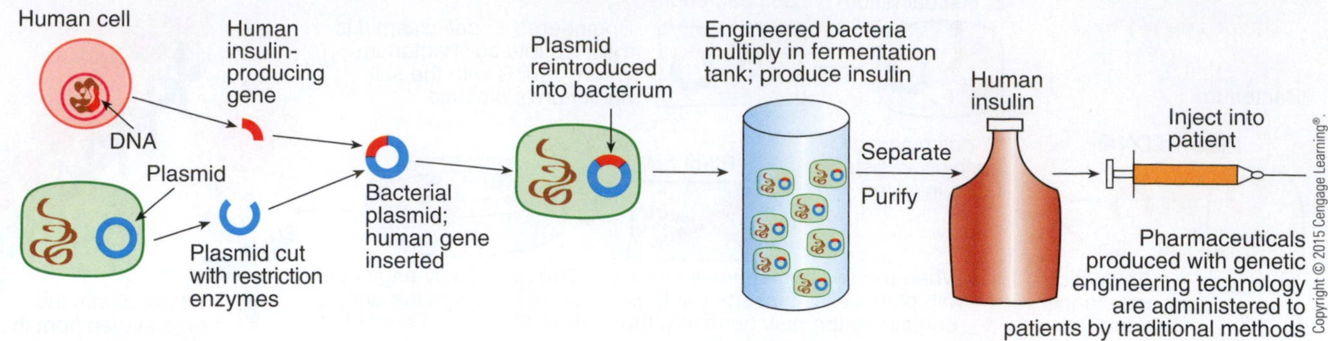

FIGURE 3-10 As a result of genetic engineering of bacteria, insulin is now readily available and relatively inexpensive.

INTERNET KEY WORDS: safe products, biotechnology

INTERNET KEY WORDS: biotechnology ethics

disposal of 40 percent or more in comparison with 1990. It is evident, however, that no matter what we do in America, waste management is an international problem that must be solved by the nations of the world.

Biotechnology can be used to help solve waste disposal problems. Genetically altered bacteria are used to feed on oil slicks and spills, transforming this serious pollutant into less harmful products. Similarly, bacteria have been developed that are capable of decomposing or deactivating dioxin, polychlorinated biphenyl (PCB), insecticides, herbicides, and other chemicals in our rivers, lakes, and streams. Bacteria are capable of converting solid waste from humans and livestock into methane fuel that is used to generate electricity or to heat buildings.

Although great progress has been made in pollution reduction in some areas, pollution is still one of the world's greatest problems. Biotechnology has brought some spectacular breakthroughs in the use or decomposition of waste materials (Figure 3-11).

FIGURE 3-11 Biotechnology is used to modify bacteria, making it possible for them to break down crude oil. These bacteria have become important in reducing the harmful effects on the environment when oil or chemical spills occur.

BIO-TECH CONNECTION CLONING OF A MULE

Idaho Gem is the first mule ever cloned. His DNA came from a male champion racing mule, and his surrogate mother is a mare.

One of the most remarkable scientific accomplishments of the twentieth century was the successful cloning of "Dolly" the sheep. Dolly was cloned in the laboratory from a single cell obtained from the udder of another sheep. A clone is an exact genetic copy of another living organism, so Dolly was genetically identical to the sheep from which the cell was obtained. Now scientists at the University of Idaho have cloned a mule named Idaho Gem. He is the first clone from the horse family. He also is a clone of a hybrid animal that is incapable of reproduction. This particular mule is cloned from the tissue of a fetus of the same parents as a champion racing mule named Taz. The DNA from his mule heritage was fused with an egg cell from a mare of the horse species, and a mare carried him through pregnancy and nurtured him as a newborn foal. In addition, two more identical clones were born later. Idaho Gem and one of his identical brothers, Idaho Star, raced successfully on the professional mule racing circuit.

What was the significance of this and other cloning events? It is now possible to produce a new generation of animals, with each animal identical to the most productive animal in the herd. Imagine having a whole herd of cows just like your best cow. Confusing, perhaps? Even their spots would be the same. You might have to give each animal a permanent microchip (or other identification) just to identify it.

SAFETY IN BIOTECHNOLOGY

Federal and state governments monitor biotechnology research and development very closely. Much anxiety has been expressed about the perceived dangers of genetically modified organisms. Therefore, appropriate policies, procedures, and laws have been developed as biotechnology has evolved. Many of these regulations have been developed by the Environmental Protection Agency (EPA). Research priorities and initiatives require discussion and interaction by scientists, government agencies, and other authorities. Products are tested in laboratories, greenhouses, and other enclosures before being approved for testing outdoors and in other less controlled environments. Even then, outdoor tests are first conducted on a small scale in remote places under careful observation. Under these conditions, the efficiency, safety, control, and environmental impact of new organisms are determined. If the new organism poses an unmanageable threat, it can be destroyed.

Customer resistance to new food products developed through biotechnology has been demonstrated since the first of these foods arrived at supermarkets. For example, some customers demanded that milk from cows treated with the biotech product BST should be labeled to distinguish it from other milk. Some believed that it should not be marketed at all. Despite assurances from the Food and Drug Administration (FDA) that no differences exist in comparisons of BST milk and milk from untreated cows, some customer resistance to BST still persists. A few milk processors initially joined the resistance movement because they were afraid that their products might be boycotted by consumers. In some instances, these processors refused milk shipments from dairy farms that treated their cows with BST. Biotechnology is rapidly becoming an important part of our daily lives. Many of its potential benefits have already been realized, and most people believe that we have only scratched the surface. With proper safeguards, we can look confidently to a bright future in this emerging field. Many other applications of biotechnology are cited throughout this text.

Customer resistance to biotech food products appears to be diminishing, both in the United States and in other nations. Some nations that refused bioengineered products are no longer doing so. However, there remains an active coalition of citizens and organizations that is actively attempting to place regulatory controls on genetically modified food products.

ETHICS IN BIOTECHNOLOGY

Ethics is a system of moral principles that defines what is right and wrong in a society. The ability to manipulate the genetics of living organisms raises important ethical questions about how the technology should be used. For example, should humans be cloned from the strongest athletes or the smartest scholars? Would it be right to sell cloned embryos to parents who are carriers for a known genetic defect so that they might have children who are free of the defect? Is it morally right or wrong to mass-produce clones of popular human beings in a laboratory? Imagine what Hitler might have done to create his "master race" if modern biotechnology methods had been known to his scientists.

It seems appropriate that a discussion of ethics should be part of the biotechnology revolution that is occurring. Such a discussion would help scientists and consumers decide how ethical issues related to biotechnology should be handled. At the very least, we can expect new laws to be passed and courtroom decisions to be rendered on the basis of ethics in biotechnology.

SCIENCE CONNECTION "FINGERPRINTING" ORGANISMS

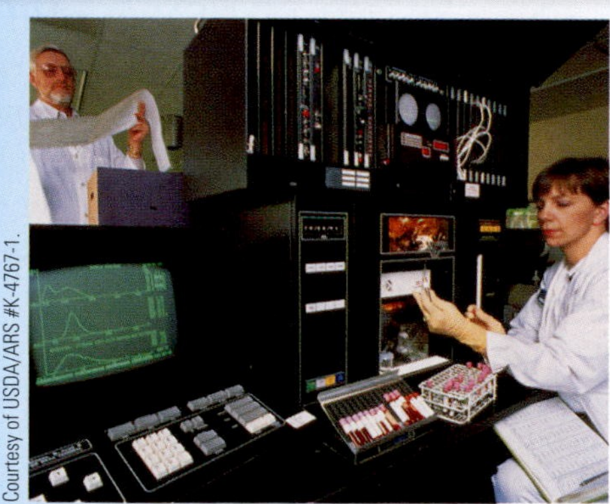
Courtesy of USDA/ARS #K-4767-1.

DNA analysis makes "fingerprinting," or individual identification, possible in all organisms.

USDA microbiologists at the National Animal Disease Center in Ames, Iowa, have been able to "crack" the mysteries of some of the worst known cases of food poisoning in North America. DNA matching, or "fingerprinting," is used to link (1) the persons who became ill, (2) the contaminated food that caused the poisoning, (3) the place where the food had been contaminated, and (4) the materials from which the food poisoning organisms had originated and spread. Food poisoning is a life-threatening condition resulting from eating food containing toxic material produced by bacteria under unsanitary conditions. *Salmonella, Campylobacter, Staphylococcus aureus, Clostridium perfringens, Vibrio parahaemolyticus, Listeria monocytogenes, Bacillus cereus,* and enteropathogenic *E. coli* are the most common bacteria to infect food.

Food-borne illnesses, though rare, have been known to be deadly. According to the U.S. Centers for Disease Control and Prevention, up to 48 million people get sick each year from contaminated food. Of these, approximately 128,000 are hospitalized, and 3,000 die from food-borne illnesses. Not all of these instances are traced back to farms and processors; a significant number occur because food is not properly refrigerated or prepared at home. Leaving food out on the counter overnight or failing to cook it adequately gives toxin-forming organisms a head start.

STUDENT ACTIVITIES

1. Write the Terms to Know and their meanings in your notebook.
2. Read an article in an encyclopedia or other reference on the process of genetic engineering.
3. Report your findings from Activity 2 to the class.
4. Make a collage depicting some important discoveries, inventions, and developments in biotechnology.
5. Form a discussion group to explore the benefits and hazards of biotechnology.
6. Arrange for a resource person to speak on the ethical and moral issues surrounding developments in biotechnology.
7. Organize a class debate on the ethical and moral issues regarding research and the use of new discoveries in biotechnology.

SELF-EVALUATION

A. MULTIPLE CHOICE

1. Bio means
   - a. a study of.
   - b. life.
   - c. three.
   - d. science.
2. An example of a fermented food is
   - a. applesauce.
   - b. bologna.
   - c. cheese.
   - d. coffee.
3. The earliest method of livestock improvement was probably by
   - a. biotechnology.
   - b. crossbreeding.
   - c. gene splicing.
   - d. selection.
4. The person providing the foundation for scientific study of heredity was
   - a. Gregor Johann Mendel.
   - b. George Washington Carver.
   - c. Joseph Glidden.
   - d. Thomas Jefferson.
5. The genetic code of life is
   - a. clone.
   - b. DNA.
   - c. progeny.
   - d. thymine.
6. Adenine, guanine, cytosine, and thymine are all
   - a. acids.
   - b. bases.
   - c. DNA.
   - d. genes.
7. Recombinant DNA technology is also known as
   - a. bovine somatotropin.
   - b. gene splicing.
   - c. porcine somatotropin.
   - d. X-Gal.
8. Genetic engineering can be done to change
   - a. animals.
   - b. microorganisms.
   - c. plants.
   - d. all of the above.
9. An important contribution of biotechnology to waste management is
   - a. bacteria that consume oil.
   - b. disease-resistant bacteria.
   - c. ice minus bacteria.
   - d. human bacteria.

10. Chemical pollutants in water that may be decomposed or deactivated by bacteria include
    a. chlorine.
    b. fluorides.
    c. iron.
    d. PCBs.

B. MATCHING

_________	1. Fruits and grains	a. Bacteria linked to food poisoning
_________	2. Yeast	b. Controls blood sugar levels
_________	3. Silage	c. Used to make alcoholic beverages
_________	4. Genetics	d. Embryo
_________	5. Genes	e. Result of dominant gene
_________	6. Tall pea plants	f. Deoxyribonucleic acid
_________	7. DNA	g. Causes bread to rise
_________	8. Fertilized cell	h. Fermented grains or forage
_________	9. Insulin	i. Heredity
_________	10. Salmonella	j. DNA and bases

SECTION 2

YOU AND AGRISCIENCE

The "agricultural industry" can use your energy in science, engineering, business and financial management, production, renewable natural resources, communications, or other challenging and rewarding careers.

Futurists predict that during the next quarter century, the most important discoveries in genetic engineering will be made in agriscience. In agriscience, you could help develop plants and animals that grow better and more efficiently, even in adverse situations. Nutritionists and food scientists study links among food, diet, and health. As one of them, you might work to ensure that new convenience foods are nutritious and healthy; or, you might design a new way of preserving, processing, or packaging food. Agriscience educators teach in high schools, colleges, and universities. They also work for land-grant universities as extension educators. The job of agriscience education is to teach new and proven scientific practices to students and farm families.

As a financial manager, you might work for a bank or a credit agency as an agricultural loan officer or for a company that sells supplies to producers and growers. As credit manager, loan officer, financial analyst, or marketing specialist, you could utilize your knowledge of business and finance, as well as production, processing, or distribution.

The farmers, ranchers, timber producers, and other growers are the foundation of the food and fiber system. To join their ranks is to play a most basic role in the future of our planet. For others who love the outdoors, a look at the work of foresters, range managers, game managers, fish and wildlife managers, park rangers, and crop scouts should be of interest. Like all parts of our world touched by science and technology, the food, fiber, and

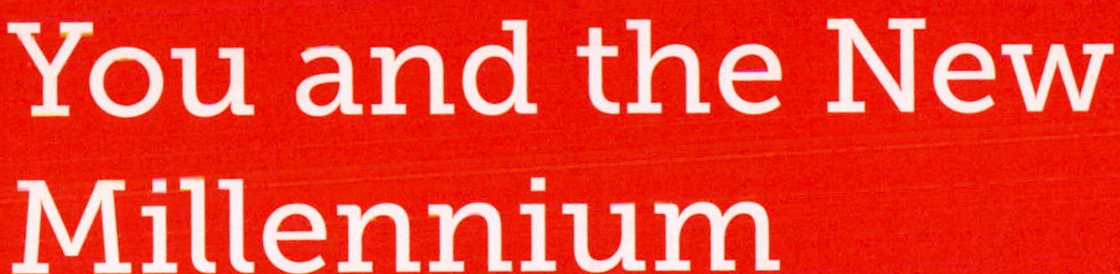

You and the New Millennium

renewable natural resources system is changing rapidly. Because it contributes to and is strongly influenced by trade and consumer lifestyles, the system must adjust continuously. If you get the appropriate training and experience, any of these careers can be yours.

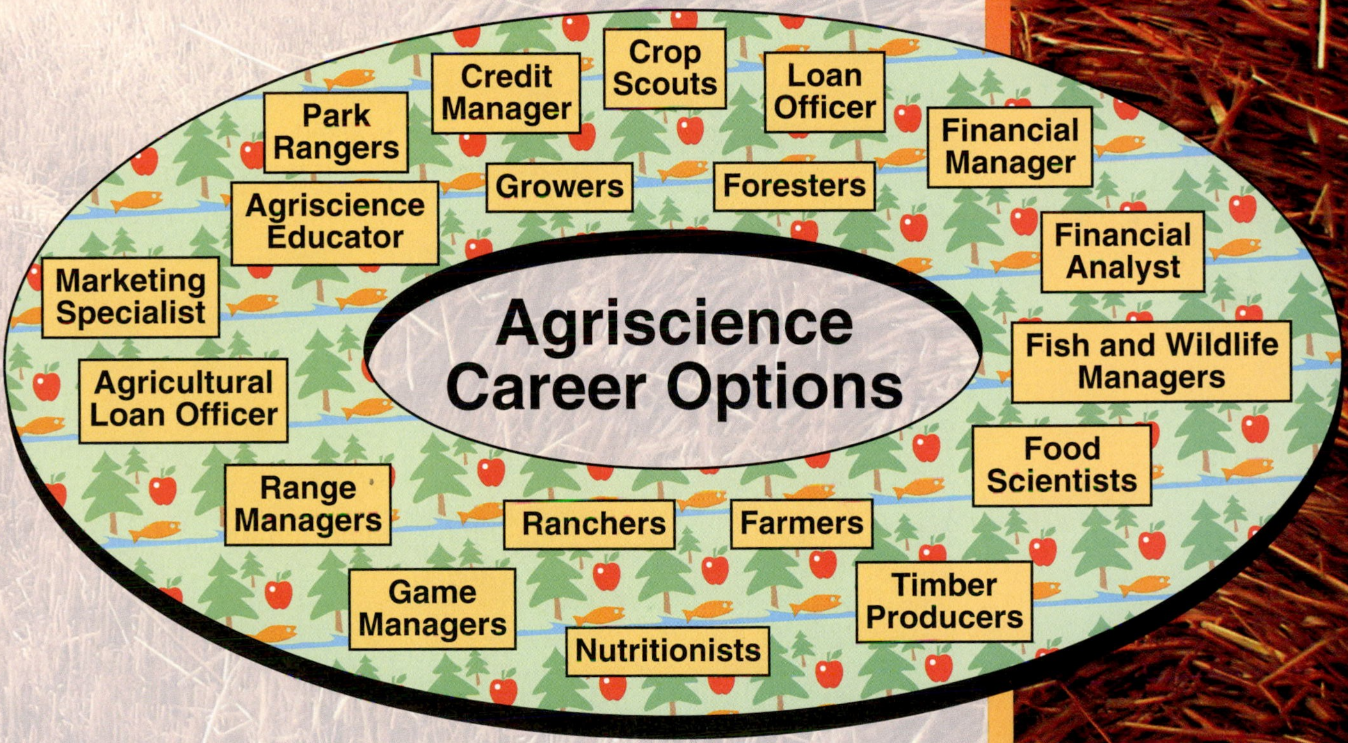

Agriscience Career Options

- Park Rangers
- Credit Manager
- Crop Scouts
- Loan Officer
- Financial Manager
- Agriscience Educator
- Growers
- Foresters
- Financial Analyst
- Marketing Specialist
- Fish and Wildlife Managers
- Agricultural Loan Officer
- Food Scientists
- Range Managers
- Ranchers
- Farmers
- Timber Producers
- Game Managers
- Nutritionists

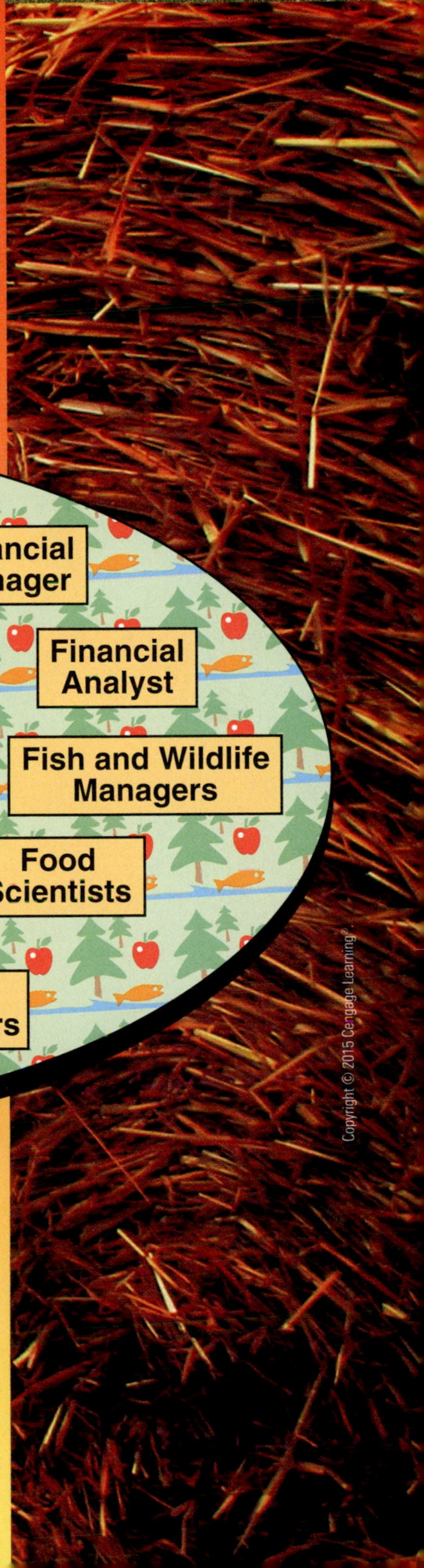

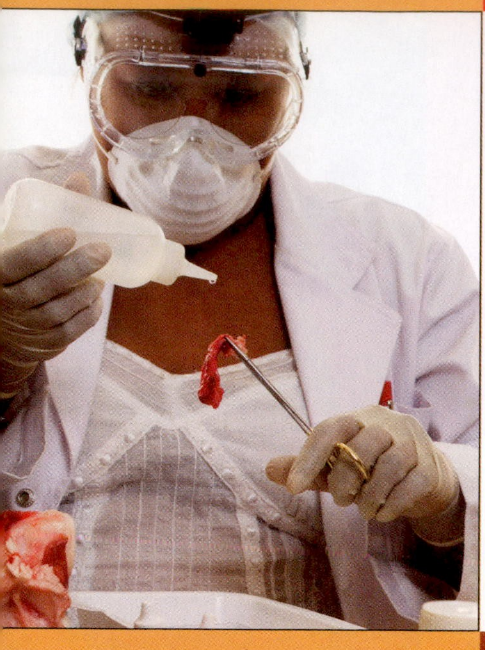

UNIT 4

Career Options in Agriscience

OBJECTIVE

To survey the variety of career opportunities in agriscience, observe how they are classified, and consider how you can prepare for careers in agriscience.

MATERIALS LIST

- paper
- pencil or pen
- bulletin board materials
- agriscience magazines and pictures
- *Occupational Outlook Handbook* or other agriscience career references
- Internet access

COMPETENCIES TO BE DEVELOPED

After studying this unit, you should be able to:

- define agriscience and its major divisions.
- describe the opportunities for careers in agriscience.
- compare the scope of job opportunities in farm and off-farm agriscience jobs.
- list activities in the middle school, high school, technical college/university to help prepare for agriscience careers.
- identify resource people for obtaining career assistance in agriscience.

SUGGESTED CLASS ACTIVITIES

1. Conduct a career day for agriscience classes. Invite agricultural professionals to present workshops about the career fields they represent. Assign class members to take notes on two or three workshops in which they are interested. Collect the notes for a class assignment and have the students keep them in their notebooks for later reference.
2. Invite the recruiter from the College of Agriculture at a state land-grant university to talk with the class about career opportunities in agriculture. Ask him or her to talk about what high school preparation is needed to succeed at the university. Encourage the recruiter to provide literature about specific career opportunities. Invite parents to attend. Repeat this exercise by inviting representatives of technical colleges to meet with students and parents.

TERMS TO KNOW

production agriculture
agricultural processing, products, and distribution
horticulture
forestry
agricultural supplies and services
agricultural mechanics
profession
agriscience professions

3. Make a list of 10 to 15 interests you currently have. As you read this unit, look for careers that may incorporate some of your interests or that are interesting to you. Write those careers next to the interests on your list and consider your future in those professions.

Life is possible without many of our modern conveniences but not without food. An adequate supply of suitable food and other products of the soil, air, and water is basic to life. This includes food for nourishment, fiber for clothing, and trees for lumber. Less obvious are alcohols for fuel and solvents, oils for home and industry, and oxygen for life itself. The industry that provides these vital basic commodities is agriculture. American agriculture is the world's largest commercial industry, with assets of nearly $1 trillion (Figure 4-1).

How Much Is One Trillion Dollars?

You can count $1 trillion ($1,000,000,000,000) by using the following procedure:

- One dollar bill every second
- Sixty bills per minute
- Thirty-six hundred bills per hour
- Eighty-six thousand per day
- Thirty-one million five hundred thirty-six thousand per year
- And continue counting for thirty-one thousand seven hundred and ten years!

FIGURE 4-1 The agricultural industry in the United States has assets of approximately $1 trillion ($1,000,000,000,000).

DEFINITION

Agriscience is a term that includes all jobs relating in some way to plants, animals, and renewable natural resources. Such jobs occur indoors and outdoors. They include people in banking and finance; radio, television, and satellite communications; engineering and design; construction and maintenance; research and education; and environmental protection. All are in the field of agriscience if their products or services are related to plants, animals, and other renewable natural resources.

PLENTY OF OPPORTUNITIES

Approximately 21 million people are employed in agriscience careers. About 400,000 people are needed each year to fill positions in this field. Of those vacancies, only 100,000 are currently being filled by people trained in agriscience (Figure 4-2). That means there are many opportunities for you. You can use what you learn today in your current job, on your farm, or in agriscience classes to go directly to full-time employment. However, if you choose to pursue a college degree in agriscience, many additional career opportunities will be open to you.

About 20 percent of the careers in agriscience require college degrees. Many professional careers in agriculture require four-year college or university degrees. These careers are in the fields of education, marketing, communications, production, social services, finance, management, science, engineering, and many others (Figure 4-3). Additional career opportunities are available to students who graduate with technical college certificates and degrees. Among these opportunities are careers as mechanics, sales representatives, field representatives, agriscience laboratory technicians, insurance adjustors, and many others. Many of these careers offer financial opportunities that are equivalent to those available to university graduates.

Careers That Help Others

Helping others is an extra bonus with a career in agriscience. Many of the jobs in processing, marketing, production, natural resources, mechanics, banking, education, writing, and other areas are people-oriented jobs. This means you have

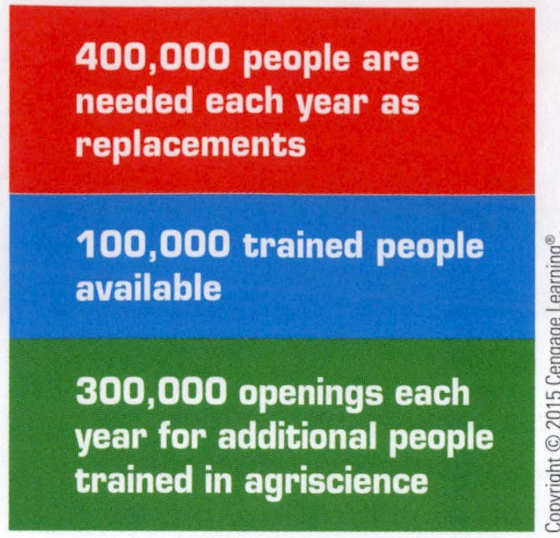

FIGURE 4-2 The employment outlook is good for people trained in agriscience.

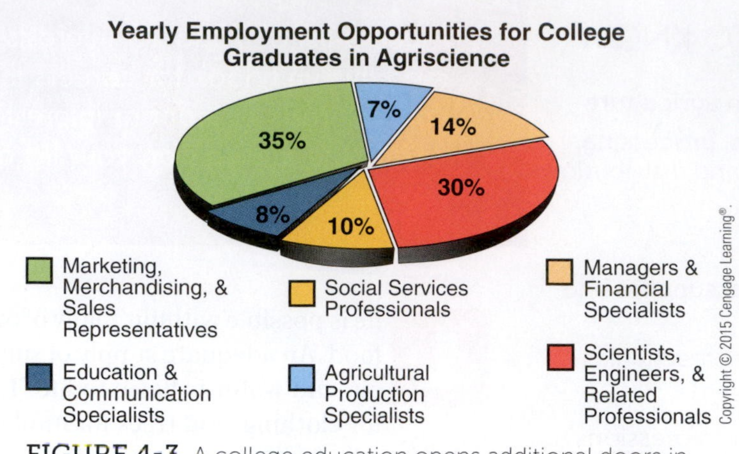

FIGURE 4-3 A college education opens additional doors in agriscience.

INTERNET KEY WORDS:
agricultural careers
SOICC

the extra benefit of being a product or process specialist and receive the special appreciation of others.

Careers That Satisfy

You might ask, "How can an agriscience career benefit me? What's in it for me?" Of course, there's the money. Salaries in agriscience vary tremendously from job to job. Generally, the better-qualified individuals will earn more. Agriscience industries employ one-fifth of all workers in the United States. And there are job openings for skilled individuals at various levels of expertise. You can be hired at the entry level as an agricultural mechanic right out of high school, or you can work for an advanced degree and be an agricultural engineer. You can do what you decide is best for you. Before you invest in a technical college or university education, it is a good idea to assess your talents and interests. What kind of work do you think you will enjoy doing? What special skills and talents do you have? How can your interests, skills, and talents be matched to a career in which your skills and talents can be developed and expressed? These are important considerations to think about before you enter advanced education and training programs.

Some valuable resources are available to help you explore your career options. One of these may be as close as the counseling center in your high school. Many high school counselors are trained to use computer programs to help identify careers that match your interests and talents. If your school does not have a computer-assisted program such as Career Information System (CIS), you may be able to identify some career choices using the Internet.

THE WHEEL OF FORTUNE

Agriscience is like a wheel with a large hub. The hub of that wheel is production agriculture, or farming and ranching. The rest of the wheel consists of the nonfarm and nonranch careers in agriscience. Because so many opportunities for rewarding careers exist in that wheel, it may be called a wheel of fortune.

Production Agriculture

Production agriculture is farming and ranching. It involves the growing and marketing of field crops and livestock. Careers in this area account for one-fifth of

FIGURE 4-4 The average U.S. farmer produces enough food and fiber for approximately 167 people.

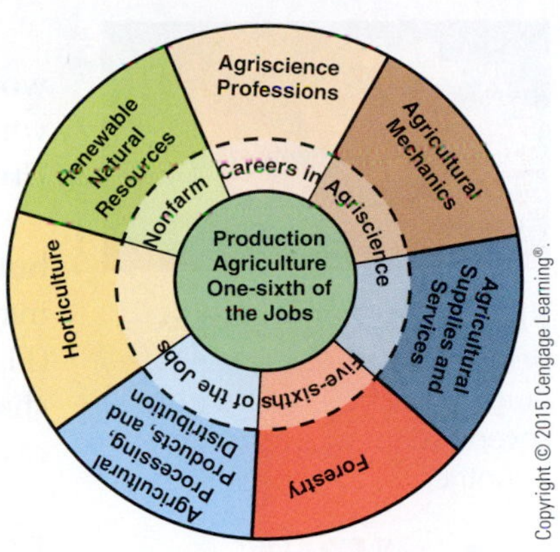

FIGURE 4-5 A wheel of fortune is a good way to illustrate agricultural careers.

all jobs in agriscience. Some estimates indicate the average U.S. farmer produces enough food and fiber for approximately 167 people (Figure 4-4). Large farm operators produce enough to feed more than 200 people.

Most other agriscience careers are involved with goods and services that flow toward or away from production agriculture. Workers in those careers permit U.S. farmers to supply goods so efficiently that U.S. consumers spent only 5.7 percent of their income on food in 2006. This is the lowest percentage in the world. Out of six workers in agriscience, five have jobs that are not on farms. The nonfarm agriscience jobs may be in rural, suburban, or urban settings. The agriscience wheel of fortune contains a hub with one-sixth of the agriscience workers in production on farms and ranches. The rest of the wheel contains the five-sixths of the agriscience jobs in nonproduction-type careers that are off the farm (Figure 4-5).

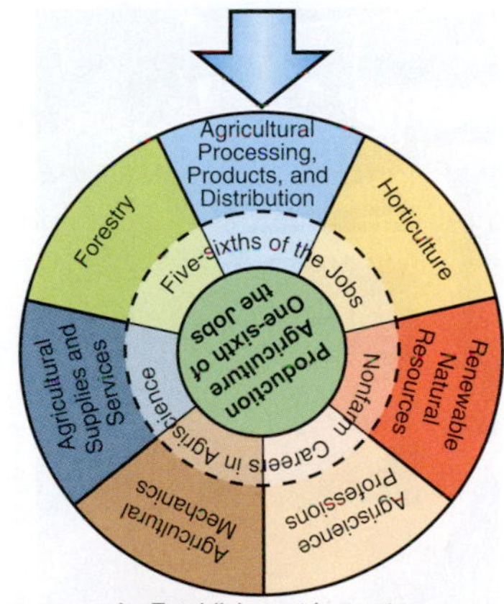

Ag Establishment Inspector
Butcher
Cattle Buyer
Christmas Tree Grader
Cotton Grader
Farm Stand Operator
Federal Grain Inspector
Food & Drug Inspector
Food Processing Supervisor
Fruit & Vegetable Grader
Food Distributor
Fruit Press Operator
Flower Grader
Grain Broker
Grain Buyer
Grain Elevator Operator
Hog Buyer
Livestock Commission Agent
Livestock Yard Supervisor
Meat Inspector
Meatcutter
Milk Plant Supervisor
Produce Buyer
Produce Commission Agent
Quality Control Supervisor
Tobacco Buyer
Weights & Measures Official
Winery Supervisor
Wood Buyer

FIGURE 4-6 Spin the wheel of fortune. What comes up for you? Agricultural processing, products, and distribution.

Agricultural Processing, Products, and Distribution

Spin the wheel of fortune! What comes up for you? Agricultural processing, products, and distribution (Figure 4-6)!

Agricultural processing, products, and distribution are those parts of the industry that haul, grade, process, package, and market commodities from production sources. Pick any item of food, clothing, or other commodity. Trace it back to its source. Except for metals and stone, most objects can be traced back to a farm, ranch, forest, greenhouse, body of water, or other agricultural production facility. If you consider a deluxe hamburger, you can trace the beef, mayonnaise, tomato, lettuce, pickle, catsup, mustard, relish, bun, and sesame seeds back to farms where they were produced (Figure 4-7). The same is true of many ingredients in soda, coffee, chocolate, or any other beverage you choose.

Courtesy of National FFA; FFA #18.

FIGURE 4-7 The components of a deluxe cheeseburger may have come from several states or even different countries.

INTERNET KEY WORDS:
food-processing careers
horticulture careers
natural resource careers
forestry careers

Check the label in your coat. Is it made of cotton, leather, vinyl, rubber, or wool? Each can be traced to a farm, ranch, or plantation. Your search may take you to a Maryland farm, a California ranch, a Colombian rubber plantation, or a Utah mink farm.

People with careers and jobs in agricultural processing, products, and distribution make it all possible. From hauling to selling, processing to merchandising, inspection, and research—the commodity moves from its source to consumption (Figure 4-8). The U.S. Department of Agriculture (USDA) reports that the producer's share of the food dollar is as low as 15.4 cents for cereal and bakery products. The rest is for handling, processing, and distribution.

Horticulture

Spin the wheel of fortune! What comes up for you? Horticulture (Figure 4-9)!

Horticulture includes producing, processing, and marketing fruits, vegetables, and ornamental plants such as turfgrass, flowers, shrubs, and trees (Figure 4-10). Horticultural production is a farming enterprise, but because it is generally done on small plots, the production of horticultural crops is classified with horticulture rather than farming. Horticultural commodities are high-labor and high-income commodities.

© Paula Cobleigh/Shutterstock.com.

A

© Richard Thornton/Shutterstock.com.

B

FIGURE 4-8 Raw farm products usually require processing before they become available to the consumer.

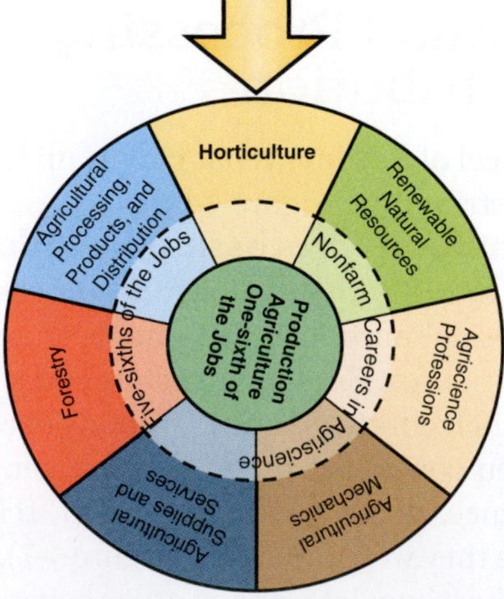

Floral Designer
Floral Shop Operator
Florist
Golf Course Superintendent
Greenhouse Manager
Greenskeeper
Horticulturist
Hydroponics Grower
Landscape Architect
Landscaper
Nursery Operator
Plant Breeder
Turf Farmer
Turf Manager

FIGURE 4-9 Spin the wheel of fortune! What comes up for you? Horticulture.

© Paul Vinten/Shutterstock.com.

FIGURE 4-10 Flowers are often imported to the United States during seasons when they are not available locally.

Courtesy of DeVere Burton.

FIGURE 4-11 Turf farming is the production of grass that is harvested, roots and all, and transplanted to provide "instant lawns."

The landscape designer, golf course superintendent, greenhouse supplier, greenhouse manager, flower wholesaler, floral market analyst, florist, strawberry grower, vegetable retailer, and turf farmer (Figure 4-11) are all horticulturists. Recent data show more than 110,000 people employed by the floral industry alone.

Forestry

Spin the wheel of fortune! What comes up for you? Forestry (Figure 4-12)!

Forestry is the industry that grows, manages, and harvests trees for lumber, poles, posts, panels, pulpwood, and many other commodities. Americans have huge appetites for wood products.

Careers in forestry range from growing tree seedlings to marketing wood products. Many jobs in forestry are outdoors and require the use of large machines to cut trees, drag logs, and load trucks (Figure 4-13). Other jobs are service oriented,

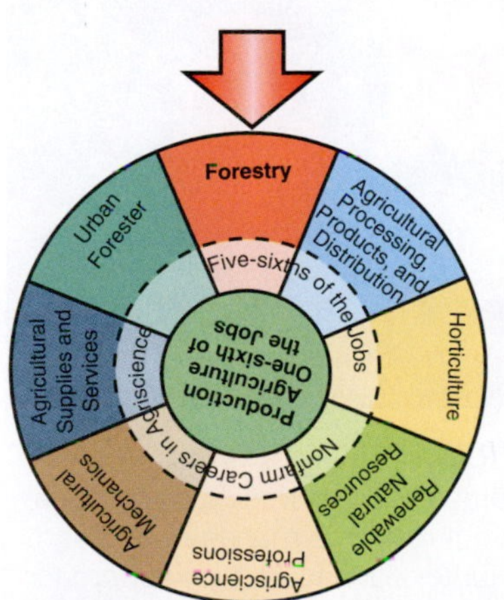

Forester
Forest Ranger
Heavy Equipment Operator
Log Grader
Logging Operations Inspector
Lumber Mill Operator
Nursery Operator
Park Ranger
Plant Breeder
Timber Manager
Tree Surgeon
Urban Forester

Copyright © 2015 Cengage Learning®.

FIGURE 4-12 Spin the wheel of fortune! What comes up for you? Forestry.

Courtesy of National FFA; FFA #187.

FIGURE 4-13 The forest industry provides many opportunities for outdoor work.

HOT TOPICS IN AGRISCIENCE A CAREER IN FOOD SCIENCE

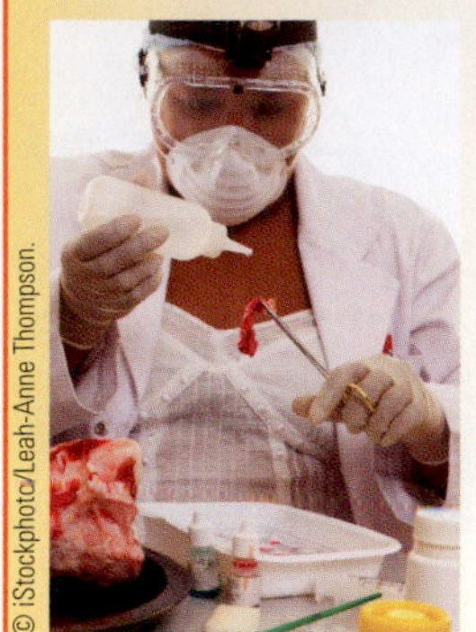

Food science has become one of the "hottest" careers in the agricultural industry.

A professional career in food science has become one of the hottest career fields available in agriculture. It combines the career fields of science and agriculture in the development of new food products. It also extends into the processing, packaging, distribution, and marketing of food products. This career requires a high degree of creativity in designing new products. Preparation for this career requires a college education with a strong emphasis in chemistry, bacteriology, and the biological sciences. Salaries in this career field rank high in comparison with most other agricultural careers, and college graduates in this field are in high demand.

such as the state or district forester whose job is to give advice and administer governmental programs. Many find enjoyable careers in forestry research, teaching, wood technology, and marketing.

Renewable Natural Resources

Spin the wheel of fortune! What comes up for you? Renewable natural resources (Figures 4-14 and 4-15)!

Renewable natural resources involve the management of wetlands, rangelands, water, fish, and wildlife. All fields require people with an appreciation for natural and scientific knowledge of plants and animals. This area of agriscience is attractive to those who enjoy working in parks, on game preserves, or with

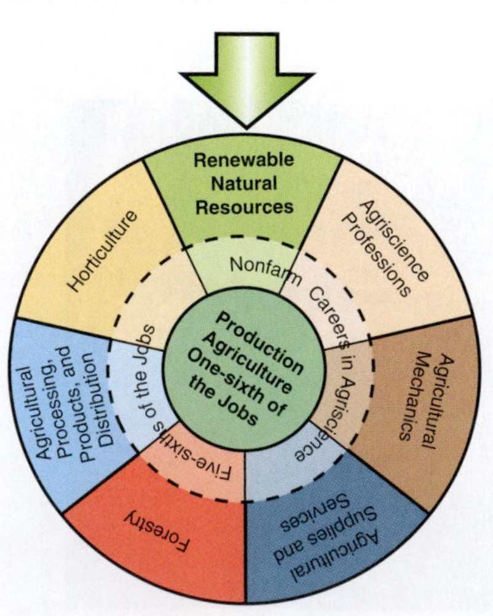

Animal Behaviorist
Animal Ecologist
Animal Taxonomist
Environmental Conservation Officer
Environmentalist
Fire Warden
Forest Fire Fighter/Warden
Forest Ranger
Game Farm Supervisor
Game Warden
Ground Water Geologist
Park Ranger
Range Conservationist
Resource Manager
Soil Conservationist
Trapper
Water Resources Manager
Wildlife Manager

FIGURE 4-14 Spin the wheel of fortune! What comes up for you? Renewable natural resources.

FIGURE 4-15 Effective management of natural resources begins with education about resource management, and care. Hunter education is one approach to teaching respect for nature's resources.

landowners to preserve and enhance natural habitat, plants, and wildlife. Water quality and soil conservation are state and regional concerns of high priority. New career opportunities in natural resource management are resulting from new efforts to save our oceans, lakes, wetlands, rivers, and bays.

Agricultural Supplies and Services

Spin the wheel of fortune! What comes up for you? Agricultural supplies and services (Figures 4-16 and 4-17)!

Agricultural supplies and services are businesses that sell supplies or provide services for people in the agricultural industry. Examples of supplies are seed, feed, fertilizer, lawn equipment, farm machinery, hardware, pesticides, and building supplies. These businesses are operated by owners, managers, mill operators, truck drivers, sales personnel, bookkeepers, field representatives, clerks, and others.

People in these jobs provide the supplies for the agricultural industry. However, many in agriscience seldom handle the commodities themselves. Instead, they provide a service. Those who provide legal assistance, write agricultural publications, advise agriculturists on money matters, or provide advice on crops, livestock, pest control, or soil fertility are working in service occupations. Such jobs are for those who are more people oriented than commodity oriented.

INTERNET KEY WORDS: agricultural sales career

Agricultural Mechanics

Spin the wheel of fortune! What comes up for you? Agricultural mechanics (Figures 4-18 and 4-19)!

Are you fascinated by tools and equipment? Are you challenged by something that does not work? Are you creative and like to build things? If so, a career in agricultural mechanics may be for you. **Agricultural mechanics** is the design, operation, maintenance, service, selling, and use of power units, machinery, equipment, structures, and utilities in agriculture.

Agricultural mechanics includes the use of hand and power tools, woodworking, metalworking, welding, electricity, plumbing, tractor and machinery

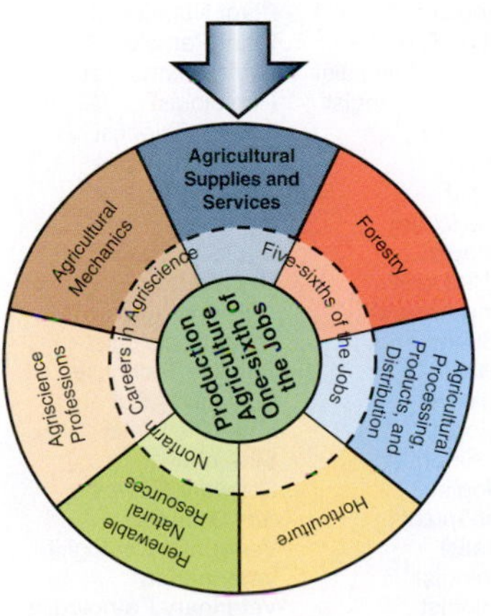

Aerial Crop Duster
Ag Aviator
Ag Chemical Dealer
Ag Equipment Dealer
Animal Groomer
Animal Health Products Distributor
Animal Inspector
Animal Keeper
Animal Trainer
Artificial Breeding Distributor
Artificial Breeding Technician
Artificial Inseminator
Biostatistician
Chemical Applicator
Chemical Distributor
Computer Analyst
Computer Operator
Computer Programmer
Computer Salesperson
Custom Operator
Dairy Management Specialist
Dog Groomer
Farm Appraiser
Farm Auctioneer
Farrier
Feed Mill Operator
Feed Ration Developer & Analyst
Fertilizer Plant Supervisor
Fiber Technologist
Field Inspector
Field Sales Representative, Agricultural Equipment
Field Sales Representative, Animal Health Products
Field Sales Representative, Crop Chemicals, Machinery
Harness Maker
Harvest Contractor
Horse Trainer
Insect & Disease Inspector
Kennel Operator
Lab Technician
Meteorological Analyst
Pest Control Technician
Pet Shop Operator
Poultry Field Service Technician
Poultry Hatchery Manager
Poultry Inseminator
Sales Manager
Salesperson
Service Technician
Sheep Shearer

FIGURE 4-16 Spin the wheel of fortune! What comes up for you? Agricultural supplies and services.

Courtesy of National FFA.

FIGURE 4-17 Agricultural supplies and services provide the vital materials and services to keep a trillion dollar industry moving.

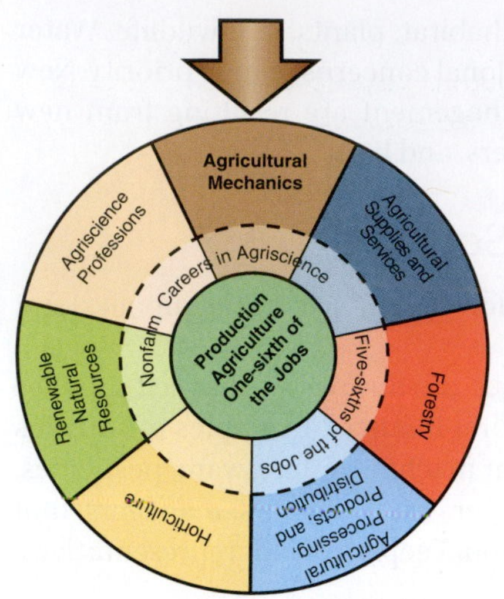

- Ag Construction Engineer
- Ag Electrician
- Ag Equipment Designer
- Ag Plumber
- Ag Safety Engineer
- Diesel Mechanic
- Equipment Operator
- Hydraulic Engineer
- Irrigation Engineer
- Land Surveyor
- Machinist
- Parts Manager
- Research Engineer
- Safety Inspector
- Soil Engineer
- Welder

FIGURE 4-18 Spin the wheel of fortune! What comes up for you? Agricultural mechanics.

FIGURE 4-19 Careers in agricultural mechanics are varied and challenging.

mechanics, hydraulics, terracing, drainage, painting, and construction. Choose your level—indoors or outdoors. Choose your role—employee, employer, or professional.

Agriscience Professions

Spin the wheel of fortune! What comes up for you? Agriscience professions (Figures 4-20 and 4-21)!

The word **profession** means an occupation requiring specialized education, especially in law, medicine, teaching, or the ministry. **Agriscience professions** are those professional jobs that deal with knowledge and understanding of agriscience. They cut across all divisions in the wheel of fortune.

Consider the agriscience teacher and the Cooperative Extension educator. Both must have a bachelor's or master's degree to be qualified. They may teach or consult about subjects in several or all divisions of agriscience.

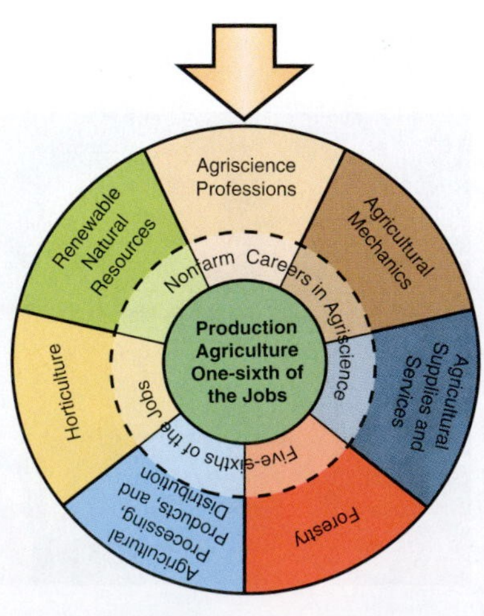

- Ag Accountant
- Ag Advertising Executive
- Ag Association Executive
- Ag Consultant
- Ag Corporation Executive
- Ag Economist
- Ag Educator
- Ag Extension Educator
- Ag Extension Specialist
- Ag Journalist
- Ag Lawyer
- Ag Loan Officer
- Ag Market Analyst
- Ag Mechanics Teacher
- Ag News Director
- Agriculture Attaché
- Agronomist
- Animal Cytologist
- Animal Geneticist
- Animal Nutritionist
- Animal Physiologist
- Animal Scientist
- Agriculturist
- Avian Veterinarian
- Bacteriologist
- Biochemist
- Bioengineer
- Biophysicist
- Botanist
- Computer Specialist
- Credit Analyst
- Dairy Nutrition Specialist
- Dendrologist
- Electronic Editor
- Embryologist
- Environmental Educator
- Entomologist
- Equine Dentist
- 4-H Youth Assistant
- Farm Appraiser
- Farm Broadcaster
- Farm Investment Manager
- Food Chemist
- Foreign Affairs Official
- Graphic Designer
- Herpetologist
- Horticulture Instructor
- Hydrologist
- Ichthyologist
- Information Director
- International Specialist
- Invertebrate Zoologist
- Land Bank Branch Manager
- Limnologist
- Magazine Writer
- Mammalogist
- Marine Biologist
- Marketing Analyst
- Media Buyer
- Microbiologist
- Mycobiologist
- Nematologist
- Organic Chemist
- Ornithologist
- Ova Transplant Specialist
- Paleobiologist
- Parasitologist
- Pharmaceutical Chemist
- Photographer
- Plant Cytologist
- Plant Ecologist
- Plant Geneticist
- Plant Nutritionist
- Plant Pathologist
- Plant Taxonomist
- Pomologist
- Poultry Scientist
- Public Relations Manager
- Publicist
- Publisher
- Reproductive Physiologist
- Rural Sociologist
- Satellite Technician
- Scientific Artist
- Scientific Writer
- Silviculturist
- Software Reviewer
- Soil Scientist
- Vertebrate Zoologist
- Veterinarian
- Veterinary Pathologist
- Virologist
- Viticulturist
- Vocational Agriculture Instructor/FFA Advisor

FIGURE 4-20 Spin the wheel of fortune! What comes up for you? Agriscience professions.

Courtesy of USDA/ARS K-3401-10.

FIGURE 4-21 Professional workers are more in demand than ever before as agriscience has embraced the information age.

Consider the veterinarian, agricultural attorney, research scientist, geneticist, or engineer—all of these professions require advanced degrees and high levels of education and skills. If you can meet the standard, then a career in an agriscience profession may be for you.

Computers in Agriscience

The use of computers is extensive in agriscience. This means there are many opportunities to combine computer skills with agriscience settings. Among the agricultural uses of computers are machinery management, farm financial records, livestock management, crop management, commodity marketing, farm/ranch inventory management, agricultural business management, taxes, irrigation management, and precision farming. In addition, tractors and combines are equipped with computer and global positioning systems and other high-technology devices that are quite sophisticated. These systems make it possible to operate the machines more efficiently. Computer skills are important in nearly every agriscience career. Every student should be certain that he or she is proficient in the use of computers.

PREPARING FOR AN AGRISCIENCE CAREER

Career education is an important part of public education today. Since the early 1970s, the career education movement has spread from occupational education programs to the general curriculum. Many school systems now emphasize career education from kindergarten through adulthood.

As you consider a career in agriscience, it is important to consider how to meet the requirements to get started in that career. Some young people have an early start on careers in agriscience. They may have grown up on a farm or ranch. Their parents may have worked in one of the other areas of agriscience such as horticulture, resource management, business, or teaching. Perhaps they have obtained jobs or worked for friends or neighbors who had agriscience businesses. Following are some suggestions for preparing for a career in agriscience.

AGRI-PROFILE AGRISCIENCE TEACHER

© Goodluz/Shutterstock.com.

Students of agriculture learn best by doing, and teachers seek ways to teach relevant skills.

Every student who has enrolled in an agricultural program has benefited from the experience of *learning to do by doing*. It is a curious thing that one of the critical worker shortages in the agricultural professions exists among teachers of agriscience programs. Too few students are enrolling as majors in university agricultural education programs, and too few graduates are entering the agricultural education profession after they graduate with their degrees. There is also a high demand that attracts experienced agriscience teachers to related agricultural careers. The result is that some high school agriculture programs have been closed due to a shortage of qualified teachers. Top students would do well to consider teaching agriscience as their first career choice. Interested students should express their interest to their teacher or a teacher educator at the university. This career requires a strong background in agriculture and related sciences and a four-year degree in agricultural education. Most agriscience educators are endorsed by their state to teach agriculture and science classes.

Agriscience Career Portfolio

An agriscience career portfolio is a collection of your best work on agriscience projects and other career-related materials that you have personally developed. The portfolio will be used to sell your skills to a prospective employer. Only your best work should be included because a mediocre portfolio is an indicator of mediocre or average job skills.

AGRI-PROFILE

CAREER AREA: AGRISCIENCE TECHNICIAN OR PROFESSIONAL?

© iStockphoto/Rubberball.

Whether you prefer to be a technician or a professional, agriscience offers a broad array of career possibilities.

An assessment of career opportunities in agriscience by the USDA revealed a bright outlook for college graduates. Other studies indicate a need for workers trained in agriscience for technical-level jobs.

The agriscience technician may be broadly trained in plant and animal sciences and employable in many fields. For the sharp individual with good work habits and a broad background in plant and animal sciences, the choices are extensive. Add agriscience mechanics skills to the package and the individual has access to dozens of career options.

Education for agriscience careers should begin at the high school level or earlier. If one plans to work at the technician level, an early start is especially helpful. The technician is expected to have firsthand experience and detailed knowledge of procedures and techniques. Such knowledge comes with experience in shops, laboratories, farms, greenhouses, fisheries, and on-the-job training situations as well as in the classroom. Technicians generally have specialized training beyond the high school level, whereas professionals are required to obtain degrees at the bachelor's, master's, or doctorate levels.

Some of the materials that you may want to include in your portfolio include the following:

- Resumé or vita
- FFA Agriscience Scholarship application
- Articles and papers written by you (published and unpublished)
- Photographs and written reports of projects that you have completed
- Documentation of participation in community and public service activities
- Personal letters and citations for services you have performed for other people
- Letters of recommendation
- Personal and career goals
- Action plan for accomplishing your goals
- Newspaper articles, video clips, and sound bytes of media coverage of your projects

Career Plan While in Middle School

- Plan and conduct science projects with plants, animals, soil, water, energy, ecology, conservation, and wildlife.
- Research and report on the projects listed earlier that interest you most.

AGRI-PROFILE WORKFORCE SKILLS: WHAT SKILLS WILL YOU BRING TO YOUR CAREER?

The SCANS report is a document that was developed by the U.S. Secretary of Labor. The report was written by the Secretary's Commission on Achieving Necessary Skills (SCANS). Commission members represented business owners, unions, workers, supervisors, and public employers. The commission developed the following five competencies that students need to master so they will be prepared for productive careers:

- **Resources: Identifies, organizes, plans, and allocates resources**
  - A. Time—selects goal-relevant activities, ranks them, allocates time, and prepares and follows schedules
  - B. Money—uses or prepares budgets, makes forecasts, keeps records, and makes adjustments to meet objectives
  - C. Material and Facilities—acquires, stores, allocates, and uses materials or space efficiently
  - D. Human Resources—assesses skills and distributes work accordingly, evaluates performance, and provides feedback
- **Interpersonal: Works with others**
  - A. Participates as a Member of a Team—contributes to group effort
  - B. Teaches Others New Skills
  - C. Serves Clients/Customers—works to satisfy customers' expectations
  - D. Exercises Leadership—communicates ideas to justify position, persuades and convinces others, responsibly brings new ideas forward to improve existing procedures and policies
  - E. Negotiates—works toward agreements involving exchange of resources, resolves divergent interests
  - F. Works with Diversity—works well with men and women from diverse backgrounds
- **Information: Acquires and uses information**
  - A. Acquires and Evaluates Information
  - B. Organizes and Maintains Information
  - C. Interprets and Communicates Information
  - D. Uses Computers to Process Information
- **Systems: Understands complex interrelationships**
  - A. Understands Systems—knows how social, organizational, and technological systems work and operates effectively with them
  - B. Monitors and Corrects Performance—distinguishes trends, predicts impacts on system operations, diagnoses deviations in systems' performance, and corrects malfunctions
  - C. Improves or Designs Systems—suggests modifications to existing systems and develops new or alternative systems to improve performance
- **Technology: Works with a variety of technologies**
  - A. Selects Technology—chooses procedures, tools, or equipment, including computers and related technologies
  - B. Applies Technology to Tasks—understands overall intent and proper procedures for setup and operation of equipment
  - C. Maintains and Troubleshoots Equipment—prevents, identifies, or solves problems with equipment, including computers and other technologies

Source: The Secretary's Commission on Achieving Necessary Skills. (June 1991). *What Work Requires of Schools, A SCANS Report for America 2000.* U.S. Department of Labor, pg. x. Retrieved from http://wdr.doleta.gov/SCANS/whatwork/whatwork.pdf

- Join 4-H or Scouts and choose agricultural projects and merit badges.
- Volunteer to work on lawn, garden, greenhouse, farm, or conservation projects.
- Enroll in agriscience or other career education programs.

SCIENCE CONNECTION WHEN I'M TWENTY-SOMETHING!

Close your eyes and relax for a minute and then think about yourself in the year 2025. How old will you be then? Twenty-something? You have finished your formal schooling and are into your career. What kind of job do you have? Are you married or single? Is life enjoyable and your work challenging? Are you staying in your hometown, or are you in some other state or nation? Are you happy with your career? If you could start again, would you make the same choices? Now read the following questions and make mental notes of the answers that are most likely to be correct for you in the year 2025.

A Day on the Job in the Year 2025

Directions: Select the response to each item that you think will best describe you and your situation in the year 2025.

1. What time do I generally wake up?
   - Early morning
   - Afternoon
   - Evening
2. What is my work environment?
   - Indoor
   - Outdoor
   - Both
3. Where am I working?
   - In an office
   - In a factory or shop
   - In the community
   - In my home
4. What type of clothes do I wear to work?
   - Dress clothes (coat and tie or suit)
   - Uniform
   - Casual (open collar)
   - Worn jeans
5. Am I married?
   - Yes
   - No
6. If married, how long?
   - Fewer than three years
   - Three years or more
7. Number of children:
   - Zero
   - One
   - Two
   - Three
   - More than three
8. I will retire when I am:
   - 54 years or younger
   - 55 to 65
   - Older than 65
9. How much education do I have? My highest level of education is:
   - High school
   - Two-year technical school degree or certificate
   - Four-year college
   - Master's degree
   - Doctorate degree
   - Other
10. Do I like keeping up with technical trends and procedures related to my work?
    - Yes
    - No
11. Where do I live?
    - Hometown
    - Home state but not hometown
    - Not home state but in the United States
    - Outside the United States
12. I am working with or for:
    - Government
    - Education
    - Myself/self-employed
    - A large company (more than 200 employees)
    - A small company (200 or less employees)
13. My income level makes me:
    - Wealthy
    - Comfortable
    - Struggling financially
    - Generally deprived
14. My amount of paid vacation is:
    - Fewer than 5 days
    - 6 to 10 days
    - 11 to 15 days
    - 16 days or more
15. The most important thing in my life is:
    - Family
    - Money and belongings
    - Prestige and status
    - Time off from work
    - The place where I live

(continues)

(*continued*)

16. My work can best be described as:
    - Producing information
    - Producing or selling a product
    - Distributing information
    - Serving people

How well do your current career preparation activities mesh with your perception of yourself in the year 2025? The material in this and other units should be helpful in determining your preferences and making plans and preparations for a successful and challenging career.

Career Plan While in High School

- Enroll in agriscience classes, including plant science, animal science, agricultural mechanics, agribusiness, and farm management.
- Enroll in college-preparatory and/or dual-enrollment courses in English, math, and science.
- Join the FFA organization and participate in leadership, citizenship, and agriscience activities.
- Develop a broad, supervised, occupational agriscience experience program.
- Acquire hands-on, skill-development experiences.
- Conduct an agriscience research project, and enter it in the FFA Agriscience Scholarship and Agriscience Fair programs.

Career Plan after High School

- Obtain an agricultural job and plan ways to get additional training while on the job.
- Enter a community college and take courses that will transfer to the college of agriculture or life sciences at your state university.
- Enter a two-year program in technical agriculture.
- Enter a college of agriculture or life sciences, and obtain a bachelor's degree (BS), a master's degree (MS), and/or a doctorate (PhD).

You can obtain information on careers, schools, and colleges from many sources. The following suggestions may be helpful to you:

High School Agriscience Teachers

- Agricultural mechanics
- General agriscience
- Animal and plant sciences
- Horticulture
- Biotechnology

High School Counselors

- Career brochures, bulletins
- Career information system
- Aptitude tests
- College/university catalogs

Cooperative Extension Service

- Listed in your phone book or found on the Internet under county or city government
- 4-H career bulletins

State Department of Education

- Specialist in agriscience, agribusiness, and renewable natural resources: scholarship, internships
- FFA State Executive Secretary: FFA career materials

HOT TOPICS IN AGRISCIENCE AGRISCIENCE CAREERS

Food Technician/Scientist
Environmental Technician
Computer Technician
Animal Technician/Scientist
Plant Technician/Scientist
Global Positioning System Technician
Biotechnology Engineer
Veterinarian/Technician
Farm/Ranch Managers
Urban Forester
Soil Technician/Scientist
Genetic Engineer

Community Colleges, Technical Colleges, and other Postsecondary Institutions

- Occupational Dean, Community College: program information, scholarships
- Director or Dean, Institute or Technical School: career/program information Typical programs include agriscience business management, animal/crop production and management, ornamental horticulture, water resources, forestry, and wildlife resources
- Dean, College of Agriculture or Life Sciences Typical programs include agricultural education, agricultural and resource economics, agronomy (crops and soils), animal sciences, food science, forestry, horticulture, natural resources management, and poultry science
- Agricultural Education Coordinator, Department of Agricultural and Extension Education, Land-Grant University: agricultural education career information

As new technologies and job opportunities emerge, so will the need for well-trained and educated new people. Agriscience is a diverse field with job opportunities available at all levels. Pick your area of interest, determine the level at which you wish to operate, obtain the appropriate education for the job, and follow through with a rewarding career!

STUDENT ACTIVITIES

1. Write the Terms to Know and their meanings in your notebook.
2. Using paper and pencil, calculate the amount of time needed to count the dollar value of agricultural assets in the United States, as suggested in Figure 4-1.

3. Develop a bulletin board that illustrates the "wheel of fortune," with its listing of the broad categories of jobs or divisions in agriscience.
4. Develop a collage that illustrates the many careers in agriscience.
5. Write the names of the divisions of agriscience, such as "Production Agriculture," "Agricultural Processing, Products, and Distribution," "Horticulture," and so on, and list five careers under each that may interest you.
6. Choose a career from the lists you developed for Activity 5 and write a one-page description for that career. Include the following sections in your description:
   a. Career title
   b. Education/training required to enter the career and advance in the field
   c. Working conditions
   d. Advantages/benefits
   e. Disadvantages of the career
   f. Salaries of beginning and advanced workers in the field
   g. Aspects of the career that you like
7. Using the career you researched for Activity 6, or another job or career area, list the things you should do during and after high school to prepare for a career in that area.
8. Find the name and address of an appropriate official of a school or college and request information from that person about educational opportunities for you in his or her institution.
9. Develop a list of agriscience careers in which computers are used.
10. At fast food restaurants, a popular order is a cheeseburger, fries, and a soda. Make a list of possible agriscience jobs that were involved in making this order a reality. For example, a wheat farmer was responsible for growing the wheat that is in the bun.
11. Make an outline of the unit. Phrases and words that are bold or in color should be included along with a brief description of each.

SELF-EVALUATION

A. MULTIPLE CHOICE

1. The industry that provides commodities that are basic to life is
   a. aerospace.
   b. agriscience.
   c. biotechnology.
   d. transportation.
2. The number of workers in agriscience in the United States is approximately
   a. 21 million.
   b. 100 million.
   c. 100,000.
   d. 400,000.
3. The percentage of total jobs in agriscience that require a college education to enter the job is
   a. 10 percent.
   b. 20 percent.
   c. 41 percent.
   d. 60 percent.
4. The products and services that are provided in most of the areas in the agriscience wheel of fortune seem to flow to or originate from
   a. agriculture processing, products, and distribution.
   b. agriscience professions.
   c. horticulture.
   d. production agriculture.

5. The management of wetlands comes under the area of
   a. agricultural processing, products, and distribution.
   b. agriscience professions.
   c. horticulture.
   d. renewable natural resources.
6. Of all agriscience jobs, the percentage that is not on farms or ranches is
   a. 20 percent.
   b. 40 percent.
   c. 60 percent.
   d. 80 percent.
7. A producer's share of each dollar spent for bread and cereals in the United States is about
   a. 11 cents.
   b. 15.4 cents.
   c. 25 cents.
   d. 75 cents.
8. The number of floral industry workers in the United States is about
   a. 110,000.
   b. 220,000.
   c. 500,000.
   d. 1,000,000.
9. A student may join 4-H or Scout groups to learn agriscience concepts as early as
   a. college.
   b. high school.
   c. middle school.
   d. none of the above.
10. Agriscience classes in high school usually include extensive instruction in
   a. plants, animals, and agribusiness.
   b. plants, animals, and social sciences.
   c. plants, mechanics, and higher math.
   d. food, fiber, and physics.

B. MATCHING

_________	1. Production	a. Teacher or veterinarian
_________	2. Processing and distribution	b. Hydraulics
_________	3. Horticulture	c. Seed, feed, or lawn supply
_________	4. Forestry	d. Farming or ranching
_________	5. Natural resources	e. Lumber
_________	6. Supplies and services	f. Ornamentals
_________	7. Mechanics	g. Grading and packaging
_________	8. Professions	h. Wildlife

UNIT 5

Supervised Agricultural Experience

OBJECTIVE

To learn the rationale for and plan a supervised agricultural experience.

MATERIALS LIST

- paper
- pencil or pen
- bulletin board materials
- Student Interest Survey form (Figure 5-12)
- Resources Inventory form (Figure 5-13)
- Selecting a Supervised Agricultural Experience form (Figure 5-14)
- Experience Inventory form (Figure 5-19)
- Placement Agreement (Figure 5-20)
- Improvement Activity Plan and Summary (Figure 5-21)
- Supplementary Agri-science Skills Plan and Record (Figure 5-22)

COMPETENCIES TO BE DEVELOPED

After studying this unit, you should be able to:

- define supervised agricultural experience (SAE) program terms.
- determine the place and purposes of SAEs in agriscience programs.
- determine the types of supervised agricultural experience activities.
- explore the opportunities for SAEs.
- set personal goals for an SAE.
- plan your personal SAE.

SUGGESTED CLASS ACTIVITIES

1. Organize a Saturday or summertime bus tour so students can see the supervised agricultural experience (SAE) programs of other students. Encourage students who show their projects to describe why they chose the project, what they liked about the project, and what they learned through the project.
2. Invite a representative from the federal Farm Service Agency (or an agricultural lender) to visit the class and share information about financial assistance that is available to young people who wish to become involved in farming or other agricultural businesses. Provide students with information about these lending programs prior to the presentation to allow them to ask appropriate questions. Invite parents to attend the presentations.

3. Spend a class period looking through agriscience magazines, old SAE project books, and any other sources your teacher can provide. After learning what projects have been done in the past, create two new ideas or ways to update an old project for an SAE you could do.

TERMS TO KNOW

real-world experience
on-the-job training
supervised agricultural experience (SAE)
FFA
project
enterprise
exploratory supervised agricultural experience
agriscience literacy
career exploration
career
agriscience research project
scientific method
entrepreneurship supervised agricultural experience
production enterprise
placement supervised agricultural experience
internship
improvement activities
resumé

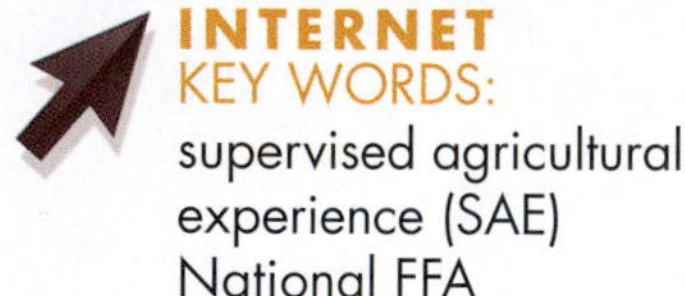
INTERNET KEY WORDS:
supervised agricultural experience (SAE)
National FFA

Agriscience programs in various schools teach basic principles and practices in plant and animal sciences, resource management, business management, agricultural mechanics, landscape design, leadership, and personal development. Such programs emphasize reading and math skill development and the application of scientific principles. Classrooms are excellent places to learn fundamentals and theory through reading, study, discussion, and planning. School laboratories, such as greenhouses, agricultural mechanics shops, school land demonstration plots, farms, and animal production facilities, help provide experiences that simulate real-world experiences. Simulate means to look or act like. **Real-world experience** means conducting the activity in the daily routine of our society. Simulation is an excellent way to learn. It imitates the real world and provides an ideal setting for many agriscience activities.

Classroom experiences and simulation, however, lack the thoroughness of real-world experiences. Also, the personal relationships, such as employer–employee, supervisor–subordinate, owner–worker, salesperson–customer, owner–government, owner–community, fellow workers, and employee competition, are rarely present in simulated activities. Therefore, a program is needed for the student to obtain real-world experiences and on-the-job training if agriscience education is to lead to a successful career. **On-the-job training** means experience obtained while working in an actual job setting. In agriscience, the method used for students to obtain real-world experiences is referred to as the **supervised agricultural experience (SAE)**.

SUPERVISED AGRICULTURAL EXPERIENCE (SAE)

An SAE consists of all the supervised agricultural experiences that are learned outside the regularly scheduled classroom or laboratory. *Supervised* means to be looked after and directed. *Agriscience* in this phrase means business, employment, or trade in agriculture, agribusiness, or renewable natural resources. *Experience* means anything and everything that is observed, done, or lived through.

PURPOSE OF SAEs

Supervised agricultural experiences provide opportunities for learning by doing. They provide the means for you to learn with the help of your teacher, parents, employer, and other adults experienced in the area of your interest. Student SAEs are an essential part of effective agriscience programs.

Some important purposes and benefits of SAEs are to:

- provide opportunities to creatively explore a variety of agriscience careers;
- provide educational and practical experiences in a specialized area of agriscience;
- provide the opportunity to become established in an agriscience occupation;
- provide opportunities for earning while learning;
- create opportunities for earning after graduation;
- develop interests in additional areas of agriscience;
- develop valuable work skills such as
    - appreciate the importance of honest work;
    - improve personal habits;
    - develop superior work habits;
    - establish good relationships with others;
    - keep effective records;
    - prepare useful reports;
    - follow instructions and regulations;
    - contribute to the advancement of your occupation;
    - contribute to your family, community, and nation;
- encourage individualization of instruction;
- assure recognition for individual achievement;
- become established in an agriscience business; and
- obtain experience as an entrepreneur.

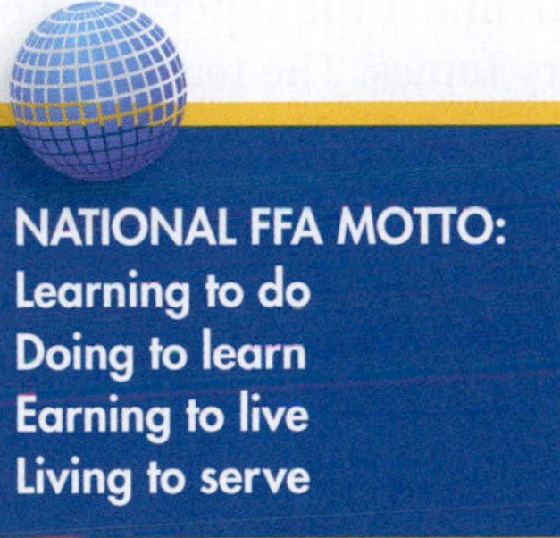

Source: Reprinted by permission of The National FFA Organization

AGRI-PROFILE CAREER AREA: SUPERVISED AGRICULTURAL EXPERIENCE

© iStockphoto/Chris Price.

Learning by doing is generally accepted as the best way to become proficient in complex procedures and psychomotor skills.

The phrase *supervised agricultural experience* has three important elements. Experience suggests hands-on or real-life activities in the workplace. Attitudes, knowledge, and skills gained here will generally be advantageous to students as they seek employment. The term *agricultural* means that the setting and skills will be in the area of plant or animal sciences and related natural resources or management, mechanics, or technologies. *Supervised* means that experienced persons such as your teacher, parents, and/or employer recognize your interest in learning and will help direct the experience.

Agricultural experience is obtained in many ways. Generally, the teacher helps the student develop an understanding of the process, and he/she helps the student identify a suitable location for a productive experience with cooperation from parents. Sometimes sympathetic parents and supportive school programs are not available. In such cases, students must develop an extra measure of determination and seek experiences on their own.

Experiences may be for pay or simply to gain experience. Many schools now provide supervised experience programs in school laboratories. Regardless of whether the student is being paid, the experiences obtained are worth the effort and give the participant an edge in the job market.

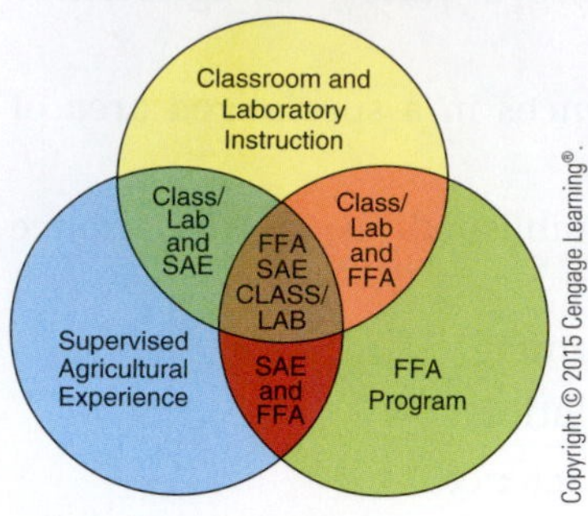

FIGURE 5-1 A comprehensive agriscience program provides students opportunities to learn through classroom/laboratory instruction, supervised agricultural experiences, and personal growth through FFA.

SAE AND THE TOTAL AGRISCIENCE PROGRAM

High school agriscience programs should be comprehensive in that they provide students with agricultural experiences, leadership development programs, and laboratory experiences together with classroom instruction. Leadership skills are developed through the FFA program, which is discussed in Unit 6. These components are integrated so they complement each other (Figure 5-1). This integration should provide the most effective program for the student and make the best use of teacher time and resources.

SAE and Classroom Instruction

The SAE is planned as part of the classroom instruction. Students use this instruction to learn how to plan an SAE, identify possible SAE choices, choose activities for their own SAEs, and make appropriate arrangements with parents, teachers, and employers. The student conducts the SAE under the supervision of the teacher, who provides instruction on the necessary topics. The teacher also arranges for small group and individual instruction (Figure 5-2). This permits the student to use the classroom and laboratory to solve problems encountered with the SAE.

SAE and the FFA

The SAE provides opportunities to use work-related skills that are studied in the classroom (Figure 5-3). It overlaps with the **FFA** (an intracurricular youth organization for students enrolled in agriscience programs). The FFA has many activities that encourage students to do more in their SAEs and provides competitive career development events that more fully develop useful skills

FIGURE 5-2 The best instruction includes hands-on activities.

© Kurhan/Shutterstock.com.

FIGURE 5-3 Supervised Agricultural Experience is based on real agricultural work opportunities.

for SAEs. The FFA provides awards and other recognition for achievements in SAEs. These frequently lead to travel experiences that greatly enrich the student's education.

FFA and Classroom Instruction

INTERNET KEY WORDS: FFA Proficiency Awards

FFA is merged with classroom and laboratory instruction in a number of ways. The teacher has instructional units on the FFA in the classroom. Instruction is provided in public speaking, parliamentary procedure, and other leadership skills. The classroom setting can become the place where FFA members polish their skills in preparation for upcoming competitive events. Therefore, the students who obtain the most benefit from their agriscience programs are those who take advantage of the opportunities provided through SAEs and the FFA.

Proficiency Awards

The FFA has an extensive system of awards for individual members. These awards provide members with incentives and rewards for excellence in leadership and agriscience achievement. These awards change as the FFA expands to meet the changing nature of agriscience. Only those awards that have an industry sponsor are distributed in any given year. Examples of proficiency awards that may be available at the chapter, state, and national levels are the following:

- Agricultural Communications
- Agricultural Education
- Agricultural Mechanics Energy Systems
- Agricultural Mechanics Fabrication and Design
- Agricultural Mechanics Repair and Maintenance – Entrepreneurship
- Agricultural Mechanics Repair and Maintenance – Placement
- Agricultural Processing
- Agricultural Sales – Entrepreneurship
- Agricultural Sales – Placement
- Agricultural Services
- Agriscience Research – Animal Systems
- Agriscience Research – Integrated Systems
- Agriscience Research – Plant Systems
- Beef Production – Entrepreneurship
- Beef Production – Placement
- Dairy Production – Entrepreneurship
- Dairy Production – Placement
- Diversified Agriculture Production
- Diversified Crop – Entrepreneurship
- Diversified Crop – Placement
- Diversified Horticulture
- Diversified Livestock Production
- Emerging Agricultural Technology
- Environmental Science and Natural Resources Management
- Equine Science – Entrepreneurship
- Equine Science – Placement
- Fiber and/or Oil Crop Production
- Food Science and Technology
- Forage Production
- Forest Management
- Fruit Production
- Goat Production
- Grain Production – Entrepreneurship
- Grain Production – Placement
- Home and/or Community Development
- Landscape Management
- Nursery Operations
- Outdoor Recreation
- Poultry Production
- Sheep Production
- Small Animal Production and Care
- Specialty Animal Production
- Specialty Crop Production
- Swine Production – Entrepreneurship
- Swine Production – Placement
- Turfgrass Management
- Vegetable Production
- Veterinary Science
- Wildlife Production and Management

FFA members may apply for the proficiency awards that fit their own situations. Medals and certificates are awarded to winners at the chapter level, whereas plaques and financial awards are given to those at the state and national levels. More information on proficiency and other award programs is found in the FFA *Student Handbook*.

American Star Awards Program

The National FFA Organization recognizes strong SAEs by selecting winners of the American Star in Agriscience, American Star in Agribusiness, American Star in Agricultural Placement, and American Star Farmer awards. These four awards are available at the chapter, district, state, and national levels. In addition, the National FFA Organization names local and national winners of Agricultural Entrepreneur awards. All of these awards are based on the quality of the students' SAEs. Students receive cash awards, and they are recognized for their achievements at state and national conventions and in local chapter banquets.

TYPES OF SAE PROGRAMS

Supervised agricultural experiences grow out of planned programs (Figure 5-4). The word **project** is used to describe a series of activities related to a single objective or enterprise, such as raising rabbits, building a porch, or improving wildlife habitats. The term **enterprise** generally refers to a business that raises animals or plants. Examples are dairy, beef, or rabbits, and corn, hay, turf, or poinsettias. Students may choose from various types of SAEs (Figure 5-5).

Exploratory SAE

An **exploratory supervised agricultural experience** conducts supervised activities to explore a variety of subjects about agriscience and careers in agriscience. Such experiences help students gain understanding of and appreciation for agriscience to satisfy personal interests and needs.

Exploratory SAEs might include investigations and experiences in small animal health, biotechnology, water rights, agriscience journalism, aquaculture, hydroponics, air pollution, crop science, tissue culture, agriscience engineering, and many other areas. The student's exploratory SAE is planned by the student under the direction of the teacher in cooperation with the parent/guardian, mentor, and others who will help the student obtain the exploratory experiences. A mentor is a person whom you admire and who has skills that you would like to learn.

Courtesy of USDA/ARS #K-3405-11.

FIGURE 5-4 Agriculture teachers and students hone specific skills prior to participating in competitive events.

Exploratory SAEs are intended for students who wish to observe and experience a variety of areas in agriscience or to explore one or more areas not covered sufficiently in class to satisfy the student's interest. Such SAEs add an exciting dimension to courses at the elementary or middle school levels where agriscience literacy and career exploration are emphasized.

TYPES OF SUPERVISED AGRICULTURAL EXPERIENCES				
Agriscience Research Project	Exploratory SAE	Entrepreneurship SAE	Placement SAE	Improvement Activities
Ag Problems/ Issues	Careers	Production Enterprise Agribusiness Enterprise	Farm/Ranch Placement (for wages and experience) Agribusiness Placement (for wages and experience) School Greenhouse or Farm	New Construction Repairs

FIGURE 5-5 Supervised agriscience experience.

Agriscience literacy means education in or understanding about agriscience. Literacy does not require the student to become proficient in a given area, but it requires general knowledge about an area. **Career exploration** means learning about occupations and jobs that could possibly become one's future **career** or life's work.

Research/Experimentation and Analysis SAE

Research is another possible area for students to obtain excellent SAEs. Research may be done in school laboratories, at home, on the job, or wherever suitable facilities may be found. Research is generally not regarded as a profit-making activity but may be conducted to answer questions that relate to increasing profits if applied in production settings. Research projects may be part of community improvement activities or state research efforts such as stream monitoring, weather watch, forest fire watch, crop scouting, insect or weed monitoring, crop reporting, and other similar projects.

The **agriscience research project** is an original research project. It consists of identifying an agriscience problem, reviewing the scientific literature, applying the **scientific method** to the problem, and reporting the results. The scientific method is sometimes referred to as the scientific method of inquiry or discovery. A real problem should be identified, and some kind of experiment or testing procedure should be used to test the hypothesis that you have developed concerning the problem. You will usually be able to find a mentor to help you design an accurate experiment or testing procedure. The mentor can be a teacher in a local high school or college, or perhaps he or she can be a science professional in a business or industry. In many cases, the mentor will help you find a laboratory to work in and equipment that can be used to conduct tests and measurements.

Research skills are valuable for solving problems and for working in interesting and well-paying careers. As a special incentive for teaching and learning research skills, the National FFA distributes awards to outstanding agriscience students and teachers annually.

The final step is to report the results of your research to the public. This is done by writing a report. Sometimes a press release to local newspapers will attract the interest of a reporter who is willing to help you. The ultimate step is to enter the National FFA Agriscience Student Scholarship competition or science fair and/or submit the report to a scientific journal.

Ownership/Entrepreneurship SAEs

Entrepreneurship supervised agricultural experience refers to supervised activities conducted by students as owners or managers for profit (Figure 5-6). Emphasis is placed on developing the skills of the job or enterprise and working in a profitable and professional manner. Students may develop and own plant, animal, recreation, or other enterprises wherein they provide services to agriculturists or grow commodities such as flowers, fruits, vegetables, field crops, turfgrass, Christmas trees, nursery stock, trees, small animals, wildlife, beef, sheep, swine, honeybees, earthworms, and other commodities. A project that produces raw materials such as crops or livestock is a **production enterprise**. Agriculture students are encouraged to own or work with a production enterprise to learn about agriculture through hands-on experiences (Figure 5-7).

Entrepreneurship activities may also be conducted in agribusiness. An agribusiness entrepreneurship enterprise is one in which the student buys and sells an

EXAMPLES OF ENTREPRENEURSHIP

A production enterprise is a crop, livestock, or agribusiness venture. The student may own or be employed on a production enterprise.

Types of Crop Enterprises

- Corn Production
- Soybean Production
- Small Grain Production
- Greenhouse Production
- Nursery Production
- Hay Production
- Vegetable Production
- Fruit Production
- Forestry Production
- Christmas Tree Production

Types of Animal Enterprises

- Commercial Cow–Calf Production
- Registered Breeding Stock Production
- Market Beef Production
- Dairy Production
- Feeder Pig Production
- Market Swine Production
- Sheep Production
- Poultry Production
- Horse Production
- Rabbit Production

Types of Agribusiness Enterprises

- Lawn Service
- Custom Farm Work
- Trapping and Pelt Sales
- Hunting Guide Service
- Tree Service
- Farm and Garden Supply Service
- Artificial Insemination Business
- Animal Care and Boarding
- Winery
- Fishing and Crabbing for Sales
- Custom

FIGURE 5-6 Agriscience students have many production projects from which to choose.

FIGURE 5-7 Agriculture students are encouraged to own or work with a productive enterprise to learn about agriculture through hands-on experience.

agricultural commodity for profit rather than raising or growing the commodity. Some examples include a pet business, florist shop, livestock sales business, game dressing service, crop scouting service, crop spraying service, feed sales, seed sales, flower vendor, auctioneer, agriscience mechanic, and trucker.

Placement SAEs

Placement supervised agricultural experiences place the student with an employer in a production unit such as a farm, ranch, greenhouse, nursery, or aquaculture facility to produce commodities for wages. The student may also be placed with an employer or mentor in an agency or agribusiness where commodities are bought and sold or where agricultural services are rendered. Some examples of agribusinesses are veterinary centers, kennels, feed or seed stores, pet shops, nursery outlets, florists, and garden centers. Some examples of agencies where students may be placed are Cooperative Extension Program, Farm Service Agency (FSA), Forest Service (FS), wildlife and environmental agencies, and school laboratories. The emphasis in placement is learning real skills and becoming proficient in a chosen agriculturally related career.

Agriscience Internship

Many agricultural businesses sponsor internships within their organizations. An **internship** may be a paid or unpaid work experience that allows a student to work in an industry. The student learns what a career in the industry is really like by experiencing a variety of jobs within the industry and gaining valuable career experiences to list on a resumé. The sponsoring organization benefits by having the opportunity to evaluate the student as a potential employee after graduation.

Improvement activities are projects that improve the appearance, convenience, efficiency, safety, or value of a home, farm, ranch, agribusiness, or other agriscience facility. For example, construction of a deck on a home is a challenging and worthwhile improvement activity (Figure 5-8). The student does not receive a wage or profit for conducting improvement activities. However, the

SCIENCE CONNECTION

AGRISCIENCE RESEARCH PROJECT

Courtesy of DeVere Burton.

Whether you prefer to be a technician or a professional, agriscience offers a broad array of career possibilities.

Agriscience research projects benefit everyone. For example, a newspaper article motivated an agriscience student in Meridian, Idaho, to conduct research on contamination of surface water in the delivery and drainage canals of a large irrigation project. The newspaper had reported that irrigation practices were causing the Snake River to become polluted with nitrates and phosphates that were dissolved in runoff water from the farm fields. Renee Burton gathered water samples throughout the irrigation season and tested them at a local university science laboratory with the help of the chemistry professor. Her research data showed that the algae problems in the river did not originate from the irrigated farmland, but from some other unknown source. Her efforts in this project resulted in being named a national agriscience student finalist by the National FFA Organization.

INTERNET KEY WORDS:
FFA Agriscience Research

Courtesy of DeVere Burton.

FIGURE 5-8 Construction of a deck is an example of an improvement activity that adds to the comfort and value of the home or other property.

student benefits by learning new skills and enjoying the benefits of the improvements. The owner of the facility should provide the materials and cover other expenses. Improvement activities provide the student with many opportunities to learn without the risks, commitment, and financial backing that is necessary for entrepreneurship activities (Figure 5-9).

EXAMPLES OF IMPROVEMENT ACTIVITIES

Soil Improvement Programs

- Liming
- Fertilizing
- Drainage
- Erosion control
- Plow under green manure
- Introduce a cropping system
- Soil sampling

Building Improvement Programs

- Painting
- Window repair
- Roof repair
- Foundation repair
- Floor repair
- Siding repair
- Door repair
- Electric wiring
- Water systems installation
- Heating
- Lighting protection
- Feeding floor construction
- Remodeling
- Home sewage system installation

Fence Improvement

- Construction of new fence
- Fence replacement and repair
- Construction of floodgates
- Construction and repair of gates

Homestead Improvement Programs

- Plan and set out a windbreak
- Seed or reseed lawn
- Plant shrubs and trees
- Clean up homestead

Orchard, Small Fruits, and Vegetables Improvement Programs

- Plan and set out a fruit tree orchard
- Plan and set out a small fruit garden
- Plan and grow a home vegetable garden
- Renovate an old orchard

Weed Control Programs

- Spray major weed areas on farm
- Mow or spray weeds in fence rows
- Pull weeds in corn
- Clip weeds in permanent pastures

Insect and Pest Control

- Rat control
- Corn borer control
- Japanese beetle control
- Livestock parasite control

Farm Management Programs

- Keep farm accounts
- Plan farm safety program
- Inventory farm equipment
- Keep checking account records

Agricultural Shop Programs

Agricultural Machine Repair and Reconditioning Programs

Livestock Improvement Programs

Crop Improvement Programs

Landscaping Improvement Programs

General Clean-up Program

- Remove dead trees or shrubs
- Remove and discard dead branches
- Remove unsightly junk, trash, and woodpiles
- Provide a specific storage area for all lawn and horticulture equipment
- Repair or remove broken lawn furniture
- Improve the grade if necessary
- Remove, replace, or repair fences, sidewalks, step railings, or porches
- Transplant trees, shrubs, or flowers
- Make or improve the driveway
- Pick up nails and broken glass
- Remove unsightly rocks

Grounds Maintenance Practices

- Mow the lawn
- Trim or prune trees and shrubs
- Fertilize the lawn, trees, shrubs, and flowers
- Apply herbicides
- Apply insecticides
- Water the lawn, trees, shrubs, and flowers
- Repair trees
- Brace trees
- Prevent sunburn of plants
- Prevent insect damage and disease
- Set up, adjust, and move a sprinkler
- Edge the lawn
- Stake trees
- Set up rain gauge
- Record precipitation from rain gauge
- Mulch trees, shrubs, and flowers
- Rake the lawn

Improvement and Beautification Activities

- Plant new trees and shrubs
- Draw a landscape plan
- Seed, plug, or sod lawn
- Plant flowers
- Relocate and replant trees and shrubs
- Renovate existing lawn
- Build a patio
- Make a window box
- Build a trellis
- Plant a windbreak

FIGURE 5-9 Improvement activities are available for agriscience students regardless of the home situation.

AGRISCIENCE SKILLS PLAN AND PROFILE

Each student is encouraged to develop an agriscience skills profile to be placed in his or her portfolio. An agriscience skills profile is a record of skills that the student has developed and a measurement of the level of competence in each. Documentation is important because it provides a record of employment skills and how well the student can perform the skills. Competent people who can document their skills are more likely to be hired in the jobs and careers they desire. Time and resources are limiting factors in preparing for a career, so every effort should be made to learn the most useful and interesting skills. Skills may be obtained in the classroom, laboratory, community, and through participating in SAEs. Lists of appropriate skills for various areas in agriscience should be helpful for the student and teacher as they develop the agriscience skills plan and profile (Figure 5-10).

EXPLORING OPPORTUNITIES FOR THE SAE

Students should use great imagination when considering the SAE in which they will participate (Figure 5-11). Some students may not have very many opportunities for meaningful SAEs. Yet other students in similar circumstances find or create opportunities for effective programs. Seek the advice of your teacher for ideas. Also, observe what successful students have done in the agriscience program and in your community. Then develop an SAE that provides the opportunity to learn and earn.

Personal Interest

Personal interest is an important factor in the success of an SAE. Consider the kinds of activities you like to do and then build on those interests. A Student Interest Survey or Inventory should help you assess your natural interests and provide some guidance in developing an SAE (Figure 5-12).

Resource Inventory

A resource inventory is a listing of the assets and sources of help that may be available for conducting SAE activities. It includes information about your home, farm, work setting, and community that might be useful in considering your SAE (Figure 5-13). Part of the inventory is a scale drawing of the property where you live or work. Making the scale drawing will help you realize what is available and it will help your teacher suggest SAE possibilities.

SELECTING AND IMPLEMENTING YOUR SAE

After completing the Personal Interest Survey and the Resources Inventory, you should arrange a conference with your teacher. The conference should include discussions of your interests, and it should take a look at the possible production enterprises, improvement projects, and supplementary skills available to you. After the conference, you should record tentative plans for the SAE (Figure 5-14). At this point, you and your teacher should discuss the plan with your parents or guardians. If an employer is involved, he or she should become a partner in the planning process.

EXAMPLES OF AGRISCIENCE SKILLS

Agribusiness

- Operate cash register
- Display merchandise
- Keep inventory records
- Write sales tickets
- Deliver products
- Set up machinery
- Assemble equipment
- Greet customers
- Close sales
- Answer telephones
- Order merchandise
- Operate adding machine
- Wrap meat
- Cut carcass into wholesale cuts
- Cut wholesale cuts into retail cuts
- Compute sales tax

Agriculture Mechanics

- Arc weld metals
- Oxy-acetylene weld metals
- Cut details with oxy-acetylene
- Operate farm machinery
- Operate wood power tools
- Operate metal power and hand tools
- Recondition and sharpen tools
- Service air cleaner
- Service electric motor
- Store machinery
- Install rings or pistons
- Grind valves
- Wire electrical convenience outlet
- Change transmission fluid
- Fasten sheet metal with rivets
- Repair flat tire
- Calibrate field sprayer
- Pour concrete
- Lay reinforcement steel
- Use farm level

Corn

- Select seed
- Plant seed
- Prepare seedbed
- Calibrate corn planter
- Adjust planter for depth
- Apply dry fertilizer
- Apply anhydrous ammonia
- Conduct corn variety test
- Check harvest losses
- Apply herbicides and insecticides
- Operate combine
- Operate corn picker
- Identify weeds
- Identify insects
- Cultivate corn
- Dry corn artificially

Forages

- Innoculate legume seeds
- Rotate pastures
- Greenchop forages
- Bale hay
- Renovate permanent pastures
- Combine grasses and legumes
- Graze pastures properly
- Rotate pastures

Horticulture

- Plant vegetable garden
- Prepare garden plot
- Plant fruit trees
- Bud-graft
- Cleft-graft
- Whip-graft
- Prepare growing medium
- Sterilize soil
- Pot plants
- Water greenhouse plants
- Root cuttings
- Harvest crop
- Process vegetables
- Store produce
- Fertilize plants
- Determine plant diseases
- Treat deficiency symptoms
- Force blooming

Small Grains

- Calibrate grain drill
- Plant small grains
- Broadcast fertilizer
- Select adapted varieties
- Clean seeds
- Harvest small grains

Soil Management

- Test soil for lime requirements
- Lime soils
- Seed grass waterway
- Rotate crops
- Plant windbreak
- Test for fertilizer
- Fertilize soils
- Lay drainage tile
- Terrace fields
- Farm on contour
- Aerate soil
- Control erosion

Soybeans

- Test for germination
- Inoculate seed
- Take soil sample
- Control weeds
- Treat seed for storage
- Market beans
- Harvest soybeans
- Check harvest losses

Beef Cattle

- Assist cow in calving
- Dehorn
- Castrate bull calves
- Disinfect navel of calves
- Select herd sire
- Select replacement heifers
- Creep feed calves
- Cull poor producers
- Ear tag
- Tattoo
- Remove warts
- Drench
- Treat for bloat
- Treat for external parasites
- Select feeders
- Trim hooves
- Vaccinate for blackleg
- Vaccinate for infectious bovine rhinotracheitis (IBR)
- Vaccinate for brucellosis
- Formulate a balanced ration
- Ring bull
- Fit animal for show or fair
- Analyze production records
- Palpate to determine pregnancy

continues

continued

Dairy Cattle

- Select replacement heifers
- Dry cows
- Control external parasites
- Assist newborn calves in getting colostrum
- Vaccinate heifers for brucellosis
- Test cows for T.B.
- Cull low producers
- Dehorn
- Production test cows
- Participate in Dairy Herd Improvement Association (DHIA)
- Treat mastitis
- Operate milking machines
- Clean facilities after milking
- Artificially inseminate cows
- Prevent milk fever
- Formulate balanced ration
- Wash udder with disinfectant
- Detect abnormal milk with strip-cup
- Feed according to production
- Clean and sterilize utensils

Poultry

- Clean and disinfect brooder
- Cull poor producers
- Select pullets
- Sex baby chicks or poults
- Prevent cannibalism
- Keep production records
- Dress broilers
- Prevent breast blisters
- Stimulate egg production
- Disinfect laying house
- Grade eggs by candling
- Size eggs
- Debeak chicks
- Vaccinate for fowl pox
- Treat for external parasites
- Worm poultry
- Provide sanitary water
- Feed balanced rations
- Provide litter

Sheep

- Select ram
- Flush ewes
- Treat for external parasites
- Assist ewes at lambing
- Creep feed lambs
- Dock lambs
- Castrate lambs
- Determine pregnancy in ewes
- Ear tag
- Shear sheep
- Tie fleece
- Cull farm flock
- Determine estrus in ewes
- Trim hooves
- Worm for internal parasites
- Keep production records

Swine

- Select herd boar
- Select replacement gilts
- Flush gilts
- Vaccinate sows for leptospirosis
- Provide farrowing stalls
- Clean facilities before farrowing
- Clean sow before farrowing
- Assist sow at farrowing
- Treat navels of baby pigs
- Clip needle teeth
- Creep feed pigs
- Ring hogs
- Treat for external parasites
- Inject iron in baby pigs
- Vaccinate for erysipelas
- Vaccinate for brucellosis
- Wean pigs at 4–6 weeks
- Weigh pigs at 56 days
- Formulate balanced ration
- Castrate boar pigs

FIGURE 5-10 Hundreds of agriscience skills are useful for employment in agriscience.

FIGURE 5-11 A good SAE plan draws on the ideas of the student, teacher, employer, and other adults who will act as mentors for you.

STUDENT INTEREST SURVEY Place an X in the blank by the tasks that you like to do or would like to learn how to do.		
Tasks Typical of Agribusiness	**Tasks Typical of Horticulture**	**Tasks Typical of Production**
____ Delivering merchandise	____ Applying pesticides	____ Applying pesticides
____ Displaying merchandise	____ Arranging flowers	____ Baling hay
____ Driving trucks	____ Balling and burlapping trees	____ Building fences and buildings
____ Keeping records	____ Building patios	____ Castrating animals
____ Mowing lawns	____ Edging flower beds	____ Cleaning animals
____ Operating cash registers	____ Identifying plants	____ Feeding animals
____ Operating equipment	____ Lifting heavy materials	____ Getting up early
____ Pricing merchandise	____ Making Christmas decorations	____ Handling manure
____ Processing meat, milk, grains	____ Making cuttings	____ Harvesting crops
____ Repairing equipment	____ Mowing lawns	____ Helping parents
____ Selling merchandise	____ Mulching beds	____ Keeping records
____ Stocking shelves	____ Operating power machinery	____ Lifting heavy materials
____ Taking customer orders	____ Planting bulbs	____ Milking cows
____ Taking inventory	____ Planting grass	____ Operating machinery
____ Taking telephone orders	____ Planting seeds	____ Painting buildings
____ Unloading trucks	____ Planting trees and shrubs	____ Planting crops
____ Working outside	____ Protecting plants from weather	____ Plowing fields
____ Working with people	____ Pruning plants	____ Repairing buildings
____ Working with plants	____ Raking leaves	____ Repairing machinery
	____ Selling plants	____ Shearing sheep
	____ Watering plants	____ Showing animals
	____ Weeding by hand	____ Taking soil samples
	____ Working in various weather conditions	____ Working in various weather conditions
	____ Working with people	____ Working with animals

FIGURE 5-12 A Student Interest Survey should be helpful to you in choosing a Supervised Agricultural Experience.

Resources Inventory

1. Name ______________________ Age ____ Class ______
2. Address ______________________ Phone ______________
3. Parent's Name __________________ Occupation ____________
4. Number in my family ________ Boys __________ Girls __________
5. I live: on farm ________ in town ________ on an acreage _______.
6. Is land available for you to rent to produce crops? ____yes ____no

   a. If yes, how many acres? ___________

   b. Which crops? ____________________

   c. Location of land? ______________________________
7. Are facilities available for you to rent to produce livestock or livestock products? __________________
   If so,

   a. What type of livestock? ____________________________

   b. Number _______________________________________

   c. Location of facilities ____________________________
8. Do you have available space for a garden? ____yes ____no
9. Do you have facilities for mechanical work? ____yes ____no
10. Do you have a greenhouse available for your use? ____yes ____no
11. Would you be interested in producing livestock or crops on the school farm?
12. Do you have an agricultural job available to you?

    _____yes _____no If so, what type? ____________________

Resources available to the student for the Supervised Agricultural Experience (SAE).

MAP OF HOME FARM AND/OR BUSINESS

Let each square represent any convenient acreage or square footage, i.e., 20, 40, 80, 100, etc.

One square =

LEGEND

Public Road
Private Drive
Farmstead
Terrace
Terrace Outlet
Natural Drain
Gullies
Stream
Pond
Trees
Fence
Railroad
Crossing

FIGURE 5-13 The Resources Inventory is especially helpful in planning production enterprises and improvement activities.

Selecting a Supervised Agriculture Experience for

____
(Name of Student)

Instructions: Use this form to tentatively decide on a beginning agriscience SAE. This information will be helpful in agriscience classes to develop detailed plans for obtaining experiences.

My interest areas in agriscience/horticulture are ______________________________

____

Based upon my interest and opportunities available to me to get practical experience in agriscience, I plan to include the following in my SAE.

1. Production or Productive enterprises (examples: beef, dairy, nursery production, Christmas trees)

____

____

2. Placement in an agribusiness (examples: supply store, florist shop, nursery, golf course, landscape contracting)

____

____

3. Improvement activities (examples: landscape your home, fertilize your lawn, plant trees)

____

____

4. Skills (examples: change spark plugs, weld, change the oil in small engines)

____

____

5. Other activities (example: projects on school facility)

____

____

FIGURE 5-14 Enterprises, improvement activities, and agriscience skills should be selected and recorded early in the school year.

Securing a Job

If placement on a farm or in an agribusiness is part of the SAE plan, you will need a brief **resumé** (a one-page summary of information about a job applicant; Figure 5-15). The prospective employer will be interested in your educational and occupational background, which will help determine your qualifications and experiences. Be sure to ask a minimum of three adults who know your character and qualifications if you may list them as references. Your prospective employer will probably want to contact them.

When your resumé is complete, be sure your teacher has approved the final version. Then you are ready to approach employers for a job that will help achieve the objectives of your SAE plan. Your approach is critical because first

Personal Resumé

Name: Jamison Ledeoux

Address: 1234 Honeylocust Drive
Frederick, Maryland 21701

Telephone: (301)555-0127

Email: JLedeoux@email.com

Education: Junior at Frederick High School

Career Interest: Landscaping

Subjects Studied:
Horticulture I
Horticulture II
Landscaping I
Landscaping II
Typing I

Student Activities:
President: FFA
Editor of school yearbook
Tennis team and softball team

Special Skills:
Ball and burlap trees, operate cash register, water plants, transplant plants, and operate a tractor.

Employment Experience:
Worked as a cashier and cook
Landscaped neighbor's yard

References:
Mr. Ralph Rece, Principal, Frederick County Vo-Tech, Frederick, MD.
Mrs. Holly Deane, Instructor, Frederick County Vo-Tech, Frederick, MD.
Mr. Bernard Rose, Manager, Hardees Fast Food, Frederick, MD.

FIGURE 5-15 A personal resumé will help you when seeking a job.

impressions are lasting impressions. It is important to dress in a businesslike manner, be courteous and confident, and conduct yourself according to standard interview procedures (Figure 5-16).

Refining the Plan

Plans should be worked out for each production enterprise. It is important to develop an accurate estimate of the anticipated expenses and income, which will help you make financial arrangements to conduct the project. Also, some goals for the enterprise should be set, including the size of the project and some efficiency factor goals (Figure 5-17).

PREPARING FOR THE JOB INTERVIEW

What Employers Look for in an Employee

- **Attitude**—The prospective employee should have a positive attitude about the job. He/she should show enthusiasm and a willingness to learn and work. Employers stress this as being one of the most important qualities they look for in prospective employees.
- **Experience**—Previous experience of the prospective employee is important. However, employers are usually willing to train the person with a positive attitude.
- **Appearance**—The prospective employee should be neat and clean, have hair combed, and be well dressed. It is better to be overdressed than underdressed for an interview.
- **Posture**—It is important to stand and sit up straight. The employer will be observing the way you carry yourself and will make judgments accordingly.
- **Mannerisms**—Mannerisms are gestures that are made that may be annoying or could be welcomed. However, one should be aware of mannerisms. Do you:
  1. Yawn a lot? If so, others will think you're bored or worse—lazy.
  2. Fidget? Squirming may indicate lack of confidence or disinterest in the job.
  3. Daydream? Give your full attention to the interviewer.
- **Handshake**—Have a firm handshake; not bone crushing and not limp.

What Questions Should be Asked?

The following questions may be asked if the information is not provided by the interviewer.

- What type of jobs or tasks are to be done?
- What are the policies and procedures for workers?
- What are the working hours?
- What is the rate of pay?
- What arrangements are needed for time off?
- If you are uncertain about something that has been discussed in the interview, you should ask the employer to clarify or explain.
- IN SUMMARY, BE POLITE AND ATTENTIVE DURING YOUR JOB INTERVIEW.

FIGURE 5-16 Preparation for the job interview will permit you to relax and give your total attention to the interviewer.

HOT TOPICS IN AGRISCIENCE SAE—INSIDE THE DAIRY BUSINESS

The Dairy Heifer Replacement Project is a program that is in place in many areas of the United States. It is designed to increase the knowledge and interest of young people in the dairy industry. The goal is to enhance life skills of its youth participants. The project begins when the participant purchases a heifer calf from a program-approved seller. For the next few months, the participant cares for the animal. Specific care must be provided, vaccinations must be given, and a magnet needs to be administered. Quality feed must be provided to ensure proper weight and good health. The participant is required to keep detailed records on the animal. In some instances, the heifer is bred to an approved sire, and the project culminates in the sale of the pregnant heifer when she is presented for show and sale. The Dairy Replacement Project is a challenging SAE that allows each participant to gain an understanding of a crucial part of the dairy industry. Contact the agriculture teacher in your high school or your local extension office for more information about such a program in your area.

SETTING GOALS FOR PRODUCTION IN THE SAE

What goals should be set for the supervised agricultural experience?

DEFINITION: A goal is the hoped-for end result of hard work and should be challenging and realistic.

- Goals should be challenging!
- Goals should be reachable!
- SAE goals should focus on scope, learning opportunities, and production efficiency factors.
- Parents, employers, agriculture teachers, and other qualified adults should help develop SAE goals.
- Goals should be recorded.
- Goals should be analyzed and evaluated periodically, and new goals should be developed.
- Goals provide direction and organization.
- Settings realistic goals should help increase profits.

What are efficiency factors?

DEFINITION: Efficiency factors are measures of production success and encourage enterprise improvement and profit. Examples of efficiency factors are as follows.

- **Size of Enterprise.** For animal weight or livestock products produced.
- **Rate of Gain and Production.**

Beef: $\text{Percent of calf crop} = \dfrac{\text{Calves born alive}}{\text{Cows bred}}$

Poultry: $\text{Percent of egg production} = \dfrac{\text{Average eggs per hen}}{\text{Number days in production}}$

Sheep: $\text{Percent of lamb crop} = \dfrac{\text{Lambs born alive}}{\text{Ewes bred}}$

Swine: $\text{Pigs farrowed per litter} = \dfrac{\text{Live pigs farrowed}}{\text{Sows bred}}$

$\text{Weight produced per litter} = \dfrac{\text{Total production lbs.}}{\text{Number of litters}}$

- **Returns and Feed Costs.** Round total income and value of feed fed to the nearest whole dollar.

$\text{Returns per \$100 feed fed} = \dfrac{\text{Total income}}{\text{Dollars worth of feed fed}} \times 100$

$\text{Returns per \$100 invested} = \dfrac{\text{Total income}}{\text{Total expenses}} \times 100$

$\text{Expense per Cwt. of Production} = \dfrac{\text{Total expenses}}{\text{Total production}} \times 100$

$\text{Average weight sold} = \dfrac{\text{Total sales weight}}{\text{Animals sold}}$

$\text{Average price received} = \dfrac{\text{Total sales value}}{\text{Units sold}}$

- **Feeding Efficiency**

Note: Convert all corn to shelled corn basis (56 lb. per bu.) beforefiguring efficiency factors.
Note: Poultry—1 unit is equal to 1 dozen eggs or 1.5 lb.
Note: Dairy—1 unit is equal to 1,000 lb. milk or 100 lb. weight.

Feed cost per Cwt. or per unit

a. $\text{For swine or beef cattle} = \dfrac{\text{Total feed cost}}{\text{lb. weight produced}} \times 100$

b. $\text{For sheep} = \dfrac{\text{Total feed cost}}{\text{lb. wool + lb. weight}} \times 100$

c. $\text{For dairy or poultry} = \dfrac{\text{Total feed cost}}{\text{Units of production}}$

Feed per Cwt. produced

a. $\text{For hogs, beef cattle, or sheep} = \dfrac{\text{lb. feed fed}}{\text{lb. weight produced}}$

(For sheep, include wool with weight as in lb. above)

b. $\text{For dairy or poultry} = \dfrac{\text{lb. of each feed fed}}{\text{Units of production}}$

- **Death Loss**

$\text{Percent death loss} = \dfrac{\text{Number of dead animals}}{\text{Total number of dead produced and purchased}}$

A low percentage of death loss means a high enterprise rating for this item.

FIGURE 5-17 Goals should state the number, size, timelines, and efficiency factors you plan to achieve.

© Goodluz/Shutterstock.com.

FIGURE 5-18 The employer acts as a mentor and supervisor to a student who participates in a placement SAE.

Employers are generally impressed with students who are eager to learn (Figure 5-18). In this regard, both the student and the employer can benefit from a carefully thought-out statement of skills to be developed on the job. You should develop the list with the guidance of your teacher and prepare an Experience Inventory to record the completion of tasks or jobs (Figure 5-19). It should be frequently updated. The experience inventory helps the student, teacher, and employer track progress in achieving goals and in developing a skills profile.

The Placement Agreement document helps finalize the plans for placement on a job (Figure 5-20). Such agreements need the signature of the student, parents or guardians, employer, and teacher. Once all parties are in agreement regarding the student's placement experiences, the chances for success will be enhanced.

Another document that will help plan and conduct the SAE is the Improvement Project Plan and Summary (Figure 5-21). This plan directs the student to describe the conditions found, plans for improvement, and estimated value of the improvement when completed. The summary is filled out as the improvement project progresses and serves as the record when finished.

Finally, a skills plan should be developed. These skills are chosen from lists that apply to the student's community and are recorded on the Agriscience Skills Plan and Record (Figure 5-22). The date completed should be added when the skill is acquired.

EXPERIENCE INVENTORY

Directions: Complete the following information sheet by listing any experiences you have had or would like to gain in the field of agriculture.

Tasks or Jobs	Can perform without help	Can perform with help	Can help perform	Cannot or have not performed	Would like to learn how to perform	How or where to obtain experience
Examples						
• Drive tractor		X				Landscaping business
• Take cuttings					X	Agriscience Class
• Keep records				X		SAE & Technical Skills
1. ________						
2. ________						
3. ________						
4. ________						
5. ________						
6. ________						

FIGURE 5-19 The Experience Inventory is a device to plan and record experiences you plan to gain. It is also a mechanism for recording how well you have learned new skills.

PLACEMENT AGREEMENT

To provide a basis of understanding and to promote business-like relationships, this memorandum is established on ______________, 20 ____ . This work will start on ___________, 20____, and will end on or about _____________ , 20 ____ , unless the arrangement becomes unsatisfactory to either party before the ending date. Person (employer) responsible for training ______________________________

The usual working hours will be as follows:

1. While attending school working hours shall be ____________________
   When not attending school working hours shall be ____________________
2. Provisions for overtime ______________________________
3. Provisions for time off ______________________________
4. Liability insurance coverage (type and amount) ____________________

Wages will be at the following rate(s): ______________________________
Trial period ______________________________
Remainder of the agreement period ______________________________
And will be paid (when?) ______________________________

A. IT IS UNDERSTOOD THAT THE EMPLOYER WILL (check the items that apply):

______ Provide the student with opportunities to learn how to do well as many jobs as possible, with particular reference to those contained in the planned program;
______ Coach the student in methods found desirable in implementing project activities and handling management problems;
______ Help the student and teacher make an honest appraisal of the student's performance;
______ Avoid subjecting the student to unnecessary hazards;
______ Notify the parents and the school immediately in case of accident or sickness and if any other serious problem arises;
______ Assign the student new responsibilities when he/she can handle them;
______ Cooperate with the teacher in arranging conferences with the student on supervisory visits; and/or
Other:
____
____

B. THE STUDENT AGREES TO (check the items that apply):

______ Do productive work, recognizing that the employer must profit from the student's labor in order to justify employment;
______ Keep the employer's interest in mind and be punctual, dependable, and loyal;
______ Follow instructions, avoid unsafe acts, and be alert to unsafe conditions;
______ Be courteous and considerate of the employer, the family, and others;
______ Keep such records of work experience and make such reports as the school may require;

PLACEMENT AGREEMENT *(cont'd.)*

______ Develop plans for management decisions with the employer and teacher; and/or
Other:
____
____

C. THE TEACHER, ON BEHALF OF THE SCHOOL, AGREES TO (check the items that apply):

______ Visit the student on the job at frequent intervals for the purpose of instruction and assurance that the student gets the most education out of the experience;
______ Show discretion in the time and circumstances of these visits, especially when the work is pressing; and/or
______ Provide appropriate job-related instruction at school; and/or
Other:
____
____

D. THE PARENTS AGREE TO (check the items that apply):

______ Assist in promoting the value of the student's experience by cooperating with the employer and the teacher of agriscience;
______ Satisfy themselves in regard to the living and working conditions made available to the student; and/or
Other:
____
____

E. ALL PARTIES AGREE TO:

______ An initial trial period of ____ working days to allow the student to adjust and prove himself/herself;
______ Discuss any issues concerning the job with the teacher before ending employment and/or
Other:
____
____

STUDENT's Signature ______________
Address ______________
____
Telephone Number ______________
Social Security No. ______________
PARENT's Signature ______________
Address ______________
____
Telephone Number ______________

EMPLOYER's Signature ______________
Address ______________
____
Telephone Number ______________
TEACHER's Signature ______________
School Address ______________
____
Telephone Number ______________
School Telephone Number ______________

FIGURE 5-20 A Placement Agreement states what each party is expected to do. It promotes good planning and reduces misunderstandings and conflicts.

IMPROVEMENT ACTIVITY PLAN AND SUMMARY

Improvement Project No. ________

A. Conditions found:

____
____
____
____
____

B. Plans for improvement (including costs):

____
____
____
____
____

C. Value of improvement when completed:

____
____
____
____
____

AGRICULTURAL IMPROVEMENT PROJECT SUMMARY

Started: ____________, _____ Completed: ____________, _____

Date	Jobs Done	Hours of Labor	Cost of Materials & Equipment

FIGURE 5-21 Improvement activities should be planned and records kept on the jobs, hours, and costs involved.

SUPPLEMENTARY AGRISCIENCE SKILLS PLAN AND RECORD

Directions: Using the list of Supplementary Practices supplied by the instructor, complete the chart below by choosing skills you would like to include as part of your SAE.

Skills, Practices, Job, or Experience	Place to obtain skill	Date planned to obtain skill	Date completed
Examples:			
• Operate Cash Register	On-Job	Sept. 16	Sept. 16
• Bud-Graft	School Farm	February	Feb. 20
1. ____________			
2. ____________			
3. ____________			
4. ____________			
5. ____________			
6. ____________			
7. ____________			

FIGURE 5-22 Agriscience skills should be selected at the beginning of the year with plans for times and places to complete each item.

SCIENCE CONNECTION

TRACKING THE DISAPPEARING GENE

Students can gain valuable research skills in supervised agriscience experience programs.

The USDA Agriculture Research Service (ARS) is on the lookout for good students who are seeking experience in research. Plant geneticist Thomas E. Devine, of the USDA-ARS plant molecular biology laboratory in Beltsville, Maryland, researched the genetic structures and characteristics of soybean plants for a significant part of his professional life. A succession of talented high school students have worked with him. These students have done work of real scientific significance and have made original contributions to science. Together, they conducted long and detailed studies of thousands of soybean plants to track down the genes responsible for disease resistance and nitrogen fixation.

Nikola Lockett, while a high school junior, started a two-summer experience program at the USDA Southern Regional Research Center. Later, while a student at Xavier University in New Orleans, she was able to continue her work with a plant physiologist as part of the Cotton Fiber Bioscience team at the Center. Similarly, a high school research apprenticeship program attracted students to the ARS Arthropod-Borne Animal Diseases Research Laboratory in Laramie, Wyoming. Students have assisted with research projects involving insects and diseases of livestock and poultry.

The USDA has numerous plant, animal, disease, insect, food, fiber, nutrition, and other research laboratories throughout the United States. For information on research assistance opportunities for students, do an online search for USDA-ARS internship opportunities.

STUDENT ACTIVITIES

1. Write the Terms to Know and their meanings in your notebook.
2. Describe the relationships between (1) classroom instruction and supervised agriscience experience and (2) the FFA program and supervised agriscience experience.
3. Construct a bulletin board showing the relationships presented in Figure 5-1.
4. Study Figure 5-6, and write five ideas for production projects or enterprises to discuss with your teacher.
5. Discuss Figure 5-9 with your parents/guardians, and select two or three improvement activities that you would like to conduct.
6. Review the examples presented in Figure 5-10, and choose 20 agriscience skills from enterprises other than your production projects and improvement activities.
7. Examine the tasks in the Student Interest Survey (Figure 5-12). Determine whether your interests are more in agribusiness, horticulture, production, or other areas of agriscience.
8. Using Figure 5-13 as a guide, draw a map (to scale) of your home, farm, or business where you can conduct an SAE. Make an inventory of the resources that may be available for you to use in conducting an SAE.
9. Talk with your teacher and parents/guardians. Write the names of the projects, activities, and skills that you definitely plan to do during the current year (Figure 5-14).
10. Prepare a personal resumé.
11. Apply and interview for a part-time job.
12. Develop an Experience Inventory (Figure 5-19).
13. Work out a Placement Agreement with an employer (Figure 5-20).
14. Set goals for production projects (Figure 5-17).
15. Make detailed plans for improvement activities (Figure 5-21).
16. Select definite agriscience skills. Write down where and when you plan to accomplish the skills (Figure 5-22).

SELF-EVALUATION

A. MULTIPLE CHOICE

1. Conducting an activity in the daily routine of our society is said to be
   a. laboratory experience.
   b. real-world experience.
   c. simulation.
   d. supervised occupational experience.
2. Which is not a purpose or benefit of SAEs?
   a. Become established in an agriscience occupation
   b. Permit early graduation
   c. Permit individualized instruction
   d. Provide educational and practical experiences
3. Which is not a major component of a comprehensive agriscience program?
   a. Classroom/laboratory instruction
   b. FFA
   c. Memorization and recitation
   d. Supervised occupational experience

4. SAEs should be planned
   a. at home with parents/guardians.
   b. in the classroom.
   c. on the job with employers.
   d. all of the above.
5. A student-drawn map of the home property is an important part of a
   a. job interview.
   b. Placement Agreement.
   c. Resources Inventory.
   d. resumé.
6. A production or productive project
   a. is the same as an improvement activity.
   b. is the same as a skill.
   c. may involve either ownership or placement for experience.
   d. must be done without pay or profit.
7. Development of agriscience skills is important because
   a. a skill inventory is part of a career portfolio.
   b. personal agriscience skills help identify career interests.
   c. skills lead to part-time jobs.
   d. all of the above.
8. Improvement activities
   a. are always connected with employment.
   b. focus primarily on leadership development.
   c. must improve some part of the instructional program.
   d. should be without pay.

B. MATCHING

_________ 1. Enterprise
_________ 2. Experience
_________ 3. SAE
_________ 4. FFA
_________ 5. Improvement Project
_________ 6. Agriscience
_________ 7. Production Enterprise
_________ 8. Program
_________ 9. Project
_________ 10. Skill

a. A project or experience in agriculture under the direction of your teacher
b. A project conducted for wages or profit
c. Plans, activities, records, and experiences related to an agricultural enterprise
d. Activities related to a single enterprise
e. The application of science to agriculture
f. A national organization for agriscience students
g. Ability to do something well
h. A business raising animals or crops
i. Activities that improve the appearance, convenience, efficiency, safety, or value of a home or other facility
j. Anything that is observed, done, or lived through

UNIT 6

Leadership Development in Agriscience

OBJECTIVE

To develop basic leadership skills.

MATERIALS LIST

- paper
- pencil or pen
- bulletin board materials
- Internet access

COMPETENCIES TO BE DEVELOPED

After studying this unit, you should be able to:

- define *leader* and *leadership.*
- explain why effective leadership is needed in agriscience.
- list some characteristics of good leaders.
- describe the opportunities for leadership development in FFA.
- demonstrate positive leadership skills.

SUGGESTED CLASS ACTIVITIES

1. Instruct class members to attend a public meeting to observe how leadership skills are used to conduct the business of the community. Include several different kinds of meetings, and assign groups of students to each meeting. Provide a worksheet to each group to be sure that students observe and report on critical leadership activities or skills. For example, you may want to have students critique the use of parliamentary procedure skills, explain how the committee process was used, or describe how community leaders used other leadership techniques.
2. View a television segment of C-Span while a congressional committee is meeting or while the Congress is in session debating a national issue. Identify and discuss the parliamentary rules used. How are they different from Robert's Rules of Order (used by the National FFA Organization)? How are they the same?

TERMS TO KNOW

leadership
plan
citizenship
integrity
knowledge
courage
tact
enthusiasm
selflessness
loyalty
Cooperative Extension System
4-H
Girl Scout
Boy Scout
extemporaneous speaking
parliamentary procedure
business meeting
presiding officer
minutes
order of business
gavel
motion
main motion
amend
refer
lay on the table
point of order
adjourn

3. In small groups, describe the characteristics of a good leader. Produce examples of people who were/are good leaders and explain what makes them a good leader. Present your ideas to the rest of the class.

What is leadership in agriscience? Agriscience has been described as a broad and diverse field. It is not just horticulture or supplies and services; not just professions or products, processing, and distribution; not just mechanics or forestry; and not just renewable natural resources or production. Agriscience is all of these. Then, what is leadership in agriscience?

LEADERSHIP DEFINED

Leadership may be defined as the capacity or ability to solve problems and to set a direction. To lead is to show the way by going in advance or guiding the actions or opinions of others. To do this in agriscience, you must have knowledge of technical information and people. You must know how to organize and manage activities. Most jobs are too big for one person. We can do only part of what needs to be done. Therefore, we need the help of others.

A leader uses the knowledge and skills of others to achieve a common goal. For instance, a quarterback on a football team uses leadership skills to coordinate the team players to achieve a touchdown. Similarly, the wise batter in baseball hits the ball in a place that not only gets the batter on first base but also permits other runners to advance around the bases. A properly placed hit supports the common goal of achieving runs. A single base hit, where everyone advances and no one gets out, may be the best for the team. Conversely, a line drive that doesn't quite make the fence may result in a third out, with no chance for other players to score.

WHY LEADERSHIP IN AGRISCIENCE?

Agriculture is a highly organized industry. It involves people and complex processes. Leadership skills are necessary whenever people are assembled. Those who teach in agriscience are part of a team of teachers, principals, supervisors, community advisory groups, and others. Those in agribusiness are typically part of teams consisting of the manager, office staff, sales representatives, field personnel, and board of directors (Figures 6-1 and 6-2). Those on farms may be the owner, manager, spouse, children, hired help, or neighbors who assist at times. The manager of a farm or business must **plan** (think through, determine procedure, assemble materials, and train staff to do a job). Once a job is planned, it may be accomplished through management. To manage is to direct people, resources, and processes to reach a goal. A manager uses leadership skills continuously in working with others on a day-to-day basis.

A landscaper is someone who plans, plants, builds, or maintains outdoor ornamental plants and landscape structures. Sometimes working alone and sometimes working with others, the landscaper is a leader

© iStockphoto/Lise Gagne.

FIGURE 6-1 A leader must learn to coordinate the work of all of the members of a team to avoid having a team member work on something that has already been completed.

© iStockphoto/Vicki Reid.

FIGURE 6-2 Agribusinesses rely on team efforts to achieve goals.

in many respects. When developing a plan for a customer, the landscaper exerts leadership. The landscaper knows the name, function, and performance of each plant. This information is used to develop an acceptable plan. Because the customer has personal ideas about landscaping, a professional must consider these ideas in the plan. It may test the landscaper's leadership skills to lead the customer to an acceptable plan with plants that survive the climate and conform to acceptable landscape practices.

The landscaper and other agriscience personnel may be called on as officers or members of professional organizations to give testimony before legislators or other public officials on the need for laws, regulations, or other actions that affect their work. It may be quite possible to end the day presiding over a meeting or taking minutes at a professional or civic meeting.

For the agriscience student, the need for leadership skills is also apparent. Having confidence to participate fully in class is important (Figure 6-3). To meet prospective employers and conduct a supervised agricultural experience program requires leadership skills. Functioning in the community, participating in group meetings, and readily making friends all require acceptable leadership skills. Good leadership skills greatly improve individual marketability in the working world.

INTERNET KEY WORDS:
leadership traits
leadership skills

Photo courtesy of the National FFA Organization.

FIGURE 6-3 Participation in class discussions is the first step toward developing the qualities of leadership.

INTERNET KEY WORDS:
National 4-H Clubs
National FFA Organization
Boy Scouts of America
Girl Scouts of the United States of America

To be a good citizen, you must earn your way in life without infringing on the rights of others. A part of effective citizenship is using leadership to promote the common good in society. **Citizenship** means functioning in society in a positive way.

TRAITS OF GOOD LEADERS

Good leaders must have **integrity** (honesty). Without it, others cannot trust an individual with the power to manage or control, even in minor things. A leader must have **knowledge**, which means familiarity, awareness, and understanding. Good leaders are dependable and have the **courage** (willingness to proceed under difficult conditions) and initiative to carry out personal and group decisions. To lead, one must demonstrate the initiative to carry out personal and group decisions. To lead, one must also communicate. This requires good speaking and listening skills.

In working with others, **tact**, or the skill of encouraging others in positive ways, is useful. Similarly, a sense of justice to ensure the rights of others is important. **Enthusiasm**, or energy to do a job and inspiration to encourage others, is useful. **Selflessness** means placing the desires and welfare of others above yourself. It, too, is an important quality for good leadership. These traits encourage **loyalty**, which results in reliable support for an individual, group, or cause. These and other traits are achieved through effective leadership development.

LEADERSHIP DEVELOPMENT OPPORTUNITIES

Modern schools provide extensive opportunities for agriscience students to develop leadership skills. Students develop leadership in school organizations, athletics, and in classroom and laboratory situations. Some become leaders at home, on the job, or in community organizations.

4-H Clubs

The **Cooperative Extension System** is an educational agency of the U.S. Department of Agriculture (USDA) and an arm of your state university. It provides educational programs for both youth and adults. Its programs include personal,

HOT TOPICS IN AGRISCIENCE THE AGRICULTURAL LOBBY

Agriculture is affected in many ways by the laws that are approved by Congress and state legislatures. In addition, state and federal agencies write regulations to implement new laws. Farmers, ranchers, and agricultural processors are subject to both the laws and the regulations. In a time when politics affects the agricultural industry in so many ways, agricultural lobby efforts have become important.

Special leadership skills are required by those who lobby on behalf of agriculture. A lobbyist must be registered in most states before he or she can actively lobby a legislature or other political organization. Most lobbyists represent a particular group of farmers or processors, such as the United Dairymen or the Wheat Commission. The lobbyist is paid to negotiate, persuade, and be persistent in promoting legislation that is favorable to the segment of the industry that he or she represents.

© Alric Bolt/Shutterstock.com.

FIGURE 6-4 Many agricultural youth experience their first leadership training as members of 4-H clubs.

home and family, community, and agriscience resources development. The Cooperative Extension System also sponsors 4-H clubs. The **4-H** network of clubs is directed by Cooperative Extension System personnel to enhance personal development and provide skill development in many areas, including agriscience (Figure 6-4). The four Hs in 4-H stand for head, heart, hands, and health. These provide the basis for the 4-H pledge, which is, "I pledge my head to clearer thinking, my heart to greater loyalty, my hands to larger service, and my health to better living for my Club, my community, my country, and my world."

Scout Organizations

Girl Scout and **Boy Scout** organizations provide opportunities for leadership development and skill development in agriscience and other areas. Scouts focus heavily on outdoor activities and provide excellent leadership development and natural resources skills. They provide recognition through a system of merit badges, which are earned by learning skills and obtaining experiences in many areas, including agriscience (Figure 6-5).

FFA

The FFA is a youth-oriented organization that was developed specifically to expand the opportunities in leadership and agriscience skill development for students in public schools. Only students under the age of 21 who are enrolled in a systematic program of agricultural education are eligible for membership in FFA.

Aim and Purposes

The FFA is part of the agriscience curriculum in most schools where agriscience programs are offered. It is an important teaching tool. It serves as a laboratory for

Courtesy of DeVere Burton.

FIGURE 6-5 Boy Scouts of America and Girls Scouts of the United States of America provide opportunities for young people to develop skills in providing leadership to other members of their organizations.

developing leadership and citizenship skills. These, in turn, are helpful in learning agriscience skills. The primary aim of the FFA is the development of agriscience leadership, cooperation, and citizenship. The specific purposes of the FFA may be paraphrased as follows:

AGRI-PROFILE

CAREER AREA: LEADERSHIP DEVELOPMENT

Courtesy of National FFA.

The art of public speaking is one of the powerful tools of a leader.

Leadership development may be a career area or specialty for teachers, consultants, personnel managers, coaches, and others. However, many agriscience positions require good leadership capabilities as a tool for everyday use. Auctioneers, salespersons, managers, entrepreneurs, corporate executives, politicians, and anyone who directs others or routinely meets the public must have strong leadership skills.

Leadership involves good planning, goal setting, and the ability to inspire others to work toward a common goal. Such skills as committee interaction, parliamentary procedure, and self-expression are important leadership techniques that are developed through study and practice. These skills are used in church, civic, and community organizations as well as in workplaces. Group projects and club activities in agriscience provide excellent opportunities for leadership development.

- to develop competent and assertive agriscience knowledge and leadership;
- to develop awareness of the global importance of agriscience and its contribution to our well-being;
- to strengthen the confidence of agriscience students in themselves and their work;
- to promote the intelligent choice and establishment of an agriscience career;
- to stimulate development and to encourage achievement in individual agriscience experience programs;
- to improve the economic, environmental, recreational, and human resources of the community;
- to develop competencies in communications, human relations, and social abilities;
- to develop character, train for useful citizenship, and foster patriotism;
- to build cooperative attitudes among agriscience students;
- to encourage wise use and management of resources;
- to encourage improvement in scholarship; and
- to provide organized recreational activities for agriscience students.

The Emblem

The FFA emblem contains five major symbols that help demonstrate the structure of the organization (Figure 6-6). They are as follows:

Eagle—The emblem is topped by the eagle and other items of our national seal. The eagle was placed in the emblem to represent the national scope of the organization. It could also represent the natural resources in agriscience.

Corn—Corn is grown in every state in the United States. It reminds us of our common interest in agriscience, regardless of where we live.

Owl—The owl represents knowledge and wisdom. Use of this symbol in the emblem recognizes the fact that people in agriscience need a good education and that education must be tempered with experience to be of greatest usefulness.

Plow—The plow has been used to represent work–labor–effort. These qualities are needed to cause things to happen and to get results in agriscience.

Rising Sun—The rising sun is a symbol of the progressive nature of agriscience. It is symbolic of the need for workers in agriscience to cooperate and work toward common goals.

Source: Reprinted by permission of The National FFA Organization.

The FFA emblem may be constructed one symbol at a time. When assembled and dissembled in this manner, it is a good device to help others understand the FFA.

Courtesy of National FFA; FFA #148.

FIGURE 6-6 The FFA emblem contains five meaningful symbols that are important to the organization.

The Colors

The official FFA colors are blue and gold. The shade of blue is national blue. The shade of gold is the yellow color of corn. Therefore, the colors are called national blue and corn gold.

Source: Reprinted by permission of The National FFA Organization.

Motto

The FFA motto contains phrases that describe the philosophy of learning and development in agriscience. The motto is:

- Learning to Do
- Doing to Learn
- Earning to Live
- Living to Serve

"Learning to Do" emphasizes the practical reasons for study and experience in agriscience. It also suggests ambition and willingness to productively use the hands as well as the mind. "Doing to Learn" describes procedures used in agriscience instruction at the doing level. Experiencing results from doing is the most permanent result of learning. "Earning to Live" suggests that FFA members intend to develop their skills and support themselves in life. And "Living to Serve" indicates an intention to help others through personal and community service.

Creed

The creed is a statement of beliefs of the members of the National FFA Organization. It is studied and memorized by agriculture students during their first year of enrollment in an agriculture program. The memorized creed is repeated word-for-word as the first public speaking event in which most students participate.

THE FFA CREED

I believe in the future of agriculture, with a faith born not of words but of deeds—achievements won by the present and past generations of agriculturists; in the promise of better days through better ways, even as the better things we now enjoy have come to us from the struggles of former years.

I believe that to live and work on a good farm, or to be engaged in other agricultural pursuits, is pleasant as well as challenging; for I know the joys and discomforts of agricultural life and hold an inborn fondness for those associations which, even in hours of discouragement, I cannot deny.

I believe in leadership from ourselves and respect from others. I believe in my own ability to work efficiently and think clearly, with such knowledge and skill as I can secure, and in the ability of progressive agriculturists to serve our own and the public interest in producing and marketing the product of our toil.

I believe in less dependence on begging and more power in bargaining; in the life abundant and enough honest wealth to help make it so—for others as well as myself; in less need for charity and more of it when needed; in being happy myself and playing square with those whose happiness depends upon me.

I believe that American agriculture can and will hold true to the best traditions of our national life and that I can exert an influence in my home and community which will stand solid for my part in that inspiring task.

The creed was written by E. M. Tiffany, and adopted at the 3rd National Convention of the FFA. It was revised at the 38th Convention and the 63rd Convention.

Salute

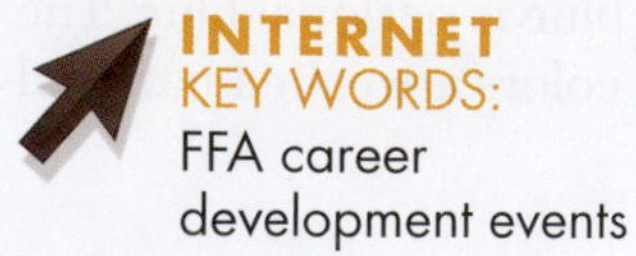

INTERNET KEY WORDS: FFA career development events

The Pledge of Allegiance to the American flag is the official FFA salute. The words of the pledge are the following: "I pledge allegiance to the flag of the United States of America and to the Republic for which it stands, one nation under God, indivisible, with liberty and justice for all."

Degree Requirements

The FFA has four degrees, each of which indicates the progress a member is making. These are Greenhand, Chapter, State, and American degrees. The Greenhand and Chapter FFA degrees are awarded by the FFA chapter in the agriscience department of the school. The State FFA degree is awarded by the State Association. The American FFA degree is awarded by the National FFA.

The Greenhand degree is so named to indicate that the member is in a learning mode. He or she is developing basic skills through FFA participation and by studying the principles of agriscience. To receive the Greenhand degree, the member must meet the requirements spelled out in the current FFA *Official Manual*. In general, the requirements for the Greenhand degree are to:

- be enrolled in an agricultural education course;
- have satisfactory plans for a supervised agricultural experience program;
- recite the FFA creed, motto, and salute;
- describe the FFA emblem, colors, and symbols;
- explain the FFA Code of Ethics and proper use of the FFA jacket;
- have satisfactory knowledge of the history of the organization and of the Chapter Constitution and Program of Activities;
- know the duties and responsibilities of members;
- own or have access to a copy of the *Official Manual* and *FFA Student Handbook*; and
- submit a written application for the Greenhand degree.

The requirements for the other three degrees help the member to learn and grow professionally from one level to the next in the organization. The *Official Manual* contains the exact requirements for all degrees, details of membership, and chapter operation for the FFA.

Career Development Events

The FFA sponsors competitive career development events for a wide range of career interests. The first level is in the local chapter at the school. The second level is the district or regional level within the state. The third level is the state association level, and the fourth is the national level. Local FFA advisors determine which events are appropriate for students in their programs. The competitions that are conducted at each level should reflect the content of the instructional programs.

The purpose of FFA career development events is to encourage agriscience students to develop technical and leadership skills and to practice these skills in friendly competition with other FFA members. These events include the following:

Agriculture Communications
Agriculture Issues
Agriculture Mechanics
Agriculture Sales
Agronomy
Creed Speaking
Dairy Cattle Evaluation
Floriculture
Food and Science Technology
Forestry
Horse Evaluation
Job Interview
Livestock Evaluation
Marketing Plan

Dairy Foods
Dairy Handlers
Environmental and Natural Resources
Extemporaneous Public Speaking
Farm Business Management
Meats Evaluation and Technology
Nursery and Landscape
Parliamentary Procedure
Poultry Evaluation
Prepared Public Speaking

Some FFA career development events require contestants to know how to grade agricultural products, such as eggs, meats, poultry, fruits, and vegetables. Other events require students to evaluate live animals, such as beef and dairy cattle, horses, poultry, sheep, and swine. Some require mechanical abilities, such as welding, plumbing, electronics, irrigation, surveying, engine troubleshooting and repair, painting, woodworking, and general tool use (Figure 6-7).

Some events, such as Floriculture, require knowledge of the art and science of floral arrangement. Some career development events include Forestry and Nursery/Landscape competitions, which require students to identify plants, plant materials, insects, and diseases. Land Judging (conducted by soil and water districts) involves evaluating the soil and land and recommending appropriate management practices. Parliamentary Procedure event teams demonstrate their ability to conduct meetings using their knowledge of correct parliamentary practices. These procedures are used to open meetings, conduct business, close meetings, and write minutes according to acceptable practice in the real world. Various types of speaking contests help individuals to sharpen their speaking skills.

INTERNET KEY WORDS:
how to give a speech

Each of these skill events is organized to ensure that the participants are competing as individuals and as teams of three or four participants. An example of a chapter achievement is the National Chapter Award. This competition involves most or all of the students in the agriscience program and encourages them to make valuable improvements within the program, school, and community.

FIGURE 6-7 Students enjoy learning by participating in a career-development event.

AGRI-PROFILE BE ALL YOU CAN BE

Courtesy of National FFA.

FIGURE 6-8 FFA members participate as delegates to state and national meetings, where they introduce, debate, and ultimately approve or disapprove proposals from members to change the way the organization is run.

On the day after Christmas in 1960, a 12-year-old boy with polio was delivered to a ranch for needy boys, together with his two brothers. Their broken home was without heat, and food was scarce. After five years of growing up on the ranch, the boy enrolled in a high school agriculture program and joined an FFA chapter.

Although love of animals attracted him to the FFA, it was recognition for his successes on the parliamentary team that spurred him on. "It was the first time I had ever won anything," was his later observation. From the parliamentary team, he advanced to area FFA president and on to state president and, eventually, he was a delegate to the national FFA convention (Figure 6-8). It was there that he presented the historic motion to open FFA membership to female members.

After graduation from college, he followed in the footsteps of his FFA advisor, teaching agriculture and inspiring young people to be all they could be. As FFA advisor, he insisted that his students could prepare themselves for any career through leadership training, talking on their feet, developing responsibility, learning to manage money, experiencing teamwork, learning respect, setting goals, and planning ahead. These experiences help the individual win in life.

His experiences in FFA, teaching, and advising helped him develop the confidence and skills he needed for a career in public life. After teaching for a while, he served eight years in the Texas Senate before going on to the U.S. House of Representatives. There, his leadership skills were soon recognized; he was elected president of the Freshman Class of Congressmen and eventually became a valued member of the House Agriculture Committee.

What advice does this highly successful statesman give to young people pursuing careers in agriscience? The Honorable Congressman Bill Sarpalius suggests the following:

1. Be in the right frame of mind. Don't dwell on your handicaps or lack of ability like in public speaking or running. Forget, "I can't."
2. Avoid negatives.
3. Stay physically and mentally sharp. Don't let yourself get lazy.
4. Develop a religious background.
5. Concentrate on doing for others—not yourself.

In summary, he asserts, "To achieve all that is possible—we must attempt the impossible. To be as much as we can be—we must dream of being more!"

PUBLIC SPEAKING

Oral communication skills are important for good leadership. Effective leaders must speak with individuals, committees, small groups, and in large forums. The ability to relax, speak clearly, and state what is pertinent to the subject at hand is useful. These skills are developed by applying basic principles of speech preparation and organization.

Speeches may be prepared or extemporaneous. **Extemporaneous speaking** is delivering a speech with little or no time for preparation. Extemporaneous speaking is a real-life skill that is used daily in agriscience careers. The ability to speak extemporaneously is enhanced by learning to deliver prepared speeches. Both prepared speeches and extemporaneous speaking are used to teach public speaking skills in FFA.

The National FFA Organization sponsors three different types of speech competitions: the Creed Speaking, Prepared Public Speaking, and Extemporaneous Public Speaking contests. Each event provides opportunities for students to stand before an audience and deliver a speech in a public setting.

Creed Speaking

The Creed Speaking event is for students who are enrolled in an agriscience class for the first time. It requires the speaker to repeat the FFA Creed from memory. The emphasis is on accuracy and delivery. When the speaker is finished, a specific statement taken from the creed is cited. The speaker explains what the statement means to him or her. Each contestant responds to the same statement.

The Creed Speaking event gets students started in speaking without undue concern over the content of the speech. It is a good way to help students succeed before they are expected to prepare the written content of a speech. Once confidence is gained, students often become motivated to participate in other speech contests.

Prepared Speaking

The Prepared Speaking contest provides an opportunity for a student to research an agricultural topic and develop his or her own ideas. The content of the speech must be original, not copied from someone else. The length of the speech should be six to eight minutes, and points are deducted if the speech is too long or too short.

Part of this speech competition is based on the quality of the written manuscript. The neatly typed manuscript (double-spaced) is given to the judges ahead of time. Questions are developed from the manuscript. Each contestant responds to the questions for five minutes following the delivery of the speech. Contestants are judged on the quality, effectiveness, and accuracy of the written manuscript, the speech delivery, and response to the questions.

Extemporaneous Speaking

Extemporaneous speaking is a valuable skill for life. This skill is used every day by most agriscience professionals. Salespeople use extemporaneous speaking skills to negotiate sales. Agricultural educators use these skills to teach classes or to teach individuals how to deal with problems and issues. Agricultural executives and administrators use extemporaneous speaking skills to convince their employees and stockholders to support their leadership and business plans. Nearly everyone can benefit from learning extemporaneous speaking skills.

Extemporaneous Speaking competitions require students to gather original documents and materials in a notebook or file, but no written preparation of a manuscript may be done before the competition. Each competitor draws for a speech topic, and he or she is allowed to prepare for 30 minutes using only the materials that were assembled earlier. The speech length is five to eight minutes, and the judges are allowed five minutes to ask questions after the speech has been delivered.

Planning the Speech

A speech should have at least three sections: the introduction, body, and conclusion. The plan should clearly identify the sections.

Introduction

The introduction indicates the need for and importance of the speech. It should be carefully planned and spoken with confidence. The introduction may be in the form of statements or questions. If the introduction is not to the point, does not fit the occasion, or is not delivered in a spirited manner, the audience may not listen to the rest of the speech. The introduction might be only a few lines long, but it must capture the attention of the audience.

Body

The body of the speech contains the majority of the information. It should consist of several major points that support one central theme or objective. Each major point is supported by additional information to explain, illustrate, or clarify the point. It is best to write the body of the speech in outline form (Figure 6-9).

An outline will help direct the thought and delivery of the speech. After outlining the speech, a carefully worded narrative may be written to help the speaker fully develop the content of the speech. However, it must be emphasized that a speech should be given from an outline to avoid the temptation to memorize the speech. Memorization of a speech has two serious pitfalls. First, there is the danger that it will sound like someone else's words and lack authenticity. Second, if

SPEECH OUTLINE

Introduction

Honorable judges, instructors, and fellow students
1. Rabbits, cows, plants, and plows—WE STILL NEED THEM!
2. When the family farm goes under, a piece of America goes under with it

Body

The 2010 census revealed . . .
1. The midsize farm is likely to be the true family farm
   - Owned and operated by the family
   - Family receives benefit from their work
2. Some believe the family farm is a relic of the past!
   a. Press fascination with bankruptcy sales
   b. Farms not in view from interstate highways
   c. Small population on farms
   d. Animal rights groups and unionizing efforts
3. The family farm endures in the United States
   a. Better managed farm businesses buy the weaker farms
   b. Disease epidemics threaten specialized operations
   c. Farm retailing is on the rise
   d. Small farms are becoming legitimate
4. Farm bankruptcy must be minimized
   a. Farm failure affects general businesses
   b. We will miss fresh farm produce
   c. We will see more pollution
   d. We will depend more on imports
5. Are there remedies? Yes!
   a. Remove politics from marketing
   b. Recognize farmers as astute businesspeople
   c. See the total industry of agriculture
   d. Tax farmand for farm use
   e. Increase agriculture land-preservation programs

Conclusion

Indeed . . .
1. General George Washington nearly lost the continental army at Valley Forge from lack of food, shelter, and clothing.
2. It can happen to us if we don't maintain a healthy farm situation in the United States today
3. Don't give away our most valuable resource—the ability to feed, clothe, and shelter ourselves!

FIGURE 6-9 An outline of a winning speech on the importance of family farms.

the line of thought is lost during delivery, the speaker may not be able to find the location in the narrative. This can greatly damage the quality of the speech.

Conclusion

The conclusion should remind the audience of the major theme or central points of the speech and briefly restate them. The conclusion should leave the audience feeling like they want to take action to implement or adopt what you have said. Some speeches call for action, whereas others call for changes in attitude or perception. The more powerful speeches move people to action. The words needed to do such a big job must be carefully planned.

Giving the Speech

Giving the speech can be fun and provide much satisfaction (Figure 6-10). However, this fun and satisfaction does not come easily. The speaker must prepare the plan well and practice the speech extensively. Practicing the speech until the content becomes familiar helps speaking become nearly automatic. The speech should be given orally to oneself several times. Then, practice in front of a mirror to observe facial expressions, posture, and gestures. Finally, give the speech in front of others and invite them to make suggestions to improve the delivery (Figure 6-11).

Books have been written on techniques to enhance the delivery of a speech. However, for the beginner, a few basic and time-tested procedures should be helpful for effective speaking. Following are some suggested procedures for giving speeches:

- Have your teacher read and make suggestions on the content of your written speech.
- Learn the content thoroughly through repeated thought and practice.

Photo courtesy of the National FFA Organization.

FIGURE 6-10 Speaking in public can be a pleasant and exciting experience once you have learned the proper way to organize and present your thoughts. Speaking is a skill that is learned only through practice as you stand and speak in front of an audience.

National Public Speaking Contest ***Judge's Score Sheet***

PART I. FOR SCORING CONTENT AND COMPOSITION

Items To Be Scored	Points Allowed	Points Awarded Contestant 1	2	3	4	5	6	7	8	9	10	11	12	13
Content of Manuscript	200													
Composition of Manuscript	100													
Score on Written Production	300													

PART II. FOR SCORING DELIVERY OF THE PRODUCTION

Items To Be Scored	Points Allowed	Points Awarded Contestant 1	2	3	4	5	6	7	8	9	10	11	12	13
Voice	100													
Stage Presence	100													
Power of Expression	200													
Response to Questions	200													
General Effect	100													
Score on Delivery	700													

PART III. FOR COMPUTING THE RESULTS OF THE CONTEST

Items To Be Scored	Points Allowed	Points Awarded Contestant 1	2	3	4	5	6	7	8	9	10	11	12	13
Score on Written Production	300													
Score on Delivery	700													
TOTALS	1000													
* Less Overtime Deductions, for each minute or major fraction thereof	20													
* Less Undertime Deductions, for each minute or major fraction thereof	20													
GRAND TOTALS														
Numerical or Final Placing														

* From the Timekeeper's record.

Explanation of Score Sheet Points

Part I-For Scoring Content and Composition

1. *Content of the manuscript* includes:
   Importance and appropriateness of the subject
   Suitability of the material used
   Accuracy of statements included
   Evidence of purpose
   Completeness and accuracy of bibliography

2. *Composition of the manuscript* includes:
   Organization of the content
   Unity of thought
   Logical development
   Language used
   Sentence structure
   Accomplishment of purpose-conclusions

Part II-For Scoring Delivery of Production

1. *Voice* includes:
   Quality
   Pitch
   Articulation
   Pronunciation
   Force

2. *Stage Presence* includes:
   Personal appearance
   Poise and body posture
   Attitude
   Confidence
   Personality
   Ease before audience

3. *Power of expression* includes:
   Fluency
   Emphasis
   Directness
   Sincerity
   Communicative ability
   Conveyance of thought and meaning

4. *Response to questions* includes:
   *Ability to answer satisfactorily the questions on the speech which are asked by the judges indicating originality, familiarity with subject and ability to think quickly.

5. *General effect* includes:
   Extent to which the speech was interesting, understandable, convincing, pleasing, and held attention.

*NOTE: Judges should meet prior to the contest to prepare and clarify the questions asked.

FIGURE 6-11 A score sheet for evaluating speeches.

Courtesy of National FFA.

- Record the speech and observe the sound, speed, power, and effectiveness of your voice. Make corrections to improve the delivery.
- Practice the speech in front of a mirror to observe posture, hand gestures, and facial expressions. Your posture should be erect and natural, with hands at your sides or resting lightly on the edges of the podium. Your hands should be used occasionally for gestures that emphasize a point, show direction, or indicate count.
- Ask your teacher for a score sheet for judging speeches. Deliver the speech in front of a trusted person who can check your delivery against the score sheet and provide suggestions for improvements. This may be a friend, relative, or teacher.
- Deliver your speech in front of your class for experience and suggestions.
- Anticipate possible questions the judges may ask and prepare for them.
- Make some statements in your speech that you are well prepared to defend. They may lead to questions from the judges that you will be ready to answer because of your preparation.
- Ask your teacher to critique your speech for final approval.
- Deliver your speech in front of civic groups and/or in FFA public-speaking competitions.
- Record a video of your speech for later critique and review.

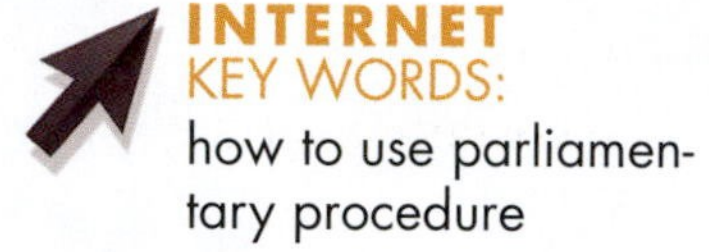

PARLIAMENTARY PROCEDURE

What is parliamentary procedure? Why are so many people familiar with it? Why is it important? Why should agriscience students be interested in learning parliamentary procedure? **Parliamentary procedure** is a system of guidelines or rules for conducting meetings. Most Americans who are influential in their communities are familiar with parliamentary procedure.

Parliamentary procedure is used to guide the meetings conducted by city councils, school boards, church groups, commissions, professional organizations, and civic organizations, such as Lions, Rotary, and Ruritan clubs. Agriscience students should be interested in learning this procedure so they can have their opinions heard and influence decisions that affect their lives. Parliamentary procedure is important because it permits a group to:

- discuss one thing at a time;
- hear everyone's opinion in a courteous atmosphere;
- protect the rights of minorities; and
- make decisions according to the wishes of the majority of the group.

Requirements for a Good Business Meeting

A good **business meeting** is a gathering of people working together to make wise decisions. Wrong decisions cause unhappiness, loss of income, inefficiency in business and social activities, injury, and other problems. Poorly run business meetings are a waste of time and accomplish little. Meetings run by groups of individuals who know and use parliamentary procedure are smooth, efficient, orderly, and focused, and such meetings accomplish much more than poorly organized meetings (Figure 6-12). Some requirements for a good business meeting are as follows:

Photo courtesy of the National FFA Organization.

FIGURE 6-12 Membership in the FFA provides opportunities for students to lead class discussions and to learn to conduct the business of an organization by taking turns as the presiding officer.

Effective Presiding Officer

A **presiding officer** is a president, vice president, or chairperson who is designated to lead a business meeting. He or she should be committed to the goals of the organization and should want to lead the group in making good group decisions. A good presiding officer must know and use proper parliamentary procedure.

Competent Secretary

A secretary is a person elected or appointed to take notes and prepare minutes of the meeting. **Minutes** is the name of the official written record of a business meeting. Minutes should include the date, time, place, presiding officer, attendance, and motions discussed at the meeting. They should be written clearly, include all actions taken by the group, and be kept in a permanent secretary's book.

Informed Members

Informed members are members who are active in the organization and want to be part of the group. They give previous thought to issues to be discussed and gather useful information about the issues. They share these thoughts with others in the meeting. This permits everyone to have the benefit of the best thinking in the group and permits the best decisions to be made. Effective members know and use parliamentary procedure to bring out important points of discussion and to advance the agenda of the meeting in an orderly manner.

A Comfortable Meeting Room

The meeting place must be comfortable and free from distractions. A moderate temperature and good lighting are essential. Members should be seated so they can hear and see each other. Seating at a table or in a circle works well for small groups. Large groups must rely on a good sound system for the presiding officer to be heard, and members must speak clearly with good volume to be heard. Good public speaking skills and thorough knowledge of parliamentary procedure help the members conduct effective meetings.

Conducting Meetings

The Order of Business

The **order of business** refers to the items and sequence of agenda items for a meeting. The order of business is usually made up by the secretary. This generally grows out of an executive meeting. An executive meeting is a meeting of the officers to conduct the business of the organization between regular meetings. They may also consider what needs to be discussed by the total membership at the regular meeting. The essential items in an order of business are:

- call to order;
- reading and approval of minutes of the previous meeting;
- treasurer's report;
- reports of other officers and committees;
- old business;
- new business; and
- adjournment.

Other items or activities that are frequently included in orders of business are programs, speakers, or entertainment.

Parliamentary Practices

Use of the Gavel

The **gavel** is a wooden mallet used by the presiding officer to direct a meeting (Figure 6-13). It is used to call the meeting to order, announce the result of votes, and adjourn the meeting. It is also used to signal the members to stand, sit down,

Photo courtesy of the National FFA Organization.

FIGURE 6-13 The presiding officer and each member should take time to learn how to use the gavel as a tool to conduct the business of an organization.

or reduce the noise level of the group. The gavel is a symbol of the authority of the office of president or chairperson, and it should be respected by all attending the meeting.

In some organizations, such as the FFA, a system of taps is used to signal the audience to do certain things. In FFA meetings, the gavel is used as follows:

- **One tap**—the outcome of or decision about the item under consideration has been announced by the presiding officer.
- **Two taps**—the meeting will come to order, members should sit down if standing, or members should be quiet except when recognized.
- **Three taps**—members should stand up.

Obtaining Recognition and Permission to Speak

For a meeting to be orderly, members must speak one at a time and in some logical and fair sequence. The presiding officer is regarded as the "traffic controller" and calls on members as they request to be recognized according to certain rules. To be recognized, the member should raise a hand to get the presiding officer's attention. The presiding officer should call the member by name; then the person should stand and address the presiding officer as Madam or Mr. Chairperson, or Madam or Mr. President. The individual should then proceed to speak. Both the presiding officer and the members should understand the correct classification of motions (Figures 6-14 and 6-15).

Presenting a Motion

A **motion** is a proposal, presented in a meeting, that is to be acted upon by the group. To present a motion, the member raises a hand and is recognized by the presiding officer. Then the member states, "Madam/Mr. President, I move that ..." (and continues with the rest of the motion). The words "I move" are important to say when beginning the motion. Otherwise, you will be regarded as incorrect in your usage of good parliamentary procedure. For a motion to be discussed by the group, at least one other member must be willing to have the motion discussed. That second individual expresses this willingness by saying, "Madam/Mr. President, I second the motion."

Some Useful Motions

There are dozens of motions, but a few basic motions are generally known and widely used. These include the following:

- **Main motion**—a basic motion used to present a proposal for the first time. The way to state it is to obtain recognition from the chairman and then say, "I move ..."
- **Amend**—a type of motion used to add to, subtract from, or strike out words in a main motion. The way to present an amendment is to say, "I move to amend the motion by ..."
- **Refer**—a motion used to refer to a committee or person for finding more information and/or taking action on the motion. The way to state a referral is to say, "I move to refer this motion to ..."
- **Lay on the table**—a motion used to stop discussion on a motion until the next meeting. The way to table a motion is to say, "I move to table the motion."

SUMMARY OF MOTIONS

Motion	Debatable	Amendable	Vote Required	Second	Reconsider
PRIVILEGED					
Fix time which to adjourn	No	Yes	Majority	Yes	Yes
Adjourn	No	No	Majority	Yes	No
Recess	No	Yes	Majority	Yes	No
Question of privilege	No	No	None	None	Yes
Call for orders of the day	No	No	None/2/3	None	No
INCIDENTAL					
Appeal	Yes	No	Majority	Yes	Yes
Point of order	No	No	None	No	No
Parliamentary inquiry	No	No	None	No	No
Suspend the rules	No	No	2/3	Yes	No
Withdraw a motion	No	No	Usually none	No	No
Object consideration of question Negative vote only	No	No	2/3	No	Yes
Division of the question	No	Yes	Majority	Yes	No
Division of the assembly	No	No	No	No	No
SUBSIDIARY					
Lay on table	No	No	Majority	Yes	No
Previous question before vote	No	No	2/3	Yes	Yes
Limit debate	No	Yes	2/3	Yes	Yes
Postpone definitely	Yes	Yes	Majority	Yes	Yes
Refer to committee	Yes	Yes	Majority	Yes	Yes
Amend	Yes	Yes	Majority	Yes	Yes
Postpone indefinitely	Yes	No	Majority	Yes	Yes vote only
Main motion	Yes	Yes	Majority	Yes	Yes
UNCLASSIFIED					
Take from table	No	No	Majority	Yes	No
Reconsider	No	No	Majority	Yes	Negative vote only

For more details on parliamentary procedure, see a parliamentary procedure book such as *Robert's Rules of Order.*

FIGURE 6-14 Parliamentary skills are useful in FFA and other organizations.

- **Point of order**—a procedure used to object to some item in or about the meeting that is not being presented properly. The procedure to use is to stand up and say, "Madam/Mr. President, I rise to a point of order!" The presiding officer should then recognize the member by saying, "State your point." The member then explains what has been done incorrectly.
- **Adjourn**—a motion used to close a meeting. The procedure is to say, "I move to adjourn."

The FFA *Student Handbook* provides a listing of additional motions and explains how to use parliamentary procedure for more effective meetings. Agriscience students are encouraged to develop effective leadership skills.

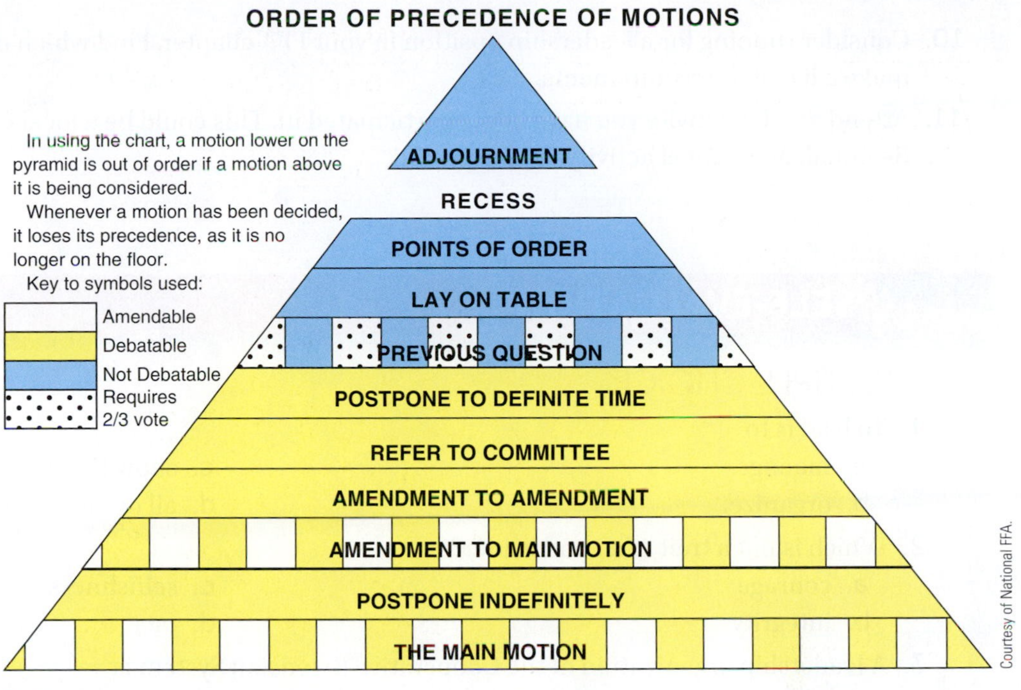

FIGURE 6-15 Correct order of precedence of motions.

The ability to work effectively as a member of a group is essential for all. The ability to function as a chairperson or officer creates more opportunities to serve and influence the communities in which we live. The development of self-confidence is essential. Self-confidence is a product of knowledge and skill. Therefore, each student should strive to learn to speak well, to function in groups through parliamentary procedure, and to use the opportunities in FFA for personal growth and development.

STUDENT ACTIVITIES

1. Write the Terms to Know and their meanings in your notebook.
2. Make a list of the many ways that you exercise leadership in your family, school, and community.
3. Develop a bulletin board showing the symbols of the FFA emblem.
4. Develop a bulletin board illustrating the purposes of the FFA. Include the FFA colors.
5. Write down five career development events and five proficiency awards for which you would like to try out. Discuss these with your classmates and teacher.
6. Prepare and present a three-minute speech on an agriscience topic to your class.
7. Ask your teacher to let you use a gavel and direct a mock class or FFA meeting to learn parliamentary skills.
8. Form a parliamentary procedure team consisting of you and your classmates, and then demonstrate various parliamentary skills to the class.
9. Under the supervision of your teacher, develop an appropriate solution for a school or district issue. Nominate and appoint a small committee to present your idea, in the form of a main motion, to the principal or school board.

10. Consider running for a leadership position in your FFA chapter. Find which offices are available and make a list of the requirements.
11. Attend an FFA activity you have never participated in. This could be a local Chapter, District, State, Regional, or National activity or convention.

SELF-EVALUATION

A. MULTIPLE CHOICE

1. To lead is to
   a. manage.
   b. organize.
   c. show the way.
   d. all of the above.
2. Which is not a trait of a good leader?
   a. courage
   b. integrity
   c. selfishness
   d. tact
3. A leadership organization of the Cooperative Extension System is
   a. Boy Scouts.
   b. FFA.
   c. Girl Scouts.
   d. 4-H.
4. Membership in FFA is limited to youth who
   a. are in the country.
   b. enroll in an agriscience program in school.
   c. plan careers in agriscience.
   d. seek leadership training.
5. Which is *not* a purpose of FFA?
   a. develop leadership
   b. intelligent choice of agriscience occupations
   c. promote scholarship
   d. promote self above others
6. The symbol that signifies that the FFA is a national organization is the
   a. corn.
   b. eagle.
   c. owl.
   d. rising sun.
7. The first line of the FFA motto is
   a. doing to learn.
   b. earning to live.
   c. learning to do.
   d. living to serve.
8. One requirement for the Greenhand degree is
   a. prepare a plan for supervised agricultural experience.
   b. $70 earned from agriscience experience.
   c. school grades of C or above.
   d. unselfish attitude in FFA activities.
9. An FFA activity not generally organized as a career-development event is
   a. dairy foods.
   b. forestry.
   c. land judging.
   d. agricultural sales.
10. One of the last items in an order of business is
    a. new business.
    b. officer reports.
    c. reading of the minutes.
    d. treasurer's report.
11. The largest part of a speech is the
    a. body.
    b. conclusion.
    c. introduction.
    d. summary.

12. The only acceptable way to start a motion is to say
    a. "I believe ..."
    b. "I make a motion that ..."
    c. "I move ..."
    d. "I think ..."

B. MATCHING

_________	1. Adjourn	a. Present a new proposal
_________	2. Amend	b. Leave it to a committee
_________	3. Lay on the table	c. Correct some procedure
_________	4. Main motion	d. Close the meeting
_________	5. Point of order	e. Consider it at the next meeting
_________	6. Refer	f. Change a motion
_________	7. Three taps of gavel	g. Prepare minutes
_________	8. Secretary	h. Members stand

SECTION 3

STEWARDS OF THE LAND

Hikers, bikers, birders, prospectors, hunters, anglers, farmers, foresters, ranchers, caretakers, and occupants all must be good stewards of the land. All are dependent on the land and rely on the soil, air, water, wildlife, and other natural resources around us. Farmers, ranchers, and foresters rely on the land to grow the crops and animals of their businesses. Similarly, hunters and other recreational users of the land and water rely on the habitat to grow and sustain the plants and wildlife. All enjoy wildlife for sport and recreation.

Land for farming and ranching is typically owned by the families who occupy the land, and they have definite property rights to grant or deny others access to their property for hunting, fishing, or other recreational pursuits. However, the good-citizen hunter or angler seeks permission of the owner to access private property and strives to protect or enhance the fish and wildlife population and habitat through legal and good stewardship practices. Both owners and good-citizen users share a love for animals and a respect for crops, pasture, and woodland. Both want to conserve the quality of land, air, and water.

Farmers and ranchers can do much to encourage growth of food, cover, and habitat for wildlife. They interact with wildlife biologists, game specialists, game officers, and other public authorities to nurture game populations and enforce game laws. Hunters and anglers (including farmers and ranchers), wildlife specialists, and the general public all help keep wildlife populations in check by harvesting excess game birds and animals. Game hunting limits are set by wildlife specialists. Scientific methods should be used in an effort to permit and encourage the removal of excess wild animals by hunting. Those animals that are not

Natural Resources Management

removed by hunting compete with young game animals for food and shelter. As a result, the young often perish by disease, predation, and starvation. This is nature's way of keeping animal and plant populations in balance.

Public lands are owned by federal, state, or local governments, and people who are educated in many specialty occupations are employed to care for them. Foresters, biologists, fish and game managers, game officers, park rangers, horticulturists, scientists, technicians, and others all contribute to the upkeep and improvement of public lands.

To be counted as good stewards of the land, the owners, managers, and users of both private and public lands must all cooperate in the use and conservation of soil, water, trees, crops, wild plants, livestock, wildlife, and wildlife habitats.

Courtesy of DeVere Burton

UNIT 7

Maintaining Air Quality

OBJECTIVE

To determine major sources of air pollution and identify procedures for maintaining and improving air quality.

MATERIALS LIST

- pencil and paper
- several aerosol cans/labels
- Internet access

COMPETENCIES TO BE DEVELOPED

After studying this unit, you should be able to:

- define the term *air* and identify its major components.
- analyze the importance of air to humans and other living organisms.
- determine the characteristics of clean air.
- describe common threats to air quality.
- describe important relationships between plant life and air quality.
- discuss the greenhouse effect and global warming.
- list practices that lead to improved air quality.

SUGGESTED CLASS ACTIVITIES

1. Identify the pollutants that are most often responsible for reducing air quality. Create a bulletin board in the classroom that illustrates the sources of these pollutants.
2. Invite an official from a government agency such as the Environmental Protection Agency (EPA) or the Department of Environmental Quality (DEQ) to do a class presentation on air quality. After the presentation, conduct a visual inspection of your community to identify potential sources of air pollution. Be sure to note that visible emissions, such as steam, do not necessarily indicate serious pollution problems. Consider ways to control those pollutants that are identified.

TERMS TO KNOW

air
water
soil
habitat
sulfur
hydrocarbon
nitrous oxide
tetraethyl lead
carbon monoxide
radon
radioactive material
chlorofluorocarbon (CFC)
ozone (O_3)
particulates
pesticide
asbestos
greenhouse effect
respiration
photosynthesis

INTERNET KEY WORDS:
air quality standards
air pollutants
air pollution organizations

3. In groups of three or four students, identify one source of air pollution. Make a suggestion for a new regulation or law, in the form of a main motion, that would improve the air quality in your community.

Life, as we know it, on our planet requires a certain balance of unpolluted air, water, and soil. **Air** is a colorless, odorless, and tasteless mixture of gases. It occurs in the atmosphere around the Earth and is composed of approximately 78 percent nitrogen, 21 percent oxygen, and a 1 percent mixture of argon, carbon dioxide, neon, helium, and other gases (Figure 7-1). **Water** is a clear, colorless, tasteless, and nearly odorless liquid. Its chemical makeup is two parts hydrogen to one part oxygen. **Soil** is the top layer of the Earth's surface that is suitable for the growth of plant life.

AIR QUALITY

Without a reasonable balance of air, water, and soil, most organisms would perish. Slight changes in the composition of air or water may favor some organisms and cause others to diminish in number or in health. Unfavorable soil conditions usually mean inadequate food, water, shelter, and other unfavorable factors related to **habitat**, the area or type of environment in which an organism or biological population normally lives (Figure 7-2).

Threats to Air Quality

The mixture of gases we call air is absolutely essential for life. The air we breathe should be healthful and life supporting. Air must contain approximately 21 percent oxygen for human survival. If a human stops breathing and no life-supporting equipment or procedures are used, the brain will die in approximately four to six minutes. Air may contain poisonous materials or organisms that can decrease the body's efficiency, cause disease, or cause death through poisoning.

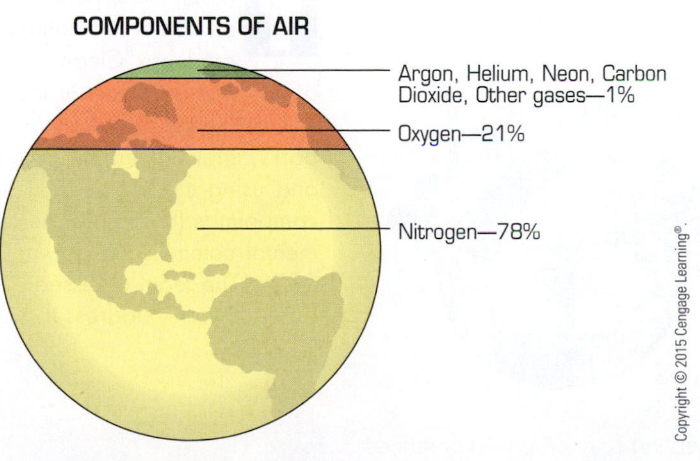

FIGURE 7-1 The atmosphere of the Earth is composed mostly of nitrogen and oxygen and small amounts of a few other gases.

© SNEHIT/Shutterstock.com.

FIGURE 7-2 Clean air, clean water, and productive soil are necessary for a good habitat for plants, animals, and humans.

Even though the Earth's circumference at the equator is 24,902 miles, the abuse of the atmosphere in one area frequently damages the environment in distant parts of the world. Air currents flow in somewhat constant patterns, and air pollution moves with them. However, when warm and cold air meet, the exact air movement is determined by the differences in temperature, the terrain, and

SCIENCE CONNECTION

ENVIRONMENTAL SUCCESS STORY: ELMHURST PARK DISTRICT

The Elmhurst Park District is an Illinois agency that manages parks and other recreational facilities. The park district was recently honored by "Clean Air Counts," an organization that promotes air quality improvements in the Chicago area. Smog-forming emissions were reduced in the region by expanding the use of native landscaping in some parks, implementing the use of B-20 biodiesel in vehicles and equipment, and using cleaning supplies and paints that are low in volatile organic compounds (VOCs). These improvements have occurred since an environmental committee, also known as the "Green Team," was organized in 2008. The team assessed environmental needs and developed strategies and policies to address the needs.

© iStockphoto/Chutima Chokkij.

Implementing science-based practices can reduce pollution of air and other basic resources.

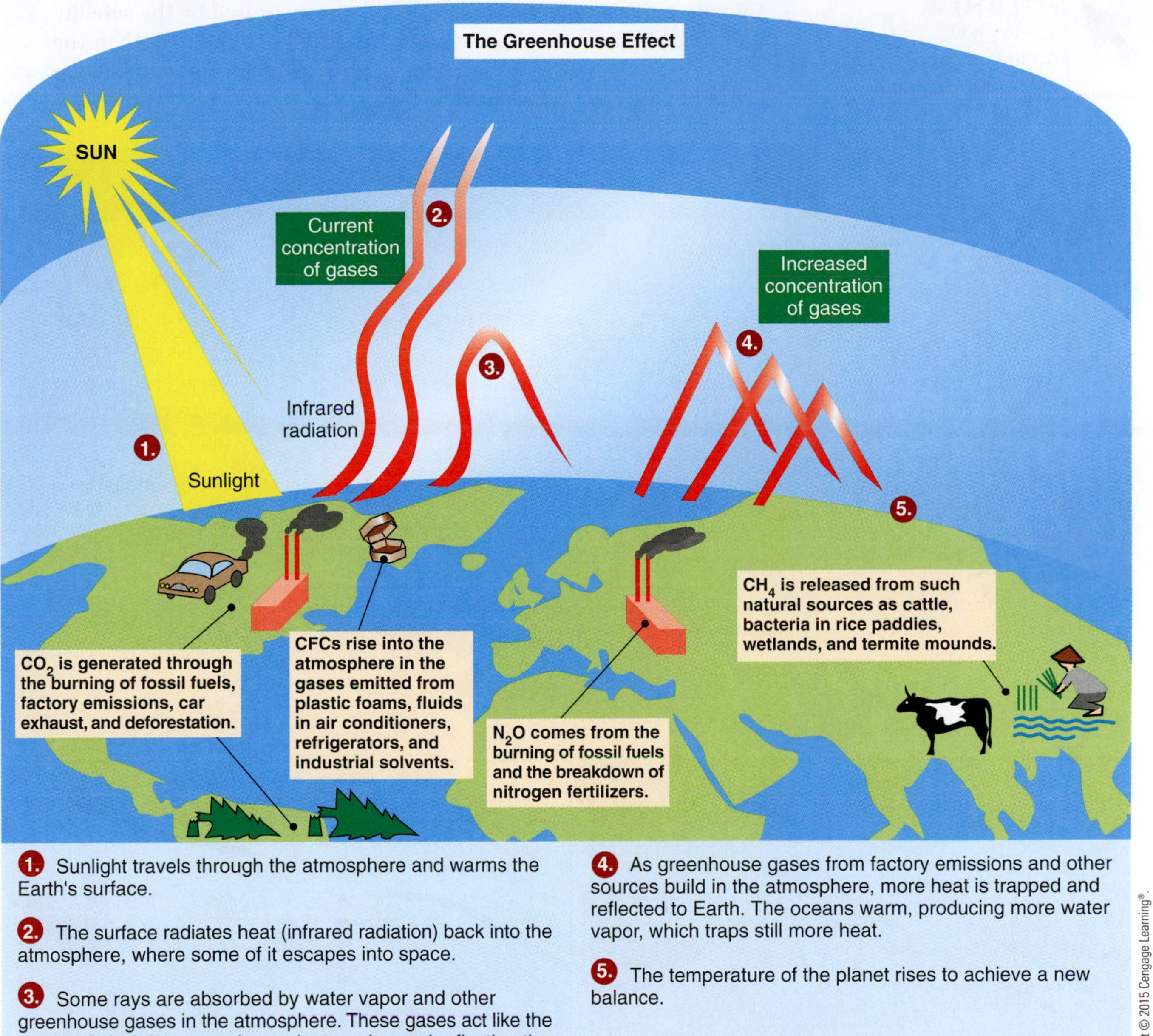

FIGURE 7-5 The greenhouse effect on land, sea, and air.

more than 100 years, and we do not have actual historical records from which to draw conclusions on climatic cycles and weather patterns of longer duration. An example of this line of thinking is evident in the following statement by Dr. Roy Spencer (1998), former Senior Scientist for Climate Studies at NASA's Marshall Space Flight Center:

> The adjusted satellite trends are still not near the expected value of global warming predicted by computer climate models. The Intergovernmental Panel on Climate Change's (IPCC) 1995 estimate of average global warming at the surface until the year 2100 is + 0.18 deg. C/decade.

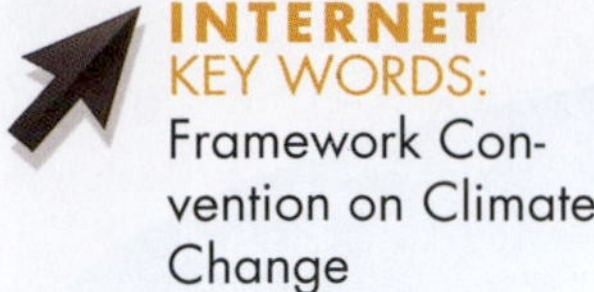

INTERNET KEY WORDS: Framework Convention on Climate Change

Climate models suggest that the deep layer measured by the satellite and weather balloons should be warming about 30% faster than the surface (+0.23 deg. C/decade). None of the satellite or weather balloon estimates are near this value.

Physical Factors	Recent Global Changes	Predictions
Temperature	+ 1.4 °F since 1880	Accelerating temperature rise
Sea Level	+ 3.3 mm/year since 1993	+ 7.1 to 23 inches in this century
Precipitation	+7 to +15%	Varied precipitation patterns across the planet
Glacial Retreat	Significant	Accelerating
Atmospheric carbon dioxide concentration	+ from 280ppm to 385ppm in the last century	Increasing concentration

Sources: Intergovernmental Panel on Climate Change; NASA's Global Climate Change website; and Schieder, S. (1990) Prudent planning for a warmer planet, New Scientist, 128(1743)

FIGURE 7-6 Evidence of Global Climate Change.

Carbon Dioxide (CO_2) The relative contribution of the greenhouse gas CO_2 to the global warming trend is expected to be about 50 percent by 2020

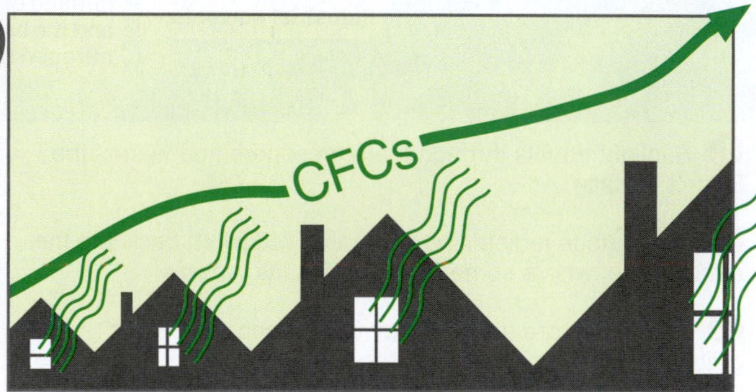

Chlorofluorocarbons (CFCs) About 25 percent of greenhouse effect by 2020

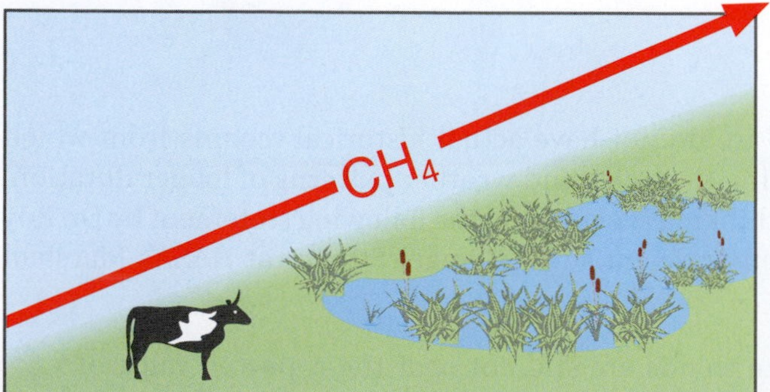

Methane (CH_4) About 15 percent of greenhouse effect by 2020

Nitrous Oxide (N_2O) About 10 percent of greenhouse effect by 2020

FIGURE 7-7 Predicted increases in the concentrations of four major pollutants and how they may affect global climate change by 2020.

In contrast with this scientific view is one group of scientists' statement on global climatic disruption that was issued in June 1997. The text of the statement follows:

> We are scientists who are familiar with the causes and effects of climatic change as summarized recently by the Intergovernmental Panel on Climate Change (IPCC). We endorse those reports and observe that the further accumulation of greenhouse gases commits the earth irreversibly to further global climatic change and consequent ecological, economic, and social disruption. The risks associated with such changes justify preventive action through reductions in emissions of greenhouse gases. In ratifying the Framework Convention on Climate Change, the United States agreed in principle to reduce its emissions. It is time for the United States, as the largest emitter of greenhouse gases, to fulfill this commitment and demonstrate leadership in a global effort.
>
> Human-induced global climatic change is under way. The IPCC (2007) concluded that global mean surface air temperature has increased by between about 0.5 and 1.1 degrees Fahrenheit in the last 100 years and anticipates a further continuing rise of 1.8 to 6.3 degrees Fahrenheit during the next century. Sea-level has risen on average 4–10 inches during the past 100 years and is expected to rise another 6 inches to 3 feet by 2100. Global warming from the increase in heat-trapping gases in the atmosphere causes an amplified hydrological cycle resulting in increased precipitation and flooding in some regions and more severe aridity in other areas. The IPCC (1996) concluded that "The balance of evidence suggests a discernible human influence on global climate" (p. 4). The warming is expected to expand the geographical ranges of malaria and dengue fever and to open large new areas to other human diseases and plant and animal pests. Effects of the disruption of climate are sufficiently complicated that it is appropriate to assume there will be effects not now anticipated.
>
> Our familiarity with the scale, severity, and costs to human welfare of the disruptions that the climatic changes threaten leads us to introduce this note of urgency and to call for early domestic action to reduce U.S. emissions via the most cost-effective means. We encourage other nations to join in similar actions with the purpose of producing a substantial and progressive global reduction in greenhouse gas emissions beginning immediately. We call attention to the fact that there are financial as well as environmental advantages to reducing emissions. More than 2000 economists observed that there are many potential policies to reduce greenhouse-gas emissions for which total benefits outweigh the total costs...."

Source: *Scientists' Statement, Global Climatic Disruption.* (June 18, 1997). Open Letter. The Woods Hole Research Center. Retrieved from http://www.whrc.org/resources/essays/pdf/1997_climate_stmt.pdf#search="statement on global climatic disruption"

Precautions against Global Warming

Reversing the perceived trend toward continued global warming and the problems it could bring may require real changes in the way we do some things. With good research, wise government, and environmentally sensitive business, we can slow down or stop the decline of our air quality brought about by human-caused

FIGURE 7-8 A lawn mower can cause 50 times more pollution per horsepower than a modern truck engine.

FIGURE 7-9 A chain saw is powered by a two-cycle engine that emits large amounts of air pollutants.

pollution. Even small-engine lawn and garden equipment is now seen as seriously contributing to pollution, and changes are needed to address this problem. Experts now observe that a gasoline lawn mower can cause 50 times more pollution per horsepower than a modern truck engine (Figure 7-8). Similarly, a lawn mower running for just 1 hour may create as much pollution as an automobile traveling 240 miles. A chain saw running for 2 hours emits as much pollution from hydrocarbons as a new car running from coast to coast across the United States (Figure 7-9).

Carbon Dioxide (CO_2)

Carbon dioxide is a major product of combustion. Our robust appetites for food, clothing, consumer goods, heated and air conditioned spaces, and transportation have led to a cultural lifestyle that requires huge volumes of fuels to be burned to raise food, manufacture goods, and move vehicles. Our highways in and around large cities are clogged with automobiles. Interstate highways are loaded with trucks, and rivers and other waterways carry heavy shipping traffic. Homes and commercial buildings use extensive volumes of electricity for light and temperature control, and farms and factories have huge machines and heating devices—all consuming fuel that expels carbon dioxide and other pollutants into the air.

Green plants can use some carbon dioxide from the air and convert it into plant food, oxygen, and water. However, in most parts of the world, the mass of green plants is being reduced even as the expulsion of pollutants increases dramatically. The practice of slash-and-burn agriculture in the jungle regions of the world threatens to remove vast areas and volumes of plant growth with little hope of replacement. Increased carbon dioxide in the atmosphere is projected to account for 50 percent of the increase in global warming by 2020.

Chlorofluorocarbons (CFCs)

Since the discovery of the damage that CFCs do to the ozone layer, other propellants have been used in pressurized spray containers, and CFCs are now recaptured from refrigeration units before they are discarded. Still, the escape of CFCs into the air and their effect on the ozone layer are projected to account for 25 percent of the increase in global warming by 2020.

Methane (CH_4)

Most methane comes from naturally decaying plant materials such as leaves and debris on the soil surfaces of forests and jungles. Some of it is a product of decaying organic matter, such as human and animal waste. Methane rises from piles, pits, and other accumulations of decaying animal manure, peat bogs, and sewage. Some large farms are now using carefully engineered systems to capture the gas from large manure-holding areas. The methane is then used as fuel for engines driving generators to provide electricity for the farm and for sale. Methane is projected to account for 15 percent of the increase in global warming by 2020.

Nitrous Oxide (N_2O)

Nitrous oxide has long been a troublesome pollutant from gasoline engines. The ever-increasing number of automobiles, trucks, tractors, heavy equipment, aircraft, chain saws, lawn motors, boats, and other engine-driven applications has

SCIENCE CONNECTION

OVERCOMING THE EFFECTS OF AIR POLLUTION

Courtesy of USDA/ARS #K-3204-1.

A biological aide uses a porometer to measure the impact of CO_2 enrichment of the air on the transpiration rate and stomata activity of a soybean leaf.

U.S. Department of Agriculture (USDA) scientists are tackling what could be the toughest conflict of the century: the battle to breathe. Although ozone depletion is a serious problem in the upper atmosphere, ozone as a product of combustion hovers just above the Earth's surface as a pollutant that decreases air quality.

Researchers have estimated that U.S. farmers experience at least 1 billion dollars per year in lost crop yields because of air pollution. Research indicates that cutting the level of ozone in the air by 40 percent would mean an extra $2.78 billion for agricultural producers.

USDA and University of Maryland scientists discovered that treatment of plants with the growth hormone ethylenediura (EDU) can reduce damage by ozone. EDU alters enzyme and membrane activity within the leaf cells where photosynthesis takes place. A single drenching of soil with EDU effectively protected some plants from damage and reduced the sensitivity of others to excessive levels of ozone in the air. Similarly, injection of EDU in the stems of shade trees in highly polluted areas could protect them from damage.

In addition to ozone, other major air pollutants that are damaging to crops include peroxyacetetyl nitrate, oxides of nitrogen, sulfur dioxide, fluorides, agricultural chemicals, and ethylene. The task of reducing the amounts of pollutants in the air and finding ways to reduce the effects of pollutants on living organisms will continue to challenge future generations.

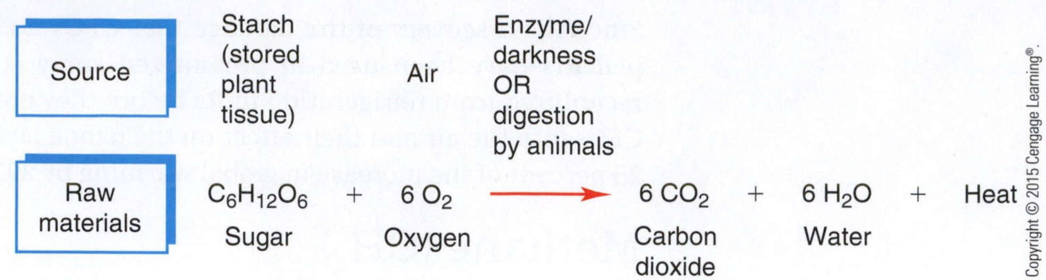

FIGURE 7-10 Respiration is the process by which plant tissues are broken down to produce heat, water, and carbon dioxide.

offset the tremendous improvements made in emission reduction from individual engines. Because nitrous oxide emissions are still increasing, it is projected that they will account for 10 percent of the increase in global warming by 2020.

Air and Living Organisms

Oxygen in the air is consumed by plants and animals during a process called **respiration** (Figure 7-10). Animals, as well as humans, use oxygen to convert food into energy and nutrients for the body. Animals breathe in or inhale to obtain oxygen. They exhale (breathe out) carbon dioxide gas. Plants release oxygen during the day. They create oxygen through the process of **photosynthesis** (a process in which chlorophyll in green plants enables them to use light, carbon dioxide, and water to make food and release oxygen) (Figure 7-11).

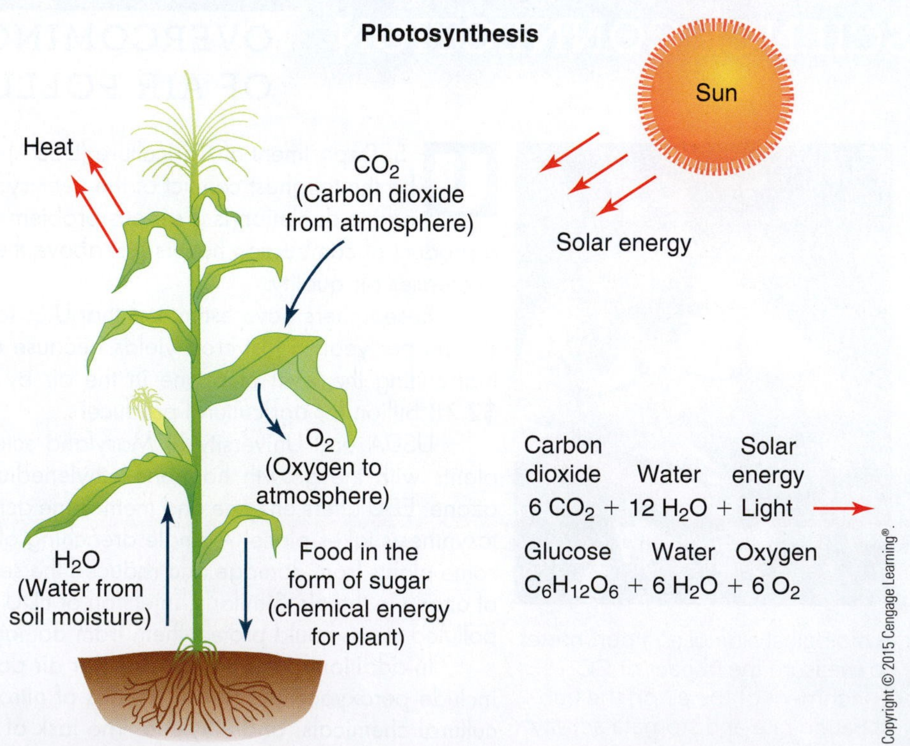

FIGURE 7-11 Photosynthesis is the process by which carbon dioxide is combined with water to store energy obtained from sunlight. Chlorophyll supports this reaction in which sugar and oxygen are produced.

Maintaining and Improving Air Quality

Air quality can be improved by reducing or avoiding the release of pollutants into the air and by removing existing pollution. Specific practices that can help reduce air pollution include:

- stopping the use of aerosol products that contain CFCs.
- providing adequate ventilation in tightly constructed and heavily insulated buildings.
- having buildings checked for the presence of radon gas.
- using exhaust fans to remove cooking oils, odors, solvents, and sprays from interior areas.
- regularly cleaning and servicing furnaces, air conditioners, and ventilation systems.
- maintaining all systems that remove sawdust, wood chips, paint spray, welding fumes, and dust to ensure that they function most efficiently.
- keeping gasoline and diesel engines properly tuned and serviced.
- keeping all emissions systems in place and properly serviced on motor vehicles.
- observing all codes and laws regarding outdoor burning.
- reporting any suspicious toxic materials or conditions to the police or appropriate authorities.
- reducing the use of pesticide sprays as much as possible.
- using pesticide spray materials strictly according to label directions.

The U.S. Congress has passed various laws to prevent the loss of air quality and resolve existing air quality problems. The first significant laws were the Clean Air Act of 1963 and the Air Quality Act of 1967. These required reductions in releases of industrial pollutants into the atmosphere. The laws have been expanded and updated several times since then.

HOT TOPICS IN AGRISCIENCE

RESEARCH TO IMPROVE PHOTOSYNTHESIS

Improving the efficiency of photosynthesis is one area of current interest in plant research. It is well documented that the processes of photosynthesis and respiration are closely related. Photosynthesis is the process by which energy from the sun is stored in molecules of sugar; respiration is a process that converts the sugars to high-energy molecules of adenosine triphosphate (ATP) during hours of darkness.

Scientists hope to be able to stimulate plant growth through more efficient photosynthesis by interrupting respiration during hours of darkness. Total plant yields will increase if science is successful in maintaining daylight gains in plant tissues.

Scientists are researching ways to modify photosynthesis to make the process more efficient.

The 1970 Clean Air Act has helped reduce air pollution. Tall smoke stacks that disperse harmful gases over larger areas instead of eliminating their release altogether are no longer legal for pollution control. The current law requires each state to develop an implementation plan that describes how the state will meet the act's requirements.

The Clean Air Acts of 1970 and 1990 are the most far-reaching of the air quality laws. From these versions, a series of clean air amendments have been approved. New federal and state agencies have been created to interpret and enforce the laws. Among these are the Environmental Protection Agency (EPA), the Office of Air Quality Planning and Standards, the Alternative Fuels Data Center, and the Commission for Environmental Cooperation. Each government office has created new regulations and standards for air quality.

Among the federal standards that have been implemented are the Clean Air Act's National Ambient Air Quality Standards, the Clean Air Act's New Source Performance Standards, the Prevention of Significant Deterioration, Air Guidance Documents from the EPA's Office of Air and Radiation, and updated air quality standards for smog (ozone) and particulate matter. Each of these air quality standards is intended to reduce air pollution and improve air quality.

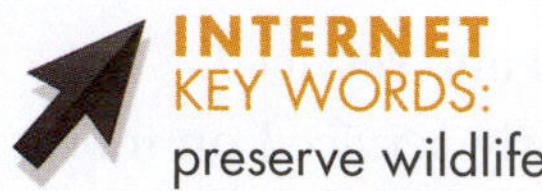

As people begin to experience the effects of pollution on the atmosphere in the form of lung and skin diseases, there will be greater motivation to solve the problems created by harmful atmospheric gases. Humans probably will not do much specifically to improve air quality for wild animals, but wild creatures will benefit when humans improve air quality for themselves.

STUDENT ACTIVITIES

1. Write the Terms to Know and their meanings in your notebook.
2. Make a pie chart illustrating the components of air.
3. Stand in a safe location about 5 feet behind and to the side of a parked truck or bus with its gasoline engine running. Notice the smell of the exhaust. Stand in the same relative position from an automobile that is fewer than 3 years old or has traveled fewer than 40,000 miles. What differences do you observe in the truck and automobile exhausts? Why? Which do you think causes more pollution of the air?
4. Examine three different aerosol cans. Which products use chlorofluorocarbons for the propellant? Why is it unwise to use such products?
5. Talk with an automobile tune-up specialist about the effect of engine adjustments on the content of exhaust gases.
6. Ask your teacher to invite an air pollution specialist to your class to discuss the problems of air pollution in your town, county, or state.
7. Do a research project on the greenhouse effect and its relationship to global warming.
8. Obtain a radon test kit, and perform a test for radon in your home.
9. Create four original drawings showing the four major pollutants. Under each of the drawings, write one sentence that explains where this pollution comes from and what can be done to improve air quality.

SELF-EVALUATION

A. MULTIPLE CHOICE

1. Air is
   a. 78 percent argon.
   b. 21 percent nitrogen.
   c. 21 percent oxygen.
   d. 10 percent carbon dioxide.
2. Pure water is
   a. a mixture of gases.
   b. metallic tasting.
   c. one part hydrogen to two parts oxygen.
   d. odorless.
3. Without proper air to breathe, a human can survive only about
   a. six minutes.
   b. twelve minutes.
   c. two hours.
   d. twelve hours.
4. Radon gas is a widespread threat to air quality
   a. on the highway.
   b. in factories.
   c. in homes.
   d. in wooded areas.
5. Radioactive dust is likely to be caused by
   a. improperly adjusted furnaces.
   b. cracks in basement floors.
   c. a damaged ozone layer.
   d. nuclear reactions.
6. Chemicals used to kill insects are called
   a. pests.
   b. pesticides.
   c. pollutants.
   d. toxic materials.
7. One ingredient not associated with photosynthesis is
   a. carbon dioxide.
   b. oxygen.
   c. radon.
   d. water.
8. Chlorofluorocarbons have been found to damage
   a. aerosol sprays.
   b. the ozone layer.
   c. refrigeration units.
   d. water pumps and equipment.
9. Poisonous gas we cannot remove from auto exhaust is
   a. carbon monoxide.
   b. hydrocarbons.
   c. nitrous oxides.
   d. radon.
10. The most reliable source of information on the use of a pesticide is
    a. experienced applicators.
    b. an extension service.
    c. personal experience.
    d. the product label.
11. The buildup of heat resulting from sunlight passing through glass and heating trapped air in the interior area is called
    a. the greenhouse effect.
    b. infrared energy.
    c. radiation.
    d. ultraviolet.
12. A product of decaying plant or animal matter is
    a. chlorofluorocarbons.
    b. methane.
    c. nitrous oxide.
    d. ozone.

B. MATCHING

_________ 1. Carbon monoxide
_________ 2. Chlorofluorocarbons
_________ 3. Hydrocarbons
_________ 4. Lead
_________ 5. Nitrous oxides
_________ 6. Ozone
_________ 7. Radon
_________ 8. Pests
_________ 9. Photosynthesis
_________ 10. Sulfur

a. Tetraethyl
b. Pale yellow
c. 5 percent of auto exhaust
d. Damage ozone layer
e. Filters ultraviolet rays
f. Diseases, insects, weeds
g. Chlorophyll, light, carbon dioxide
h. Causes death from auto exhaust
i. Leaks into houses
j. Pollutant from autos and factories

UNIT 8

Water and Soil Conservation

OBJECTIVE

To determine the relationships between water and soil in our environment and the recommended practices for conserving these resources.

MATERIALS LIST

- three growing plants in 4- to 6-inch pots for Student Activity 5
- kitchen scale or laboratory balance scale
- plant watering containers
- plant growing area
- Internet access

COMPETENCIES TO BE DEVELOPED

After studying this unit, you should be able to:

- define water, soil, and related terms.
- cite important relationships between land characteristics and water quality.
- discuss some major threats to water quality.
- describe types of soil and water and their relationships to plant growth.
- cite examples of enormous erosion problems worldwide.
- describe key factors affecting soil erosion by wind and water.
- list important soil and water conservation practices.

SUGGESTED CLASS ACTIVITIES

1. Make a bulletin board display in the school or classroom about conservation of water and soil. Illustrate causes of soil and water losses, and illustrate ways to prevent losses. Include some statistics about the value of lost agricultural production caused by soil and water losses.
2. Prepare a class presentation and demonstration on soil and water conservation. Make arrangements with elementary school teachers to present the demonstration to elementary school students. Include a model demonstration on the effects of slope or soil type on soil loss caused by erosion by water. Assign teams of class members to give short presentations on conservation topics. Topics are easily identified on the Internet using key words.

TERMS TO KNOW

potable
fresh water
tidewater
food chain
water cycle
irrigation
watershed
water table
free water
gravitational water
capillary water
hygroscopic water
no-till
contour
erosion
aquifer
sheet erosion
mulch
conservation tillage
contour farming
strip cropping
crop rotation
lime
fertilizer
grass waterway
terrace

3. Add large amounts of three different pollutants to carefully labeled gallons of water. Leave another gallon of water unpolluted; this will be your control for comparison with the polluted treatments. Purchase four inexpensive houseplants. Apply one-half cup of a specific water treatment to each plant daily. Be sure to label each plant with the water pollutant that each has been given. Observe and discuss the short-term appearance and health of the plants after a few days, and then after a few weeks (long-term).

Flying over the vast continents of the Americas, Europe, Asia, and Africa, we might feel that the landmass of the Earth is an endless resource (Figure 8-1). The great oceans of the world, however, combine to provide an even larger area. Even so, both land and water resources have become limited, and there is genuine concern that we are rapidly depleting them. In developed countries, a safe and adequate water supply generally faces only temporary shortages and mild inconveniences, such as restricted water for nonessential tasks like washing cars and watering lawns. In developing countries, however, a safe water supply is a luxury. Although it seems that most areas of the world have a sufficient volume of water, supplies of safe water are sometimes insufficient because of misuse, poor management, wastage, and pollution.

Similarly, productive land also is becoming a scarce commodity, and ownership is expensive. Individuals seek good land for homes, farms, and recreation. It is also needed by businesses for banks, stores, warehouses, car lots, and other uses. Industry needs land for factories and storage areas. Governments need land for roads, bridges, buildings, parks, recreation, and military facilities. Most parties look for the best land—land that is level, with deep and productive soil. Such land is in great demand because it

Courtesy of USDA/ARS #K5051-5. Photo by Scott Bauer.

FIGURE 8-1 It is difficult to view our land as a resource that is in short supply because the amount of farmland seems so vast. However, it is important to realize that we are already farming most of the land that is suited to producing crops.

© EugeneF/Shutterstock.com.

Courtesy Elmer Cooper/Copyright © 2015 Cengage Learning®.

Courtesy Elmer Cooper/Copyright © 2015 Cengage Learning®.

© Ruud Morijn Photographer/Shutterstock.com.

FIGURE 8-2 Good land is in high demand for (A) housing developments; (B) apartments, condominiums, business, and industry; (C) roads and bridges; and (D) farmland.

provides a firm foundation for roads and buildings; fertile soil produces good crops; soil also supports trees and shrubs, and it can be modified to support the desired use (Figure 8-2).

THE NATURE OF WATER AND SOIL

Water

We live on the water planet! Most of the Earth's surface is covered with water (Figure 8-3), and the oceans and lakes are vast. Most people around the world, including in the United States, live near an ocean, river, lake, or stream. Those who do not live near such bodies of water must have access to water from

© StudioSmart/Shutterstock.com.

FIGURE 8-3 Most of the Earth's surface is covered by water.

Courtesy of Wendy Troeger.

FIGURE 8-4 Clean, fresh water serves many life-supporting functions, including nutrient carrier, waste transporter, coolant, home for aquatic plants and animals, oxygen for fish, and cleanser for humans and animals.

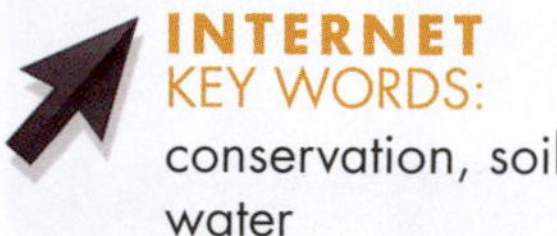

INTERNET KEY WORDS: conservation, soil, water

INTERNET KEY WORDS: American Farmland Trust

deep wells. Like the tissues of plants and other animals, our bodies are about 90 percent water, so we can survive only a few days if our supply of **potable** water—that is, drinkable and free from harmful microorganisms or chemicals—is cut off.

Water is essential for all plant and animal life (Figure 8-4). It dissolves and transports nutrients to living cells and carries away waste products. Evaporating water cools the surfaces of leaves and plants and the bodies of animals and humans when the temperature is uncomfortably high. Water serves so many useful functions that a sufficient supply of potable water is one of the first considerations for a healthy community.

Fresh versus Salt Water

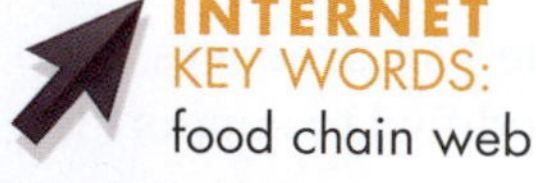

INTERNET KEY WORDS: food chain web

Most of the water on Earth is saltwater, not fresh water, and except for transportation, it is not suitable for human use. **Fresh water** refers to water that flows from the land to oceans and contains little or no salt. The water in our oceans contains heavy concentrations of salt. Similarly, our bays and tidewater rivers contain too much salt for domestic or household use. **Tidewater** refers to the water that flows up the mouth of a river with rising or in-flowing ocean tides; saltwater is not fit for land-based animal consumption or for plant irrigation.

Aquatic Food Chains and Webs

A **food chain** is made up of a sequence of living organisms that eat and are eaten by other organisms living in the community (Figure 8-5). Each member of the chain feeds on lower-ranking members of the chain. The general organization of an aquatic food chain moves from organisms known as producers (food plants) to herbivores (plant-eating water insects). Herbivores are eaten, in turn, by carnivores (meat-eating animals).

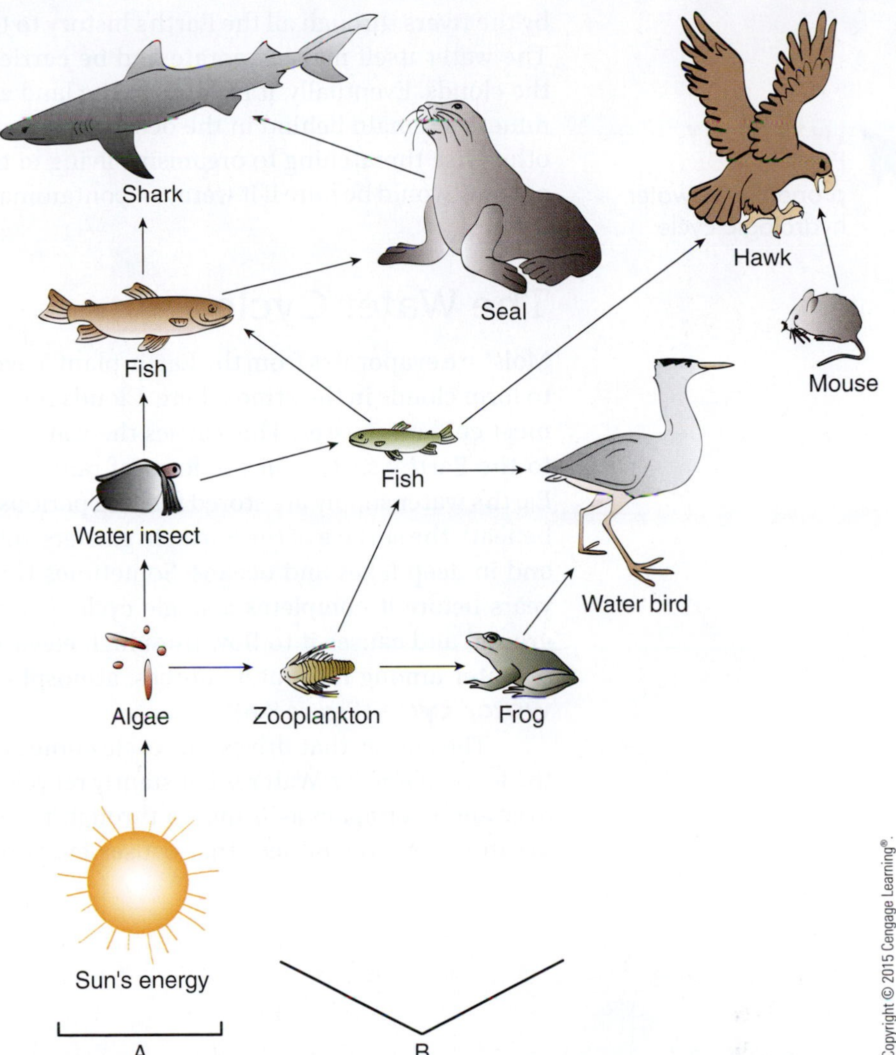

FIGURE 8-5 (A) A food chain model showing nutritional energy flowing in one direction. (B) A food web model showing the nutritional energy flow throughout the ecosystem.

The Universal Solvent

Water has been described as the universal solvent (a substance that dissolves or otherwise changes most other materials). Nearly every material will rust, corrode, decompose, dissolve, or otherwise yield to the presence of water. Therefore, water is seldom seen in its pure form. It generally has something in it.

Some minerals in water are healthful and give the water a desirable flavor. However, water sometimes carries toxic or undesirable chemicals or minerals. Water may also contain decayed plant or animal remains, disease-causing organisms, or poisons. Ocean water may be described as a thin soup. It is like our blood. Also, like blood, it gathers and transports nutrients, and it is the habitat for microorganisms. It carries life-supporting oxygen, and it neutralizes and removes wastes.

Scientists tell us that the purity levels of our rivers, bays, and seas are in trouble. People are polluting air, water, and land faster than nature can cleanse and purify these resources. The sea contains all of the dissolved elements carried by the rivers through all the Earth's history to the low places on the planet's crust. The water itself may evaporate and be carried back to the land as moisture in the clouds. Eventually, it may fall to the land again as rain or snow. However, the minerals remain behind in the ocean. Many of those substances may be toxic or otherwise threatening to organisms living in the sea. The water that falls as rain or snow would be pure if it were not contaminated by pollutants as it falls through the air.

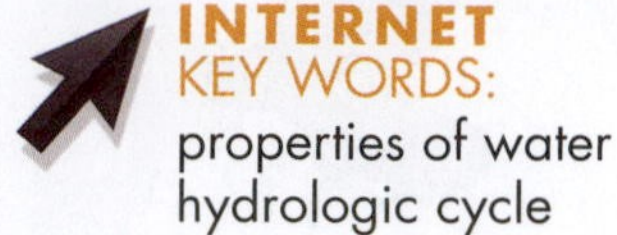

The Water Cycle

Moisture evaporates from the Earth, plant leaves, freshwater sources, and the seas to form clouds in the atmosphere. Clouds remain in the air until warm air masses meet cold air masses. This causes the water vapor to change to a liquid and fall to the Earth's surface in the form of rain, sleet, or snow. Large amounts of the Earth's water supply are stored for long periods of time. Storage occurs in aquifers beneath the surface of the Earth, in glaciers and polar ice caps, in the atmosphere, and in deep lakes and oceans. Sometimes this water is stored for thousands of years before it completes a single cycle. Gravity draws the water back into the ground and causes it to flow from high elevations to low elevations. The cycling of water among the water sources, atmosphere, and surface areas is called the **water cycle** (Figure 8-6).

The energy that drives this cycle comes from two sources: solar energy and the force of gravity. Water is constantly recycled. A molecule of water can be used over and over again as it moves through the cycle. Solar energy trapped by the ocean is a source of heat that causes evaporation. Additional water enters the

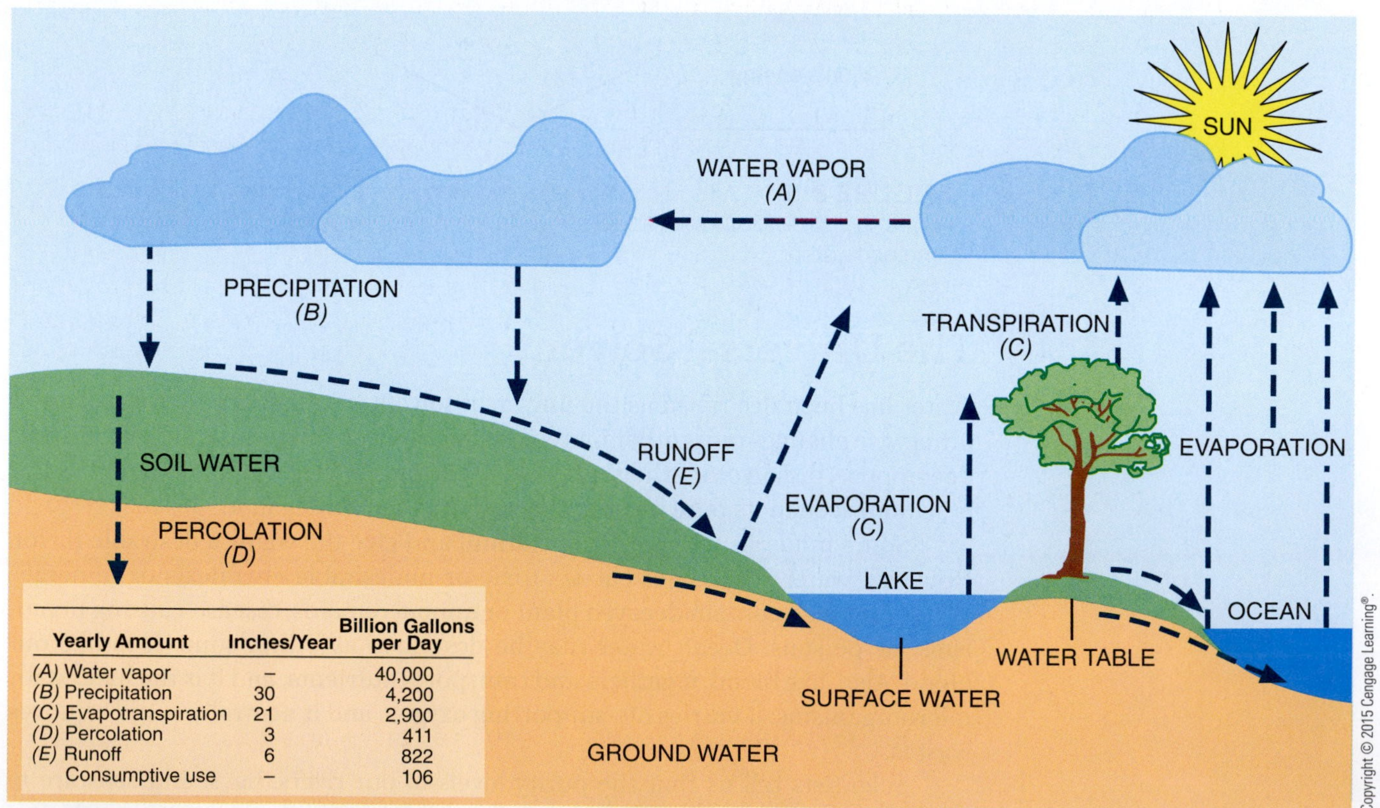

Yearly Amount	Inches/Year	Billion Gallons per Day
(A) Water vapor	–	40,000
(B) Precipitation	30	4,200
(C) Evapotranspiration	21	2,900
(D) Percolation	3	411
(E) Runoff	6	822
Consumptive use	–	106

FIGURE 8-6 The water cycle is a natural process by which water moves in a circular flow from oceans to land and back to the oceans.

SCIENCE CONNECTION FOOD CHAINS AND WEBS

Bays and oceans provide excellent support for the growth of algae, which in turn become major food sources for water-dwelling insects. Insects provide nutritional energy for shellfish. Shellfish are eaten by larger fish, which are finally consumed by top-level predators such as sharks. In this example, energy is passed from one organism to another in the form of food. This mode, referred to as a *food chain,* describes the interdependence of plants and animals for nutrition. All food chains begin with the sun, which is the primary energy source for most living organisms.

Producers at the base of the food chain capture solar energy and convert it into nutrients such as sugars and proteins. Plants and algae are good examples of producers. The next step in the food chain takes place when a consumer eats a producer. Animals that depend on producers or other animals for food are called *consumers.* In a simple food chain, energy is passed from the sun to the producers and then to consumers in one straight line. In most ecosystems, however, feeding relationships are much more complex than a single food chain suggests. *Food webs,* which are made up of many different overlapping food chains, represent the sum of all feeding relationships in an ecosystem. Using the preceding example, it is easy to see the involvement of other organisms. For example, birds will also eat fish, as do sharks. Water insects may eat aquatic plants other than algae. Both foods chains and food webs explain the feeding relationships in given ecosystems and can be observed on land and in water.

atmosphere by evaporating from soil and plant surfaces, especially in areas of hot temperatures and high precipitation.

Land

Land provides us solid foundations for buildings. It also provides nutrition and support for plants and space for work and play. Its aquifers provide storage for groundwater. Landmass also serves as a heat and compression chamber, converting organic material into coal and oil.

Soil is an important component of land. Productive soil is made up of correct proportions of soil particles and has the correct balance of nutrients. It also contains at least some organic matter and has adequate moisture. Much of the Earth's crust is too rocky or has an incorrect balance of nutrients for crop production, and much of the Earth's landmass is covered by only a thin layer of productive soil or is too steep to permit cultivation. Where there is some useful soil, however, trees are often able to survive. Forest lands provide lumber, poles, paper, and other products. Forest lands also provide areas where humans can benefit from the pleasures of wildlife and recreation.

Large areas of Earth have soil with a usable balance of soil particles and minerals for plant growth, but large areas have insufficient water. Areas with continuous, severe water shortages are called deserts. Some desert areas have become productive through the use of modern irrigation practices. **Irrigation** is the addition of water to the land to supplement the water provided for crop production by rain or snow. Many nations in desert regions do not have the money, technical knowledge, or water to make their deserts productive. Many nations of the world have such limited land resources that all of their land must be used to its greatest capacity.

During the 1930s, a large area of western Oklahoma and neighboring states became known as the *dust bowl* because lack of rainfall and increased wind shifted the dry soil and ruined the productivity of the land. Later, the productivity returned as weather patterns changed. Soil and water conservation practices were implemented to help prevent similar problems from recurring in the future.

Relationships of Land and Water

Precipitation

Land and water are related to each other in many ways. Land in cold regions or high altitudes retains moisture on its surface in the form of snow. This moisture is then released gradually to feed the streams and rivers after the precipitation (moisture from rain and snow) has stopped falling. Precipitation is caused by the change of water in the air from a gaseous state to a liquid state. It then falls to the land or bodies of water. Moisture-laden, warm-air clouds contact cold-air masses in the atmosphere and the result is precipitation. Clouds are formed by water changing from a liquid to a gas when it is evaporated by air movement over land and water. Evaporation means changing from a liquid to a vapor or gas.

INTERNET KEY WORDS: irrigation systems science, watershed

A **watershed** is a large land area in which water from rain or melting snow is absorbed and from which water drains as it emerges from springs and moves into streams, rivers, ponds, and lakes. The watershed acts as a storage system by absorbing excess water and releasing it slowly throughout the year (Figure 8-7).

Land as a Reservoir

In more ways than one, land serves as a container or reservoir for water. Where water soaks down into the soil, it forms a **water table**. Below that level, soil is saturated or filled with water. Water held below the water table may run out onto the Earth's surface at a lower elevation in the form of springs. Springs feed streams, which in turn form rivers and flow into lakes, bays, and oceans. Since ancient times, people have been known to dig wells below the water table to extract water for human needs.

FIGURE 8-7 Soil acts like a huge sponge that soaks up excess water in times when a surplus exists and releases water in a somewhat uniform flow from springs and wells throughout the year.

Even above the water table, excess water, held in the soil, is taken up by plant roots, from which water travels throughout the plant. Much of it evaporates from the leaves through transpiration to contribute to the moisture supply in the atmosphere.

Water helps soil by improving its physical structure. It is also essential for microorganisms that live in the soil and contributes to the soil's fertility (the amount and type of nutrients in the soil).

Types of Groundwater

Soil is saturated when water fills all the spaces or pores in the soil. If soil remains saturated for too long, plants will die from lack of air around their roots. The water that drains out of soil after it has been wetted is called **free water**, or **gravitational water**. Gravitational water feeds wells and springs. When gravitational water leaves the soil, some moisture—**capillary water**—remains, and this can be taken up by plant roots. Water that is held too tightly for plant roots to absorb is **hygroscopic water** (Figure 8-8).

Groundwater is easily polluted by chemicals and manure through abandoned wells and other depressions. In addition, the constant irrigation can increase salt concentration. These and other problems are being addressed by soil conservation and water-management districts.

Benefits of Living Organisms

Plants break the fall of raindrops and reduce damage to the soil from the impact of water. When plants drop their leaves, plant materials accumulate, and they provide a rain-absorbing layer on the soil surface. Worms, insects, bacteria, and other small and even microscopic plants and animals contribute to soil by decomposing dead plant and animal matter. As these materials decay, they add

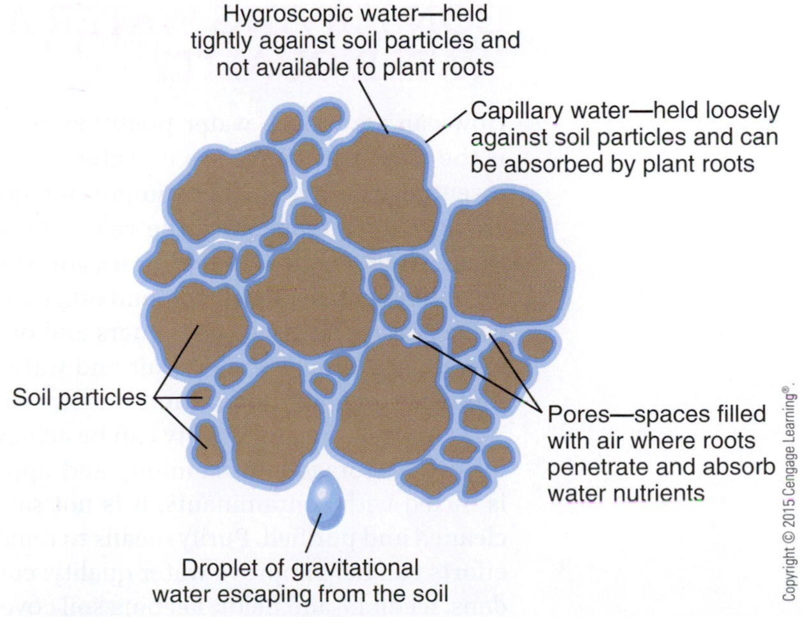

FIGURE 8-8 Plants use only capillary water. However, hygroscopic water contributes to soil structure, and gravitational water is held in reserve for future use by plants and animals.

AGRI-PROFILE CAREER AREA: SOIL CONSERVATION HYDROLOGY

Courtesy of USDA/ARS #K-1918-10.

Salt-laden wastewater flows from irrigated land in California's Imperial Valley into the Salton Sea.

Some major challenges in soil and water conservation include the following: (1) keeping rain water on the land where it falls, (2) keeping soil in place and stabilized with plants, (3) minimizing pollution of fresh water, and (4) wisely managing land and water to maintain ecological balance. For instance, soil and water scientists at the University of California are developing ways to reuse salt-laden irrigation water. As the human population increases in populous states such as California, the competition for fresh water becomes more intense. Imagine what it would mean to the world if the ocean waters could be purified and used to irrigate crops and supply the needs of the manufacturing and processing industries.

Soil and water conservation career options provide opportunities for indoor and outdoor work. Typical job titles are conservation technician, farm planner, soil scientist, and soil mapmaker. Work is available in offices of the Federal Farm Service Agency. One may work in the field served by such offices in nearly any county in the United States.

The United States Department of Interior, state departments of natural resources, city and county governments, industry, and private agencies hire people with soil and water conservation expertise. These professionals manage water resources for recreation, conservation, and consumption. They may work as consultants, law enforcement officers, technicians, administrators, heavy equipment operators, and other emerging careers.

organic matter to the soil, improving its structure. Soil organisms, therefore, contribute to water absorption and reduce soil erosion. Decayed plant and animal matter contributes substantially to the nutrient content of the soil. Plants assist in distributing water by taking it up through their roots and releasing it in the atmosphere. Both soil and water benefit from the living organisms that live in the soil.

CONSERVING WATER AND IMPROVING WATER QUALITY

How can we reduce water pollution? How can soil erosion be reduced? What is the most productive use of water and soil that does not pollute or lose these essential resources? These important questions deserve answers now. Farmers have long appreciated the value of these resources and generally have used them wisely. Economic conditions, governmental policies, production costs, farm income, personal knowledge, and other factors, however, also influence the use of conservation practices by farmers and other land users. Every citizen, business, agency, and industry affects air and water quality. Similarly, all of us have some influence on how our land and water resources are used.

Improved water quality can be achieved by proper land management, careful water storage and handling, and appropriate water use. Once clean water is mixed with contaminants, it is not safe for use for other purposes until it is cleaned and purified. Purify means to remove all foreign material. Several general efforts can help improve water quality: controlling water runoff from lawns, gardens, feedlots, and fields; keeping soil covered with plants; constructing livestock facilities so manure can be collected and spread on fields; and where feasible, using no-till or minimum tillage practices to produce crops.

SCIENCE CONNECTION FROM MANURE TO METHANE

© muzsy/Shutterstock.com.

Technology makes it possible to convert animal manure to heating gas and electricity.

The size of livestock enterprises has increased dramatically in recent years, with thousands of animals concentrated on large farms. Manure disposal from such farms has become challenging to many owners, particularly when they do not have enough cropland on which to spread it. Too much manure on the land is likely to contaminate the groundwater beneath the soil surface.

What to do? Treat the manure as a profitable commodity by converting it to methane gas. A manure digester is capable of providing a reliable supply of gas that can be (1) used to heat buildings on the farm, (2) used to generate electricity, or (3) sold as gas to local utilities for distribution to their customers for heating water and homes.

Some practices that help reduce water pollution are the following:

- Save clean water. Whenever we permit a faucet to drip, leave water running while we brush our teeth, or flush toilets excessively, we are wasting water. Taking long showers, leaving water running while we wash automobiles or livestock, and using excessive water for pesticide washup wastes clean water.
- Dispose of household products carefully. Many products under the kitchen sink, in the basement, or in the garage are threats to clean water (Figure 8-9). Never pour paints, wood preservatives, brush cleaners, or solvents down the drain. Eventually, they enter the water supply, rivers, or oceans. Use all products sparingly and completely. Put solvents into closed containers and permit the suspended materials to settle out before reusing the solvent. Fill empty containers with newspaper to absorb all liquids before discarding. Then send the containers to approved disposal sites.
- Maintain lawns, gardens, and farmland carefully. Improve soil by adding organic matter. Mulch lawn and garden plants as shown in Figure 8-10. Use proper amounts and types of lime, fertilizer, and other chemicals. Leave soil untilled if it is likely to erode excessively. Cover exposed soil with a new crop immediately. Apply water to the soil only when it is excessively dry, and continue to irrigate until the soil is soaked to a depth of 4–6 inches.
- Practice sensible pest control. Many insecticides kill all insects—both harmful and beneficial. Incorrect applications of insecticides also pollute water. Therefore, use cultural practices, such as crop rotations and resistant varieties, instead of insecticides, whenever possible. Encourage beneficial insects and insect-eating birds by improving the habitat around lawn, garden, and fields. Drain pools and containers of stagnant water to prevent mosquitoes from laying eggs resulting in larvae populations. Keep untreated wood away from soil to avoid damage from termites, other insects, and rot. Follow all pesticide label instructions exactly.

© iStockphoto/Alistair Forrester Shankie.

FIGURE 8-9 Many products around the home are hazardous and require correct disposal to avoid pollution of groundwater.

Courtesy of DeVere Burton.

FIGURE 8-10 A mulching system conserves water and allows plant vegetation from lawns and gardens to be added back to the soil as a nutritious organic mulch.

- Control water runoff from lawns, gardens, feedlots, and fields. Keep the soil surface covered with plants. Construct livestock facilities in such a manner that manure can be collected and spread on fields. Where it is feasible, use no-till cropping. **No-till** means planting crops without plowing or disking the soil. Plant alternate strips of close-growing crops (such as small grains or hay) with row crops (such as corn or soybeans). Farm on the **contour** (following the level of the land around a hill). Plant cover crops where regular crops do not protect the soil throughout the year. A cover crop is a close-growing crop planted to temporarily protect the soil surface.
- Control soil erosion. Reduce the volume of rainwater runoff by minimizing the amount of blacktop or concrete surface constructed. Use grass waterways in areas of fields where runoff water tends to flow. Add manure and other organic matter to soil to increase water-holding capacity. Construct terraces on long or steep slopes (Figure 8-11). Leave steep areas in tree production and excessive sloping areas in close-growing crops.
- Avoid spillage or dumping of petroleum products such as gasoline, fuel, or oil on the ground or in storm drains. Recycle used petroleum products.
- Prevent chemical spills from running or seeping away. Do not flush chemicals away. Chemicals will damage lawns, trees, gardens, fields, and groundwater. Sprinkle spills with an absorbent material such as soil, kitty litter, or sawdust. Remove the contaminated material and place it in a strong plastic bag to be discarded according to local recommendations.
- Properly maintain your septic system (if your home has one). Avoid using excessive water. Do not flush inappropriate materials down the toilet. Avoid planting trees where their roots might penetrate or interfere with sewer lines, septic tanks, or field drains. Do not run tractors and other heavy equipment over field drains.

INTERNET KEY WORDS:
wetlands, water

By following these recommendations, we can all help decrease water resource pollution.

Courtesy of DeVere Burton.

FIGURE 8-11 Construction of terraces on sloping land distributes water precipitation along the terrace and helps to prevent soil erosion.

SCIENCE CONNECTION INDICATOR SPECIES

Courtesy of DeVere Burton.

One way scientists can tell that water has been polluted is by studying *indicator species,* or sensitive organisms that show the state of an environment's overall health. The four categories of indicator species in water are fish, aquatic invertebrates (including worms, snails, and aquatic insects), algae, and aquatic plants. A pollutant in water affects these organisms in different ways. Depending on the type of pollution, the organisms can die off, have their numbers reduced, or increase their numbers. An algal bloom is an example of how an indicator species is affected by pollution.

Runoff water from farmlands on which fertilizer is too heavily applied may contain excess fertilizer. As this fertilizer enters a lake or river, it can stimulate algae growth. The result is an overgrowth called an *algal bloom.* These are often seen late in the growing season and indicate to scientists that a fertilizer pollutant is in the water. Specific problems may arise from such a bloom. Algae can produce toxins that are harmful to fish, animals, humans, and other organisms. When the algae die and decompose, oxygen is consumed and oxygen levels in the water fall, often killing other organisms that depend on the oxygen. Clean and safe water is needed for drinking, watering crops, and maintaining healthy living environments. Using indicator species helps scientists determine the safety of the water supply.

LAND EROSION AND SOIL CONSERVATION

The Problem

Land Erosion—A Worldwide Problem

Land **erosion**, the process of wearing away of soil, is a serious problem worldwide (Figure 8-12). Both wind and water are capable of eroding soil. The food and fiber production capabilities of large nations are being compromised because of extensive damage from soil erosion. Numerous cases can be cited from history when soil erosion caused enormous problems. Yet, in the twenty-first century, the people of the world are making many of the same mistakes that caused extensive soil losses in the past.

Courtesy of DeVere Burton.

FIGURE 8-12 Erosion of soil means loss of productive soil, damage to machinery, additional costs of production, and pollution of streams, rivers, lakes, bays, and oceans.

Consider, for example, the port of St. Mary's, Maryland, an early European colony established in 1634 on the eastern coast of what is now the United States. Like all ports, it was located along the shoreline, where a natural harbor made it possible for ships to load and unload cargo. By 1706, the port had been replaced by another town where the central port for the colony was located. The harbor area of the original port rapidly filled with soil eroded from the surrounding area. The water was no longer deep enough for the ships to enter safely.

Consider the massive size of the Mississippi River Delta. This land area was created from the soil deposits carried by the river. The Mississippi River Delta is more than 15,000 square miles in area. The Delta was formed as the river deposited soil on the bottom of the Gulf of Mexico, continuing until it reached the surface of the water. The soil gets into the river from smaller rivers and streams receiving runoff water from land areas in the heartland of the United States. Therefore, one must conclude that the Mississippi River Delta was built from the bottom to the surface of the Gulf of Mexico with the best topsoil in the United States. Yet, in these modern times, it is estimated that the amount of soil being dumped by the river into the Gulf each day would fill a freight train 150 miles long. It has been estimated that the amount deposited at the mouth of the Mississippi River in one year is sufficient to cover the entire state of Connecticut with soil 1 inch deep, or more than 8 feet in 100 years. Similar deltas exist at the mouth of the Nile River and most other great rivers of the world.

INTERNET KEY WORDS:
soil conservation
stabilization

During the late 1960s and early 1970s in the People's Republic of China, much of the forested land was cleared. It soon became apparent that much of the land depended on the trees to prevent soil erosion. Due to the extent of damage from soil erosion, the country has responded with a national policy of rapid reforestation. China has more than 1.3 billion people, and all are expected to contribute to the tree-planting effort. However, it will take many years for newly planted trees to be mature enough to protect the soil again. Unfortunately, the soil that was washed away during the absence of trees can never be returned to the land. Soil scientists report that it takes 300–500 years for nature to develop 1 inch of topsoil from bedrock.

Another serious threat to large areas of the world is the farming system known as "slash and burn." This is used in the tropical rain forests of South America, Africa, and Indonesia. In these areas, impoverished farmers cut or slash the jungle growth and burn the plant residues. The extensive burning causes serious air pollution and destroys useful fuel. These farms produce crops for several years until the land's fertility is depleted and soil erosion takes its toll. The land is then abandoned and the farmers move on to forested land, where the cycle is repeated. Once the land is cleared, native plant growth is not easily reestablished and the land is often left permanently damaged. The loss of the rain forest of the Amazon has slowed since the early 1990s, but it still declines by approximately 7,250 square miles each year. In some regions, 80 percent of the rainfall is attributed to transpiration from trees. When the trees are removed, such areas may become subject to drought.

National Problems

Each year, about 1.6 billion tons of soil are worn away from 417 million acres of U.S. farmland and deposited into lakes, rivers, and reservoirs. One ton equals 2,000 pounds. Although some soils are deep and can tolerate a certain amount of erosion, many fragile soils cannot. The U.S. Department of Agriculture (USDA) National Resources Inventory indicates 41 million acres (or about 10 percent) of our nation's cropland are highly erodible at rates of 50 or more tons per acre per year (Figure 8-13).

Of growing concern is the contamination of groundwater under large areas of the United States. Groundwater pollution emerged as a public issue in the late 1970s. The first reports documented sources of contamination associated with the disposal of manufacturing wastes. By the early 1980s, several incidences of groundwater contamination by pesticides used on field crops were confirmed. Groundwater contamination can threaten the health of large populations. For instance, in New York State, the vast **aquifer**, a water-bearing rock formation that underlies Long Island, represents the only supply of drinking water for more than 3 million people. In the United States, major aquifers underlie areas that contain thousands of square miles of land, often encompassing several states. Contamination of the aquifer in one region usually results in contamination of larger areas.

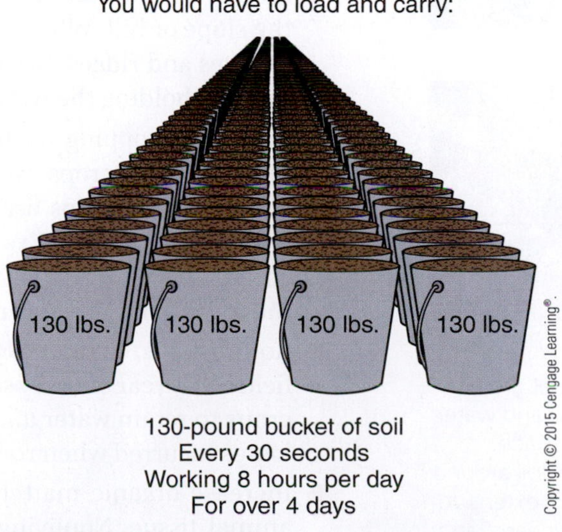

FIGURE 8-13 Imagine how much work it would require to replace the soil that is lost each year from one acre of highly erodible land.

Soil Conservation

Preventing and Reducing Soil Erosion

Soil erosion can be reduced or stopped by good land management. Fortunately, management practices that reduce soil erosion also increase water retention (Figure 8-14). This helps stabilize the supply of fresh water available for crops, livestock, wildlife, and people. It also reduces airborne soil particles, which are threats to good air quality. Most soil conservation methods are based on (1) reducing raindrop impact, (2) reducing or slowing the speed of wind or water moving across the land, (3) securing the soil with plant roots, (4) increasing absorption of water, or (5) carrying runoff water safely away. If water is free to run down a hill, it increases in volume and speed, picks up and pushes soil particles, and carries these particles off the land. This results in sheet and gully erosion and deposits soil, nutrients, and chemicals into streams. **Sheet erosion** occurs as layers of soil are removed from the land, whereas gully erosion is severe soil loss that leaves trenches in the land surface.

Courtesy of USDA/ARS #K-3226-6.

A

Courtesy of USDA/ARS #K-3213-1.

B

Courtesy of USDA/ ARS #K-5210-18.

C

FIGURE 8-14 Plant growth holds the key to soil and water conservation. (A) Surface mulch protects the soil around cotton seedlings, (B) extensive root systems of soybeans, and (C) revegetation of sandbars with willow trees.

Some recommended practices to reduce or prevent wind and water erosion include the following:

- Keep soil covered with growing plants. Plants reduce the destructive impact of raindrops, reduce the speed of wind and water movement across the land, and hold soil in place when threatened by wind or moving water.
- Cover the soil with mulch. **Mulch** is a material placed on the soil surface to break the fall of raindrops, prevent weeds from growing, and/or improve the appearance of the area.
- Utilize conservation tillage methods. **Conservation tillage** means using techniques of soil preparation, planting, and cultivation that disturb the soil the least, leaving the maximum amount of plant residue on the surface. Plant residue is the plant material that remains when a plant dies or is harvested.
- Use contour practices in farming, nursery production, and gardening. **Contour farming** means conducting all operations, such as plowing, disking, planting, cultivating, and harvesting, across the slope and on the level. This way, any ruts or ridges created by machinery occur around the slope or hill. When water tries to run down the hill, it encounters the grooves and ridges. Because these are level, the water tends to soak into the soil, holding the water for future use.
- Use strip cropping on hilly land. **Strip cropping** means alternating strips of row crops with strips of close-growing crops. Examples of close-growing crops are hay, pasture, and small grains, such as wheat, barley, oats, and rye. Examples of row crops are corn, soybeans, and most vegetables. The strips of close-growing crops intercept runoff water from the row crops and prevent it from entering streams.
- Rotate crops. **Crop rotation** is the planting of different crops in a given field every year or every several years. Crop rotation permits close-growing crops to retain water and soil, and it tends to rebuild the soil, countering losses incurred when row crops occupied the land.
- Increase organic matter in the soil. Organic matter is dead plant and animal tissue. Nonliving plant leaves, stalks, branches, bark, and roots decay and become organic matter. Similarly, animal manure, dead insects, worms, and animal carcasses decompose to make organic matter. The

HOT TOPICS IN AGRISCIENCE

CONSTRUCTED AND NATURAL WETLANDS

Constructed wetlands are human-made marshlands that can be used to treat polluted water. They contain plants that tolerate standing water and provide habitat for bacteria that are capable of breaking down materials that cause water pollution. As water passes slowly through the marsh, these bacteria and other microbes cleanse the water by consuming pollutants and changing them to harmless compounds. Some water plants also take up pollutants.

A natural wetland performs the same function, removing impurities from the water. Many of the natural wetlands, however, have been drained to prepare land for farming. Currently, we are beginning to recognize the value of the wetlands. Constructed wetlands usually consist of rectangular plots arranged in a series that are filled with gravel or porous soil and lined to prevent pollutants from leaching into the groundwater. Plants are added to the plots to simulate a natural marshland. Constructed marshlands do a good job of mimicking the advantages of natural marshlands. They also help create excellent wildlife habitat.

decomposed organic matter forms a gel-like substance that holds soil particles in absorbent granules called aggregates. An aggregated soil is a water-absorbing and nutrient-holding soil. Organic matter also releases nutrients to growing plants.

- Provide the correct balance of lime and fertilizer. **Lime** is a material that reduces the acid content of soil. It also supplies nutrients such as calcium and magnesium to improve plant growth. **Fertilizer** is any material that supplies nutrients for plants.
- Establish permanent grass waterways. A **grass waterway** is a strip of grass growing in an area of a field where flowing water tends to erode the soil surface (Figure 8-15).

Courtesy of DeVere Burton.

FIGURE 8-15 Soil that is most vulnerable to erosion should be planted to permanent vegetation.

SCIENCE CONNECTION SAFEGUARDING GROUNDWATER

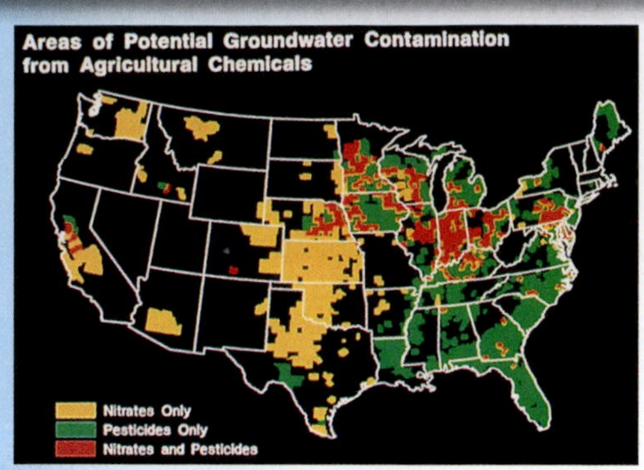

Groundwater contamination by agricultural chemicals is considered a risk in many areas.

Source: Nielsen, E. G. and Lee, L.K. October 1987. The Magnitude and Costs of Groundwater Contamination from Agricultural Chemicals: A national perspective. Resources and Technology Division, Economic Research Service, U.S. Department of Agriculture. Agricultural Economic Report No. 576. Figure 8. Retrieved from http://naldc.nal.usda.gov/download/CAT88907300/PDF

Authorities estimate that nearly 330,000 tons of pesticides are applied yearly on U.S. crops. Added to this are the fertilizers that are used on farms, yards, gardens, parks, and so on, and the chemicals used to control fleas, flies, mosquitoes, ticks, termites, roaches, and other pests of livestock and humans. Although agricultural chemicals have enabled us to feed and clothe ourselves and many others in the world, pollution from these chemicals has become a threat to our air, water, and land.

Applying these pesticides where they are intended and keeping them there until they change or biodegrade into harmless products is a major goal in protecting our water supplies. To help safeguard our surface water and groundwater from pesticide pollution, the Agricultural Research Service of the USDA has developed a specific strategy that has helped shape research priorities. The ultimate goals of the plan are to (1) provide U.S. farmers with cost-effective best-management practices that will ensure ample supplies of food and fiber at a reasonable cost, while reducing pesticide movement into the groundwater; (2) identify the factors that accelerate or retard pesticide movement; and (3) provide computer models that will quickly and accurately predict contamination.

USDA water research is being concentrated in three areas to protect against contamination of groundwater: (1) agricultural watershed management, (2) irrigation and drainage management, and (3) water quality protection and management.

The computer models being used and developed are decision-enhancing tools that use a database and computer program to help select management practices. Such practices may include pesticide application procedures that are most likely to restrict pesticide movement in various soils due to climatic and other environmental conditions. These models integrate data from many sources that are frequently not available to decision makers. New technologies, such as advances in slow-release formulations, improved pesticide application scheduling, and selective placement, are also under study. Today, groundwater contamination by agricultural chemicals is considered a risk in many of the major crop and livestock production areas of the United States.

- Construct terraces. A **terrace** is a soil or wall structure built across a slope to capture water and move it safely to areas where it is not likely to cause erosion.
- Avoid overgrazing. Overgrazing results when too much of a plant is eaten, reducing its ability to recover after grazing. This reduces the ability of the plant roots to hold soil in place.
- Follow a soil conservation plan. A conservation plan is a plan developed by soil and water conservation specialists to use land for its maximum production and water conservation without unacceptable damage to the land.

The USDA's Natural Resources Soil Conservation Service (NRCS) and Farm Service Agency (FSA) provide advice, technical assistance, and funds to assist landowners with soil and water conservation practices. The Environmental Protection Agency (EPA) monitors groundwater quality and enforces point and nonpoint source pollution laws to help protect our land and water.

There are many careers in the field of water and soil conservation. The material in this unit discusses some of the problems created by humans as they use natural resources to provide food, water, shelter, recreation, and other resources for living. The health, wealth, peace, and general welfare of humanity depend on our skill in conserving these most basic resources.

Constructed wetlands are a growing trend in agriculture. Farmers understand that water is priceless. Some have chosen to set aside a fraction of their property as a wetland to help purify water and ensure its safety.

STUDENT ACTIVITIES

1. Write the Terms to Know and their meanings in your notebook.
2. Collect a sample of drinking water from each of five sources as follows: a safe spring or well; bottled pure water; and faucets attached to (a) galvanized pipe, (b) plastic pipe, and (c) copper pipe. (Do not run off the water before obtaining the samples from the faucets.) Taste a small amount of each sample and describe the taste. Do the samples have different tastes? If so, what causes the differences?
3. Study the eating habits of one species of birds in your community. Describe the food chain that accounts for the survival of that species. What effect did the use of DDT as an insecticide have on that species of birds before DDT was banned from use in the United States?
4. Obtain a rain gauge and record the precipitation on a daily basis for several months. What variations did you observe from week to week? How do you explain the variations? What effects did these variations have on the agricultural activities of the community?
5. Conduct an experiment to determine the effect of soil water on plant growth. Obtain three inexpensive pots of healthy flowers or other plants. Each pot must be the same size and type, have the same amount and type of soil, and contain the same size and number of plants. Use the following procedure:
   a. Mark the pots "1," "2," and "3."
   b. Plug the holes in the bottoms of pots 2 and 3 so water cannot drain from the pots. Leave pot 1 unplugged so it has good drainage.
   c. Add water to pot 1 as needed to keep the plant healthy for several weeks. Use it as a comparison specimen (called the "control").
   d. Add water slowly to pot 2 until water has filled the soil and the water is just level with the surface of the soil. All pores of the soil are now filled and the soil is "saturated." Weigh the pot and record the weight as "A."
   e. Remove the plug from the bottom of pot 2 and permit the water to drain out. After water has stopped flowing from the drain hole, immediately weigh the pot again. Record the weight as "B." The water that flowed from pot 2 when the plug was removed is the free or gravitational water. Weight of the gravitational water, "D," should be calculated and recorded using the formula A – B = D. Do not add any more water to pot 2 for two weeks.
   f. With the hole plugged in pot 3, add water slowly until the water is just level with the top of the soil. Do not remove the plug, and keep pot 3 filled with water to the saturation point for two weeks.
   g. Keep all three pots in a good growing environment for two weeks and record all observations.
   h. When the plant in pot 2 wilts badly due to lack of water, weigh the pot and record the weight as "C."

      Make the following calculations:

      Weight of the gravitational water (D) = A – B

      Weight of the capillary water (E) = B – C

      The weight of the hygroscopic water can be determined only by driving the remaining water from the soil by heating the soil in an oven.

6. Ask your teacher to help you design and conduct a project that demonstrates some or all of the following:
   a. effect of the force of raindrops on soil
   b. effect of soil aggregation on absorption
   c. effect of slope on erosion
   d. effect of living grass on erosion control
   e. effect of plant residue on erosion control
7. Consider some feeding relationships in your area. Illustrate and label a food web or food chain that consists of five or more organisms. Write one paragraph to explain your drawing.

SELF-EVALUATION

A. MULTIPLE CHOICE

1. Most of the Earth's surface is covered with
   a. crops.
   b. farms.
   c. trees.
   d. water.
2. The bodies of plants, animals, and humans consist of about what percentage water?
   a. 10 percent
   b. 40 percent
   c. 70 percent
   d. 90 percent
3. The universal solvent is
   a. gasoline.
   b. paint thinner.
   c. varsol.
   d. water.
4. The content of ocean water may be likened to
   a. fresh water.
   b. pure water.
   c. thin soup.
   d. varsol.
5. The interdependence of plants and animals on each other for food is known as
   a. domestic.
   b. the food chain.
   c. symbiosis.
   d. a universal relationship.
6. The land serving as a heat and compression chamber gives us
   a. building foundations.
   b. coal and oil.
   c. crops.
   d. wildlife habitat.
7. Groundwater that is available for plant root absorption is called
   a. capillary.
   b. free.
   c. gravitational.
   d. hygroscopic.
8. Improvement of water quality can be achieved by
   a. appropriate use of water.
   b. careful water storage and handling
   c. proper land management.
   d. all of the above.
9. The problem of land erosion is
   a. characteristic of developed countries only.
   b. found worldwide.
   c. mostly found in poor countries.
   d. not a substantial problem in view of food surpluses.

10. A conservation plan
    a. conserves soil and water.
    b. is prepared by professionals.
    c. maximizes productivity of land.
    d. all of the above.

B. MATCHING

________	1. Contour	a.	Amount and type of nutrients
________	2. Delta	b.	Without plowing or disking
________	3. Evaporation	c.	Town with a harbor
________	4. Fertility	d.	Removal of foreign material
________	5. Irrigation	e.	Soil deposited by water
________	6. No-till	f.	Top of saturated soil
________	7. Port	g.	Rain and snow
________	8. Precipitation	h.	Supplemental water
________	9. Purify	i.	On the level
________	10. Water table	j.	Change from liquid to gas

C. COMPLETION

1. An _______ is a water-bearing rock formation.
2. The process of creating narrow and deep trenches eroded in soil is known as _______.
3. Material placed on soil to break the fall of raindrops is called _______.
4. Alternating row crops with close-growing crops is known as _______.
5. _______ is used to reduce acid in soil.
6. One inch of topsoil may be formed from bedrock in about _______ years.
7. _______ is dead plant and animal material in soil.
8. A _______ may be planted to temporarily protect soil from erosion.
9. In the United States, about _______ tons of soil are worn away from farmland each year.
10. About _______ acres, or 10 percent, of the U.S. cropland is highly erodible.

UNIT 9

Soils and Hydroponics Management

OBJECTIVE

To determine the origin and classification of soils and to identify effective procedures for soils and hydroponics management.

MATERIALS LIST

- writing materials
- newspapers
- topography map
- Internet access

COMPETENCIES TO BE DEVELOPED

After studying this unit, you should be able to:

- define terms in soils, hydroponics, and other plant-growing media management.
- identify types of plant-growing media.
- describe the origin and composition of soils.
- discuss the principles of soil classification.
- determine appropriate amendments for soil and hydroponics media.
- discuss fundamentals of fertilizing and liming materials.
- identify requirements for hydroponics plant production.
- describe types of hydroponics systems.

SUGGESTED CLASS ACTIVITIES

1. Assign class members to fill a 1-gallon container with soil and bring it to school. This will provide a wide range of soil types that can be used as teaching materials. Each student should measure his or her soil sample to determine the soil texture. Students should work in pairs, and each student should also test his or her partner's soil sample as a check for accuracy.
2. Conduct drainage tests on the soil samples. Begin by placing a filter paper at the bottom of a funnel. An acceptable funnel can be made by cutting off the top third of a plastic beverage bottle and inverting it. Next, place a measured amount of soil inside the funnel and on top of the filter paper. Make sure the soil is directly on the filter paper. Then carefully add a measured amount of water to the soil surface while being

TERMS TO KNOW

medium
hydroponics
compost
sphagnum
peat moss
perlite
vermiculite
horizon
soil profile
residual soil
alluvial deposit
lacustrine deposit
loess deposit
colluvial deposit
glacial deposit
percolation
capability class
O horizon
A horizon
clay
silt
sand
topsoil
B horizon
subsoil
C horizon
bedrock
loam
soil structure
decomposer
soil amendment
pH
acidity
alkalinity
neutral
fertilizer grade
active ingredient
starter solution
nitrate
nitrogen fixation
aeroponics

careful not to disturb the placement of the soil. Observe the length of time that is required for the water to move through the soil to the container below the funnel. This is a good time to line up the soil samples from most permeable to least permeable. Have students check the texture of the soil with their fingers according to the chart that is provided in this unit.

3. Invite a local expert to come and introduce hydroponics to the class. Students have the ability to do new solution culture experiments at little cost by using plants that have never been grown in solution culture before, such as most houseplants and bedding plants. Students could do experiments in the areas of nutrient deficiencies, toxicities, carbon dioxide and oxygen deficiencies, pH, fertilizer testing, growth regulators, nitrogen fixation, shoot-to-root ratio, bulb forcing, and others. The possibilities for hydroponics projects are nearly endless.

The roles of plants in our environment and their importance in our lives have been discussed in previous units. Plants are necessary to nourish the animals of the world and maintain the balance of oxygen in our atmosphere. However, they depend on a medium of soil, water, or, in some cases, air for a supportive living environment. A **medium** is a material or surrounding environment in which a living organism functions and lives.

PLANT-GROWING MEDIA

For discussion in this unit, the word *media* is used to mean the material that provides plants with nourishment and support through their root systems.

Types of Media

Media comes in many forms. The oceans, rivers, land, and man-made mixtures of various materials are the principal types of media for plant growth. Seaweed, kelp, plankton, and many other plants depend on water for their nutrients and support. It is only recently that we have become aware of the tremendous amount of plant life in the sea. The plant life in oceans and rivers is important for feeding the animals found in water.

It has long been known that water could be used to promote new root formation on the stems of certain green plants, and it can completely support plant growth for a short time. Recently, it has been found that food crops can be grown efficiently without soil. This is done using structures where plant roots are submerged in or sprayed with solutions of water and nutrients (Figure 9-1). These solutions feed the plants, whereas mechanical structures provide physical support. The practice of growing plants without soil is called **hydroponics**. Hydroponics has become an important commercial method of growing some plants.

Courtesy of USDA/ARS.

FIGURE 9-1 Plants can live and grow without any soil by spraying or submerging their roots in water in which the nutrients that are necessary for plants to live are dissolved.

Soil

Soil is defined as the top layer of the Earth's surface, which is suitable for the growth of plant life. It has long been the predominant medium for cultivated plants (Figure 9-2). In early years, humans accepted the soil as it existed. They planted seeds using primitive tools and did not know how to modify or enhance the soil to improve its plant-supporting performance. Ancient civilizations discovered that plant-growing conditions were improved on some land where deposits remained

© J. Helgason/Shutterstock.com.

FIGURE 9-2 Soil is the most important plant medium for most crops.

after river waters flowed over the land during flood season. Similarly, other land was ruined by floodwaters. Therefore, early efforts to improve plant-growing media were a matter of moving to better soil. Obviously, good soil was a valuable asset and, therefore, possession of it was the cause of intense personal disputes and wars among nations.

Other Media

In addition to water and soil, certain other materials will retain water and support plant growth. Fortunately, some of the best non-soil and non-water plant-growing media are partially decomposed (decayed) plant materials. Hence, plants tend to improve the environment where they grow. One common material available around most homes is leaf mold and compost. Leaf mold is partially decomposed plant leaves. **Compost** is a mixture of partially decayed organic matter such as leaves, manure, and household plant wastes. Decaying plant matter should be mixed with lime and fertilizer in correct proportions to support plant growth (Figures 9-3 and 9-4). There is a group of pale or ashy mosses called **sphagnum**. These are used extensively in horticulture as a medium for encouraging root growth and growing plants under certain conditions. **Peat moss** consists of partially decomposed mosses that have accumulated in waterlogged areas called bogs that are saturated with water. Both sphagnum and peat moss have excellent air- and water-holding qualities. Many other sources of plant and animal residues may be used as plant-growing media. For instance, a fence post or a log may rot on its top and hold moisture from rainfall. Horse manure mixed with straw is used extensively as a medium for growing mushrooms. In this instance, both animal residue (manure) and plant residue (straw) combine to make an effective medium.

Some mineral matter can also become plant-growing media. For instance, volcanic lava and ash eventually accumulate soil particles on their surface. Seeds settle into cracks, and moisture causes the seeds to germinate. Roots then penetrate and break up the volcanic residue. As time passes, the area that once was only lava and ash becomes covered with plant life. Horticulturists use certain mineral materials in plant-growing areas, too. **Perlite** is a natural volcanic glass material that has water-holding capabilities. Perlite is used extensively for starting new plants. **Vermiculite**, a mineral matter from a group of mica-type materials, is also used for starting plant seeds and cuttings.

Layers should be turned and mixed together every several weeks as the materials decay into a fertile mass. Depending upon moisture, temperature, content, and frequency of turning, it should take 3–12 months to make compost.

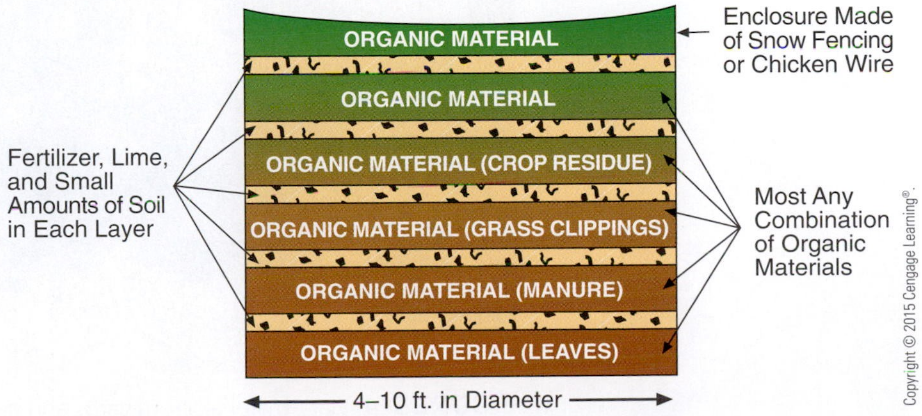

FIGURE 9-3 Compost is excellent organic matter. Most mineral soils can be improved by adding compost.

HOT TOPICS IN AGRISCIENCE PRODUCING "HOTHOUSE" VEGETABLES

Commercial production of vegetables is extended into winter months using hydroponic practices in heated greenhouses.

The production of "hothouse" vegetables is done in a controlled greenhouse environment. In many cases, these vegetable production units are located near natural hot water springs to take advantage of the heat source during the winter months. The hot water is often piped into the greenhouse where it is an inexpensive source of supplemental heat. Most of these vegetable farms use some variation of hydroponics rather than using soil as a medium for plant growth. The plants are supported as they grow by tying them up using wires and strings. Tomatoes and other vegetable products are harvested frequently at their peak of quality and are shipped immediately to markets. Most "hothouse" vegetables are produced during seasons when field production is limited or impossible because of climate restrictions. As a result, these vegetables command high prices in the markets.

INTERNET KEY WORDS:
soil formation
soil formation and weathering
soil formation and topography
soil formation and climate
soil formation and bedrock materials

ORIGIN AND COMPOSITION OF SOILS

Factors Affecting Soil Formation

Productive soils develop on the Earth's surface as the atmosphere, sunlight, water, and living things meet and interact with the mineral world. If soil is suitable for plant growth to a depth of 36 inches or more, the soil is regarded as "deep." Many soils of the Earth are much shallower than this. Plants attach themselves to the soil by their roots, where they grow, manufacture food, and give off oxygen. Plants

FIGURE 9-4 Plant materials from yards and gardens can be converted to valuable compost by placing the materials in an environment with moderate temperatures where bacteria known as decomposers have adequate moisture and oxygen.

Factors Affecting Soil Formation

Climate/location	Affects rate of weathering
Living organisms	Cause decay of organic material
Parent material	Influences fertility and texture
Topography	Affects distribution of soil particles and water
Temperature	Influences rate of weathering
Weathering	Causes soils to develop, mature, and age

FIGURE 9-5 Soil formation depends on a number of natural factors.

and animals of various sizes live on and in the soil, using carbon dioxide, oxygen, water, mineral matter, and products of decomposition.

Soils vary in temperature, organic matter, and the amount of air and water they contain. The kinds of soils formed at a specific site are determined by the forces of climate, living organisms, parent soil material, topography, and time (Figure 9-5).

soil microorganisms

Climate and Location

Climatic factors, such as temperature and rainfall, greatly affect the rate of weathering. When temperature increases, the rate of chemical reactions increases and the growth of fungi, organisms (such as bacteria), and plants increases. The rate and amount of rainfall in a locality greatly affect the soil. In areas of high rainfall, the soils are usually leached and somewhat acidic. Leached means that certain contents have been removed from the soil by water. If the land is covered by trees, the action of high temperatures and moisture on leaf residues generally creates an acidic soil. Rainfall during cold weather has less effect on the soil than during warm or hot weather.

Slope and location of a field affect soil erosion and drainage, thus influencing soil formation. Moreover, free water in the soil carries fine particles to the deeper layers and tends to produce "layering." Too much water prevents or retards microbial growth and may exclude air by waterlogging the soil. Water and temperature also have the effect of swelling and contracting soil particles.

FIGURE 9-6 Living organisms like earthworms play important roles in breaking down organic matter such as leaves, stubble, and grass.

Living Organisms

Living organisms, such as microbes, plants, insects, animals, and humans, exert considerable influence on the formation of soil. Certain types of soil bacteria and fungi aid in soil formation by causing decay or breakdown of the plant and animal residues in the soil. Carbon dioxide and other compounds essential to soil formation are released by microbe activity. Microbes are microscopic plants and animals. Without soil microbes, organic materials would not decay.

Numerous insects, worms, and animals contribute to the formation of soil by mixing the various soil materials (Figure 9-6). Earthworms consume and digest certain soil substances and discharge body wastes. This aids decomposition and soil mixing. All such dead organisms add to soil organic matter.

FIGURE 9-7 Clearing natural vegetation from land usually contributes to soil loss at a much faster rate than soil is formed.

Human activity also influences soil formation. Cultivation, bulldozing, and construction projects all disturb the surface layer. Clearing of land removes native plant life and greatly modifies soil-forming activities (Figure 9-7).

Parent Material

Parent material is the horizon of unconsolidated material from which a soil develops. **Horizon** means layer. Parent materials compose the C horizon of a typical soil profile. **Soil profile** means a cross-sectional view of soil (Figures 9-8 and 9-9).

Parent materials formed in place are called **residual soils**. Other soils are transported and deposited by water, wind, gravity, or ice. **Alluvial deposits** are transported by streams, and **lacustrine deposits** are left by lakes. **Loess deposits** are left by wind, **colluvial deposits** by gravity, and **glacial deposits** by ice.

The kind of parent material from which soil is formed influences the many characteristics of that soil. Natural fertility and texture are influenced greatly by the parent material of a profile.

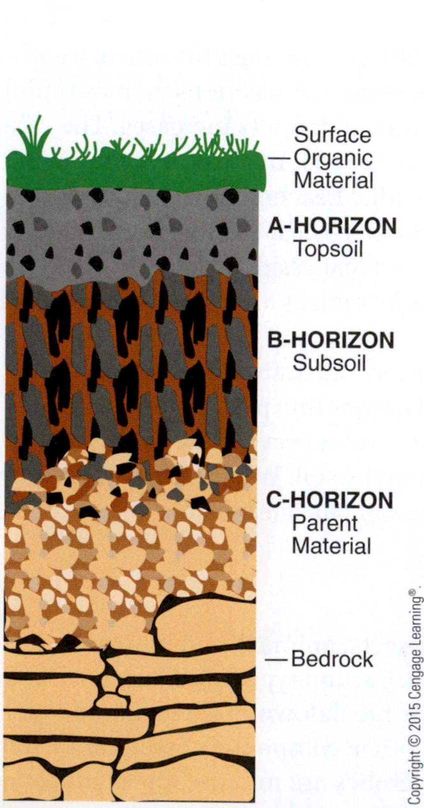

FIGURE 9-8 A soil profile consists of a cross-sectional view of the different layers of materials, beginning at the surface and going down to bedrock.

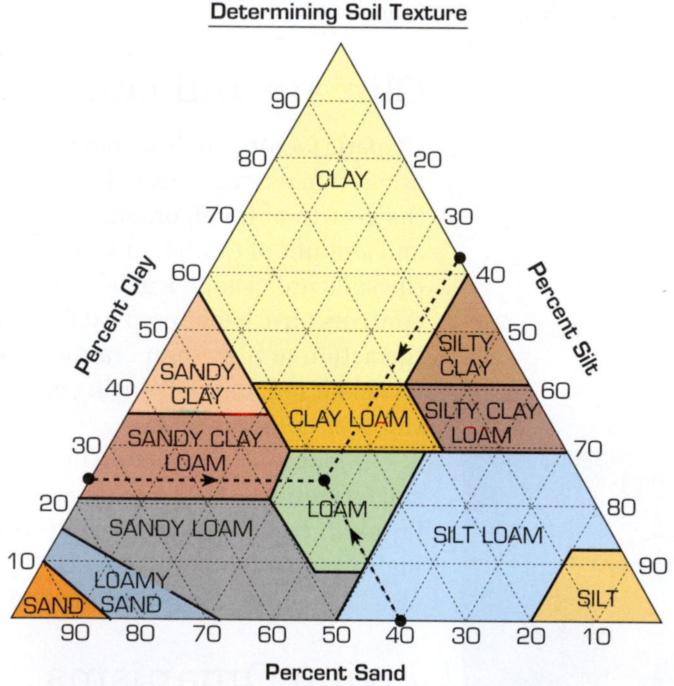

Example: Identify a soil that is 40% sand, 22% clay, and 38% silt.

1. Find 40 on the side for sand.
2. Draw a line in the direction of the arrow.
3. Do the same for clay (22%) and silt (38%).
4. The spot where the three come together is the soil texture. In this case, the soil is a loam.

A textural name may include a prefix naming the dominant sand size, as in "coarse sandy loam."

FIGURE 9-9 Soil texture is determined by the percentages of sand, silt, and clay found in a soil sample. The soil triangle is a tool that is used to determine texture.

Topography

Slope and drainage affect soil formation both directly and indirectly. On a steep slope, loose material is moved downward by runoff water, gravity, and movement of humans and animals. This movement not only breaks up soil materials and adds them to the lower levels but it also exposes subsoil materials along the upper slopes. The movement of soil materials has a pulverizing effect on the material being moved as well as on the material left behind.

Slope affects the distribution of water that falls on the Earth's surface. On level areas, the water soaks in and moves through the soil in a process called **percolation**. On sloping land, the water tends to run off and it moves some surface soil with it. Soils that develop on level land at low elevations tend to be poorly drained, whereas soils on gentle slopes tend to be better drained and more productive (Figure 9-10).

Drainage (or lack of it) affects the water table in a particular field or area. The water table has a direct bearing on soil formation, especially if it is near the surface. When a soil is saturated with water, little or no air can penetrate it. The lack of air reduces the action of fungi, bacteria, and other soil-forming activities in the soil.

A wet soil is, therefore, a slow-forming soil and is usually low in productivity. Because of the lack of air, undecomposed organic matter accumulates in a wet soil. This organic matter generally causes the soil to be a blackish color. Poor drainage, accompanied by free water in the soil, reduces or retards plant growth and affects soil formation.

Time

Soils are formed by the chemical and physical weathering of parent material over time, as affected by climate, living organisms, and topography. Therefore, time itself is regarded as a factor in soil formation. Chemical weathering is the result of the chemical reactions of water, oxygen, carbon dioxide, and other substances that act on the rocks, minerals, organic matter, and life that compose the soil. The leaching action of water hastens the weathering process by removing soluble materials, and chemicals react with each other to form new chemicals in the soil.

FIGURE 9-10 Topography, or slope of the land, influences the formation of soil and how well excess water drains from it.

Weathering

Weathering refers to mechanical forces caused by temperature change such as heating, cooling, freezing, and thawing. As these processes occur, rocks, minerals, organic matter, and other soil-forming materials are broken into smaller and smaller particles until soil is formed. Soils at different stages of weathering will differ widely.

Weathering causes soils to develop, mature, and age much as people do. Soils develop rapidly, mature, and then develop certain characteristics of age. Plant nutrients are released quickly from the minerals, plant growth increases, and organic matter accumulates. Soils age more slowly during the later stages of weathering.

Eventually, nutrients in the soil are depleted. Water moving through the soil leaches away many soluble materials. At this stage, many soils are acidic because the limestone originally found in them is gone. As the supply of nutrients in the soil decreases, the amount of plant growth is reduced to the point where the organic matter decomposes faster than it is produced. When soils become acidic and have lost their native fertility, they require expensive amendments to keep them productive.

In permeable soils that permit water movement, the fine clay particles tend to move downward from the surface soil into the subsoil during the weathering process. This movement, together with further breakdown of the rock material, accounts for the fact that many soil types have a greater percentage of clay in the subsoil than in the surface soil.

SOIL CLASSIFICATION

Soil scientists have developed a system for mapping soils according to the physical, chemical, and topographical aspects of the land. Such maps have lines showing the outline of soil types, and they provide numerical codes keyed to large amounts of information about the land. The experienced soil technician can obtain a wealth of information about the land by consulting a soils map and the accompanying material.

Land Capability Maps

Soil mapping and land classification have been priorities of the U.S. Department of Agriculture (USDA) throughout most of the twentieth century. Much of the original work of mapping was completed by the Soil Conservation Service (SCS). However, in recent years, this agency has been incorporated into the Natural Resources Conservation Service (NRCS). Now the NRCS can provide maps and classification information for almost any area in the United States, including small areas on individual farms. Agency personnel work with local farmers to develop farm plans. These provide recommendations for land use, cropping systems, crop production practices, and livestock systems. These plans are designed to maximize production while controlling soil erosion and enhancing productivity.

land capability classes

Land capability maps indicate (1) capability class, (2) capability subclass, and (3) capability unit.

Capability Class

Capability classes are the broadest classifications on a soils map and are designated by Roman numerals I through VIII (Figure 9-11). The numerals indicate progressively greater limitations and narrower choices for practical use of land as follows.

Land Capability Classes

USDA	SCS
Suitable for cultivation	No cultivation-pasture, hay, woodland, and wildlife
I Requires good soil management practices only	V Requires good soil management practices only
II Moderate conservation practices necessary	VI Moderate conservation practices necessary
III Intensive conservation practices necessary	VII Intensive conservation practices necessary
IV Perennial vegetation infrequent cultivation	VIII Perennial vegetation infrequent cultivation

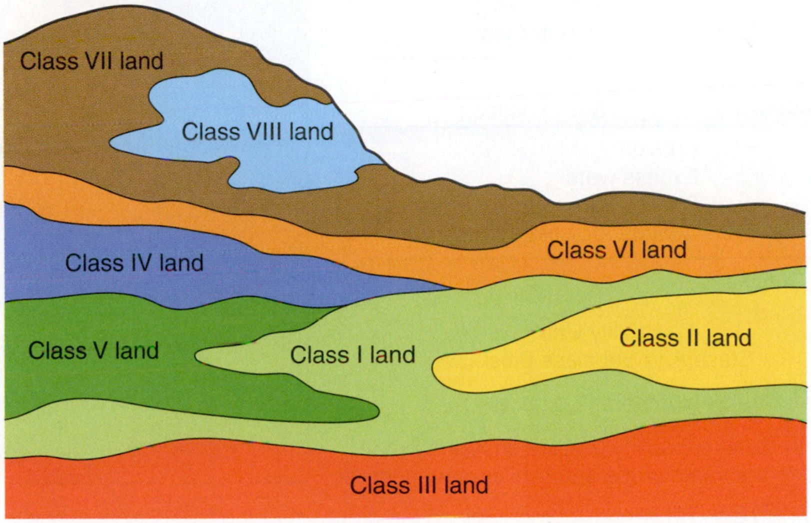

Source: USDA/NRCS Soil Conservation Service.

FIGURE 9-11 The eight classes of land in the United States.

Class I—Soils have few limitations that restrict their use.

Class II—Soils have moderate limitations that reduce the choice of plants or require moderate conservation practices.

Class III—Soils have severe limitations that reduce the choice of plants, require special conservation practices, or both.

Class IV—Soils have severe limitations that reduce the choice of plants, require careful management, or both.

Class V—Soils are not likely to erode but have other limitations that are impractical to remove and limit their use.

Class VI—Soils have severe limitations that make them generally unsuitable for cultivation.

Class VII—Soils have severe limitations that make them unsuitable for cultivation.

Class VIII—Soils and geologic features have limitations that nearly always prevent their use for agricultural production except light grazing.

Land capability maps are usually color-coded for ease in differentiating capability classes.

Capability Subclasses

Capability subclasses are soil groups within one class. They are designated by adding a lowercase letter *e*, *w*, *s*, or *c* to the class numeral (e.g., IIe).

Capability Class

Numeral		Color Code
I	—	Light green
II	—	Yellow
III	—	Red
IV	—	Blue
V	—	Dark green
VI	—	Orange
VII	—	Brown
VIII	—	Purple

Capability Class		Subclass
Plus Lower Case Letter		Soil Limitation
e	—	Erosion
w	—	Excess water
s	—	Shallow, droughty, or stony soil
c	—	Climate too cold or too dry

Capability Unit
Class And Subclass Plus:

Arabic Numbers (1, 2, 3, etc.)

Soils with same capability unit are enough alike to be suited to the same crops and similar management

Land Capability Map

Source: USDA/NRCS Soil Conservation Service.

FIGURE 9-12 Land capability map. A land capability map is used to identify soil locations with similar characteristics. Similar land capability classes indicate that similar crops and management practices may be applied.

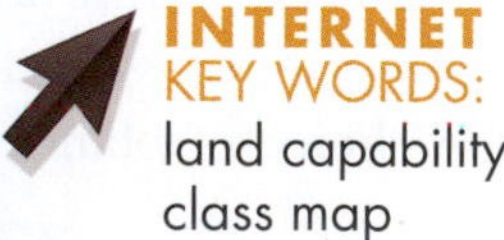

INTERNET KEY WORDS: land capability class map

The letter *e* indicates the main limitation is risk for erosion unless close-growing plant cover is maintained; *w* indicates that weather in or on the soil interferes with plant growth or cultivation; *s* indicates the soil is limited mainly because it is shallow, droughty, or stony; and *c*, used in only some parts of the United States, indicates the chief limitation is climate—too cold or too dry (Figure 9-12).

In Class I, there are no subclasses because the soils of this class have few limitations. Conversely, Class V contains only the subclasses indicated by *w, s*, or *c* because the soils in Class V are subject to little or no erosion. However, they have other limitations that restrict their use to pasture, range, woodland, wildlife habitat, or recreation.

Capability Units

The soils in one capability unit (soil groups within the subclasses) are enough alike to be suited to the same crops and pasture plants and require similar management. They have similar productivity capabilities and other responses to management. The capability unit is a convenient grouping for making many statements about the management of soil.

Capability units are generally designated by adding numbers (0–9) to the subclass symbol (e.g., IIIe4 or IIw2). Thus, in one symbol, the Roman numeral designates the capability class or degree of limitation; the lowercase letter indicates the subclass or kind of limitation; and the Arabic numeral specifically identifies the capability unit within each subclass. A map legend for each soil grouping is included with the soil and land capability map. This type of map is also an example of how symbols can be effectively used. They make it possible to place much information in a small space on the map by use of a code.

Courtesy of USDA #K-5218-03.

FIGURE 9-13 Natural Resources personnel advise on the many uses of land. Satellite imagery and high-altitude photography help document changing conditions on the soil surface.

Use of Maps

When the landowner and the soil conservationist start planning for the most intensive use of a farm, they need a soil and land capability map. This will help them prepare a conservation plan or proposed land-use map. The soil conservationist and the landowner, through the use of the soil and land capability map, discuss the kinds of soil on the farm. The current and original land uses are also discussed and noted on the map. In developing a proposed land-use map, the soil conservationist must know the personal goals or objectives of the landowner and the plans for developing the land.

Many things can be involved in reaching land-use decisions, field by field. Field boundaries may need to be changed so that all the soil in each field is suited for the same purpose and management practices. The desired balance among cropland, pasture, woodland, and other land uses needs to be considered. Appropriate livestock enterprises should also be considered to match the land's characteristics and potential. If there is a good potential for income-producing recreation enterprises in the community, an area may be used for hunting, campsites, or fishing.

The landowner and soil conservationists must consider how to treat each field to get the desired results. The NRCS conservationist can give many good suggestions, but the landowner must decide what to do, when to do it, and how to do it. As planning decisions are made, the conservationist will record them in narrative form and make them part of the plan map. This becomes the farm conservation plan. It is the guide for the farming operation in the years ahead. A conservation plan is required to obtain federal support to encourage good farming and conservation practices.

INTERNET KEY WORDS:
soil profiles, horizons

Workers from the NRCS are also available to give on-site technical assistance in applying and maintaining the farm conservation plan. They provide management advice and services to non-farm landowners, developers, strip-mining operations, and other activities where soil is involved (Figure 9-13).

PHYSICAL, CHEMICAL, AND BIOLOGICAL CHARACTERISTICS

Soil Profile

Undisturbed soil will have four or more horizons in its profile. These are designated by the capital letters *O*, *A*, *B*, and *C* (Figure 9-14). The **O horizon** is on the surface and is composed of organic matter and a small amount of mineral matter. Organic matter originates from living sources such as plants, animals, insects, and microbes. Mineral matter is derived from non-living sources such as rock materials.

The **A Horizon** is located near the surface and consists of mineral matter and organic matter. It contains desirable proportions of organic matter, fine mineral particles called **clay**, medium-sized mineral particles called **silt**, and larger mineral particles called **sand**. The appropriate proportion of these soil components creates soil that is tillable, or workable with tools and equipment. With the presence of desirable plant nutrients, chemicals, and living organisms, the A horizon generally supports good plant growth. The A horizon is frequently called **topsoil**.

Horizon	Name	Colors	Structure	Processes Occurring
O	Organic	Black, dark brown	Loose, crumbly, well broken up	Decomposition
A	Topsoil	Dark brown to yellow	Generally loose, crumbly, well broken up	Zone of leaching
B	Subsoil	Brown, red, yellow, or gray	Generally larger chunks, may be dense or crumbly, can be cement-like	Zone of accumulation
C	Parent material (slightly weathered material)	Variable—depending on parent material	Loose to dense	Weathering, disintegration of parent material or rock

FIGURE 9-14 Characteristics of soil horizons.

The **B horizon** is below the A horizon and is generally referred to as **subsoil**. The mineral content is similar to the A horizon, but the particle sizes and properties differ. Because organic matter comes from decayed plant and animal materials, the amount naturally decreases as distance from the surface increases.

The **C horizon** is below the B horizon and is composed mostly of parent material. C Horizon is important for storing and releasing water to the upper layers of soil, but it does not contribute much to plant nutrition. It is likely to contain larger soil particles and may have substantial amounts of gravel and large rocks. The area below the C horizon is called **bedrock**.

INTERNET KEY WORDS: soil texture, soil structure

Texture

Texture refers to the proportion and size of soil particles (Figures 9-15 and 9-16). Texture can be determined accurately in the laboratory by mechanical analysis (Figure 9-17). The feel of the soil can also be used to analyze soil as follows:

1. Make a stiff mud ball.
2. Rub the mud ball between your thumb and forefinger.
3. Note the degree of coarseness and grittiness caused by the sand particles.
4. Squeeze the mud between your fingers, and then pull your thumb and fingers apart.
5. Note the degree of stickiness caused by the clay particles.
6. Make the soil slightly more moist and note that the clay leaves a "slick" surface on your thumb and fingers.

The outstanding physical characteristics of the important textural grades, as determined by the "feel" of the soil, are discussed next (Figure 9-18).

Coarse-textured (sandy) soil is loose and single grained. The individual grains can be seen readily or felt. Squeezed in the hand when dry, it will fall apart when the pressure is released. Squeezed when moist, it will form a cast, but will crumble when touched.

Medium-textured soil is known as **loam**. It has a relatively even mixture of sand, silt, and clay. However, the clay content is less than 20 percent. (The

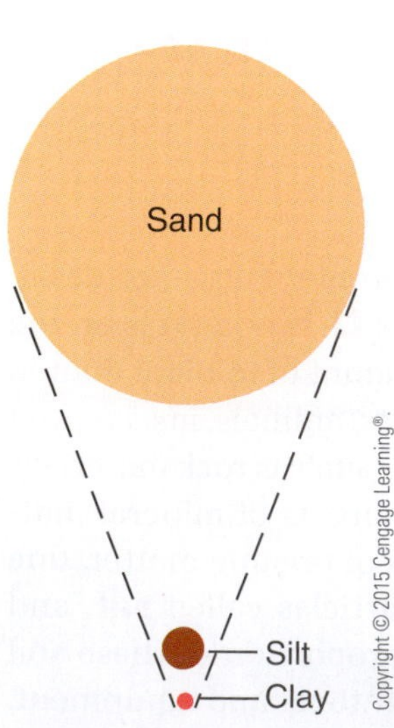

FIGURE 9-15 Relative size of sand, silt, and clay particles.

SIZES OF SOIL PARTICLES

Name	Size, Diameter in Millimeters
Fine gravel	2–1
Coarse sand	1.00–0.50
Medium sand	0.50–0.25
Fine sand	0.25–0.10
Very fine sand	0.10–0.05
Silt	0.05–0.002
Clay	less than 0.002

FIGURE 9-16 Range of sizes of soil particles.

characteristic properties of clay are more pronounced than those of sand.) A loam is mellow with a somewhat gritty feel, yet fairly smooth and highly plastic. Squeezed when moist, it will form a cast that can be handled quite freely without breaking.

Fine-textured (clay) soil usually forms hard lumps or clods when dry. It is usually very sticky when wet and is quite plastic. When the moist soil is pinched between the thumb and fingers, it will form a long, flexible "ribbon." A clay soil leaves a "slick" surface on the thumb and fingers when rubbed together with a long stroke and firm pressure. The clay tends to hold the thumb and fingers together because of its stickiness.

Structure

Soil structure refers to the tendency of soil particles to cluster together and function as soil units called aggregates. Aggregates, or crumbs, contain mostly clay, silt, and sand particles held together by a gel-type substance formed from organic matter.

Aggregates absorb and hold water better than individual particles. They also hold plant nutrients and influence chemical reactions in the soil. Another major benefit of a well-aggregated soil or a soil with good structure is its resistance to damage by falling raindrops. When hit by falling rain, the aggregate stays together as a water-absorbing unit, rather than separating into individual particles. When aggregates on the surface of soil dry out, they remain in a crumbly form and permit good air movement. Dispersed soil particles run together when dry and form a crust on the soil surface. The crust prevents air exchange between the soil and the atmosphere and decreases plant growth. The process and benefits of aggregation apply mostly to fine- and medium-textured soils.

INTERNET KEY WORDS:
soil organic matter, soil texture, soil classification

DEMONSTRATION TO SHOW SAND, SILT, AND CLAY

MATERIALS: A quart jar with lid
1/2 teaspoon of Calgon (dispersing agent)
A pint of medium-textured soil

PROCEDURE: Fill the jar half-full of soil.
Add enough water to make the jar 3/4 full.
Add 1/2 teaspoon of Calgon.
Shake for 5 minutes.
Set jar aside and allow to settle undisturbed.
Note that the soil separates, settling into layers.

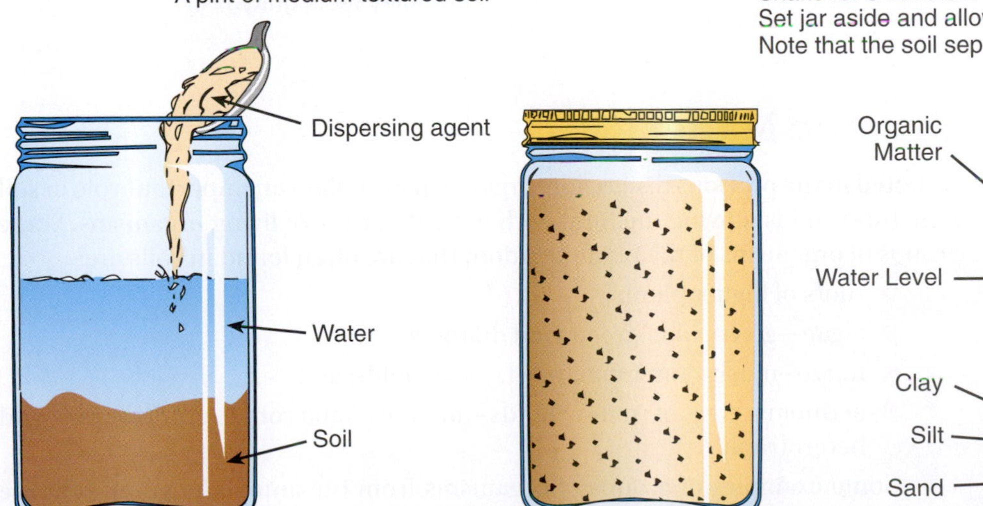

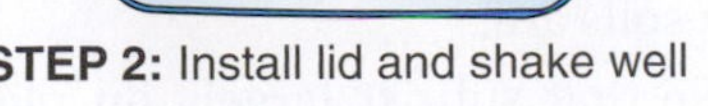

FIGURE 9-17 Mechanical analysis of soil to accurately determine the soil texture.

SOIL TEXTURAL CLASSES

SAND

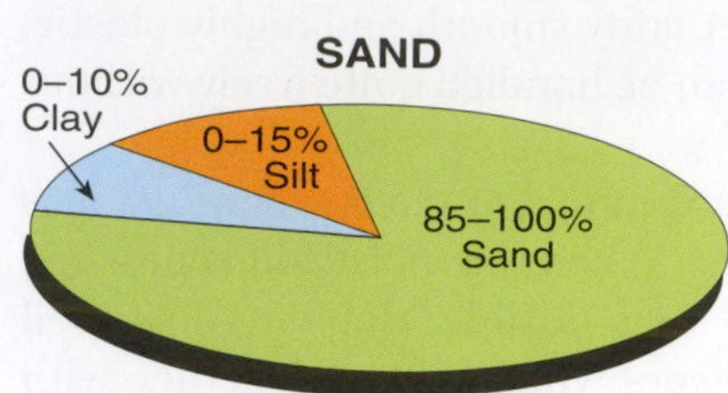

Dry: loose and single grained; feels gritty.
Moist: will form a ball that crumbles very easily.

LOAMY SAND

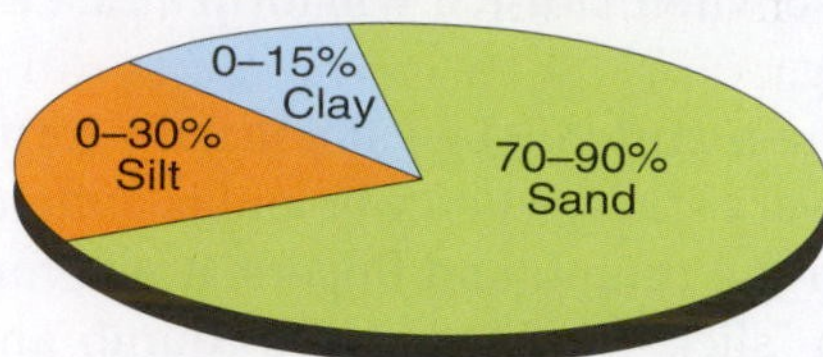

Dry: silt and clay may mask sand; feels loose, gritty.
Moist: feels gritty; forms a ball that crumbles easily; stains fingers slightly.

SANDY LOAM

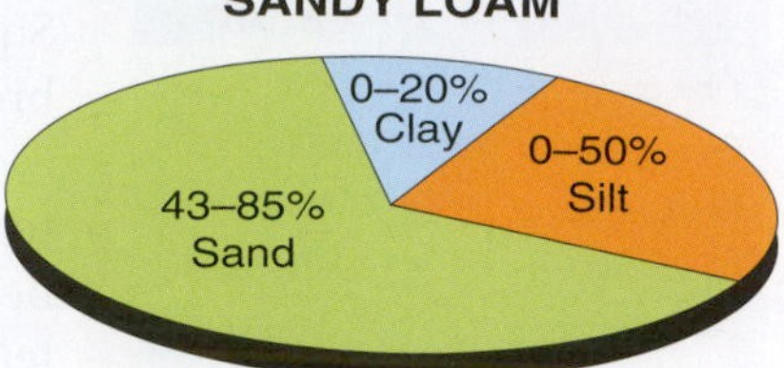

Dry: clods easily broken; sand can be seen and felt.
Moist: moderately gritty; forms a ball that can stand careful handling; definitely stains fingers.

LOAM

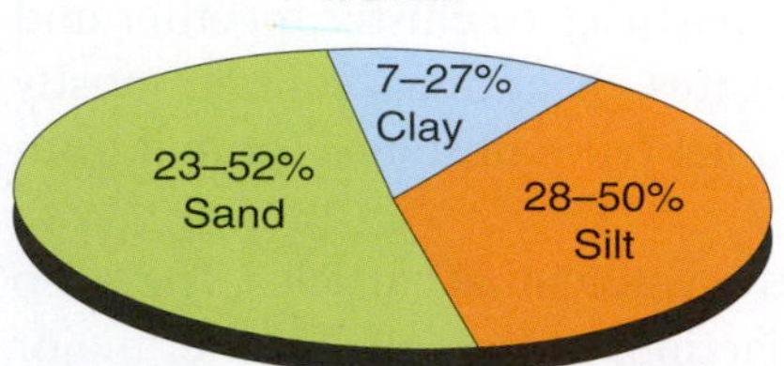

Dry: clods are moderately difficult to break; mellow, somewhat gritty.
Moist: neither very gritty nor very smooth; forms a firm ball; stains fingers.

SILT LOAM

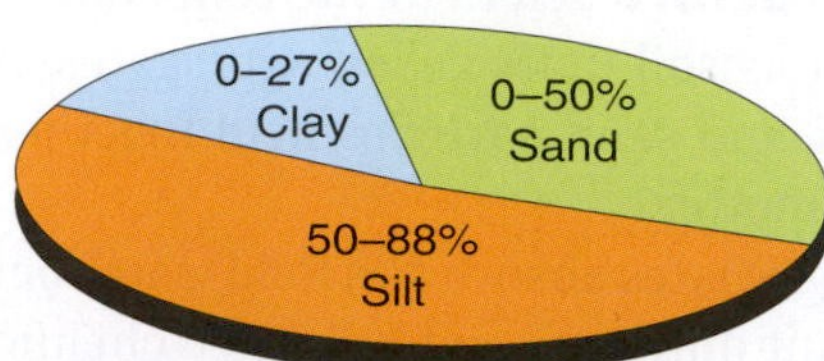

Dry: clods are difficult to break; feels smooth, soft, and floury when pulverized; shows fingerprints.
Moist: has smooth or slick "buttery" or "velvety" feel; stains fingers.

CLAY LOAM

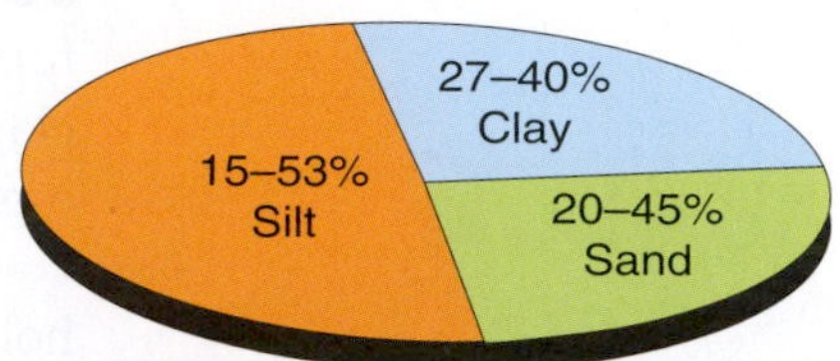

Dry: clods very difficult to break with fingers.
Moist: has slightly gritty feel; stains fingers; ribbons fairly well.

SILTY CLAY LOAM

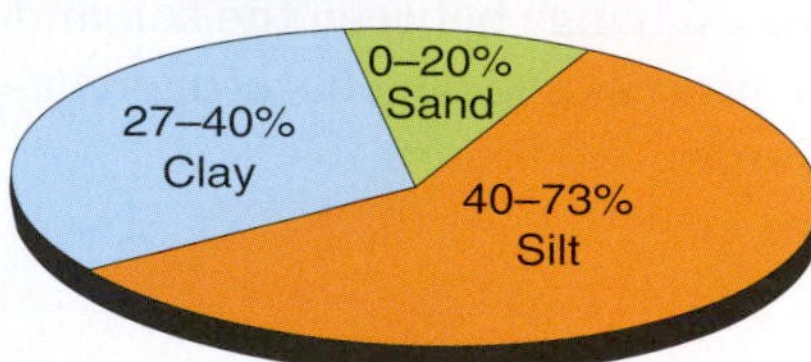

Same as **CLAY LOAM** but very smooth.

SANDY CLAY LOAM

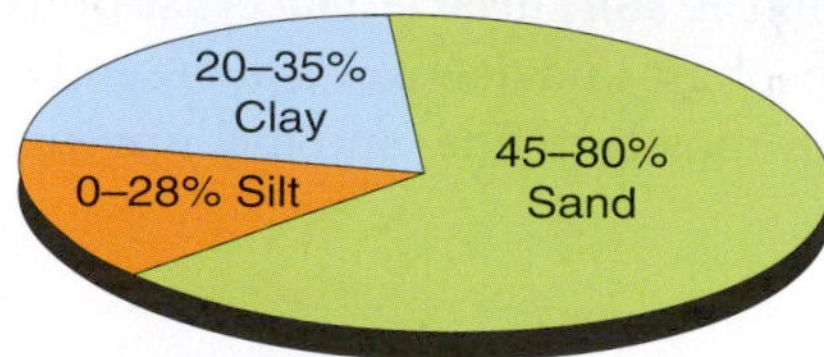

Same as **CLAY LOAM.**

CLAY

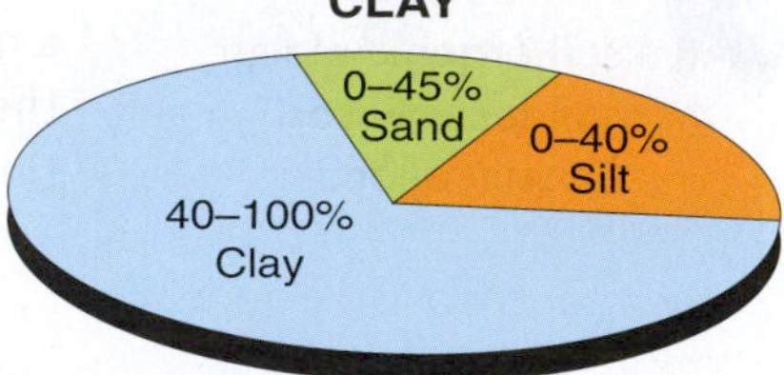

Dry: clods cannot be broken with fingers without extreme pressure.
Moist: quite plastic and usually sticky when wet; stains fingers. (A silty clay feels smooth; a sandy clay feels gritty.)

FIGURE 9-18 Major soil textural classes.

Organic Matter

As noted in the previous discussion, organic matter plays an important role in soil structure. Soil is a living medium with a great variety of living organisms. Some groups of organisms of the plant kingdom that are often found in soils are:

- roots of higher plants;
- algae—green, blue-green, and diatoms;
- fungi—mushroom fungi, yeasts, and molds; and
- actinomycetes of many kinds—aerobic, anaerobic, autotrophic, and heterotrophic.

Some examples of groups of organisms from the animal kingdom that are prevalent in soils are:

- those that subsist largely on plant material—small mammals, insects, millipedes, sow bugs (wood lice), mites, slugs, snails, earthworms;

- those that are largely predators—snakes, moles, insects, mites, centipedes, and spiders; and
- microanimals that are predatory, parasitic, and live on plant tissues—nematodes, protozoans, and rotifers.

Living Organisms

Living organisms excrete cell or body wastes that become part of the organic content of soil. The microbes of the soil and the remains of larger plants and animals decompose or decay into soil-building materials and nutrients.

People who grow indoor plants at home, raise gardens, farm, produce greenhouse crops, or grow nursery stock generally find it useful to add organic materials to the soil. Popular sources of organic matter for soil amendments are peat moss, leaf mold, compost, livestock manure, and sawdust.

Some important benefits or functions of organic matter in soil include:

- making the soil porous;
- supplying nitrogen and other nutrients to the growing plant;
- holding water for future plant use;

SCIENCE CONNECTION DIGGING LIFE

Soil is the most diverse ecosystem on Earth. The number of organisms per acre in soil far exceeds the concentration of organisms of any other place in the world. Most of us do not think about the abundant life under our feet as we cross a lawn. But just one square inch of this soil is teeming with busy creatures, most of which cannot be seen. Gardens and fields consist of fertile soil filled with living organisms. One gram of this soil contains approximately the following organisms:

3,000,000–500,000,000	Bacteria—A group of one-celled microscopic organisms
1,000,000–20,000,000	Actinomycetes—Microscopic organisms that resemble both fungi and bacteria
5,000–1,000,000	Fungi—Nonmicroscopic organisms that get their food from dead material
1,000–100,000	Yeast—Single-celled fungi. Many are used in producing food.
1,000–500,000	Protozoa—Small single-celled organisms. An example is an amoeba.
1,000–500,000	Algae—One-celled organisms that contain chlorophyll
1–500	Nematodes—Nonsegmented roundworms

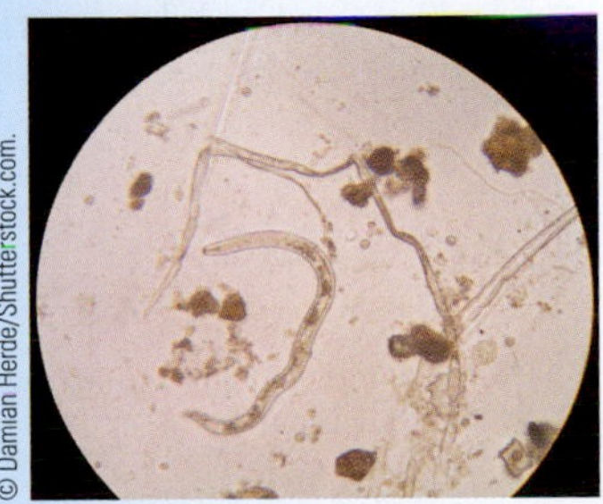

A nematode is a microscopic worm found in the soil.

Large numbers of slime molds, viruses, insects, and earthworms are also present. Some of these organisms are harmful to plants and animals. Most are decomposers. A **decomposer** is an organism that breaks down material that was once living. They change things that are dead into the rich organic substances that add to the fertility of the soil.

The role these tiny creatures play in an ecosystem is irreplaceable. Imagine a world without decomposers. Leaves, dead animal carcasses, and huge amounts of other dead organic matter would pile up. In a short period, dead and undecomposed material would crowd out all life. Soil is a unique substance. It provides living space for billions of organisms. It is the medium in which plants and other producers grow. Soil is also a place where once-living things are broken down and changed into the fertile organic material of the soil.

- aiding in managing soil moisture content;
- furnishing food for soil organisms;
- serving as a storehouse for nutrients;
- minimizing leaching;
- serving as a source of nitrogen and growth-promoting substance; and
- stabilizing soil structure.

Other Properties of Soils

Soil scientists, managers, technicians, and operators must be aware of numerous other properties of soils in their work. Some of these are external factors such as land position, slope, and stoniness. Soil color, depth, drainage, permeability, and erosion are important considerations.

MAKING AMENDMENTS TO PLANT-GROWING MEDIA

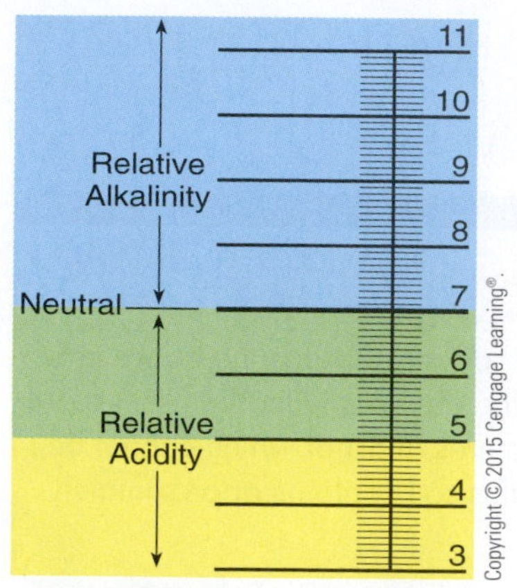

FIGURE 9-19 The pH is neutral in the middle. It gets progressively more acidic from 7 down to 3 and progressively more alkaline from 7 up to 11.

The term **soil amendment** is used here to mean addition to or change. Most soil amendments are made to add organic matter, add specific nutrients, or modify soil pH. The **pH** is a measure of the degree of acidity or alkalinity. **Acidity** is sometimes referred to as sourness, and **alkalinity** is referred to as sweetness. The pH scale ranges from 0 (maximum acidity) to 14 (maximum alkalinity). The midpoint of the scale is 7, which is **neutral**, meaning neither acidic nor alkaline (Figure 9-19).

Crops grow best in media with a narrow pH range unique to that plant species (Figure 9-20). Most plants require a pH somewhere between 5.0 and 7.5. Some crops, such as potatoes, prefer a soil pH around 5.5. Alfalfa responds best to a pH of 7.0 to 8.0.

Liming

Areas that were historically covered by trees, such as the northeastern, western, and northwestern parts of the United States, develop acidic soils. When cleared of trees, such soils require additional lime to increase the pH for the efficient production of most farm crops.

A pH test can be performed using a pH test kit, or soil samples can be sent to a university or commercial laboratory for analysis. Laboratory analyses generally include an analysis of phosphorus, potash, and magnesium, as well as pH. Liming and fertilizing recommendations may also be provided by testing laboratories and universities (Figure 9-21).

How to Take a Soil Sample in Your Field, Lawn, or Garden

1. Select an appropriate sampling tool (spade, auger, or soil tube) (Figure 9-22).
2. Make a sketch dividing the area into sampling areas—for example, front lawn, garden, flower bed, slope, and back lawn. Appropriately label each area (see Figure 9-21A).

collecting a soil sample

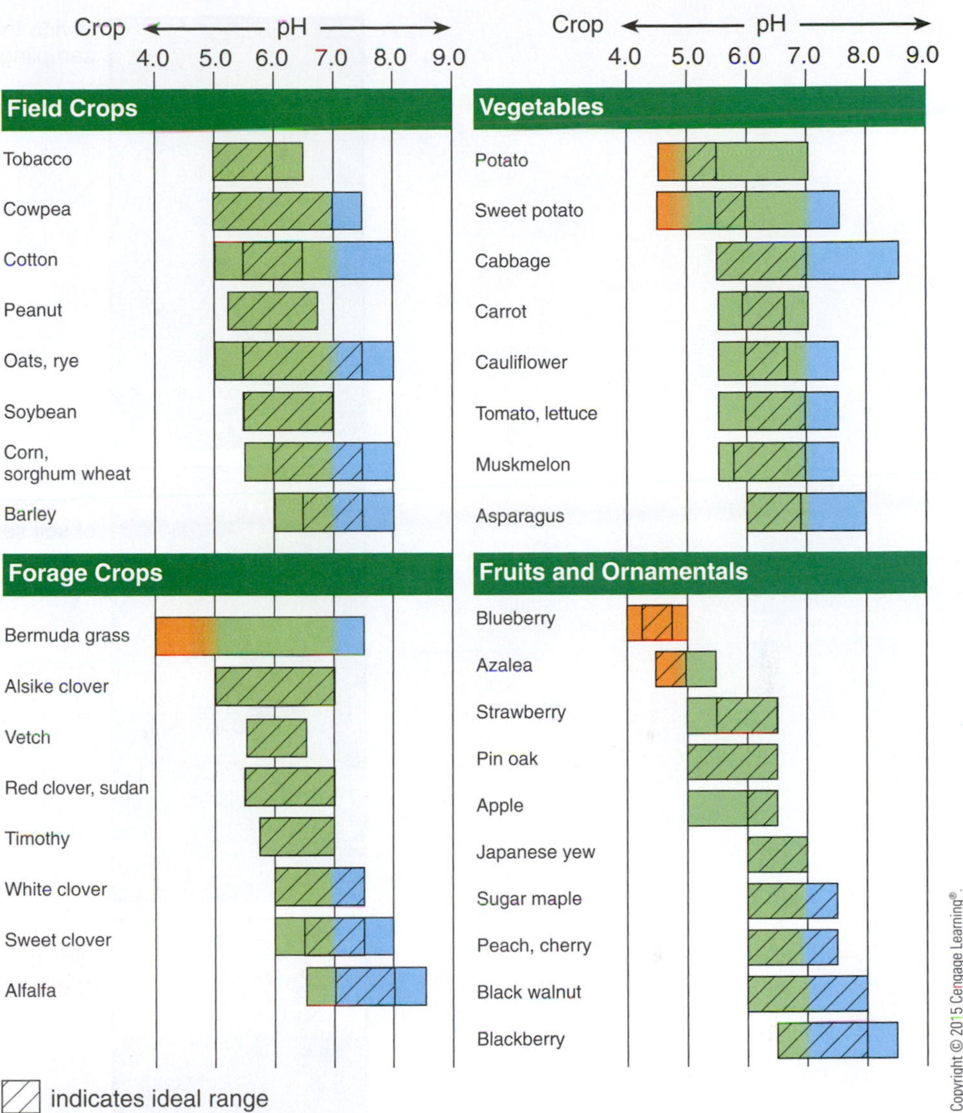

FIGURE 9-20 Plants grow in pH ranges from approximately 4 (very acidic) to 8.5 (very alkaline), but most plants function best within a specific pH range.

3. When taking samples, avoid wet or bare spots (Figure 9-21B). Soils that are substantially different in plant growth or past treatment should be sampled separately, provided their size and nature make it feasible to fertilize or lime each area separately.
4. After removing surface litter, take a sample from the correct depth. This is 2 inches for established lawns and about 6 inches for gardens, flower beds, farm crop land, and other areas to be tilled.
5. Submit a separate composite sample for each significantly different area—for example, front lawn, back lawn, and flower bed. Your composite sample for each area should include a small amount of soil taken from each of 10 to 20 randomly selected locations in the area represented by each sample.
6. When using a spade, first make a V-shaped cut. Then remove a 1-inch slice from one side of the cut (see Figure 9-21C). Then take a 1-inch strip from the middle of this slice (see Figure 9-21D). This represents the soil from one spot in the sample.

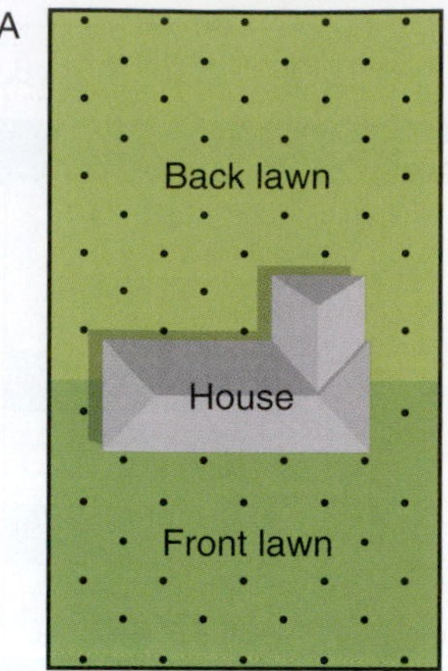

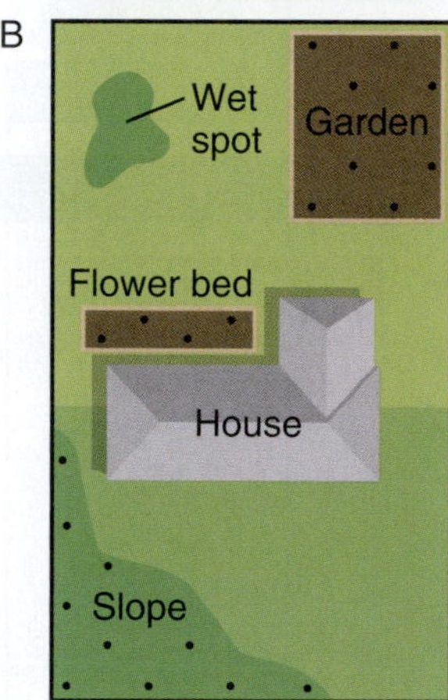

FIGURE 9-21 A procedure for collecting a dependable soil sample is illustrated in these views.

FIGURE 9-22 A soil sample that is collected with a soil probe collects a sample that is uniform in volume at any given soil depth.

7. Air-dry the soil; do not use heat. Mix the soil from a composite in a clean bucket. Place about 1 pint of this mixture into the sample box. Use a separate box for each composite. Fill in the blanks on the box or information sheet for each box.
8. Send soil sample(s) and information sheet(s) to the soil test laboratory.

Soil samples are collected routinely to help determine appropriate management of soil amendments. The same soil sampling procedure that was described for lawns and gardens is used for farm fields, golf courses, and greenhouses. It is critical that adequate amounts of fertilizers and other soil amendments are provided to a crop without applying excessive amounts. Adding too much of a chemical to the soil is not only expensive but it also may leach into the groundwater and surface water.

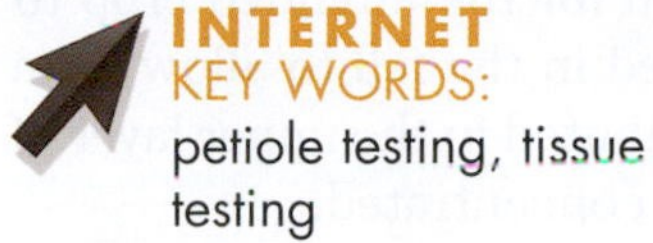

INTERNET KEY WORDS: petiole testing, tissue testing

Note: The Cooperative Extension System office in most counties and/or cities can arrange for soil testing.

The Petiole Test

Some crops are expensive to raise because they require large amounts of soil amendments. For example, potatoes require relatively high applications of nitrogen. When the nitrogen supply in the soil becomes depleted, the newest leaves on the plant will also have a nitrogen deficiency. By sampling and testing the petiole (young stem and leaves) of the potato plant, the deficiency can be identified and corrected before the crop yield is affected. The petiole test is effective when it is used to manage soil nutrients during the growing season.

The pH Test

The amount of agricultural lime required to raise soil to the desired pH level is indicated by a pH test. The same pH value may require different amounts of agricultural lime. This is because soils contain varying amounts of organic matter, clay, silt, and sand. The greater the organic matter and clay content, the greater the amount of lime required to correct the acidity.

Even though all the lime required by your soil is applied, do not expect the pH to increase quickly. It will generally require 2 to 3 years for all of the lime to be used and the desired pH to be reached.

Mix Well

The soil on 1 acre that is 6 inches in depth weighs about 2 million pounds. Therefore, it takes a lot of mixing to distribute a relatively small amount of lime with the soil. Liming recommendations are based on a specific plowing depth, such as 6 or 7 inches. Application rates may need to be adjusted for deep or shallow tilling.

Standard Ground Limestone

Lime recommendations are based on standard ground limestone, which should contain a minimum of 50 percent lime oxides (calcium oxide plus magnesium oxide). About 98 percent should pass through a 20-mesh sieve, with a minimum of 40 percent passing through a 100-mesh sieve. The higher the percentage of limestone passing through a 100-mesh screen, the faster this limestone will correct soil acidity. The best limestone will have the greatest calcium and/or magnesium content and will be ground to a small particle size. It is more important to finely grind a high-magnesium stone than a high calcium stone. A high-magnesium or dolomitic limestone should always be used when a magnesium deficiency is indicated by a soil test.

One ton (2,000 lbs.) standard ground limestone is approximately equivalent to 1,500 lbs. of hydrated lime or 1,100 lbs. of ground burned limestone.

Correcting Excessive Alkalinity

A reduction of soil alkalinity is desirable where soils have a high pH and the alkaline condition causes unsatisfactory crop growth. In most cases, this condition will exist where heavy applications of lime are made at one time or where lime has been applied to a soil with a high pH.

Gypsum is a soil amendment that is often used to reduce the alkalinity of agricultural soils. It is added when soil pH is too high for the intended crop to grow and thrive. In most instances, it should be added in the fall to allow time for it to be effective. For best results, it should be distributed in the upper layer of topsoil within the area where the roots of the crop are concentrated.

Another method that may be used to decrease the pH value of soil is to use sulfur or aluminum sulfate. Sulfur at 1.5 lbs. per 100 square feet or aluminum sulfate at 5 lbs. per 100 square feet will decrease the alkalinity, under most conditions, by 0.5 pH. Aluminum sulfate acts rapidly, producing acidification in 10 to 14 days. Sulfur requires 3 to 6 months in some soils before it forms into compounds the plants can use. Broadcast the material over the surface and thoroughly work it into the soil. For full benefit, sulfur should be applied in the fall, after garden crops are harvested. Aluminum sulfate may be applied in early spring.

Use chemicals cautiously to decrease alkalinity. Before these measures are taken, you should consider other factors that may be responsible for poor growth (such as drought, insect and disease injury, and fertilizer burning).

Fertilizers and Fertilizing

Essential plant nutrients are discussed in Unit 16, "Plant Physiology." However, it should be noted that nitrogen, phosphorus, and potassium are known as the three primary nutrients. These three ingredients must be present for a fertilizer to be called a complete fertilizer.

The proportions of nitrogen, phosphorus, and potassium are known as the **fertilizer grade**, expressed on a fertilizer container as percentages of the contents of the container by weight. Therefore, a 100-lb. container of fertilizer with a grade of 10-10-10 contains 10 percent nitrogen, 10 percent phosphorus, and 10 percent potassium.

If the total amount of fertilizer (100 lbs.) is multiplied by the percentage of each ingredient, the pounds of each ingredient may be calculated. Therefore, 0.10 (percent of nitrogen) × 100 (total lb. of fertilizer) = 10 lbs. nitrogen. The amount of phosphorus and potassium is also 10 lbs. each. The other 70 lbs. consist of inert filler material.

Fertilizer is frequently shipped in 80-lbs. bags. Therefore, one bag of 10-10-10 fertilizer would have 8 lbs. nitrogen, 8 lbs. phosphorus, and 8 lbs. potassium for a total of 24 lbs. of **active ingredients** (components that achieve one or more purposes of the mixture).

Some popular grades of fertilizer are 5-10-5, 5-10-10, 10-10-10, 6-10-4, 0-15-30, 0-20-20, 8-16-8, and 8-24-8. These grades are formulated to meet the needs of a variety of crops on various soils. The amount and grade of fertilizer to apply are determined by (1) the specific crop to be grown, (2) the potential yield or performance of the crop, (3) fertility of the soil, (4) physical properties of the soil, (5) previous crop, and (6) type and amount of manure applied.

Therefore, decisions on rate of application must be made on a local basis. Soil tests, tissue tests, and plant observations are useful techniques for determining fertility needs (Figure 9-23).

Organic Fertilizers

Organic fertilizers include animal manures and compost made with plant or animal products. Organic commercial fertilizers include dried and pulverized manures, bone meal, slaughterhouse tankage, blood meal, dried and ground sewage sludge, cottonseed meal, and soybean meal.

FIGURE 9-23 Corn plants are good indicators of nutrient deficiencies and other factors that impact plant health and productivity.

Organic fertilizers have certain definite characteristics. First, nitrogen is usually the predominant nutrient, with lesser quantities of phosphorus and potassium. One exception to this is bone meal, in which phosphorus predominates and nitrogen is a minor ingredient. Second, the nutrients are only made available to plants as the material decays in the soil, so they are slow acting and long lasting. Third, organic materials alone are not balanced sources of plant nutrients, and their analysis in terms of the three major nutrients is generally low.

Inorganic Fertilizers

Various mineral salts, which contain plant nutrients in combination with other elements, are called inorganic fertilizers. Their characteristics are different from organic fertilizers. First, the nutrients are in soluble form and are quickly available to plants. Second, the soluble nutrients make them caustic to growing plants and can cause injury. Care must be used in applying inorganic fertilizers to growing plants. They should not come in contact with the roots or remain on plant foliage for any length of time. Third, the analysis of chemical fertilizers is relatively high in terms of the nutrients they contain.

Fertilizers are needed to replenish mineral nutrients depleted from a soil by crop removal or by such natural means as leaching. Some soils with high fertility may need only nitrogen or manure. Use of fertilizers that also contain small amounts of copper, zinc, manganese, boron, and other minor elements is not considered necessary for most soils but may be needed in certain soils and for certain crops.

There are also unmixed fertilizers that carry only one element (Figure 9-24). Most important of these unmixed materials are the nitrogen and phosphate carriers. Nitrogen carriers vary from 16 to 45 percent nitrogen. Be careful in using nitrogen materials. Too much may cause excessive and soft growth.

Superphosphate fertilizers carry only phosphorus. Phosphorus promotes flower, fruit, and seed development. It also firms up stem growth and stimulates root growth. Superphosphate may be added to manure to give a better balance of nutrients for plants (100 lbs. to each ton of horse or cow manure and 100 lbs. to each half-ton of sheep manure).

Fertilizer Applications

There are many ways to apply fertilizers. For the home lawn, the most likely method is broadcasting (spreading evenly over the entire surface). In the case of gardens or fields, broadcasted fertilizer may be incorporated or mixed into the soil by spading, tilling, plowing, or disking.

Band application places fertilizers about 2 inches to one side of and slightly below the seed. This method is used extensively for row crops in gardens and fields.

PLANT NUTRIENTS IN FERTILIZER MATERIALS

Materials	Nitrogen N%	Phosphorus P_2O_5%	Potassium K_2O%	Calcium %	Magnesium %	Sulfur %
Ammonium Nitrate	33.5	—	—	—	—	—
Ammonium Sulfate	21	—	—	—	0.3	23.7
Urea	45	—	—	1.5	0.7	0.02
Sodium Nitrate	16	—	0.2	0.1	0.05	0.07
Calcium Nitrate	15	—	—	19.4	1.5	0.02
Calcium Cyanamide	21	—	—	38.5	0.06	0.3
Anhydrous Ammonia	82	—	—	—	—	—
Superphosphate	—	20	0.2	20.4	0.2	11.9
Liquid Phosphoric Acid	—	58	—	—	—	—
Ammonium Phosphate	11	48	0.2	1.1	0.3	2.2
Potassium Chloride	—	—	62	—	0.1	0.1
Potassium Sulphate	—	—	53	0.5	0.7	17.6

FIGURE 9-24 Plant nutrients in common fertilizer materials.

Side-dressing is done by placing fertilizer in bands about 8 inches from the row of growing plants. This method is popular for row crops such as corn and soybeans.

Top-dressing is a procedure where fertilizer is broadcast lightly over close-growing plants. Top-dressing is used for adding nitrogen to small grain, hay, and turf crops after they are established.

Starter solutions are diluted mixtures of fertilizer used when plants are transplanted. Their purpose is to provide small amounts of nutrients that will not burn the tender roots of young plants.

Other methods of applying fertilizers include the application of foliar sprays directly onto the leaves of plants and knife application of anhydrous ammonia gas into the soil. The practice of adding liquid fertilizer to irrigation water is also used extensively in the United States.

Using Manure

Animal manure is a valuable product when handled properly. Its content of plant nutrients makes it a valuable fertilizer material. In addition, the organic matter aids in developing and maintaining structure in soils (Figure 9-25).

To obtain the most nutrient value from manures, the following practices should be followed:

- Use adequate bedding to absorb all of the liquids.
- Balance the phosphorus in cow manure by adding superphosphate to the fields.
- Spread manure evenly over fields. An 8- to 12-ton application per acre is recommended. (Fifty bushels of manure and litter weigh about 1 ton.)
- When possible, incorporate manure into the soil immediately after spreading.
- Do not spread on steep slopes when the ground is frozen.
- When storing manure, keep it compact and under cover.
- Prevent liquid runoff from escaping from manure holding areas.
- Apply manure to crops that will give best response, such as corn, sorghums, potatoes, and tobacco.

When applying manure, the amount of commercial fertilizer should be reduced accordingly (Figure 9-26).

Legume Crops

Legumes are plants in which specialized bacteria use nitrogen gas from the air and convert it to **nitrate** (the form of nitrogen used by plants). Some examples of

APPROXIMATE AMOUNTS OF PLANT NUTRIENTS AVAILABLE FROM ONE TON OF MANURE

	Cattle	Sheep	Swine	Poultry	Horse
Nitrogen (lb.)	10	28	10	30	14
Phosphorus (lb.)	5	10	5	20	5
Potassium (lb.)	10	25	10	10	14

FIGURE 9-25 Animal manure is excellent fertilizer for most crops, and it also improves the soil by adding organic matter to the topsoil.

	N NITROGEN	P PHOSPHORUS	K POTASSIUM
Suppose the fertilizer recommendations per acre for a certain crop are:	150 lbs.	100 lbs.	100 lbs.
If well-managed cow manure is applied at the rate of 10 tons per acre, it can be determined from Figure 9-32 that the 10-ton application would provide:	100 lbs.	50 lbs.	100 lbs.
Therefore, the amount of nutrients that must be provided per acre through commercial fertilizer is:	50 lbs.	50 lbs.	0 lbs.
The shortfall of ingredients listed above could be made up by applying 500 lbs. of 10-10-0 commercial fertilizer per acre. NOTE: Caution must be exercised when estimating the nutrient values of manure due to variations in liquid and solid content captured from the animal, amount and type of bedding, and procedures used to handle and store the manure.			

FIGURE 9-26 When manure is applied to soil, the amount of commercial fertilizer used should be adjusted downward, taking into account the nutrients in the manure.

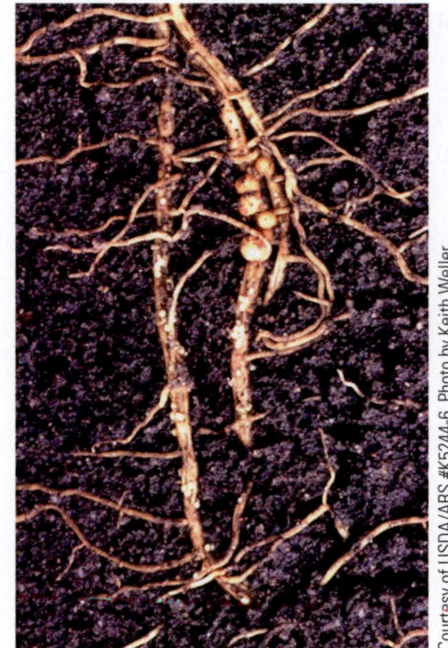

FIGURE 9-27 Nitrogen-fixing nodules on the roots of legume plants.

legumes are beans, peas, clovers, and alfalfa. The process of converting nitrogen gas to nitrates by bacteria in the roots of legumes is called **nitrogen fixation** (Figure 9-27).

Nitrogen fixation reduces or eliminates the need for adding expensive nitrogen fertilizer to legume crops. When the roots of legume plants decay, large amounts of nitrates are left behind for the next crop. Crops such as corn, which requires large amounts of nitrogen, should follow alfalfa or clover in a field because the legumes leave unused nitrogen behind.

Rotation Fertilization

Many crop rotations start with a small-grain crop. The preparation of the seedbed for small grains provides an excellent opportunity to put lime and fertilizer into the feeding zone for roots of perennial forage crops that produce for more than a year. Lime and phosphorus move slowly in the soil, and unless the roots contact these elements, they cannot provide for the high requirements of some seedlings for phosphorus in the critical first year of growth.

A properly limed and fertilized forage crop is the backbone of a successful crop rotation. The growth of nutrient-enriched grass and legume roots throughout the soil provides a most favorable medium for natural soil-building processes. The addition of organic matter, the movement of fertilizer elements by the roots into the soil, and the production of root channels by sod roots produce soils that absorb water better, erode less, are easier to work, and produce higher-yielding and better-quality crops.

Phosphorus and potassium added to sod will benefit the sod and, in addition, will be placed in the best location and be in the best form to supply these elements to the long-season row crops that follow. Nitrogen produced by the legumes or added to grass will be present in organic form and will be released in the best possible form for the row crops and small-grain crops.

The composition of soils is complex and depends on many factors. Soil is dynamic and changing all the time. Nature has provided many cycles to help provide for soil renewal. The scientific management of soils makes them more productive and helps to ensure productive soils capable of efficient production for future generations.

INTERNET KEY WORDS: hydroponics

HYDROPONICS

The term *hydroponics* refers to a number of systems used for growing plants without soil (Figure 9-28). Some major systems are:

- aggregate culture—in which a material such as sand, gravel, or marbles supports the plant roots;
- water culture, solution culture, or nutriculture—the plant roots are immersed in water containing dissolved nutrients;
- **aeroponic**—in which the plant roots hang in the air and are misted regularly with a nutrient solution; and
- continuous-flow systems—in which the nutrient solution flows constantly over the plant roots. This system is the one most commonly used for commercial production.

Hydroponics is growing in importance as a means of producing vegetables and other high-income plants (Figure 9-29). In areas where soil is lacking or unsuitable for growth, hydroponics offers an alternative production system. Equally good crops can generally be produced in a greenhouse in conventional soil or bench systems.

When plants are grown hydroponically, their roots are either immersed in or coated with a carefully controlled nutrient solution. The nutrients and water are

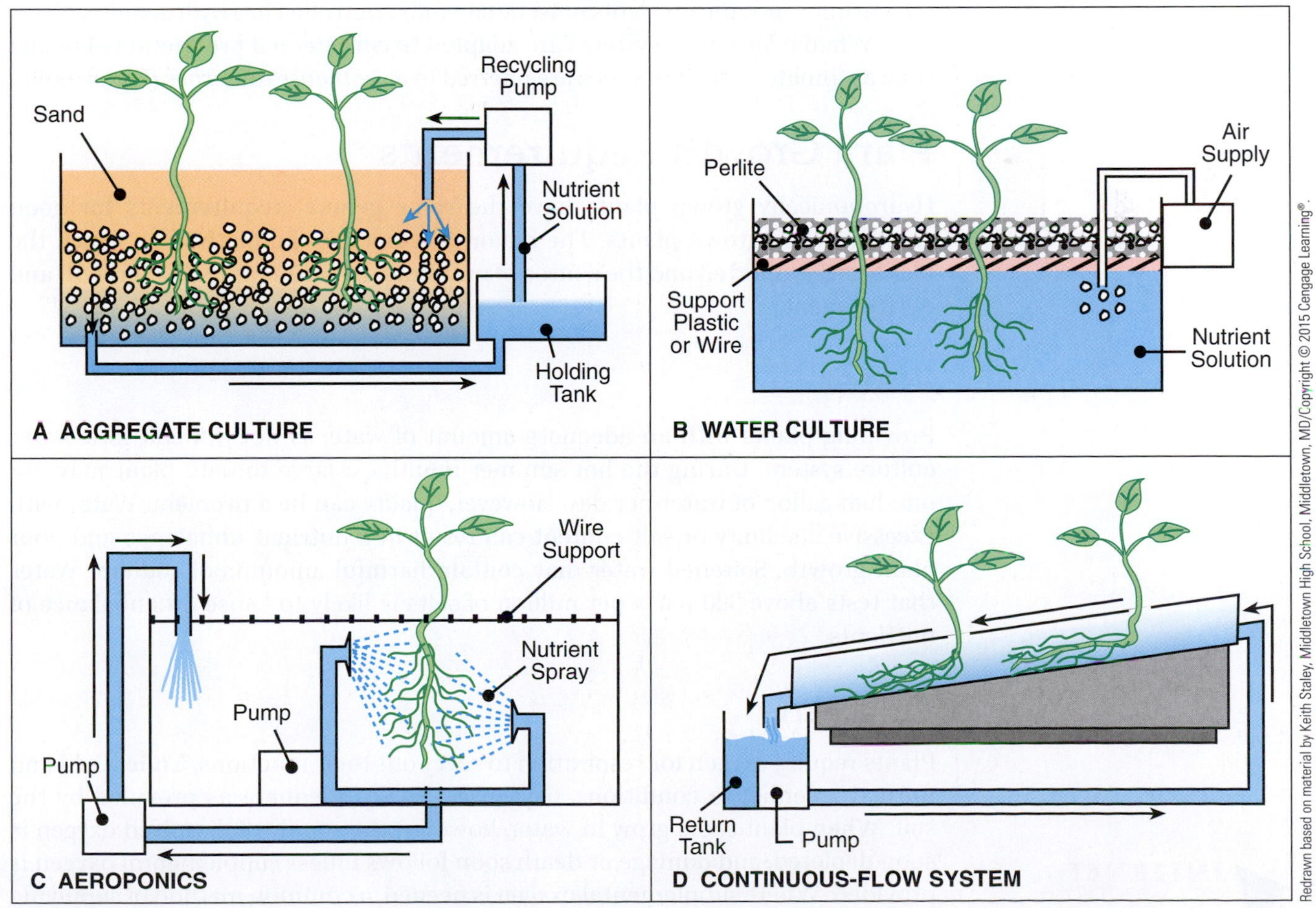

FIGURE 9-28 There are many types of hydroponics systems, including (A) aggregate culture, (B) water culture, (C) aeroponics, and (D) continuous-flow systems.

FIGURE 9-29 High-value crops such as tomatoes are ideal for hydroponic production methods.

FIGURE 9-30 "Hothouse" production of high-value crops is a commercial use of hydroponics on a large scale.

supplied by the solution alone and not by aggregates or other inert materials that support the roots. Inert means inactive. Without the presence of soil to absorb and release nutrients, the nutrients must be carefully controlled in a hydroponic system.

When hydroponic systems are adapted to commercial production of plants such as tomatoes, the tomatoes are referred to as *hothouse* tomatoes (Figure 9-30).

Plant Growth Requirements

Hydroponically grown plants have the same general requirements for good growth as soil-grown plants. The major difference is the method by which the plants are supported and the source of nutrients that are supplied for growth and development.

Water

Providing plants with an adequate amount of water is not difficult in a water culture system. During the hot summer months, a large tomato plant may use one-half gallon of water per day. However, quality can be a problem. Water with excessive alkalinity or salt content can result in a nutrient imbalance and poor plant growth. Softened water may contain harmful amounts of sodium. Water that tests above 320 parts per million of salts is likely to cause an imbalance of nutrients.

Oxygen

Plants require oxygen for respiration to carry out their functions. Under field and normal greenhouse conditions, oxygen is usually adequate as provided by the soil. When plant roots grow in water, however, the supply of dissolved oxygen is soon depleted, and damage or death soon follows unless supplemental oxygen is provided. Where supplemental oxygen is needed, a common method of supplying oxygen is to bubble air through the water. It is not usually necessary to provide supplementary oxygen in aeroponic or continuous-flow systems.

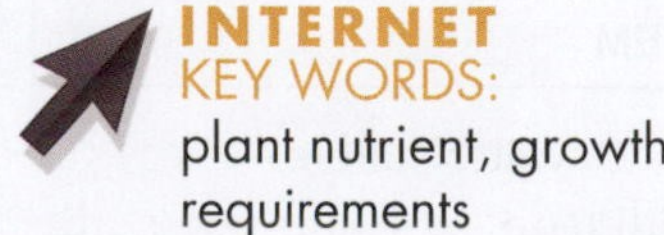

INTERNET KEY WORDS:
plant nutrient, growth requirements

Mineral Nutrients

Green plants must absorb certain minerals through their roots to survive. These minerals are supplied by soil, organic matter, or soil solutions. The elements needed in large quantities are nitrogen, phosphorus, potassium, calcium, magnesium, and sulfur. The nutrients needed in small amounts are iron, manganese, boron, zinc, copper, molybdenum, and chlorine. An oversupply of any nutrient is toxic or detrimental to plants. In hydroponic systems, all nutrients normally supplied by soil must be included in the water to form the solution or media.

Light

All vegetable plants and many flowers require large amounts of sunlight. Hydroponically grown vegetables, such as those grown in a garden, need at least 8 to 10 hours of direct sunlight each day to produce well. Electrical lighting is a poor substitute for sunshine because most indoor lights do not provide enough intensity to produce a crop. Incandescent lamps supplemented with sunshine or special plant growth lamps can be used to grow transplants, but they are not adequate to grow the crop to maturity. High-intensity lamps, such as high-pressure sodium lamps, can provide more than 1,000 foot-candles of light. They may be used successfully in small areas where sunlight is inadequate. However, these lights are too expensive for most commercial operations.

Spacing

Adequate spacing between plants will permit each plant to receive sufficient light in the greenhouse. Tomato plants, pruned to a single stem, should be allowed 4 square feet per plant. European seedless cucumbers should be allowed 7 to 9 square feet, and seeded cucumbers need about 7 square feet. Leaf lettuce plants should be spaced 7 to 9 inches apart within the row and 9 inches between rows. Most other vegetables and flowers should be grown at the same spacing as recommended for a garden.

Greenhouse vegetables will not do as well during the winter as in the summer. Shorter days and cloudy weather reduce the light intensity, thus limiting production.

Temperature

Plants grow well only within a limited temperature range. Temperatures that are too high or too low will result in abnormal development and reduced production. Warm-season vegetables and most flowers grow best between 60° and 80° F. Cool-season vegetables, such as lettuce and spinach, should be grown between 50° and 70° F.

Support

In the garden or field, plants are supported by roots anchored in soil. A hydroponically grown plant must be artificially supported with string, stakes, or other means.

Hydroponics in the Classroom and Laboratory

Hydroponics has become an important teaching area in agriscience programs. The use of common plastic bottles and low-cost aquarium supplies permits

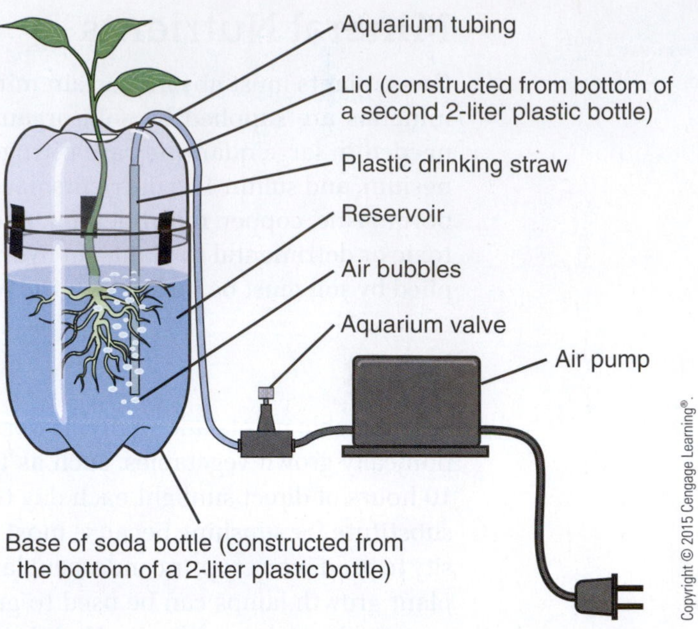

FIGURE 9-31 A static solution culture system built from two plastic soda bottles.

students and teachers to set up and perform numerous research and demonstration projects in the classroom or laboratory.

According to Dr. David R. Hershey (1994), biology education consultant and author, a solution culture system (Figure 9-31) may be constructed as follows:

1. Fill two dark 2-liter plastic soda bottles with hot water to loosen the glued label and base. Remove the labels from both bottles and the base from one bottle; keep the base for future use.
2. Using a fine-point felt-tip marker, draw a line around the bottle 9 inches up from the bottom.
3. Using short-bladed scissors, cut on the line made in Step 2 and remove the upper portion of the bottle. This will be the reservoir.
4. Using a cork-borer, drill, or scissors, make a hole in the center of the bottle base approximately 0.5 inches inches in diameter to accommodate a plant stem. Close all except one of the existing (prepunched) holes in the base with black vinyl electrician's tape to prevent light passage. Then place the base on the reservoir as a dome-shaped lid.
5. Insert about 8 inches of 2-foot-long aquarium tubing into the open, prepunched hole. To keep the tubing in the reservoir rigid, place a plastic drinking straw over the tubing. Attach the free end of the aquarium tubing to an aquarium air pump with an air control valve.

Nutrient Solutions

There are many nutrient-solution recipes, but Hoagland Solution No. 1 is used widely and can be modified to create nutrient deficiencies (Figure 9-32). Often, Hoagland solution is used at less than full strength. Nutrient stock solutions are prepared using a balance to weigh out the salts and a graduated cylinder or volumetric flask to measure the water. Plastic soda bottles can be used to store

Stock Solution			milliliters of stock solution per liter of nutrient solution							
Formula	**Name**	**grams /liter**	**Complete**	**–N**	**–P**	**–K**	**–Ca**	**–Mg**	**–S**	**–Fe**
$Ca(NO_3)_2{\bullet}4H_2O$	Calcium nitrate, 4-hydrate	236	5	–	4	5	–	4	4	5
KNO_3	Potassium nitrate	101	5	–	6	–	5	6	6	5
KH_2PO_4	Monopotassium phosphate	136	1	–	–	–	1	1	1	1
$MgSO_4{\bullet}7H_2O$	Magnesium sulfate, 7-hydrate	246	2	2	2	2	2	–	–	2
FeNa EDTA[z]	Iron EDTA	18.4	1	1	1	1	1	1	1	–
Micronutrients[y]		–	1	1	1	1	1	1	1	1
K_2SO_4	Potassium sulfate	87	–	5	–	–	–	3	–	–
$CaCl_2{\bullet}2H_2O$	Calcium chloride, dihydrate	147	–	2	–	–	–	–	–	–
$Ca(H_2PO_4)_2{\bullet}H_2O$	Calcium phosphate, monobasic	12.6	–	10	–	10	–	–	–	–
$Mg(NO_3)_2{\bullet}6H_2O$	Magnesium nitrate, 6-hydrate	256	–	–	–	–	–	–	2	–

[z]Ferric-sodium salt of ethylene-diamine tetraacetic acid. Differs from original Hoagland recipe, which used Fe tartrate.

[y]Contains the following in grams/liter:

2.86 H_3BO_3, (boric acid)

1.81 $MnCl_2{\bullet}4H_2O$, (manganese chloride, 4-hydrate)

0.22 $ZnSO_4{\bullet}7H_2O$, (zinc sulfate, 7-hydrate)

0.08 $CuSO_4{\bullet}5H_2O$, (copper sulfate, 5-hydrate)

0.02 $H_2MoO_4{\bullet}H_2O$, (85% molybdic acid).

[x]For Mn-, B-, Cu-, Zn-, and Mo-deficient solutions, substitute micronutrient stock solutions missing one of the five salts in the regular micronutrient stock solution. For Cl-deficient micronutrient stock solution, substitute 1.55 $MnSO_4{\bullet}H_2O$ (manganese sulfate, monohydrate) for 1.81 $MnCl_2{\bullet}2H_2O$.

Source: Hoagland, D. R. and Arnon, D. I. 1950. The Water-Culture Method for Growing Plants Without Soil. California Agricultural Experiment Station Circular 347, revised 1950.

FIGURE 9-32 Preparation of modified Hoagland nutrient solutions for nutrient deficiency system development.

stock solutions. When stored in a dark place at room temperature, stock solutions should last for many years. To prepare the nutrient solution, add measured volumes of stock solutions to a measured volume of water. Stock solutions can be dispensed using pipettes or graduated cylinders. Complete hydroponic salt mixtures can be purchased, and they greatly simplify nutrient solution preparation.

Aeration

Solution cultures are typically aerated using an aquarium air pump and aquarium air tubing. In a soda bottle system, the aquarium air tubing is inserted into the reservoir through one of the prepunched holes, and the tubing in the reservoir is made rigid by slipping a plastic soda straw onto the tubing. To prevent clogging, a piece of cotton or aquarium filter floss is placed in the aquarium tubing as it exits the pump. The filter is necessary to either trap dirt from the air or pieces of the pump diaphragm. Adjust the aeration to a gentle rate of one to three bubbles per second. Some plants do not benefit from aeration but most do. An alternative to pump aeration is to let the top third of the root system remain uncovered by solution in the humid air in the top of the reservoir.

Maintenance

Water must be added to static solution cultures every few days or so to replace water lost by transpiration and evaporation. Nutrient solutions are typically replaced with fresh solutions on a schedule, such as every 1 or 2 weeks. Frequency of solution replenishment depends on the rate of plant growth relative to the volume of

AGRI-PROFILE CAREER AREAS: AGRONOMY/HYDROPONICS

Courtesy of USDA/ARS #K-5050-3.

Soil scientist Ronald Schnabert (left) and hydrologist technician Earl Jacoby study natural riparian zone processes that lessen the impact of upstream agriculture on water quality.

Electronic devices have become tools of the trade for soil and water research and management. Career options in soil and hydroponics management overlap those in soil and water conservation to a certain extent, and additional opportunities occur in management of farms and hydroponics greenhouses. The practice of hydroponics is not new, but hydroponics for commercial production has captured the imagination of the world. The recent popularity of hydroponics operations provides new career opportunities, especially in urban areas.

Soil and water management specialists are in demand on the global scene. Progressive developing countries need help in policy development, education, and project management. They hope to leap from their primitive agriculture of the past to the agriscience of the present in a few short years.

nutrient solution. An electrical conductivity (EC) meter is useful for determining when a nutrient solution has been depleted. To prevent interruption of growth, replace the solution when the solution EC is about half the original value.

Plants

Beans, corn, sunflowers, and tomatoes are often used for teaching hydroponics because they grow rapidly from seed; but they are often difficult to handle because of their large size, high light requirements, and need for staking. Wisconsin Fast Plants, which are becoming a standard plant for teaching use, are excellent for solution culture. Fast plants go from seed to flower in 2 weeks under a bank of six 4-foot, 40-watt fluorescent lamps and remain less than a foot in height.

Houseplants, such as piggyback (*Tolmiea menziesii*), wandering Jew (*Tradescantia* species), evergreen euonymus (*Euonymus japonica*), coleus, common philodendron, and pothos (*Epipremnum aureum*), are also excellent in teaching hydroponics. They thrive with low light levels, are readily available, and root quickly in solution. Radish and lettuce are easily grown from seed. A carrot placed in tap water in the 2-liter soda bottle system described earlier will sprout lateral roots, produce a shoot, and flower. Pineapple fruit tops easily root in solution.

Germination

Seeds for hydroponics are conveniently germinated in containers of perlite, which is inert and is easily removed from roots before transferring the plants to solution. Small seeds, such as fast plants or coleus, can be germinated on a paper towel in a

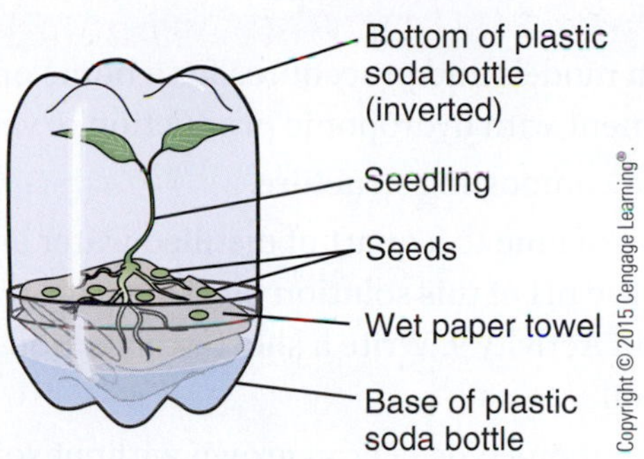

FIGURE 9-33 A germination chamber may be constructed with the bottom cut from a 2-liter soda bottle inverted over a soda bottle base. The dome section is lined with a wet paper towel, and seeds are pushed against the towel to absorb moisture and germinate before transplanting.

seed germinator built from a 2-liter soda bottle. The base is removed from a bottle and a 4-inch bottom section of the bottle is inverted over the base to complete the germinator. A wet paper towel is used to line the domed section, and small seeds are pressed into the towel. The seeds remain stuck until the seed germinates and the seedling is then transplanted (Figure 9-33).

Future of Hydroponics

Hydroponics is increasing in use commercially and undoubtedly will become more important in the future. Research is expanding, and new techniques are being developed. The use of nutrient solutions as media for growing plants will be an important part of agriscience in the future.

STUDENT ACTIVITIES

1. Write the Terms to Know and their meanings in your notebook.
2. Observe soils and other media used in flowerpots, trays of vegetable seedlings, greenhouses, gardens, road banks, construction sites, and other places.
3. Examine the living organisms in a shovelful of soil taken from an outdoor area that is damp, moist, and high in organic matter.
4. Invite a Soil Conservation Service professional to the classroom to discuss soil mapping and land-use planning.
5. Obtain a land-use map from a Farm Service Agency (FSA) for your home, farm, or other area. Study the material and report your findings to the class.
6. Observe a soil profile at a cut in a road bank, stream bank, or hole. Identify the O, A, B, and C horizons.
7. Do a mechanical analysis of a sample of soil using a fruit jar, water, and a dispersing agent such as Calgon (see Figure 9-17).
8. Observe the feel of some of the soil used in Activity 7.
9. Obtain a pH test kit and test samples of soil from around your home.

10. Do research on models and procedures for a home or school hydroponics unit. Plan, build, and use the unit to experiment with hydroponic production of various plants.
11. Build and use a composting structure.
12. Add 1 teaspoon of lime to a quart of distilled water and mix well. Be sure to wear gloves and protective eyewear. Find the pH of this solution by using a pH test kit. Compare the pH of the sample to the pH of the soil taken in Activity 9. Write a short explanation describing the pH differences and what impact lime may have on soil.
13. Learn more about one type of crop grown without soil using the Internet or other agriscience resources. In one or two paragraphs, tell which crop you learned about, in which non-soil media it was grown, and what procedure the grower used to produce crops.

SELF-EVALUATION

A. MULTIPLE CHOICE

1. An example of plant-growing media is
   a. soil.
   b. water.
   c. perlite.
   d. all of the above.
2. Which is *not* organic matter?
   a. leaf mold
   b. peat moss
   c. sphagnum moss
   d. vermiculite
3. Decay of organic matter is caused by
   a. large animals.
   b. microbes.
   c. rodents.
   d. water.
4. Which is *not* a factor affecting soil formation?
   a. hydroponics
   b. gravity
   c. ice
   d. water
5. Humans affect soil formation by
   a. landscaping.
   b. irrigating.
   c. bulldozing.
   d. weathering.
6. The land class with the fewest limitations is
   a. Class I.
   b. Class III.
   c. Class VI.
   d. Class VIII.
7. The land classes suitable for field crop production are
   a. I, II, IV, and VI.
   b. I, II, III, and IV.
   c. I, IV, V, and VIII.
   d. I, II, VI, and VII.
8. The horizon that is most supportive of plant growth is
   a. horizon O.
   b. horizon A.
   c. horizon B.
   d. horizon C.
9. The smallest soil particle is
   a. clay.
   b. gravel.
   c. sand.
   d. silt.

10. A term related to soil structure is
    a. aggregate.
    b. loam.
    c. silt.
    d. topsoil.
11. Soil pH is generally increased by adding
    a. sulfur.
    b. nitrogen.
    c. lime.
    d. complete fertilizer.
12. Hydroponics refers to
    a. aggregate culture.
    b. aeroponics.
    c. nutriculture.
    d. all of the above.
13. The importance and use of hydroponics is
    a. increasing.
    b. decreasing.
    c. about the same.
    d. a new field.
14. Hydroponically grown plants differ principally from soil-grown plants by
    a. light requirements.
    b. nutrient requirements.
    c. oxygen requirements.
    d. support mechanisms.
15. The best and most economical light source for growing plants hydroponically is
    a. incandescent light.
    b. fluorescent light.
    c. sodium.
    d. sunlight.

B. MATCHING (GROUP I)

________	1. Colluvial	a. Deposited by ice
________	2. Alluvial	b. Deposited by lakes
________	3. Glacial	c. Formed in place
________	4. Lacustrine	d. Deposited by gravity
________	5. Loess	e. Class II land with erosion problem
________	6. Leaching	f. Cross section of soil
________	7. Organic matter	g. Makes soil dark in color
________	8. Residual	h. Removal of soluble materials
________	9. IIe	i. Deposited by streams
________	10. Profile	j. Deposited by wind

C. MATCHING (GROUP II)

________	1. Horizon A	a. Mostly organic matter
________	2. Horizon B	b. Parent material
________	3. Horizon C	c. Topsoil
________	4. Horizon O	d. Subsoil
________	5. Coarse	e. 5-10-5
________	6. Medium texture	f. Equal parts of nitrogen, phosphorus, and potassium
________	7. Fine texture	g. Sulfur
________	8. 10-10-10	h. Loamy soils
________	9. Complete fertilizer	i. Clay soils
________	10. Decreases pH	j. Sandy soils

D. COMPLETION

11. Soil is defined as the top layer of the Earth's surface suitable for the growth of _______.
12. Three groups of plants found in soil are _______, _______, and _______.
13. Three groups or types of animals that live in the soil are _______, _______, and _______.
14. Five important benefits of organic matter in soil are _______, _______, _______, _______, and _______.
15. In solution culture, about one-third of the roots must be exposed to air, or air must be bubbled through the solution to prevent damage or death from lack of the element _______.
16. The six mineral elements taken up by roots and needed in large quantities by plants are _______, _______, _______, _______, _______, and _______.
17. The seven mineral elements taken up by roots and needed in small quantities by plants are _______, _______, _______, _______, _______, _______, and _______.

UNIT 10

Forest Management

OBJECTIVE

To determine the relationship of forests to the environment and the recommended practices for using forest resources.

MATERIALS LIST

- bulletin board materials
- samples of forest and wood products
- tree leaf collection
- forestry reference materials
- Internet access

COMPETENCIES TO BE DEVELOPED

After studying this unit, you should be able to:

- define forest terms.
- describe the forest regions of the United States.
- discuss important relationships among forests, wildlife, and water resources.
- identify important types and species of trees.
- describe how a tree grows.
- discuss important properties of wood.
- apply principles of good woodlot management.
- describe procedures for seasoning lumber.

SUGGESTED CLASS ACTIVITIES

1. Use a chainsaw to cut some 3- to 4-inch crosscut sections from a large log. The grain of the tree should be distinct. Assign class members to determine the age of the tree by counting the annual rings. Identify the dates of important historical events by pinning a label to the appropriate annual ring. Seek an opportunity for students to present a mini-history lesson to an elementary school class using the crosscut visual to identify important dates.
2. Invite a forestry professional to instruct the class on the kinds of forest management activities he or she encounters. The forestry professional might be employed by a federal or state agency or may be self-employed in a privately owned forest operation. In addition, many towns and cities employ urban foresters to care for trees and shrubs in parks and along streets. Prepare the class for the visit by compiling a list of questions that the students might ask. Have students write a short report of the forester's presentation.

TERMS TO KNOW

forest land
timberland
forest
tree
shrub
lumber
board foot
evergreen
conifer
softwood
deciduous
hardwood
pulpwood
clear cut
plywood
veneer
cambium
annual ring
xylem layer
sapwood
phloem
inner bark
heartwood
hardness
shrinkage
warp
woodlot
silviculture
arboricultur
seedling

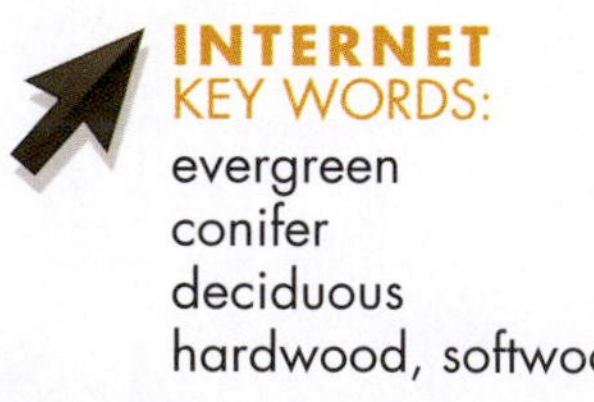

INTERNET KEY WORDS:
evergreen
conifer
deciduous
hardwood, softwood

3. Take a walk around the schoolyard. Collect leaves or needles from the trees. Identify as many trees as you can using the illustrations contained in this unit.

According to the U.S. EPA's Web page on Forestry (www.epa.gov/agriculture/forestry.html), the United States has 490 million acres of timberland and 246 million acres of other forest land, for a total of 736 million acres. This represents about one-third of the total land in the United States. **Forest land** is defined by the U.S. Department of Agriculture (USDA) as land at least 10 percent stocked by forest trees of any size. **Timberland** is defined as forest land that is capable of producing in excess of 20 cubic feet per acre per year of industrial wood and that has not been withdrawn from timber utilization by statute or administrative regulation.

A **forest** is a complex association of trees, shrubs, and plants that all contribute to the life of the community (Figure 10-1). A **tree** is a woody perennial plant with a single stem that develops many branches. Trees vary greatly in size, but by definition, they grow to more than 10 feet in height. A **shrub** is a woody plant with a bushy growth pattern and multiple stems. A productive forest is one that is growing trees for lumber or other wood products on a continuous basis. **Lumber** consists of boards that are sawed from trees. It is bought and sold by the board foot. A **board foot** is a unit of measurement for lumber that is equal to 1 × 12 × 12 inches (Figure 10-2).

Forest land may include parks, wilderness land, national monuments, game refuges, and some areas where harvesting of trees is not permitted.

FIGURE 10-1 A forest contains trees, shrubs, and other plants as well as animal life.

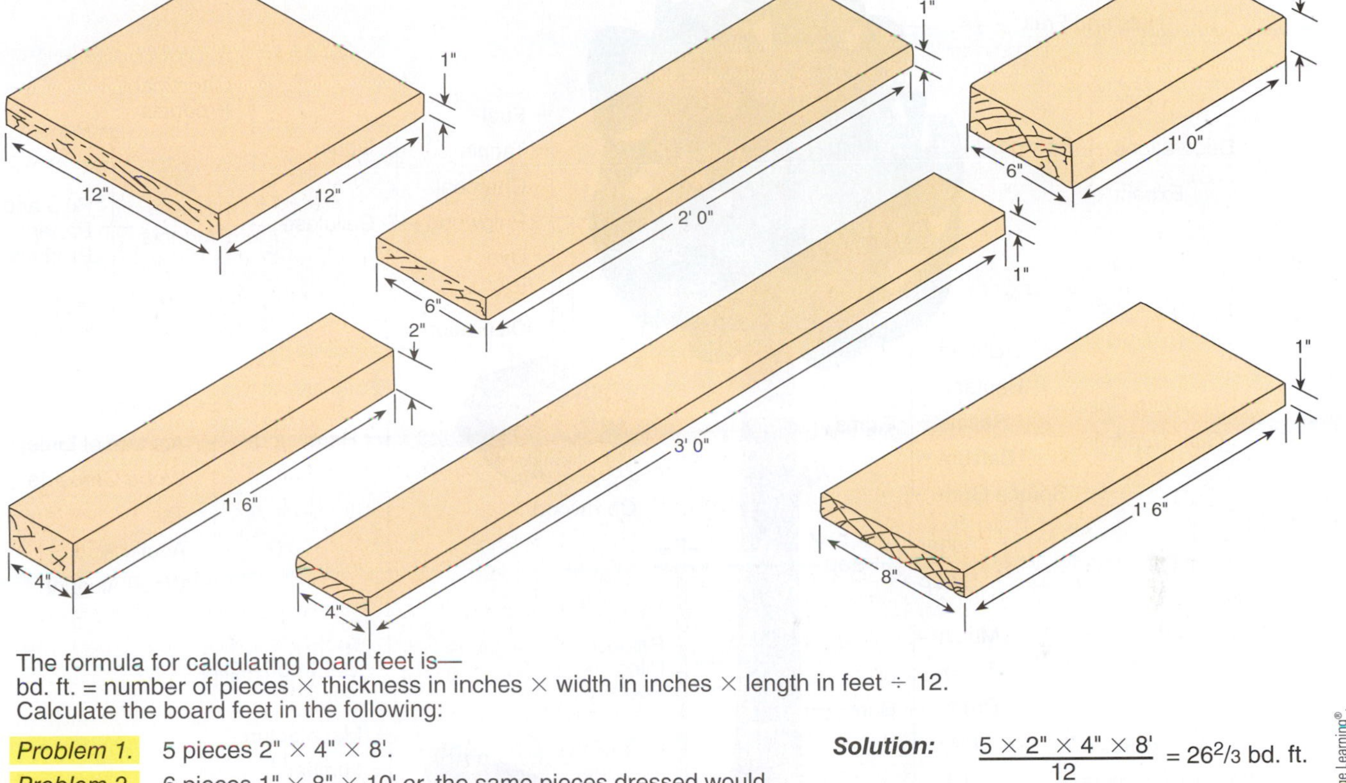

The formula for calculating board feet is—
bd. ft. = number of pieces × thickness in inches × width in inches × length in feet ÷ 12.
Calculate the board feet in the following:

Problem 1. 5 pieces 2" × 4" × 8'.

Solution: $\frac{5 \times 2" \times 4" \times 8'}{12} = 26^2/_3$ bd. ft.

Problem 2. 6 pieces 1" × 8" × 10' *or*, the same pieces dressed would be 6 pieces $^3/_4$" × $7^1/_2$" × 10'. (Fractions from $^1/_2$" to 1" are considered as 1".)

Solution: $\frac{6 \times 1" \times 8" \times 10'}{12} = 40$ bd. ft.

Problem 3. 8 pieces 2" × 6" × 38".
(If the length is in inches, divide the product by 144 instead of 12. Why?)

Solution: $\frac{8 \times 2" \times 6" \times 38"}{144} = 25^1/_3$" bd. ft.

FIGURE 10-2 Lumber is bought and sold by the board foot. One board foot has a volume of 144 cubic inches. Any combination of dimensions that equals 144 cubic inches is 1 board foot.

When you consider that there are 860 species of trees in the United States, it is evident that forestry is an important part of the economy. Forestry is the management of forests. Our citizens depend on a great variety of products that come from trees (Figure 10-3).

Trees in the forests of the United States are divided into two general classifications: evergreen and deciduous. **Evergreen** trees do not shed their leaves on a yearly basis. Evergreen trees of commercial importance are mostly conifers. **Conifers** are evergreen trees that produce seeds in cones, have needle-like leaves, and produce lumber called **softwood**. **Deciduous** trees shed their needles or leaves every year and produce lumber called **hardwood**.

FOREST REGIONS OF NORTH AMERICA

Eight major forest regions are generally recognized by forestry educators and other forestry professionals (Figure 10-4). They include the Northern Coniferous Forest, Northern Hardwoods Forest, Central Broad-leaved Forest, Southern Forest, Bottomland Hardwoods Forest, Pacific Coast Forest, Rocky Mountain Forest, and Tropical Forest. In addition, some experts make reference to the Wet Forest and the Dry Forest in Hawaii (Figure 10-5).

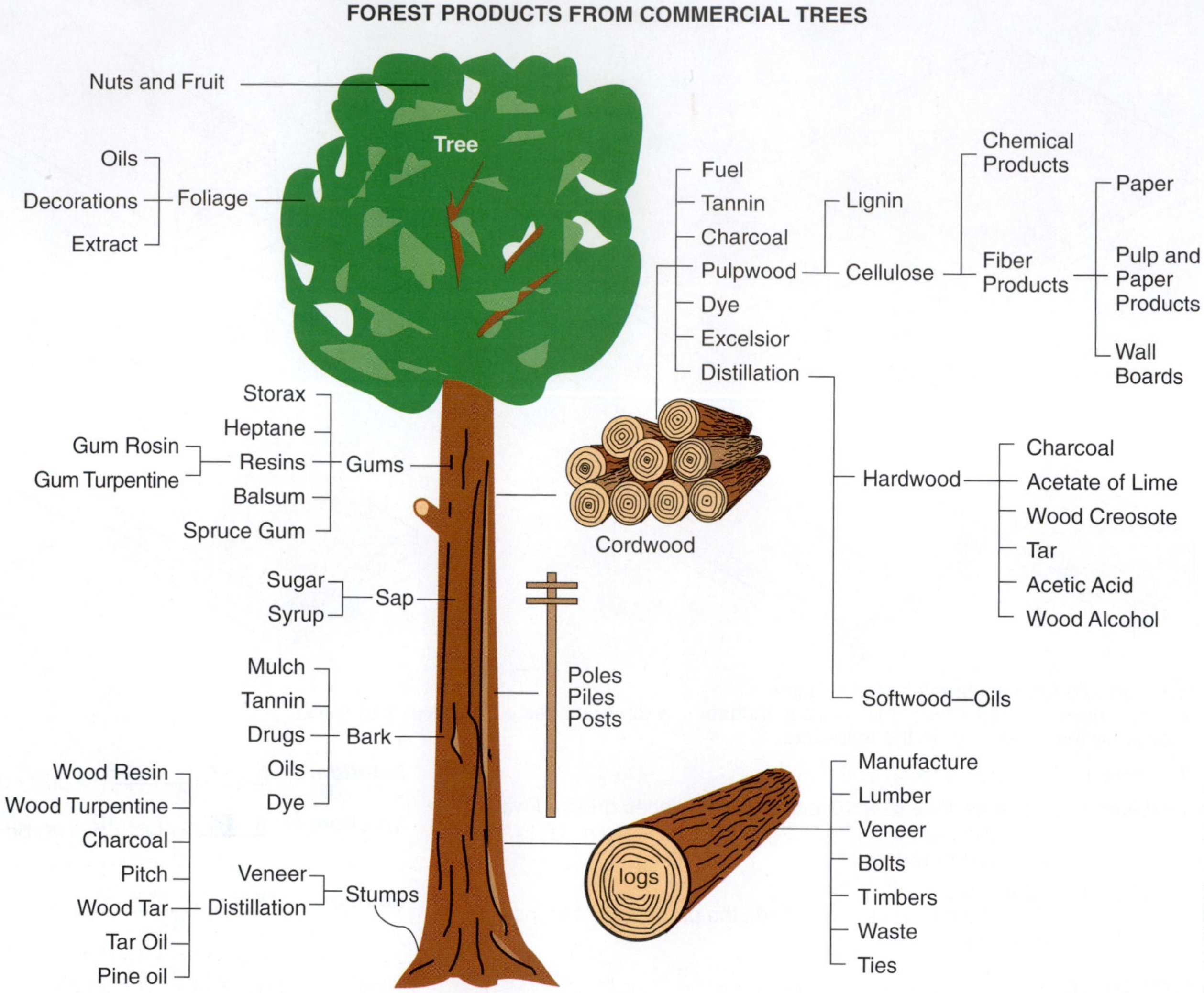

FIGURE 10-3 Many kinds of commercial products are obtained from trees and wood by-products.

Northern Coniferous Forest

The Northern Coniferous Forest region contains vast regions of softwoods. Some areas along the border between Canada and the United States contain a mixture of softwoods and hardwoods. The region is characterized by swamps, marshes, rivers, and lakes. The climate is cold. This forest is the largest region in North America, extending across Canada and Alaska. The most dominant type of tree is the evergreen, and large amounts of **pulpwood** are harvested in this region. The most important species include the white spruce, Sitka spruce, black spruce, jack pine, black pine, tamarack, and western hemlock.

INTERNET KEY WORDS:
pulpwood production

Northern Hardwoods Forest

The Northern Hardwoods Forest region reaches from southeastern Canada through New England to the northern Appalachian Mountains. It blends with the Northern Coniferous Forest on the northern border and the Central Broad-leaved Forest on the south. The region extends westward beyond the Great Lakes region, and it is populated by a number of important hardwood species, including beech, maple, hemlock, and birch trees.

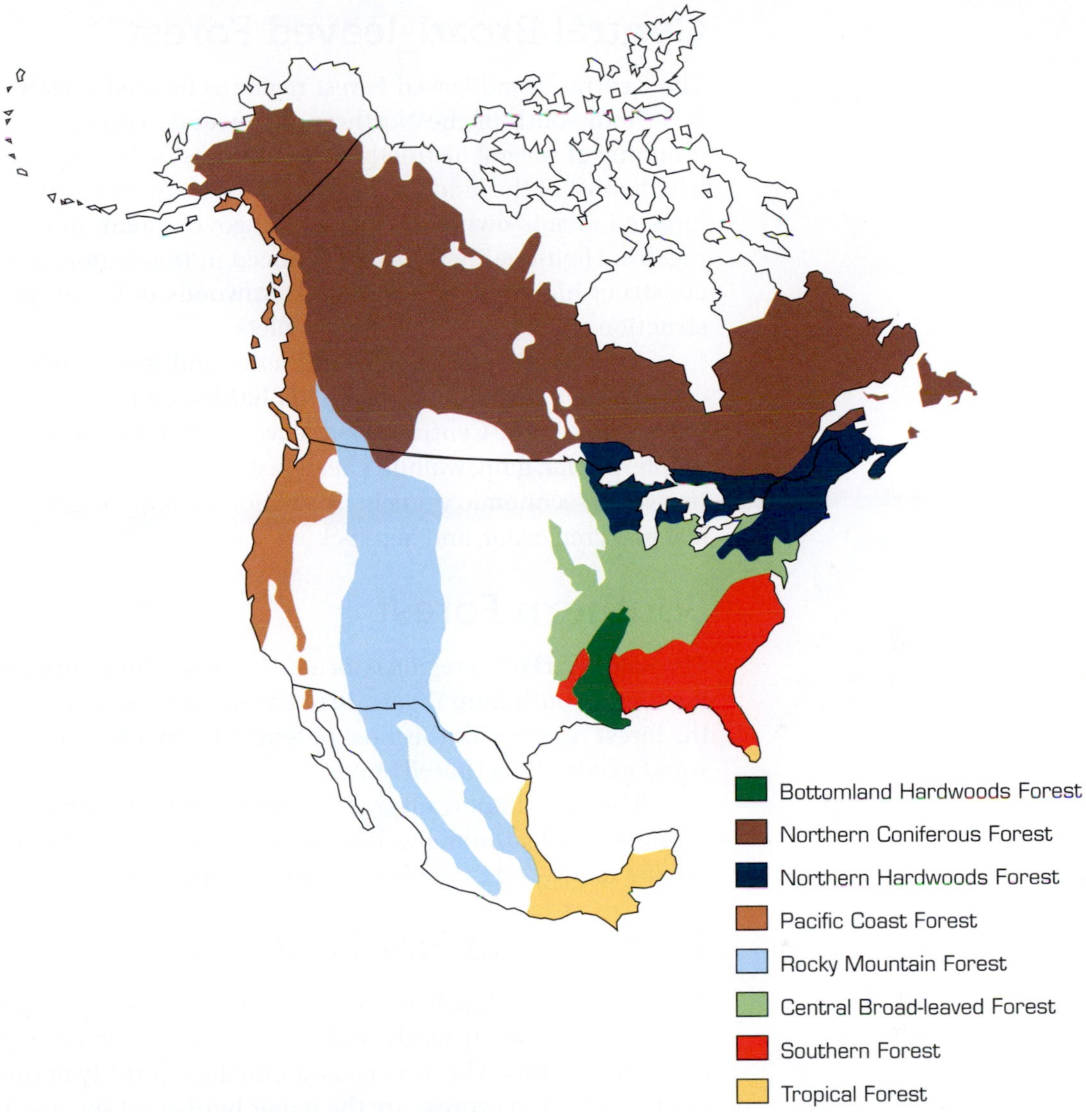

Bottomland Hardwoods Forest
Northern Coniferous Forest
Northern Hardwoods Forest
Pacific Coast Forest
Rocky Mountain Forest
Central Broad-leaved Forest
Southern Forest
Tropical Forest

FIGURE 10-4 Forest regions of North America.

FIGURE 10-5 Forest regions of Hawaii.

Central Broad-leaved Forest

The Central Broad-leaved Forest region is located mostly east of the Mississippi River and south of the Northern Hardwoods Forest. It consists of an arbitrary grouping of several distinctly different forest subgroups. It is a farming region in which most of the land has been cleared to produce cultivated crops. Little of the forested area is owned by the federal government, in contrast with some other regions. High-quality wood is produced in this region, and much of it is used to construct high-quality furniture. Hardwoods of lesser quality are used for construction and to make industrial pallets.

This forest contains more varieties and species of trees than any other forest region. It is composed mostly of hardwood trees. Hardwoods of commercial importance in the Central Broad-leaved Forest region include oak, hickory, beech, maple, poplar, gum, walnut, cherry, ash, cottonwood, and sycamore. The conifers that are of economic value in this region include Virginia pine, pitch pine, shortleaf pine, red cedar, and hemlock.

Southern Forest

The Southern Forest region is located in the southeastern part of the United States. It extends south from Delaware to Florida, and west to Texas and Oklahoma. It is the forest region with the most potential for meeting the future lumber and pulpwood needs of the United States.

The most important trees in the Southern Forest are conifers. They include Virginia, longleaf, loblolly, shortleaf, and slash pines. Oak, poplar, maple, and walnut are hardwood trees of economic importance.

Bottomland Hardwoods Forest

The Bottomland Hardwoods forests occur mostly along the Mississippi River. They contain mostly hardwood trees and are often among the most productive of the U.S. forests. This is because of the high fertility of the soil in this area. Oak, gum, tupelo, and cypress are the major hardwood species found here.

Pacific Coast Forest

The Pacific Coast Forest region is located in Northern California, Oregon, and Washington. It is the most productive of the forest regions in the United States and has some of the largest trees in the world. Approximately 48 million acres of Pacific Coast Forest provide more than 25 percent of the annual lumber production in the United States. About 19 percent of the pulpwood and 75 percent of the plywood produced in the United States comes from trees grown in the Pacific Coast Forest region.

water cycle, forest relationships

Trees in the Pacific Coast forests include 300-foot-tall redwoods and giant Sequoias that may be as much as 30 feet in diameter. Douglas fir, ponderosa pine, hemlock, western red cedar, Sitka spruce, sugar pine, lodgepole pine, noble fir, and white fir are conifers that are important in this region. Important hardwood species include oak, cottonwood, maple, and alder.

Rocky Mountain Forest

The forests of the Rocky Mountain Forest region are much less productive than those of the Pacific Coast region. This region is divided into many small areas and extends from Canada to Mexico. About 27 percent of the lumber produced in the United States comes from the 73 million acres of this forest region.

Most of the trees of commercial value in the Rocky Mountain forests are the western pines: western white pine, ponderosa pine, and lodgepole pine. Spruce, fir, larch, western red cedar, and hemlock also grow there in small quantities. Aspen is the only hardwood of commercial importance in the Rocky Mountain Forest region.

Tropical Forest

The tropical or subtropical forests of the continental United States are located in southern Florida and in southeastern Texas. These compose the smallest forest region in the United States. The major trees in this region are mahogany, mangrove, and bay, which are unimportant commercially. However, they are very important ecologically.

Hawaiian Forest

The Wet Forest region of Hawaii produces ohia, boa, tree fern, kukui, tropical ash, mamani, and eucalyptus. Most of these woods are used in the production of furniture and novelties.

The Dry Forest region of Hawaii produces koa, haole, algaroba, monkey pod, and wiliwili. None of these is of commercial value.

RELATIONSHIPS BETWEEN FORESTS AND OTHER NATIONAL RESOURCES

The relationships between forests and other natural resources, such as water and wildlife, are important to the overall well-being of the ecological system.

Forests play important roles in the water cycle. As water circulates between the oceans and the land areas, forests reduce the impact of falling rain on the soils and serve as storage areas for vast amounts of water. The stored water is released slowly from a forest watershed, allowing streams to flow throughout the year. In this manner, a forest regulates the flow of water, making it possible for fish and other aquatic plants and animals to survive.

Forests filter rain as it falls and help reduce erosion of the soil. They trap soil sediment and help maintain water quality. Trees and shrubs of the forest are also instrumental in removing many pollutant materials from the air and from water runoff. Similarly, the roots of trees filter excess nutrients from water runoff. They also help reduce the harmful effects of excess fertilizer nutrients that enter streams and underground water systems.

The relationships between forests and many types of wildlife are numerous. Algae, fungi, mosses, and numerous other plant and animal forms make their homes in forests. Forests provide food, shelter, protection, and nesting sites for many species of birds and animals. They also help maintain the quality of streams so that fish and other types of aquatic life can live and thrive. The shade provided by forests helps maintain proper water temperatures for the growth and reproduction of aquatic life.

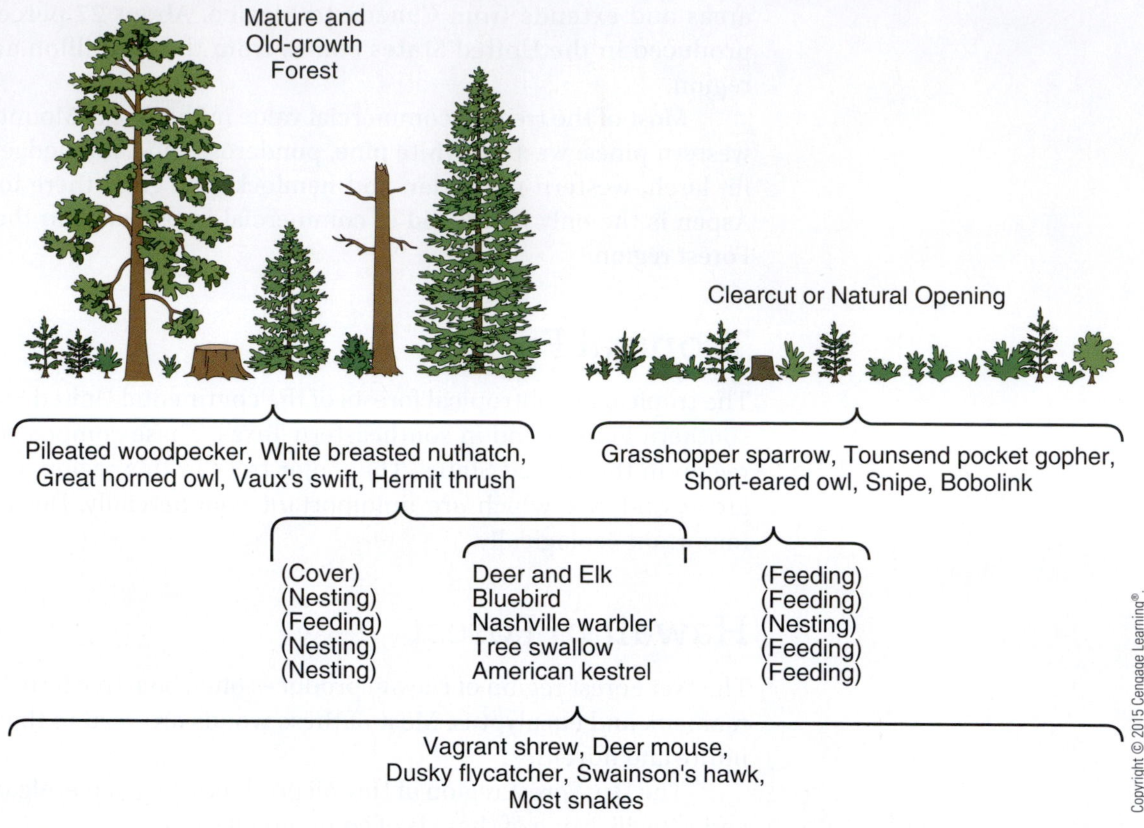

FIGURE 10-6 Preference of 20 wildlife species for different forest habitats.

INTERNET KEY WORDS:
forest relationships
water cycle forest relationships
wildlife
tree species

The wildlife found in the different types of forests varies considerably (Figure 10-6). Some species must have the open areas that occur naturally or that have been clear cut by people. A forest that has been **clear cut** has had all of the marketable trees removed from it. Other wild birds and animals depend to a greater extent on mature forests to provide their needs in life. As a forest changes naturally or as a result of harvesting, the types of wildlife that inhabit it also change.

SOME IMPORTANT TYPES AND SPECIES OF TREES IN THE UNITED STATES

Trees may be described in terms of the lumber they produce. Characteristics such as hardness, weight, tendency to shrink and warp, nail- and paint-holding capacity, decay resistance, strength, and surface qualities are used to evaluate the usefulness of lumber.

INTERNET KEY WORDS:
Douglas fir
balsam fir
hemlock
cedar

Softwoods

The softwoods, or needle-type evergreens, that are important commercially in the United States include Douglas fir, balsam fir, hemlock, white pine, cedar, southern pine, ponderosa pine, and Sitka spruce (Figure 10-7).

© Christopher Barrett/Shutterstock.com.

A Douglas Fir

© Helder Almeida/Shutterstock.com.

B Balsam Fir

© R.L. Hausdorf/Shutterstock.com.

C Western Hemlock

© Noah Strycker/Shutterstock.com.

D Western Red Cedar

© Candia Baxter/Shutterstock.com.

E Eastern White Pine

© IrinaK/Shutterstock.com.

F Loblolly Pine

© Tom Grundy/Shutterstock.com.

G Ponderosa Pine

Courtesy of DeVere Burton.

H Sitka Spruce

FIGURE 10-7 Some species of softwood trees that are commercially important for wood products.

Douglas Fir

Douglas fir is probably the most important species of tree in the United States today. It grows to a height of more than 300 feet and a diameter of more than 10 feet. About 20 percent of the timber harvested in the United States each year is Douglas fir. One-hundred-year-old stands of Douglas fir can produce 170,000 board feet of lumber per acre. This is five to six times the production of most other softwood species. Douglas fir is popular as construction lumber and for the manufacture of plywood. **Plywood** is a construction material made of thin layers of wood glued together.

Balsam Fir

Found in the forests of the Northeast, the lumber from balsam fir is used mostly for framing buildings. Balsam fir trees have soft, dark green needles and a classic triangular shape when grown at low densities. They are often used as Christmas trees.

Hemlock: Eastern and Western

Eastern hemlock is strong and is often used for building material. However, at times, it can be brittle and difficult to work with. Eastern hemlock grows over most of the Northern Coniferous Forest range.

Western hemlock grows in the Pacific Coast Forest region, where yearly rainfall averages 70 inches. Western hemlock lumber is very strong and is one of the most important sources of construction-grade lumber. It is also important for pulpwood.

Cedar: Eastern Red, Eastern White, and Western Red

Eastern red cedar is used for fence posts because it is resistant to decay. It is also used for lining chests and closets because its odor repels many insects. White cedar is a swamp tree with decay-resistant wood that is often used for shingles and log homes. Western red cedar resembles redwood in appearance. It is used where decay resistance, rather than strength, is important.

White Pine

White pine lumber is soft, light, and straight-grained. It has less strength than spruce or hemlock, and it is more popular as a wood for cabinetmaking. Eastern white pine grows from Maine to Georgia. Western white pine is found in the Rocky Mountain Forest region.

Southern Pine

Included in the category of southern pine is longleaf pine, shortleaf pine, loblolly pine, and slash pine. The southern pines grow in the southern and south Atlantic states. Lumber from southern pine is used for construction, pulpwood, and plywood.

Ponderosa Pine

The ponderosa pine is a large tree that grows up to 130 feet in height and 4 feet in diameter. It is widely distributed in the western United States. The wood is heavy,

and it can be brittle; however, it is reasonably free of knots and other defects. Its most valuable use is for construction of wooden windowpanes and doors.

Sitka Spruce

Sitka spruce trees grow from California to Alaska, attaining a height of 300 feet and a diameter of 18 feet. Lumber from Sitka spruce is of very high quality. It is strong, straight, and even-grained. Sitka spruce is also used in large quantities for pulpwood.

Hardwoods

Hardwoods come from deciduous trees. Commercially important species of hardwoods in the United States include birch, maple, poplar, sweetgum, oak, aspen, ash, beech, cherry, hickory, sycamore, walnut, and willow (Figure 10-8).

Birch

Easily recognized by their white bark, birch trees grow in areas where summer temperatures seldom exceed 70° F. Birch lumber is dense and fine-textured. It is used for furniture, plywood, paneling, boxes, baskets, and veneer, as well as for many small novelty items. **Veneer** is a very thin sheet of wood glued to a cheaper species of wood that is used in paneling and furniture making.

Maple

Maple lumber is classified as both hard and soft. Hard maple lumber is heavy, strong, and hard. It is used for butcher blocks, workbench tops, flooring, veneer, and furniture. Soft maple is only about 60 percent as strong as hard maple, and it is used in many of the same applications. Some species, such as the sugar maple, produce sweet sap that is made into maple syrup.

Poplar

Poplar grows over most of the eastern United States. It is classified as a hardwood because of its deciduous structure. However, lumber from poplar trees is soft, light, and usually knot-free. Poplar lumber may be white, yellow, green, or purple in color, and can be stained to resemble most of the fine hardwoods. Poplar is used for furniture, baskets, boxes, pallets, and building timbers.

Sweetgum

The sweetgum tree is easily recognized by its star-shaped leaves and distinctive ball-shaped fruit. Sweetgum trees grow to as much as 120 feet in height and 3–5 feet in diameter. Lumber from sweetgum trees has interlocking grain and is used for house trim, furniture, pallets, railroad ties, boxes, and crates. The gum that comes from wounds in trees can be used as natural chewing gum or as a flavoring or perfume.

Oak: White and Red

The two general types of oak in the United States are white and red. White oak lumber is hard, heavy, and strong. Its pores are plugged with membranes that

A Paper Birch

B Sugar Maple

C Yellow Poplar

D Sweetgum

E White Oak

F Quaking Aspen

FIGURE 10-8 Some species of hardwood trees that are commercially important for wood products.

© ukmooney/Shutterstock.com.

G White Ash

© Maksym Bondarchuk/Shutterstock.com.

H American Beech

© bjonesphotography/Shutterstock.com.

I Black Cherry

© Cynthia Burkhardt/Shutterstock.com.

J Bitternut Hickory

© BONNIE WATTON/Shutterstock.com.

K American Sycamore

© Bernd Schmidt/Shutterstock.com.

L Black Walnut

© sandra zuerlein/Shutterstock.com.

M Black Willow

FIGURE 10-8 Continued

make it nearly waterproof. It is used for structural timbers, flooring, furniture, fencing, pallets, and other uses where wood strength is a necessity.

Red oak is similar to white oak, except that it is very porous. It is not very resistant to decay and must be treated with wood preservatives when used outside. Chief uses of red oak include furniture, veneer, and flooring.

Aspen

Aspen trees grow in the Northeast, Great Lake states, and the Rocky Mountains. Aspen is rapid growing, but the lumber tends to be weaker than most construction-grade timber. It is also used for pulpwood.

Ash

Ash lumber is heavy, hard, stiff, and has a high resistance to shock. It also has excellent bending qualities. It is popular for use in handles, baseball bats, boat oars, and furniture. It resembles oak in appearance.

Beech

Beech is grown in the eastern United States. It is heavy and hard, and it is noted for its shock resistance. It is hard to work with and is prone to decay. Beech is used in veneer for plywood and for flooring, handles, and containers.

Cherry

Cherry can be found from southern Canada through the eastern United States. Cherry wood is dense and stable after drying. It is desirable and popular in the production of fine-quality furniture. It is expensive and in limited supply; therefore, it is used mostly for veneer and paneling. It may, however, be used for other woodworking purposes.

Hickory

Hickory grows best in the eastern United States. Hickory lumber is hard, heavy, tough, and strong. It is somewhat stronger than Douglas fir when used as construction lumber. Other uses for hickory include handles, dowel rods, and poles. Hickory is also popular for use as firewood and for smoking meat.

Sycamore

Sycamore wood is used for flooring in barns, trucks, and wagons because of its strength and shock resistance. It grows from Maine to Florida and west to Texas and Nebraska. Boxes, pallets, baskets, and paneling are other uses for sycamore.

Black Walnut

A premier wood for the manufacture of fine furniture, black walnut grows from Vermont to Texas. The wood has straight grain and is easily machined with woodworking tools. Because walnut is slow-growing and demand for it is high, it is often made into veneer to get more use from its chocolate brown heartwood. It is also the source of black walnut nutmeats.

Black Willow

Most of the black willow of commercial value is grown in the Mississippi River Valley. It is soft and light and has a uniform texture. Willow is used mostly in construction for subflooring, sheathing, and studs. Some willow is also used for pallets and for interior components of furniture. Black willow is sometimes a low-cost substitute for black walnut because it has a similar brown appearance when finished. Several other domestic softwoods and hardwoods grow in the United States and many are important in local areas. The types discussed here are but a sampling, rather than a definitive list, of the commercially important trees in the United States.

tree, age, annual rings

AGRI-PROFILE

CAREER AREAS: DENDROLOGY/SILVICULTURE/FORESTRY

© iStockphoto/Tadeusz Przybyl.

Many challenging careers are available to college graduates in the forest industry.

Forestry is known as a career area for rugged individuals who prefer the outdoors and like to work in relative isolation. However, many jobs in forestry are in urban areas and involve much indoor work. The United States Forest Service hires large numbers of forestry technicians and managers. Many forestry jobs do involve an extensive amount of outdoor work, but most jobs provide a desirable mix of outdoor and indoor work.

Forestry includes the work of the dendrologist engaging in the study of trees, the silviculturist specializing in the care of trees, the forestry consultant advising private forest land owners, lumber industry workers, government foresters, loggers, national and state forest rangers, and firefighters. A relatively new position is that of the urban forester, who is responsible for the health and well-being of the millions of trees found in parks, along streets, and in other areas of our cities.

The arborist is an urban forester whose work may include planting, transplanting, pruning, fertilizing, or tree removal.

TREE GROWTH AND PHYSIOLOGY

Trees use carbon dioxide (carbon and oxygen) from the air and water (hydrogen and oxygen) from the soil to manufacture simple sugars in their leaves. The leaves then use additional carbon, hydrogen, and oxygen to convert simple sugars into complex sugars and starches. Nitrogen and minerals from the soil are then used to manufacture proteins, the building blocks for growth and reproduction.

A tree typically starts from a seed. For instance, oak trees grow from seeds called acorns, pine trees start from seeds in pine cones, and beech trees grow from beech seeds. Trees may also sprout and grow from stumps or other tree parts. When a seed germinates, a shoot grows upward to form the top growth, and roots grow downward and outward to form the root system.

Both roots and shoots extend themselves by growth at the tips through cell division and elongation. At the same time, tree roots, stems, and trunks grow in diameter by adding cell layers near their outer surfaces (Figure 10-9). This growth layer in a tree root, trunk, or limb is called the **cambium**. The outward growth of the cambium in 1 year creates an **annual ring**, as seen in the cross section of a root, trunk, or limb. Water and minerals are taken in by the roots and transported up to the leaves through a layer of cells called the **xylem layer**, or **sapwood**. The xylem is located just inside the cambium layer. Just outside the cambium is another layer of cells called the **phloem**, or **inner bark**, which carries food manufactured in the leaves to the stems, trunk, and roots. Each year, the tree grows new cambium, xylem, and phloem tissues, and the older sapwood becomes heartwood. **Heartwood** is the inactive core that gives a tree strength and rigidity.

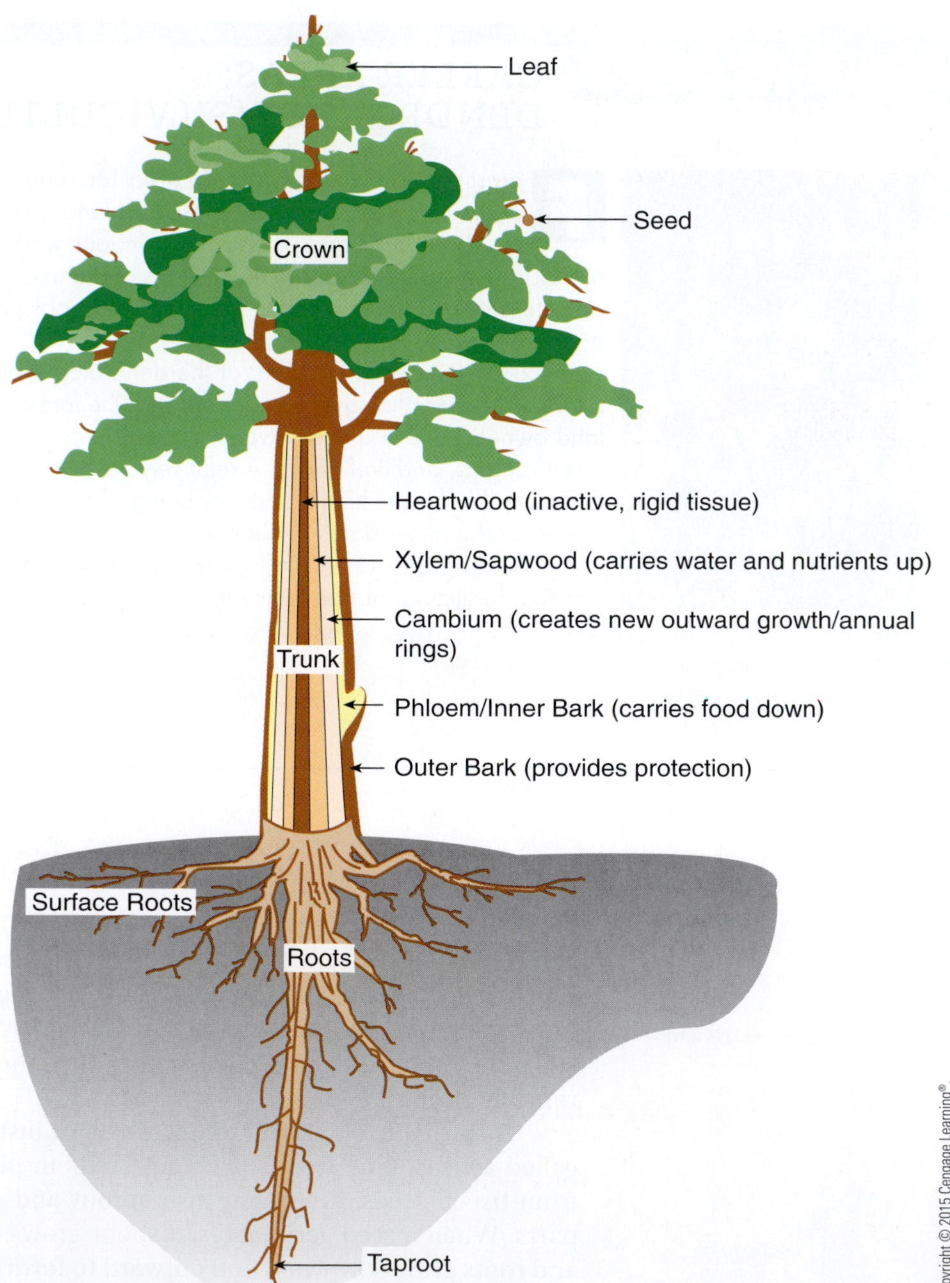

FIGURE 10-9 Major parts of a tree.

PROPERTIES OF WOOD

The various properties of wood determine the uses for which it is best adapted. It should be noted that there are wide variations within a specific species of trees, and the properties discussed here are general in nature.

Hardness

The property of **hardness** refers to a wood's resistance to compression. Hardness determines how well it wears. It is also a factor in determining the ease of working the wood with tools. Splitting and difficulty in nailing are problems that occur when wood is too hard.

Weight

The weight of wood is a good indication of its strength. In general, heavy wood is stronger than light wood. The moisture content affects the weight of wood; therefore, comparisons should always be made between woods with similar amounts of moisture.

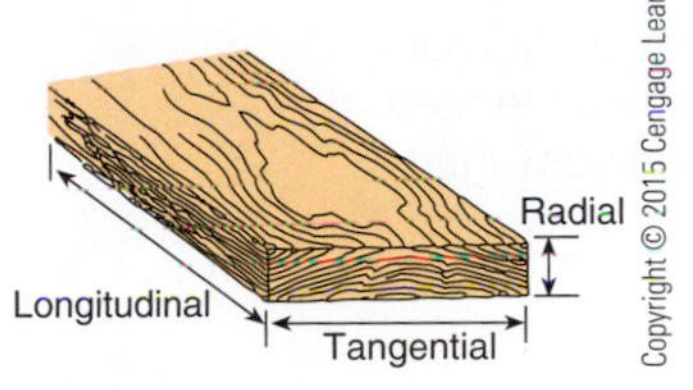

FIGURE 10-10 Dimensions of wood shrinkage.

Shrinkage

The change in dimensions of a piece of wood as it reacts to reduced humidity or temperature is referred to as **shrinkage**. The amount of expansion and contraction of wood greatly affects techniques used in building construction or manufacturing items from wood (Figure 10-10).

Warp

Warp refers to the tendency of wood to bend permanently because of moisture change. The tendency to warp is more of a problem in some types of wood than in others. Warping is caused by uneven drying of wood across its three dimensions.

Ease of Working

In wood, ease of working refers to the level of difficulty in cutting, shaping, nailing, and finishing the wood. It is influenced by the hardness of the wood and the characteristics of the grain. The grain of wood is formed by alternating hard and soft layers of new wood resulting in annual growth rings (Figures 10-11 and 10-12). In general, soft wood is easier to work with than hard, dense wood.

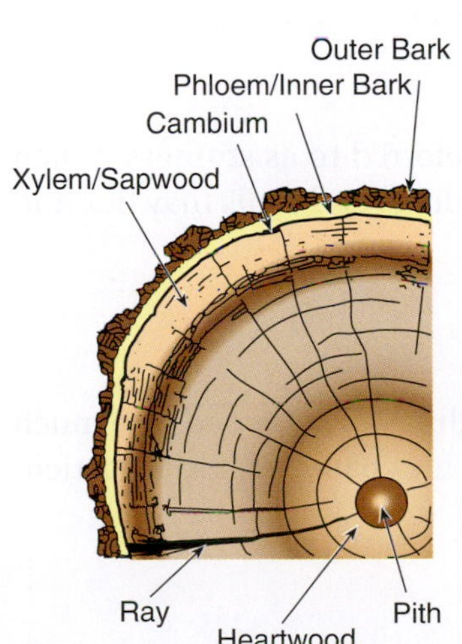

FIGURE 10-11 The structures of a tree illustrated in a cross section of a log or stump.

Paint Holding

The ability of wood to hold paint may be determined by the type of paint, surface conditions, and methods of application as well as characteristics of the wood itself. Moisture content, amount of pitch in the wood, and the presence of knots all affect how well paint adheres to wood.

Nail Holding

Nail-holding capacity refers to the resistance of wood to the removal of nails. In general, the harder and denser the wood, the better it will hold nails.

FIGURE 10-12 The annual rings in the cross section of a log or stump reflect the conditions of growth experienced by the tree on a year-to-year basis. These conditions influence the appearance of the grain.

Decay Resistance

The resistance of wood to microorganisms that cause decay is a chief factor in selecting wood. Natural resistance to insects that live in, tunnel through, or devour wood also affects the choice of species for a given application.

Bending Strength

The ability of wood to carry a load without breaking is a measure of bending strength. Bending strength is important when determining types and sizes of lumber to use for rafters, beams, joists, and other building applications.

Stiffness

The resistance of wood to bending under a load is referred to as stiffness. When wood used in construction is not sufficiently stiff, ceilings and walls may flex and buckle, and the wallboards will crack under the load.

Toughness

The ability of wood to withstand blows is called toughness. Hardwoods are much more resistant to shock and blows than softwoods. This characteristic is particularly important in woods used for tool handles.

Surface Characteristics

The appearance of the wood's surface is sometimes the determining factor in its use. The pattern of the grain greatly influences the staining characteristics and beauty of the finished product. The number and type of knots and pitch pockets are surface characteristics that need to be considered when selecting lumber for a project.

WOODLOT MANAGEMENT

The proper management of a wooded area or woodlot involves more than just the harvest of trees and the removal of unwanted species. A **woodlot** is a small, privately owned forest. The production of trees for harvest is a long-term investment, and mistakes in management take a long time to correct.

Some of the factors that need to be considered in the management of woodlots include soil, water, light, type of trees, condition of trees, available markets, methods of harvesting, and replanting. Using scientific methods in the management of forests is called **silviculture**. The care and management of trees for ornamental purposes is called **arboriculture**.

Restocking a Woodlot

The least expensive method of replacing trees harvested from a woodlot is natural seeding. Sources of seed for the desired species must be available in the forest, and conditions must be right for seed germination for natural seeding to take place. If seed from natural sources is not available, seeds from other sources may be planted on the forest site.

A surer method of restocking a woodlot is to plant trees of the desired species, rather than relying on seeds to do the reforestation. In most cases, **seedlings** (young trees started from seeds) are planted during late winter and early spring, before the new season's growth begins. Woodlots can be planted with one species of tree or a mixture of several compatible species (Figure 10-13).

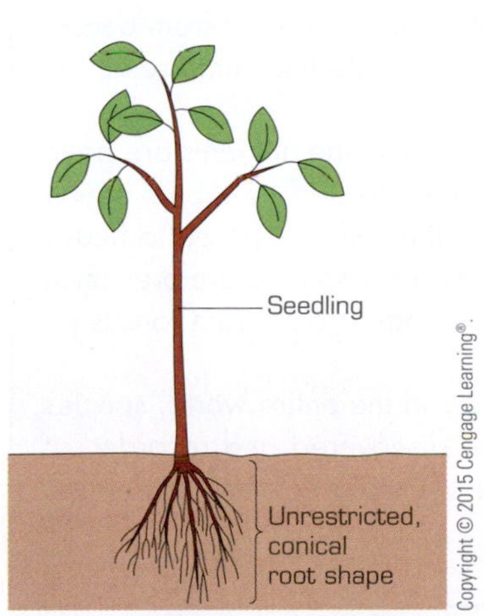

FIGURE 10-13 It is important to the survival of the seedling to maintain a conical root shape as the young tree is transplanted into the soil.

Management of a Growing Woodlot

Management of a woodlot is much more involved than just sitting back and watching it grow. Proper care and management are important if the forestry enterprise is to be successful.

Trees that are of no commercial value should be removed as soon as possible to eliminate competition for light, moisture, and nutrients. Because these "weed" trees are removed when they are small, there is seldom a market for them, and they are left on the woodlot floor to decay. If weed trees are of sufficient size, however, they may be used for firewood. When all the trees of a woodlot are nearly the same age (typically 15–30 years old), they often need to be thinned. Trees should be thinned any time that the crowns or branches occupy less than one-third of the height of the trees. Usually, about one-fourth to one-third of the trees in a woodlot are removed during thinning.

Trees that are being grown for lumber are often pruned of side branches to produce a better-quality log. Prompt removal of side branches helps keep logs free of knots. Only rapidly growing trees should be pruned. Branches should be pruned flush with the trunk of the tree. Pruning is usually done during the fall and winter, when the trees are dormant.

Planning a Harvest Cutting

A woodlot should never be harvested without a plan. A harvesting plan can maximize the income from the woodlot over many years. The use of a forester's

SCIENCE CONNECTION CONSERVING OUR BIODIVERSITY

© Dr. Morley Read/Shutterstock.com.

Rain forests are our richest source of plant and animal species and biodiversity.

Our biodiversity is at risk! Not far from the shores of Florida and Texas, the tropical rain forests of Central and South America, as well as parts of Africa and Southeast Asia, are disappearing from the face of the Earth. Commonly called a jungle, a tropical rain forest is a hot, wet, green place characterized by an enormous diversity of life and a huge mass of living matter. Here, biologists estimate that more than half of the plant and animal species of the world are found. Unfortunately, many of these species are not found anywhere except in the tropical rain forests. The rain forests occupy only 7 percent of the land area of the Earth, and their total area is diminishing at an alarming rate!

Abundant moisture and warm temperatures encourage tremendous plant growth year round. Plants and moisture then provide an excellent environment for animals and microorganisms to grow and develop. Rain forests are covered by a dense canopy formed by the crowns of trees up to 150 feet in height. Under the shade of this canopy are smaller trees and shrubs creating a second layer called the understory. The two layers are typically woven together by strong woody vines, which may exceed 700 feet in length.

The third layer is the forest floor. Palms, herbs, and ferns dot the forest floor. The jungle floor, with its shrubs and trees, is home to a myriad of insects and animals. Annual rainfall may reach 400 inches, and the daily rate exceeds the rate of evaporation. Life is plentiful, however, and the soil beneath the jungle is shallow, and nutrients from decaying leaves are quickly used up, leaving few residual nutrients. As human populations expand into the rain forests, the foliage is cut and burned to make way for logging, grazing, and cropping.

Sadly, the land will generally support grazing or crop production for only a few years until the nutrients are gone and the soil is depleted of minerals. The occupants then typically move on to virgin jungle land to slash and burn a new tract and repeat the cycle. Once the jungle is destroyed, crops extract the scant nutrients, the soil is exposed to heavy rainfall and is soon eroded, and the land has too few nutrients to support the resurgence of jungle. Therefore, once cleared, the jungle area is often lost forever. It is estimated that one-half of the world's original tropical rain forests no longer exist.

Because the rain forests are the storehouse for more than half of the genetic diversity in the entire world, species are being lost at an alarming rate. Many species are becoming extinct before they are discovered and recorded. It is estimated that scientists have discovered and named only a sixth of all the plant and animal species in rainforests. Earth's biodiversity is indeed threatened by destruction of the tropical rain forests.

services is usually wise in developing a harvesting plan. A forester is a person who studies and manages forests.

INTERNET KEY WORDS: trees, salvage, harvest

Several systems can be used to harvest a woodlot. They include clear cutting, seed-tree cutting, shelterwood cutting, diameter limit cutting, and selection cutting.

Clear Cutting

Clear cutting is a system of cutting timber where all the trees in an area are removed. It was used extensively in cutting the virgin forests (those that have never been harvested) of the United States when there was little thought of reforestation of the cut area.

Clear cutting is usually done in small patches, ranging in size from one-half acre to 50 acres. Prompt reforestation with desirable species of trees on the clear-cut areas is essential. This prevents erosion and the loss of fragile forest soil.

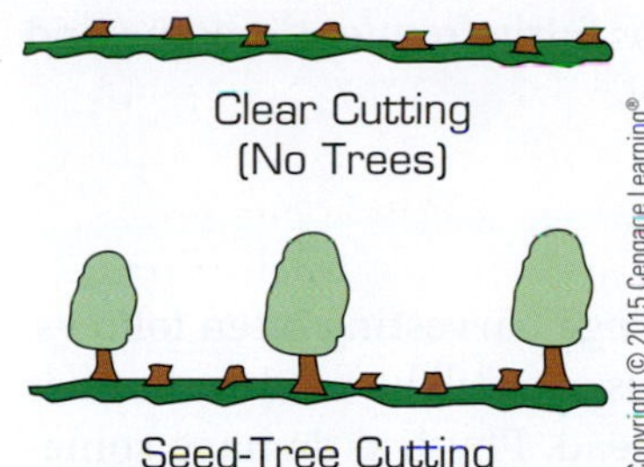

FIGURE 10-14 Clear cutting is generally done in small patches, but seed-tree cutting is an economical way to harvest and reseed large areas.

Seed-Tree Cutting

Harvesting trees by the seed-tree cutting method is very similar to clear cutting. The primary difference is that enough seed-bearing trees are left uncut to provide seeds for reforestation (Figure 10-14). The trees left to produce seeds should be representative of the desired species and free from insect, disease, or mechanical damage.

The removal of the seed trees usually takes place after the cut area is repopulated with desirable seedlings.

Shelterwood Cutting

In shelterwood cutting, enough trees are left standing after harvesting to provide for reseeding of the woodlot. They will also protect the area until the young trees are well established. After the young trees become established, the residual trees are harvested.

Diameter Limit Cutting

When harvesting trees by the diameter limit method, all trees above a certain diameter are cut. Slow-growing or diseased trees are often left standing when harvesting is done using this method. Rapid-growing trees of desirable species may be cut before they reach their production potential. Often this method leaves a woodlot with only slow-growing trees that may be undesirable. Diameter limit cutting is used advantageously to remove trees left from previous harvesting, which reduces competition with younger trees.

Selection Cutting

Selection cutting is the usual method of harvest used when the trees in a woodlot are of different ages. Selecting the trees to harvest can be a difficult task for an uninformed person, and obtaining the services of an experienced forester

HOT TOPICS IN AGRISCIENCE

INTEGRATED PEST MANAGEMENT

Photo courtesy of Boise National Forest.

Wood-boring insects cause greater economic damage to forests than any other group of insects.

An insect management plan that uses a variety of control methods is the most acceptable form of pest control because it does minimal damage to insect species for which the control method is not intended. Integrated pest management is a concept for controlling harmful insects or other pests, while providing protection for useful organisms. It involves the use of some chemical pesticides in emergency situations, but it also relies on natural enemies and other biological control strategies to control harmful pests. Integrated pest management is a control program that does not attempt to kill all of the harmful insects because insect control of this kind also kills the natural enemies of the pest. To survive, natural insect enemies must have a small population of the harmful pest upon which they can depend for food.

is usually wise. This system of harvesting allows for fairly frequent income and maintains the woodlot's aesthetic value.

Salvage Harvesting

Natural disasters occur regularly in forests, and salvage harvesting often follows. Trees that are dying from insect damage and diseases can still be used for lumber products if harvested before they are completely dead. Physical damage sometimes occurs to trees. Examples include charred trunks from forest fires and broken trunks and limbs caused by high winds. Once damage has occurred to the trees, they need to be harvested as soon as possible. Waiting longer than 2 or 3 years to harvest damaged trees will result in poor-quality lumber.

Protecting a Woodlot

A woodlot must be protected from fire, pests, and domestic animals if it is to yield a consistent harvest. It is estimated that more than 13 billion board feet of timber are destroyed each year by pests. This represents nearly 25 percent of the estimated net growth of forests and woodlots each year.

Fire

Millions of dollars worth of timber are destroyed by fire each year (Figure 10-15). In addition to actually killing trees, fire slows the growth of others and damages some so that insects and diseases may destroy them. Fire also burns the organic matter on the forest floor, which takes nutrients away from the trees and exposes the soil to erosion.

To help prevent fires, debris should be removed from around trees. Weeds, brush, and other trash around the edges of a woodlot should be removed. Conducting controlled burns at regular intervals under the supervision of professional fire personnel is an acceptable practice when it is correctly done. Limiting human use of the area during dry periods may also reduce the potential for fire. The construction of permanent firebreaks is useful for fire control.

A plan for dealing with fire is important in minimizing damage should a fire occur. State and county foresters can assist in developing fire prevention and control plans. These strategies include planning for water storage and setting procedures for notifying proper authorities and obtaining appropriate equipment and help.

Courtesy of Boise National Forest.

FIGURE 10-15 Fire can cause major damage to timber resources when it gets out of control. It is important to plan for fire protection, especially in high-value forests.

Pests

Insects and diseases cause more damage to existing forests than fires (Figure 10-16). Pests cause trees to be weak and deformed. They consume leaves, damage bark, and retard the growth of trees. They kill billions of trees each year. However, control of diseases and insects in forest lands is difficult and expensive.

Removal of dead, damaged, or weak trees may help reduce disease and insect problems. Not only are weak trees attacked but healthy ones also are sometimes favorite targets of pests. Prompt action in dealing with outbreaks of insect infestations and diseases will do much to ensure a profitable harvest.

Domestic Animals

The grazing of woodlots by cattle and sheep usually results in the destruction of all small seedlings in the forest. It also eliminates some of the woodlot floor

Courtesy of DeVere Burton.

FIGURE 10-16 Insects and diseases sometimes become epidemic in a forest, killing many trees during periods of drought or stress.

coverage. Livestock may also strip the bark from trees when grazing is inadequate, causing the trees to die. Woodlots do not provide much food for grazing animals, so it is usually wise to exclude livestock from them.

Harvesting a Woodlot

Before timber is harvested, a market for it must be found. Various methods of marketing the timber should be explored to determine the most profitable alternatives. There are many uses for forest products, and trees should be marketed to maximize their value.

HOT TOPICS IN AGRISCIENCE

TO FIGHT OR NOT TO FIGHT

Photo courtesy of Boise Interagency Fire Center.

Once a decision has been made to fight a fire, resources, such as fire-retardant chemicals, are often used to gain control and keep it from spreading.

Before humans began managing our forests, natural processes prevailed. Fire was a natural occurrence. When it erupted, it would burn underbrush, ground litter, and trees. Often, these fires would not kill the larger mature trees. Fire would kill the smaller trees that were in competition with these older trees for nutrients. The giant sequoia or ancient redwoods that stand today in some parts of the country would not have grown as tall or as thick as they have if fire had not thinned competing trees from the population.

Some types of trees need fire to reproduce. The seeds of the Lodge Pole Pine tree cannot germinate without first going through the intense heat of fire. In the early 1900s, people in the United States began fighting forest fires. For almost 100 years, few forest fires have been allowed to burn. The result is thicker, more densely populated forests than were present a century ago. There are more trees, and they are much closer together in today's forests. Now, when fires start, the damage is much more severe. Fires burn hotter, faster, and with more intensity than in the past.

(Continued)

HOT TOPICS IN AGRISCIENCE TO FIGHT OR NOT TO FIGHT (*Continued*)

Forest fires demolish forests, kill wildlife, destroy ecosystems, burn homes, and threaten animal and human life. Although fire was once nature's way of cleaning and thinning forests, it now can wipe out all life in its path. The job of keeping our forest healthy now belongs to the forest industry. Some types of logging practices thin trees; the wood is then sold for construction, papermaking, and many other uses.

Some people and organizations are strongly opposed to logging and fighting fire. They would prefer that whole forests burn and lives of forest creatures be lost than fight fires or thin crowded trees by logging. More moderate conservationists recognize that we must thin trees and remove forest debris using all of the proven practices for doing so. This may include use of controlled burns, logging dead or diseased trees, and clear-cutting small areas to remove the most dangerous fuels. Individuals who are truly concerned with the health of our forests understand that thinning our forests by logging and seeking to control dangerous forest fires is the best way to preserve this natural resource for generations to come.

The harvest of a woodlot can be done by the owner or by a contractor skilled in timber harvest. Another alternative is to sell the standing timber to a company that harvests and markets it. Regardless of how the timber is harvested, it is extremely important that contracts covering all pertinent harvest details be drawn up and signed by all parties involved before the harvest begins.

SEASONING LUMBER

The proper seasoning of lumber is essential to protect it from damage. As soon as a tree is cut, it starts to lose moisture. If the wood dries too slowly, it may be subject to rot, stain, or insect damage. If it is allowed to dry too quickly, lumber may twist and warp or split. This may make it unsuitable for most uses.

Wood that is sawed for lumber should be stacked immediately after sawing to allow for even drying (Figure 10-17). The stacking should take place on a

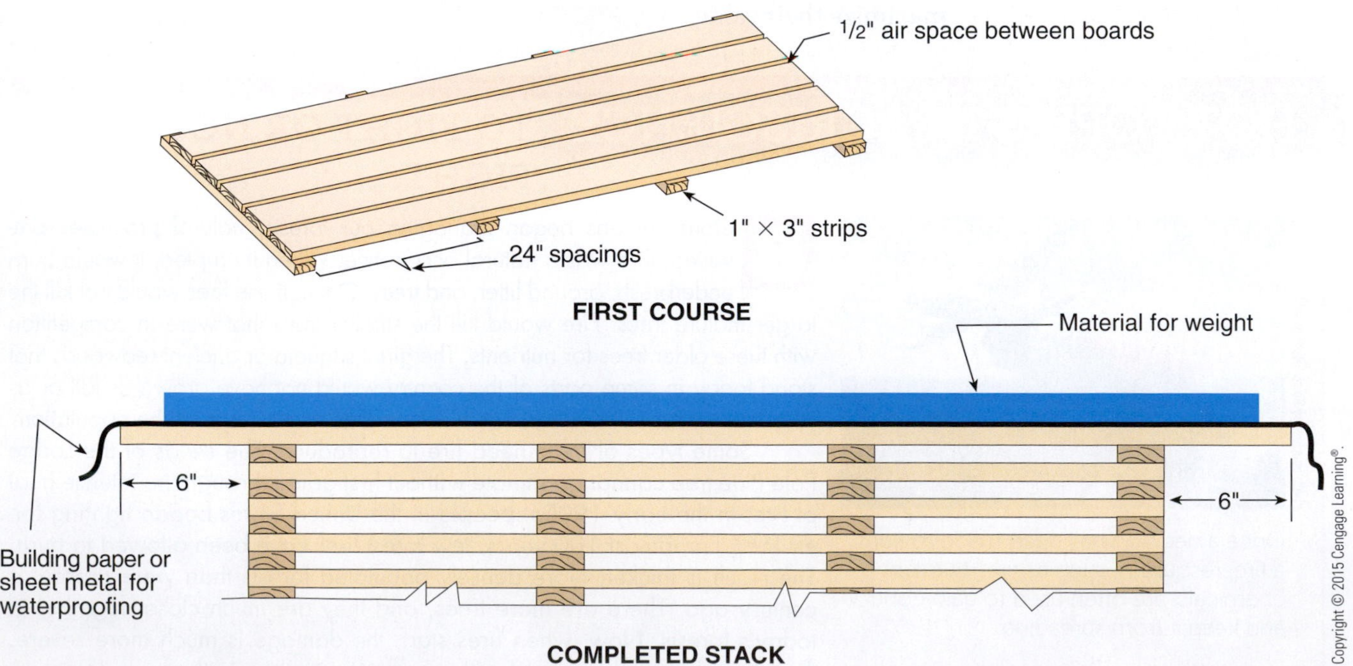

FIGURE 10-17 Freshly sawed lumber must be properly stacked to prevent warping and end splitting as the lumber dries.

level, well-drained location. Wood should be off the ground and stacked so that air can circulate freely around each board. The stacked lumber should also be protected from weather. The amount of time required for lumber to dry depends on the thickness of the lumber and the species of tree from which it was cut. Drying times usually range from 30 to 200 days.

Lumber can be seasoned quickly by placing the stacked lumber in a heated kiln for a few days to dry it out. This procedure will keep the lumber from becoming warped as it dries. Under the controlled conditions that exist inside a heated kiln, most of the moisture is removed from the wood within a few hours.

The forestry industry in the United States produces about 16.3 billion cubic feet of wood products each year. The demand for wood products, such as lumber, pulp wood, posts, and pilings, is almost 19 billion cubic feet per year. Only by carefully managing our forest resources and replacing our trees can the United States fulfill its need for wood and wood products in the future.

STUDENT ACTIVITIES

1. Write the Terms to Know and their meanings in your notebook.
2. Make a collection of forest products.
3. Make a collection of tree leaves, bark, or twigs that are of economic importance in your area.
4. Make a bulletin board showing the many uses of forests and forest products.
5. Write a report on a species of trees or a type of forest product of interest to you.
6. Visit a local forest and identify the types of trees growing there.
7. Have a forester visit the class and speak about forest management.
8. Visit a wood-processing plant operation to learn how trees are processed into other products.
9. Write a report on a career in the forest or wood products industry.
10. Study Figure 10-2 and learn how to calculate board feet (BF). Determine the number of board feet in the following:
    a. a board that is 1" × 12" × 12' =___ BF
    b. a plank that is 2" × 8" × 16' = ___ BF
    c. a 4"× 4" × 8' board = ___ BF
    d. six 2"× 4" × 8' boards = ___ BF
    e. twenty 2" × 10" × 16' boards = ___ BF
11. What is the total board feet in Activity 10?
12. If the items in Activity 10 were rough-sawed lumber and were priced at $0.20 per board foot, what would be the total cost of the order? $ ___
13. Look up "rain forests" on the Internet or using the school library. Find one organism that is used to benefit humans (besides building products). Write a one-page report describing the location the organism is found, its importance to humans, and its survival expectations if forests are destroyed.
14. Obtain a coring tool from your teacher and find a large tree. Core the tree by following your teacher's instructions. Then, count the rings in the core to determine the age of the tree.

SELF-EVALUATION

A. MULTIPLE CHOICE

1. Trees that are used for making paper are called
   a. timber.
   b. lumber.
   c. pulpwood.
   d. veneer.
2. The scientific management of forests is
   a. silviculture.
   b. arboriculture.
   c. pomology.
   d. olericulture.
3. The continental United States has about ___ acres of productive forests.
   a. 105 million
   b. 235 million
   c. 500 million
   d. 735 million
4. The most important commercial species of trees in the United States are
   a. oak.
   b. Douglas fir.
   c. redwood.
   d. walnut.
5. About 75 percent of the wood for plywood is harvested from the ___ Forest region.
   a. Northern Coniferous
   b. Central Broad-leaved
   c. Hawaiian
   d. Pacific Coast
6. A forest that has never been harvested is called
   a. virgin.
   b. hardwood.
   c. clear cut.
   d. seedling.
7. The seed-tree method of harvesting
   a. cuts all trees over a certain diameter.
   b. cuts all trees under a certain diameter.
   c. cuts about one-third of the trees in a woodlot.
   d. cuts all but a few trees left for seed.
8. Which of the following was not stated as a property of wood?
   a. nail-holding capacity
   b. bending strength
   c. color
   d. surface characteristics
9. The yearly demand for wood products in the United States is about ___ cubic feet.
   a. 19 billion
   b. 23 billion
   c. 190 billion
   d. 253 million
10. In a woodlot with even-aged trees, the lot usually needs to be thinned when the trees are ___ years old.
    a. 1 to 5
    b. 5 to 10
    c. 10 to 15
    d. 15 to 30

B. MATCHING

__________	1. Tree	a. Thin layers of wood glued together
__________	2. Shrub	b. Wood from conifers
__________	3. Woodlot	c. Small, multistemmed plant
__________	4. Veneer	d. Management of wooded land
__________	5. Plywood	e. Wood from deciduous trees
__________	6. Lumber	f. Thin slices of wood
__________	7. Forestry	g. Small forest
__________	8. Softwood	h. Tree with needle-like leaves
__________	9. Hardwood	i. Woody, single-stem plant
__________	10. Conifer	j. Boards sawed from trees

C. COMPLETION

1. The air drying of lumber is referred to as _______.
2. Nearly _______ percent of the estimated net growth of forests is lost each year to fire and pests.
3. _______ is a type of harvesting where every tree over a certain size is cut.
4. Two general types of oak are _______ and _______ lumber.
5. Douglas fir is an example of an _______ or _______ tree.
6. Cutting all of the trees in an area is called _______.
7. The _______ Forest region is located along the Mississippi River.
8. The forest region with the most potential for meeting future needs for forest products is the _______ Forest region.

UNIT 11
Wildlife Management

OBJECTIVE

To determine the relationship between wildlife and the environment and approved practices in managing wildlife enterprises.

MATERIALS LIST

- bulletin board materials
- reference materials on wildlife, wildlife management, and pollution
- Internet access

COMPETENCIES TO BE DEVELOPED

After studying this unit, you should be able to:

- define wildlife terms.
- identify characteristics of wildlife.
- describee relationships between types of wildlife.
- understand the relationships between wildlife and humans.
- describe classifications of wildlife management.
- identify approved practices in wildlife management.
- discuss the future of wildlife in the United States.

SUGGESTED CLASS ACTIVITIES

1. Conduct a wildlife survey of the area where you live. Create a master list in the classroom that is updated each day as students report sightings of different species of birds, mammals, reptiles, and amphibians. Conclude the activity by writing a press release that details the findings for a local newspaper.
2. Identify a local wildlife expert, for example, a professional who works with a fish or wildlife agency or a local bird-watcher. Invite this person to tell the class about the local species of wildlife and where they may be viewed. Encourage the expert to bring photos or other materials to illustrate the presentation.
3. A week before beginning this unit, encourage the class to find and bring in newspaper or magazine clippings that focus on wildlife issues in your area. To introduce the topic, a few students may want to share some of the articles and direct a discussion pertaining to their articles.

TERMS TO KNOW

wildlife
vertebrate
predator
prey
parasitism
mutualism
predation
commensalism
competition
wetlands

Wildlife has been part of the life of humans since the beginning of time. **Wildlife** includes animals that are adapted to live in a natural environment without the help of humans. Early humans followed herds of wild animals and killed what they needed to survive. They observed what the animals did and what they ate to determine what was safe for human consumption. Early humans also used wildlife as models for their artwork and in many of their ceremonial rites.

As settlers came to the New World and moved westward, wildlife often provided the bulk of the available food until food production systems could be developed. Supplies of wildlife seemed to be inexhaustible as the skies were blackened with the flight of millions of passenger pigeons, and herds of bison created dust storms as they migrated on the vast prairie (Figure 11-1). Unfortunately, supplies of wildlife were not and are not unlimited. Human activities have damaged or destroyed wildlife habitat (the area where a plant or animal normally lives and grows). Humans have polluted the air and water supplies, killed wildlife in tremendous numbers, and in some instances, generally disregarded the needs of wildlife. As a result, many species of wildlife in the United States now require some degree of management.

Courtesy of U.S. Fish and Wildlife Service.

FIGURE 11-1 Many wildlife populations declined in the early 1900s, but management efforts have been successful in restoring many of them, including several species of wild sheep.

Fortunately, humans also have the ability to restore wildlife habitat and manage many wildlife species successfully. The Great Plains may never sustain vast herds of roving bison, but the bison have been brought back from the brink of extinction. Populations of large game animals such as deer, elk, moose, and black bear have expanded to fill the available habitat. Many smaller species such as eagles have also experienced population growth. A few species, however, such as the whooping crane, have not responded well to management efforts. These species of wild creatures require our best efforts to accommodate their survival needs.

CHARACTERISTICS OF WILDLIFE

All **vertebrate** animals (animals with backbones), except humans, are included in the classification of wildlife. They have many of the same characteristics as humans. Growth processes, laws of heredity, and general cell structure are common to both humans and animals. When populations become too dense, disease outbreaks occur, populations suffer from starvation, and disposal of waste becomes a problem.

The wildness of an animal itself is a characteristic that allows the animal to survive without interference or help from humans. The animal's wildness often contributes to the interest that humans have in wildlife. Characteristics identified as wildness are what attract hunters to hunting and fishermen to fishing. Birdwatching and wildlife photography would be far less fascinating if wildlife were less wild and wary of humans (Figure 11-2).

With few exceptions, wildlife species live in environments over which they have no control. Wildlife must be able to adapt to whatever they are presented in terms of food and environment, or they will perish. They must also possess natural senses that allow them to avoid predators and other dangers. A **predator** is an animal that feeds on other animals. The animal being eaten by the predator is the **prey**.

The ability to avoid overpopulation is a characteristic of many groups of wildlife. Establishing and defending territories is one way that wildlife may

FIGURE 11-2 The nature of many wild animals is to avoid humans. It is human nature to be curious about elusive wild animals, but we must be careful to make sure that our interests do not intrude on their needs.

naturally avoid overpopulation. The stress of overpopulation causes some animals to slow their rate of reproduction or completely stop reproducing.

WILDLIFE RELATIONSHIPS

Every type of wildlife is part of a community of plants and animals where all individuals are dependent on others. Any attempt to manage wildlife must take into account the natural relationships that exist. This is because relationships within the wildlife community are constantly changing, and it is difficult to set standard procedures for their management.

The balance of nature is actually a myth because wildlife communities are seldom in a state of equilibrium. The numbers of various species of wildlife are constantly increasing and decreasing in response to each other and to many external factors such as natural disasters. These include fires, droughts, and disease outbreaks. Interference of humans often upsets sensitive relationships in nature. Some of the natural relationships that exist in the wildlife community include parasitism, mutualism, predation, commensalism, and competition.

Parasitism

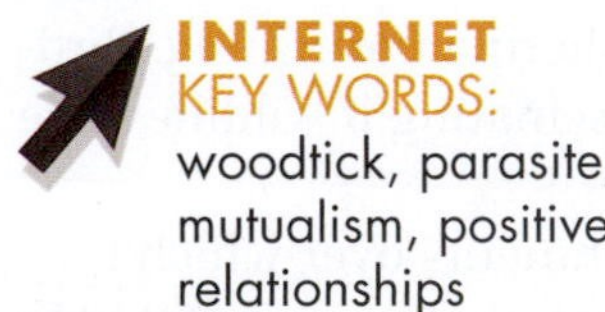
INTERNET KEY WORDS: woodtick, parasite, mutualism, positive relationships

The relationship between two organisms, either plants or animals, in which one feeds on the other without killing it is called **parasitism**. Parasites may be either internal or external. An example of a parasitic relationship is the wood tick, which lives on almost any species of warm-blooded animal. Warm-blooded animals have the ability to regulate their body temperatures.

Mutualism

Mutualism refers to two types of animals that live together for mutual benefit. There are many examples of mutualism in the wildlife community. Tick pickers are birds that remove and eat ticks from many of the wild animals in Africa, to the mutual benefit of both. The wild animals have parasites removed from them, and the birds receive nourishment from the ticks. A moth that lives only on a certain

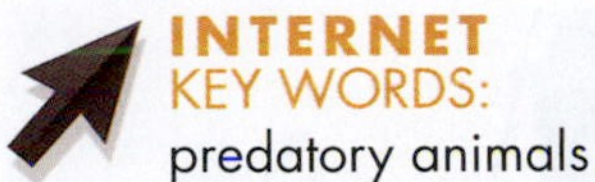

INTERNET KEY WORDS:
predatory animals

plant is also the only pollinator of that plant in several relationships. Some plant seeds will germinate only after having passed through the digestive tract of a specific bird or animal.

Predation

When one animal eats another animal, the relationship is called **predation** (Figure 11-3). Predators are often very important in controlling populations of wildlife. Foxes are necessary to keep populations of rodents and other small animals under control. Populations of predators and prey tend to fluctuate widely. When predators are in abundance, prey becomes scarce because of overfeeding. When prey becomes scarce, predators may starve or move to other areas. This permits the population of the prey species to increase again.

Commensalism

Commensalism refers to a plant or animal that lives in, on, or with another, sharing its food, but not helping or harming it. One species is helped, but the other is neither helped nor harmed. Vultures waiting to feed on the leftovers from a cougar's kill is an example of commensalism.

Competition

When different species of wildlife compete for the same food supply, cover, nesting sites, or breeding sites, **competition** exists. Competition may exist between two or more species that share the same resources. It also exists among members of the same species, especially when food or shelter is in short supply or during the mating season (Figure 11-4). When competition exists, one species may increase in number, whereas the other declines. Often, the numbers of both

Courtesy of U.S. Fish and Wildlife Service; photo by Pedro Ramirez, Jr.

FIGURE 11-3 Predatory animals play an important role in nature by keeping populations of rodents, birds, and other animals from expanding beyond the capacity of their environments to provide food and shelter for them.

Courtesy of U.S. Fish and Wildlife Service; photo by John D. Wendler.

FIGURE 11-4 Competition among wildlife species helps keep animal populations in balance. Deer have been known to battle to the death of one or both combatants.

Courtesy of Chesapeake Bay Foundation.
A

Courtesy of U.S. Fish and Wildlife Service; Photo by William Janus.
B

Courtesy of U.S. Fish and Wildlife Service.
C

FIGURE 11-5 Ecology is the branch of biology that describes relationships between living organisms and their living environments. Shown in their natural habitats are the (A) blue heron; (B) eastern cottontail; and (C) arctic hare.

species decrease as a result of competition. For example, owls and foxes compete for the available supply of rodents and other small animals.

The various relationships that exist among species of wildlife make it necessary to consider more than just one species anytime management is contemplated. Understanding the relationships that exist in the entire wildlife community is essential if wildlife management programs are to be successful.

RELATIONSHIPS BETWEEN HUMANS AND WILDLIFE

Relationships between humans and wildlife may be biological, ecological, or economic. Biological relationships exist because humans are similar to wildlife in the biological processes that control life. Relationships may be ecological because humans are but one species among nearly 1 million species of creatures that inhabit Earth, sharing its resources and environments (Figure 11-5).

Courtesy of U.S. Fish and Wildlife Service; Photo by Richard Baldes.

FIGURE 11-6 Hunting and fishing are popular recreational uses of wildlife.

The economic relationship that exists between humans and wildlife is important. Originally, humans were dependent on wildlife for food, clothing, and shelter. Today, wildlife relationships with humans have six positive values: commercial, recreational, biological, aesthetic, scientific, and social.

The harvesting and sale of wildlife and/or wildlife products is an example of the commercial relationship between humans and wildlife. Raising wild animals for use in hunting, fishing, or other purposes also falls into this category.

Hunting and fishing, as well as watching and photographing wildlife, are examples of recreational relationships (Figure 11-6). Although it is estimated that more than $2 billion are spent each year on hunting and fishing and at least another $2 billion on other recreational uses of wildlife, many of the recreational values of wildlife are intangible.

The value of the biological relationship between humans and wildlife is difficult to measure. Examples of the biological relationship include pollination of

HOT TOPICS IN AGRISCIENCE SUPPLEMENTAL FEEDING OF WILDLIFE

© Irene Pearcey/Shutterstock.com.

Supplemental feeding has made it possible for more elk to survive during the cold winter months.

Every year, as the snow starts to fall in big game country, a major controversy begins. During especially difficult winters, it is common for elk, deer, and other big game animals to die of starvation. The question of feeding the animals is debated again each time large numbers of animals are found dead or dying. In mild years when conditions are not as difficult, most big game animals survive to the next season. Wildlife managers are concerned with stabilizing animal populations. When a dramatic decline in the number of animals occurs during the winter, managers become concerned that the animals will not be able to recover on their own. Sometimes, the decision is made to feed these animals during periods of extreme food shortages. Some individuals believe that when humans become involved in this way, the animals are no longer self-sufficient, and that they are therefore less wild. This is a concern of the Fish and Game Departments in some states. It becomes a balancing act to keep wildlife wild and to keep wildlife alive. For this reason, many factors must be considered before a decision is made to provide supplemental feed for big game animals. Some of these factors include whether the animals constitute a threat to private property. Are excess animals a threat to public safety? Will excess deaths affect recovery of the animal population? Will saving the animals create a bigger problem later because of a limited or unavailable supply of food? Even though wildlife managers make every effort to limit supplemental feeding practices, the controversy continues. Will supplemental feeding take the "wild" out of "wildlife"?

Courtesy of USDA/ARS #K-4716-1.

FIGURE 11-7 Biological values of wildlife and human relationships include the pollination of crops by honeybees.

crops, soil improvement, water conservation, and control of harmful diseases and parasites (Figure 11-7).

Aesthetic value refers to beauty. Watching a butterfly sipping nectar from a flower, a fawn grazing beside its mother, or a trout rising to a hatch of mayflies are all examples of the aesthetic value of wildlife. Wildlife also provides the inspiration for much artwork. Even though the aesthetic value of wildlife is not measurable in economic terms, wildlife can contribute greatly to the mental well-being of the human race.

Using wildlife for scientific studies often benefits humans. The scientific relationship between humans and wildlife has existed from the beginning of time when early humans watched wild animals to determine which plants and berries were safe to eat.

The value of the social relationship between humans and wildlife is also difficult to measure. However, wildlife species have the ability to enhance the value of their surroundings simply by their presence (Figures 11-8 and 11-9). They provide humans the opportunity for variety in outdoor recreation, hobbies, and adventure. They also make leisure time much more enjoyable for humans.

CLASSIFICATIONS OF WILDLIFE MANAGEMENT

Wildlife management can be divided into several classifications for ease in developing management plans. Techniques for management vary tremendously according to classification. Some wildlife management classifications are farm, forest, wetlands, streams, and lakes and ponds. The management of farm wildlife is probably the most visible wildlife management classification. The development of fence rows, minimum tillage practices, improvement of

© gracious_tiger/Shutterstock.com.

FIGURE 11-8 The presence of birds can be encouraged by providing them with feeding stations.

Courtesy of U.S. Fish and Wildlife Service. Photo by Bob Ballou.

FIGURE 11-9 Wildlife sanctuaries are created to provide safe habitats for animals. They are also attractive to humans who go there to observe and enjoy the birds and other animals.

SCIENCE CONNECTION POPULATION DYNAMICS

© Christopher Kolaczan/Shutterstock.com.

Northwest wildlife managers are studying Bighorn sheep to identify causes for their declining population.

A population is a group of individuals of a species that live in a specific area. A wolf pack in Yellowstone National Park is an example of a population. "Dynamics" refers to the factors that cause changes in a population. Biologists are interested in population dynamics that affect animal behavior and the number of animals in a population. Wildlife biologists spend a lot of time observing, counting, and calculating the information they gather from studying a population. The goal of this work is to predict what changes may take place within a population over time. Factors that can affect a population's size include changes in the number of births or the number of deaths. The number of animals that move into and out of the group affects the population, as do such factors as predation, disease, and competition for resources.

Another factor that must be considered by scientists is the carrying capacity of the environment. Carrying capacity is the maximum number of individuals the resources of an ecosystem can support. This is critical information. If the carrying capacity of the area is not determined, then it would be difficult to see what effect the other factors may have. The size of the Bighorn Sheep population in the Northwest has been on the decline in recent years. Wildlife biologists have been trying to determine what has been causing the sheep to die. Through observations, counting, blood sampling, and using radio collars, they hope to identify what is responsible for the decline. Once these things are understood, they hope to be able to predict the future for the Bighorn Sheep more accurately.

petroleum remediation, bioremediation

woodlots, and controlled hunting are all techniques that have long been used to manage farm wildlife. Rabbits, quails, pheasants, doves, and deer are the types of wildlife that are normally managed in this category.

Forest wildlife is difficult to manage. Plans should be developed so that timber and wildlife can exist in populations large enough to be sustained and possibly harvested. Management of forest wildlife may include population controls to prevent destruction of habitat. Deer, grouse, squirrels, and rabbits are wildlife species that are usually included in forestry wildlife management programs (Figure 11-10).

FIGURE 11-10 Management of wildlife in forest environments is often difficult because many species of plants and animals occupy the living environment. Interactions among all of the organisms in the environment must be considered in developing the management plan.

The most productive wildlife management areas are wetlands. **Wetlands** include all areas between dry upland and open water. Marshes, swamps, and bogs are all wetland areas. Because these areas are sensitive to changes in environmental conditions, careful management of them for wildlife is essential. The wetlands provide homes to ducks, geese, beaver, muskrats, raccoons, deer, pheasants, grouse, woodcock, fish, frogs, and many other species of wildlife (Figure 11-11).

The management of running water or streams is often a difficult task. Water pollution and the need for clean water for a growing human population continue to increase at a rapid pace. Potential damage to the wildlife in streams from chemical pollution, the building of dams and roads, home construction, and the drainage of swampland are critical considerations for the stream wildlife manager.

FIGURE 11-11 Canada geese populations have benefitted from man-made wetlands that provide resting areas during migration and favorable habitats during other seasons.

Management of wildlife in lakes and ponds is normally somewhat easier than it is in streams because water is standing rather than running. Population levels of pond wildlife, oxygen levels, pollutants, and the availability of food resources are all concerns of the pond and lakes wildlife manager.

APPROVED PRACTICES IN WILDLIFE MANAGEMENT

Farm Wildlife

Management of wildlife on most farms is usually a by-product of farming or ranching. It is often given little attention by the farmer or rancher, except when wild animals cause crop damage and financial loss. However, there are many wildlife proponents in the agricultural sector. Many of them plant food plots on their property to support deer and other wildlife populations.

Much of the management of farm wildlife involves providing a suitable habitat for living, growth, and reproduction. This may involve leaving some unharvested areas in the corners of fields, planting fence rows with shrubs and grasses that provide winter feed and cover, or leaving brush piles when harvesting wood lots.

The timing of various farming operations is also important in a farm wildlife management program. Crop residues that are left standing over the winter will provide food and cover. Planting crops that are attractive to wildlife on areas that are less desirable as cropland is an excellent farm wildlife management practice. Providing water supplies for wildlife during dry periods is often necessary to maximize the numbers of farm wildlife on the area being managed (Figure 11-12).

Harvesting farm and ranch wildlife by hunting has been shown, by extensive research, to have little impact on spring breeding populations. Excess populations of farm and ranch wildlife that are not harvested by humans usually die during the winter. Even heavy hunting pressures seldom result in severe damage to wildlife populations. The sale of hunting rights to hunters is a way to increase the income of many farms and ranches. In addition, it often means the difference between profit and loss in the farming enterprise.

Courtesy of Wendy Troeger.

FIGURE 11-12 Maintaining clean water for wildlife is an important wildlife management practice.

Management of wildlife on game preserves or farms set up specifically for hunting often differs drastically from other wildlife management programs. Species of animals and birds that are not native to the area are sometimes raised and released on the preserve. It is important to note, however, that introduction of nonnative species may violate laws and regulations of state wildlife agencies, and special permits are usually required. In some instances, native wildlife species are also raised in pens and released to the farm or preserve expressly for harvest by hunters.

HOT TOPICS IN AGRISCIENCE

ENVIRONMENTAL CLEANUP: BIOTIGER™

© iStockphoto/Danny Hooks.

BioTiger, a product consisting of a collection of oil-ingesting microbes, is used to restore soil and water environments that have been contaminated with oil.

One of the remarkable scientific developments in recent years is identification of microbes for special uses. They have been developed for a variety of purposes, one of which is breaking down accidental spills of crude oil or other petroleum products. Spills from ships and tanker trucks have caused serious environmental damage, especially to marine animals. *BioTiger™* is a collection of microbes found in a century-old waste lagoon in Poland. The microbes break down oil-based materials to carbon dioxide and other harmless products, and they are becoming important tools in reducing environmental damage caused by oil spills. They can also be used to clean oil spills from driveways, contaminated soil, and construction materials. In addition, they are useful in oil production to separate oil from tar sands in the oil fields.

Forest Wildlife

The types and numbers of forest wildlife in any specific woodland are dependent on many factors. These include type and age of the trees in the forest, density of the trees, natural forest openings, types of vegetation on the forest floor, and the presence of natural predators.

Management of forest wildlife is usually geared toward establishing and maintaining stable populations of desired species of wildlife. If desired populations of wildlife are present, the management goal is usually to sustain those populations. Sometimes, numbers of certain species of forest wildlife increase to the point where destruction of habitat occurs. When this happens, control measures may have to be instituted to restore proper balance. The steps in developing a forest wildlife management plan should include taking an inventory of the types and numbers of wildlife living in the forest area to be managed. Goals for the use of the forest and the wildlife living in it need to be developed. The third step in the development of a forest wildlife management plan is determining the types and populations of wildlife that the forest area can support and how best to manage the forest so that required habitat is available.

The requirements for forest wildlife include food, water, and cover. These necessities must be readily available to the desired species of forest wildlife at all times. Management practices that meet these requirements include making clearings in the forest so that new growth will make twigs available for deer to feed on. Another practice is selective harvesting so that trees of various ages exist

AGRI-PROFILE CAREER AREAS: WILDLIFE BIOLOGIST/MANAGER/OFFICER

Courtesy of USDA/ARS #K-5213-3. Photo by Scott Bauer.

Conducting game counts and doing habitat analyses provide information for wildlife managers.

Wildlife biologists work with fish and game species living in habitats such as land, freshwater streams and lakes, tidal marshes, bays, seas, and oceans. Wildlife biologists generally have master's or doctorate degrees in biology. They use the basic sciences in their work.

Wildlife managers typically have associate's or bachelor's degrees. They work in government agencies, advising landowners and managing game populations on public lands. Their work frequently requires the use of helicopters, small planes, snowmobiles, all-terrain vehicles, horses, and Land Rovers, as well as time in the wild on foot or horseback.

Wildlife officers interact continuously with the hunting and fishing community. They advise governments in establishing fish and game laws and programs for habitat improvement. Wildlife officers have the backing of strict laws and stiff penalties for offenders. However, much of their time is spent educating the public and obtaining private assistance in improving habitats and maintaining game populations.

in the forest to make a more suitable habitat for squirrels and many other species of forest wildlife. Leaving piles of brush for food and cover is also a management practice that leads to increased production of forest wildlife. Care in managing harvests of forest products so that existing supplies of water are not contaminated is also important in good wildlife management.

Deer, grouse, squirrels, and rabbits are the forest wildlife species that are usually targeted for management because they are valuable for recreational purposes, especially hunting (Figures 11-13 and 11-14). They may also be managed to prevent the destruction of valuable forest trees and other products.

Courtesy of DeVere Burton.

FIGURE 11-13 Habitat improvements, such as developing water sources for wildlife and livestock, are usually joint projects that are developed by ranchers and forest managers.

© Alan and Sandy Carey/Photodisc/Getty Images.

FIGURE 11-14 Deer management is a key activity of fish and game agencies. Management units are studied carefully to determine the size of the deer population and to determine appropriate hunting season regulations.

Notably, during times of overpopulation of forest animals, especially deer, it is seldom a good idea to provide supplemental food. Natural losses should nearly always be allowed to occur, including starvation of excess animals or allowance of heavier-than-normal hunting pressures. Artificial feeding of wildlife populations may result in further population increases and an expansion of the problem.

Wetlands Wildlife

No area of U.S. land is more important to wildlife than the wetlands. Wetlands include any land that is poorly drained—swamps, bogs, marshes, and even shallow areas of standing water (Figure 11-15). The wetlands are constantly changing as wet areas fill in with mud and decaying vegetation. They eventually become dry land that contains forests.

Wetlands provide food, nesting sites, and cover for many species of wildlife. Ducks and geese are probably the most economically important type of wildlife that depends on the wetlands for survival (Figure 11-16). Other types of wildlife found in the wetlands include woodcock, pheasants, deer, bears, mink, muskrats, raccoons, and many other lesser-known species.

INTERNET KEY WORDS: wetlands, wildlife

Management of wetlands for wildlife may include impounding water. Open water areas should occupy about one-third of the wetlands for optimum use by wildlife. The most useful wetlands include shallow, standing water, not more than about 18 inches in depth. This allows for growth of reeds and other aquatic plants and feeding by shore birds.

The management of the plant life in the wetlands is also important. This may include cutting trees to open up the wetland area. Many species of wildlife require large, open areas in order to thrive. Care must be taken not to remove hollow trees that are used as nesting sites for some species of wildlife. Wetland areas can also be opened up by killing excess trees rather than cutting them. This provides resting areas for many types of wetlands wildlife.

Establishing open, grassy areas around wetlands and planting millet, wild rice, and other aquatic plants in the wetlands also helps attract many types of wildlife to the area.

One hazard to wildlife in wetland habitats is excessive pollution. Pollution of water flowing into the wetlands may come from agriculture, industry, or the disposal of domestic wastes. Because pollutants such as heavy metals are trapped in

Courtesy of Chesapeake Bay Foundation.

FIGURE 11-15 No area of U.S. land is more important to wildlife than the wetlands.

Courtesy of U.S. Fish and Wildlife Service.

FIGURE 11-16 Wetlands are important resting areas for migratory birds. They also provide nesting areas for millions of ducks and geese in the United States and Canada.

the mud and silt of the wetlands, the effects of these pollutants are often long-term. However, wetlands are sometimes constructed precisely for the purpose of removing such pollutants as nitrates and phosphates from the water. Wetlands tend to recycle some substances that would otherwise pollute the surface water.

In areas lacking natural nesting sites, populations of some wildlife species can be greatly increased by providing artificial nesting sites. Wood duck boxes, platforms, and islands surrounded by open water provide safe nesting sites for many species of wetlands wildlife (Figure 11-17).

Raising certain species of ducks in captivity and later releasing them in wetlands areas has helped maintain viable duck populations. This has been important as more and more natural duck nesting areas have been destroyed to meet the needs of people.

Stream Wildlife

Stream wildlife can be divided into two general categories: warm-water and cold-water wildlife. These categories are based on the water temperatures at which the wildlife, primarily fish, can best grow and thrive. There is little or no difference in

Courtesy of Cameron Waite. Photo taken by Claire Waite.

FIGURE 11-17 Artificial nesting sites are beneficial to many kinds of birds. Wood ducks would benefit from the artificial nesting site that is illustrated here.

TERMS TO KNOW

aquaculturist
gradient
amphibian
estuary
saltwater marsh
brackish water
fry
spawn
osmosis
adaptation
gills
shellfish
crustacean
molt
ppm
buffer
turbidity
ammonia/nitrite/nitrate
TAN
toxin
salmonid
ppt
larvae

Water covers three quarters of the surface of the earth. This resource produces both plants and animals that are used to feed the world. Aquaculture is the management of this and other water environments to increase the harvest of usable plant and animal products. An **aquaculturist** is a person trained in production of plants and animals in water environments. He or she must understand where and how organisms live, eat, grow, and reproduce in water. The manipulation of these factors determines the success of the aquaculture system. Aquaculture systems are becoming important to our food system because natural fisheries have reached their maximum capacities to produce more fish (Figure 12-1).

Aquaculture production systems are part of an integrated industry that requires specialized products and services. Aquaculturists include nutritionists, feed mill operators, pathologists, fish hatchery managers, processing managers, researchers, and growers. Their services are used to produce fresh and processed seafoods, freshwater fish, shellfish, and ornamental fish and plants. Natural fisheries are fish production areas that occur in nature without human intervention. They include the oceans, continental shelves, reefs, bays, lakes, and rivers. These fisheries are currently fished by sophisticated fishing fleets that are so efficient that yearly catches (yields) have leveled off or are decreasing because of insufficient supplies of fish. The population of the world continues to increase; therefore, aquaculture must be used to produce more aquatic plants and animals for food. Understanding the aquatic environment, the biology of the organisms, and how to control the production of aquatic plants and animals is essential.

THE AQUATIC ENVIRONMENT

The oceans represent the largest expanse of natural resources in the world (Figure 12-2). They are filled with water containing soluble nutrients and materials washed from the land. Over time, the evaporation of water into the atmosphere has increased the concentration of these nutrients until the salinity, or mineral content, of the water is high. We call this seawater. The concentrations of these nutrients and salts are so high that land plants and animals are unable to survive if this water is used for irrigation or drinking.

Courtesy of Chesapeake Bay Foundation.

FIGURE 12-1 Natural fisheries have reached their capacities to produce more fish.

FIGURE 12-2 The oceans represent some of our most abundant and most promising future resources.

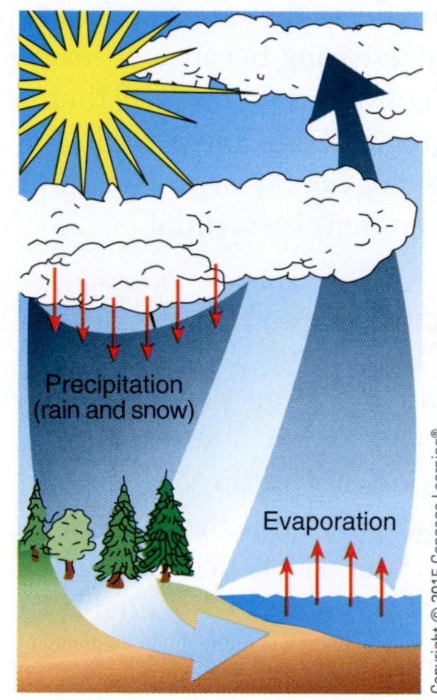

FIGURE 12-3 The natural water cycle removes water from the oceans in the form of precipitation that falls on the landmasses. The water then gradually returns to the ocean through springs, streams, lakes, and rivers.

Rain contains only small amounts of salts. Therefore, accumulation of water on land and its flow into the oceans generates a **gradient**, or measurable change over time or distance, in salinity. The water cycle is the means by which the increase in salinity has occurred in ocean water (Figure 12-3). This affects the types of organisms that can flourish in an ocean environment.

The Salinity Gradient

Freshwater wetlands, such as marshes, ponds, and streams, are generated by rainwater (Figure 12-4). Here, natural rainfall accumulates and provides aquatic environments that represent the transition between aquatic (water) and terrestrial (land) plants and animals. **Amphibians**, such as frogs, turtles, and reptiles, are animals that live part of their lives in freshwater and the remaining period on land. Several plant species, such as cattails (*Typha* sp.), watercress (*Nasturtium officinale*), water spinach (*Ipomoea reptans*), and rice, also require a transitional period of flooding and drainage to flourish.

As water flows from fields and urban areas into large streams and lakes, this runoff accumulates more soluble nutrients, and the salinity increases. The profiles of plants and animals change as other organisms that are more adapted to this changed environment displace the original residents. Flows accumulate into rivers that empty into bays, estuaries, and saltwater marshes. *Bays* are open waters along coastlines where freshwater and saltwater mix. **Estuaries** are ecological systems influenced by brackish or salty water. The **saltwater marshes** are lowland areas that are covered by seawater when the tide comes in.

Coastal areas near locations where freshwater rivers and streams flow into the ocean are different from either aquatic or terrestrial environments. In those areas, freshwater mixes with seawater to create **brackish water**, a mixture of freshwater and saltwater that fluctuates with the tide, flow of the rivers, and weather conditions. These areas provide unique growing conditions for various fish, shellfish, and aquatic plants.

Several types of seaweeds are commercially produced in seawater in East Asia. These include types of red, green, and brown algae. Similar plant growth of algae must be maintained in intensive fish systems to feed newly hatched fish called **fry**.

FIGURE 12-4 Freshwater marshes serve as transition areas where some species of animals and plants are able to adjust while they make the transition between the aquatic and terrestrial phases in their life cycles.

The migration of saltwater salmon to upstream locations to **spawn** (lay eggs) in freshwater streams illustrates the gradient effect on the life cycle of aquatic organisms. The body of the salmon must adjust from expelling salts and other minerals to retaining salts and expelling excess water that builds up in its tissues in the freshwater environment. At the same time, its body adjusts from drinking large amounts of sea water to drinking almost no fresh water. When the salmon entered the ocean as a smolt, these body processes were exactly opposite to those described earlier. Oceangoing fish and other aquatic animals are indeed able to accommodate tremendous changes in their living environments.

Saltwater and Freshwater Fish

Osmosis is the process by which water moves from an area of high concentration to an area of low concentration through a selectively permeable membrane. A selectively permeable membrane will allow some molecules to pass through, while other molecules (e.g., salts) cannot. Figure 12-5 illustrates water movement from side A (pure water), which has a greater water concentration, to side B (sugar solution), which has a lower water concentration. This happens because water flows across membranes to make both sides of the membrane equal in concentration.

The skin of a fish and the cells of its gills can act as selectively permeable membranes that control water movement. In a marine environment, water would naturally move from the inside of the fish to the outside of the fish in an attempt to dilute the saltwater outside the body. If this took place without the membranes of the fish reacting to control water loss, the fish would become dehydrated and die. Conversely, in a freshwater environment, where there is more salt in a fish's body than outside it, water would move inside in an attempt to dilute the salt concentration in the body. This would result in a waterlogged fish, which would also cause severe problems and death.

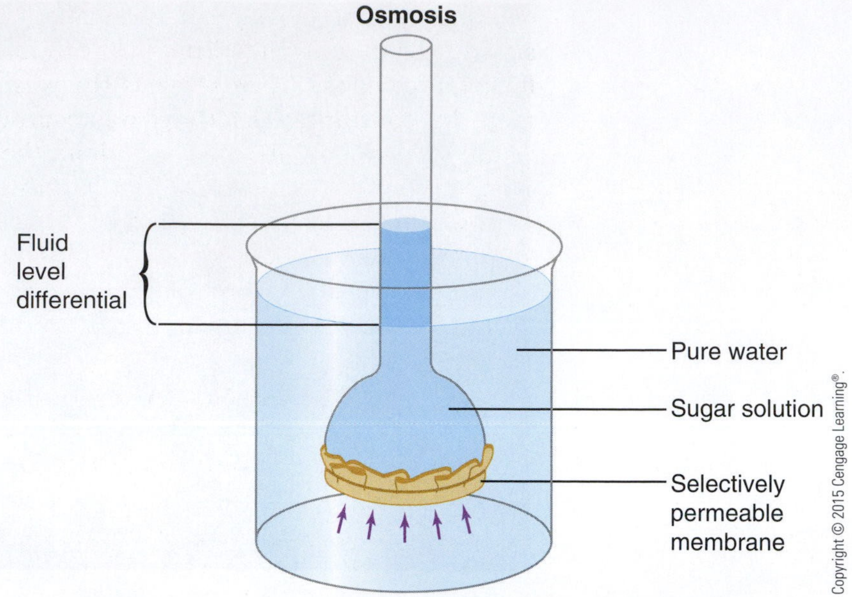

FIGURE 12-5 Osmosis is a process by which water moves across a membrane from an area of high concentration to one of lower concentration.

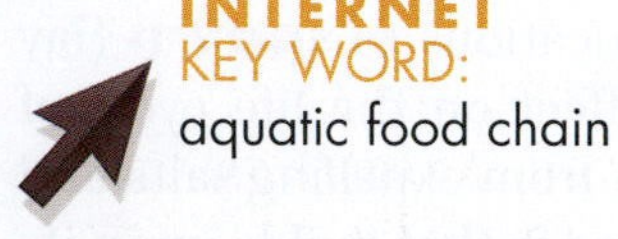

INTERNET KEY WORD: aquatic food chain

Adaptation occurs when heritable traits that increase an organism's chances for survival are passed from one generation to the next. One adaptation that fish have made over time is the ability to live in freshwater, saltwater (marine), or both environments. Marine fish drink water to make up for water loss. They also have specialized "chloride" cells that rid the body of salt. Freshwater fish, in contrast, have no need to drink water. They also have chloride cells in their gills. In freshwater fish, these cells work conversely to those in their marine relatives. Their chloride cells absorb salt from the water. They also excrete a large amount of very diluted urine. Some fish, such as salmon, have adapted even further and the function of their chloride cells changes as they move between freshwater and marine environments.

The Aquatic Food Chain

The aquatic environment constantly changes to maintain a balance of organisms that function in the food chain or system. A simple illustration is the makeup of a freshwater pond. The food chain is fueled by sunlight. Green plants and algae use this energy to grow and use nutrients they absorb from the water. These plants are eaten by fish and other animals that become, in turn, the prey of larger animals (fish, reptiles, and others) and sometimes humans. These large animals return nutrients to the water as waste, or carrion, that is reabsorbed by plants for growth. The maintenance of this food chain supports life (Figure 12-6).

The aquaculturist must understand the effect of any management activity on this cycle and make adjustments using technology or design to maintain an aquaculture production system.

FIGURE 12-6 The aquatic food chain in a freshwater pond.

General Biology

The biology of aquatic plants and animals is similar to that of terrestrial plants and animals. Both assimilate nutrients, grow, reproduce, and interact with the environment. Green plants harvest energy from the sun through photosynthesis and absorb nutrients from water to manufacture carbohydrates, proteins, fats, and cellulose. They serve as the waste recyclers in the aquatic environment by constantly absorbing waste products (nutrients) and contributing to the food chain. Like any land plant, they respond to fertilization, shading, competition, insects, disease, and weather. Aquatic plants are composed of many parts, as discussed in Unit 15, "Plant Structures and Taxonomy." Certain parts help them compete within the aquatic environment. Green plants absorb carbon dioxide from water and release oxygen during photosynthesis. During the hours of darkness, plants reabsorb a smaller amount of oxygen and release carbon dioxide.

Aquatic animals, particularly fish, complement the relationship with plants by generating carbon dioxide during respiration and releasing soluble nutrients through waste products and decay. Figure 12-7 illustrates the anatomy of a fish and shows the specialized **gills** that exchange gases by absorbing oxygen from water and releasing carbon dioxide into the water.

In this competitive environment of predator and prey, plants provide the shelter and food that are essential to the life cycles of these animals.

Shellfish are aquatic animals with a shell or shell-like extensions. Some adult shellfish are nonmotile—that is, they cannot move. They include clams, mussels, oysters, and others, and they occupy a unique niche in the aquatic environment. Located on the bottom of a body of water, these organisms have developed an efficient pumping mechanism that filters great quantities of water. It filters out edible microscopic plants and animals known as *plankton.*

Crustaceans are a group of aquatic organisms with exoskeletons. These organisms **molt**, or replace, their outer shells as they grow. Saltwater lobsters (*Homarus* sp.), crawfish (*Procambarus* sp.), and the various crabs, shrimps, and prawns are important crustaceans. These mobile organisms are characterized

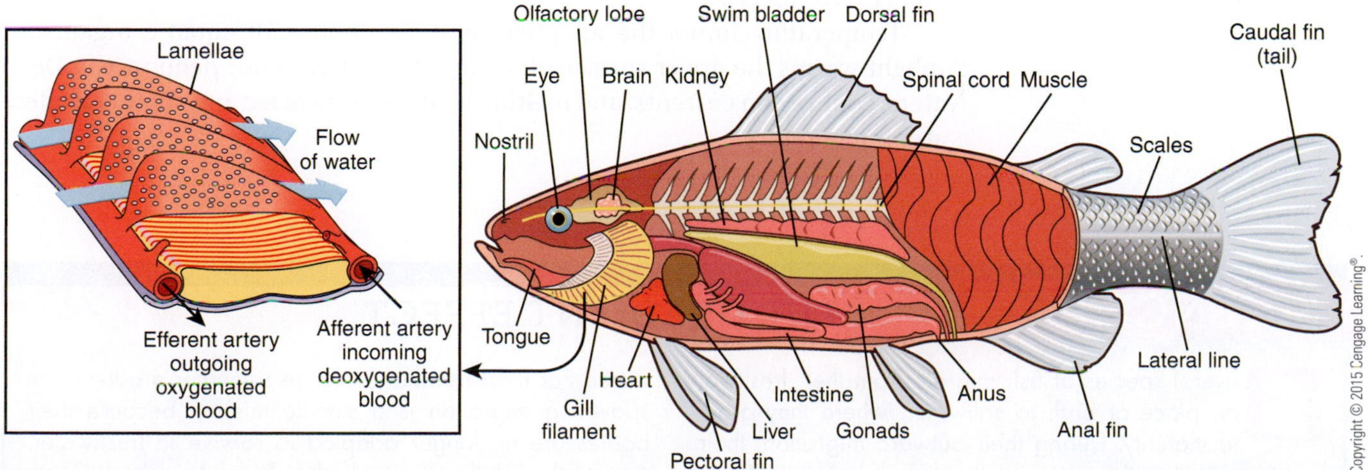

FIGURE 12-7 Fish are equipped with specialized organs such as gills and swim bladders that make it possible for them to survive in a water environment.

FIGURE 12-8 Water quality must be monitored closely to maintain an environment in which aquatic plants and animals can live.

by hard exoskeletons that must soften and split as the animals molt and secrete larger shells.

Aquaculture Production

The aquaculturist, in an attempt to increase the production of any aquatic organism, must monitor and maintain the optimum water quality. Water quality has several chemical and physical characteristics that interact within the water. These characteristics must be measured and maintained within a narrow range to promote growth and development of aquatic plants and animals (Figure 12-8).

Several different water-quality test kits are available to test the characteristics discussed in the following paragraphs.

The concentration of dissolved oxygen (oxygen in water) depends on the temperature and pressure of the water and the concentration of atmospheric oxygen. As water is cooled, the pressure increases. The more contact with atmospheric oxygen, the greater and faster oxygen can be dissolved in water. Measured by oxygen probes or chemical tests, the results are reported as 0 to 10 **ppm** (parts per million). Water at 85° F (30° C) is saturated at about 8 ppm. Most fish can survive at levels as low as 3 ppm but quickly become stressed and succumb to other problems. Rainbow trout must have excellent, or high, levels of dissolved oxygen and can be cultured only in water that is saturated with oxygen (Figures 12-9 and 12-10).

The measurement of acidity or alkalinity in water is the pH. This factor affects the toxicity of soluble nutrients in the water. This measurement is recorded, using a pH meter or litmus paper tape, as a number from 1 to 14. Readings less than 7 are acidic solutions, 7 is neutral, and numbers greater than 7 are alkaline. Most aquatic plants and animals grow best in water with a pH between 7 and 8.

Water hardness is measured by chemical analysis and is expressed as ppm calcium. This element is essential in the development of the exoskeletons of shellfish and crustaceans. It also serves as a chemical **buffer** that stabilizes rapid shifts in pH.

INTERNET KEY WORDS:
water, hardness water, turbidity water, nitrogen, TAN

Turbidity in water is caused by the presence of suspended matter. High turbidity limits photosynthesis and visibility. A simple method for estimating pond turbidity uses a white Secchi disc that is lowered into the water. When visibility is impaired, the depth is recorded. The greater the depth, the less turbid the water (Figure 12-11).

Temperature limits the adaptive range of almost all aquatic organisms. Sunlight warms the upper surface of open water but does not penetrate it. Deep waters, cool-region currents, and melting winter covers of ice and snow can affect water temperature.

SCIENCE PROFILE THE GRADIENT EFFECT

Several species of fish migrate from their freshwater place of birth to saltwater, where they grow to maturity. During their outward migration, their bodies adapt to thrive in a saltwater environment. Eventually, they migrate back to freshwater to spawn. This is necessary because the young fry must hatch in low-salinity water. For most fish that cross the saltwater gradient, the spawning migration is a suicide mission because their bodies are no longer adapted to survive in freshwater. Among the fish that live on both sides of the saltwater gradient are eels, salmon, steelhead trout, and some species of bass.

FIGURE 12-9 Oxygen content of water is increased by churning and mixing water through air.

FIGURE 12-10 Trout require high levels of dissolved oxygen and a constant environment of cold water to survive and grow in an aquaculture setting. The water environment is controlled carefully to provide adequate oxygen and to eliminate waste.

FIGURE 12-11 The turbidity of water is measured with a Secchi disc. It is lowered in the water until visibility is impaired. The depth at which this occurs is used to calculate the turbidity of the water.

Ammonia/nitrite/nitrate compose a group of nitrogen compounds generated by aquatic animals, first as urea and ammonia. These waste products are converted first to nitrite by microscopic organisms in the water, and then to nitrate. They are ultimately converted to nitrogen gas or absorbed by plants. The accumulation of both ammonia and nitrite is toxic to fish and often limits commercial production. Total ammonial nitrogen, or **TAN**, is recorded by chemical assay in ppm. This does not reflect the toxicity of the measured amount because the toxicity of ammonia is dependent on the pH. Generally, levels of TAN are maintained at less than 1 ppm.

Toxins represent a host of materials that act as poisons, adversely affecting the growth and development of aquatic plants and animals. These include agricultural chemicals, pesticides, municipal wastes, and industrial sludges. Chemical analyses are difficult and often inconclusive.

Selection of Aquaculture Crops

The actual selection of aquatic crops that may be grown is dependent on the resources and experience of the aquaculturist. Like terrestrial crops, each species of aquatic crop has a particular set of water-quality standards to ensure survival and reproduction. A discussion of a few of the well-known aquatic crops should indicate the diversity of this commercial industry. Characteristics will also vary between species. Information on ideal conditions should be requested from your local county extension office and local aquaculturists. Trout and **salmonids** are high-quality fish products in high demand (Figure 12-12). The trout flourishes in high-quality water. Dissolved oxygen must be kept at greater than 5 ppm. Salmonids also require low salinity, cool temperature(60° F/15° C), and low turbidity. The TAN must be maintained at less than 0.1 ppm.

Catfish (*Ictalurus* sp.) farming represents one of the fastest-growing aquaculture industries in the United States. Current figures project that more than 130,000 acres of ponds are producing catfish annually. Catfish thrive at 75° to 79° F (24–26° C), a pH between 6.6 and 7.5, a water hardness of 10 ppm, and dissolved oxygen greater than 4 ppm.

FIGURE 12-12 Trout are produced on fish farms in large numbers. A high-quality source of cool water is required to raise them successfully.

FIGURE 12-13 The highly prized blue crab of the Chesapeake Bay.

Crawfish (*Procambarus* sp.) thrive in freshwater lakes and streams. Good growth occurs at 70° to 84° F (21–29° C), a pH close to neutral, a water hardness of 50 to 200 ppm, salinity up to 6 **ppt** (parts per thousand), and dissolved oxygen greater than 3 ppm.

Clams, crabs, and oysters are cultivated in bays and estuaries that are subject to tidal flows (Figure 12-13). Good growing conditions vary greatly with species but include approximately 6 to 20 ppt salinity, 15° to 30° C, dissolved oxygen greater than 1 ppm, and adequate amounts of microscopic organisms. Production should improve on cultivated sites when improved spat, or young oysters, are developed for stocking.

Shrimp and prawn are cultured in brackish-water ponds and estuaries. Conditions for good growth are temperatures greater than 25° C, a salinity of 20 percent, high levels of microorganisms, and dissolved oxygen greater than 4 ppm. Hatchery techniques are complex and involve several distinct growing stages.

Production Systems

The cultivation of any aquatic organism by the aquaculturist integrates the necessary cultural requirements with existing resources. This blend has developed three general production programs.

Open ponds, rivers, and bays are stocked with natural or cultured young and are maintained with densities that are balanced with the existing ecosystem. Competing species are controlled, and natural recycling techniques are encouraged. This form of aquaculture can use both natural and constructed ponds (Figure 12-14). Some care must be taken to prevent any peaks in TAN. Over-fertilization stimulates rapid algal blooms, with high levels of dissolved oxygen during photosynthesis, but very low levels during early morning hours. The stress that results often leads to high levels of mortality. The aquaculturist must be careful to monitor incoming water for toxins and other suspended materials. Harvesting must involve draining or seining the entire production area. Seining is the removal of fish with nets.

Caged Culture

Caged culture represents a more capital-intensive program. Aquatic animals or plants are contained in a small area, and waste products are removed by the flushing action of flowing waters (Figure 12-15). Confining growing fish in floating pen

Courtesy of R. O. Parker.

FIGURE 12-14 Constructed ponds made of concrete are used by aquaculturists to produce large numbers of fish and other freshwater species in controlled water environments.

cages or shellfish on suspended float tables is a technique for managing increasing densities of aquatic organisms. As is true on a cattle feedlot, the aquaculturist confines the animals in a limited area and provides the necessary feed and cultural management. The natural ebb and flow of the water removes waste products and replenishes dissolved oxygen.

Cage culture can be designed for both natural waters and newly constructed ponds. Aquaculturists have a better idea of growth rates and can adjust feeding ratios more economically. Some growers have reported problems when a pond rolls over (i.e., changes in water quality occur suddenly during certain weather

© iStockphoto/Scott Hailstone.

FIGURE 12-15 Some aquaculture businesses raise fish and other aquatic animals in pens or cages that sit on the ocean bottom or are suspended in water.

conditions, bringing the less-oxygenated water from the lower levels of a pond or lake to the surface). Fish in cages are unable to move and can be stressed or killed. In this intensive production system, the aquaculturist must ensure adequate nutrition, disease control, predator control, and physical maintenance. The young stock must be legally caught from natural waters or produced in controlled hatcheries. Successful operations include the production of Atlantic salmon off the coast of Norway, Nova Scotia, and Maine. Hybrids of striped bass have been cultured in cages in Maryland and California. Trout have been cultured in net pens suspended in mountain streams and ponds.

INTERNET KEY WORDS:
cage fish culture production

SCIENCE PROFILE SALINITY

Salinity is the measurement of total mineral solids in water. It is measured by either electrical conductivity (ohms/cm) or is calculated against known standards and converted to ppt. As the concentration of salts rises, the conductivity reading becomes greater. A seawater standard can be made with artificial sea salts or by dissolving 29.674 g NaCl in 1 liter H_2O. Measurements of salinity usually range from 0 to 32 ppt (freshwater to saltwater).

INTERNET KEY WORDS:
aquaculture, cage culture aquaculture, hatchery

Shellfish growers in Japan and the United States have demonstrated the production of oysters, clams, and mussels on suspended float tables or ropes.

Recirculating Tanks

Many areas of the world lack sufficient water resources to maintain a viable aquaculture industry. Recirculating systems must circulate the waste water through a biological purifier and return it to the growing tank (Figure 12-16). This complicated process is similar to that in a city waste-treatment plant. The system must

AGRI-PROFILE CAREER AREAS: AQUACULTURE RESEARCHER

Courtesy of USDA/ARS #K1-5325.

Gentle pressure is applied to cause a female catfish to release her eggs for artificial fertilization.

Catfish farming and other aquaculture enterprises are big business in many nations of the world. Similarly, catfish farming is one of the fastest growing food production enterprises in the United States. Cultured seafood production is rapidly approaching the volume of that taken from natural waters.

The rapid growth in aquaculture has spurred research-and-development activities. These, in turn, have stimulated career opportunities in animal science, nutrition, genetics, physiology, aquaculture construction, facility maintenance, pollution control, fish management, harvesting, marketing, and other areas.

As productive land becomes scarce among the global resources, food production will become more concentrated on the ocean environment. Aquaculture will play an increasing role in providing high-quality food at affordable prices.

© iStockphoto/defun.

FIGURE 12-16 New technology is making fish production profitable in tanks located in sheds, warehouses, greenhouses, and other enclosures.

remove the solid fish wastes, soluble ammonia/nitrite/nitrates, and carbon dioxide, and it must replace depleted oxygen. The pH must be maintained and integrated into the biological needs of the bacteria that populate the biological filters. As in any pond, if any single parameter is ignored, the organisms become stressed and production is decreased.

SCIENCE CONNECTION WHIRLING DISEASE

Courtesy of Sascha Hallett, OSU.

A tiny parasite is responsible for a serious disease in trout and salmon called whirling disease.

A major problem for trout and salmon is a small parasite called *Myxobolus cerebralis*. This harmful organism is native to Europe and Asia, but it was accidentally transported to the United States in 1955. It was not until the early 1990s that scientists discovered that a major epidemic was occurring in the waters of many states. Both wild fish and hatchery fish were affected.

The parasite is virtually indestructible. The spores can live in water systems for as many as 30 years. Eventually, a common aquatic worm ingests these spores. Inside the worm, the second life stage develops. This is the free-floating stage that moves through the water until it comes in contact with a young fish called a fry. The parasite then attaches to the host fish and burrows into the head and spinal cartilage of the fry. Here, it begins to multiply very quickly. This puts pressure on the fry's brain, causing the fish to swim in circles. Other symptoms of this disease include a deformed jaw and spine and a tail that shrivels and turns black. The fish appear to be whirling or swimming in circles. The fry that are affected have a hard time eating and difficulty avoiding predators. "Whirling disease," as it is now called, is a concern of sportsmen, aquaculturists, fish farmers, and scientists. Whirling disease is a major focus of research in aquaculture as attempts are made to control it.

BIO-TECH CONNECTION AQUACULTURE RESEARCH IS BIG TIME!

Courtesy of USDA/ARS #K-4248-2.

Activity (left) and solids (right) tests are among the many tests needed periodically to keep fish growing in this 7400-gallon recycled water aquaculture system.

The Aquaculture Research Center in Baltimore, Maryland, was established to provide combined research facilities for the University of Maryland Center of Marine Biotechnology, the $160 million Columbus Center for Marine Sciences, and the National Aquarium in Baltimore. The center was established by converting a shipbuilding structure located beside the harbor into a state-of-the-art aquaculture research facility.

Scientists are growing high densities of valuable fish, such as striped bass, in closed systems using biological filters for cleaning the water to make the facility environmentally friendly. Tools of biotechnology are being used to acquire knowledge about native species and to develop novel solutions to problems associated with commercial finfish and shellfish culture. Studies in the environmental, hormonal, and molecular regulation of spawning will enable growers to have access to seedstocks whenever they need them, rather than being dependent on the normal spawning cycles.

The study of fish genes and hormones that control fish growth and the study of fish nutrition will decrease the time and cost of raising fish to market weight. Of primary importance is the study of devastating diseases that have nearly eliminated the commercial oyster population and certain other species in many natural waters. The worldwide demand for seafood is projected to double by 2030 (UMBC, n.d.). Yet the world's natural fisheries are already stressed beyond sustainable limits. To meet the increased demand for seafood with decreasing harvests from natural waters, it is believed that aquaculture will have to increase its output three- to fourfold over the next 25 years (UMBC, n.d.).

As a working, participatory science laboratory with public spaces for observation, the Aquaculture Research Center is an international focal point for scientists. It includes a major research center and an educational center.

Extensive research in aquaculture has been done at the University of Wisconsin with yellow perch; the University of Maryland with striped bass, trout, and tilapia; the Walt Disney World EPCOT Center with numerous species; Mississippi State University with catfish; University of Idaho with trout; and many other institutions in the United States.

Hatcheries

The development of more intensive aquaculture systems will depend on a constant supply of high-quality, young organisms (Figures 12-17A and 12-17B). The natural fisheries are threatened by overfishing, pollution, and habitat destruction. Hatcheries are investigating the parameters that affect fish breeding habits, induced spawning, and fry or larvae production. **Larvae** are mobile aquatic organisms that develop into nonmotile adults. These advances in fish and shellfish management allow the industry to develop improved breeds and hybrids to support improved production. Improved genetic lines of trout, catfish, and salmon are already commercially available.

Aquaculture and Resource Management

The demand for aquaculture products will continue to increase. This demand will stimulate a tremendous growth in commercial aquaculture. During the

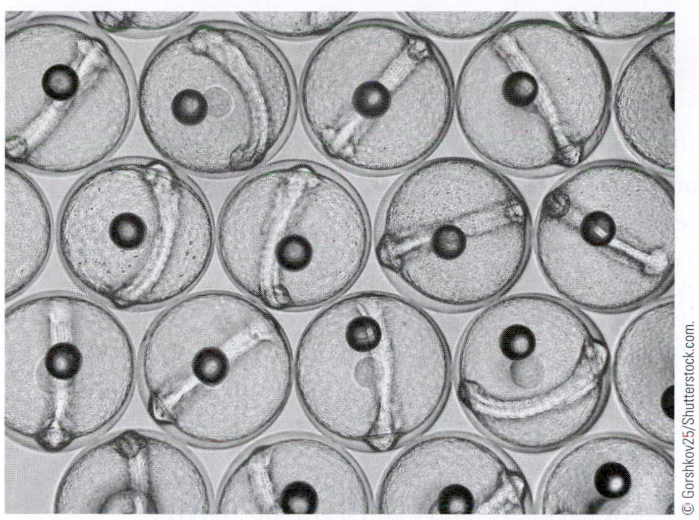
© Gorshkov25/Shutterstock.com.

FIGURE 12-17A Healthy fertilized eggs are required in order for aquatic animals such as fish to be raised in captive living environments.

Courtesy of U.S. Fish and Wildlife Service.

FIGURE 12-17B Young fry require a living environment with a constant temperature and adequate dissolved oxygen.

INTERNET KEY WORDS:
Fish hatchery practices culture

expansion of the commercial aquaculture industry, sources of clean, pure water will have to be located or developed. Conflicts for scarce natural resources and clean water, the impact on recreational areas, and the potential pollution effects will need to be resolved by trained specialists.

SCIENCE PROFILE AQUACULTURE—A SOURCE OF FISH AND SEAFOOD

It is well known that the harvest of fish and other seafood from the oceans of the world is not likely to increase greater than current production levels. At the same time, the demand for seafood such as fish and shellfish is continuing to increase. These two trends have created ideal economic conditions for expanding the aquaculture industry. The key to success will be to produce and market high-quality fish and other seafood products economically to take advantage of this opportunity. Obtaining access to enough pure water to support aquaculture production is the first challenge in establishing this kind of farming enterprise. The second big challenge is to return the water to streams and rivers without polluting them. Aquaculture appears destined to be a growth industry to support the demand for fish and other seafood products.

© leungchopan/Shutterstock.com.

© Vladislav Gajic/Shutterstock.com.

Aquaculture is the practice of raising fish in captivity. Methods have been developed for conducting aquaculture in both freshwater and saltwater environments.

STUDENT ACTIVITIES

1. Write the Terms to Know and their meanings in your notebook.
2. Visit a local supermarket or seafood market and list the seafood products. Classify them as fish, shellfish, or crustaceans; freshwater or saltwater products; or imported products.
3. Describe why some of the seafood in Activity 2 might not be produced in your area.
4. Locate three local aquaculturists and discuss their production systems. How do they maintain water quality?
5. Visit a local pet store. Describe the various parts of a freshwater aquarium. How is the water quality maintained in this recirculating system?
6. Make a bulletin board illustrating the food chain of a freshwater pond.
7. Set up a class aquarium and discuss the balance between plants and animals in the system.
8. Make saltwater and place a fresh cucumber in the water overnight. Observe any changes. In a few sentences, explain what changes you observed.
9. Research three aquaculture careers that interest you. In a short paragraph, describe the jobs and the training they require.
10. In your own words, describe the eight factors that affect water quality.

SELF-EVALUATION

A. MULTIPLE CHOICE

1. The aquaculturist must understand how aquatic organisms
   a. eat.
   b. reproduce.
   c. live.
   d. all of the above.
2. The yearly catch of fish from natural waters is
   a. increasing.
   b. holding constant or decreasing.
   c. difficult to determine.
   d. mostly catfish.
3. The highest salinity level is measured in
   a. pond water.
   b. irrigation water.
   c. creeks.
   d. ocean water.
4. The accumulation of salts in water occurs most often when
   a. water runs across agricultural land.
   b. water settles in a pond.
   c. water collects in a drainage ditch.
   d. water is lost through evaporation.
5. Brackish water is
   a. colored black.
   b. located in tidal areas.
   c. collected from small creeks and branches.
   d. mostly high in salinity (20–34 ppt).
6. Water quality is least affected by which of the following factors?
   a. fish density
   b. weather
   c. chemical runoff
   d. fish species

7. The greater the density of fish in a system, the
   a. smaller the tank.
   b. larger the fish.
   c. the more difficult the management.
   d. the greater the temperature.
8. A fish death can occur when a pond "rolls over"
   a. because of the temperature shock.
   b. because the cages sink to the bottom.
   c. because of low levels of dissolved oxygen.
   d. because the fish turn upside down.

B. MATCHING

_________	1. Salinity	a. Measured in ppm calcium
_________	2. Dissolved oxygen	b. Recorded as degrees F or C
_________	3. Turbidity	c. Measured as ppm or percent
_________	4. Temperature	d. Depth of visual penetration
_________	5. Ammonia	e. TAN
_________	6. Hardness	f. Conductivity or ppt

C. COMPLETION

1. Both plants and animals are part of the food ______.
2. The ______ are where freshwater and seawater mix.
3. The ______ of a fish absorb oxygen from the water.
4. Between 32 and 24 ppt is the salinity of ______.
5. Trout need a ______ dissolved oxygen concentration than clams to grow and mature.
6. Crawfish must ______, or break out of their exoskeletons to grow.
7. Salmon must return to ______ water to spawn and complete their life cycles.
8. Pond culture relies mostly on ______ recycling of fish waste products.
9. The recirculating production systems must treat the fish wastes with ______ filters.

SECTION 4

A BALANCING ACT!

Honeybees are a major player in the production of food, fiber, flowers, and ornamentals in the United States. We rely on them as the only method of pollinating certain plants and count on them to do some of the pollination of nearly all species of plants. Bees enter flowers of plants to gather nectar and pollen for their own food and nourishment of their young. Their service to humans and animals in pollinating plants, and thereby producing seeds and fruit, is one we cannot do without. Most plants could not reproduce and survive without producing seeds. Keeping bees safe from pesticides used to control harmful insects is always at the top of the agenda for entomologists. Honeybees are numbered among the many insects that are beneficial to humans. Though they can sting if threatened, honeybees in the United States are of the European type and are predictable. Except for the inexperienced person approaching a beehive, honeybee stings are generally a single sting by a single bee. And, except for the relatively few individuals who have life-threatening allergic reactions to bee stings, honeybees pose little threat because they act individually when they are aggravated enough to sting.

Enter the Africanized honeybee—a hybrid cousin of the domestic pollinator, famous for its aggressiveness and dubbed "killer bee"!

Africanized honeybees resulted when bees were imported from Africa to Brazil by a Brazilian scientist in 1957. The plan was to experiment by crossbreeding them with the domestic European bees prevalent in the Americas to develop a better strain of honeybees for the tropics. Unfortunately, some African bees were inadvertently released in the countryside and promptly interbred with the domestic bees. The new hybrids and their descendents are known as Africanized honeybees. They have migrated as far south as Argentina and as far north as the United States.

On October 15, 1990, the first Africanized honeybee swarm to migrate naturally to the United States was identified by entomologists near Brownsville, Texas. The swarm was promptly destroyed using standard procedure. By 2004, Africanized bees had been detected and verified in at least 10 states. See nationalatlas.gov for more information about Africanized honeybees.

Unfortunately, the dispositions of Africanized honeybees are different from the domestic bees of the United States. They tend to defend their colonies more vigorously, stinging in greater numbers and with less provocation. One bee is likely to inflict many stings. Therefore, there is greater danger in an encounter with the Africanized bees. U.S. agriculture and the beekeeping industry fear that domestic bees interbred with Africanized bees may become harder to manage as pollinators of crops and may not be as efficient as honey producers.

The challenge of observing, detecting, and stopping the northward migration of Africanized honeybees will be a top priority of government inspectors, entomologists, beekeepers, farmers, and citizens at large. At the same time, animal behaviorists will study the bees' habits and look for ways to manage them. Geneticists will study the bees' genes and look for ways to genetically engineer future bees so as to decrease their objectionable habits and enhance their abilities as pollinators and honey makers.

Courtesy of USDA/ARS #K3653-12.

Honeybees are friends we cannot do without. However, the Africanized hybrids can be dangerous and threaten to decrease bee productivity in the United States.

UNIT 13

Biological, Cultural, and Chemical Control of Pests

OBJECTIVE

To develop an understanding of the major pest groups and some elements of effective pest management programs.

MATERIALS LIST

- insect net
- killing jar
- insect-mounting pins
- insect specimen labels
- pictures of pests
- Internet access

COMPETENCIES TO BE DEVELOPED

After studying this unit, you should be able to:

- define pest, disease, insect, weed, biological, cultural, chemical, and other terms associated with integrated pest management.
- understand how the major pest groups adversely affect agriscience activities.
- describe weeds based on their life cycles.
- describe both the beneficial and detrimental roles that insects play.
- recognize the major components and the causal agents of disease.
- understand and explain the concept of integrated pest management.

SUGGESTED CLASS ACTIVITIES

1. Contact the agricultural extension office that serves your county or parish to obtain literature about pests that are problems in the area near your school. Ask the agent to talk with your class about the best way to control the pests that have been identified. Make a chart with the following elements: pest name, kind of damage, and methods of control.
2. Design and construct a public display that tells the story of integrated pest management. Prepare teams of students to explain this concept for controlling pests, and help them set appointments to present what they have learned to civic clubs, community groups, and other interested parties.
3. Have each student bring five weeds, including the roots, to class. Spend a class period identifying the weeds using field guides. Discuss the type of damage each of the weeds is known to inflict on the environments of other plants, animals, and humans.

TERMS TO KNOW

vector
arachnid
pathogen
annual weed
biennial weed
perennial weed
rhizome
node
stolon
meristematic tissue
noxious weed
exoskeleton
metamorphosis
instars
entomophagous
disease triangle
abiotic (nonliving) disease
biotic disease
fungi
hyphae
mycelium
bacteria
virus
mosaic
nematode
economic threshold level
quarantine
targeted pest
eradication
pheromone
cultivar
biological control
cultural control
chemical control
pesticide resistance
pest resurgence

The ability to control pests by chemical, cultural, or natural control methods has afforded people in the United States an unprecedented standard of living. We often take for granted an unlimited food supply, good health, a stable economy, and an aesthetically pleasing environment. Without effective pest control strategies, our standard of living would decrease.

Good pest management practices have resulted in dramatic yield increases for every major crop. A single U.S. farmer in 1850 could only support himself and four people; currently, a farmer can provide food and fiber for approximately 167 people. The ability to control plant and animal diseases or disorders vectored by insects and arachnids has reduced the incidence of malaria, typhus, West Nile virus, and Rocky Mountain spotted fever. A **vector** is a living organism that transmits or carries a disease organism. An insect is a six-legged animal, such as a mosquito, with three body segments. An **arachnid** is an eight-legged animal, such as a spider or a mite.

The impact of pest management in maintaining a stable economy can be seen on a regional and national basis. The regional economy suffered shortly after the cotton boll weevil's introduction into the United States in 1892. The weevil devastated much of the cotton crop in the early 1900s (Figure 13-1). Similarly, the potato blight disease in Ireland caused famine and mass migration of Irish people to other parts of the world in 1845. Today, blights are still serious threats to our crops (Figure 13-2).

TYPES OF PESTS

The word *pest* is a general name for any organism that may adversely affect human activities. We may think of an agricultural pest as one that competes with crops for nutrients and water, tends to defoliate plants by eating the leaves, or carries and transmits plant or animal diseases. The major agricultural pests include weeds, insects, nematodes, and plant diseases. However, other types of pests exist. Some examples, and the classes of pesticides or chemicals used for killing them, are included in the following list:

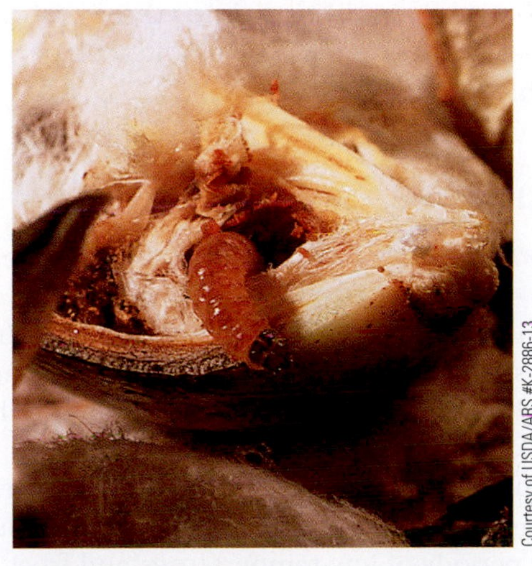
Courtesy of USDA/ARS #K-2886-13.

FIGURE 13-1 A sliced-open cotton boll showing a pink bollworm and the damage it has done.

Courtesy of USDA/ARS #K-5300-1.

FIGURE 13-2 Pear fruits yellowed and shriveled when the fire blight disease cut off the flow of nutrients from the tree to the fruit.

Type of Pest	Class of Pesticide
mites, ticks	acaricide
birds	avicide
fungi	fungicide
weeds	herbicide
insects	insecticide
nematodes	nematacide
rodents	rodenticide

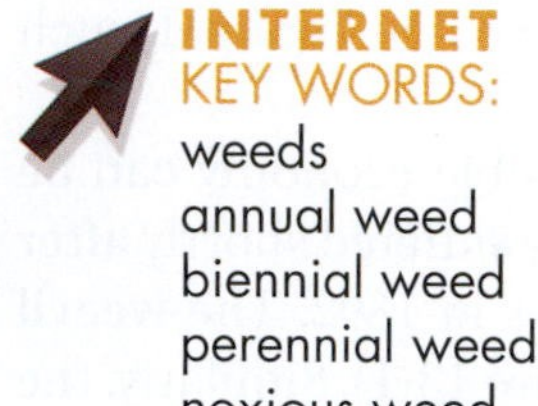

Damage by pests to agricultural crops in the United States has been estimated to be one-third of the total crop-production potential. Therefore, an understanding of the major pest groups and their biology is required to ensure success in reducing crop losses caused by pests.

Weeds

Weeds are plants that are considered to be growing out of place (Figure 13-3). Such plants are undesirable because they interfere with plants grown for crops. The word *weed* is therefore a relative term. Corn plants growing in a soybean field or white clover growing in a field of turfgrass are examples of weeds, just as crabgrass is considered a weed when it grows in a yard or garden.

Weeds can be considered undesirable for any of the following reasons:

- They compete for water, nutrients, light, and space, resulting in reduced crop yields.
- They decrease crop quality.
- They reduce aesthetic value.
- They interfere with maintenance along rights-of-way.
- They harbor insects and disease **pathogens** (organisms that cause disease).

Weeds can be divided into three categories—annual, biennial, and perennial—based on their life spans and their periods of vegetative and reproductive growth.

Annual Weeds

An **annual weed** is a plant that completes its life cycle within 1 year (Figure 13-4). Two types of annual weeds occur, depending on the time of year they germinate. A winter annual germinates in the fall and actively grows until late spring. It will then produce seed and die during periods of heat and drought stress. Examples of winter annuals are chickweed, henbit, and yellow rocket.

A summer annual germinates in the late spring, with vigorous growth during the summer months. Seeds are produced by late summer, and the plant will die during periods of low temperatures and frost. Examples of summer annuals are crabgrass, spotted spurge, and fall panicum.

© Darla Hallmark/Shutterstock.com.

A Crabgrass (Summer Annual)

© Cheryl Casey/Shutterstock.com.

B Johnson Grass (Perennial)

© Rob Byron/Shutterstock.com.

C Henbit (Winter Annual)

© Martin Fowler/Shutterstock.com.

D Mullein (Biennial)

FIGURE 13-3 A plant growing out of place is considered to be a weed. This illustration identifies different types of plants considered to be weeds.

© Karin Hildebrand Lau/Shutterstock.com.

FIGURE 13-4 An annual weed is a plant that completes its life cycle within 1 year. Yellow mustard is a common example of this kind of weed.

Biennial Weeds

A **biennial weed** is a plant that will live for 2 years (Figure 13-5). In the first year, the plant produces only vegetative growth, such as leaf, stem, and root tissue. By the end of the second year, the plant will produce flowers and seeds. This is referred to as reproductive growth. After the seed is produced, the plant will die. Only a few plants are considered biennials. Some examples are bull thistle, burdock, and wild carrot.

Perennial Weeds

A **perennial weed** can live for more than 2 years and may reproduce by seed and/or vegetative growth (Figure 13-6). By producing rhizomes, stolons, and an extensive rootstock, perennial plants reproduce vegetatively. A **rhizome** is a stem that runs underground and gives rise to new plants at each joint, or **node**. A **stolon** is a stem that runs on the surface of the ground and gives rise to new plants at each node. These plant parts have **meristematic tissue** (tissue capable of starting new plant growth). Examples of perennial weeds are dandelion, Bermuda grass, Canada thistle, and nutsedge.

Courtesy of DeVere Burton.

FIGURE 13-5 A biennial weed is a plant that completes its life cycle in 2 years. The bull thistle is a widespread example of this type of weed.

© ArjaKo's/Shutterstock.com.

FIGURE 13-6 A perennial weed can live for more than 2 years and may reproduce by seed and/or vegetative growth. An example of a perennial weed is hoary cress, also known as white top.

SCIENCE CONNECTION

A THORN IN YOUR SOCK

Courtesy of DeVere Burton.

Cheat grass has a competitive advantage over most other plants in its native range because it germinates early and uses up much of the available moisture before other plants begin to compete for it.

Bromus tectorum is a noxious weed that was introduced to the United States from Europe in the 1850s. It has subsequently invaded every state in the country. Early pioneers nicknamed the weed *cheatgrass* because they believed the weed was cheating them out of greater wheat yields. Other common names include downy brome, downy chess, bronco grass, cheat, and 6-weeks grass. In early spring, range animals take advantage of it as a nutritious source of food. Once the plant dies in late spring, its prickly seeds become a potential danger to range animals by irritating the mouth and throat, resulting in sliver-like painful sores.

B. tectorum is a successful weed. A single plant can produce up to 5000 seeds under favorable growing conditions. Wind, animal fur, and human clothing easily transport the seeds. Most people who spend time outdoors have collected these seeds in their shoes and socks. Cheatgrass germinates earlier than most plants and quickly uses most of the available water in the soil. When other plants begin to germinate, they often die because of lack of water. This is why cheatgrass is so harmful to crops. Cheatgrass and other noxious weeds must be aggressively dealt with using proven integrated pest management techniques.

Noxious Weeds

A **noxious weed** is a plant that causes great harm to other organisms by weakening those around it. Most states have developed lists of noxious weeds, and great effort is directed to control or eradicate them. Most noxious weeds are difficult to control, and they require extended periods of treatment followed by close monitoring. Noxious weeds should be handled carefully to avoid spreading seeds to unaffected areas. Noxious weeds are often spread when seeds become airborne, fall into flowing streams, become attached to the hair of an animal or to human clothing, or are eaten and distributed intact by birds.

Insects

INTERNET KEY WORDS:
cotton boll weevil insects
beneficial, useful insects
harmful, destructive pests

Insects have successfully adapted to nearly every environment on the earth. There are more species of insects than any other class of organism. Part of their success is due to the large numbers of offspring they are capable of producing and the short time they require to reach physical maturity. The human race is dependent on insects in many ways, and insects provide great service to us. Some of the most beneficial insects are the ladybug, praying mantis, parasitic wasps, and honeybees. Insects also cause great losses to crops, livestock, and people by injuring them or infecting them with diseases or parasites.

Insect Pests

When compared with the total number of insect species, relatively few species cause economic loss. However, it is estimated that in the Florida citrus industry alone, pest control costs $300 million annually and makes up approximately 40 percent of the cost of production.

Insects can cause economic loss by feeding on forests, cultivated crops, and stored products (Figure 13-7). They also vector plant and animal diseases, inflict painful stings or bites, and act as nuisance pests.

Insect Anatomy

Insects are considered to be one of the most successful groups of animals present on Earth. Their success in numbers and species is attributed to several characteristics, including their anatomy, reproductive potential, and developmental diversity.

Insects are in the class Insecta and are characterized by the following similarities (Figure 13-8):

- Each insect has an **exoskeleton**, which is the body wall of the insect. It provides protection and support for the insect.
- The exoskeleton is divided into three regions: head, thorax, and abdomen.
- The segmented appendages on the head are called antennae and act as sensory organs.
- Three pairs of legs are attached to the thorax of the body.
- Wings are present (one or two pairs) in the majority of species. This permits mobility and greater use of habitat.

Courtesy of Boise National Forest.

FIGURE 13-7 Some insects injure or kill plants by feeding on the leaves and stems. Such insects are called defoliators.

Feeding Damage

Insects have either chewing or sucking mouthparts. Damage symptoms caused by chewing insects are leaf defoliation, leaf mining, stem boring, and root feeding. Insects with sucking mouth parts produce distorted plant growth, leaf spotting, and leaf burn (Figure 13-9).

Development

As an insect matures from an egg to an adult, it passes through several growth stages. This growth process is known as **metamorphosis**. The two types of metamorphosis are gradual and complete.

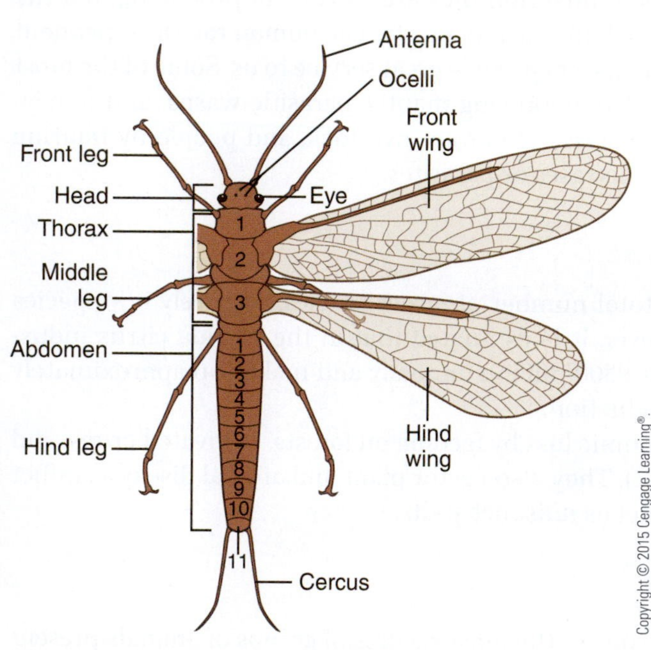

FIGURE 13-8 Diagram of an adult insect.

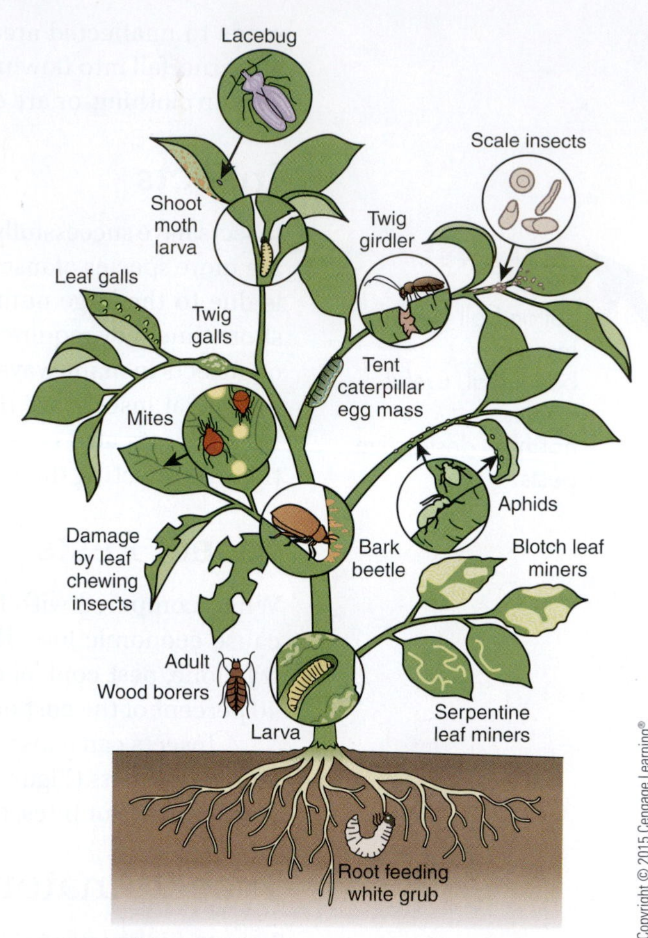

FIGURE 13-9 The different types of insect damage.

INTERNET KEY WORDS:
insect metamorphosis

Gradual metamorphosis consists of three life stages: egg, nymph, and adult (Figure 13-10). As a nymph, the insect will grow and pass through several **instars** (the stage of the insect between molts). Each time the insect sheds its exoskeleton, or molts, it passes into the next instar phase. For example, chinch bugs have five instars before they reach adult form but will vary in size, color, wing formation, and reproductive ability. When the insect reaches the adult stage, no further growth will occur.

SCIENCE PROFILE BENEFICIAL INSECTS

Honeybees' service to humans and animals in pollinating plants is a service we cannot do without.

Scientists estimate that more than 1 million species of insects inhabit the Earth. A majority of them are beneficial, or helpful, to humans. For example, insects are necessary for plant pollination. According to the Natural Resources Defense Council (March 2011), in the United States, it is estimated that bees pollinate more than $15 billion worth of fruit, vegetable, and legume crops per year. Honey, beeswax, shellac, silk, and dyes are just a few of the commercial products produced by insects. Many insects are entomophagous and help in the natural control of their insect species. **Entomophagous** insects feed on other insects. Insects that inhabit the soil, act as scavengers, or feed on undesirable plants all play important roles. These insects increase soil tilth, contribute to nutrient recycling, and act as biological weed-control agents. Insects are at the lower levels of the food chain. Thus, they support higher life forms, such as fish, birds, animals, and humans.

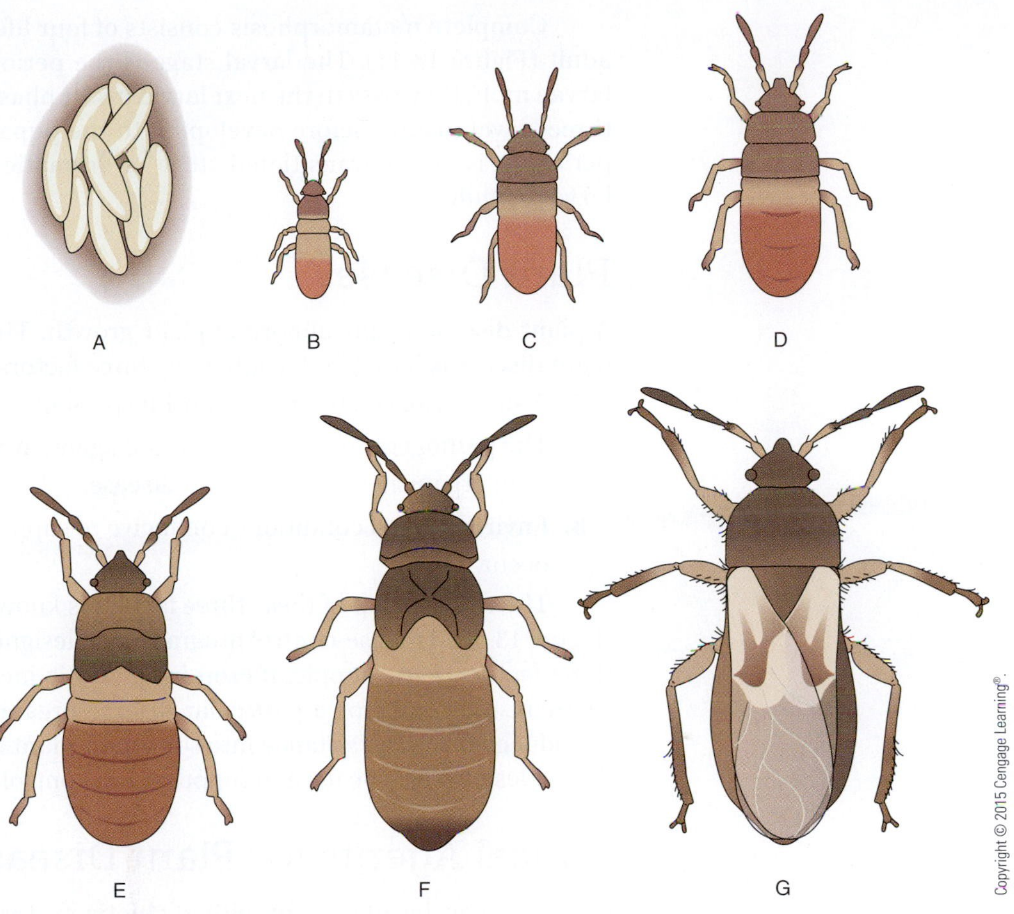

FIGURE 13-10 Gradual metamorphosis of the chinch bug: (A) egg, (B–F) first to fifth instars, and (G) adult.

AGRI-PROFILE CAREER AREA: ENTOMOLOGY/PLANT PATHOLOGY

Plant pathologists observe a tree damaged by fire blight disease (foreground) and healthy, fire blight–resistant trees (background).

The work of entomologists and plant pathologists is never done. They do battle in the laboratory and in the field against insects and diseases that consume or ruin much of what we produce. The advice and service of a specialist is sought to control diseases and damaging insects, as well as to encourage beneficial ones.

Entomologists and plant pathologists attempt to control or reduce the buildup of damaging insect and disease populations. Such work may include assessing damage; attracting, trapping, counting, and observing insects; and advising, directing, and assisting those who attempt to control insects and plant diseases. The mysteries of some insects are so great that scientists must specialize in just a few insects to be truly knowledgeable about them. Chemicals no longer can be our only means of controlling insects and diseases. Rather, we now use a variety of techniques collectively known as "integrated pest management."

Career opportunities exist for field and laboratory technicians as well as for degree-holding specialists. Neighborhood jobs may include termite control, scouting, spraying, crop dusting, inspecting, monitoring, selling, and managing field research projects. Honeybee specialists may manage hives to pollinate crops for improved seed and fruit production.

Complete metamorphosis consists of four life stages: egg, larva, pupa, and adult (Figure 13-11). The larval stage is the period when the insect grows. As larvae molt, they pass to the next larval instar phase. A Japanese beetle will have three larval instars before developing to the pupa stage. The pupa is a resting period. It is also a transitional stage of dramatic morphological change from larva to adult.

Plant Diseases

A plant disease is any abnormal plant growth. The occurrence and severity of plant disease is based on the following three factors:

1. A susceptible plant or host must be present.
2. The pathogenic organism, or causal agent, must be present. A causal agent is an organism that produces a disease.
3. Environmental conditions conducive to support of the causal agent must occur.

The relationship of these three factors is known as the **disease triangle** (Figure 13-12). Disease-control programs are designed to affect each or all of these three factors. For example, if crop irrigation is increased, a less favorable environment may exist for a particular disease organism. Breeding programs have introduced disease resistance into some new plant lines for many different crops. Pesticides may also be used to suppress and control disease organisms.

Causal Agents for Plant Disease

Diseases may be incited by either abiotic factors or biotic agents. **Abiotic (nonliving) diseases** are caused by environmental or man-made stress. Examples of abiotic diseases are nutrient deficiencies, salt damage, air pollution, chemical damage, and temperature and moisture extremes.

Biotic means living. **Biotic diseases** are caused by living organisms (Figure 13-13). Examples of causal agents or organisms are fungi, bacteria, viruses, nematodes, and parasitic plants. Organisms are parasites if they derive their nutrients from other living organisms. Examples and discussions of causal agents for plant diseases follow.

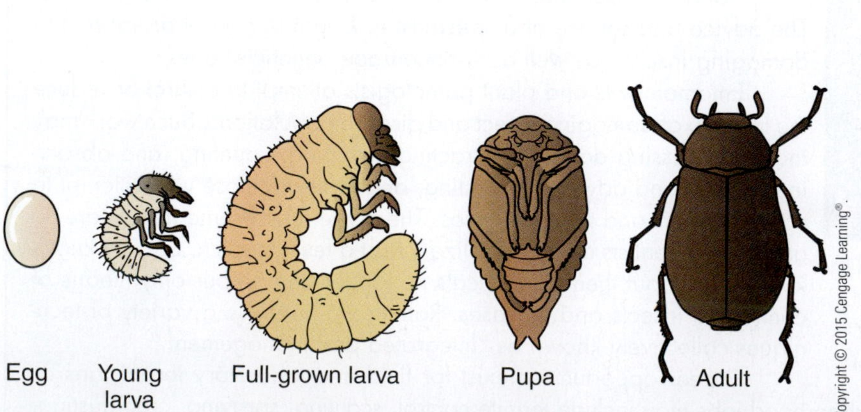

FIGURE 13-11 Complete metamorphosis of the June beetle.

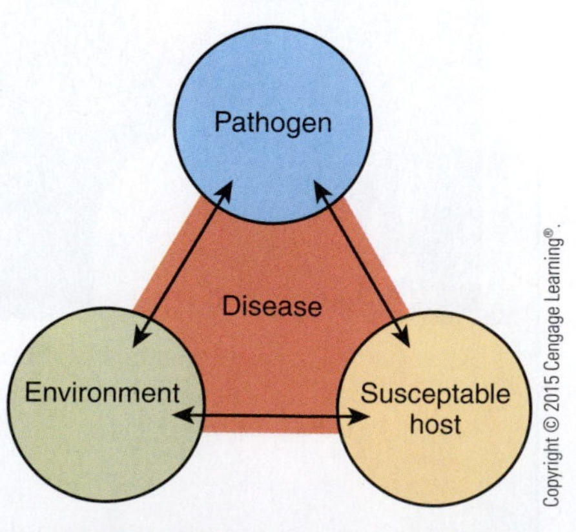

FIGURE 13-12 Components of the disease triangle.

Courtesy of DeVere Burton.

FIGURE 13-13 Biotic diseases occur when living organisms such as fungi, bacteria, viruses, nematodes, and others cause damage to growing plants. In this instance, the roots have been killed.

© Copit/Shutterstock.com.

FIGURE 13-14 Powdery Mildew is a fungal disease that spreads from plant to plant through airborne spores.

Fungi

Fungi (plural for fungus) are the principal causes of plant diseases. **Fungi** are plants that lack chlorophyll. Their bodies consist of threadlike vegetative structures known as **hyphae** (Figure 13-14). When hyphae are grouped together, they are called **mycelium**. Fungi can reproduce and cause disease by producing spores or mycelia. Spores can be produced asexually or sexually by the fungus. For example, a mushroom produces millions of sexual spores under its cap. These spores can be dispersed by wind, water, insects, and humans.

Bacteria

Bacteria are one-celled or unicellular microscopic plants. Relatively few bacteria are considered plant pathogens. Being unicellular, bacteria are among the smallest living organisms. Bacteria can enter a plant only through wounds or natural openings. Bacteria can be scattered in ways similar to fungi. Some important bacterial diseases are fire blight of apples and pears and bacterial soft rot of vegetables.

Viruses

Plant viruses are pathogenic, or disease-causing, organisms. **Viruses** are composed of nucleic acids surrounded by protein sheaths. They are capable of altering a plant's metabolism by affecting protein synthesis. Plant viruses are transmitted by seeds, insects, nematodes, fungi, grafting, and mechanical means, including sap contact. Viral diseases produce several well-known symptoms. A symptom is the visible change to the host caused by a disease. These symptoms are ring spots, stunting, malformations, and mosaics. A **mosaic** symptom is a light- and dark-green leaf pattern.

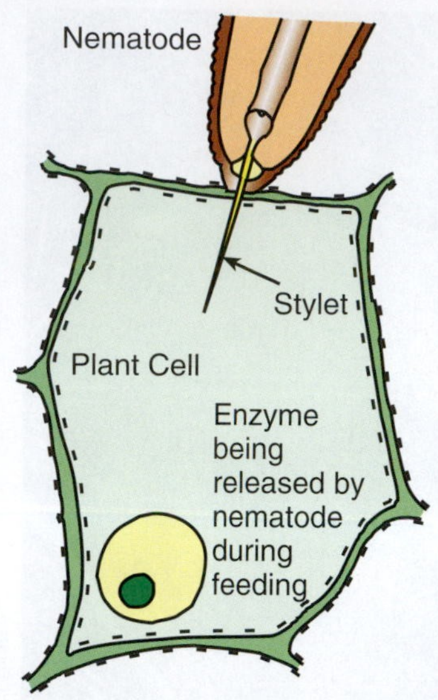

FIGURE 13-15 A nematode feeding on a plant cell. The stylet is a tiny needle-like feeding structure.

Nematodes

Nematodes are tiny roundworms that live in the soil or water, within insects, or as parasites of plants or animals. Plant parasitic nematodes are quite small, often less than a quarter inch (4 mm). They produce damage to plants by feeding on stem or leaf tissue (Figure 13-15). The main symptom of nematode damage is poor plant growth, resulting from nematodes feeding on the roots. The major plant parasitic nematodes are included in one of three groups: root-knot, stunt, or root-lesion.

INTEGRATED PEST MANAGEMENT

History

Integrated pest management (IPM) is a pest-control strategy that relies on multiple control practices. It establishes the amount of damage that will be tolerated before control actions are taken. The concept of integrated control is not new. Entomologists had developed an array of cultural and natural controls for the boll weevil and other insect pests by the early 1900s.

However, our approach to pest management during the period from 1940 to 1972 moved to a major reliance on chemical pesticides. Alternate control strategies were deemphasized because chemical control gave excellent results at a low cost.

It was not until 1972 that a major change in policy occurred in the United States to encourage other pest-control strategies. Natural, biological, and cultural control programs began to be introduced as alternatives to chemical pest control.

The more recent trend toward reduced use of chemicals for pest control was triggered by a book published in 1962 entitled *Silent Spring*, by biologist Rachel Carson. After 1962, heavy reliance on chemical pest management began to be questioned. Carson's book created a public awareness of the environmental pollution that results from the overuse of pesticides. Adverse effects from misuse and/or overuse of pesticides were beginning to occur as well. These effects included pest resurgence, resistance to pesticides, and concern over human health from exposure to pesticides.

INTERNET KEY WORDS: integrated pest management

Since the 1970s, great strides have been made in the development and implementation of IPM programs. The end result has been to reduce dependency on chemical use, while still achieving acceptable pest control.

Principles and Concepts of Integrated Pest Management

The following concepts or principles are important in understanding how IPM programs should function.

Key Pests

A key pest is one that occurs on a regular basis for a given crop (Figure 13-16). It is important to be able to identify key pests and to know their biological characteristics (Figures 13-17 and 13-18). The weak link in each pest's biology must be found if management of the pest is to be successful.

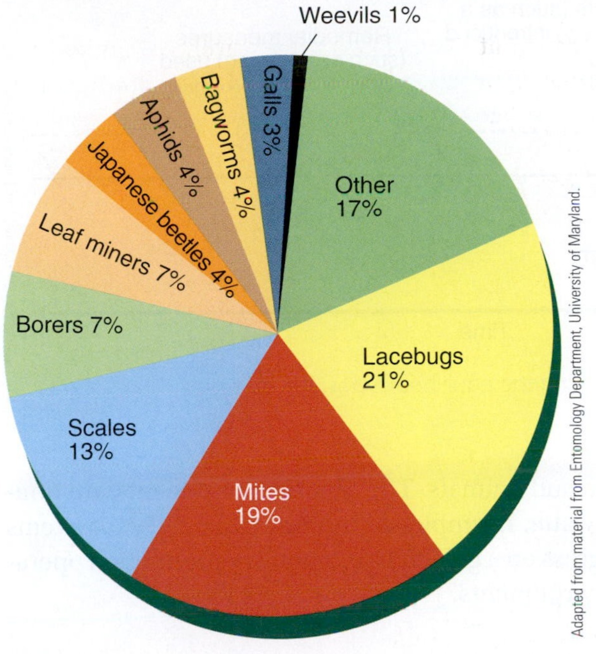

Adapted from material from Entomology Department, University of Maryland.

FIGURE 13-16 Arthropod pests and their percentages in six Maryland communities.

Courtesy of DeVere Burton.

FIGURE 13-17 Grasshoppers, locusts, and crickets are capable of crop destruction to such an extent that resulting famines have afflicted people throughout recorded history.

FIGURE 13-18 Vast forests are afflicted with severe insect damage throughout North America. Insects account for the death of many trees every year, and they contribute to the devastating fires that occur in unhealthy forests.

Damage to crops:	weevils	aphids
	beetles	leafhoppers
	grasshoppers	crickets
	moths	locusts
	bugs	cockroaches
	nematodes	
Damage to trees:	sawflies	cone beetles
	cone maggots	seed bugs
	cone worms	cone borers
	leafminers	needleminers
	moths	mites
	aphids	psyllids
	carpenter worms	termites
Damage to animals and people:	flies	mosquitos
	ticks	keds
	lice	mites
	fleas	Screwworms

Crop and Biology Ecosystem

The integrated pest manager must learn the biology of the crop and its ecosystem. The ecosystem of the crop consists of the biotic and abiotic influences in the living environment of the crop. The biotic components of the ecosystem are the

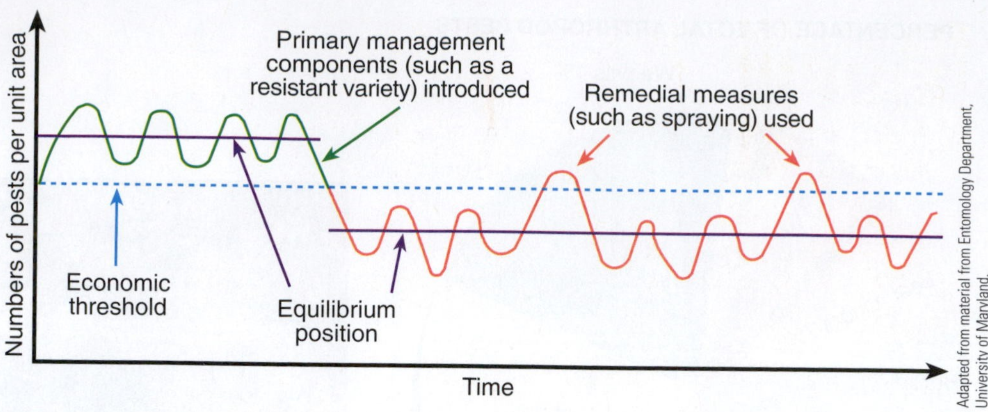

FIGURE 13-19 The effect of lowering the equilibrium position of a pest.

living organisms, such as plants and animals. The abiotic components are nonliving factors, such as soil and water. Examples of human-managed ecosystems include a field of soybeans, turfgrass on a golf course, a poultry production operation, and many other similar environments.

Ecosystem Manipulation

With IPM, the attempt is made to understand the influence of ecosystem manipulation on reducing pest populations (Figure 13-19). To illustrate this concept, the manager must ask, "What would happen to the pest population equilibrium if a disease-resistant plant were introduced?" Pest population equilibrium occurs when the number of pests stabilizes or remains steady. The introduction of disease-resistant plants should decrease the pest population to less than the economic threshold level. The **economic threshold level** is the point where pest damage is great enough to justify the cost of additional pest-control measures. Until the pest population increases to a high enough level that the cost of controlling the pest is less than the cost of the losses that the pest causes, no control actions are taken.

Threshold Levels

The level of a pest population is important. For instance, the mere presence of a pest may not warrant any control measures. But, at some point, the damage created by insects may be great enough to warrant control measures. Various threshold levels are developed to determine if and when a control measure should be implemented. This prevents excessive economic loss of plants to pest damage, while minimizing the use of pesticides (Figure 13-20). Economic threshold levels are determined by first developing a pest-damage index (Figure 13-21). It is crucial in the decision-making process to know the level of pest infestation that will cause a given yield reduction. Pest populations are measured in several different ways. They can be counted in number of pests per plant or plant part, number of pests per crop row, or number of pests per sweep with a net above the crop.

Monitoring

For IPM to be successful, a monitoring (checking) or scouting procedure must be performed. Different sampling procedures have been developed for various crops

Economic Injury Threshold for Alfalfa Weevil; Number of Larvae from 30-Stem Sample

How to use table below:

1. Use plant height category that fits the field.
2. Estimate the value of crop in dollars per ton of hay equivalent and the cost to spray an acre.
3. From monitoring the field, find the number of alfalfa weevil larvae from a sample of 30 stems.
4. The number in each small box indicates the number of larvae per 30-stem sample that is required before a spray application would be profitable under these conditions.

EXAMPLE:

Plants in the field are 20 inches high (use Category II), hay is valued at $80 per ton, cost to spray is $8.00 per acre, and you collected 40 larvae from the sample of 30 stems. The number in the box common to $80 and $8 is 75. This means that under these conditions, 75 larvae are needed before a spray would be profitable. Since you collected only 40 larvae, a spray at this time will not be profitable.

Category I plant height 12 to 18 inches

Value of hay per ton	$8	10	12	14	16	20
$ 60	91	114	137	160	183	225
$ 80	68	85	102	119	136	171
$100	54	68	81	95	108	137
$120	45	57	68	79	91	114
$140	39	49	59	68	77	99
$160	34	43	51	60	68	86

Category II plant height 18 to 24 inches

Value of hay per ton	$8	10	12	14	16	20
$ 60	99	124	149	174	199	240
$ 80	75	94	113	131	150	186
$100	62	75	90	105	120	149
$120	50	62	75	87	100	124
$140	43	54	64	75	86	107
$160	37	47	56	65	75	93

Category III plant height 24 to 30 inches

Value of hay per ton	$8	10	12	14	16	20
$ 60	104	130	156	182	209	260
$ 80	78	97	117	137	157	195
$100	63	78	94	110	126	156
$120	52	65	78	91	105	130
$140	45	56	67	78	90	112
$160	39	49	58	68	79	98

Cost of insecticide application per acre

FIGURE 13-20 A chart to determine economic threshold level for the alfalfa weevil.

Source: Tooker, J. (May 2009). Entomological Notes, Alfalfa Weevil. [Insect Fact Sheet]. Penn State College of Agricultural Sciences, Cooperative Extension, Department of Entomology. Retrieved from http://ento.psu.edu/extension/factsheets/pdf/Alfalfa%20Weevil.pdf

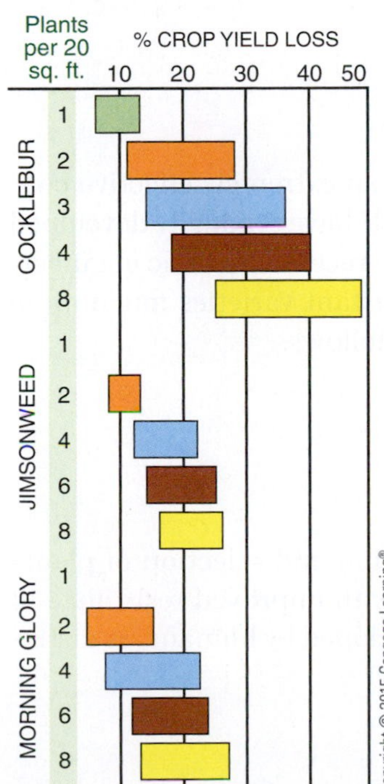

FIGURE 13-21 The pest-damage index for several weeds in soybeans.

and pest problems (Figure 13-22). The presence or absence of the pest, amount of damage, and stage of development of the pest are several visual estimates a scout must make. The method used must be speedy and accurate.

Scouts are people who monitor fields to determine pest activity. They must be well trained in entomology, pathology, agronomy, and horticulture.

PEST-CONTROL STRATEGIES

Pest-control programs can be grouped into several broad categories. These include regulatory, biological, cultural, physical/mechanical, and chemical.

Regulatory Control

Federal and state governments have created laws that prevent the entry or spread of known pests into uninfested areas. Regulatory agencies also attempt to contain or eradicate certain types of pest infestations. The Plant Quarantine Act of 1912 provides for inspection at ports of entry. Plant or animal quarantines are implemented if shipments are infested with targeted pests. A **quarantine** is the isolation of pest-infested material. A **targeted pest** is a pest that, if introduced in a new area, poses a major economic threat.

If a targeted pest becomes established, an eradication program will be started. **Eradication** means total removal or destruction of a pest. This type

SCIENCE PROFILE IPM: BIOLOGICAL PEST CONTROL IN NATIONAL PARKS

Hungry beetles provide biological control of purple loosestrife in national parks.

National Parks manage pest control issues with a philosophy that "less is best." This philosophy is in line with the science-based decision making of IPM. Park managers consider IPM to be the most appropriate and cost-effective approach to control specific pests. For example, the invasive weed, purple loosestrife, is known to completely choke meadows, lakes, and ponds with its aggressive foliage. Under the IPM model, such a weed infestation is approached in different ways. When a small, isolated plant population is located, it may be sprayed with a pesticide to eradicate it before it spreads. However, a well-established population of this invasive weed is more likely to be controlled by establishing a population of hungry beetles that feed almost exclusively on purple loosestrife. Two beetle species have been approved by the USDA to control this pest by eating the foliage. Over time, the beetles are expected to provide biological control over this particular plant pest.

FIGURE 13-22 Random sampling of a plant stem and leaf to determine pest populations and damage.

of pest control is extremely difficult and expensive to administer. In California, the Mediterranean fruit fly was eradicated at a cost of $100 million in 1982. It has recurred since then, and the cost is high each time it is eradicated. This program relies on chemical spraying, sanitation, sterile male releases, and pheromone traps to ensure complete eradication. A **pheromone** is a chemical secreted by an organism to cause a specific reaction by other organisms of the same species, for example, seeking mates.

Pest Resistance

The development of plants having pest resistance is an extremely effective control practice. New genetic strains of plants and animals have gradually developed genetic resistance to some diseases and insects. More recently, genetic engineering techniques have made it possible to develop resistant varieties much more rapidly. Some advantages of resistant varieties are as follows:

- minimal cost of pest control
- no adverse effect on the environment
- a significant reduction in pest damage
- ability to fit into any IPM program

Breeding programs have focused on identification and selection of plants with pest resistance. Currently, new plant cultivars with improved resistance to pests are released annually. A **cultivar** is a plant developed by humans, as distinguished from a variety as it is found in nature.

INTERNET KEY WORDS: pests, biological control

Biological Control

Biological control means control by natural agents. Such agents include predators, parasites, and pathogens. A predator is an animal that feeds on a smaller or

HOT TOPICS IN AGRISCIENCE INSECT CONTROL USING STERILE MALE INSECTS

Imagine a laboratory that raises millions of insects that are harmful to agricultural crops. Once the insects are mature, they are irradiated. The treatment does not kill them, but it eliminates their ability to reproduce. The treated insects are released in huge numbers into areas where the same species of insect pest is found. Because the treated insects vastly outnumber the untreated insects, the odds of treated insects mating with the local insect population are high, but no offspring are produced from these matings. By repeating the releases of sterile insects over a 2- or 3-year period, the original population of harmful insects is reduced to manageable levels or, in some instances, it is completely eradicated.

weaker organism. An example of a predator is the lady beetle. Aphids are the lady beetle's principal prey. Parasites are organisms that live in or on another organism. The braconid wasp is parasitic on the caterpillars of many moths and butterflies. Pathogens are organisms that produce diseases within their hosts. For example, the bacterium *Bacillus popilliae* is a pathogen because it causes the milky spore disease in Japanese beetle grubs.

Successful biological control programs reduce pest populations to less than economic thresholds and keep the pests in check. Such programs require a thorough understanding of the biology and ecology of the beneficial organism along with that of the pest. Careful research can even match desirable plant pathogens against undesirable weeds (Figure 13-23).

Cultural Control

Cultural control is achieved when the crop environment is altered to prevent or reduce pest damage. It may include such agricultural practices as soil tillage, crop rotation, adjustment of harvest or planting dates, and irrigation schemes. Other practices that are considered cultural control are clean culture and trap crops. Clean culture refers to any practice that removes breeding or overwintering sites of a pest.

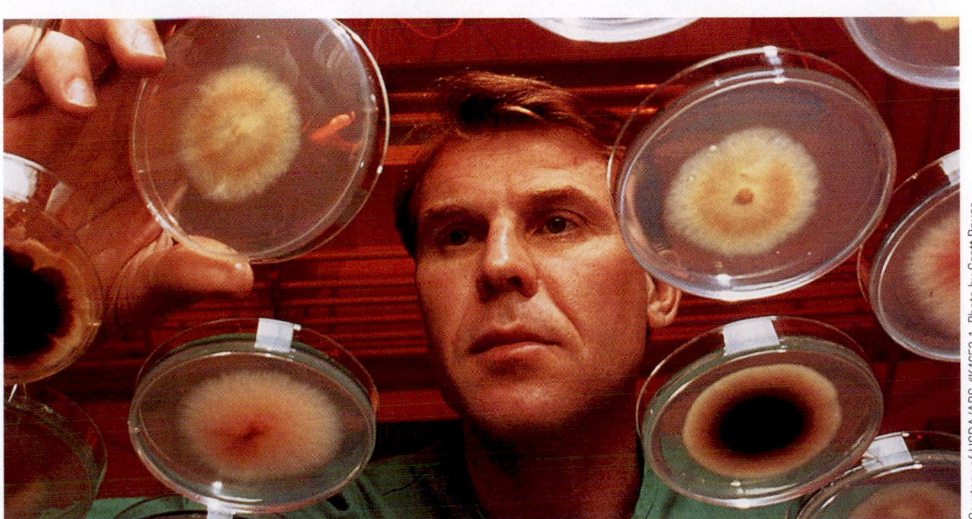

Courtesy of USDA/ARS #K4652-1. Photo by Scott Bauer.

FIGURE 13-23 A plant pathologist examines fungi that may be used for biological control of weeds.

BIO-TECH CONNECTION BIOENGINEERING A SAFER MOSQUITO!

The common mosquito is a scourge from the Alaskan tundra to the equatorial marshes. The biological need of the female mosquito to obtain a blood meal to provide protein before laying her eggs causes the mosquito to be an annoyance and a threat to humans and animals alike. The small amount of blood extracted during a "bite" is not the problem. First, the bite causes discomfort. However, the biggest problem lies in the life-threatening yellow fever or malaria-causing microorganisms that certain mosquitos transfer to humans while taking their blood meals. Currently, the threat of West Nile virus is a serious mosquito-borne disease that affects birds, mammals, and humans. The search for methods to rid the environment of mosquitos is continuous.

Is there any such thing as a "good" mosquito? Yes, if the mosquito is a non-malaria mosquito that replaces a malaria carrier. Geneticist Andrew Cockburn has developed a genetic engineering technique that may be useful in creating a malaria-resistant mosquito. Such mosquitos could theoretically thin out or replace their protozoan-carrying cousins and reduce or eliminate the malaria disease in humans.

A promising technique is the use of tiny needles, new genes, and insect eggs spun in an ordinary laboratory vortex in saline solution. The needles gently pierce the eggs and allow new material to enter. Cockburn has transferred either a test gene or dye into eggs of house flies, fruit flies, and stable flies. So far, the mosquito eggs have proven to be too tough for penetration. However, Cockburn believes he eventually will find a combination of needle number and vortex speed that will permit him to slip material into mosquito eggs.

When the technique for egg penetration is worked out, perhaps a gene for malaria resistance will be found and a malaria-resistant population of mosquitos will fill the niche now occupied by mosquitos that host the organism carrying malaria. You would still have biting mosquitos, but you would not get malaria!

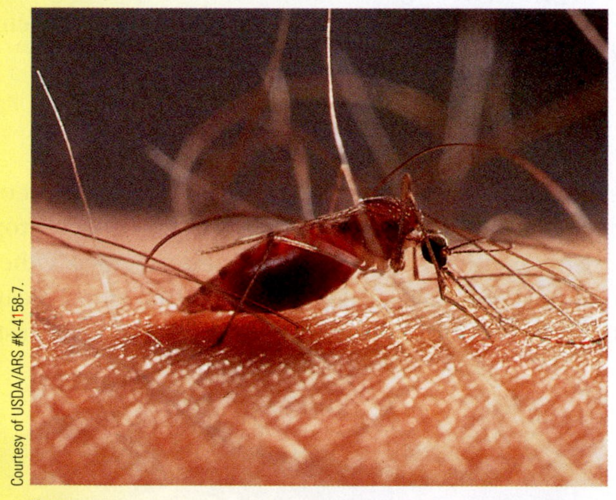

Courtesy of USDA/ARS #K-4158-7.

A

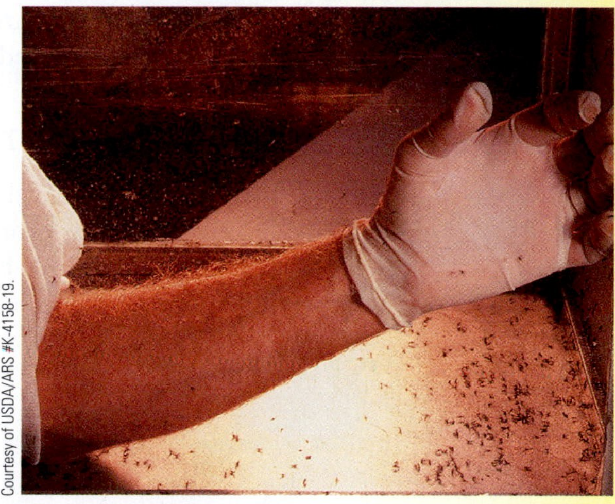

Courtesy of USDA/ARS #K-4158-19.

B

Scientists have tried for decades to control the dreaded mosquito. (A) Mosquito in blood-sucking position. (B) An arm treated with a modern mosquito repellent.

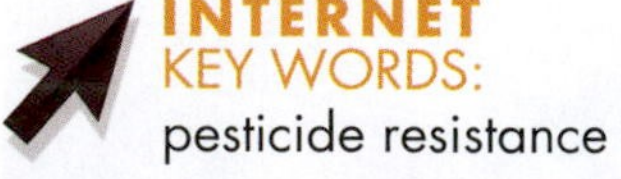

INTERNET KEY WORDS:
pesticide resistance

This may include removal of crop leaves and stems, destruction of alternate hosts, or pruning of infested parts. A trap crop is a susceptible crop planted to attract a pest to a localized area. The trap crop is then either destroyed or treated with a pesticide.

Physical and Mechanical Control

Physical and mechanical control programs use direct measures to destroy pests. Examples of such practices are insect-proof containers, steam sterilization, hand removal, cold storage, and light traps. Implementation of these control practices is costly and provides pest-control results that are not always reliable.

Chemical Control

Chemical control is the use of pesticides to reduce pest populations. Chemical-control programs have become very cost effective. However, various problems occur if this practice is misused or overused. Problems that can develop are environmental pollution, pesticide resistance, and pest resurgence. **Pesticide resistance** is the ability of an organism to tolerate a lethal level of a pesticide. **Pest resurgence** refers to a pest's ability to repopulate after control measures have been eliminated or reduced.

Integrated pest management seems to be our best defense against pests. Biological and cultural controls are favored when they are effective. However, we cannot control certain pests without the use of chemical pesticides. Under such circumstances, it is important to use chemical pesticides safely.

STUDENT ACTIVITIES

1. Write the Terms to Know and their meanings in your notebook.
2. Use reference materials to find and list 10 examples of beneficial insects and 10 examples of insect pests.
3. Name five insects that are beneficial some of the time but are pests at other times.
4. Make an insect collection, with the insects properly identified and named.
5. Make a drawing of an insect and label the various body parts, appendages, and mouth parts.
6. Research a major crop in your community, and discuss its key pests and measures recommended to control those key pests.
7. Develop a collage showing pests of plants in your community.
8. Make a weed collection and identify each sample. Your instructor may have a weed identification key for this purpose.
9. Using the Internet, library, magazines, or other sources, prepare a two-minute class presentation on a local pest. Be sure to include the type of harm the pest is responsible for causing and the integrated pest management techniques used to control it.
10. Make an outline of the unit. Phrases and words that are bold should be included, together with a brief description of key concepts.

SELF-EVALUATION

A. MULTIPLE CHOICE

1. A biennial weed will live for
   a. 1 year.
   b. 2 years.
   c. 3 years.
   d. more than 3 years.
2. The major causal agent of plant disease is
   a. nematodes.
   b. bacteria.
   c. viruses.
   d. fungi.

3. The number of insect species in the world is estimated to be
   a. 100,000.
   b. 500,000.
   c. 1 million.
   d. none of the above.
4. The term *instar* refers to the development stage of
   a. plants.
   b. fungi.
   c. bacteria.
   d. insects.
5. Plant diseases are vectored by
   a. wind.
   b. rain.
   c. insects.
   d. none of the above.
6. A nematode is a type of
   a. fungus.
   b. roundworm.
   c. annual plant.
   d. insect.
7. A type of regulatory control is
   a. plant quarantine.
   b. sanitation.
   c. crop rotation.
   d. soil tillage.
8. The control practice that relies on the introduction of parasites and predators is
   a. cultural.
   b. chemical.
   c. biological.
   d. host resistance.
9. A threshold level is also known as the
   a. pesticide residues.
   b. control program.
   c. degree of pest control.
   d. pest concentration.
10. A pesticide used to control diseases is a/an
    a. fungicide.
    b. nematacide.
    c. insecticide.
    d. acaricide.

B. MATCHING

________	1. Biotic	a. The insect stage between molts
________	2. Pest	b. A chemical used to control mites and spiders
________	3. Summer annual	c. Integrated pest management
________	4. Acaricide	d. A visible change to the host caused by pests
________	5. Instar	e. Adversely affects human activities
________	6. Fungi	f. Diseases caused by living organisms
________	7. Abiotic	g. A plant that germinates in the summer and lives for only 1 year
________	8. Entomophagous	h. Thread-like plants that lack chlorophyll and spread plant diseases
________	9. Symptom	i. Nonliving factors
________	10. IPM	j. Insects on which other insects feed

C. COMPLETION

1. Soil tillage is an example of _______ control.
2. An _______ will control weeds.
3. A _______ lives in or on another organism.
4. Insects have _______ pairs of legs.
5. Complete metamorphosis consists of the following stages: egg, _______, _______, and adult.
6. _______ resistance is when a plant has developed its own defensive response to pests.
7. _______ control will use quarantine practices.
8. Monitoring is essential for _______ programs to work successfully.
9. _______ pests occur on a regular basis for a given crop.
10. Causal agents of disease are _______, _______, _______, and _______.

UNIT 14

Safe Use of Pesticides

OBJECTIVE

To determine the nature of chemicals used to control pests, to know important terms regarding chemical safety, and to practice the safe use of pesticides.

MATERIALS LIST

- writing materials, encyclopedias
- assorted pesticide labels
- Internet access

COMPETENCIES TO BE DEVELOPED

After studying this unit, you should be able to:

- describe the previous and current trends of pesticide use in the United States.
- recognize some popular classes of chemicals used for pest management and their roles in pest control.
- read and interpret information on pesticide labels.
- state the components of protective clothing for individuals handling pesticides.
- describe the environmental and health concerns relating to pesticide use.

SUGGESTED CLASS ACTIVITIES

1. Divide the class into as many work teams as you have computer workstations with access to the Internet. Assign the students to search for Web sites that (1) identify new pesticides and (2) describe licensing requirements of different states for pesticide consultants and applicators. Have the teams report their findings to the class.
2. Conduct a class debate on pesticide safety. Have students work in teams of two on an assignment to debate food safety. Each team will debate in both the affirmative and the negative positions on the issue of using pesticides on food crops. Identify a champion debate team, and give the students appropriate recognition.
3. Invite a local organic farmer to come and discuss the advantages and disadvantages of organic farming. Invite the guest speaker to discuss what alternate methods he or she uses to keep yields high and pest infestations at bay.

TERMS TO KNOW

element
compound
inorganic compound
organic compound
contact herbicide
systemic herbicide
phloem
preemergence herbicide
postemergence herbicide
photodecomposition
insecticide
protectant fungicide
eradicant fungicide
formulation
signal word
symbol
LD_{50}
LC_{50}
toxicity
acute toxicity
chronic toxicity
carcinogen

The development and use of pesticides has provided many benefits for both the producer of agricultural commodities and the consumer. However, there are also risks associated with pesticide use. These risks are reduced with proper pesticide application, storage, and disposal. Improper use of pesticides will increase the risk of environmental contamination and adverse effects on human health.

Currently, pesticide use is a controversial issue in the United States. It is important to objectively balance the benefits and the risks associated with pesticide use. The Environmental Protection Agency (EPA) conducts benefit–risk assessments on each pesticide that is registered. When a pesticide is registered for use and is applied according to label directions, the benefits greatly outweigh the risks.

Some benefits of pesticide use are summarized as follows:

- increased yields of food and fiber;
- reduced loss of stored products;
- increased crop quality;
- economic stability;
- better health; and
- environmental protection and conservation.

The increase in crop and animal yields and improved quality of crops and livestock products resulting from the use of pesticides have been adequately documented (Figure 14-1). It has been estimated that the average total income spent on food in the United States would increase to 30 percent or more without the protection offered by pesticides. Benefits to human health are best illustrated where insecticides are used to control insects that carry and spread malaria and typhus diseases.

Minimum-tillage or no-tillage practices reduce soil erosion, but these practices would not be possible without the use of pesticides. Many substantial benefits to the quality of life result from the proper use of pesticides.

HISTORY OF PESTICIDE USE

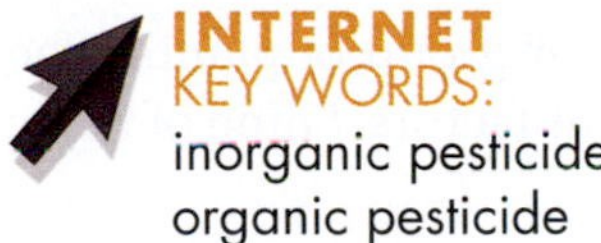

The use of chemicals to control pests is not new. Elements such as sulfur and arsenic were among the first chemicals used for this purpose. An **element** is a uniform substance that cannot be further decomposed by ordinary chemical means. Homer, in about 1000 BC, wrote that sulfur could be used for pest-control purposes. The Chinese, in about AD 900, discovered the insecticide potential of arsenic sulfide, a chemical **compound** that is composed of more than one element.

Until the late 1930s, pest-control chemicals or pesticides were mainly limited to **inorganic compounds** (any compounds that do not contain carbon). Examples of other inorganic pesticides include mixtures of mercury and Bordeaux. Bordeaux mixture is a combination of copper sulfate and lime. It is used for plant disease control. A majority of currently used pesticides are synthetically produced **organic compounds** (compounds that contain carbon). The organic chemistry involved in pesticide production is often complex and extremely diverse. However, a classification system for pesticides that is based on the type of pest to be treated is useful (Figure 14-2). The major pesticide groups are herbicides, insecticides, and fungicides.

In the United States, the amount of pesticide used each year totals 5.1 billion pounds (U.S. EPA, January 2012). Currently, the EPA has registered more than 600 chemicals that are formulated into some 30,000 products for pest control. Of the three major pesticide categories, the largest volume was for herbicides, followed by insecticides and fungicides.

Crop	Estimated Percent Increase	Major Pests Controlled
Corn	25	Weeds, rootworms, corn borers, seedling blights
Cotton	100	Pink bollworms, boll weevils, nematodes, boll rots
Alfalfa seed	160	Weeds, alfalfa weevils
Potatoes	35	Tuber rots, blackleg, soft rots, blights
Onions	140	Botrytis blights, neck rot, smut, onion maggots

FIGURE 14-1 Yield increases from pesticide applications.

Source: Bohmont, B. L. (1983). The New Pesticide User's Guide. Reston Publishing Company.

Pesticide Type	Targeted Pests
acaricide	mites, ticks
algaecide	algae
attractant	insects, birds, other vertebrates
avicide	birds
bactericide	bacteria
defoliant	unwanted plant leaves
desiccant	unwanted plant tops
fungicide	fungi
growth regulator	insect and plant growth
herbicide	weeds
insecticide	insects
miticide	mites
molluscicide	snails, slugs
nematicide	nematodes
piscicide	fish
predacide	vertebrates
repellents	insects, birds, other vertebrates
rodenticide	rodents
silvicide	trees and woody vegetation
slimicide	slime molds
sterilants	insects, vertebrates

FIGURE 14-2 Classifications of pesticides based on the target pests.

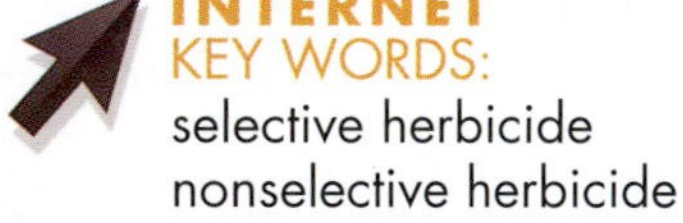

HERBICIDES

Herbicides are grouped into several major categories based on application method, type of control, and chemical structure. The terminology for herbicide use, type of control, and chemical family follows.

Selective Herbicides

A selective herbicide kills or affects only a certain type or group of plants. The selectivity of an herbicide can be caused by many different factors. Some of these factors include:

- differences in herbicide chemistry, formulation, and concentration;
- differences in plant age, morphology, growth rate, and plant physiology; and
- environmental differences such as temperature, rainfall, and soil type.

Nonselective Herbicides

A nonselective herbicide controls or kills all plants. These herbicides are used for many different purposes. Examples of their use are for railroad rights-of-way, industrial areas, fence rows, irrigation and drainage ditch banks, and renovation programs.

Contact Herbicides

Contact herbicides will not move or translocate within the plant. They affect only that part of the plant with which they come in contact. In addition, they are often used for controlling annual weeds.

Systemic Herbicides

A **systemic herbicide** is absorbed by the plant and is then translocated in either xylem or phloem tissue to other parts of the plant. Xylem tissue transports water and dissolved minerals from the roots to the leaves. **Phloem** tissue is responsible for transporting carbohydrates from the leaves and stems to the plant roots.

Preemergence and Postemergence Herbicides

INTERNET KEY WORDS:
acetanilides
Lasso, Bullet, Gardian, Harness, Dual
dinitroanilines
Surflan, Stomp, Balan, Prowl
phenoxys
2, 4-D; 2, 4, 5-T
triazines
Atrazine, Simazine, Cyanazine

A **preemergence herbicide** is applied before weed or crop seeds germinate. A **postemergence herbicide** is applied after the weed or crop is actively growing.

Chemical Families of Herbicides

Herbicides are chemicals that are used to control weeds. These compounds can affect plant growth in many different ways. Currently, more than 23 chemical families of herbicides have been developed. Each one has a unique chemical structure and a different site or mode of action. *Mode of action* is a phrase used to describe the way herbicides adversely affect plant growth. Several of the more important chemical herbicide families and their characteristics are described in this section.

Acetanilides

Acetanilides interfere with cell division and protein synthesis. They are applied either as preemergence or as preplant applications for control of annual grasses and some annual broadleaf weeds. Popular herbicides within this category are Lasso, Bullet, Guardian, Harness, and Dual. They are used for weed control in corn and soybeans.

BIO-TECH CONNECTION: PEST RESISTANCE USING GENETICALLY ENGINEERED PLANTS

Courtesy of DeVere Burton.

A Colorado potato beetle that takes a bite out of a beetle-resistant plant has a big surprise coming. The juice in the foliage is deadly to the beetle, but harmless to those who eat the tubers produced by the plant.

Scientists have identified the genes that make some plants resistant to a number of specific pests. Once such a gene is located on the chromosome of a pest-resistant plant, it can be isolated and transferred to other plants. This process is used to develop new plant varieties that are resistant to specific insects. For example, new potato varieties have been developed that have genetically engineered resistance to the Colorado potato beetle. This beetle is capable of destroying entire fields of potatoes by eating all of the leaves and part of the stems of potato plants. It is a devastating pest to potato plants. The beetle-resistant potato plants have a new gene that produces a natural selective pesticide in the juices of the plant. Potato beetles that eat the foliage of these plants are poisoned without causing risk to other insects or the human population who eat the potato tubers.

Dinitroanilines

Dinitroanilines act on root tissue, preventing root development in seedling plants. They are preemergence herbicides applied to prevent weed germination and should be incorporated into the soil. The dinitroanilines are deactivated quickly by **photodecomposition** (chemical breakdown caused by exposure to light), volatilization (changing to gases), and other chemical processes. Examples of these herbicides are Surflan, Stomp, Balan, and Prowl. They are used for control of annual grasses and broadleaf weeds in many different crops.

Phenoxys

Phenoxy herbicides affect plants by overstimulating growth. They perform best when applied as postemergence foliar sprays. They are selective herbicides that affect broadleaf weeds in grass crops. The herbicide 2, 4-D was first used in 1942 and is still widely used. The herbicide 2, 4, 5-T is another product that is widely used.

Triazines

Triazines are photosynthetic inhibitors that interfere with the process of photosynthesis. They are preemergence herbicides used to control both annual and broadleaf weeds in grass crops. Atrazine, Simazine, and Cyanazine are examples of these herbicides. They are primarily used in control programs for grasses.

AGRI-PROFILE

CAREER AREAS: PESTICIDE APPLICATOR/PESTICIDE SPECIALIST/CHEMIST

Courtesy of USDA/ARS #K-3663-15.

Aerial applicators are important members of the teams that apply chemical and biological pesticides to large crop, range, and forest areas.

Tens of thousands of chemical formulations have been developed in recent years to control insects, diseases, weeds, nematodes, rodents, and other pests. We must use chemicals to help control pests. However, these materials are likely to be hazardous to the operator, plants, animals, or the environment if not properly used.

Pesticide applicators are used by farmers and ranchers, lawn service companies, farm and garden supply firms, termite control companies, highway departments, and railroads as well as self-employed individuals. Special training and licensing are required for handling most pesticides.

State agencies such as state departments of agriculture are responsible for developing the regulations for pesticide consultants and applicators. In most states, people who apply for licenses must attend classes and seminars to learn how to handle pesticides properly. The training usually includes pesticide storage, mixing, application, and disposal procedures. Those who attend these classes and seminars may be required to show evidence of attendance before they are allowed to take the licensing tests that are given. Many states use these tests to establish that the candidates are qualified to prescribe and use pesticides.

Pesticide applicators may work in homes, buildings, fields, forests, and even the holds of ships. They may dust, spray, bait, or fumigate, depending on the setting and the pest. Pesticide applicators must always use protective clothing and safety devices to protect against accidental poisoning. Their tools may be as simple as aerosol cans or as complex as orchard spray rigs or specially equipped helicopters.

Many companies throughout the world are continually researching new and existing chemicals to determine what characteristics they may exhibit, such as pesticide qualities, toxicity levels, persistence in the environment, risk to crops and animal life, and effectiveness in controlling pests. A tremendous amount of research is required to develop and gain government approval for a single pesticide. The cost is high, but the reward for developing an effective pesticide product is also high. A single product that is safe and effective has the potential to earn huge profits for the manufacturer. For this reason, chemical companies are willing to pay for the research to develop and gain government approval for a new product.

INSECTICIDES

Insecticides are chemicals used to control insects. They can affect the insect in many different ways. The classification system of insecticides may be based on chemical structure and/or mode of action, as shown in Figure 14-3.

The botanical, inorganic, and oil insecticides are some of the original chemicals used for insect control. Sulfur, for example, may be used to control mites. Its effectiveness as an insecticide was discovered thousands of years ago.

Rotenone is a chemical present in the roots of certain species of legume plants. It affects insects by inhibiting respiratory metabolism and nerve transmission. Rotenone, first used in 1848, is a botanical insecticide still used today. It is a contact- and stomach-poison insecticide. Rotenone is principally used in vegetables for controlling fleas, beetles, loopers, Japanese beetles, and other insects (Figure 14-4).

Superior oils are highly refined oils and are applied to ornamentals and citrus crops to control scale insects, mites, and other soft-bodied insects. Oils act by excluding oxygen from the insect, thus causing suffocation. It is also believed that oils may destroy cell membrane function. Oils were first used in the early 1900s to control San Jose scale in fruit orchards (Figure 14-5).

Chemical Group	Mode of Action	Common/Trade Names
Inorganics	Protoplasmic Poisons Physical Poisons	Sulfur Silica Aerogel
Oils	Physical Poisons	Superior Oils
Botanicals	Metabolic Inhibitors	Pyrethrum Rotenone
Synthetic Organics		
Chlorinated Hydrocarbons	Nerve Poison	Lindane Endosulfan-Thiodan
Organophosphates	Nerve Poison	Diazinon-Spectracide Parathion-Thiophos
Carbamates	Nerve Poison	Carbofuran-Furadan Carbaryl-Sevin
Pyrethroids	Nerve Poison	Permethrin-Ambush Fluvalinate-Mavrik
Insect Growth Regulators (IGRs)	Alter Insect Growth	Methoprene-Altosid Kinoprene-Enstar
Biorational/Microbial Insecticides	Wide Range of Activity	*Bacillus thuringiensis*

FIGURE 14-3 Insecticide classification system.

Courtesy of DeVere Burton.

FIGURE 14-4 Rotenone is a chemical insecticide that is used to control beetles and other insects.

Courtesy of DeVere Burton.

FIGURE 14-5 Superior oils are insecticides that are used to control insects such as mites, scale, and soft-bodied insects by blocking their supply of oxygen.

The synthetic organic group of insecticides was principally developed after 1940. This group includes the chlorinated hydrocarbons, organophosphates, carbamates, pyrethroids, insect growth regulators, and other minor classes. A majority of these insecticides adversely affect nerve transmission. The chlorinated hydrocarbons were the first to be synthesized (man-made). Released in 1940, DDT (dicloro-diphenyl-trichloroethane) was the first and most popular chlorinated hydrocarbon. This chemical had excellent insecticidal properties and was used extensively. However, environmental and health problems were linked to DDT and other insecticides within this group. Because of these risks, many of the chlorinated hydrocarbons (such as DDT, aldrin, dieldrin, and chlordane) are banned from use in the United States.

The organophosphate and carbamate insecticides are the principal insecticides currently used to control insects. Approximately 60 percent of all insecticides produced in the United States are from these two groups. They control insects by affecting the nervous system. The potential for pesticide poisoning of people and livestock is high for these insecticides. They will affect the nervous systems of humans and animals in a manner similar to the way they affect the insects they are designed to kill. The dose, or the amount, of insecticide applied is the discriminating factor with respect to insect control or human poisoning.

INTERNET KEY WORD: fungicide

FUNGICIDES

Fungicides are chemicals used to control plant diseases caused by fungi. Chemical control of plant diseases is more difficult than it is for weed and insect control. Fungi are plants without chlorophyll. They are parasites of other plants. The fungicide must be selective enough to control the fungus but must not adversely affect the host. Also, fungi have many generations each growing season. Therefore, reapplication of the fungicide is required to provide effective control.

Kinds of Fungicide

Protectant Fungicide

A **protectant fungicide** is applied before disease infection. This will provide a chemical barrier between the host and the germinating spores. However, this

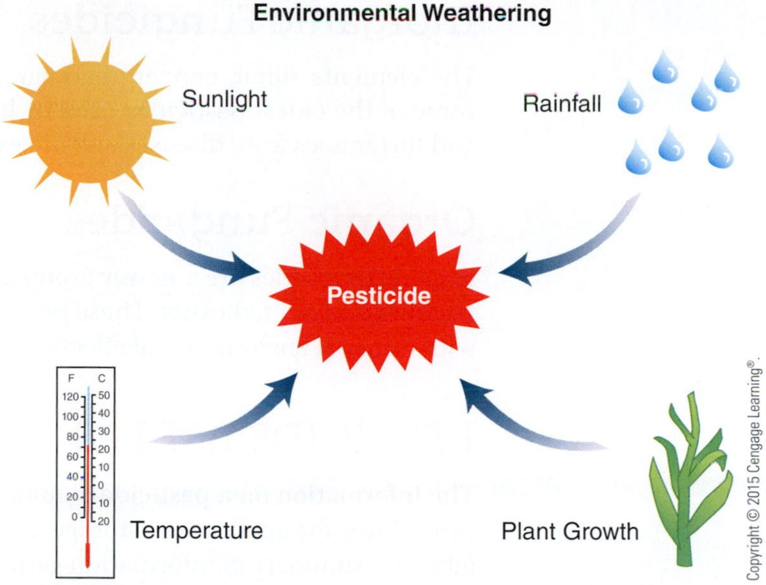

FIGURE 14-6 Pesticides are broken down by environmental weathering, which is caused by sunlight, rainfall, temperature, and plant growth.

barrier will be broken down by environmental weathering of the fungicide. Rainfall, sunlight, temperature, and plant growth are the major causes of this breakdown (Figure 14-6). The fungicide must be reapplied if adequate disease control is expected.

Eradicant Fungicide

An **eradicant fungicide** can be applied after disease infection has occurred. These fungicides act systemically and are translocated by the plant to the site of infection. They offer a longer control period than do protectant fungicides because they are not so prone to environmental weathering.

Chemical Structure

Fungicides, like insecticides and herbicides, can also be classified into different chemical families. Figure 14-7 lists some of the major groups of fungicides.

Chemical Family	Fungicide Activity	Common/Trade Name
Benzimidazoles	Eradicant	Benomyl-Tersan 1991
Dicarboximides	Protectant	Captan-Captane
Dithiocarbamates	Protectant	Mancozeb-Dithane M45
Oxathiins	Eradicant	Carboxin-Vitavax

FIGURE 14-7 Fungicide classification system.

Inorganic Fungicides

The elements sulfur, copper, mercury, and cadmium, or mixtures of them, are some of the oldest pesticides used by humans. They protect many ornamentals and turfgrasses from diseases and are examples of inorganic pesticides.

Organic Fungicides

Organic fungicides are a newer group of fungicides. These are used both as protectants and as eradicants. These products are used for effective control of fungi with minimal environmental effects.

PESTICIDE LABELS

The information on a pesticide container label instructs the user on the correct procedures for application, storage, and disposal of the pesticide. The pesticide label is a summary of information gathered by the pesticide manufacturer and is required for product registration. The registration process is estimated to take 8 to 10 years and costs millions of dollars per pesticide. The multiple studies required to meet federal standards are summarized as follows:

- Chemical and physical properties: water solubility, volatility, movement in soils, stability to heat and light, and other factors affecting environmental stability
- Toxicology studies: determine acute oral, dermal, and inhalation toxicity to various animals; evaluate chronic toxicity, including any effect on reproduction or the tendency of the pesticide to be a carcinogen
- Residue analysis: determine the amount of pesticide residues at the time of harvest; develop safe tolerance levels for any pesticide residue
- Metabolism studies: determine application exposure, determine consumer exposure to pesticide residues, and establish safety practices that minimize exposure

The label is a legal document indicating proper and safe use of the product. Pesticide use that differs from that specified on the label is a misuse. Such use is illegal, and the offender can be charged with civil or criminal penalties. When improperly applied, a pesticide can pose danger to the applicator, the environment, animals, and other people. Therefore, it is important to read, understand, and follow the information on the pesticide label.

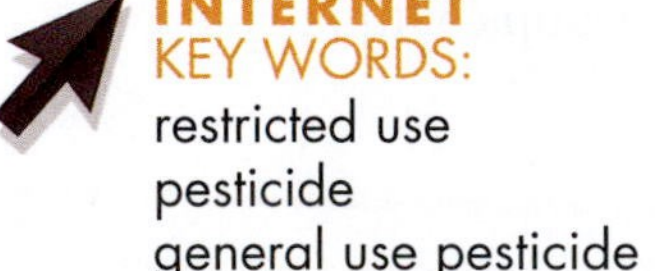

INTERNET KEY WORDS:
restricted use pesticide
general use pesticide

The pesticide label and other labeling information must meet federal standards. The label consists of a front panel and a back panel on the product (Figures 14-8 and 14-9). If there is insufficient room on the panels, additional labeling information will be attached, in booklet form, to the product. The following sections discuss the parts of a label.

Use Classification

Pesticides are classified as either general use or restricted use. A general-use pesticide poses minimal risk when applied according to label directions.

Restricted-use pesticides pose a greater risk to humans and the environment. Therefore, anyone applying a restricted-use pesticide must be properly trained and certified. Applicator certification is administered by each state. An applicator must meet a minimum set of standards and is usually evaluated by testing to be certified.

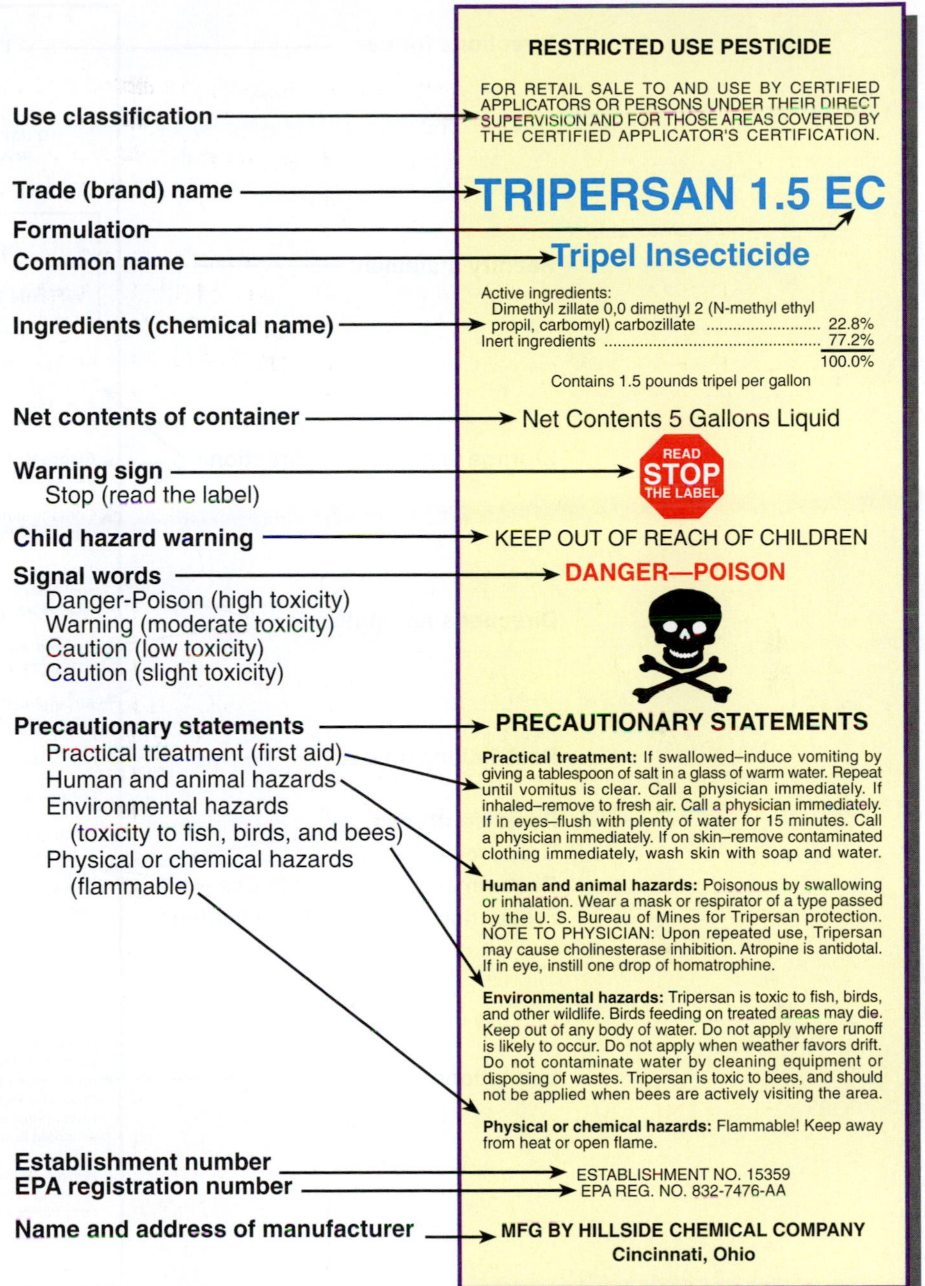

FIGURE 14-8 Front panel of a sample pesticide label.

Trade Name

A trade name is the manufacturer's name for its product. It appears on the label as the most conspicuous item. The manufacturer will use the name in all of its promotional campaigns. The same chemical may have several different trade names, depending on the type of formulation and patent rights.

Formulation

Formulation refers to the physical properties of the pesticide. The pesticide chemical or active ingredient will often have to be modified to allow for field

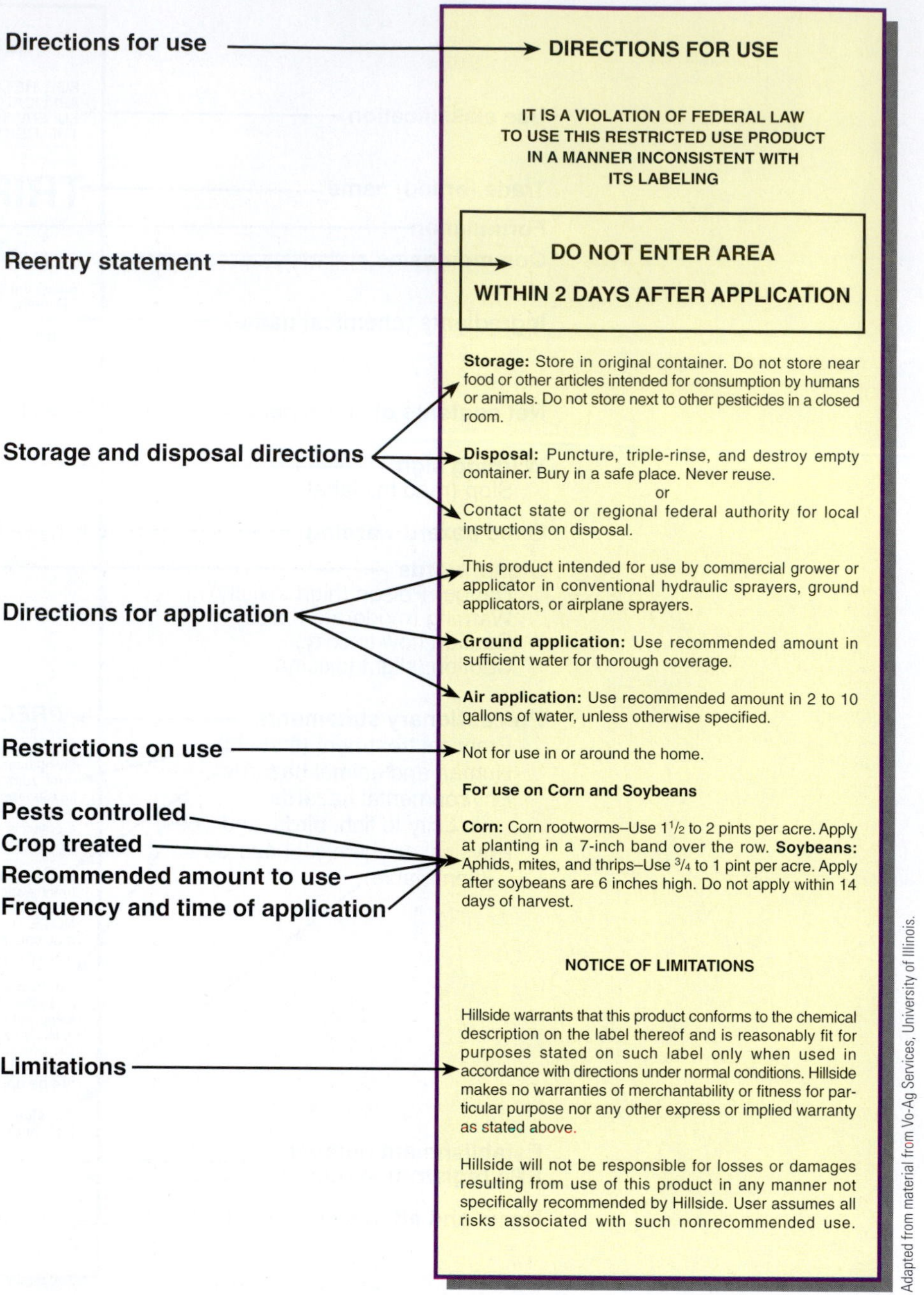

FIGURE 14-9 Back panel of a sample pesticide label.

use. These modifications may include adding inert ingredients, such as solvents, wetting agents, powders, or granules, to the pesticide. This will result in different formulations. Some examples of the different types of formulations and their abbreviated label names are as follows:

Granules—G

Solutions—S

Wettable Powders—WP or W

Soluble Powders—SP

Dry Flowables—DF

Emulsifiable Concentrates—EC or E

Dusts—D

Common Name

The common name is given to a pesticide by a recognized authority on pesticide nomenclature. A pesticide is identified by a trade name, common name, and chemical name. It may have several trade names but will have only one common and one chemical name. The common name, or generic name, identifies the active pesticide ingredient and can be used for comparison shopping.

Ingredients

The percentages by weight of both the active and inert ingredients are stated on the label. The active ingredient is identified by its chemical name. This is the name of the chemical structure of that pesticide. The inert, or inactive, ingredients do not have to be listed by their chemical names. The label must state only the total percentage of inert material.

Net Contents

The label will state the amount of product in the container. This is referred to as net contents. This quantity can be expressed in gallons, quarts, pints, or pounds.

Signal Words and Symbols

The **signal words** and **symbols** describe acute toxicity of the pesticide. The different categories are based on LD_{50} (lethal dose) and LC_{50} (lethal concentration) values and on skin and eye irritation. The signal words used on pesticide labels are (1) danger—poison, (2) warning, and (3) caution. These words are used to alert the person handling or using the pesticide to the poisoning effect of contact with the chemical.

Pesticides with high toxicity—only a few drops to 1 teaspoon will kill a 150-lb. person—are labeled "DANGER—POISON." Pesticides with moderate toxicity—1 to 2 tablespoons will kill a 150-lb. person—are labeled "WARNING." Those restricted-use pesticides requiring more than 2 tablespoons of the chemical to kill a 150-lb. person are labeled "CAUTION." Obviously, even the pesticides with "CAUTION" as the signal word must be handled with extreme care.

Acute toxicity is measured by determining LD_{50} values when the pesticide is absorbed through the skin or is ingested orally. These values are determined by inhalation studies. **LD_{50}** is the amount or dose of the pesticide that is required to kill 50 percent of test populations. It is expressed in milligrams (mg) of pesticide per kilogram (kg) of body weight. The lower the LD_{50} value, the more toxic a pesticide is rated.

LC_{50} is the lethal concentration of the pesticide in the air that is required to kill 50 percent of test populations. It is expressed in micrograms per liter (μg/l) or in parts per million (ppm). The lower the LC_{50} value, the more toxic the pesticide is rated.

All labels must bear the statement, "KEEP OUT OF REACH OF CHILDREN," regardless of pesticide toxicity. Figure 14-10 shows the toxicity ratings for the various signal words.

Toxicity Rating	Label Signal Words	Lethal Oral Dose, 150 lb. Person	Oral LD_{50} (mg/kg)	Dermal LD_{50} (mg/kg)	Inhalation LC_{50} (μg/l or ppm)
High	Danger–Poison	few drops to 1 teaspoon	0–50	0–200	0–2,000
Moderate	Warning	1 teaspoon to 1 ounce (2 tablespoons)	50–500	200–2,000	2,000–20,000
Low	Caution	1 ounce to 1 pint+ or 2 lbs.	500–5,000	2,000–20,000	200,000+
Very low	Caution	1 pint+ or 2 lbs.+	5,000+	20,000+	—

FIGURE 14-10 Toxicity ratings and signal words for pesticides.

Precautionary Statements

Precautionary statements on the label will list any known hazards to humans, animals, and the environment. They will advise the user how to minimize any adverse effect that the pesticide may have. The categories that are normally listed are as follows:

- Hazards to Humans and Domestic Animals
- Statement of Practical Treatment
- Environmental Hazards
- Physical and Chemical Hazards

Establishment and EPA Registration Numbers

The establishment number identifies the manufacturer. The EPA registration number indicates that the product has passed the review process imposed by the EPA.

Name and Address of Manufacturer

All pesticide labels must contain the name and address of the company that manufactures and distributes the pesticide.

Directions for Use

The correct amount, timing, and mixing of the pesticide are given under the "Directions for Use" section of the label. The label will also list the different pests that are controlled, the application technique, and any other specific directions for optimum control.

Misuse Statement

The misuse statement appears on the label to remind the user to apply the pesticide according to label directions. Problems associated with pesticides, whether they involve environmental pollution or human poisoning, usually occur because of pesticide misuse.

Reentry Information

Specific directions on reentering a treated area appear under this heading. Only a few pesticides require reentry times of more than a day after application. However,

BIO-TECH CONNECTION SUBSTITUTING BIOLOGICAL CONTROL FOR CHEMICAL PESTICIDES

Biological control of pests includes the use of predatory insects and microbes that destroy the target pest during a phase of its life cycle when it is most vulnerable. The eggs that are attached to this pest will hatch, and the larvae will eat their host.

The sound of a roaring cyclone sprayer or the sight of an aircraft fogging a field or treetops creates an image of chemical pesticide application in the minds of most people. However, more and more such sprays are the new, user-friendly, environmentally safe, biological-control pesticides. These materials generally contain bacteria, fungi, viruses, or other microbes that attack the targeted host but are harmless to other living organisms and the environment.

The newer, safer pesticides increase the chances of eradicating or permanently subduing some of our most troubling and costly pests. For instance, the efforts made to control cotton bollworms and tobacco budworms on the Mississippi Delta cost an estimated $50 million per year using chemical pesticides. Marion Bell, formerly an Agriculture Research Service entomologist at Stoneville, Mississippi, believed he had a better way—the *Heliothis* nuclear polyhedrosis virus. *Heliothis* has been found to be very specific to the bollworm and tobacco budworm because it works only on insect larvae that have an alkaline midgut.

Bell specialized in microbial control of insects. He and other authorities determined that treating the entire 4.7 million acres of the Mississippi Delta with *Heliothis* would cost about $7 million. However, the team was aware of the natural fears of the populace with such a widespread, general spraying program. Farmers are concerned about the effect of any general spraying program on their crops, aquaculturalists are concerned that such material could contaminate their catfish ponds, and the general public is concerned about any spray that may be damaging to the environment.

Therefore, they opted for an extensive educational program coupled with a spray program on small areas until the system was proven and more widely accepted by the public. The educational program included informational releases and face-to-face contacts with farmers, congressional representatives, extension personnel, environmental and health officials, physicians, civic clubs, and private interest groups. The message was, "The approach is solid, and the virus is harmless to every living thing except the target pests."

Before a test site was sprayed, brochures were given to the people who were affected, and catfish farmers were contacted to get permission to spray the virus around their ponds. The first year, spray planes were used to apply the virus mixture at the rate of 2 ounces of virus mix in 2 gallons of water per acre. Spray trucks were used to spray certain areas not accessible to aircraft. The researchers found that the virus application does work, and they have fine-tuned the procedure in subsequent years.

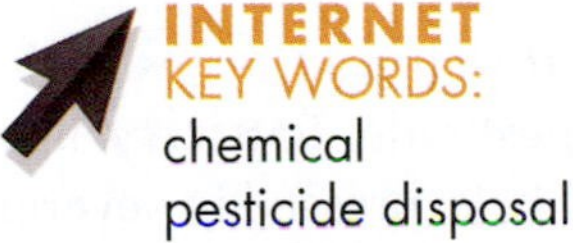

INTERNET KEY WORDS: chemical pesticide disposal

even if the pesticide label does not contain specific restrictions, no one should ever be allowed to enter a treated area until the pesticide has dried.

Storage and Disposal Directions

This section describes the proper storage and disposal of the pesticide. It is recommended that you purchase only the amount of pesticides needed for the current season. Stockpiling them will only increase storage risks and, ultimately, the problem of pesticide disposal, if they no longer can be used.

Many states have enacted laws and established regulations that control the disposal of unused chemicals and chemical containers. It is never wise to bury these kinds of materials because they can ultimately pollute our groundwater and soil. It may also be illegal. The best choice is to take the containers and chemicals to designated sites where hazardous materials are collected so that they can be disposed of properly (Figure 14-11). Chemicals that are left over should be applied to another area or crop where they will not pollute the environment.

Courtesy of DeVere Burton.

FIGURE 14-11 Dispose of leftover chemicals by taking the material to a hazardous waste station.

Notice of Limitations

The manufacturer guarantees that the product will perform as the label states. The company also conveys inherent risks to the user if the pesticide is applied in a manner inconsistent with the label. The manufacturer will limit its liability in case of lawsuits stemming from misuse of the pesticide.

RISK ASSESSMENT AND MANAGEMENT

Risk Measurement

An experienced pesticide applicator who understands the hazards of pesticides will take steps to reduce risks. The risk, or hazard, of pesticide applications has been expressed as the following:

Risk (Hazard) = Toxicity × Exposure

The hazard, or risk, is the relationship between the toxicity of the pesticide and the length and degree of exposure while using the pesticide. **Toxicity** is a measure of how poisonous a chemical is. These data may be expressed in several ways. **Acute toxicity** describes the immediate effects (within 24 hours) of a single exposure to a chemical. Acute toxicity data based on dermal (skin), oral (by mouth), and inhalation (breathing) exposure routes have been determined. Signal words on pesticide labels indicate acute toxicity values. **Chronic toxicity** measures the effect of a chemical over a long period. To determine this information, the chemical is administered at low levels, with repeated exposures to the test animals. The effect of the chemical on reproduction or as a potential carcinogen and any other adverse effects are evaluated.

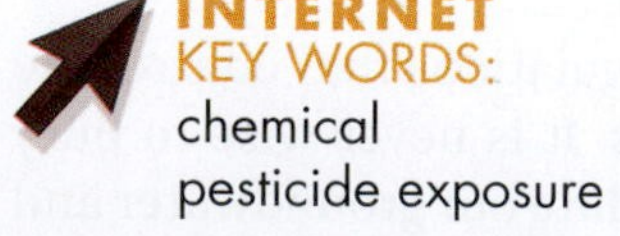

INTERNET KEY WORDS:
chemical pesticide exposure

Limiting Exposure

A pesticide's toxicity cannot be changed, but risk can be managed by addressing the exposure component of the formula. Many things can be done to

FIGURE 14-12 Proper clothing items and appropriate safety equipment are essential in the safe use of pesticides.

reduce exposure. Examples of practices that can reduce exposure include the following:

- Select a pesticide formulation with a lower exposure potential. For example, granule formulations have a much lower exposure potential than do emulsifiable concentrates.
- Use protective clothing and other safety equipment during the time of pesticide mixing, application, and disposal.
- Apply pesticides during weather conditions that will not cause pesticide drift and that provide for the most effective control.
- Check all application equipment for proper working condition before applying pesticides.
- Store pesticides and application equipment properly.

The pesticide label will provide guidance concerning acceptable protective clothing or gear for the application of the pesticide (Figure 14-12). Recommended protective clothing and gear will vary according to the toxicity of the pesticide. Even if no special gear is required, it is best to minimize your exposure to all pesticides by selecting appropriate clothing.

Appropriate clothing includes long pants, a long-sleeved shirt, nonabsorbent shoes, and socks. Avoid the use of any leather clothing, particularly shoes, because leather absorbs pesticides. Use of heavy denim clothing provides good repellency to any pesticide and it can be washed to remove any pesticide residue.

Special Gear

Gloves and Boots

Unlined rubber or neoprene gloves and boots significantly reduce pesticide exposure. Any type of cloth-lined or leather boot or glove will only increase exposure if the lining becomes contaminated. Gloves should be tucked inside sleeves if you are working below the waist. They should be left outside your sleeves if you are working above the waist. Pants legs should be placed over boots. By following these rules, you can prevent material from entering the inside of protective clothing or gear.

Hat and Coveralls

Absorption of pesticides through the skin and into the body is greatest in the scrotal area, ear canal, forehead, and scalp (Figure 14-13). The use of an appropriate hat and coveralls minimizes exposure to pesticides in these areas. Lightweight, one-piece, repellent coveralls with hoods are available, and they provide excellent protection.

Apron

During the mixing and loading operations, the applicator is exposed to pesticide concentrates. The use of a rubber or neoprene apron at this time will prevent pesticide concentrates from splashing onto the chest, waist, and legs of the applicator.

Goggles and Face Shield

Eyes are extremely sensitive to many pesticides. The use of goggles and face shields is recommended when mixing pesticide concentrates or working with a spray, dust, or fog.

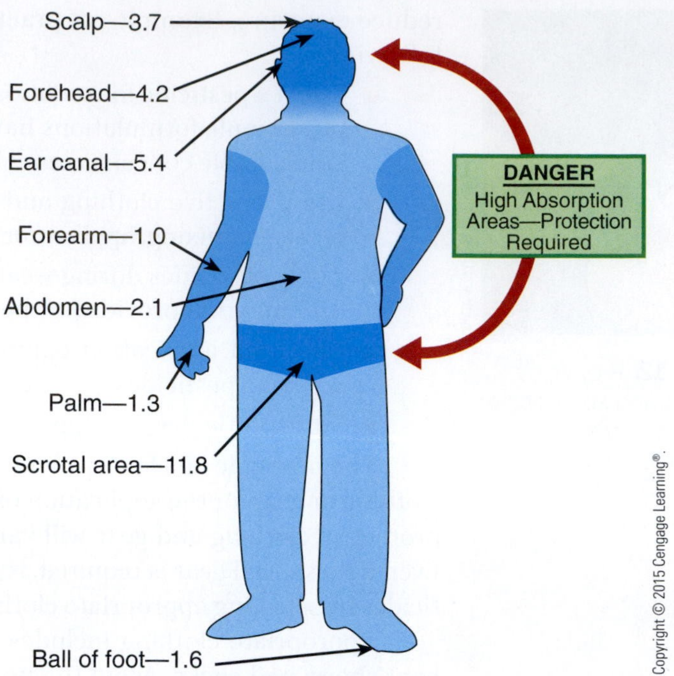

FIGURE 14-13 Dermal absorption sites and rates of pesticide absorption into the body. Comparison is based on forearm absorption rates.

Respirators

Respirators reduce the inhalation of pesticide fumes and/or dust. A recommended respirator for pesticide applications is a cartridge respirator that will absorb toxic fumes and vapors and filter any dust particles in the air. These respirators are often used during the mixing and application of a pesticide.

Personal Hygiene

Dermal exposure is the principal method of pesticide entry for the applicator. Personal hygiene can drastically reduce this type of exposure. Washing or showering at the end of the workday will remove any pesticide residue on the body. In case of a pesticide spill or splash at the work site, water can be used to immediately remove the material from the skin. After pesticide use, washing your hands before eating or using the restroom will further decrease pesticide exposure. Cleaning protective gear and clothing should also be done to prevent any residual exposure to pesticides on these objects.

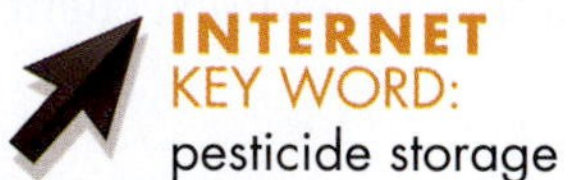
INTERNET KEY WORD: pesticide storage

PESTICIDE STORAGE

Improper storage of pesticides can pose as much danger to the applicator, other people, animals, and the environment as misapplication of pesticides. Some important considerations in selecting a storage facility are the site location and building specifications.

Ideally, the site should be separate from other equipment or material storage facilities. This will reduce risk by decreasing exposure of individuals not involved in pesticide applications. The building should not be located on a floodplain,

FIGURE 14-14 Pesticides must be stored in secure storage units approved for chemical storage.

where flooding will introduce pesticides into surface water. It should be built to prevent any runoff or drainage from the site onto sensitive areas. Spill and drainage containment for large storage facilities is highly recommended. Containment systems would trap the pesticides and aid in emergency situations to minimize any environmental damage if the pesticides were to move from the site.

A well-planned storage building should be well ventilated, have a source of heat and water, be fireproof, have a secure locking system, and have sufficient storage area. The storage area should be well marked with placards indicating the presence of pesticides (Figure 14-14). The arrangement of the pesticides within the storage area should allow for ease in handling and safety. Tips to provide good storage conditions are the following:

- Separate each pesticide class for storage on its own shelf.
- Keep products off the floor.
- Store containers so the labels remain in good condition and the containers remain orderly.
- Practice good housekeeping.

HEALTH AND ENVIRONMENTAL CONCERNS

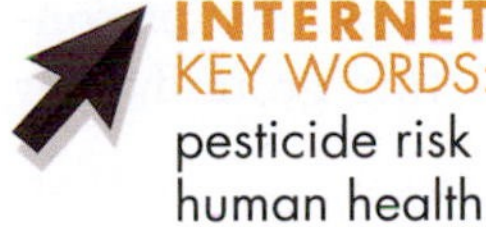
INTERNET KEY WORDS:
pesticide risk
human health

The current use pattern of pesticides has caused a heightened awareness of their risks. Human health and environmental quality are the major issues in assessing the hazards of pesticide use. The EPA must conduct a benefit–risk assessment for each pesticide that is registered or reregistered for use. This is an extremely controversial issue. The EPA presently defines acceptable risk to the public at one death per million caused by pesticide exposure.

Human Health

The number of lethal pesticide-related poisonings in the United States has decreased over the years. In one year, the total number of accidental deaths from all causes in the United States was 92,000, whereas deaths attributed to pesticide poisoning numbered only 27. However, more than 100,000 nonfatal pesticide poisoning cases per year have been estimated. A majority of the reported deaths were children involved in accidental ingestion of pesticides in the home. The causes have been traced to improper handling and storage practices by homeowners and other consumers.

The residues of a few pesticides used on food crops can pose potential health problems as carcinogens. A **carcinogen** is a material capable of producing a cancerous tumor. The National Academy of Science estimated that certain types of pesticide residues in food crops "may" cause up to 20,000 cancer cases per year. This estimate was based on a "worst-case" scenario. It assumes that exposure to the pesticide would be continuous over 70 years, with the maximum allowable pesticide residue on the food when eaten. Obviously, careful handling and preparation of food eliminates most of this risk.

HOT TOPICS IN AGRISCIENCE ORGANIC FOOD

There is a consumer group that has demonstrated a preference for organic food in contrast to foods produced traditionally. Some farmers are converting their farms into organic farms to take advantage of this market niche. The aim of organic farming is to produce safe, high-quality fruits and vegetables without the use of synthetic fertilizers and pesticides. Biological controls are used to keep insects at acceptable and manageable levels. To boost production, animal manures, legumes, and green manures are used as a natural alternative to chemical fertilizers. The U.S. Department of Agriculture has set strict standards for the production, handling, and labeling of "organic food." In a grocery store, you may see food labels with four different organic claims. "100% Organic"-labeled foods contain 100 percent organically produced ingredients. Labels with an "Organic" claim must contain at least 95 percent organic ingredients. Food "Made with Organic Ingredients" has to be composed of at least 70 percent organic ingredients; if a product claims that it "Has Some Organic Ingredients," it may have less than 70 percent organic ingredients. Whether a person prefers foods grown organically or traditionally, it is important to offer consumers a choice in the kind of food they buy.

Environmental Concerns

After a pesticide is applied, not all of the pesticide reaches or remains in the target area (Figure 14-15). When this happens, the pesticide is often considered an environmental pollutant.

The movement of a pesticide from the designated area may occur in several ways. Drift, soil leaching, runoff, improper disposal and storage, and improper application are some of the major causes of a pesticide becoming an environmental pollutant. Natural resources that can be contaminated are groundwater, surface water, soil, air, fish, and wildlife. Surveys have shown that more than 50 percent of the counties in the United States have potential groundwater contamination from agrichemicals. The three main factors affecting groundwater contamination by agrichemicals are:

- soil type and other geological characteristics;
- the pesticide's persistence and mobility within the soil; and
- the production and application methods of pesticide users.

Pesticide drift is a major cause of soil and air contamination. Drift is the movement of a pesticide through the air to nontarget sites. It will occur at the time of pesticide application when small spray particles are moved by air currents

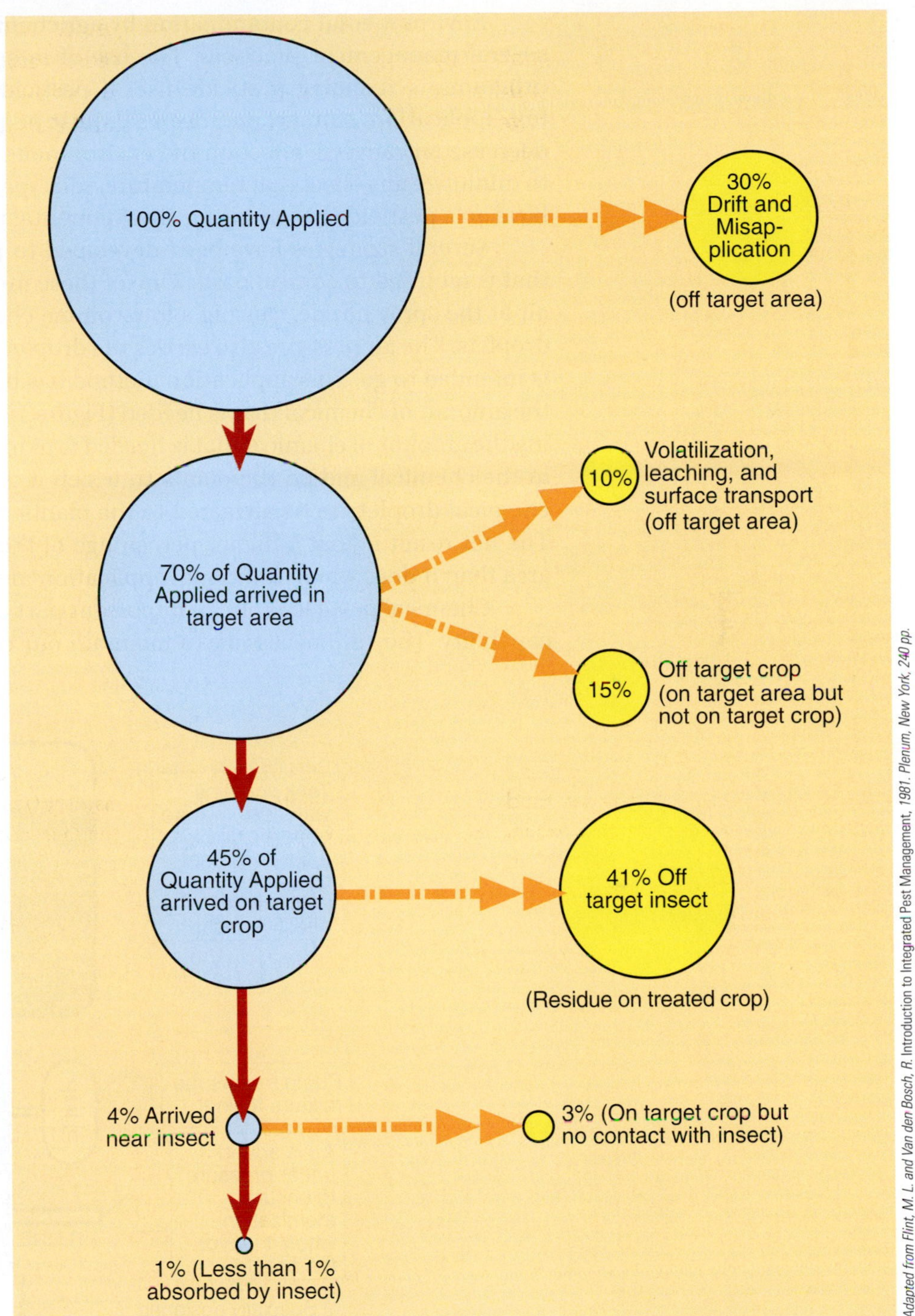

FIGURE 14-15 Aerial applications of pesticides result in very small amounts actually being absorbed by the insect that is targeted.

to nontargeted areas. Also, vapor drift of a pesticide may occur after an application. Vapor drift is movement of pesticide vapors because of chemical volatilization of the product.

The adverse effects of pesticides on fish and wildlife may directly result in animal mortality. Pesticides may also indirectly influence animal feeding or reproduction. Pesticide labeling will indicate any potential harm to wildlife, and this information should be heeded to minimize risk. Fish, birds, bees, and other animals will be affected when pesticides reach them or their habitats.

Environmental contamination by agrichemicals can be decreased through several management practices. The use of integrated pest management (IPM) programs is reducing pesticide use. If pesticides are used, then proper mixing, application, storage, and disposal must be performed. These practices will decrease any adverse effect on the environment. An attempt must also be made to minimize any effect that temperature, soil type, rainfall, and wind patterns may have on a pesticide becoming an environmental pollutant.

Several strategies have been developed to reduce the amount of chemical that is required to control pests. One of these methods is to inject high-pressure air at the spray nozzle, causing a low-volume chemical mix to separate into tiny droplets. The air pressure also carries the droplets to the area where the chemical is intended to go. This application method has been proven effective in reducing the amount of chemical that is needed (Figure 14-16). Another method for reducing the amount of chemical that is needed is to induce opposite electrical charges in the chemical and on the plants to which it is being applied. This causes the chemical droplets to be attracted to the plants to which they are being applied. The net result is that a higher percentage of the chemical reaches the targeted area than it does when traditional application methods are used.

Chemical pesticides are an important part of our food and fiber production capability. They are necessary to maintain our current standard of living in the

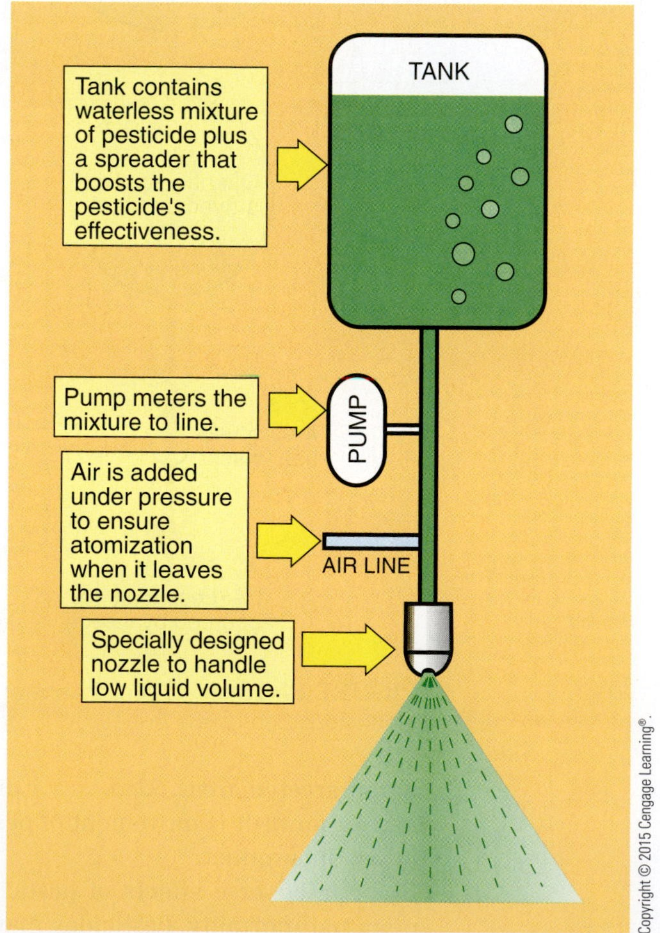

FIGURE 14-16 New technology permits pesticide application equipment to use less water and less pesticide to achieve equal or better control of pests.

United States and the world. However, they do create risks if used by improperly trained or careless individuals. Therefore, it is important that those using pesticides be properly informed and follow label instructions (Figure 14-17).

It also is important that consumers wash and handle food products to minimize the intake of pesticide residues on food (Figure 14-18). Although the government does extensive research to arrive at safe pesticide-residue levels, it is in the best interest of the consumer to wash fruits, vegetables, and all plant parts that may contain pesticide residues (Figure 14-19). This precaution will further reduce the hazard of pesticides.

Pesticide Safety
Choose the right pesticide for the job.
Carefully read the label and apply the chemical according to the directions for its use.
Mix the chemical properly to assure that the correct amount is applied.
Make sure that the applicator wears appropriate protective clothing.
Mix only the amount of chemical that is needed, and apply chemicals that are left over to a crop where it will not affect the environment in negative ways.
To avoid chemical drift, consider wind and weather conditions before making the decision to apply a chemical.
Clean the sprayer and carefully dispose of the wash water to avoid contamination to the environment.
Avoid inhaling or ingesting the chemicals and wash dangerous chemicals off the skin immediately.
Store chemicals in locked areas in the original containers away from people, animals, and feeds.
Dispose of chemical waste and empty containers according to the rules for handling hazardous waste materials.

FIGURE 14-17 Pesticide safety.

FIGURE 14-18 The government does extensive research to arrive at safe pesticide residue levels.

FIGURE 14-19 Fresh fruits and vegetables should always be carefully washed to remove any possible pesticide residue.

Similarly, the wise person will become familiar with the uses of pesticides. Such uses include pesticides to control roaches, termites, flies, insects, and diseases of lawn and garden plants; insect repellants and insecticides used for outdoor camping and recreation; and other everyday common practices. The benefits of pesticides far outweigh the risks.

STUDENT ACTIVITIES

1. Write the Terms to Know and their meanings in your notebook.
2. Ask your instructor to show examples of an approved pesticide applicator's respirator, goggles, gloves, boots, clothing, and other protective items.
3. Prepare and present a class demonstration on the proper use of one or more items of protective clothing and devices for safe handling of pesticides.
4. Do research and prepare to defend the position that pesticide residue on food in the United States is or is not a serious problem. Arrange for at least three classmates to prepare for a debate presenting various issues regarding the use of pesticides.
5. Ask a professional pesticide applicator with a special interest in pesticide safety to demonstrate safe pesticide-application principles to the class.
6. Collect newspaper and magazine articles on accidents involving pesticides. Study the articles and determine how each accident could have been avoided. Report your conclusions to the class. Note: Many pesticide accidents occur in or on the home, lawn, and garden. Do not overlook such cases.
7. Conduct a survey of pesticide storage-and-use practices in your home, farm, and place of employment. Correct all safety violations.
8. List the classes of chemicals used for pest management. Give a brief description of each, and explain how they work to prevent or control pests.

SELF-EVALUATION

A. MULTIPLE CHOICE

1. An example of an inorganic pesticide is
   a. pyrethrum.
   b. rotenone
   c. Bordeaux mixture.
   d. organophosphate.
2. The total amount of pesticides used annually in the United States is
   a. 820 million pounds.
   b. 420 million pounds.
   c. 620 million pounds.
   d. 5.1 billion pounds.
3. A preemergence herbicide is applied
   a. after the weed or crop is present.
   b. before the weed or crop is present.
   c. at the time of planting.
   d. none of the above
4. Which herbicide family inhibits photosynthesis?
   a. acetanilines
   b. dinitroanilines
   c. phenoxys
   d. triazines

5. An example of botanical insecticide is
   a. 2, 4-D.
   b. Rotenone
   c. diazinon.
   d. sulfur.
6. Pesticide registration often takes
   a. between 1 and 2 years.
   b. between 3 and 4 years.
   c. between 8 and 10 years.
   d. 15 years.
7. Pesticide risk can be decreased by
   a. proper pesticide exposure.
   b. reading the label.
   c. minimizing pesticide exposure.
   d. all of the above.
8. The signal word(s) for a highly toxic pesticide is
   a. CAUTION.
   b. WARNING.
   c. DANGER—POISON.
   d. none of the above.
9. An example of protective clothing or gear that will minimize inhalation of a pesticide is
   a. a respirator.
   b. boots.
   c. gloves.
   d. Coveralls.
10. The number of lethal pesticide poisoning cases per year is
   a. aless than 20.
   b. between 20 and 30.
   c. between 40 and 60.
   d. more than 100.

B. MATCHING

______	1. Carcinogen	a. The effect of a single exposure to a pesticide
______	2. Chronic toxicity	b. Movement of a pesticide through the air to nontarget sites
______	3. Water	c. Repeated exposures to low doses of a pesticide
______	4. DF	d. A soluble powder pesticide formulation
______	5. Signal word	e. A material capable of producing a tumor
______	6. Acute toxicity	f. A fungicide used after disease infection
______	7. SP	g. Used to remove pesticides from the body
______	8. Drift	h. A dry, flowable pesticide formulation
______	9. Protectant	i. Describes the acute toxicity of a pesticide
______	10. Eradicant	j. A fungicide used before disease infection

C. COMPLETION

1. An ______ will control insects.
2. A ______ herbicide will control all types of plants.
3. A systemic herbicide will be translocated in the ______ and ______ tissue.
4. The amount of money spent on pesticides in 1999 was ______ billion dollars.
5. A pesticide will have only one ______ and one ______ name.
6. To reduce exposure to pesticides, the use of ______ is recommended for clothing and boots.
7. Environmental and health hazards of a pesticide are listed under the ______ of the label.
8. The ______ number identifies where the pesticide was manufactured.
9. LD refers to the ______ of a pesticide.
10. The ______ name is the manufacturer's name for its product.

SECTION 5

Scientists, growers, propagators, breeders—everyone gets into the act. How can we grow plants that are different; plants to provide materials for new medicines and industrial products; plants that provide more beautiful flowers and ornamentals; plants that resist diseases, insects, drought, cold, or heat? Where can we find trees that produce wood faster? These questions all drive our quests in plant science.

During World War II, the USDA sent botanist Richard Schultes to South America in search of rubber trees resistant to diseases. Such trees were discovered, which opened the way for rubber plantations in the Western Hemisphere. These and other plants, which USDA explorers collected, changed agriculture around the world, as noted in the following examples:

- Curare vines provide a muscle relaxant used in surgery.
- Ucu uba trees in Colombia have bark that provides a highly desired appetite suppressant and skin medicine.
- Rootstock from China strengthens peach trees.
- Navel oranges from Brazil have strengthened the California citrus industry.
- Durum wheats from Russia set the standard for U.S. varieties for years.
- A peanut from Peru has genes that are resistant to two major diseases that impacted the U.S. peanut industry.
- Wild oats have resulted in one of the most disease-resistant oat varieties ever developed.
- California's avocado industry was started with germ plasm from Mexico.
- Sorghum, dates, tung oil, and numerous forage grasses from around the world are now grown in the United States.

More than 1600 species of plants are used for medicinal purposes by people of the Colombian Amazon. However, many of these have yet to be studied by scientists. The USDA Chief of the Agriculture Research Service coordinates about 10 trips a year in

The Quest for More and Better Plants!

search of new plants and plant materials. The need for plant collecting, cataloging, and preserving becomes more urgent each year. The genetic base for many of the crops we take for granted is narrow. Furthermore, the encroachment of modern civilization on plants in remote places of the world is wiping out some important sources of new genes. Such plants could carry the genetic material needed to help important species survive. They may also be the source of genetic material that is needed to develop and introduce new plant species with new uses.

Small samples of some of the diversity of plant materials are brought back to the United States. The following guidelines are used by modern plant hunters and collectors when exploring for or making new discoveries:

PLANT SCIENCES

- Trips are to be organized as collaborations between the United States and the host country.
- All collected material is divided at least equally with the host country.
- All germ plasm collected with the support of the USDA is deposited in the National Plant Germ Plasm System and is available to all valid users.
- All collection must "be done with a conservation ethic in mind."
- Collection must not endanger natural plant populations, and enough must be left behind so that the plant population can regenerate naturally.

Scientists believe that genetic diversity can be better preserved when new plants are cataloged and most of them are left and protected where they are found.

Centers of origin for some major fruits, vegetables, grains, and oil crops.

UNIT 15

Plant Structures and Taxonomy

OBJECTIVE

To identify major parts of plants and state the important functions of each.

MATERIALS LIST

- plant collection materials
- bulletin board materials
- Internet access

COMPETENCIES TO BE DEVELOPED

After studying this unit, you should be able to:

- draw and label the major parts of plants.
- describe the major functions of roots, stems, fruits, and leaves.
- draw and label the parts of a typical root, stem, flower, fruit, and leaf.
- explain some of the variations found in the structures of root systems, stems, flowers, fruits, and leaves.
- describe the relationship of plant parts to fruits, nuts, vegetables, and crops.

SUGGESTED CLASS ACTIVITIES

1. Gather plant materials that can be used to illustrate as many of the Terms to Know as possible. Assign students to work in teams, and have them label as many of the plant parts as they can identify (reference books are encouraged). Follow up with a review to reinforce correct labels and to correct mistakes.
2. Collect, prepare, and laminate samples of plants that are found in your community. Accomplish this during the summer months before the students are in class. Add additional samples to the collection as students bring them in. Have class members memorize the common names of each of the samples.
3. As a class, brainstorm ways to remember the taxonomic order: kingdom, phylum, class, order, family, genus, and species, for example, "King Phillip Came Over For Good Soup."

TERMS TO KNOW

taproot
fibrous root
adventitious root
root cap
xylem
root hair
herbaceous
bulb
corm
tuber
internode
axil
terminal bud
vegetative bud
flowering bud
phototropism
margin
simple leaf
compound leaf
leaf blade
petiole
cuticle
epidermis
chloroplast
stoma
guard cell
flower
stamen
filament
anther
pollen
pistil
stigma
style
ovary
ovule
perfect flower
pollination
petal
sepal
calyx
fruit
taxonomy
genus
species
binomial

Plants are a basic part of the food chain. Without plants, the Web of life cannot exist, and most animals and humans would die. Knowledge of plant growth is essential. To have a better understanding of plants, it is necessary to identify the parts that make up plants. The casual observer sees stems, branches, leaves, and possibly flowers and some nuts or fruits. The agriscience technician, however, sees a series of interconnected tissues and organs that depend on each other to function. The technician knows that all of the organs do not need to be present at one time but is aware that each cell or organ has an important role in the successful growth of the plant. Plant technicians and scientists are concerned with the efficient growth of plants. A plant may become stressed if one or more parts are absent or not functioning properly.

The basic industry of agriculture is dependent on the proper functioning of plants. The animal grower needs many kinds of plants to feed livestock and poultry. The plant industry also needs superior plants for feed and seed production. The horticulture industry needs plants for seeds and cuttings and for food, such as fruits, vegetables, and nuts. Plants are also needed for landscaping the inside and outside of homes and office buildings.

To be successful with plants, one must have knowledge of plant parts and how they function. Such knowledge is essential, whether you are growing, selling, or using plants.

THE PLANT

Plants are composed of many parts. Each part is important in the overall life and function of a plant. The root system is normally under the ground and is responsible for anchoring the plant and supplying water and nutrients. The stem, or trunk, is normally above the ground and functions as a support system for the rest of the aboveground parts. Leaves constitute the food-manufacturing parts of plants. Flowers come in many sizes, colors, and shapes and function as the seed-producing parts of the plant. Healthy plants produce seeds, nuts, fruits, and vegetables. These parts are popular foods for animals and humans. They are also used for reproduction of the plant (Figure 15-1).

Basic Necessities of Plant Life

For a plant to survive, its basic needs must be met. These needs include light, water, air, and minerals (Figure 15-2). Animals must capture their food to have sufficient energy. Plants must also capture an energy source. Some plants can thrive in quite shady areas, whereas others need direct sunlight. The sun's light is required by plants to perform photosynthesis. In this process, plants convert the sun's energy into food. This process is discussed in Unit 16. Another requirement for healthy plants is water. The amount of water needed varies from plant to plant. A desert cactus needs far less water than a tropical fern, but no matter the amount, all plants need it. Most water is taken in through the roots; however, many plants can absorb small quantities through the leaves. Plants also need air. Oxygen is used during plant respiration, whereas carbon dioxide is required for photosynthesis. Minerals are necessary to supply nutrients for plants. Many of these minerals are found free in the soil, but others must be supplemented by providing fertilizer.

FIGURE 15-1 Seeds, nuts, and fruits are plant parts commonly used for food.

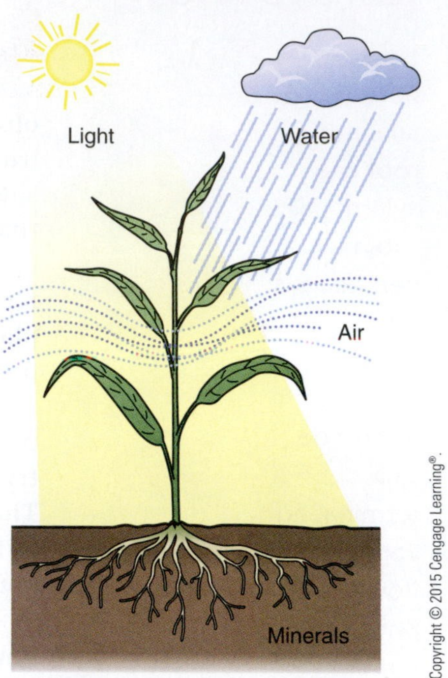

FIGURE 15-2 A plant draws all of its needs from the environment in which it lives.

ROOTS

Root Systems

The largest part of the plant is often the root system. Roots take up more space in the soil than does the top part of the plant seen in the air above the ground. In fact, some roots will go down into the soil 6, 8, or even 10 feet. Some plants, such as squash, have massive root systems.

Root systems are generally either taproot or fibrous roots (Figure 15-3). Knowledge of these two types of root systems can be of value in caring for and

FIGURE 15-3 Corn plants showing extensive fibrous root systems at an early age.

handling plants. The **taproot** is the main root of a plant and generally grows straight down from the stem. It is a heavy, thick root that does not have many side, or lateral, branches. Taproots are often used for human and livestock consumption because they are food-storage organs. Carrots (*Daucus carota*) and sugar beets (*Beta vulgaris*) are examples (Figures 15-4A, B, C, and D). Some plants with taproots are used for ornamental purposes. An ornamental plant is used to improve the appearance of a structure or area.

Plants that have taproot systems have the ability to survive periods of drought. Because they grow deep into the soil and have few fine secondary roots, taproots do not stabilize the soil well. **Fibrous roots** are generally thin, somewhat hair-like, and numerous. The fibrous root system is normally shallow. Grasses, corn, and many ornamentals, such as begonia, are good examples of

FIGURE 15-4 Taproots of carrots (A), radishes (B), turnips (C), and beets (D) are healthy sources of food.

plants with fibrous root systems. Many small, thin-branched roots are present in this type of system. The result is that they are able to hold soil much better than taproot systems. However, fibrous root systems dry out more easily. They cannot tolerate drought conditions.

Although many roots are in the soil, other types are seen above the ground that may not be considered roots. Some plants, such as poison ivy (*Rhus radicans*) and English ivy (*Hedera helix*), have roots that help them climb trees, walls, and sides of buildings. These are called adventitious roots. **Adventitious roots** appear where roots are not normally expected (Figure 15-5). Adventitious roots also prop up plants such as corn, strengthening them against the wind.

The mistletoe, a popular Christmas plant, has roots that penetrate the bark of trees in the upper branches, or crown (Figure 15-6). These roots grow into the xylem and phloem tissues of the host plant and extract nutrients that originate in the soil. The dodder plant (*Cuscuta campestris*) has soil roots that die off as the plant gains a foothold in a plant. The dodder plant forms root-like attachments that penetrate the stem of the host plant and extract nutrients from that plant.

Root Tissues

Although there are different systems, all roots look similar when they are examined on the inside (Figure 15-7). The parts of the root have very specific functions in the plant. Knowledge of these parts is helpful in diagnosing diseases and other dysfunctions of plants.

The Root Cap

The **root cap** is the outermost part of the root. It protects the tender growing tip as the root penetrates the soil. The root cap is a tough set of cells that is able to withstand the coarse conditions that the root encounters as it pushes its way through soil with rock and small sand particles. As the root cap wears away, the cells are replaced by more cells that develop at the root tip. This portion of the root is known as the area of cell division.

Courtesy of DeVere Burton.

FIGURE 15-5 Adventitious roots are prop roots that are located above the soil surface. Their purpose is to provide a strong base that aids in keeping a plant in an upright position, especially when soils are wet and conditions are windy.

Courtesy of Boise National Forest.

FIGURE 15-6 Mistletoe is a parasitic plant with roots that grow through the bark and into the tissue of the tree branch.

AGRI-PROFILE

CAREER AREAS: BOTANY/BIOLOGY/TAXONOMY

Courtesy of USDA/ARS.

Plant scientists have studied individual plant specimens and devised a system of classification of plants based on characteristics of plant parts.

Everyone relies heavily on our system of plant classification. Without it, we could not order, buy, sell, or instruct others about a plant unless the plant was in the presence of both parties at the same time. Taxonomists devote their careers to identifying, classifying, and teaching others about plants. An important and exciting part of their work is that of discovering, studying, and naming new plants in their appropriate places in the classification system. Consider the excitement of discovering plants in faraway lands or even at home that have not been observed before or recognized even by the most knowledgeable specialists.

Botany is the study of plants, and biology is the study of both plants and animals. Knowledge of plant structures is critical to both. Consumers of plants for food, ornamentation, medicine, shade, wood, fuel, fiber, and other uses should have some knowledge of plant structures. Plant structures affect plant nutrition, functions, disease susceptibility, adaptation, and use.

Area of Cell Division and Elongation

The area of cell division provides new cells that allow the root to grow longer. The cells in this area multiply in two directions. Small, tough cells are produced on the front edge of this region. They replace cells of the root cap that are worn off or destroyed as the tip pushes its way through the soil. Small, tender cells are produced on the back of this area. They are used as the root tip grows longer. The area of cell division is actually quite thin, perhaps as thin as a strand of hair.

The next area, moving toward the base of the plant, is the area of cell elongation. In this area, the cells start to become longer and specialized. They also begin to look like the older cells and will start to perform their specific functions.

Xylem and Phloem

There are many types of cells in the root. Perhaps the most important ones are the xylem and the phloem cells. The **xylem** cells are responsible for carrying the water and nutrients in the soil to the upper portion of the plant. The phloem cells function as the pipeline to carry the manufactured food down from the leaves to other plant parts, including the roots, where it is used or stored. There are other specialized cells in the root; some are discussed later in this unit.

INTERNET KEY WORDS:
plant parts functions

Area of Cell Maturation

The area of cell maturation is where cells become fully developed. This is also where the root hairs emerge. **Root hairs** are small microscopic roots. They will

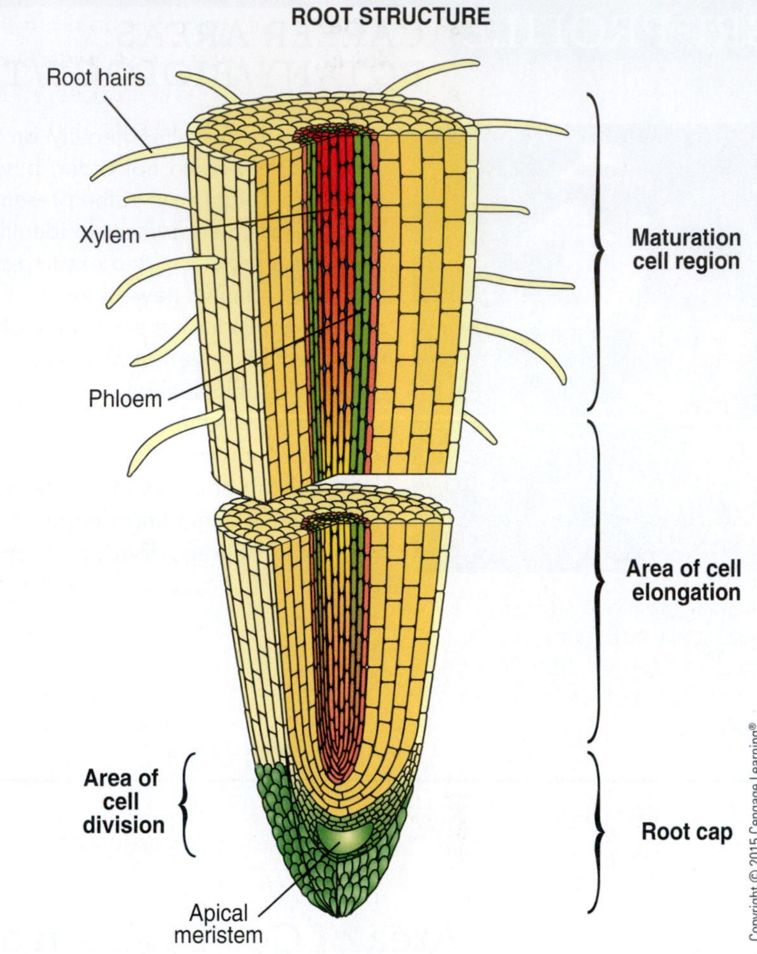

FIGURE 15-7 Root structure.

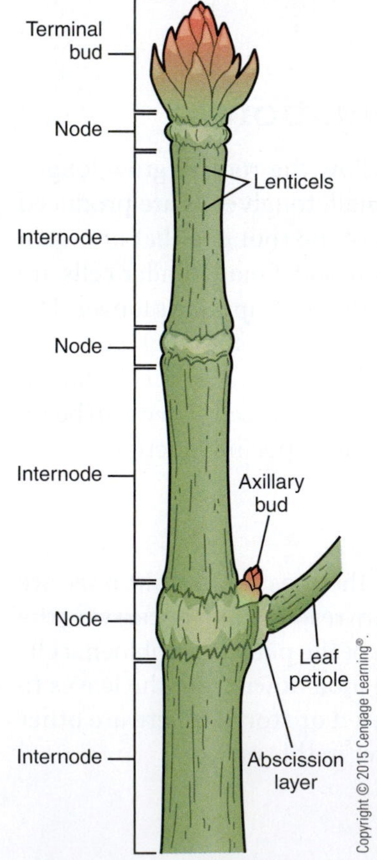

FIGURE 15-8 Important parts of stems.

rise from existing cells located on the surface of the root. It is the job of root hairs to absorb water and nutrients. Water and nutrients move into the root hairs, enter the xylem, and move to the upper portions of the plant. Root hairs are small and tender. They will break off very easily. This means plants must be handled very carefully when transplanting. Once the root hairs are broken off, they cannot grow again or be replaced.

Although roots are normally hidden from view, they are very important parts of the plant. Roots need the same care and consideration as the other parts for plants to grow well.

STEMS

Stems are among the first things seen by the casual observer when looking at plants (Figure 15-8). Stems and branches are noticeable in the winter when the leaves are gone. They are easily seen as a plant grows. Stems support the leaves, flowers, and fruit.

Types of Stems

Two types of stems grow above the ground: woody and herbaceous. Woody stems are composed of tough material. They often have bark around them.

Herbaceous stems are succulent and somewhat tender. They usually do not survive the winter season in cold climates.

Not all stems are erect structures growing above the ground. Some grow along the ground or even underground. Some stems have specialized jobs to perform. Such stems are referred to as modified stems. Examples of modified stems are bulbs, corms, rhizomes, and tubers. **Bulbs** are short stems that are surrounded by modified leaves called scales. Some examples of bulbs are Easter lilies (*Lillium longiflorum*) and onions (*Allium* sp.; Figure 15-9). **Corms** are thickened, compact, fleshy stems. An example of a corm is the gladiola (*Gladiola* sp.; Figure 15-10). Rhizomes are thick stems that run below the ground. Johnson grass and the iris (*Iris germanica*) are examples of plants with rhizomes (Figure 15-11). **Tubers** are thickened, underground stems that store carbohydrates. We often eat an example of this type of stem, the Irish potato (*Solanum tuberosum*) (Figure 15-12).

Courtesy of USDA/ARS.

FIGURE 15-9 Bulbs, such as Easter lilies and onions, are shortened stems surrounded by modified leaves called scales.

© nimblewit/Shutterstock.com.

FIGURE 15-10 Corms, such as the gladiola, are stems that are thick, compact, and fleshy.

© iStockphoto/Thomas Arbour.

FIGURE 15-11 Rhizomes are thick stems that grow underground near the surface and give rise to new plants at each node.

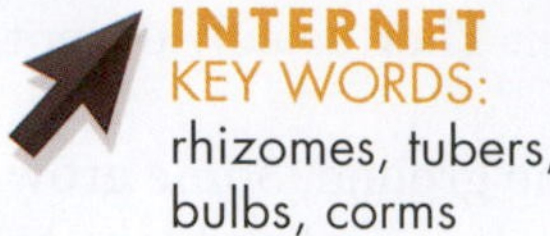

INTERNET KEY WORDS:
rhizomes, tubers, bulbs, corms

Parts of Stems

Stems have some of the same internal parts as roots. The xylem and the phloem continue to run the length of the stem and into all the branches of the plant. In a subclass of plants called dicotyledons, the xylem and phloem occur together in tissues called vascular bundles. In another important subclass called monocotyledons, the xylem and phloem occur in separate areas (Figure 15-13).

Courtesy of USDA/ARS #K-4016-5.

FIGURE 15-12 The potato is really a specialized stem called a tuber.

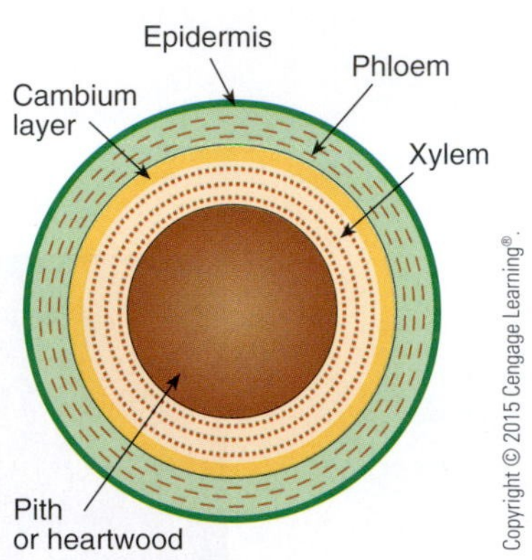

FIGURE 15-13 This illustration of the cross section of a stem shows the xylem and phloem cells that make up the vascular system of a plant.

BIO-TECH CONNECTION

INSIDE A CELL: PLANT BIOTECHNOLOGY AT WORK

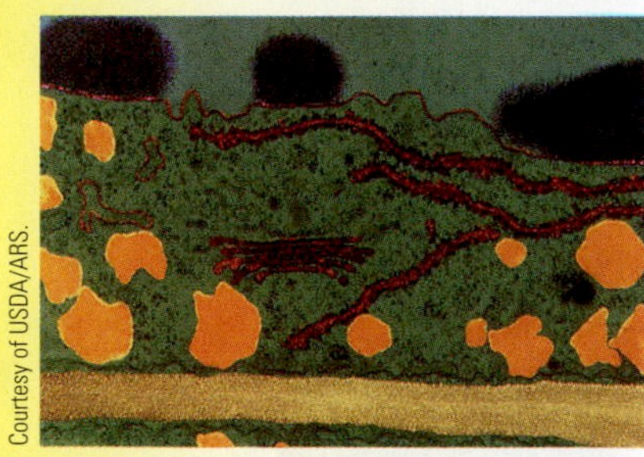

Courtesy of USDA/ARS.

A computer-enhanced view of the interior of a soybean cell.

This electron micrograph shows some of the components in a soybean seed cell. For better visibility, computer processing has colored seed storage proteins purple and stored oil yellow. Red areas are cellular compartments where synthesis of the stored protein and oil occurs. The electron micrograph was taken by Eliot Herman of the Agriculture Research Service, and the color enhancement was performed by Terry Yoo of the Science and Technology Center for Computer Graphics and Scientific Visualization at the University of North Carolina.

Through the use of standard and electron microscopes, together with other modern technology, scientists can peer into the most minute parts of plant structures. They can improve the images seen through various instruments via computer enhancement. The fields of molecular biology, cellular biology, and genetics search the mysteries of the cell and its components and seek to manipulate its biology to bring about desired changes. Larger parts of plants are observed with hand glasses or the naked eye.

Plant structures permit the plant to function as a whole. If one part becomes diseased, it will limit the performance of the rest of the plant and may lead to death of the plant. If the roots cannot obtain sufficient water and nutrients, the leaves cannot manufacture food for the plant. If the stem grows too fast without developing strength, the plant may topple and die. If seeds do not form and mature properly, the plant may not be able to reproduce and perpetuate itself. The whole is dependent on the health of its parts.

HOT TOPICS IN AGRISCIENCE SHORTER STRAW—A STRONGER WHEAT PLANT

A significant improvement has occurred in some varieties of wheat because of the efforts of scientists. They have discovered ways to improve the wheat plants by reducing the length of the stem and increasing the thickness to make it stronger. This change in the structure of the wheat plant results in fewer problems caused by lodging. Lodging occurs when wind and moist conditions combine to cause the wheat plants to fall over and remain flat on the surface of the soil. This condition makes it difficult to harvest the wheat, and some of the grain is usually left in the field because the grain combine cannot pick it up. The short, thick straw of some of the new wheat varieties makes them much less susceptible to this problem.

Some important external parts of plant stems include the node, internode, axillary bud, lenticels, and terminal bud. The node is the portion of the stem that is swollen or slightly enlarged where buds and leaves originate. The **internode** is the area between the nodes. The **axil** is the angle above a leaf stem or flower stem and the stalk. The axillary bud grows out of the axil. The function of the axillary bud is to develop into a leaf or branch. Lenticels are pores in the stem that allow the passage of gases in and out of the plant. The **terminal bud** is located on the tip or top of the stem or its branches. It may be either a vegetative or flowering bud. The **vegetative bud** produces the stem and leaf growth of the plant. The **flowering bud** produces flowers.

INTERNET KEY WORDS:
plants, phototropism
internal leaf structure

LEAVES

The plant leaf has an important function. It manufactures food for the plant by using light energy.

A plant leaf is capable of adjusting its angle of exposure to the sun. The leaves of some plants turn to allow full sunlight to shine on the leaf surfaces as the position of the sun changes during the day. The process by which this occurs is called **phototropism**. Without this important plant reaction to sunlight, plant growth would be reduced.

Leaf Margins

Plants may be identified by the edges, shape, and arrangement of the leaves. The leaf edges are known as **margins**. Leaf margins are named or described according to the toothed pattern on each leaf edge (Figure 15-14).

Leaf Shape and Form

Leaves vary in shape and form according to their species. Therefore, knowledge of the name given to each leaf shape and form is useful in identifying the plant (Figures 15-15 and 15-16).

Types of Leaves

Leaf types vary according to the species. Therefore, leaf type is also used to identify plant species. A single leaf arising from a stem is called a **simple leaf**.

FIGURE 15-14 Leaf margin characteristics are helpful in identifying plants.

Two or more leaves arising from a common point on the stem are referred to as **compound leaves** (Figure 15-17).

Leaf Parts

A leaf consists of a petiole and blade. These are the most familiar parts of a leaf. The **leaf blade** is the wide portion. It may be of many shapes and

FIGURE 15-15 Names of various leaf shapes.

sizes. The **petiole** is the stem of the leaf. It may be almost absent or may be very long.

Internal Structure

The leaf is the food-manufacturing unit for the plant. The process of manufacturing food is called photosynthesis. Food that is created through photosynthesis enables

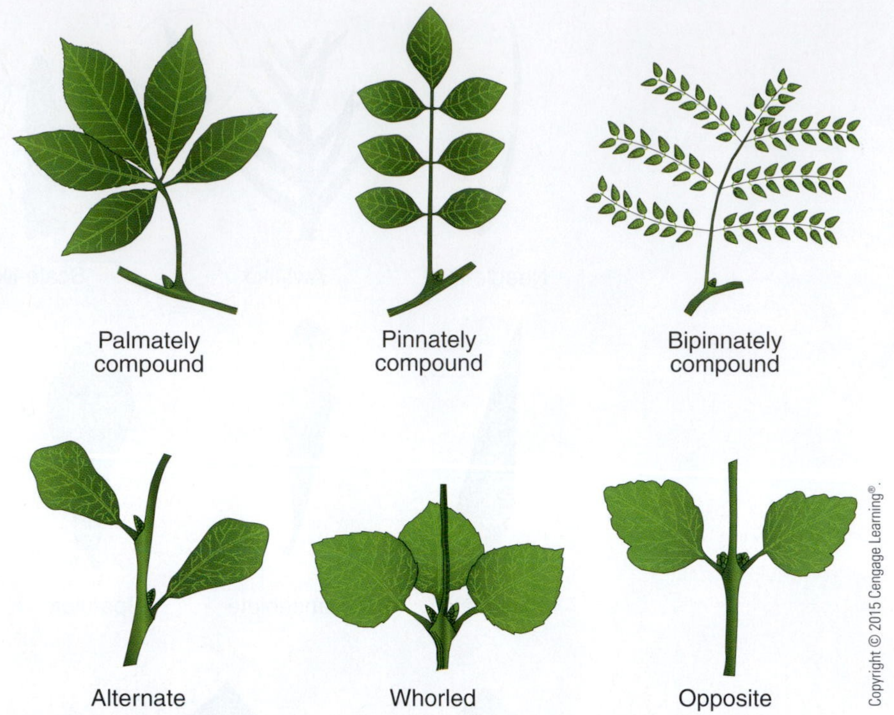

FIGURE 15-16 Examples of various leaf arrangements.

the plant to grow. The process of photosynthesis is illustrated in Figure 15-18 and is discussed in greater detail in Unit 16.

The **cuticle** is the topmost layer of the leaf. It is waxy and functions as a protective covering for the rest of the leaf. The **epidermis** is the surface layer on the lower and upper sides of the leaf. The epidermis protects the inner leaf in many ways. Elongated, vertical palisade cells give the leaf strength and are the sites for the food-manufacturing process. These cells, as well as the lower

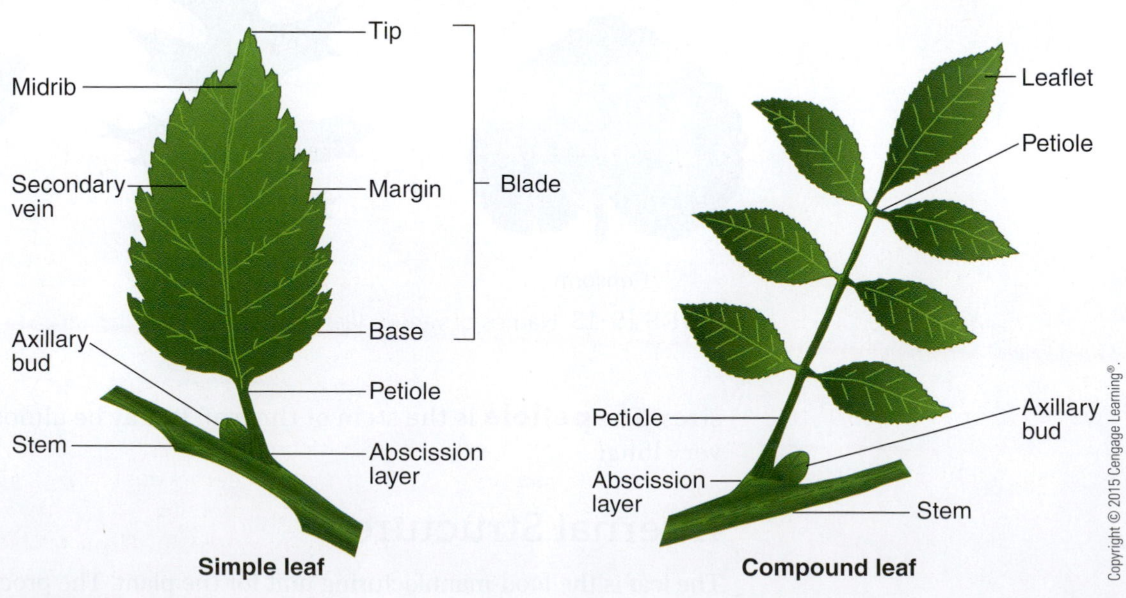

FIGURE 15-17 Parts of a leaf.

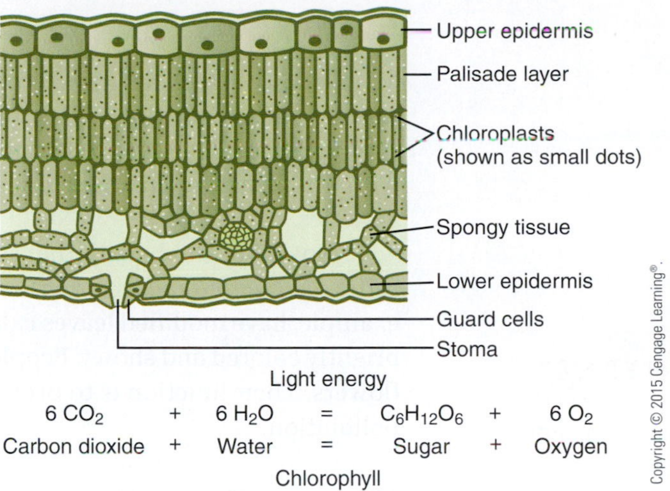

FIGURE 15-18 The leaf is the food-manufacturing part of a plant.

spongy tissue layer, contain chloroplasts. A **chloroplast** is the part of the cell that contains chlorophyll, a structure that is necessary for photosynthesis to occur. The lower layer is irregular and allows the veins, or vascular bundle, to extend into the leaf. The palisade cell layer and the spongy tissue are often referred to as the mesophyll.

The vascular bundles contain the xylem and the phloem. These are extensions of the same tissues that are located in the root. They extend through the stem to the leaves. The xylem brings the water and minerals from the root. The phloem carries the manufactured food from the leaf to the various parts of the plant to nourish plant tissue or to be stored. The lower epidermis contains some special cells called **stomas**. These openings allow for the exchange of carbon dioxide and oxygen as well as some water. The stomas are surrounded by **guard cells** that open and close the stoma. If the plant is stressed by the lack of water or by a low light level, the guard cells will close the stoma. The result is that the plant cannot manufacture food because it will not have all the necessary ingredients.

SCIENCE PROFILE THE DIVERSITY OF PLANTS

There are four different plant types in the Kingdom Planta. Bryophytes are plants that lack a vascular system. They do not produce seeds, and they must depend on water for reproduction. A vascular system is a system of vessels in a plant, each of which carries water and nutrients independently from each other throughout the plant. Moss is an example of a Bryophyte. Seedless vascular plants have distinct vascular systems. They also need water environments to reproduce, but unlike the Bryophytes, they have true roots, leaves, veins, and stems. Ferns are seedless vascular plants.

Seed plants include both the gymnosperms and the angiosperms. Both groups of plants produce seeds. In contrast to plants mentioned earlier, they do not require water environments to reproduce. Both plant types produce seeds and have extensive vascular systems. Gymnosperms bear their seeds in cones, whereas angiosperm seeds are protected in a thick tissue called fruit. Examples of gymnosperms are pine trees, ginkgo trees, and all conifers. Angiosperms, which are flowering plants, are the most abundant of all of the plants. Grasses, corn, petunias, and most crop plants are angiosperms.

FLOWERS

Many people see plants mostly for the beauty of the flower. Others see only a fruit to eat. Fruit production is only part of the job of the flower. The **flower** has as its primary function the production of seeds needed to continue the species. It is with this structure that the plant scientist will work to produce new and different varieties.

Not all of the beauty that is seen as flowers is actually flowers. The poinsettia (*Euphorbia pulcherrima*) and the flowering dogwood (*Cornus florida*), for example, have modified leaves called bracts. A bract is a modified leaf that is often brightly colored and showy. People often see the red or white bracts and call them flowers. Their function is to protect the flower parts as well as attract insects for pollination.

INTERNET KEY WORDS: flower structure

Flower Structure

Flowers are composed of many parts, including the filament, anther, pollen, stigma, style, ovary, petals, and sepals (Figure 15-19). These are the most important parts of the flowers.

The male part of the flower is the **stamen**. It consists of the filament, anther, and pollen. The **filament** supports the anther. The **anther** manufactures the pollen. The **pollen** is the male sexual reproductive cell. The female part of the flower, the **pistil**, is made up of the stigma, style, and ovary. The **stigma** receives the pollen. The pollen travels down the **style** and into the **ovary**, which

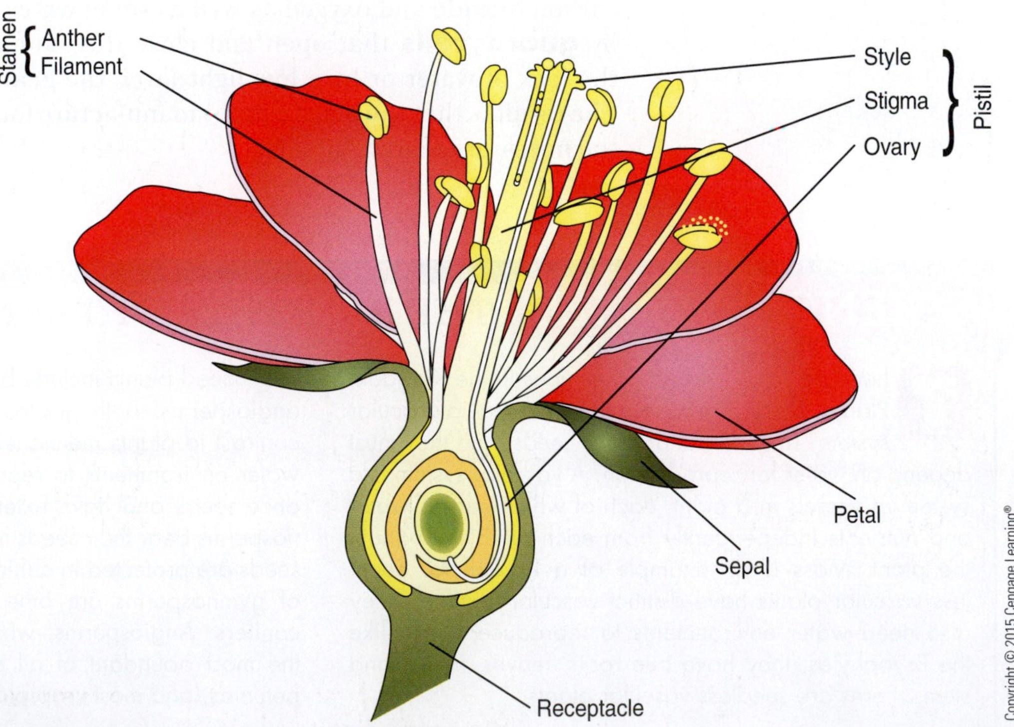

FIGURE 15-19 Major parts of a flower.

contains **ovules**. Ovules are the eggs, which are the female reproductive cells. When the eggs are fertilized by the pollen, they will ripen into seeds.

If a flower contains all the parts just mentioned, it is a **perfect flower**. If one or more of the parts is missing, it is considered an imperfect flower. In some plants, particularly the flowering plants used in horticulture, it is desirable to remove the anther sacs before the pollen ripens. This prevents pollination and stains from the pollen on the petals of the flower. **Pollination** means the union of the pollen with the stigma. In some cases, orchids, for example, the unfertilized flower will last for many months.

In plant breeding, the anther sac is removed from the plant to prevent natural pollination. It may be destroyed, or it may be used to pollinate another flower to create a new variety. Many hybrids are created in this way.

The colored **petals** attract insects or other natural pollinators. The flower petals are collectively called the corolla. The **sepals** function together as a protective device for the developing flower. Collectively, the sepals compose the **calyx**.

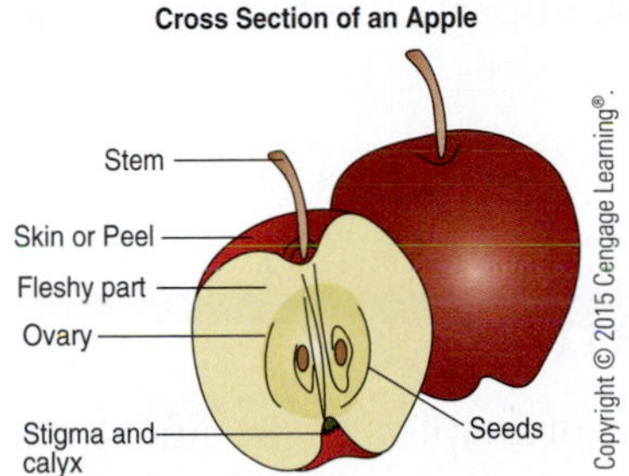

FIGURE 15-20 The ovary of a flower matures into a fruit that surrounds the seeds. When the fruits are eaten, the seeds are scattered to new locations.

Fruits, Nuts, and Vegetables

After fertilization, the ripening seed develops in the pistil. The pistil then enlarges and becomes the **fruit** (Figure 15-20). The fruit may be of many different shapes and sizes (Figure 15-21). The true fruit consists of the seeds that carry the male and female genetic characteristics of the plant. However, the fleshy material surrounding the mature seed, and the seed itself, is commonly called the fruit. The

A

B

FIGURE 15-21 Fruits from different plants vary in size, shape, and taste. (A) Pineapple growing in Hawaii and (B) peaches in Georgia.

purpose of the fleshy part of the fruit is to attract animals and humans to the seed to help spread it over wide areas. This helps in the reproduction of the plant. Entire fruits or just the seeds may be moved by wind, water, animals, or humans. Often, the fruit is eaten by animals and humans as a source of food. Then the seeds may be discarded, where they can take root and grow into new plants. People assist greatly in spreading seeds and starting new plants when they plant seeds for crops.

There are many kinds of fruits and vegetables. The two terms are sometimes used incorrectly. A vegetable can be any part of a plant that is grown for its edible parts. This can be a root, stem, leaf, or ripened flower. However, fruit, which is a ripened or mature ovary, is a specific plant part. A nut is also a type of fruit.

INTERNET KEY WORDS:
plant, taxonomy, binomial system

PLANT TAXONOMY

Importance of Classifying Plants

Taxonomy is defined as the science, laws, and principles of classification. In biology, taxonomy provides the means for classifying organisms into established categories according to characteristics they exhibit. Classification makes it easier to understand and remember plants and animals by the similarities and differences found in their structures and parts. Living organisms are given Latin names to help scientists and technicians around the world communicate better. Latin is regarded as the universal language for those in professions dealing with the biological sciences. Agriscience has its origin in the biological sciences.

About 300,000 species of plants have been identified and classified. The plant names are based on Latin descriptions and must be approved by a special committee of plant scientists. Carl Linnaeus, a Swedish botanist, developed the current system of plant classification in 1753.

Without the botanical classification system to identify and classify plants, many different species would carry one common name. For example, all clovers would be identified as clover, even though crimson clover is a winter annual, sweet clover is a biennial, and white clover is a perennial.

Complete classifications of field corn and petunia are shown in Figure 15-22.

	Yellow Corn	Petunia
Common Name:	Corn	Petunia
Kingdom:	Plant	Plant
Phylum:	Spermatophyta (seed plants)	Embryophyta
Subphylum:	Angiosperm (seed in fruit)	Angiosperm
Class:	Monocotyledonae (single leaf seed)	Dicotyledonae (two-seed leaf)
Order:	Graminales (grasslike families)	Tubiflorea
Family:	Gramineae (grass family)	Solanaceae
Genus:	*Zea* (the corns)	*Petunia*
Species:	*Mays* (dent corns)	*Hybridea*
Variety:	Reid's yellow dent	Blue Moon

FIGURE 15-22 Example of a field crop (yellow corn) and an ornamental plant (petunia) showing their complete botanical classifications.

Binomial System Used in Classifying Plants

When identifying a plant by its scientific name, it is not necessary to give its entire classification. Rather, a specific plant can be identified by using the genus and species only because the genus and species name is not used in combination for any other plant or animal. For example, grain sorghum is *Sorghum* (genus) *vulgare* (species). **Genus** is the taxonomic category between family and species. It is customarily capitalized when written together with a species name. **Species** is the subgroup under genus. A species name is generally not capitalized when written in combination with its genus. It is appropriate for both the genus and the species names to be printed in italics. To compare the name of a plant or animal with that of a human, the species corresponds to the person's first name and the genus to the person's last name. The system of using genus and species in combination is referred to as a binomial system of classification. **Binomial** means consisting of two names.

Some species are broken down into varieties. A variety is a subgroup of plants developed and maintained in production, as opposed to a species that originates in the wild. Variety is a rank within a species. When writing a plant variety name, it is generally capitalized. An example is Triumph wheat.

STUDENT ACTIVITIES

1. Write the Terms to Know and their meanings in your notebook.
2. Observe the plants that are commonly grown in your area. Classify them by the type of root system they have.
3. Make a chart of the plants listed in Activity 2. Indicate their common and scientific names, and discuss their responses to drought or their water maintenance requirements.
4. Make a collection of different kinds of leaves. Classify each according to its shape and type of margin.
5. Sketch the parts of roots, stems, and leaves. Label all items.
6. Make a bulletin board showing the major parts of plants.
7. Ask your teacher to provide a microscope and slides of plant tissue. Diagram the plant parts and label the cells and other structures that you see.
8. Record the common and scientific names of plants listed in this unit and learn the correct spelling of each.
9. List the kingdom, phylum, class, order, family, genus, and species for three plants, using the Internet, library, and other resources. Be sure to use proper punctuation.

SELF-EVALUATION

A. MULTIPLE CHOICE

1. The primary function of the root is to
   a. make sure that the plant will grow.
   b. anchor the plant and supply water and nutrients.
   c. ensure that the plant can be propagated.
   d. hold up the stem of the plant and provide propagation material.

2. The portion of the root that takes in the water and plant nutrients is the
   a. root cap.
   b. area of root division.
   c. root hair.
   d. area of cell maturation.
3. The major types of root systems are
   a. area of cell division and fibrous.
   b. fibrous and root cap.
   c. cuttings and root hairs.
   d. fibrous and taproot.
4. The area of cell division is
   a. responsible for the production of new cells on the tip of the root.
   b. where the cells will start to specialize.
   c. located in the area where the root hairs start to erupt from the wall of the epidermal cell.
   d. Where the roots drop off on special plants such as the dodder.
5. The phloem
   a. is the pipeline that carries the water and nutrients from the soil to the leaves.
   b. is the part of the stem that gives support to the node.
   c. is the part of the leaf that holds it to the stem.
   d. carries the manufactured food from the leaves to the roots.
6. Herbaceous stems
   a. are tough and have bark around them.
   b. come from herbs.
   c. are green and are not winter hardy.
   d. are part of the bulb.
7. The node
   a. is the part of the stem that supports the flower.
   b. is the part of the stem where the leaf is attached.
   c. is the part of the stem that carries the nutrients.
   d. will become detached when dry weather sets in.

B. MATCHING

________	1. Root cap	a. Located on the tip of the stem
________	2. Terminal bud	b. The wide portion of the leaf
________	3. Leaves	c. Protects the root tip as it moves in soil
________	4. Cuticle	d. Manufacture food for the plant
________	5. Blade	e. The topmost layer on the leaf
________	6. Guard cells	f. Surround the stoma

C. COMPLETION

1. The roots are responsible for _______ the plant.
2. An _______ plant is a plant used to improve the appearance of an area.
3. The area of _______ _______ is where the cells start to become specialized.
4. _______ are thick stems that run below the ground.
5. _______ are pores in the stem that allow the gases to pass through the stem.
6. _______ _______ are cells that give the leaf strength.
7. The _______ is the male part of the flower.
8. When a flower contains the stamen, pistil, petals, and sepals, it is considered a _______ flower.
9. The _______ is an enlargement that results after fertilization.
10. A _______ can be any part of a plant that is grown for its edible parts.

D. TRUE OR FALSE

1. Plants with many thin, hair-like roots have fibrous root systems.
2. Plants with taproot systems are more likely to survive in a dry period.
3. The root cap protects the young growing tip of the root.
4. The xylem carries the nutrients and water down the stem of the plant.
5. Herbaceous stems are tough and winter hardy.
6. Bulbs are stems that are thick and compact.
7. The vascular bundle is only in the leaf of the plant.
8. The leaf consists of the petiole and the blade.
9. The stomas allow for the passage of gases only through the leaf surface.
10. The bract is the colored petal located in the flower.

UNIT 16

Plant Physiology

OBJECTIVE

To determine how plants make food and to describe the relationships among air, soil, water, and essential plant nutrients for optimum plant growth.

MATERIALS LIST

- writing materials
- encyclopedias
- corn or bean plant
- Internet access

COMPETENCIES TO BE DEVELOPED

After studying this unit, you should be able to:

- explain how plants make food.
- describe the roles of air, water, light, and media in relation to plant growth.
- trace the movement of minerals, water, and nutrients in plants.
- describe the ways that various plants store food for future use.
- compare the activity in a plant during exposure to light with periods of darkness.
- explain how plants protect themselves from disease, insects, and predators.

SUGGESTED CLASS ACTIVITIES

1. Meet as a class in the science laboratory or obtain sufficient microscopes to conduct a plant science laboratory class in another location. Stain some of the thin tissue from an onion bulb with iodine and observe the plant cells under the microscope. Observe under both low and high power. Identify the parts of a plant cell as they appear under the microscope.
2. Obtain two small potted plants. For one of the plants, fashion a covering from aluminum foil that completely blocks out sunlight to the plant. Place both plants in a window or other sunny location and wait one week. At the end of the week, remove the covering from the plant and examine the two plants to observe differences. Discuss the differences with the class.
3. Examine pondweed under a microscope on high magnification. Observe the plant cells. Make a drawing and label all visible structures.

TERMS TO KNOW

physiology
chlorophyll
glucose
light intensity
respiration
turgor
transpiration
osmosis
semipermeable membrane
pore
plant nutrition
primary nutrient
secondary nutrient
macronutrient
micronutrient
ion
anion
cation
precipitate
acid
alkaline
chlorosis

The life of a plant from its beginning to its maturity is a complex process. Many factors influence and directly control how a plant grows and what it produces. Growth, as in all living organisms, occurs by the division of cells and their enlargement as the plant increases in size (Figure 16-1). As the plant grows to maturity, cells are produced, divide, grow, and become specialized organs. These specialized organs are stems, leaves, roots, flowers, fruits, and seeds (Figure 16-2). The study of how these organs function and the complex chemical processes that permit the plant to live, grow, and reproduce is **physiology**. An understanding of the processes of germination, photosynthesis, respiration, absorption of water and nutrients, translocation, and transpiration enable the agriscience technician to maximize production of plants. The technician who works with interior and other ornamental plants must understand the environment of the plant because it is not in its native habitat.

PHOTOSYNTHESIS

The most important life-sustaining process is photosynthesis. Without this chemical process, maintenance of life on this planet would not be possible. Plants need carbon dioxide to manufacture food. Animals need oxygen to live. The complex chemical process of photosynthesis permits both plants and animals to live and support each other.

Photosynthesis is a series of processes in which light energy is converted to a simple sugar. Chlorophyll and chloroplasts are also essential in this process. **Chlorophyll** is the green material inside the leaves and stems of the plant. It is the substance that gives the green color to plant leaves. Chloroplasts are small, membrane-bound bodies inside cells that contain the green chlorophyll pigments. The chloroplasts are located in the mesophyll of the leaf. They

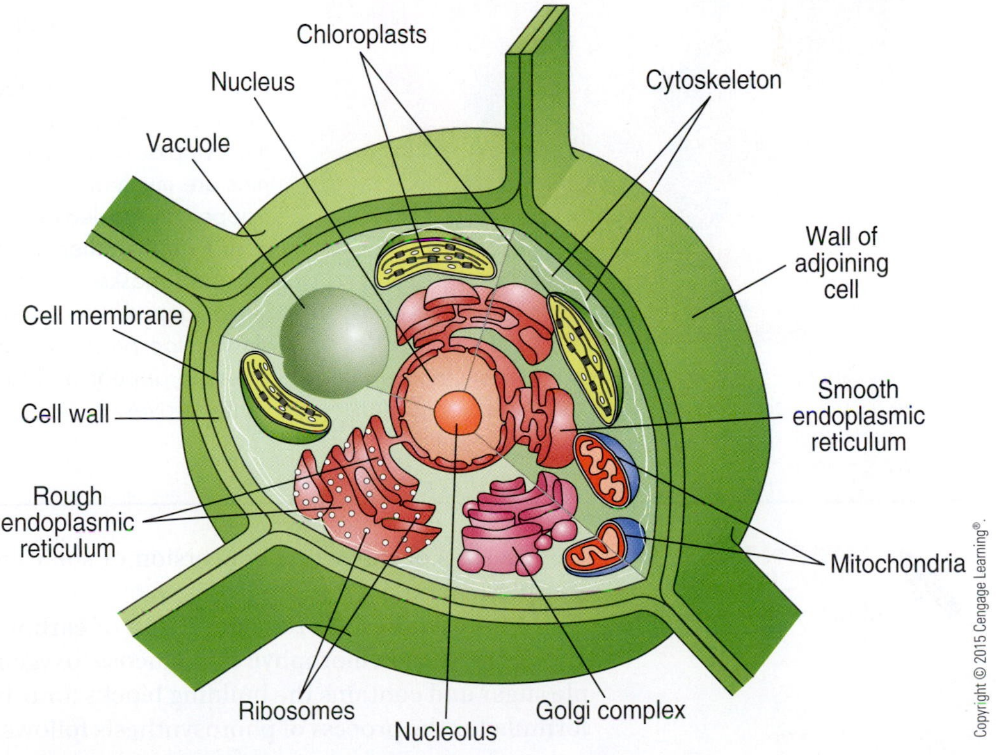

FIGURE 16-1 Major parts of a plant cell.

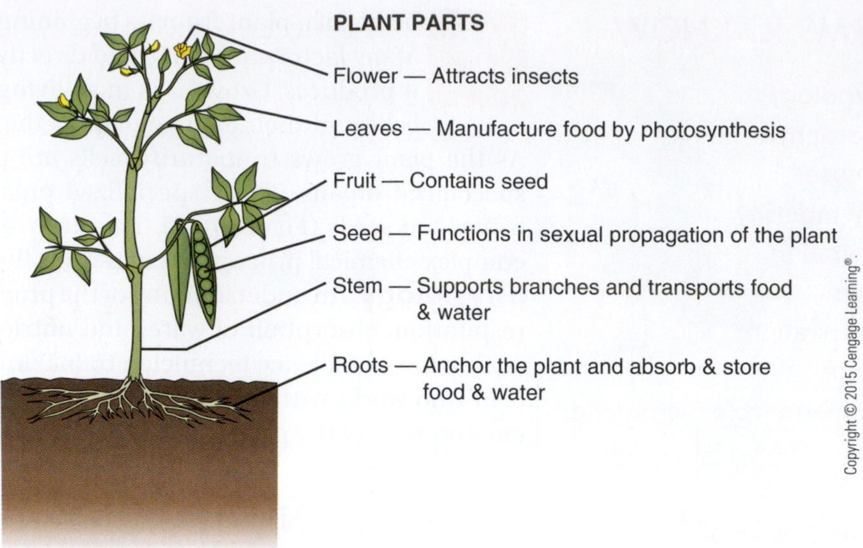

FIGURE 16-2 Major parts of a typical plant.

SCIENCE PROFILE WHY TREES CHANGE COLOR

The pigments that are responsible for the changes in leaf color become visible due to a reduction in plant chlorophyll that occurs as temperatures decline in the fall season.

In the fall of every year, trees put on a display of magnificent beauty. Leaves turn from green to orange, red, yellow, brown, crimson, purple, and scarlet. The change is a beautiful reminder that the cold temperatures of winter will soon arrive. Although the change may appear to be art, science is responsible. During the spring and summer months, the leaves of trees appear green. Chlorophyll, the green light-capturing pigment, also contains pigments called carotenoids. They are responsible for the yellow, brown, and orange colors found in bananas, rutabagas, and carrots. Anthocyanins are the red pigments that are found in the fluid within the cytoplasm of a plant cell. During the fall, anthocyanins are much more plentiful in response to sugars being trapped in the leaves. Chlorophyll is much more abundant than the other two pigments in the spring and summer and masks their colors until the fall when the day length gets shorter. As the days become shorter, plants slowly stop producing chlorophyll. This allows the other colors to appear and to produce a colorful show before winter arrives.

INTERNET KEY WORDS:
photosynthesis, plants

are the sites of the actual conversion of solar energy (light) into stored energy (simple sugars).

Photosynthesis is the conversion of carbon dioxide and water in the presence of light and chlorophyll into glucose, oxygen, and water. **Glucose** is a simple sugar and contains the building blocks for other nutrients. A simple chemical formula for the process of photosynthesis follows:

$$6CO_2 + 12H_2O \rightarrow C_6H_{12}O_6 + 6O_2 + 6H_2O$$

The rate at which the food-making process occurs depends on the light intensity, temperature, and concentration of carbon dioxide in the atmosphere. **Light intensity** is also known as the quality of light, or the brightness of light. Light must be present with sufficient brightness for the process to be successful. Some plants are able to adapt to various levels of light. Knowledge of the level of light required for plants to grow well is essential, particularly for indoor plant production. Temperature is also an important factor in the process of food manufacturing in the leaf. Photosynthesis occurs best in a temperature range of 65° to 85° F (18° to 27° C). Extremes of temperatures slow down or may completely stop the process of photosynthesis. A lack of carbon dioxide also will affect photosynthesis. Carbon dioxide is especially important in the beginning of the process. Under normal outdoor conditions, its availability is not a problem. However, in enclosed conditions such as those found in a greenhouse, carbon dioxide shortage could be a limiting factor. To correct this problem, a carbon dioxide generator is sometimes used (Figure 16-3).

RESPIRATION

All living cells carry on the process of respiration. **Respiration** is a process by which living cells (plant or animal) take in oxygen and give off carbon dioxide. Unlike photosynthesis, which occurs only in the light, respiration occurs both day and night. It is not easily measured during the day because the presence of photosynthesis will mask or obscure the occurrence of respiration. Respiration is a breaking-down process. It uses the sugars and starches produced by photosynthesis, converting them into energy. The chemical equation for respiration follows:

$$C_6H_{12}O_6 + 6O_2 \rightarrow 6CO_2 + 6H_2O + \text{heat (energy)}$$

A comparison of the activities that occur during photosynthesis and respiration may be helpful in understanding the two processes (Figure 16-4).

TRANSPIRATION

Water saturates all of the spaces between the cells throughout the plant. About 10 percent of the water that enters from the roots is used in chemical processes and

Courtesy of USDA/ARS #K3750-12. Photo by Jack Dykinga.

FIGURE 16-3 A technician measures the effect of carbon dioxide enrichment of the atmosphere on the transpiration rate and level of activity of the stomata that are structures in the plant leaves.

Photosynthesis	Respiration
1. Food is produced.	1. Food is used for plant energy.
2. Energy is stored.	2. Energy is released.
3. It occurs in cells that contain chloroplasts.	3. It occurs in all cells.
4. Oxygen is released.	4. Oxygen is used.
5. Water is used.	5. Water is produced.
6. Carbon dioxide is used.	6. Carbon dioxide is produced.
7. It occurs in sunlight.	7. It occurs in dark as well as light.

FIGURE 16-4 Comparison of the activities that occur during photosynthesis and respiration.

FIGURE 16-5 When a plant is unable to obtain enough moisture from the soil to replace moisture that is lost to the atmosphere, it loses turgor pressure and becomes wilted.

in the plant tissues. Functions of this water include transporting minerals throughout the plant, cooling the plant, moving sugars and plant chemicals, and maintaining turgor pressure. **Turgor** is a swollen or stiffened condition in stems and leaves as a result of the plant cells being filled with liquid. When the plant does not have enough water, turgor pressure is lost, and the plant becomes wilted (Figure 16-5).

The exchange of gases is important to the plant because air is needed for photosynthesis to occur, and water vapor must exit the plant to draw more dissolved nutrients into the roots. Both of these important functions occur through the tiny openings in the leaf called stomas. The stomas are surrounded by specialized cells called guard cells. The guard cells control the size of the opening in the surface of the leaf, depending on the amount of water available to the plant and other conditions in the environment surrounding the plant.

Transpiration is the process by which a plant gives up water vapor to the atmosphere (Figure 16-6). Transpiration takes place primarily through the stoma, which open to allow water vapor and air to be exchanged by the leaf. Most plants transpire about 90 percent of the water that enters through the roots.

Transpiration is greatly influenced by humidity, temperature, wind, and other air movement (Figure 16-7). As humidity in the air around the plant

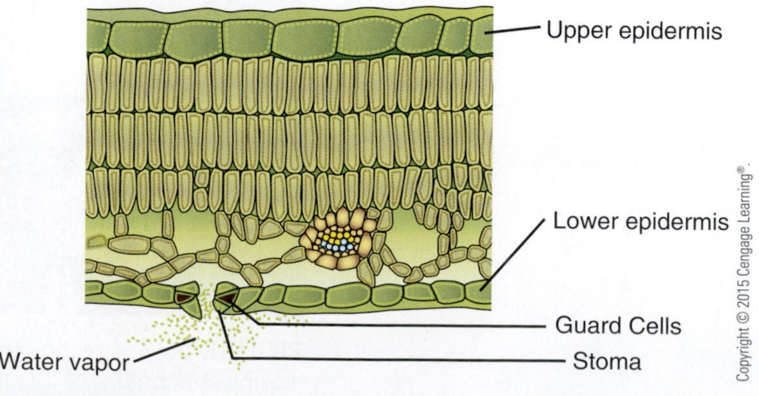

FIGURE 16-6 The process of transpiration occurs as water vapor and air are exchanged through the stoma.

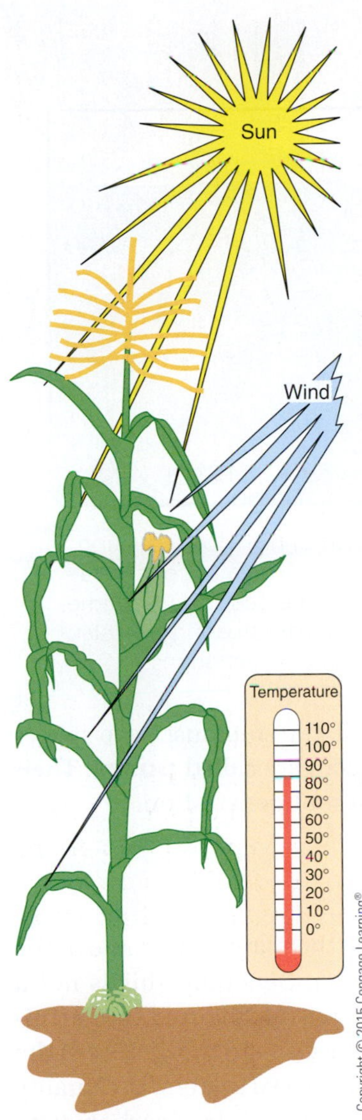

FIGURE 16-7 Factors that influence transpiration.

increases, the rate of transpiration decreases. Conversely, as humidity decreases, the rate of transpiration increases. Increased air movement around the plant increases the rate of transpiration due to evaporation that is accelerated by air movement. Similarly, as temperature increases, the rate of transpiration increases.

Often, during dry weather or when plants are not adequately watered, transpiration causes plants to lose water faster than it can be replaced by the root system. When this occurs, the guard cells will close the stomata in the leaves, thus slowing down the rate of transpiration. This mechanism enables the plant to preserve the water it contains. If there is water in the soil, the plant may wilt slightly, but it will recover. However, if there is insufficient moisture in the soil, the plant may not be able to recover.

SOIL

Productive soil provides a natural environment for the root zone. It provides air, water, and nutrients for the plant. Root hairs penetrate the pore spaces in the soil and absorb dissolved nutrients (Figure 16-8). A process called osmosis is used to get water and nutrients into root cells so they can be transported to the remainder of the plant. **Osmosis** is the process by which water moves from an area of high concentration to an area of low concentration through a semipermeable membrane that separates two solutions (Figure 16-9). A **semipermeable membrane** will allow certain substances to pass through, whereas others cannot. The epidermis or outside cell layer of a root hair is a semipermeable membrane. The root allows substances such as water, minerals, and nutrients to enter the plant. The movement of water and dissolved minerals tends to concentrate minerals and nutrients inside the plant at greater levels than in the soil. Soil moisture containing a lower concentration of dissolved nutrients and a greater concentration of water is able to move into the root hairs. Once inside the root-hair cell, nutrients can be transported to other parts of the plant as needed.

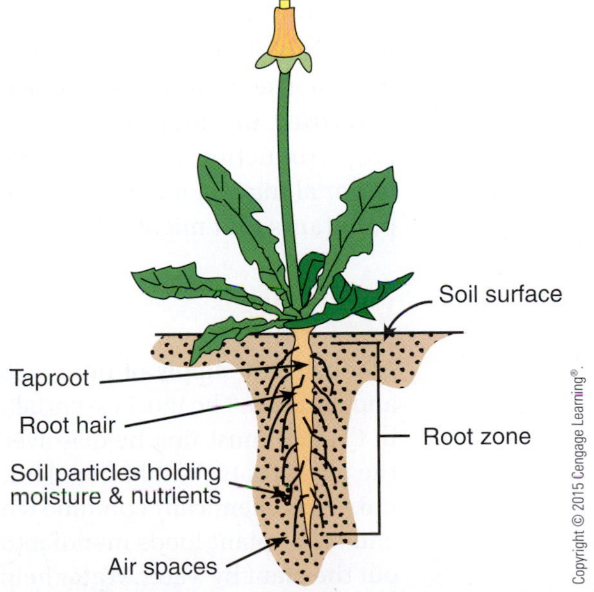

FIGURE 16-8 Root hairs extend from the main root into the pore spaces in the soil from which they absorb water and dissolved nutrients.

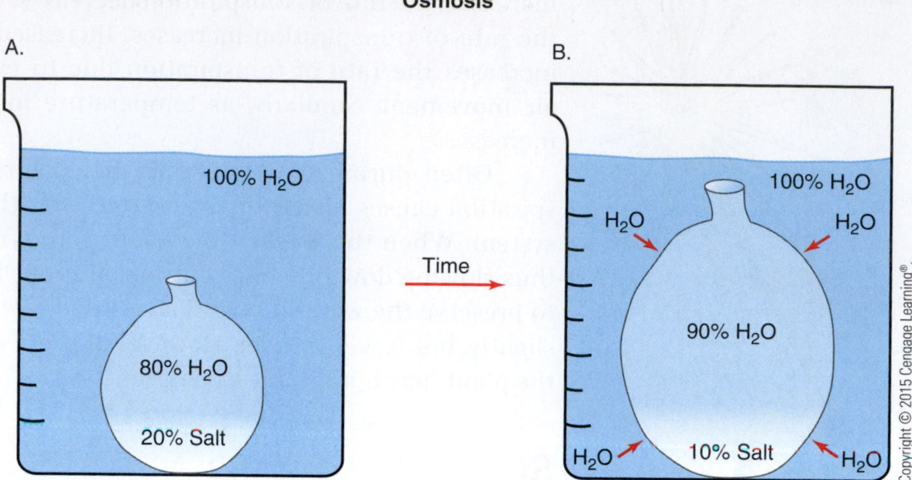

FIGURE 16-9 During osmosis, water moves from areas of high concentration to low concentration. (A) The water outside the balloon has a greater concentration at 100 percent than the water inside the balloon at 80 percent. (B) Over time, some of the water from outside the balloon has moved inside, crossing the selectively permeable membrane, in an attempt to balance the concentrations.

INTERNET KEY WORDS:
osmosis, semipermeable membrane
plants, carbon dioxide
plants, water

To allow the root hairs to move through the soil, the soil must have spaces between the particles of sand, silt, and clay. Such spaces are called **pores**. Their role is to store air, water, and nutrients and to permit root penetration.

AIR

The air or atmosphere that surrounds the portion of the plant that is above the ground must supply carbon dioxide as well as oxygen. Generally, this is not a problem when plants grow outdoors. When plants are transplanted into artificial or unnatural environments, consideration must be given to the quality of the air surrounding the plant. In a greenhouse or other enclosed system, the quality of the atmosphere should be monitored. The presence and levels of carbon dioxide and pollutants must be understood for maximum production to occur in a greenhouse environment. With certain crops in greenhouses, such as carnations and roses, the addition of some carbon dioxide might be desirable to increase crop production. In areas where crops are growing near industrial plants or cities, or along major highways, the technician must be aware of the many types of pollutants that might affect production or severely damage the plants.

WATER

A consistent supply of pure water is absolutely necessary for growth of plants and animals. The most essential ingredient for all living things is water. Nutrients in the soil must first be dissolved in water before they can be absorbed through the plant roots. Within the plant, water carries the nutrients to the leaves. These nutrients chemically combine with water in the process of photosynthesis. Sugars and other plant foods manufactured in the leaves are then transported throughout the plant by water. Water helps control the temperatures in and around plants through transpiration. Water gives the plant support by maintaining rigidity in the cells. It is important, therefore, that water used for plant production be of good quality and in adequate supply.

AGRI-PROFILE CAREER AREA: PLANT PHYSIOLOGY

Courtesy of USDA/ARS #K4191-5

Plant physiologists Marcia Holden and Douglas Luster inspect tomato plants for iron deficiency.

Physiology refers to the many functions that occur inside plants. These include familiar activities such as osmosis, nutrient uptake, translocation, respiration, photosynthesis, food movement, and food storage.

Plant physiologists work closely with technicians and scientists in other fields of plant science. They may be consultants to or collaborators with specialists in agronomy and horticulture. The work of plant physiologists is typically done by college or university faculty, employees of state or national research institutes, or specialists with agriscience corporations developing and selling seeds and plant materials.

INTERNET KEY WORDS:
plant nutrient deficiency symptoms
plant micronutrients
plant macronutrients

PLANT NUTRITION

Plant nutrition is often confused with plant fertilization. There is a difference. **Plant nutrition** refers to availability and type of basic chemical elements in the plant. Plant fertilization is the process of adding nutrients to the soil or leaves so they become available in the growing environment of the plant. Before chemicals that are supplied as fertilizer can be taken up and used by plants, they generally undergo changes.

Essential Nutrients

Sixteen elements are essential for normal plant growth. They are required in various amounts by plants and must be available in the relative proportions needed if the plants are to produce well. Three elements are used in huge amounts and are obtained from the atmosphere and water around the plant. They are carbon (C), hydrogen (H), and oxygen (O). Three more elements are used in relatively large amounts, and they are known as **primary nutrients**. These nutrients are nitrogen (N), phosphorus (P), and potassium (K). Other nutrients that are required in smaller amounts include calcium (Ca), magnesium (Mg), and sulfur (S). They are **secondary nutrients**. The primary and secondary nutrients, considered together, are called **macronutrients**, and all of them are obtained from the soil.

An additional seven elements are used in small quantities. They are called **micronutrients** (trace elements). The micronutrients are also obtained from the soil. They are boron (B), copper (Cu), chlorine (Cl), iron (Fe), manganese (Mn), molybdenum (Mo), and zinc (Zn). For a plant to grow at maximum efficiency, it must have all the essential plant nutrients (Figure 16-10). The absence of any one of these nutrients will cause the plant to grow poorly or show some signs of poor health.

Remembering the 16 Plant Nutrients

Various schemes have been devised to help you remember the names of the 16 plant nutrients. One technique is to first learn the chemical symbols. Use the

Courtesy of USDA/ARS #K3694-5

FIGURE 16-10 Soil and nutrient management specialists do cooperative research to determine the best rates of fertilizer to optimize corn growth.

symbols to make a logical string of words that are easy to remember. One such string of words that uses the symbols of most of the nutrients is "C. Hopkin's cafe, mighty good." By remembering this phrase, you can recall the symbols of 10 of the 16 nutrients as follows: C HOPKNS CaFe Mg (carbon, hydrogen, oxygen, phosphorus, potassium, nitrogen, sulfur, calcium, iron, and magnesium). The remaining ones are boron, copper, chlorine, manganese, molybdenum, and zinc. Can you devise a string of words to help you remember the symbols of these micronutrients?

Ions

Plant nutrients are absorbed from the soil–water solution that surrounds the root hairs of the plant. In fact, 98 percent of the nutrients obtained from the soil are absorbed in solution, whereas the other 2 percent is extracted by the root directly from soil particles. Most of the nutrients are absorbed as charged ions. An **ion** is an atom that has an electrical charge.

Negatively charged ions are called **anions**. Positively charged ions are called **cations**. The electrical charges in the soil are paired so that the overall effect in the soil is not changed. These ions compete and interact with each other according to their relative charges. For example, nitrogen, in its nitrate form, has a negative charge and chemical formula NO_3^-. Therefore, nitrates are anions with negative charges.

Conversely, potassium has a positive charge and is an example of a cation (K^+). Potassium nitrate, $K^+NO_3^-$, is a combination of potassium and nitrate consisting of one nitrate ion and one potassium ion. Calcium nitrate, $Ca^{++}(NO_3^-)_2$, has two nitrate ions and one calcium ion because the calcium cation has two positive charges. As you might guess, this could be confusing, but it illustrates the need to understand chemistry to manipulate plant fertility if conditions are not ideal in the natural environment.

The balance of ions is important and must be carefully monitored for good plant growth. Opposite charges attract each other, but ions with similar charges compete for chemical reactions and interactions in the soil–water environment. Some ions are more active than others and might be able to compete better in the soil. Further study would be needed to thoroughly understand why soil tests may indicate the presence of a certain element in sufficient amounts for plant growth, yet the plants may show deficiency symptoms (Figure 16-11). Deficiency means a shortage of a given nutrient

FIGURE 16-11 Nutrient deficiencies decrease plant health, vigor, and growth. Deficiency of specific nutrients can be determined by observing the plant.

that is available for plant use. A good example of a nutrient deficiency symptom is blossom-end rot of tomato. This is common in gardens and occurs when there is not enough water to dissolve and carry calcium to the plant in sufficient quantities. The end opposite the stem is called the blossom end. The calcium deficiency produces a tomato that looks good from the top, but, when picked, the bottom end is rotten.

Soil Acidity and Alkalinity

The chemistry of plant elements in the soil can be affected by pH. Soil pH is a measurement of acidity (sourness) and alkalinity (sweetness) (Figure 16-12). Many of the nutrients in soil form complex combinations and are capable of precipitating out of solution, where they are unavailable to the plant. **Precipitate** occurs when a solid is dropped out of solution. If the soil pH is **acid**, or extremely low, some micronutrients become too soluble and occur in concentrations great enough to harm the plants (Figure 16-13).

In contrast, when soil pH is high, in the **alkaline** range, many of the nutrients can be precipitated out and not be available to the plants. The pH of soils can be determined with low-cost test kits. Fortunately, soil pH can be corrected by adding lime if the pH needs to be increased or by adding sulfur or gypsum if pH needs to be decreased. Such practices are common because soil pH is often less than ideal for the crop being grown (Figure 16-14).

INTERNET KEY WORDS:
plant nutrients

Plant Nutrient Functions

The importance of carbon, hydrogen, and oxygen has already been discussed under the topic of photosynthesis. The other nutrients have very specific functions and must be available in the appropriate form and correct amounts. The effects of plant nutrients may be likened to a chain—the weakest link will

HOT TOPICS IN AGRISCIENCE

NEW PLANT VARIETIES—KEY TO AN ABUNDANT FOOD SUPPLY

© Seregam/Shutterstock.com

New varieties of low-protein wheat are of great benefit to food processors that manufacture pasta products.

Agricultural crop production is becoming more efficient all the time. Yields and quality are both increasing. These improvements are partially because of the development of better plants. Scientists are continually seeking ways to improve our food crops. Most of our highly productive crop plants are resistant to one or more diseases or pests. Some of them are able to withstand drought conditions or other environmental conditions that are known to reduce crop production.

Some new plants have been modified to produce food with nutrient levels that are different from those of the parent varieties. For example, some new wheat varieties contain less protein in the grain than the same quantity of grain produced by the parent stock. This is important to the industries that make pasta and crackers. The quality of these products is improved by using low-protein wheat. New plant varieties will always be in demand because the world will always be seeking greater crop yields and higher quality.

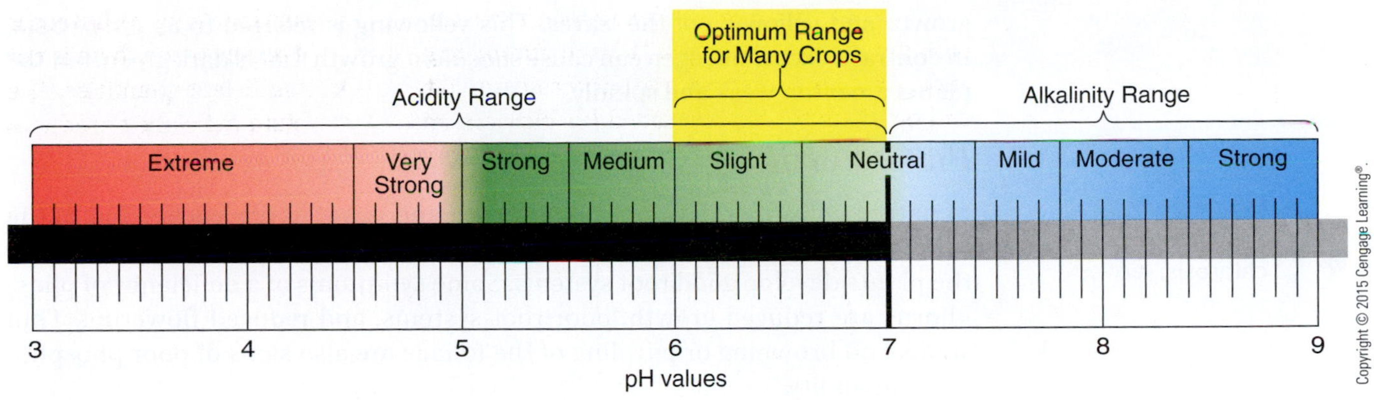

FIGURE 16-12 The pH scale measures the balance of positive and negative ions in a solution or in the soil.

determine how much the chain will pull. Similarly, the nutrient in shortest supply will determine the maximum growth that can be achieved by the plant.

Nitrogen

Nitrogen is present in the atmosphere as a gas. It is added to the soil in some fertilizers. Because it exists in nature as a gas, it is easily leached (washed out of the soil). Nitrogen is responsible for the vegetative growth of the plant and its dark green color. When nitrogen is lacking, the deficiency symptoms are reduced

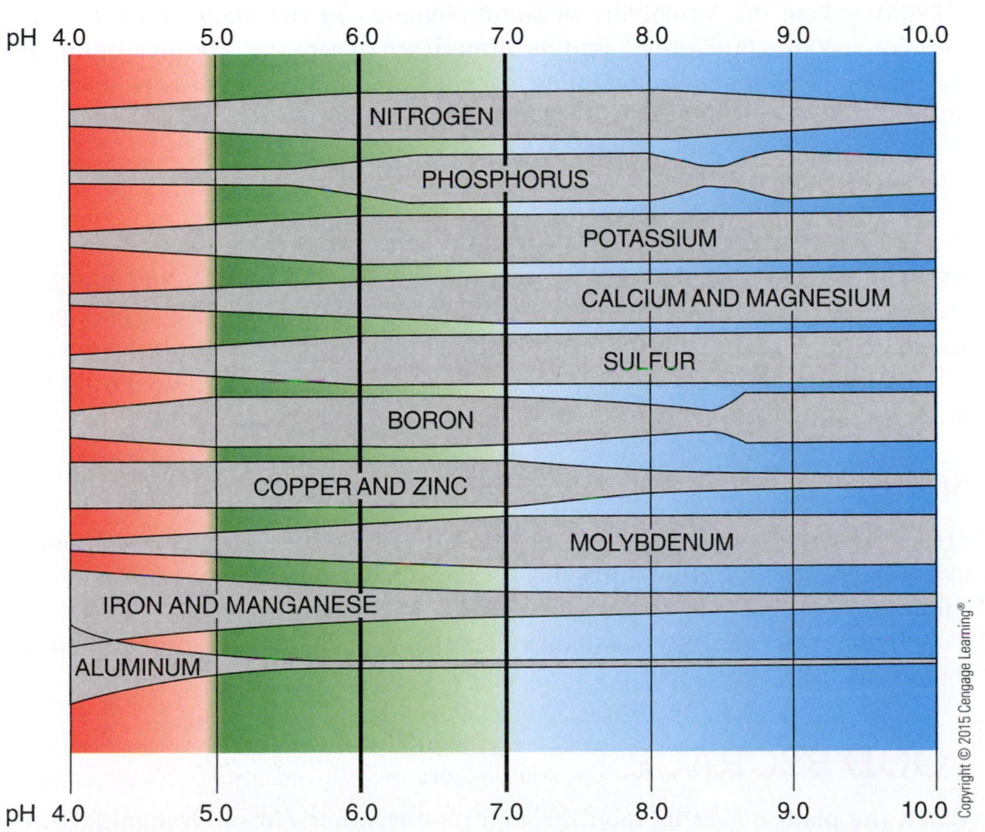

FIGURE 16-13 This chart illustrates the effect of soil pH on the availability of plant nutrients. A wide section indicates high availability of the nutrient, and a narrow section indicates that the nutrient is not available to a plant.

FIGURE 16-14 Soil scientist Charles Foy compares barley plants grown in soils having different pH levels.

growth and yellowing of the leaves. This yellowing is referred to as **chlorosis**. In contrast, excess nitrogen can cause succulent growth that is dark green, but the plants are often weak and spindly.

Phosphorus

In nature, phosphorus is present as a rock and is not easily leached out of the soil. It is important in the growth of seedlings and young plants, and it helps the plants develop good root systems. Some symptoms of a deficiency of phosphorus are reduced growth, poor root systems, and reduced flowering. Thin stems and browning or purpling of the foliage are also signs of poor phosphorus availability.

Potassium

Potassium is mined as a rock and made into fertilizer, but it can be leached from the soil. If too much potassium is present, it can cause a nitrogen deficiency. A lack of potassium will appear as reduced growth or shortened internodes and sometimes as marginal burn or scorching (brown leaf edges). Dead spots in the leaf and plants that wilt easily are also indications of a potassium deficiency.

Calcium

Calcium is often supplied by adding lime to the soil. It can be leached out and does not move easily throughout the plant. Too much calcium can cause a high pH and reduce the availability of some elements to the plant. A lack of this element can stop bud growth and result in death of root tips, cupping of mature leaves, and blossom-end rot of many fruits. Pits on root vegetables are also signs of calcium deficiency.

Magnesium

Magnesium can be added by using high-magnesium lime. It can also be leached from soil. If magnesium is lacking, some reduction of growth and marginal chlorosis can be noticed. In some plants, even interveinal chlorosis can be seen. Cupped leaves and a reduction in seed production can also be symptoms of magnesium deficiency. Foliage plants are most likely to be deficient in this nutrient.

Sulfur

Present in the atmosphere as a result of combustion, sulfur is often an impurity in fertilizers carrying other nutrients. As a result, it is rarely deficient. However, if sulfur is deficient, a yellowing of the entire plant may result.

Deficiencies and excesses of nutrients cause predictable symptoms in plants (Figure 16-15).

FOOD STORAGE

When the plant makes its food through photosynthesis, it often manufactures more than what it needs to maintain itself. This excess is stored in the plant for future use. Such food may be stored in roots, stems, seeds, or fruits.

Nutrient	Excess	Deficiency
Iron	Rare	Interveinal chlorosis, especially on young growth
Zinc	Might appear as an iron deficiency	Interveinal chlorosis, reduction in leaf size, short internodes
Molybdenum	Not known	Interveinal chlorosis on older leaves; may also affect leaves in the middle of the plant
Boron	A blackening or death of tissue between veins	Failure to set seed; death of tip buds
Copper	Might occur in low pH; will appear as an iron deficiency	New growth small and misshapen, wilted
Manganese	Brown spotting on leaves; reduced growth	Interveinal chlorosis of the leaves and brown spotting; checkered effect possible

FIGURE 16-15 An excess or deficiency of most nutrients will cause predictable symptoms in plants.

Roots

The most common type of root that serves as a storage organ is the taproot. Some common examples of plants with extensive storage capacity are sugar beets, carrots, radishes, and turnips. The sugars and carbohydrates are transported down the phloem and into the root cells. They are held there as the root enlarges. Most of the time, this type of plant is a short-term crop that does not take long to mature. This type of root system is easy to dig or harvest.

Stems

The stems of plants usually contain cells that are necessary for plant support. Some specialized stems, however, are excellent food storage organs (Figure 16-16). Some are used for propagation and some for food. The most common specialized stem used for food is the tuber. It is an enlarged portion of a stem containing all of the parts of a normal stem. Nodes, internodes, and buds can be identified in tubers. The Irish potato is an example of a tuber.

Corms and bulbs are other examples of specialized stems that contain large amounts of food manufactured by the plant. A rhizome is yet another example; however, its purpose is for propagation.

Seeds

As the ovule of a plant matures, it stores food for the young embryo to start its growth when it germinates. Seeds tend to be high in carbohydrates, fats, and oils. These are concentrated food deposits, and both humans and animals use seeds as major food sources. They help the plant by spreading the seed to new locations, increasing the ability of the plant population to survive.

SCIENCE CONNECTION DIAGNOSING AN EMINENT KILLER

Courtesy of USDA/ARS #K3323-3

Pale green, chlorite leaves and reduced leaf size typical of citrus blight are displayed by plant pathologist Michael Bausher.

A subtle and mysterious killer, citrus blight, has eluded plant pathologists and other scientists for more than a century. Recognized by citrus growers and the federal government as a deadly disease of citrus trees, the U.S. Horticultural Research Laboratory was established in Eustis, Florida, in 1892 to research this problem. Citrus blight is a mysterious disorder that renders a tree worthless for fruit production. It is most likely to attack young trees that are just beginning to produce fruit, but it also kills trees 20, 30, or even 50 years old. No cure or prevention is known for the citrus blight. First diagnosed in Florida in 1874, it causes physiological changes in the tree and first leads to a yellowing of leaves and eventually to leaf wilt. With no cure and no chance of recovery, diseased trees are destroyed as soon as the disorder is observed. Authorities estimate that a half million trees valued at $60 million are lost each year to the blight (Futch, S. H, Derrick, K. S. and Brlansky, R. H., March 2011).

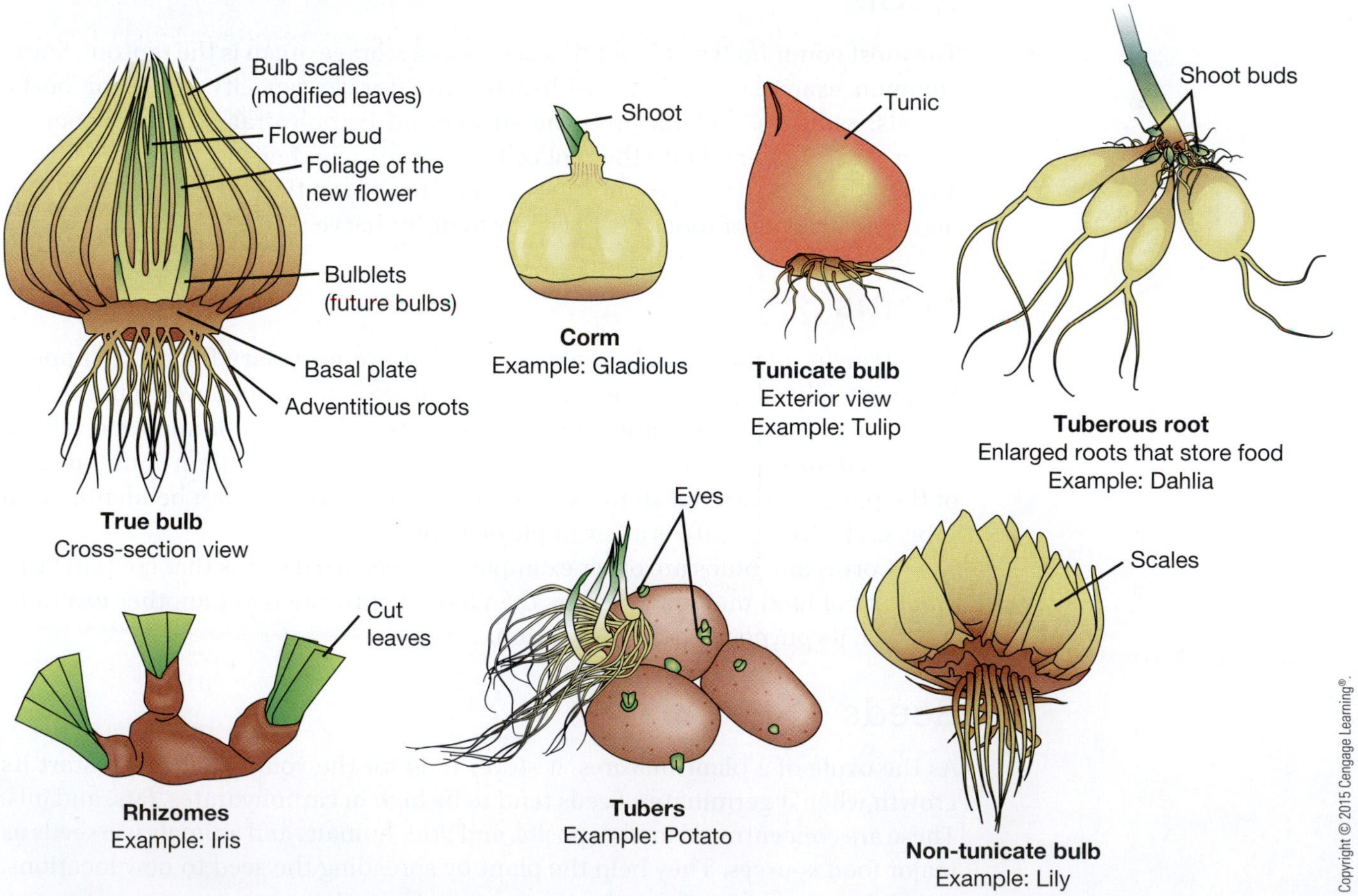

Bulb scales (modified leaves)
Flower bud
Foliage of the new flower
Bulblets (future bulbs)
Basal plate
Adventitious roots
True bulb
Cross-section view

Shoot
Corm
Example: Gladiolus

Tunic
Tunicate bulb
Exterior view
Example: Tulip

Shoot buds
Tuberous root
Enlarged roots that store food
Example: Dahlia

Cut leaves
Rhizomes
Example: Iris

Eyes
Tubers
Example: Potato

Scales
Non-tunicate bulb
Example: Lily

FIGURE 16-16 Examples of stems that are major food storage organs for the plant.

U.S. Department of Agriculture (USDA) plant pathologist Michael Bausher has found a way that leads to early diagnosis, and thus reduces losses from the disorder. At the U.S. Horticultural Research Lab in Orlando, Bausher discovered unique proteins in the leaves of diseased trees that are not present in healthy trees or in trees with other diseases or stress problems. Bausher prepared, freeze-dried, and partially purified an antigen derived from ground-up leaves of blighted trees. When the antigen was injected into rabbits, he discovered the rabbits produced a unique antisera. The rabbit-produced antisera was then found to react positively with the unique proteins from the blighted trees. However, the antisera reacted negatively when exposed to proteins derived from tissue of healthy trees and trees with other disorders or diseases.

Growers usually remove trees at the first indication of citrus blight to cut the losses from the disease. Unfortunately, other disorders may cause yellowing or mottling of leaves, which are the first observable symptoms of the blight. Therefore, it is hard to tell how many trees with correctable disorders are sacrificed. It is hoped that a process can be developed wherein proteins can be used as biological markers to help identify trees with citrus blight before the visual symptoms appear. This early detection would permit growers to remove only trees that really have the citrus blight and spare those that have other correctable disorders. Biological markers may help researchers develop trees that are resistant to the blight.

INTERNET KEY WORDS:
plants, new varieties
biology, plants, seeds

This unit has addressed some basic principles of plant physiology. Physiology is complex and must be studied in great depth to gain understanding. Plant physiologists typically have master's or doctorate degrees. However, most technicians and scientists in the plant sciences will have some training in plant physiology. A basic knowledge of soils and how plants use nutrients and function in general will greatly help in the successful production and management of plants. Farmers, ranchers, and growers generally have access to sophisticated technology (Figure 16-17).

Courtesy of USDA/ARS #K4914-2

FIGURE 16-17 Missouri farmer Bill Holmes and specialists from the Space Remote Sensing Center examine soil fertility variations in Holmes's cropland.

STUDENT ACTIVITIES

1. Write the Terms to Know and their meanings in your notebook.
2. Make a bulletin board showing the cross section of a leaf. Label the various cells and leaf parts. Include the formula for photosynthesis.
3. Write an article for a newspaper or magazine that explains the importance of photosynthesis.
4. Collect plants or pictures to make a display that could be used by others to help identify nutrient deficiencies.
5. Select a crop that interests you and conduct research to determine the optimum nutrient requirements.
6. Set up a demonstration to explain osmosis.
7. Make a list of all the chemical symbols of plant nutrients. Write a sentence or story to help you remember them.
8. In your own words, explain the process of photosynthesis. Be sure to include all of the steps in the process.
9. In your own words, explain respiration. Be sure to include all of the conditions that are necessary for the steps to occur.

SELF-EVALUATION

A. MULTIPLE CHOICE

1. The study of functions and the complex chemical processes that allow plants to grow is known as
   a. plant taxonomy.
   b. plant physiology.
   c. plant nutrition.
   d. photosynthesis.
2. Chlorophyll is important in plants because it
   a. creates an atmosphere where it can determine the osmotic pressure.
   b. allows the plant to make good xylem tissue.
   c. makes it possible for plants to grow.
   d. is also known as the chloroplasts.
3. The rate at which photosynthesis is carried out depends on
   a. the amount of fertilizer in the water.
   b. the amount of oxygen in the atmosphere.
   c. the amount of respiration carried on during the daylight hours.
   d. the light intensity, temperature, and concentration of carbon dioxide.
4. Photosynthesis will work best in which temperature range?
   a. 50° to 60° F
   b. 60° to 70° F
   c. 65° to 85° F
   d. 85° to 95° F
5. Respiration
   a. uses food for plant energy.
   b. stores energy.
   c. occurs in cells that contain chlorophyll.
   d. uses carbon dioxide.
6. Plant nutrition is
   a. plant food added to the plant pot.
   b. use of basic chemical elements in the plant.
   c. chemical processes providing plants with elements for growth.
   d. the measurement of acidity (sourness) and alkalinity (sweetness).

B. MATCHING

_______	1. pH	a. Movement through a semipermeable membrane
_______	2. Osmosis	b. Addition of nutrients to the plant-growing environment
_______	3. Corm	c. Storage organ for excess plant food
_______	4. Root	d. Site of photosynthesis
_______	5. Sulfur	e. Measurement of acidity and alkalinity
_______	6. Leaves	f. Macronutrient
_______	7. Fertilization	g. Specialized stem

C. COMPLETION

1. Extremes of temperature will slow down or completely stop _______.
2. Respiration will occur only in the _______.
3. When the temperature increases, the rate of transpiration _______.
4. Soil provides a natural environment for the _______ _______.
5. The spaces between the soil particles, where the soil water is found, are called the _______.

D. TRUE OR FALSE

1. Respiration is a building process that uses sunlight to work.
2. The process by which a plant loses water is perspiration.
3. All plant elements perform the same function in the plant.
4. Humidity refers to the amount of water in the atmosphere.

UNIT 17
Plant Reproduction

OBJECTIVE

To determine the methods used by plants to reproduce themselves and to explore new propagation technology.

MATERIALS LIST

- seed catalog
- grafting knife
- grafting rubber
- grafting wax
- cutting knife
- rooting hormone
- stock plants
- rooting media
- tissue-culture tubes
- tissue-culture media
- scalpel
- razor blade (single edge)
- tweezers
- 50 percent alcohol solution
- sanitary work area
- Internet access

COMPETENCIES TO BE DEVELOPED

After studying this unit, you should be able to:

- distinguish between sexual and asexual reproduction.
- explain the relationship between reproduction and plant improvement.
- draw and label the reproductive parts of flowers and seeds.
- state the primary methods of asexual reproduction and give examples of plants typically propagated by each method.
- explain the procedures used to propagate plants via tissue culture.

SUGGESTED CLASS ACTIVITIES

1. Assign each class member to conduct an experiment to determine how long it takes for several kinds of seeds to germinate. Place three to five seeds of a single species of plant in damp tissue paper. Place the tissue in a section of an egg carton. Keep the tissue damp without excess water. Label the seeds by writing on the carton. Check the seeds daily and add water as necessary. Keep a record of the time that is required for germination to occur.
2. Conduct a cloning exercise using African violets or potatoes as parent stock. The procedure to be followed is described later in this unit. Have students work alone or in pairs. A written report should always be part of a classroom science experiment. It should describe the process that was followed, supplies and equipment that were used, and results that were attained.
3. Spend the first half of a class period collecting pollen from different plants. This can be done outdoors during the

TERMS TO KNOW

propagation
sexual reproduction
asexual reproduction
hybrid vigor
germinate
dormant
imbibition
scarify
cutting
fungicide
rooting hormone
stem tip cutting
stem section cutting
cane cutting
heel cutting
single-eye cutting
double-eye cutting
leaf cutting
leaf petiole cutting
leaf section cutting
split-vein cutting
root cutting
simple layering
tip layering
air layering
grafting
scion
rootstock
graft union
bud grafting
T-budding
tissue culture
aseptic
agar

appropriate seasons. If outdoor collection is not possible, check with the local flower shops for discarded flowers. Next, set up a number of microscopes and examine the pollen. As a class, discuss the similarities and differences that can be seen. *Caution*: Check with your students ahead of time to see if any of them suffer from uncontrolled pollen allergies.

Plant **propagation**, or reproduction, is simply the process of increasing the numbers of a species, or perpetuating a species. The two types of plant propagation are sexual and asexual. **Sexual reproduction** is the union of an egg (ovule) and sperm (pollen grain), resulting in a seed. Two parents creating a third individual is referred to as sexual propagation. In plants, it involves the floral parts. It may involve one or two plants. **Asexual reproduction** uses a part or parts of only one parent plant. The purpose is to cause the parent plant to produce a duplicate of itself. The new plant is a clone (exact duplication) of its parent. Because this type of reproduction uses the vegetative parts of the plant, namely the stem, root, or leaf, it is known as vegetative propagation.

Sexual propagation has some distinct advantages. It is often less expensive and quicker than some other methods. It is the only way to obtain new varieties and also to capture hybrid vigor. A hybrid is a plant obtained by crossbreeding. **Hybrid vigor** refers to the tendency of hybrid plants to be stronger and survive better than plants of a pure variety. Sexual propagation also is a good way to avoid passing on some diseases. In some plants, sexual propagation is the only way they can reproduce.

Asexual propagation has many advantages as well (Figure 17-1). With some plant species, it is easier and less expensive to obtain the next generation of plants this way. In some species or cultivars, it is the only way they can be propagated.

SEXUAL PROPAGATION

A seed is made up of the seed coat, endosperm, and embryo (Figure 17-2). The seed coat functions as a protector for the seed. Sometimes it is thin and soft, or it may be hard and impervious to water or moisture. The endosperm functions as a food reserve. It will supply the new plant with nourishment for the first few days of life. The embryo is the young plant itself. When a seed is fertilized and matures,

PLANT PROPAGATION	
Sexual Propagation Advantages	**Asexual Propagation Advantages**
• Less expensive • Many plants can be produced quickly • Crosses result in hybrid vigor • Avoids passing on some diseases	• Less time is required to produce a salable plant • Plants are genetically identical • The only way to reproduce some plant varieties

FIGURE 17-1 Two methods of plant reproduction, sexual and asexual propagation, are in common use in the plant industry. Each method has distinct advantages over the other.

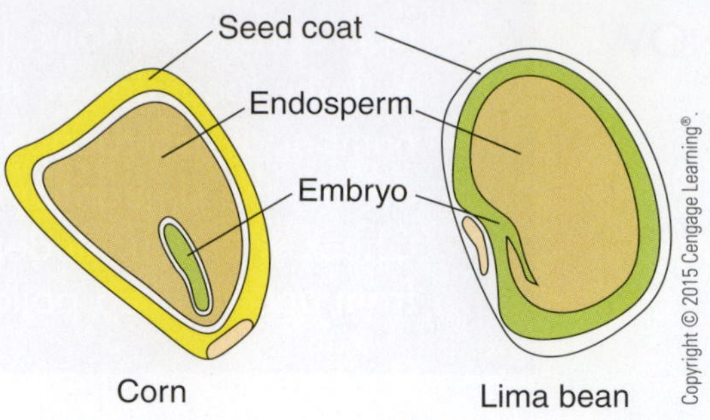

FIGURE 17-2 Parts of a seed.

it will be dormant. When it is subjected to favorable growth conditions, it will **germinate**, meaning that the seed will sprout and begin to grow.

INTERNET KEY WORDS: pollen, sperm, ovule, egg

Seed propagation starts with quality seed. Crop production by sexual reproduction allows consideration to be given to the type of plant that is needed or the variety that is best adapted to a particular area or purpose. Hybrid plants are developed by cross-pollinating two different varieties. Many varieties in production today are the result of hybridization or crossbreeding. Seeds of hybrid plants cost more than open-pollinated varieties. However, the increased quality and vigor of the plants generally offset the increased cost of the seed. New varieties are constantly being developed for disease and insect resistance. It is natural to expect the seeds of such improved varieties to cost more than standard or regular seeds. In addition, some varieties have unusual cultural or product characteristics (Figure 17-3).

Generally, seeds collected from plants used for commercial production will not save money in the long run. Seeds from such plants are often small. They may be poorly managed and improperly handled and stored (Figure 17-4). It is recommended that only certified seed be used. Seed saved from season to season should be stored in a sealed jar at 40° F (4.4° C) and at low humidity.

Germination

When seed is harvested, or collected, it is normally mature and in a **dormant**, or resting, state. To germinate and start to grow, it must be placed in certain

SCIENCE CONNECTION PLANT REPRODUCTION

It takes two chromatids to make up one chromosome. One chromatid comes from the male and one from the female parent. A haploid cell has half the normal amount of genetic material because only a single chromatid from one of the parents is present. A diploid cell has two chromatids, one from the female and one from the male parent. These two chromatids, when paired, make one chromosome. The number of chromosomes in an organism depends on its species. Genes from both parents are represented in a diploid cell.

Plants sexually reproduce by a process called alternation of generation. In their life cycles, plants change back and forth between producing haploid cells and producing diploid cells. During the haploid generation, the plant produces sperm, eggs, and, in some cases, both. In flowering plants, the sperm is called pollen, and the egg is called an ovum. Once the sperm and the egg are fused, their chromatids become paired. This starts the diploid generation. After fertilization, the membrane around the ovum hardens and becomes a seed with a developing embryo inside. Under the right growing conditions, the seed will develop into a plant. At this point, the alternation of generation is complete. Once the plant matures, the cycle begins all over.

Courtesy of USDA/ARS #K-1657-15. Photo by Jack Dykinga

FIGURE 17-3 The National Seed Storage Laboratory in Fort Collins, Colorado, preserves seeds of all varieties. They are packaged in flexible, moisture-proof bags.

Courtesy of DeVere Burton

FIGURE 17-4 Hybrid seeds are produced by pollinating one variety with pollen from a different variety. Note the taller rows of corn that will pollinate the shorter variety. All of the pollen-bearing tassels will be removed from the shorter plants to ensure that cross-pollination occurs.

favorable conditions (Figure 17-5). The four environmental factors that must be right for effective germination are water, air, light, and temperature.

Water

INTERNET KEY WORDS:
seed, germination

Imbibition is the absorption of water. It is the first step in the germination process. The seed, in its dormant stage, contains little water. The imbibition process allows the seed to fill all its cells with water. If other conditions are favorable, the seed then breaks its dormant stage and germinates.

A good germination medium is important. The medium must not be too wet or too dry. An adequate and continuous supply of water must be available. This is often difficult to control with crops directly seeded in the field. It is much easier to control in crops that are started in the greenhouse for transplanting at a later date.

Agricultural crops should usually be planted when soil moisture conditions are favorable for germination. However, this is not always possible, particularly when large acreages need to be planted at about the same time. In regions where irrigation is practiced, planting can be timed to follow applications of water to the fields. In many regions, however, timely precipitation is depended on to germinate the seeds.

A dry period during the germination process will result in the death of the young embryo. Too much water will result in the young seed rotting. In some species, the seed coat is very hard, and water cannot penetrate to the endosperm. In these cases, it is necessary to scarify the seed.

FIGURE 17-5 Seeds require water, air, light, and favorable temperature to germinate and begin to grow.

A common way to **scarify** seed is to nick the seed coat with a knife or a file. Another method is to soak the seeds in concentrated sulfuric acid. This requires special care and experience because sulfuric acid is a dangerous material. Another technique is to place seeds in hot water—180° to 212° F (82.2° to 100° C)—and allow them to soak as the water cools. This process takes 12 to 24 hours. A warm, moist scarification process may be used by simply placing the seeds in warm, damp containers and letting the seed coat decay over time. Commercial methods for seed scarification are used by suppliers to provide seed that is ready to plant.

Air

Respiration takes place in all viable seed. Viable seed is alive and capable of germinating. Oxygen is required. Even in nongerminating seeds, a small amount of oxygen is required even though respiration is low. As germination starts, the respiration rate increases. It is important that the seed be placed in good soil or media that is loose and well drained. If the oxygen supply is limited or reduced during the germination process, germination will be reduced or inhibited.

Light

Some seeds are stimulated to grow by light, while others are inhibited by the presence of light. It is necessary to have some knowledge of the presence or absence of special light requirements for some specialty crops. Many of the agronomic crops do not require light for germination. Ornamental bedding plants are more likely to require light for germination. Some crops requiring light for germination are ageratum, begonia, impatiens, and petunia. Lettuce also requires light for successful germination. Seeds of these plants are often deposited on the surface of the soil by nature, and the grower should follow the same procedure for successful germination.

Temperature

Heat is another important requirement for germination. The germination rate, or percentage of seed that germinates, is affected by the availability of favorable

Plant	Time to Seed Before Last Frost (Weeks)	Germination Time (Days)	Germination Temperature Requirements (Degrees F)	Germination Light Requirements Light (L) Dark (D)
Ageratum	8	5–10	70	L
Aster	6	5–10	70	–
Begonia	12+	10–15	70	L
Coleus	8	5–10	65	–
Cucumber	4	5–10	85	–
Eggplant	8	5–10	80	–
Marigold	6	5–10	70	–
Pepper	8	5–10	80	–
Portulaca	10	5–10	70	D
Snapdragon	10	5–10	65	L
Tomato	6	5–10	80	–
Watermelon	4	5–10	85	–
Zinnia	6	5–10	70	–

FIGURE 17-6 Time, temperature, and light requirements for germination of some common flower and vegetable plants.

ambient (surrounding) temperatures. Some seeds will germinate over a wide range of temperatures, whereas others have more narrow limits. In the agronomic crops that are directly seeded in the field, the only way to control heat is to plant when the ground is warm. In horticultural crops, particularly bedding plants and perennials, knowledge of a plant's specific heat requirements for germination should result in more efficient production. The germination requirements usually correspond to zone maps and recommended planting dates listed in seed catalogs (Figure 17-6).

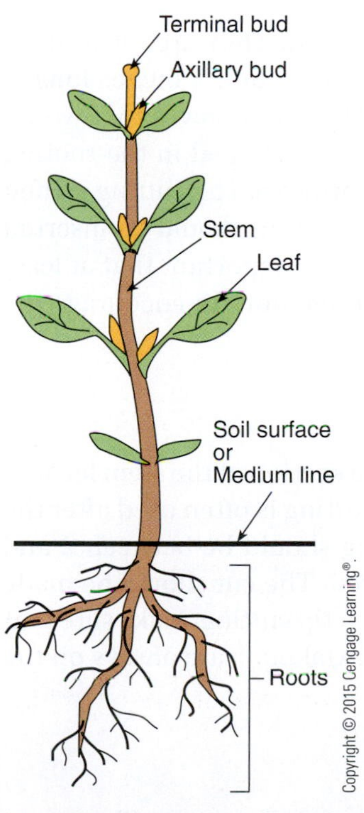

FIGURE 17-7 Vegetative parts of plants.

ASEXUAL PROPAGATION

As stated previously, asexual propagation is using the vegetative parts of the plant to increase the number of plants (Figure 17-7). Its primary advantages are economy, time, and plants that are identical to the parents. The primary methods of asexual propagation are cuttings, layering, division, grafting, and tissue culture.

Stem Cuttings

Herbaceous and woody plants are often propagated by **cuttings** (vegetative plant parts that are used to generate new plants). Types of cuttings are named for the parts of the plant from which they are obtained. There are stem tip cuttings, stem cuttings, cane cuttings, leaf cuttings, leaf petiole cuttings, and root cuttings.

The procedure for taking cuttings is relatively simple. The tools needed consist of a sharp knife or a single-edge razor blade. Sharp equipment will make the job easier and reduce injury to the parent plant. To prevent the possibility of diseases spreading, the cutting tool should be immersed in bleach water made with one part bleach to nine parts water. The tool can also be dipped in rubbing alcohol.

Flowers and flower buds should be removed from all cuttings. This allows the cutting to use its energy and stored food reserves for root formation instead of flower and fruit development. A rooting hormone containing a fungicide is used to stimulate root development. A **fungicide** is a pesticide that helps prevent diseases.

Rooting hormone is a chemical that will react with the newly formed cells and encourage the plant to develop roots faster. The proper way to use a rooting hormone is to put a small amount in a separate container and work from that container. This procedure will ensure that the rooting hormone does not become contaminated with disease organisms. Do not put the unused hormone back in the original container.

Cuttings are normally placed in a medium consisting of coarse sand, perlite, soil, a mixture of peat and perlite, or vermiculite. It is best to use the correct medium for a specific plant to obtain the most efficient production in the shortest possible time. The rooting medium should always be sterile and well drained, but it should also be capable of moisture retention to prevent the medium from drying out. The medium should be moistened before inserting the cuttings. It should then be kept continuously and evenly moist while the cuttings are forming roots and new shoots. This is accomplished through regular and frequent applications of small amounts of water. Stem and leaf cuttings do best in bright, but indirect, light. However, root cuttings are often kept in the dark until new shoots are formed and start to grow.

For most plants, the most popular method of making cuttings is by stem cutting. On herbaceous plants, stem cuttings may be made almost any time of the year as long as other growing conditions are maintained. However, stem cuttings of many woody plants such as grapes or ornamental bushes are normally taken in the fall or the dormant season or both.

Stem Tip Cuttings

Stem tip cuttings normally include the terminal bud. They are taken from the end of the stem or branch. A piece of stem between 2 and 4 inches long is selected, and the cut is made just below the node. The lower leaves that would be in contact with the medium are removed. The stem is dipped in the rooting hormone and is gently tapped to remove the excess hormone. The cutting is then inserted into the rooting medium (Figure 17-8). The cutting should be inserted deep enough so the plant material will support itself. It is important that at least one node be located below the surface of the growth medium to encourage the growth of new roots from it.

FIGURE 17-8 Stem tip cuttings include the terminal bud and are 2 to 4 inches in length.

Stem Section Cuttings

Stem section cuttings are prepared by selecting a section of the stem located in the middle or behind the tip cutting. This type of cutting is often used after the tip cuttings are removed from the plant. The cutting should be between 2 and 4 inches long, and the lower leaves should be removed. The cut should be made just beyond a node on both ends. It is then handled as a tip cutting. Make sure that the cutting is positioned with the right end up. The axial buds are always on the tops of the leaves.

Cane Cuttings

Some plants, such as the dumb cane (*Dieffenbachia* sp.), have cane-like stems. These stems are cut into sections that have one or two eyes, or nodes, to make

AGRI-PROFILE

CAREER AREAS: PLANT BREEDING/PLANT PROPAGATION/ CROP IMPROVEMENT/TISSUE CULTURE

The plant breeder or plant geneticist searches for plants, transfers genes, or crossbreeds plants from various sources in an effort to develop new varieties with desired characteristics.

Plant breeders' objectives might include developing plants that are faster growing, disease resistant, drought tolerant, insect resistant, wind resistant, frost tolerant, more beautiful, or better flavored, depending on the uses of the plants. Much plant breeding occurs in greenhouses, but outdoor plots are often used with new crop varieties. Some plants have small flowers that require the use of magnifying glasses and tweezers to transfer pollen or to remove reproductive parts. Generally, such research is followed by field trials and seed production, which gives the plant breeder some variety in the settings where work is done.

Asexual reproduction involves rooting, budding, grafting, layering, and other procedures in addition to pollination. Tissue culture, a procedure developed in biotechnology, permits the production of thousands of new plants identical to a single superior parent plant. The procedure is relatively cheap and easy, though exacting, and it is used extensively to reproduce ornamental plants. Many jobs are available in the area of plant reproduction.

cane cuttings. The ends are dusted with activated charcoal or a fungicide. It is best to allow the cane to dry in the open air for one or two hours. The cutting is then placed in a horizontal position with half of the cane above the surface of the medium. The eyes, or nodes, should be facing upward. This type of cutting is usually potted when the roots and new shoots appear (Figure 17-9).

Heel Cuttings

Heel cuttings are used with woody-stem plants. A shield-shaped cut is made about halfway through the wood around the leaf and the axial bud. Rooting hormone may be used in the same manner as in the other types of cuttings. The cutting is inserted horizontally into the medium (Figure 17-10).

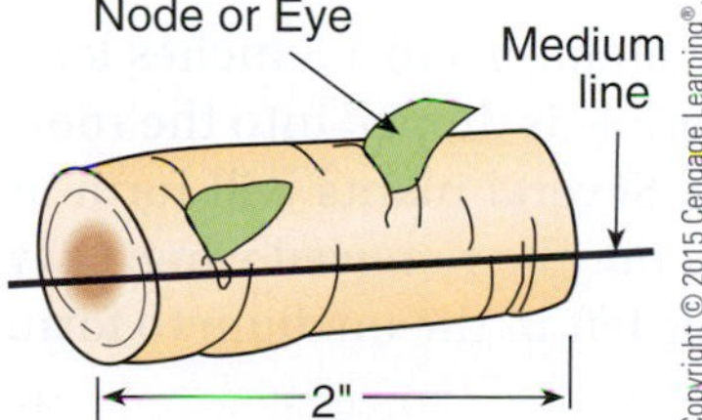

FIGURE 17-9 A cane cutting is made from the stem of a plant with a cane-like growth structure. One to two adjacent nodes are selected, and the cut is made to include the node or nodes that are desired.

Single-eye Cuttings

When the plant has alternate leaves, **single-eye cuttings** are used. The eye refers to the node. The stem is cut about a half inch above and below the same node (Figure 17-11). The cutting may be dipped in rooting hormone and then placed either vertically or horizontally in the medium.

Double-eye Cuttings

When plants have opposite leaves, **double-eye cutting** is the preferred type of cutting. It is often used when the stock material is limited. A single node is selected, and the stem is cut a half inch above and below the node with a sharp tool. The cutting should be inserted vertically in the soil medium (Figure 17-12).

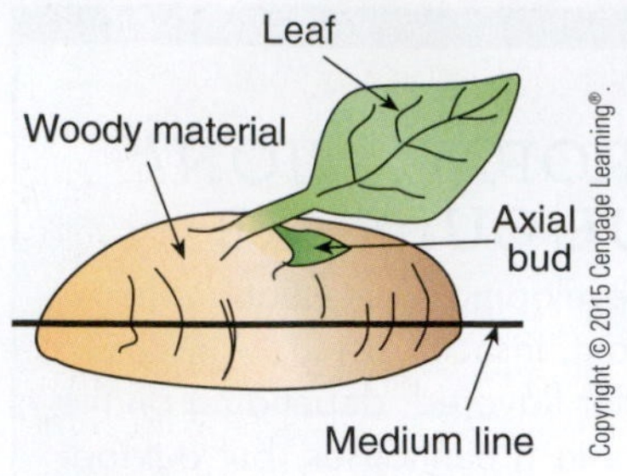

FIGURE 17-10 To make a heel cutting, a shield-shaped cut is made about halfway through the wood around a leaf and axial bud. The shield is placed horizontally into the growth medium with the leaf and axial bud above the medium line.

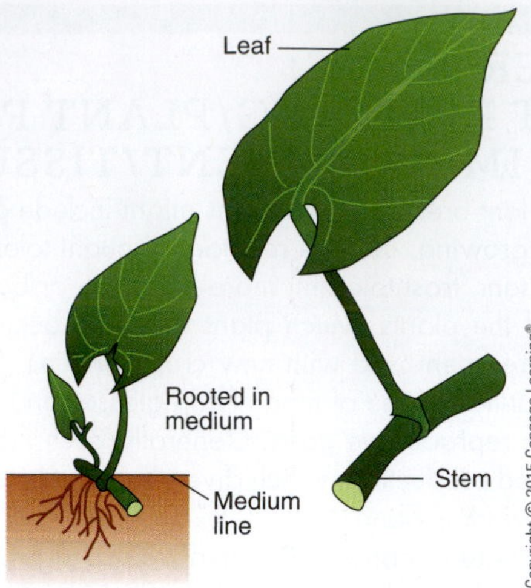

FIGURE 17-11 A single-eye cutting is made by cutting a section of stem above and below a single node and placing the cutting either vertically or horizontally into the growth medium.

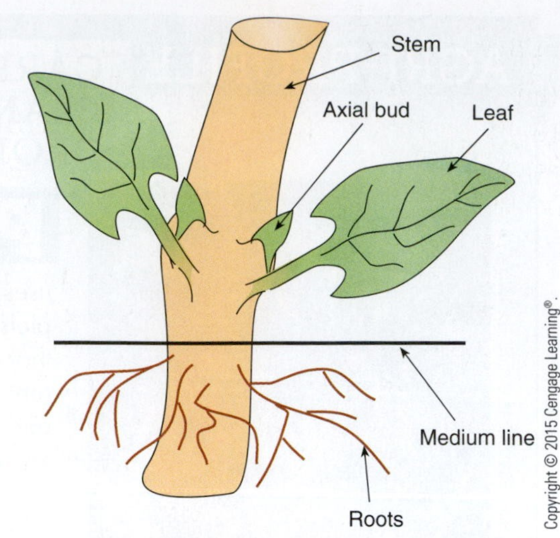

FIGURE 17-12 A double-eye cutting is used for propagating plants with an opposite leaf pattern. The stem is cut on both sides of a single node, and it is inserted vertically into the growth medium.

INTERNET KEY WORDS:
propagation, heel cutting
propagation, single-eye cutting
propagation, double-eye cutting
propagation, leaf cutting
propagation, petiole cutting
propagation, leaf section cutting
propagation, split-vein cutting

Leaf-type Cuttings

For many of the indoor herbaceous plants, a leaf-type cutting will produce plants quickly and efficiently. This type of cutting will not normally work for woody plants.

Leaf Cuttings

A cutting made from a leaf with a petiole cut to less than a half inch is referred to as a **leaf cutting**. To prepare a leaf cutting, detach the leaf from the plant with a clean cut and dip the leaf into the rooting hormone. Place the leaf cutting vertically into the medium. New plants will form at the base of the leaf and may be removed when they have formed their own roots (Figure 17-13).

Leaf Petiole Cuttings

For **leaf petiole cuttings**, a leaf with a petiole about 0.5 to 1.5 inches long is detached from the plant. The lower end of the petiole is dipped into the rooting medium and is then placed into the medium. Several plants will form at the base of the petiole (Figure 17-14). These plants may be removed when they have developed their own roots. The cutting may be left in the medium to form new plants.

Leaf Section Cuttings

Fibrous-rooted begonias are frequently propagated using **leaf section cuttings**. The begonia leaves are cut into wedges, each containing at least one vein (Figure 17-15). The sections are then placed into the medium.

New plants will form at the vein that is in contact with the medium. A section-type leaf cutting is made with the snake plant (*Sanseveria* sp.). The leaf is cut

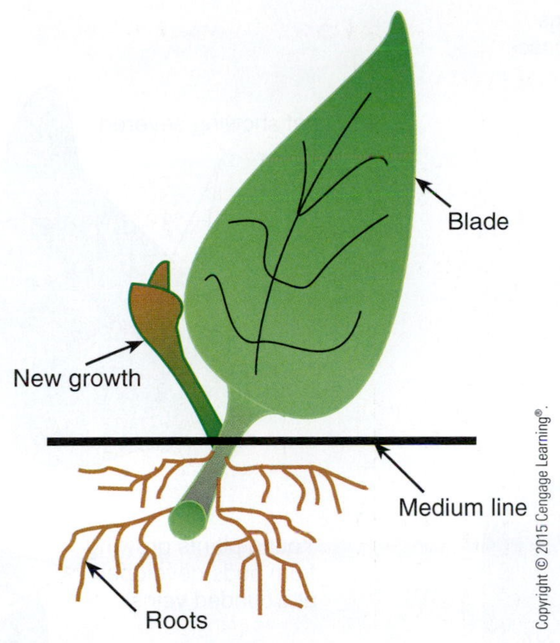

FIGURE 17-13 A leaf cutting is prepared by cutting a single leaf from a plant, dipping it in rooting hormone, and planting it vertically in the growth medium.

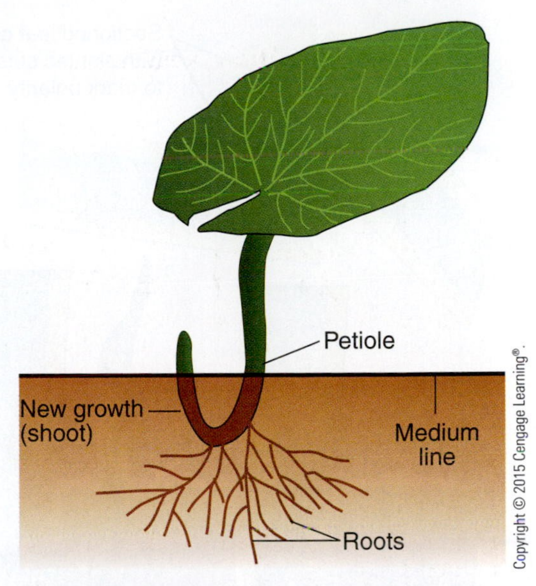

FIGURE 17-14 A leaf petiole cutting uses a leaf with an attached petiole that is 0.5 to 1.5 inches in length. The lower end of the petiole is dipped in the rooting compound and planted in the growth medium.

into sections, 2 to 3 inches long. It is a good practice to make the bottom of the cutting on a slant and the top straight. This is done so you can tell the top from the bottom (Figure 17-16). The sections are placed in the medium vertically. Roots will form reasonably soon, and new plants will start to appear. These are to be cut off from the cutting as they develop root systems. The original cutting may be left in the medium for more plants to develop.

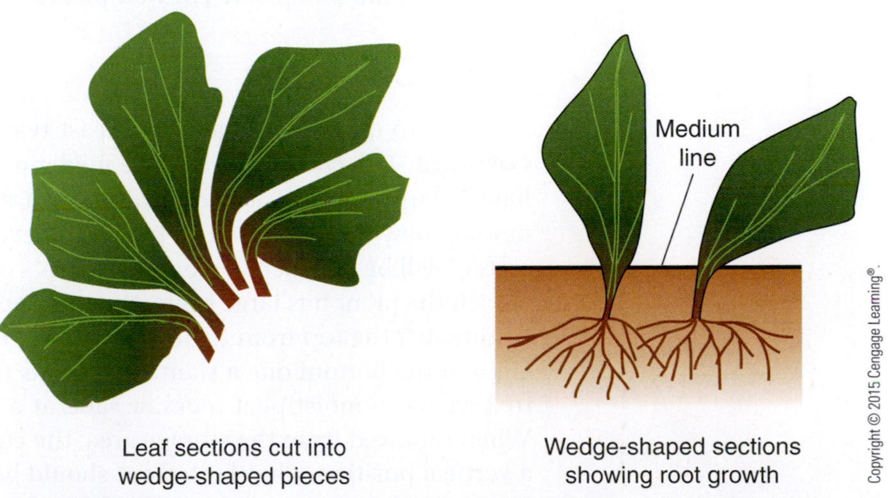

FIGURE 17-15 A leaf section cutting is obtained by cutting a single leaf into strips. The lower end of each leaf section is dipped in the rooting compound, followed by planting in the growth medium.

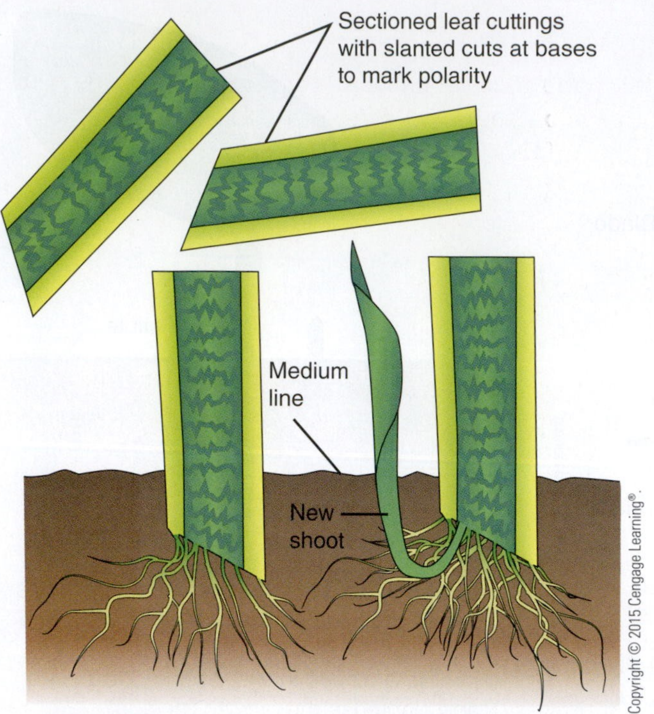

FIGURE 17-16 A snake plant leaf is cut into sections, 2 to 3 inches in length, with the bottom cut at a slant and the top straight. Each section is placed vertically in the growth medium.

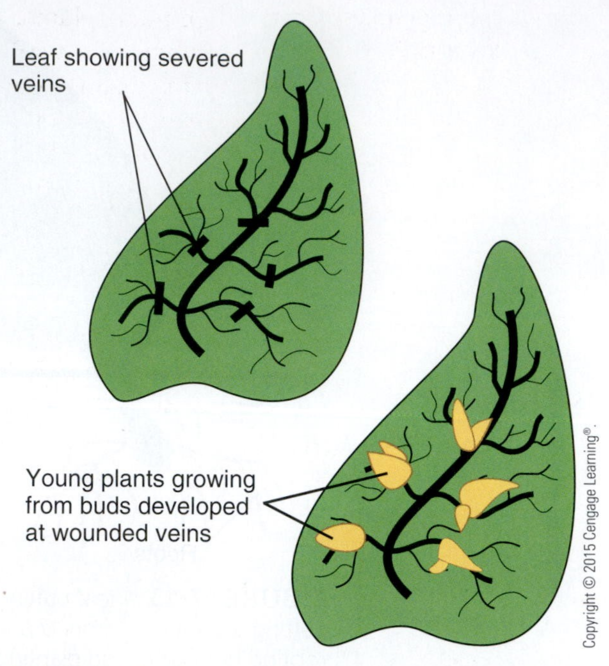

FIGURE 17-17 A split-vein cutting is prepared by making slits across the veins on the underside of the leaf. The cutting is then secured on the growth medium with the lower side down.

Split-vein Cuttings

Split-vein cuttings are often used with large leaf types, such as begonias and other large-leaf plants. With split-vein cuttings, the leaf is removed from the stock plant, and the veins are slit on the lower surface of the leaf (Figure 17-17). The cutting is then placed on the rooting medium with the lower side down. It might be necessary to secure the leaf to make it lie flat on the surface. A good method is to use small pieces of wire, bending them like hairpins and pushing them through the leaf to hold it in place. The new plants will form at each slit in the leaf.

Root Cuttings

It is best to use plants that are at least two to three years old for making **root cuttings**. The cuttings should be made in the dormant season when the roots have a large reserve of carbohydrates. In some species, the root cuttings will develop new shoots, which, in turn, will develop root systems. In others, the root system will be produced before new shoots develop.

If the plant has large roots, the root section should be 4 to 6 inches long. To distinguish the top from the bottom of the root, make the top cutting a straight cut and the bottom one a slanted cut. This type of cutting should be stored for 2 to 3 weeks in moist peat moss or sand at a temperature of about 40° F (4.4° C). When removed from the storage area, the cutting is inserted into the medium in a vertical position. The slanted cut should be down, and the top straight cut just level with the top of the medium. If the plant typically has small roots, a section 1 to 2 inches long is used. The cutting is placed horizontally a half inch below the surface of the medium.

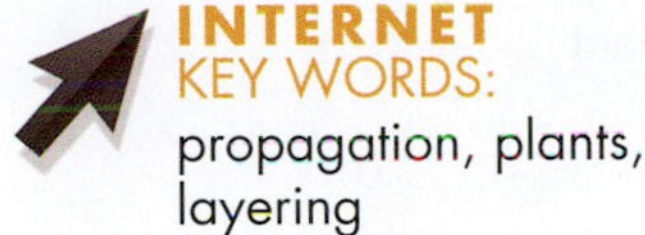

LAYERING

In many plants, stems will develop roots in any area that is in contact with the media while still attached to the parent plant. After roots form, shoots develop at the same point. An advantage of this type of vegetative propagation is that the plant does not experience water stress, and sufficient carbohydrates are supplied to the new plant that is forming. The following sections discuss some of the more common methods of layering.

Simple Layering

Simple layering is an easy method that can be used on azaleas, rhododendrons, and other plants. A stem is bent to the ground and is covered with medium. It is advantageous to wound the lower side of the stem to the cambium layer. The last 6 to 10 inches of the stem are left exposed (Figure 17-18).

Tip Layering

Raspberries and blackberries are propagated using **tip layering**. With this method, a hole is made in the medium, and the tip of a shoot is placed in the hole and covered. The tip will start to grow downward and will then turn to grow upward. Roots will form at the bend. When the new tip appears above the medium, a new plant is ready to be transplanted. It will be necessary to separate the new plant from the parent by cutting the stem just before it enters the medium (Figure 17-19).

Air Layering

Many foliage plants are propagated using **air layering**. Some ornamental trees, such as dogwood, can also be reproduced by this type of layering. The stem is girdled with two cuts about 1 inch apart. The bark is removed. The wound is dusted with a rooting hormone and is surrounded with damp sphagnum moss (Figure 17-20). Plastic is wrapped around the moss-packed wound and tied at

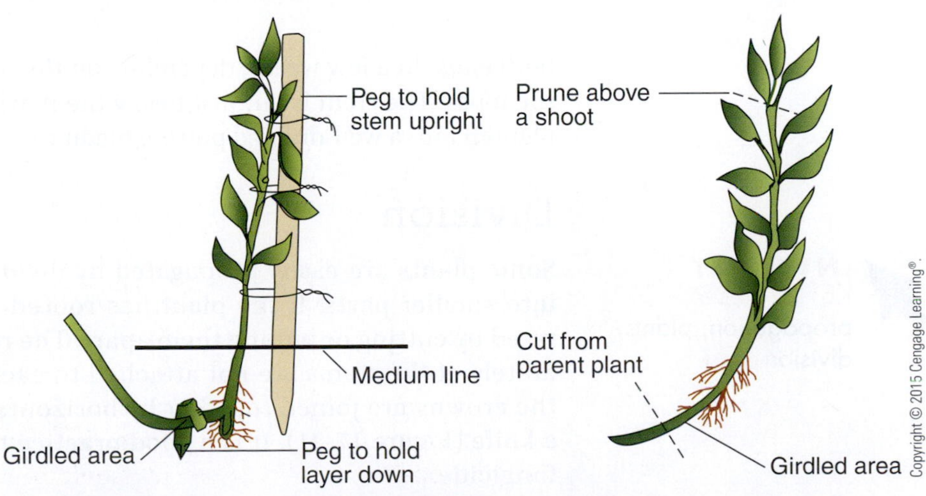

FIGURE 17-18 Layering is a method of promoting root growth by placing an attached stem or a cut stem beneath the soil partway along its length, with the last 6 to 10 inches of the stem exposed to sunlight.

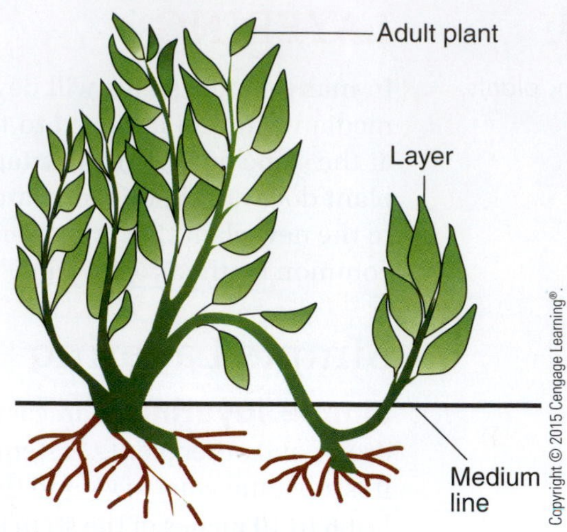

FIGURE 17-19 Tip layering is accomplished by making a hole in the medium next to a growing plant and burying the tip of a plant shoot in it to promote the growth of new roots and shoots.

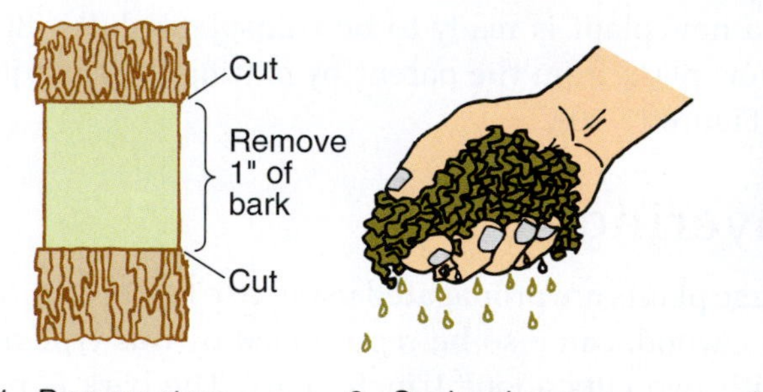

FIGURE 17-20 Air layering.

both ends. In a few weeks, depending on the plant, roots will appear throughout the moss. The stem is cut just below the newly formed root ball, and the ball is planted into a well-drained potting medium.

Division

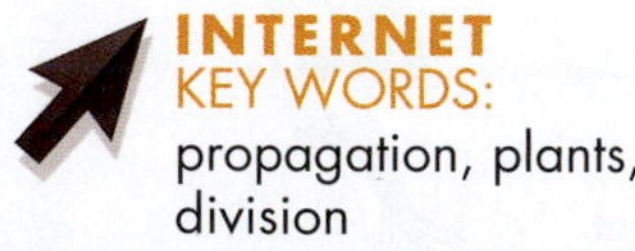

Some plants are easily propagated by dividing or separating the main part into smaller parts. If the plant has rooted crowns, these crowns are separated by cutting or pulling them apart. The resulting clumps are planted separately. If the stems are not attached to each other, they are pulled apart. If the crowns are joined together by horizontal stems, they are cut apart with a knife (Figure 17-21). It is a good practice to dust the divided plants with a fungicide.

Some plants that grow from bulbs or corms form little bulblets or cormels at their bases. To produce more plants from this type, simply separate the newly formed plant part and place it in a good medium (Figure 17-22).

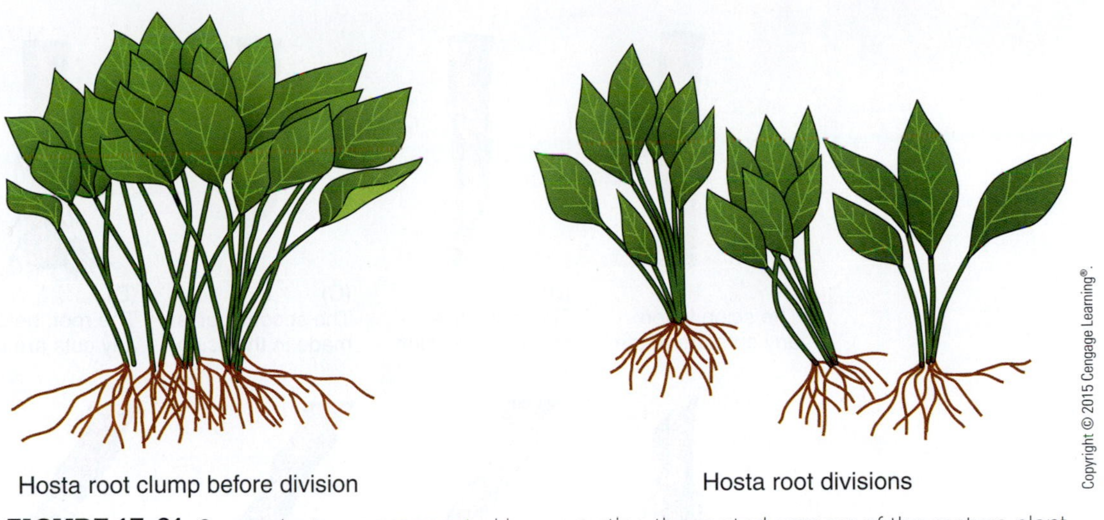

FIGURE 17-21 Some plants are propagated by separating the rooted crowns of the mature plant.

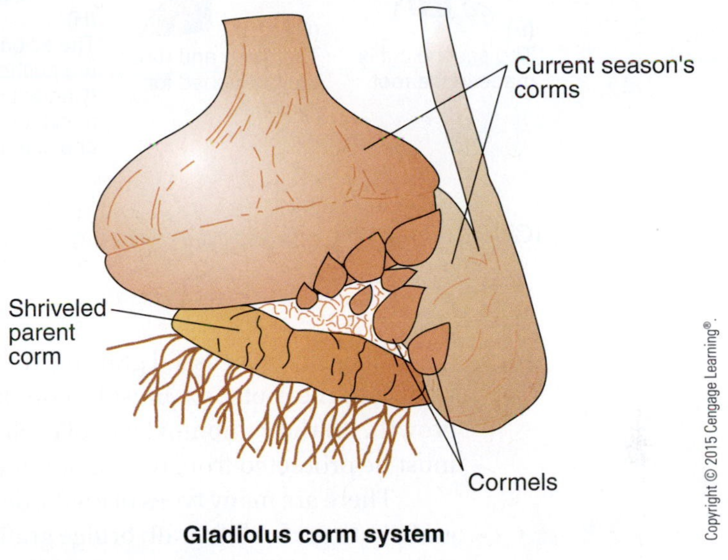

FIGURE 17-22 Plants that grow from bulbs or corms may be separated by pulling the small bulblets or cormels from the base of the mature bulb or corm.

Grafting

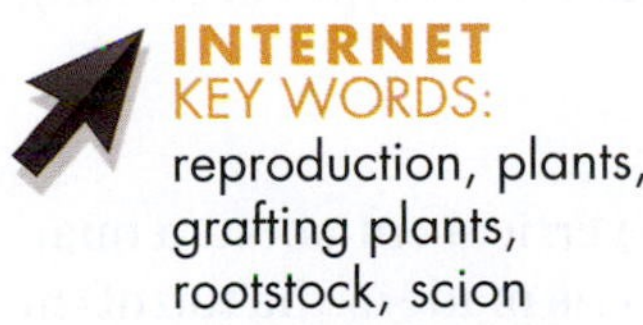

INTERNET KEY WORDS:
reproduction, plants, grafting plants, rootstock, scion

Grafting is a procedure for joining two plant parts together so they grow as one. This method of asexual propagation is used when plants do not root well as cuttings, or when the root system is inadequate to support the plant for good growth. Grafting will allow the production of some unusual combinations of plants. For instance, several varieties of apples can be grown on one tree. Some nut trees can be induced to grow varieties other than their own. Some unusual foliage plants can also be made by grafting. Finally, dwarf fruit trees are created by grafting regular varieties on dwarfing root stock obtained from related trees with similar, yet different genetics.

The top part of the plant that is to be propagated is called the **scion**. The **rootstock**, or stock, will provide the new plant's root system and will supply the nutrients and water. The **graft union** is where the two parts meet.

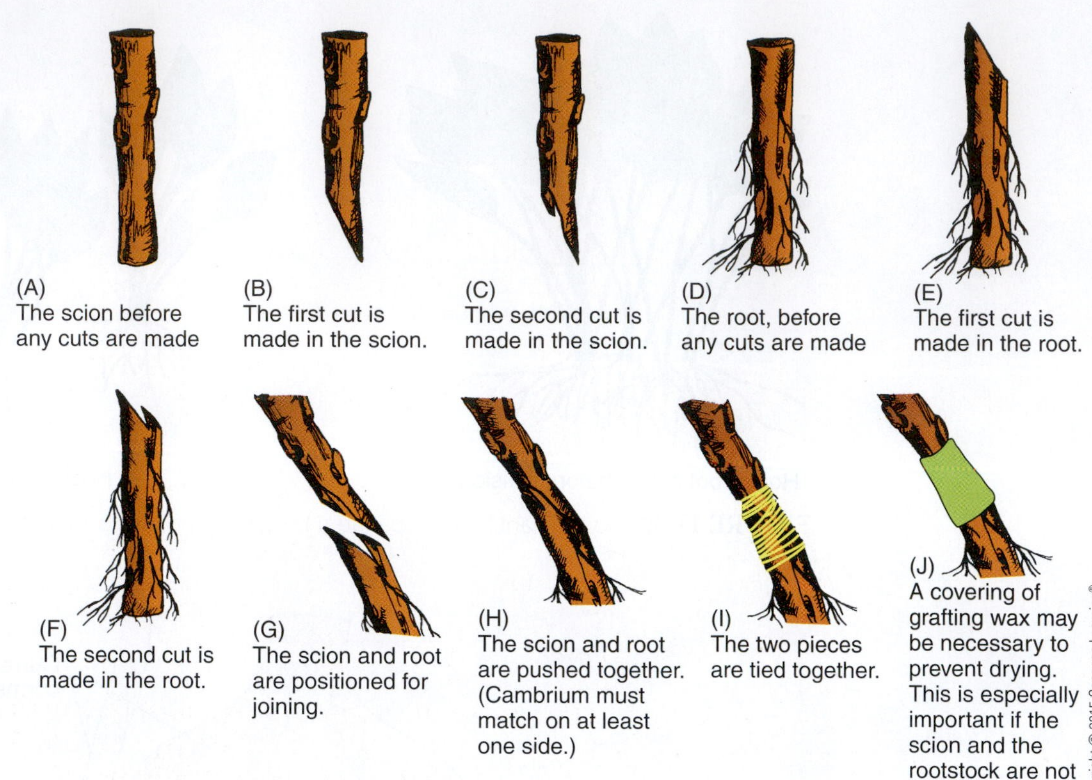

FIGURE 17-23 The process for performing a whip graft.

To ensure successful grafting, the following conditions are necessary: (1) the scion and the rootstock must be compatible, (2) each must be at the right stage of growth, (3) the cambium layer of each section must meet, and (4) the graft union must be protected from drying out until the wound has healed (Figure 17-23).

There are many types of grafts. Some common grafts are the whip, or tongue, graft; bark graft; cleft graft; bridge graft; and bud graft. Each type is used for a special purpose. The most commonly used and the easiest to perform is the bud graft.

Bud Grafting

The union of a small piece of bark with a bud and a rootstock is called **bud grafting**. It is most useful when the scion material is in short supply. This type of grafting is faster, and it will make a stronger union than other types of grafting.

T-Budding

T-budding is a popular type of bud graft in which a vertical cut about a quarter-inch long is made on the rootstock. A horizontal cut is made at the top of the vertical cut. The result is a T-shape. The bark is loosened by twisting the point of a knife at the top of the T. A small, shield-shaped piece of the scion, including a bud, bark, and a thin section of the wood, is prepared. The bud is pushed under the loosened bark of the stock plant. The union is wrapped with a piece of rubber band called a budding rubber. The bud remains exposed. After the bud starts to grow, the remainder of the rootstock plant is cut off above the bud graft (Figure 17-24)

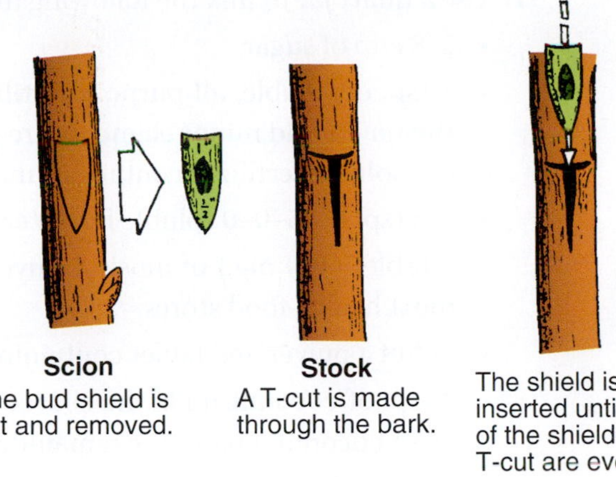

FIGURE 17-24 The process for performing a T-bud graft.

TISSUE CULTURE

A relatively new method of plant propagation is micropropagation, or **tissue culture**. Instead of using a large part of the plant, as in other types of vegetative, or asexual, propagation, a small and actively growing part of the plant is used (Figure 17-25). The result is that many new plantlets may be obtained from a section of a leaf. The process must be done in a very clean atmosphere, and it is not successful in the greenhouse or other traditional propagation areas. Tissue culture requires an aseptic environment. **Aseptic** means sterile or free from microorganisms. This kind of environment must be maintained for tissue culture to be successful. Currently, many commercial tissue-culture laboratories are producing a large variety of plants.

Advantages over Traditional Methods

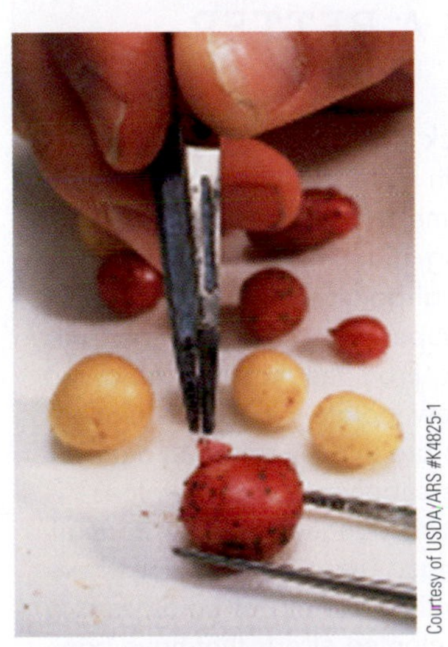

FIGURE 17-25 Only small amounts of tissue are needed to produce an exact clone of a plant when propagating through tissue culture.

The greatest advantage of tissue-culture propagation is that numerous plants can be propagated from a single disease-free plant. Plants can be propagated more efficiently and economically than with traditional methods of asexual reproduction. The main disadvantage is that the work area must be very clean. All of the equipment must be sterile. In commercial production by tissue culture, there is more expense in equipment and facilities than with traditional methods of propagation.

The materials necessary for tissue culture are (1) a clean, sterile area in which to work, (2) clean plant tissue, (3) a multiplication medium, (4) a transplanting medium, (5) sterile glassware, (6) sterile tools, (7) a scalpel, razor, or an X-acto® knife, and (8) tweezers.

Preparing Sterile Media

The first step in preparing for tissue culture is to prepare the medium in which the tissue will grow. The medium is called **agar**, which is available commercially from many of the scientific supply houses. The Virginia Cooperative Extension Service offers the following formula and procedure for preparing media for experimentation in tissue culture on a small basis:

1. Use a quart jar to mix the following materials:
    - 1/8 cup of sugar
    - 1 tsp. of soluble, all-purpose fertilizer. The label will indicate that all of the major and minor elements are present. It is especially important that the soluble fertilizer contain ammonium nitrate.
    - 1/3 tsp. of 35–0–0 soluble fertilizer
    - 1 tablet (100 mg.) of inositol (myo-inositol). This can be obtained from most health-food stores.
    - 1/4 of a pulverized tablet containing 1 to 2 mg. of thiamine
    - 4 tsp. of coconut milk, the source of cytokinin. This is obtained from a fresh coconut. Freeze the remainder for later use
    - 3 to 4 grains of a rooting hormone containing 0.1 active ingredient Indole-3-butyric acid (IBA)
2. Fill the jar with purified, distilled, or de-ionized water.
3. Shake the jar to dissolve all materials.

After the medium is dissolved, prepare the culture tubes using either test tubes with lids or other suitable glass containers. Fill the culture tubes one quarter of the way with sterile cotton balls. Use one or two per tube. They do not need to be packed tightly. Pour the prepared medium into the culture tubes to just below the top levels of the cotton. Place the lids on loosely.

After all medium is placed in culture tubes, it is ready to be sterilized. Sterilization may be done in two ways: (1) heat in a pressure cooker for 30 minutes, or (2) heat in an oven for 4.5 hours at 320° F. After the culture tubes are removed, place them in a clean area and allow them to cool (Figure 17-26). If several days will go by before using all of the tubes, wrap them in small groups in plastic wrap or foil before sterilizing.

HOT TOPICS IN AGRISCIENCE

CLONING A BETTER POTATO

© iStockphoto/annedde

Tissue culture is a relatively new method for propagating large numbers of plants from a limited amount of plant material.

Much of the seed stock that is used to produce potatoes is replaced with better cultivars over a period of a few years. Farmers who produce "seed potatoes" are constantly seeking plants that are resistant to diseases and pests. Individual potato plants that are identified in the field as being superior to other potato plants are selected as parent stock. From the materials obtained from these plants, many new potato plants are cloned. These valuable young plants are initially raised in a greenhouse environment, and the supply of "seed potatoes" is expanded in the field until an adequate supply is available for commercial plantings. The "seed potatoes" are then cut into pieces for planting. A field of potatoes is a perfect example of massive cloning efforts that have been advanced to a commercial scale of operation.

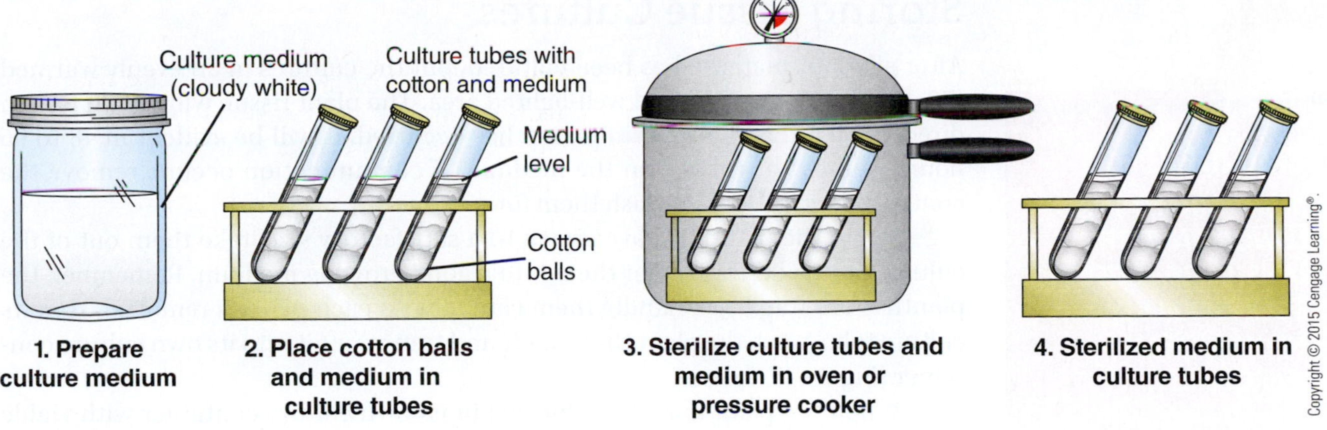

FIGURE 17-26 Procedures for sterilizing tissue-culturing medium.

Sterilizing Equipment and Work Areas

The tools and equipment used for tissue culture must also be sterilized. This can be done as the medium is sterilized by placing the tweezers, razor blade scalpel, or knife in the pressure cooker or oven. After the initial sterilization, they may be cleaned by dipping them in alcohol before and after each use. The work area must be thoroughly cleaned and sterilized. Wash the area with a disinfectant. Keep a mist bottle filled with a mixture of 50 percent alcohol and sterilized water to spray work areas and tools as well as the hands and arms of the propagator.

Preparing and Placing Plant Tissue in the Culture Tube

After the growing medium is properly prepared and cooled and the work area properly cleaned, the next step is to prepare the plant tissue. Various parts of the growing plant may be tissue cultured. For the production of vigorous plantlets, use only actively growing plant parts. With some species of plants, only a small, quarter-inch-square section of the leaf is used, whereas for others, a half inch of the shoot tip is used. With ferns, a quarter inch of the tip of the rhizome is used.

Remove the part of the plant to be used and discard the excess plant material. Submerge the plant part in a solution of one part commercial bleach and nine parts water for about 10 minutes. This will disinfect the plant tissue. Remove the tissue with sterile tweezers and rinse the material in sterile water. When the plant part has been disinfected in the bleach solution, it can be handled only with sterile tweezers and must not touch any nonsterile surface.

When the plant material has been disinfected and rinsed, remove any damaged tissue with a sterilized scalpel or razor blade. Remove the lid from a properly prepared culture tube or jar and place the plant material on the agar. Take care that the plant material is not completely submerged (Figure 17-27). Recap quickly to avoid contamination from the air. It is best if the material is placed in front of the propagator so that work is not done over the uncapped culture tube.

FIGURE 17-27 Plant material in culture tube.

REMINDER!

IT IS IMPORTANT THAT ALL WORK AND THE TRANSFER OF MATERIALS BE DONE QUICKLY AND IN A CLEAN ENVIRONMENT. Scrub all areas with disinfectant, and clean all tools with a disinfectant solution. Any contamination may lead to unsuccessful work. Bacteria and fungus will grow in unclean culture tubes and will overtake the new plant growth.

Storing Tissue Cultures

After all plant material has been cultured, put the cultures in an evenly warmed (70–75° F/21.1–23.9° C) and well-lighted area. The plant tissue will *not* do well in direct sunlight. If any contamination has occurred, it will be evident in 48 to 96 hours as mold or rotting on the medium. If contamination occurs, remove the contaminated tubes and wash them for reuse.

When the plantlets have grown to a satisfactory size, take them out of the culture tubes and transplant them into a good growing medium. Remember, the plantlets are fragile, so handle them carefully. As each plant is removed from its culture tube, wash the plant thoroughly and transplant it into its own culture container (Figure 17-28).

When the plant is well established in its own culture container with viable roots and top growth, transplant the plant again into a suitable potting mix. Place the plant inside a protected area with high humidity. The plants are coming out of a well-protected environment with plenty of humidity and light. After they have adapted to the pot and are growing well, they may be treated like any other growing plant. This process will take about 3 to 6 weeks from the beginning to a successfully growing plant.

Courtesy of USDA/ARS #K3279-11

FIGURE 17-28 A young peach tree after being moved from the original culture tube and allowed to grow in sterile medium in a sterile container.

Laboratory: Cloning of African Violets

Plant cloning by tissue culture is one of the most widely used biotechnologies. Most potatoes and many houseplants are propagated by cloning. Cloning generates multiple, genetically identical offspring from the nonsexual tissues of a parent plant.

In theory, cloning is simple: Cut a leaf off a plant, disinfect it, cut it into fragments, and then plant the fragments in nutrient agar. This may take 30 minutes.

In practice, contaminants from the air, hands, and tools quickly take over. Instead of healthy clones, you get colorful molds and bacteria. You can minimize contamination by using a simple hood and aseptic handling techniques.

Cloning African violets in the classroom is a long-term project, but it can be done within a few class periods (Figure 17-29). The first stage takes about 30 minutes. The violets can be transplanted to a mini-greenhouse 6 to 8 weeks later. In another few weeks, the plant will be ready for repotting.

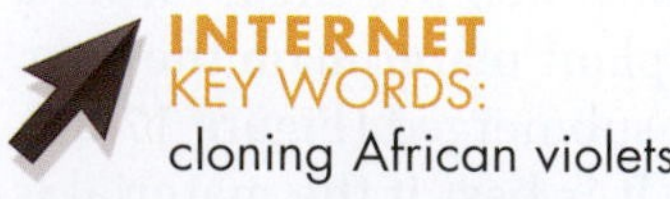

INTERNET KEY WORDS:
cloning African violets

Aseptic Technique

Aseptic handling is critical for successful plant tissue culture. The culture vessel is a battleground between rapidly growing microbes and slowly regenerating plant fragments. Five simple techniques of aseptic handling will minimize transfer of contaminants, and thus growth of molds and bacteria:

1. Wash hands thoroughly and scrub nails using regular soap and paper towels. Do not touch your face or other objects or put your hands in your pockets. Such practices put contaminants back on your hands and can re-contaminate your plants, equipment, work area, or medium.
2. Keep your hands from passing over open vessels.
3. Touch vessels far from the rim, neck, and similar areas. Keep the caps on when not in use.
4. Grasp tools as far from the working ends as possible.
5. Use sterile materials.

FIGURE 17-29 African violets can be readily propagated through tissue culture.

Practice Activities

1. Make thumbprints on bacteriologic nutrient agar plates before and after handwashing. Incubate overnight in a warm (not greater than 99° F/37° C) place. Bacterial colonies will appear on both plates. These bacteria are normal for humans but will hinder cloning.
2. Study violet leaf fragments using a hand lens or dissecting microscope. Note the hairs protruding from the leaf's upper surface. Now look at your fingers. Note the many ridges. Where do contaminants hide in each case? How does washing the hand and leaf change how each looks?
3. Practice aseptic handling of tools. How do you pass scissors at home? How would you do this aseptically? Why might a scalpel be better for cloning work than a single-edge razor when both cut just fine?
4. Conduct a dry run of the cloning procedure using a spinach leaf. A spinach leaf bruises as easily as a violet leaf, so it readily shows how gently it has been handled.

A Hood for Cloning

Materials:

60- by 80-inch sheet of 2-, 3-, or 4-mil clear plastic (painter's tarp), bulldog clips (2 to 3 inches in length), support frame for hanging file folders

Assembly:

1. Place a file folder frame on the table with its arms facing you.
2. Fold the plastic sheet so it is two layers thick.
3. Drape the folded sheet with the fold line in front, so it overhangs the arms by about 2 inches.
4. Clamp the sheeting to the top of the arms to form a flat roof.
5. Straighten the sheeting to minimize creases.
6. Spray the inside of the hood and the work surface with 70 percent ethanol. Dry only the work surface, not the plastic.

The hood will look like a lean-to with a short curtain valance in front. This fits well on a student's desk. The overhanging plastic on the sides and back creates a larger work area than just the frame.

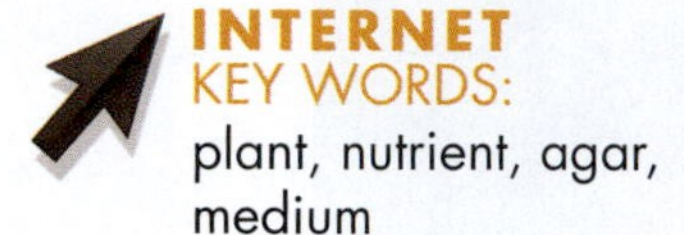

plant, nutrient, agar, medium

When working, stand over the hood. Do not breathe onto the cloning materials or work area.

Materials Preparation

To sterilize materials, autoclave for 15 minutes at 15 to 21 lbs. of pressure. Open only in a hood that has been surface sterilized with ethanol.

1. Mix and sterilize plant nutrient agar. One type is Murashige African violet/gloxinia multiplication medium, available from companies that supply biological products. One pack makes 30 to 40 plates. Stir one pack into 1 liter distilled water. Add 30 g. sucrose and 15 g. agar. The agar will not dissolve until it is heated. Loosely cap the flask with aluminum foil and autoclave or otherwise heat for 15 minutes. Cool in hot tap water of about 122° F (50° C). Pour 25 to 30 mL agar solution into each plastic petri dish (20 $\times$ 100 mm size). Allow gel to set up at room temperature, and then store in a refrigerator.
2. Sterilize a 100-mL beaker to hold disinfected leaves. Cap with a 4-inch square of heavy-duty aluminum foil.
3. Sterilize 150 mL tap water in a foil-capped, 250-mL Erlenmeyer flask.
4. Wrap the glass petri plate (cover and bottom assembled) in a double layer of white T-shirt rag, and then autoclave. The inside of the cover will be used as the cutting surface, and the rim of the bottom will be used to support the tools as they drain.
5. Assemble but do not sterilize the following:
    - 500-mL beaker for waste liquid
    - 20$\times$150-mm test tube filled to the brim with 70 percent ethanol (support in a 250-mL Erlenmeyer flask)
    - Curved 8-inch forceps and a 6-inch scalpel handle with #11 blade
    - 200-mL beaker containing 100 mL of 70 percent ethanol to dip the leaf
    - Single-edge razor to cut the leaf from the plant
    - 100-mL beaker to hold the leaf during disinfection

6. Prepare the disinfectant. Mix a 20 percent solution of liquid chlorine bleach and add one drop of Joy dish detergent per 500 mL. Swirl gently to mix; too many bubbles will inhibit wetting of the leaf surface. (Remember those little hairs.)
7. Soak the forceps and scalpel in the ethanol tube for at least 5 minutes before use. Do not store in the ethanol because the blade will rust.
8. Place the following in the hood:
   - Plant nutrient agar plate
   - Sterile H_2O flask
   - Sterile beaker
   - Sterile glass petri dish
   - 500-mL waste beaker
   - Test tube of ethanol
   - Forceps and scalpel
   - 200-mL beaker with ethanol

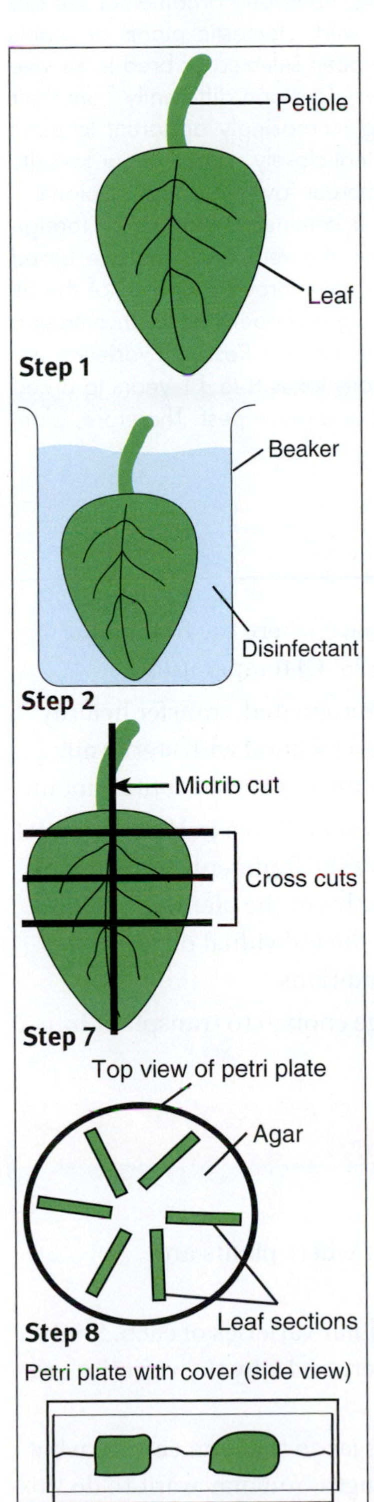

FIGURE 17-30 Cloning Procedure: Illustrated steps.

Cloning Procedure

As you work through the cloning procedure, review Figure 17-30 for illustrations related to specific steps in the process.

1. Cut a young leaf so the petiole (stem) remains attached.
2. Put the leaf and petiole in the nonsterile beaker with disinfectant to remove dirt, mites, or other vermin. Leave it there for 10 minutes, but swirl it occasionally.
3. Work under the plastic cover, using the alcohol-soaked forceps to transfer the leaf, by the petiole, to the small sterile beaker. Pour the disinfectant into the waste beaker and remove the nonsterile beaker.
4. Rinse the leaf with 50 mL sterile water. Swirl and pour the water into the waste beaker. Gently hold the leaf in the small beaker with the forceps.
5. Wash again.
6. Disinfect the leaf by dipping the leaf in the ethanol leaf soak, count to 10, and remove the leaf. Meanwhile, open the glass petri plate. Use the bottom plate to drain the forceps and blade. (Resoak the forceps before cutting.)
7. Cut the leaf into fragments. Use the lid of the glass petri plate as the operating table.
   a. Cut off the petiole. Do not plant it.
   b. Cut down the midrib firmly.
   c. Cut across each half into four pieces, being sure to cut through a branching vein.
8. Plant each fragment in the plant nutrient agar so each piece is in, not just on, the agar. Plant in a spoke arrangement to minimize spread of contaminants.
9. Put the lid back on and cover the plastic petri plate with plastic cling wrap to keep moisture in.

SCIENCE CONNECTION THE SEARCH FOR PERFECT PLANTS

Courtesy of USDA/ARS #K2492-13

Plant geneticist Keith Schertz examines grain sorghum bred for tropical climates. Bags prevent the sorghum flowers from cross-pollinating, so the plant breeder can control which plants provide the male pollen to fertilize the female part of any given plant.

The plants we use today for food, clothing, fiber, and ornamental use are quite different from those found in the wild. Domestic plants or plants grown for a specific use have generally been selected or bred to survive better, grow faster, look different, or in some way perform differently from their ancestors in the wild. However, it is becoming increasingly apparent to plant breeders that we must have wild plants that are not closely related to our favorite domestic species to inject new characteristics into our favored domestic plants.

Pest resistance is an area that requires a continuous reserve of foreign genetic sources. This is to be expected because the very pests that we breed plants to resist are constantly adapting to our plants through survival of the fittest among their kind. Insects and disease-causing pathogens have an amazing capacity to adapt to and eventually break crop resistance. Resistant varieties usually become obsolete in 3 to 10 years. It generally takes 8 to 11 years to breed a new variety to resist the changing individuals of a given pest. Therefore, plant

10. Store in the dark for about 1 week, and then place where the fragments have a daily light cycle and 68° to 77° F (20° to 25° C) temperatures.
11. Observe weekly. If part of a plate becomes contaminated, transfer healthy fragments to a fresh plate. Even with a commercial hood with sterile air, only about 50 percent of the plates remain completely free of contaminants.
12. Plantlets will appear on some fragments after about 8 weeks. When plantlet leaves grow to about 0.5 cm, aseptically remove the fragment. Separate its plantlets and return the plate to incubate. Gently cut the plantlets, being sure to have some root and some shoot. Place the individual plantlets in growing medium to develop under aseptic conditions.
13. After about a month, the plants should be large enough to transplant into a pot in sterile potting medium.

STUDENT ACTIVITIES

1. Write the Terms to Know and their meanings in your notebook.
2. List the crops that are grown commercially in your area. Visit a site or sites where plants are propagated, and ask the grower to discuss the propagation methods used.
3. List the prevalent crops that are grown from seed in your area and the popular varieties of each. Study a seed catalog and determine the requirements for germination of each variety and why the varieties are popular in your community.
4. Plant a seed in a jar filled with medium. Place the seed near the edge of the jar so that you can see what is happening. Keep a journal of daily observations as the seed or plant changes. You may want to do this with several seeds in different jars with different amounts of water, light, air, or temperature in each jar. Note your observations and the conditions regarding each seed. Write your conclusions about what is best for maximum germination results.

breeding programs must function on a continuous basis to maintain our current capability to feed, clothe, and otherwise supply the needs of our population.

U.S. farmers grow more than 200 varieties of wheat, 85 varieties of cotton, 200 varieties of soybeans, and many varieties of fruit, vegetables, and ornamental crops. Disease and insect resistance must be bred into each of these varieties. Many of them thrive only in specific, limited growing areas, such as one part of one state. Keeping up with known and persistent pests is a relatively manageable process, as long as our state and federal experiment stations are reasonably well funded to research evolving problems. However, the sudden appearance or introduction of a pest without resistant varieties or natural biological enemies can be devastating.

For instance, the Russian wheat aphid first appeared in the United States in 1986 and has cost wheat growers hundreds of millions of dollars since its arrival. U.S. wheat varieties have little resistance to the Russian wheat aphid. Therefore, chemical insecticides have to be used until either resistance can be bred into our domestic wheat varieties or biocontrol agents can be found or developed. Entomologist Robert Burton, now deceased, of the Plant Science Laboratory in Stillwater, Oklahoma, believed the answer would be found by introducing selected genes from wheat varieties from Southwest Asia. Burton earned the respect of scientists throughout the world for his work in developing Russian aphid–resistant wheat and barley plants.

5. Experiment with different kinds of plants and various kinds of cuttings. Keep a journal to determine the best type of cutting for specific plants. Keep notes on different media, temperatures, light, and rooting hormones.
6. Practice making each type of graft discussed in this unit under the supervision of your instructor.
7. Research additional grafting methods.
8. Conduct an experiment with the tissue-culture method of propagation. Keep notes on the different kinds of plants used, the time required for root formation, and the observations regarding the benefits of propagation by tissue culture.
9. Prepare a statement about the propagation methods best suited to your purpose.

SELF-EVALUATION

A. MULTIPLE CHOICE

1. Propagation is defined as
   a. the union of an egg and sperm.
   b. the process of increasing the numbers of a species.
   c. a cheaper method of propagation than with seeds.
   d. the only way to reproduce some species and cultivars.

2. A seed consists of
   a. a root, stem, and flower.
   b. a root, seed coat, and endosperm.
   c. a seed coat, endosperm, and embryo.
   d. an embryo, cotyledons, and new plant.

3. A type of stem cutting used where stock material is limited and has alternate leaves is a
   a. stem tip cutting.
   b. cane cutting.
   c. simple layering.
   d. single-eye cutting.
4. A cutting that is usually made from a large-leaf plant with the veins split is
   a. split-vein cutting.
   b. leaf petiole cutting.
   c. terminal tip cutting.
   d. tissue propagation.
5. Grafting is
   a. a type of sexual propagation.
   b. a type of hybridization.
   c. a method by which two plants are propagated.
   d. a method of joining two parts of two different plants.
6. The most common type of grafting is
   a. T-budding.
   b. simple layering.
   c. stem-tip propagation.
   d. scion cut out of the stock plant.
7. Tissue culture may be used for
   a. cloning.
   b. disinfecting.
   c. sexual reproduction.
   d. sterilization.
8. In tissue culture, contamination may be a problem from
   a. air.
   b. hands.
   c. tools.
   d. all of the above.
9. The first stage of cloning African violets takes approximately
   a. 1 to 3 weeks.
   b. 4 to 5 weeks.
   c. 6 to 8 weeks.
   d. 10 to 12 weeks.
10. To make a very inexpensive hood for tissue culturing, the recommended covering is
   a. aluminum.
   b. plastic.
   c. sheet steel.
   d. wood.

B. MATCHING

________	1. Outer seed coat	a. A cutting from an end of a branch containing a terminal bud
________	2. Germinate	b. Best taken when the plant is dormant
________	3. Imbibition	c. The absorption of water into the young seed
________	4. Tip cutting	d. Functions as a protector for the seed
________	5. Root cutting	e. When plants are separated and then replanted
________	6. Division	f. When a seed starts to sprout

C. COMPLETION

1. ________ propagation uses a part or parts of one parent plant.
2. The ________ will supply food to the young seedling until it is able to make its own.
3. A ________ might contain fungicide and is used to help plants produce roots more quickly.
4. ________ cuttings are normally made of a section containing one or two nodes.
5. Tissue culture is also known as ________.
6. A major aspect of tissue culture is that the area to be worked in must be ________ and ________.

D. TRUE OR FALSE

1. A clone is almost like the parents.
2. The embryo is actually the young plant.
3. All seeds need light to germinate.
4. An advantage of tissue culture is that only one plant can be made from a disease-free plant.
5. Double-eye cuttings are often used when plants have opposite leaves.
6. Herbaceous plants are propagated by cuttings.

SECTION 6

CROPS IN SPACE!

As humans push back the frontiers of outer space and consider the likelihood of space travel taking days or even years, we face the age-old dilemma—how do we feed the crew? Early manned space voyages were accommodated by freeze-dried food and food in tubes. Now scientists are gearing up for food production in space. Not only is there a need to produce food but there also is a need to generate oxygen, eliminate carbon dioxide, and dispose of organic waste. The solution is a controlled ecological life support system (CELSS). A CELSS will provide the basic components to sustain life without requiring inputs from external sources.

Consider a crew in space. For basic survival, they need food, water, and oxygen. The food could come from fruits, vegetables, grains, meat, milk, or eggs. Without a source of oxygen to replenish what is breathed in, the crew would soon die. In addition, if carbon dioxide were permitted to build up in the air, the crew would soon be poisoned. Similarly, if human waste could not be decomposed, its accumulation would become unbearable.

It has become clear that an effective CELSS is needed. Through photosynthesis, plants can take up nutrients from soil or water and use carbon dioxide from the air to manufacture food for themselves as well as for animals. They give off oxygen as a by-product of this process. Fish and poultry are excellent sources of protein for humans and are exceptionally efficient converters of plant material into essential proteins. They are small, grow fast, and would be excellent candidates for a space farm. Like humans, they give off carbon dioxide for plants to use. Also, like humans, they produce organic wastes that can be decomposed to replace the nutrients in the soil and water, where plants are growing. Microorganisms, worms, insects, and other forms of life would round out the system. Heat and combustible gases would be energy sources generated in the process.

Could it work as a closed system? Scientists believe it will. However, many questions must be answered, and problems must be overcome first. Steven Britz, a plant physiologist in the Plant

Crop Science

Photobiology Laboratory in Beltsville, Maryland, has been trying to ask the right questions for the scientific community to answer to create a CELSS in space. Some questions raised so far are as follows:

1. What is critical to producing a harvest when you must supply every need in a closed environment?
2. What kind of and how much light is needed, and how can it be supplied?
3. How much room is needed for plant roots, and what are the effects of root area restriction?
4. What type of medium is needed—soil, water, or another medium?
5. What compounds will plants release in a closed environment, and are these useful or toxic?
6. What trace metals are picked up and transported in recirculated water?
7. What are the effects of little or no gravity on plant and animal growth and reproduction?
8. What light/dark cycles are best to optimize plant performance?
9. What plant varieties respond best to the light conditions found in space?

Many more questions need to be answered before we grow plants and animals in space. Can you suggest some questions? Answers to questions about functioning in a space environment are helping us raise better plants today. A delegation to the United States from the Chinese Agricultural Ministry was interested in the details of raising maximum yields in extremely confined spaces. In China and other parts of the world, raising hydroponic vegetables for the hotel trade is a growing industry. Using controlled environments that are not natural to plants allows us to discover the optimum conditions for the functioning of a particular plant.

Controlled environment facilities are gaining acceptance as a viable means of commercially growing high-value crops under scheduled production management in an environment free from harmful insects and pollution.

Courtesy of NASA #89-HC-130

At Kennedy Space Center in Florida, scientists use a pressure chamber from a previous space craft to develop a controlled ecological life support system, where plants recycle air, water, and waste to produce food.

UNIT 18

Home Gardening

OBJECTIVE

To plan, plant, and manage a home garden.

MATERIALS LIST

- grid-type paper
- pencil and eraser
- seed catalogs
- Internet access

COMPETENCIES TO BE DEVELOPED

After studying this unit, you should be able to:

- analyze family needs for homegrown fruits, vegetables, and flowers.
- determine the best location for a garden.
- plan a garden to meet family needs.
- establish perennial garden crops.
- prepare soil and plant annual garden crops.
- list recommended cultural practices for selected garden crops.
- protect the garden from excessive damage caused by drought and pests.
- harvest and store garden produce.
- describe the use of cold frames, hotbeds, and greenhouses for home production.

SUGGESTED CLASS ACTIVITIES

1. Obtain some potting soil and plant containers. Instruct the class on proper procedures for planting vegetable seeds such as tomatoes, cabbage, melons, flowers, and others to transplant into home gardens. Have each student plant some vegetable and flower seeds and care for the young plants as the seeds sprout and grow. Send the plants home with the students when it is time to transplant them outside in the garden.
2. Visit a commercial greenhouse to see what kinds of plants are available for sale. Note the kind of potting soil that is used, and make a list of the varieties of plants that are offered. Ask the manager to explain why these varieties have been chosen. Inquire about the methods of controlling insects and diseases

TERMS TO KNOW

seasonal
loam
clod
furrow
climate
annual
biennial
perennial
cultivation
herbicide
cold frame
hotbed

and the sources of the greenhouse's trees, shrubs, and ornamental plants. Have class members prepare written reports on the field trip covering these topics and other points that come up during the visit.

3. Provide students with tasting samples of produce that can be grown in a home garden. Include some foods that have been properly preserved by canning or dehydrating. As a class, discuss the advantages and disadvantages of growing a garden.

Gardening is an activity that can be enjoyed by all members of the family. It provides fresh fruits, vegetables, and flowers for immediate use or to sell for profit. Gardening is both an art and a science. It is demanding of the gardener, both in skill and creativity. A garden is alive and changing every day, presenting new challenges to the gardener (Figure 18-1).

Courtesy of USDA/ARS #K5134-04

FIGURE 18-1 Some people seem to have a natural talent for gardening; however, the growth of plants and the procedures for obtaining best results are based on scientific principles.

ANALYZING A FAMILY'S GARDENING NEEDS

First, one must decide what vegetables and flowers the family likes. It would not make sense to plant sweet potatoes, green beans, and marigolds if no one in the family cared for these vegetables and flowers. The home gardener can provide a seasonal and continuous variety of vegetables and flowers. **Seasonal** pertains to a certain season of the year. There are garden fruits and vegetables that are available and productive despite the subtle changes that occur during the spring, summer, fall, and winter (southern latitudes) gardening seasons. Fresh vegetables are important to everyone's diet. Those who are fortunate to have access to a garden plot would do well to plan to have plenty of produce available during the growing season with some to store for future use. Vegetables should be selected that meet the family needs, both in yield and quality (Figure 18-2).

When it has been decided what vegetables and flowers the family likes, it is time to determine how much ground will be needed. It is better to have a small, well-cared-for garden than to have a garden that goes to waste because it becomes

Courtesy of USDA/ARS #K2282-7. Photo by Tim McCabe

FIGURE 18-2 The garden plan should provide for high-yielding fruits, vegetables, and flowers that meet the needs and preferences of the family.

AGRI-PROFILE COMMUNITY GARDENS

A popular idea that has emerged in urban areas is to establish community gardens. A plot of land is prepared by measuring it into moderate-sized plots with clearly marked boundaries. Each plot is then rented to an interested person who enjoys gardening but does not have access to a garden plot. The plot owner establishes the rules that each gardener must follow. They might include weeding, use of chemicals, observing established work times, taking water turns, and respecting the privacy of other garden plots, among others.

Some community gardens are planted and tended together with no personal ownership of the produce. At harvest time, all of the produce is divided among the shareholders who paid the membership fee and helped do their share of the work. Whatever the arrangement, community gardens are appreciated by those who enjoy working with others and eating fresh produce.

too much to handle. A good rule of thumb for four grown people is to start with a plot 10 feet wide and 26 feet long, or 260 square feet.

THE GARDEN PLAN

A prospective gardener should make a sketch on paper detailing the amount and placement of the various crops. At this stage, it is important to consider successive plantings. These crops follow each other in the season so that the ground is occupied throughout the growing season. Fall crops can follow spring and summer crops in many areas. This can be done easily by planting perennial crops and different varieties of specific annual crops. Allow adequate space between the rows for cultivation.

LOCATING THE GARDEN

Depending on where you live—in the city, suburbs, or a rural area—the location of the garden is an important consideration. The garden should be convenient to the house. It should also be accessible to a water supply; be on loamy, well-drained soil; be in a sunny area; and be visible from the home, if possible. **Loam** is granular soil with a balance of sand, silt, and clay particles.

One must visualize where to locate the garden. What happens to the proposed spot when there is a heavy rain? What happens when it is very dry? From the chosen spot, look up to determine whether trees or branches will cause problems by excessively shading the garden. When planting flower beds around the house, remember that along the south and west sides, the heat will be reflected onto these beds, so they may require extra water. Select the best site you can for both vegetable and flower gardens.

PREPARING THE SOIL

Conditioning the Soil

Garden soil should be loose and well drained. The ideal soil type should be granular—like coffee grounds—so that water will soak in rapidly. However, the soil also needs to contain enough organic matter to retain water within the root zone. Few soils are originally found this way. Soil-building practices can improve the soil over time, and the gardener must prepare the soil to achieve the best results.

INTERNET KEY WORDS: garden soil preparation, garden seeds

If the proposed site has not been previously used as a garden, it is a good idea to add organic matter and plow, spade, or rototill it into the soil. The decayed organic matter and the freezing and thawing of the soil during the winter months will help improve the soil's physical condition. The application of fine organic matter in the spring can also improve soil condition and fertility.

Materials such as composted leaves and grasses, peat moss, composted sawdust, and sterilized, weed-free manure are good soil conditioners. Peat moss is a soil conditioner made from sphagnum moss. A good rate of application for organic matter is 1 lb. dry material per square foot of surface area.

Preparing to Plant

If the selected site has never been planted, it is best to remove the sod before tilling the soil; then spread the organic matter over the soil. When moisture conditions are favorable, turn the soil with a shovel, spade, plow, or rototiller (Figure 18-3). When turning the soil, it is important to break up all clods. A **clod** is a lump or mass of soil.

After the spading (turning of the soil) is completed, make the planting beds. A good procedure is to heap the soil to make raised rows with **furrows** between them. Another method is to prepare raised beds that are 4 to 8 feet wide with sunken walks between them (Figure 18-4). The furrows or sunken walkways can deliver irrigation water or drain off excess rain.

Next, level the raised beds with a rake, but do not push the soil back into the furrows. Then, walk over the beds or tamp them with the head of a garden rake to firm them and to eliminate air pockets (Figure 18-5). Finally, use the hoe and back of the garden rake to push and pull the soil to level it and to break any remaining clods into fine particles (Figure 18-6). The soil is now ready for planting.

FIGURE 18-3 Soil preparation is one of the most important aspects of gardening. Soil is tilled to kill weeds, to mix materials into the soil, and to break up large chunks of soil into a granular texture.

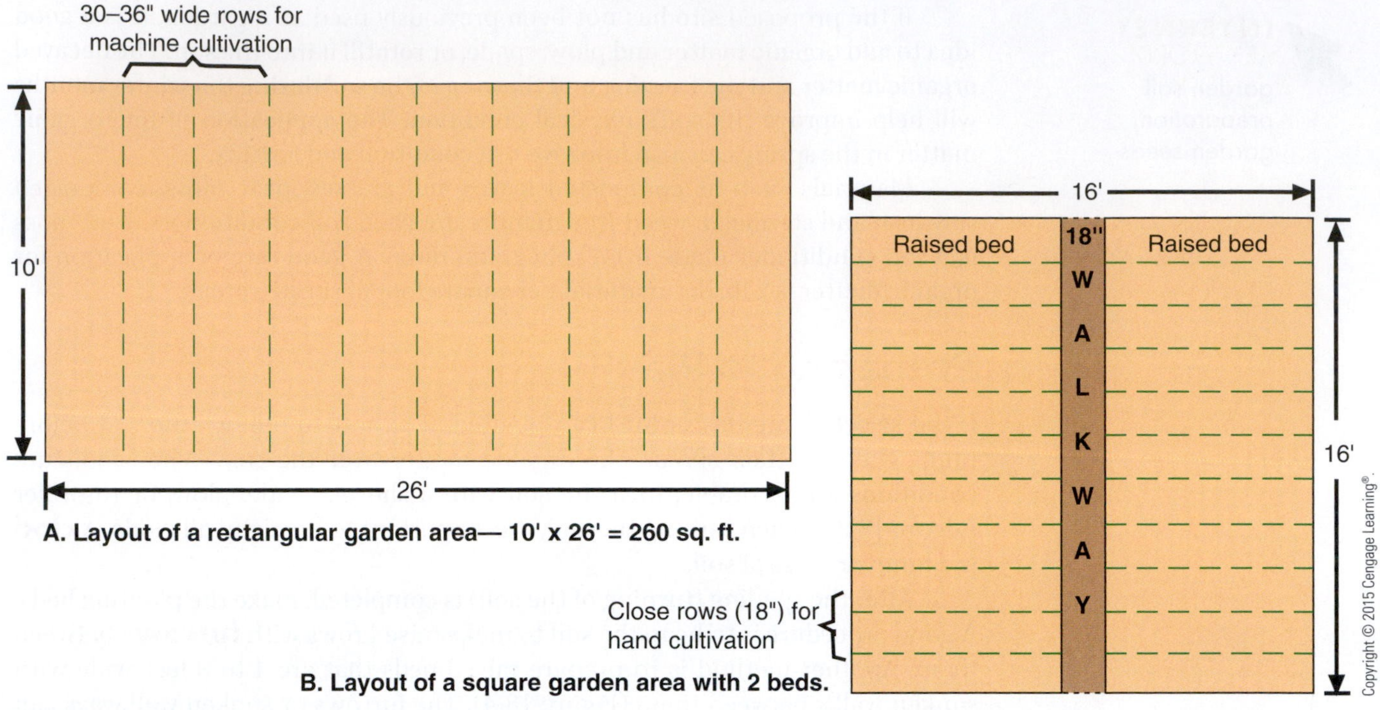

FIGURE 18-4 Layouts for rectangular (A) and square (B) gardens for a family of four.

A

B

FIGURE 18-5 Raised beds in a garden allow the soil to warm up earlier in the spring (A); they also drain away excess moisture (B).

FIGURE 18-6 The shovel, rake, and hoe are garden tools that are used to break up dirt clods and loosen, smooth, and level the soil.

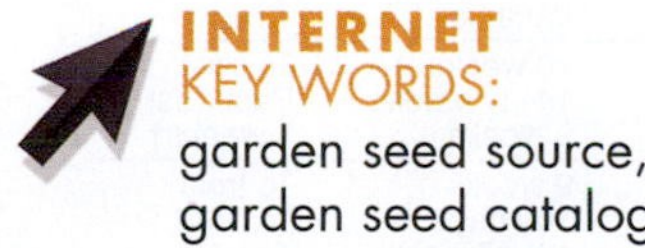

INTERNET KEY WORDS:
garden seed source, garden seed catalog

COMMON GARDEN CROPS AND VARIETIES

Consider the Climate

When choosing the varieties of vegetables and flowers to plant in the home garden, consider the climate. **Climate** refers to the weather conditions of a specific region. Do not plant crops outdoors until all danger of frost is gone. Except for a few mountainous areas and the northern tier of states, almost all areas of the United States are frost-free from June through mid-September (Figure 18-7).

Plants grow rapidly in frost-free weather. In some areas, there is not enough time for long-season crops, such as eggplant, cantaloupe, and watermelon, to

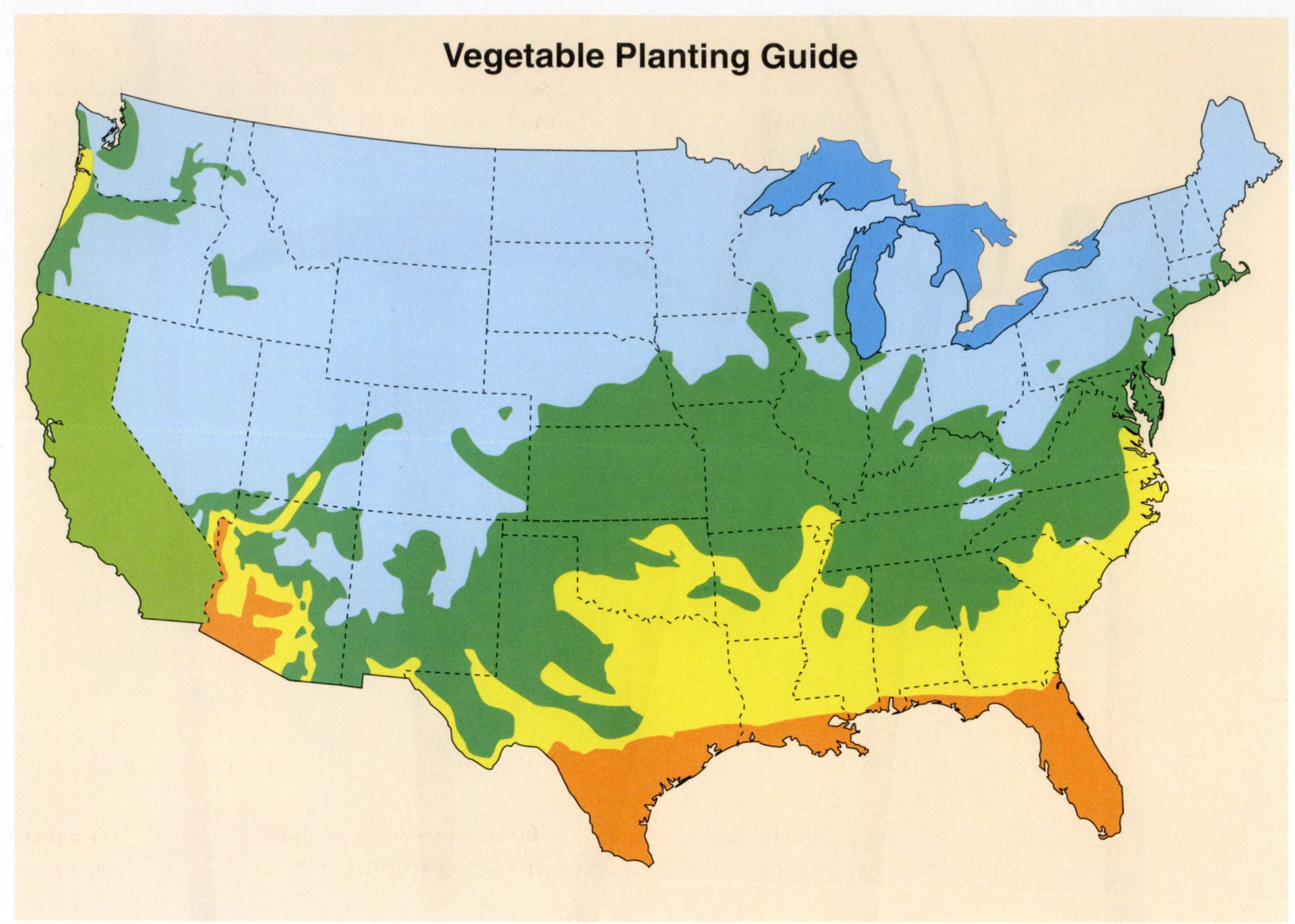

This map and accompanying chart are based largely on U.S. Department of Agriculture records showing the average dates of the last killing frosts in various parts of the country. Use this information to plan the planting and harvesting periods for your garden, but remember that these are averages, and individual conditions may vary.

	SUBTROPIC	WARM	MILD	COOL	CALIF.	PLANTING DEPTH	ROW SPACING	FIRST HARVEST	HARVEST LASTS
Beans	Apr.–Aug.	Apr.–June	May–June	May–June	Mar.–Aug.	1"	2–3 ft.	8 weeks	Until frost
Beets	Jan.–Dec.	Feb.–Oct.	Mar.–July	Apr.–July	Jan.–Dec.	1/2"	15–18"	50–78 days	6 weeks†
Broccoli	July–Oct.	Feb.–Mar	Mar.–Apr	Mar.–Apr*	Sep.–Feb.	1/4"	3 ft.	65–70 days	To frost
Brussels sprouts	Feb.–May	Feb.–Apr	Mar.–Apr*	Mar.–Apr*	Sep.–Feb.	1/2"	3 ft.	14–20 weeks.	Past frost
Carrots	Jan.–Dec.	Jan.–Mar	Mar.–June	Apr.–June	Sep.–May	1/4"	1 1/2–2 ft.	8 weeks	8 weeks†
Chard	Jan.–Dec.	Feb.–Sep.	Mar.–Aug.	Apr.–July	Jan.–Dec.	3/4"	1 1/2–2 ft.	8 weeks	To frost
Chives	Feb.–May	Mar.–May	Mar.–May	Apr.–June	Mar.–May	1/4"	1 1/2 ft.	8 weeks	To frost
Corn	Apr.–June	Mar.–June	May–July	May–July	Mar.–Aug.	1 1/2"	3 ft.	9–12 weeks	10 days
Cucumbers	Apr.–June	Apr.–June	Apr.–June	May–June	Mar.–Aug.	1/2"	6 ft.	7 weeks	5 weeks
Eggplant	Feb.–Mar	Feb.–Apr	Mar.–May*	Apr.–May*	Mar.–May	1/4"	3–4 ft.	10–14 weeks.	To frost
Endive	July–Sep.	Aug.–Sep.	Mar.–May	Apr.–June	Jan.–Dec.	1/4"	2–3 ft.	10–12 weeks	7 weeks
Lettuce	Jan.–Dec.	Aug.–May	Mar.–June	Apr.–June	Sep.–May	1/4"	15"	6 weeks	6 weeks†
Mustard	Feb.–May	Feb.–May	Mar.–June	May–July	Mar.–Aug.	1/4"	1 1/2–2 ft.	40–50 days	2 weeks
Okra	Apr.–June	Apr.–June	Apr.–June	May–June*	Mar.–May	1/2–3/4"	3 ft.	55–60 days	To frost
Onions	Dec.–Mar	Dec.–Apr	Feb.–May	Mar.–June	Sep.–May	1/2"	18"	Variable	—
Parsley	Jan.–Dec.	Jan.–June	Feb.–June	Mar.–June	Jan.–Dec.	1/4"	1 1/2–2 ft.	10 weeks	To frost
Parsnips	Mar.–June	Feb.–June	Apr.–June	May–June	Dec.–May	1/2"	1 1/2–2 ft.	14–17 weeks.	Past frost
Peas	Jan.–May	Jan.–Apr	Feb.–May	Mar.–June	Sep.–May	1 1/2"	2–3 ft.	9 weeks	2 weeks†
Peppers	Feb.–Mar	Feb.–Apr	Mar.–May*	Mar.–May*	Mar.–May*	1/4"	3 ft.	9 weeks	To frost
Potatoes	Jan.–Dec.	Feb.–Oct.	Mar.–Apr	Apr.–May	Jan.–Dec.	5"	2 1/2–3 ft.	100 days	To frost
Pumpkin	Apr.–June	Apr.–June	Apr.–June	May–June	Mar.–May	1"	8–10 ft.	14–17 weeks.	To frost
Radishes	Jan.–Dec.	Feb.–Oct.	Mar.–Aug.	Apr.–July	Sep.–May	1/4"	1–2 ft.	3–6 weeks	1–2 weeks†
Squash, Summer	Apr.–June	Apr.–June	Apr.–June	May–June	Mar.–Aug.	1"	4–6 ft.	60 days	To frost
Squash, Winter	"	"	"	"	"	1"	6–8 ft.	100 days	"
Tomatoes	Jan.–Mar	Feb.–Mar	Mar.–May*	Mar.–May*	Mar.–May*	1/4"	3–4 ft.	9–12 weeks	To frost
Turnips	Feb.–Mar	Jan.–Mar	Feb.–Apr	Mar.–May	Jan.–Dec.	1/2"	1–2 ft.	6–10 weeks	3 weeks†

*Transplants recommended.
†Following harvest, space may be used for late planting of carrots, beets, or bush beans.

Adapted from USDA

FIGURE 18-7 Planting times, planting depths, row spacing, and harvest times for common garden vegetables.

mature reliably. Some regions of the United States have intermediate growing seasons, with 5 to 6 months of frost-free weather. Long-season crops can be grown in these areas consistently. Growing seasons of 7 to 10 months are common across the mid-South, South, and low-elevation regions of the Southwest and West. In these regions, gardeners can grow many varieties of both spring and fall crops. However, these same regions are susceptible to summers that are so hot or dry that some crop varieties have difficulty surviving the intense summer heat.

Consider the Variety

There are both warm-season and cool-season crops. Certain flowers and vegetables must have continuous cool weather to do well. Heat will quickly make the plants dry up or go to seed. Where the growing season is 5 months or longer, outdoor seeding in the late summer usually results in an excellent harvest during the cool fall season. The growing season can be extended by planting varieties of the cool-season plants after other crops are harvested in late summer. Most cool-season crops can withstand light frosts with little damage.

There are two types of warm-season flowers and vegetables: those that mature quickly and those that require 4 months or more from planting to maturity. The quick-maturing types are almost always started in the garden when the soil is warm. The later-maturing types are usually started indoors and are transplanted to the garden after all danger of frost is past.

Most packets of home garden seeds are labeled to indicate the number of days that are required for the plants to mature. For example, the Blue Lake pole bean variety is labeled to mature in 60 to 65 days. Seed packet labeling is the most accurate information we have with regard to how long a plant will require to

HOT TOPICS IN AGRISCIENCE

CHOOSE CORRECT GARDEN VARIETIES: PLANT HARDINESS MAP

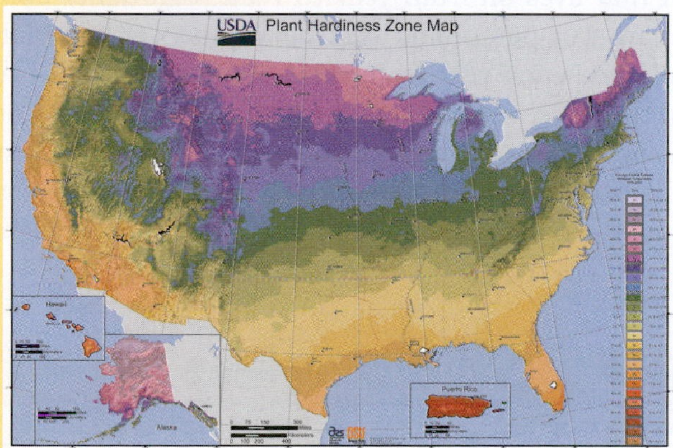

Source: USDA Plant Hardiness Zone Map, 2012. Agricultural Research Service, U.S. Department of Agriculture, Accessed from http://planthardiness.ars.usda.gov/.

A new interactive-GIS plant hardiness map was released by USDA in 2012 to replace the widely recognized USDA map that has been in use since the 1990s. The new map is based on average annual minimum winter temperatures that are divided into zones for each temperature difference of 10° F. Each zone is further divided into 5° F subzones.

Plant hardiness depends on the extreme minimum temperature it can tolerate. The new map was developed using temperature data recorded over 30 years at each of the nation's weather stations. The plant hardiness zone map is only intended to be a guide to gardeners and farm producers in selection of plant varieties. Extreme temperatures may be recorded in the future that may exceed those on which the map is based. However, the makers of this map acknowledge that microclimates exist within larger zones where temperatures are affected by large bodies of water, large paved areas, or changes in elevation.

An important feature of the map is that it can be localized by selecting a state or a ZIP Code at the following Web site: www.planthardiness.ars.usda.gov.

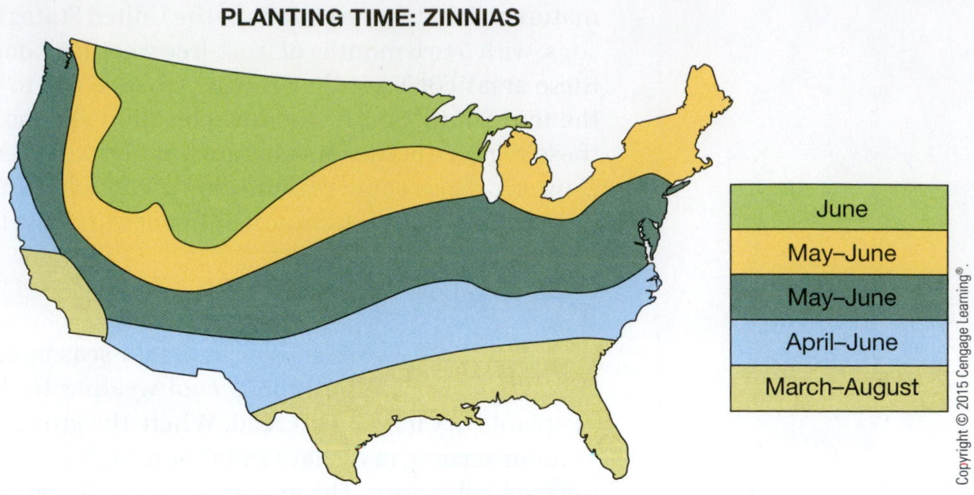

FIGURE 18-8 The best time to plant a specific plant can be determined by considering the length of the growing season, climatic zone maps, and the days to maturity as listed on the seed packet.

reach maturity (Figure 18-8). There are exceptions, however, such as the extreme northern latitudes, such as Alaska, where the sun goes down only briefly during the summer and the number of hours of daylight are much greater than in other areas of the country.

Annuals, Biennials, and Perennials

Flowers and vegetables are classified as annuals, biennials, or perennials (Figure 18-9). An **annual** is a plant whose life cycle is completed in one growing season. Growth is rapid. Practically all vegetables, except asparagus, rhubarb, and parsley, are annuals.

A **biennial** is a plant that takes two growing seasons or 2 years from seed to complete its life cycle. Some biennials bloom very little or not at all the first year, but they come into full bloom the second year and then go to seed.

A **perennial** is a plant that lives on from year to year. A gardener commonly treats some plants that are true perennials as either annuals or biennials. This is done because, when planted each year from seed, some perennials produce higher-quality blooms in their first year than older plants that remain from year to year. Perennial crops, such as asparagus, artichokes, rhubarb, and some herbs and flowers, should be planted in one section of the garden, separate from the annuals. By such separation, the perennials do not interfere with the cultivation of the annuals.

Annual	Biennial	Perennial
Lives only one year	Lives two years	Lives more than two years
Germinates, matures, and reproduces in one growing season	Germinates and grows roots, a short stem, and a cluster of leaves in the first year	Germinates, matures, and reproduces in the first year
Produces flowers, fruits and seeds in same growing season	Produces flowers, fruits and seeds in second growing season	Produces new shoot systems, flowers, fruits, and seeds each growing season

FIGURE 18-9 Characteristics of annual, biennial, and perennial plants.

Vegetables or Flowers

The kinds of vegetables and flowers to plant depend on the individual tastes of the members of your family. Certain top-ranking vegetables are popular everywhere. Others, such as okra, are popular only within certain regions of the country. The most popular vegetables are tomatoes, snap beans, onions, cucumbers, peppers, radishes, lettuce, carrots, corn, beets, cabbage, squash, and peas. Favorites of specific regions are artichokes in Louisiana and California, southern peas in warm southern climates, and melons in warm summer regions.

The gardener has a wide selection of flowers from which to choose. Generally, flower varieties are chosen because of their characteristics. Various varieties may survive well in poor soil, have a wonderful fragrance, or grow well in the shade. A few of the flowers that do well in poor soils are alyssum, cactus, cosmos, marigolds, petunias, and phlox. Flowers that are particularly fragrant are alyssum, carnations, petunias, sweet peas, and sweet William. Some that do well in partial shade are ageratums, begonias, coleus, impatiens, and pansies.

HOT TOPICS IN AGRISCIENCE

MASTER GARDENER PROGRAMS

A Master Gardener is a local expert on gardening who advises community members on garden issues.

The Cooperative Extension System associated with land-grant universities has developed a program for Master Gardeners. Under this program, people who are expert gardeners are available as consultants to answer questions about home gardening problems. This service is not available in all counties in the United States, but information about many of these programs is available on the Internet using the key words *Master Gardener*. People who want to participate by becoming Master Gardeners must fill out applications, enroll in training programs, and commit to work as a volunteer to consult with people in the community on gardening problems.

CULTURAL PRACTICES FOR GARDENS

Cultivating

Cultivation is the act of preparing and working the soil. Never put off cultivation. It is easier to do a little each day than to let weeds get ahead or the soil get too hard. Once this happens, it is difficult to get the garden back into good condition. The main purpose of cultivation is to control weeds and to loosen and aerate the soil. A sharp hoe is a necessary tool for any gardener. Buy one that can fit between plants for easy cultivation. Begin cultivation as soon as weeds or grass break through the soil. Do not wait until the weeds and grasses are tightly rooted

Courtesy of USDA/ARS #K4175-13

FIGURE 18-10 Plant materials, newspaper, and plastic films all make excellent mulches that help control weeds and conserve moisture.

and threaten to take over the garden. When cultivating, take care not to damage the roots of the vegetables and flowers. Use short, shallow scraping motions that cut the tops of the weeds from their roots instead of chopping deeply into the soil. Deep hoeing or cultivation will destroy desirable plant roots.

Weeding

Begin weeding as soon as weeds appear. When weeds grow near the plants, their roots can become intertwined with the crop. Therefore, take care to avoid pulling out both the weed and the plant. When weeding, select a time of day when the soil and plants are fairly dry, so the weeds wither when pulled. Be careful to shake the soil from weed roots so the weeds will die quickly. Remove persistent weeds such as purslane, crabgrasses, and Bermuda grasses from the garden area to prevent them from taking root again or reseeding a new crop of weeds. Mulches of various types provide weed barriers, and using them is an important practice for controlling weeds (Figure 18-10).

Weed-control chemicals called **herbicides** can be used in the home garden, but they should always be handled with care. Usually, the use of herbicides is best suited to large gardens or for the purpose of controlling established stands of persistent perennial weeds. Chemicals are specific in their actions in certain types of plants, so they must be used with caution to prevent unintended injury to susceptible vegetables and flowers. Before using any herbicides, obtain detailed information from an extension educator or an informed garden center salesperson. Always read the label on the container and use the chemical strictly in accordance with label instructions.

AGRI-PROFILE

CAREER AREAS: GARDENER/CARETAKER/ HOMEOWNER/MARKET GARDENER

Courtesy of USDA/ARS #K4173-2

Home gardens, pick-your-own farms, roadside markets, and farmers' market have all developed because people like to eat farm-fresh fruits and vegetables.

Fruit and vegetable gardening has long been an important enterprise for providing fresh and wholesome food for rural and suburban families. Similarly, flower gardening provides a rewarding pastime and greatly increases the beauty of homes in urban, as well as rural and suburban, settings. Career opportunities exist for gardeners on estates, institutions, colonial farms, truck farms, and residential neighborhoods.

Nationally, there is a resurgence of small farms and truck gardens where roadside stands and farmers' markets have created new interest in farm-fresh fruits and vegetables. Pick-your-own operations provide opportunities for people to harvest their own field-fresh produce and save the producer the cost of labor.

People have a new awareness of the benefits of fresh food and pay premium prices when freshness is ensured. This has created new markets for garden produce and has opened new career opportunities in production, processing, and marketing.

Watering

During a growing season, there will probably be some days or weeks of dry weather when the garden must be watered. Most soils require at least 1 inch of water per week, either through rain or irrigation. If water is needed, make it a practice to soak the soil to a depth of 6 inches. Frequent light watering tends to promote shallow root development and should be avoided.

When watering, run the water on the soil near the plants. A good sprinkler head, soaker hose, or water breaker is needed to ensure even distribution of water. Let the garden hose run for 15 to 30 minutes to water 100 square feet at a depth of several inches. You should apply water when the soil lacks moisture 1 to 2 inches below the surface.

Protection from Pests

Courtesy of USDA/ARS #K4179-19

FIGURE 18-11 Lady beetles feed on pea aphids and help keep the damaging insect population under control.

Gardens can be damaged from a variety of insects and diseases. To help prevent severe damage, the following practices should be helpful:

- Rotate crops so that the same or a related crop does not occupy the same area every year. This helps control soil-borne diseases.
- Watch closely for insects. Pick them off by hand or knock them off with a hard stream of water. Introduce predatory insects (Figure 18-11).
- When using sprinklers, water early in the day so the foliage can dry before nightfall. Diseases of leaves and fruit prosper in damp conditions.
- Keep weeds out of the garden. Weeds may harbor diseases or insects and will interfere with spraying and dusting of crops.
- Use correct amounts of fertilizer and lime to promote vigorous growth.

HARVEST AND STORAGE OF GARDEN PRODUCE

Harvesting

Harvest the vegetables and flowers at the peak of quality or at the stage of maturity preferred by the users (Figure 18-12). Some vegetables, such as cucumbers, will lose their quality within a day or two, whereas others hold it for a week or more.

For the best results with flowers, pick them in the early morning or late afternoon. Immediately place the cut flowers in lukewarm water for a short time. Display or store the flowers in a cool place.

INTERNET KEY WORDS:
vegetables, post-harvest storage fruits, post-harvest storage flowers, post-harvest storage

Storage

Many vegetables that are not processed or frozen for later use can be stored successfully if proper temperature and moisture conditions are met. Only vegetables that are of good quality and at the proper stage of maturity should be stored.

Warm Storage

Vegetables that tolerate moderate storage temperatures are squash, pumpkins, and sweet potatoes. These vegetables may be stored on shelves in an upstairs storage area. Damp basement areas are not recommended. Squash and pumpkins should be kept in a heated room, with a temperature between 75° and

85° F (23.9° and 29.4° C) for 2 weeks to harden the shells. Long-term storage temperature should then be reduced to 50° to 55° F (10° to 13° C).

Cool Storage

Most vegetables require cool temperatures and relatively high humidity for successful storage. The storage area should be cool, dark, and ventilated. The room should be protected from frost, heat from a furnace, or high outdoor temperatures.

Vegetable	Maturity Level	Vegetable	Maturity Level
Asparagus	From seeds—3rd year. From roots—after 1st year. Cut when spears are 6" to 10" high, in spring.	**Lettuce**	Leaf—When tender and desired size. Head—When round and firm.
Beans, Pole	When pods are nearly full size.	**Muskmelons**	When stem separates easily from fruit.
Beans, Bush	When pods are nearly full size.	**Mustard Greens**	When large leaves are still tender.
Beans, Bush Lima	When tender, pods are nearly full size.	**Okra**	Cut pods when about 2" or 3" long.
Beets	When bulbs are 1 1/4" to 2" in diameter.	**Onions**	For fresh use—When 1/4" to 1 1/2" in diameter. For storage—When tops shrivel at the bulb and fall over.
Swiss Chard	Outer leaves can be harvested anytime.	**Parsley**	Any time the outer leaves are desired size.
Broccoli	Before green clusters begin to open.	**Parsnips**	After hard frost. Can leave in ground all winter for spring use.
Brussels Sprouts	When sprouts are firm, pick from bottom up on stalks.	**Peas**	When pods are well-filled, before seeds are largest.
Cabbage	When heads are solid, before splitting.	**Peppers**	When solid and nearly full size.
Cauliflower	After blanching, when firm curds are 2" to 3" in diameter.	**Pumpkins**	When skin is hard and not easily punctured. Cut with some stem on.
Carrots	When top of root is 1" to 1 1/2" in diameter.	**Radishes**	When desired size.
Celery	When about 2/3 mature, harvest as needed.	**Rhubarb**	When stems are of desired size, twist off near base of plant and discard leaves. Do not pick more than 1/3 of plant during a season.
Collards	When leaves are large but tender. Can harvest up to winter.	**Spinach**	When outer leaves are large enough.
Corn	When kernels are filled out and milky. The silk at the tip of the ear should be dry and brown.	**Squash**	Summer—When skin is soft. Winter—When skin is hard, cut with part of stem on, before frost.
Cucumbers	When slender and dark green.	**Tomatoes**	When uniformly red and firm.
Eggplant	When half grown, glossy, and bright.	**Turnips**	When 2" to 3" in diameter.
Endive	When leaves are tender and desired size. About 15" in diameter.	**Watermelons**	When underside is yellow and thumping produces a muffled sound.
Kale	When young and tender.		
Kohlrabi	When 2" to 3" in diameter.		

FIGURE 18-12 Levels of maturity for harvesting vegetables for top quality.

SCIENCE CONNECTION PRESERVING FOOD

The produce from a garden can be enjoyed all year when it is preserved properly.

During the growing season, a gardener spends hours weeding, watering, fertilizing, and finally harvesting. With the amount of time and energy that is put into a home garden, one would not want to see any of the produce go to waste. That is why many gardeners consider food preservation the final step in home gardening.

Within the fertile soil of a home garden, millions of microorganisms thrive. Many of these creatures are beneficial to the plants as they grow. Once eaten, however, they can cause disease. During the food preservation process, care must be taken to ensure that food is free from living microorganisms.

Common organisms that are found in improperly preserved foods include fungi and bacteria. If left in food, the food will be unsafe to eat. Fungi include molds and yeasts. Bacteria are plentiful on freshly harvested foods. Many bacteria that are commonly found in the soil can cause serious illnesses if ingested.

Local agricultural extension offices can provide material, classes, and advice on safe and germ-free food preservation. Canning, freezing, and dehydrating are all methods that can keep garden produce safe and delicious for long periods. Molds and some bacteria are destroyed between 190° and 212° F. The boiling point of water at sea level is 212° F. By boiling canned food and blanching food that will be dehydrated or frozen, the chances that any microorganisms will remain in the food are slight. However, it takes temperatures of 240° F to kill some species of bacteria. Natural acids found in some foods can kill most bacteria. In produce with low acid levels, the food must be heated to 240° F using a pressure cooker. This will kill any remaining organisms. Although a lot of care must be taken to preserve the foods grown in a home garden, the benefits can be enjoyed all year long.

The suggested temperature and relative humidity for storage of crops are shown in Figure 18-13. This list will provide you with a general idea of the type of storage area that each vegetable requires.

COLD FRAMES, HOTBEDS, AND GREENHOUSES FOR HOME PRODUCTION

Cold Frames

Cold frames are very useful to the gardener. A **cold frame** is a bottomless wooden box with a sloping glass or clear plastic top (Figure 18-14). It can be constructed by using plywood or common lumber and a window sash. The size of the sash should determine the dimensions of the box. The length and width should be 2 inches smaller than the overall dimensions of the cover sash.

The following procedures should be helpful in constructing a cold frame:

1. Make the front of the box approximately 8 inches high and the back of the box about 12 inches high.
2. Cut the two sides on an angle, from 12 inches at the back to 8 inches in the front.
3. Nail the four sides together.

Crop	Temperature °F	Relative Humidity Percent
Asparagus	32	85–90
Beans, snap	45–50	85–90
Beans, lima	32	85–90
Beets	32	90–95
Broccoli	32	90–95
Brussels sprouts	32	90–95
Cabbage	32	90–95
Carrots	32	90–95
Cauliflower	32	85–90
Corn	31–32	85–90
Cucumbers	45–50	85–95
Eggplants	45–50	85–90
Lettuce	32	90–95
Cantaloupes	40–45	85–90
Onions	32	70–75
Parsnips	32	90–95
Peas, green	32	85–90
Peppers, sweet	45–50	85–90
Potatoes	38–40	85–90
Pumpkins	50–55	70–75
Rhubarb	32	90–95
Rutabagas	32	90–95
Spinach	32	90–95
Squash, summer	32–40	85–95
Squash, winter	50–55	70–75
Sweet potatoes	55–60	85–90
Tomatoes, ripe	50	85–90
Tomatoes, mature green	55–70	85–90
Turnips	32	90–95

FIGURE 18-13 Recommended temperatures and relative humidity for storage of fresh vegetables.

FIGURE 18-14 Cold frames are structures with glass or plastic covers used for starting garden plants.

4. Try the sash or top for size.
5. Hinge the sash to the back of the frame for easy use. The hinge allows you to prop the cover open slightly in warm weather for ventilation.
6. Place the cold frame in a southern exposure, with good protection from the wind and with proximity to a water supply.

When the cold frame is built, it can serve three purposes:

1. It can be used as a protective home for seedlings that have been started indoors. The seedlings can continue to grow inside the cold frame and become "hardened off" before they are transplanted to the garden.
2. You can start plants in the cold frames. Seeds can be planted directly into the soil in the cold frame. After they are 4 to 6 inches tall, they can be transplanted straight into the garden.
3. Vegetables, such as lettuce and endive, grow well into the fall. Sow seeds in early autumn. Cover the cold frame with a blanket or tarp during extremely cold nights.

Hotbeds

A **hotbed** is simply a cold frame with a heat source. In many areas, a cold frame is usually adequate. However, in colder areas, some type of artificial heat is necessary. The hotbed should be located on well-drained land.

Electricity is a convenient means of heating hotbeds. Use either lead or plastic-coated electric heating cables or 25-watt frosted light bulbs. Temperature can be controlled with a thermostat. When using light bulbs, attach the electrical fixtures to strips of lumber and suspend the bulbs 10 to 12 inches above the soil surface. Allow one 25-watt bulb per 2 square feet of space. When using a heating cable, a standard 60-foot length will heat a 6 × 6- or a 6 × 8-foot bed. Lay cable loops about 8 inches apart for uniform heating. Lay the cable directly on the floor of the bed, unless drainage material is needed.

Greenhouses

Growing flowers and vegetables in a greenhouse can be enjoyable as well as profitable. The conventional greenhouse is designed primarily to capture light and control temperature. It can be freestanding, but most often, it is attached to a building

SCIENCE CONNECTION

FINDING AND PROMOTING THE GOOD ONES

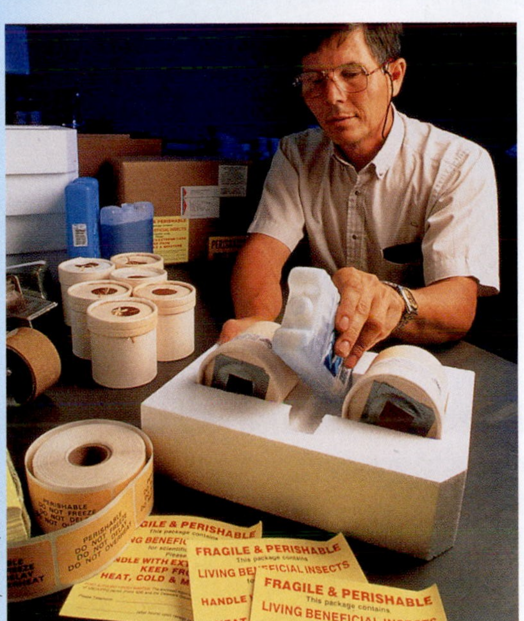

Courtesy of USDA/ARS #K4183-12

Entomologist and quarantine officer Larry Ertle at the Beneficial Insects Introduction Research Laboratory in Newark, Delaware, gently places beneficial insects in custom-designed packaging for shipment.

Newark, Delaware, is the site of an unusual facility specializing in foreign imports—insects! The imports are flies, wasps, beetles, and other insects that parasitize, or feed on, the insects that devour our crops, devastate our gardens, cause our lawns to turn brown, and infest our houseplants. These insects have been found doing their "good deeds" in countries all over the world. Most of our recent imports have come from Argentina, Brazil, Canada, Chile, China, France, Germany, India, Indonesia, New Zealand, South Korea, and Russia and other previously Soviet nations.

The facility is called the U.S. Department of Agriculture (USDA) Beneficial Insects Introduction Research Laboratory. Its job is to see that insects brought into the United States are, indeed, beneficial and that the eggs, larvae, pupae, or adults of harmful insect species do not hitchhike along with the beneficial ones. The work of the entomologists at the Newark Laboratory includes picking up incoming shipments of beneficial insects at the Philadelphia airport; transporting them to the nearby Newark laboratory; taking them through a special receiving process; disinfecting all shipping materials that arrive with the insects; isolating the insects in special refrigeration units; exercising them daily; examining each insect under magnification to confirm identity; separating eggs, larvae, pupae, and adults for rearing purposes; increasing their numbers through reproduction; and shipping them under carefully controlled conditions to USDA and state research facilities throughout the United States and cooperating foreign nations.

Three sets of heavy steel doors lead into the quarantined main work area. There, a large sink and autoclave are used to clean and sterilize materials used in the laboratory. Sterilization is essential before discarding materials to avoid accidental release of both beneficial insects and "hitchhiking intruders." It is interesting to note that insects must receive regular exercise to remain healthy, just like humans! Once a day, even on weekends and holidays, batches of insects are removed from cool storage and released in restricted, warm, and lighted areas to warm up, stretch, soak up light, and move around in the greater space.

Many innovations in packaging and handling have evolved and become standardized over the years. One of the more unusual ones is the parasite pill, or Trichocap. The parasite pill is a shipping capsule made of soft, gray cardboard and is about the size of an oyster cracker. It is used to ship insect eggs. One parasite pill holds about 500 eggs of the Mediterranean flour moth. Inside each egg, there is one future parasitic wasp used as a biocontrol against the devastating European corn borer. The discovery and introduction of parasitic wasps have helped control some of our most damaging insects, while reducing our reliance on chemical sprays.

to provide convenient access, simplified construction, and a potential source of supplemental heat if needed (Figure 18-15). The greenhouse can provide an environment for starting plants, hardening them off, or completely growing the plants.

In summary, gardening is both an art and a science. You can keep a garden for fun, reduce the cost of food for the family, generate income, or improve the beauty of the surroundings. If gardening is to be profitable, the operator must read extensively, get help occasionally with disease and insect problems, and perform garden chores in a timely fashion. For many people, the garden is a real source of satisfaction and creates a feeling of self-sufficiency and achievement.

FIGURE 18-15 A home greenhouse can be attached to the house for convenience and efficiency.

STUDENT ACTIVITIES

1. Write the Terms to Know and their meanings in your notebook.
2. Discuss with family members the types of vegetables and flowers they want in the home garden.
3. Select vegetable and flower varieties for the home garden from seed catalogs.
4. Sketch a garden plot containing both flowers and vegetables that are to be grown.
5. Make a calendar indicating the dates to plant and harvest the various crops in the garden.
6. Start and manage your own garden area. Use the flowers and vegetables at home or sell them for profit.
7. Learn to calculate area in square feet. A square foot is an area that is 1 foot long and 1 foot wide. An area that is 1 foot long and 2 feet wide contains 2 square feet. Area (A) in square feet is calculated by multiplying length (L) in feet times width (W) in feet. Therefore, $A = L \times W$, or $A = LW$.
   a. How many square feet are in a garden that is 10 ft. long × 5 ft. wide?
      A = ________ sq. ft.
   b. How many square feet are in a lawn that is 40 × 70 ft.?
      A = ________ sq. ft.
   c. What is the area of a building lot that is 109 × 150 ft.?
      A = ________ sq. ft.

d. Suppose the lot in item c (above) is covered with lawn except for the house, which is 28 × 42 ft.
What is the area of the house?
A = ________ sq. ft.
What is the area of the lawn?
A = ________ sq. ft.

8. Using the produce you grew during Activity 6, dehydrate, freeze, and/or can some of it for future use. Be sure to follow approved techniques. Contact your local extension education office for more information.

SELF-EVALUATION

A. MULTIPLE CHOICE

1. Gardening is both a
   a. science and an art.
   b. chore and hard work.
   c. science and hobby.
   d. hobby and job.
2. A good rule of thumb for planning a garden for four people is to start with a plot that is
   a. 40 × 60 feet.
   b. 15 × 25 feet.
   c. 10 × 26 feet.
   d. 3 × 7 feet.
3. The south and west sides of a house may not be the best locations for a flower garden because they
   a. are not warm enough.
   b. reflect heat on the planting.
   c. have poor exposure to the sun.
   d. collect rainwater.
4. The furrow in a garden is a
   a. sunken walkway.
   b. hole.
   c. place to plant seeds.
   d. pile of weeds.
5. An annual is a plant whose life cycle is completed in
   a. two growing seasons.
   b. one growing season.
   c. four growing seasons.
   d. three growing seasons.
6. When cultivating, a gardener should use a ________ for best results.
   a. rake
   b. rototiller
   c. shovel
   d. hoe
7. The technical name for weed-control chemicals is
   a. pesticides.
   b. fungicides.
   c. herbicides.
   d. weed killers.
8. After cutting flowers from the garden, it is best to immediately put them in
   a. hot water.
   b. cold water.
   c. salty water.
   d. warm water.
9. To heat a hotbed, frosted light bulbs or ________ are commonly used for heat sources.
   a. a furnace
   b. lead or plastic-coated heating cables
   c. a fire
   d. a small stove
10. The conventional greenhouse is designed to
   a. add more living space to the home.
   b. keep plants warm.
   c. capture light and control temperature.
   d. protect plants from pests and diseases.

B. MATCHING

_____ 1. Perennial
_____ 2. Peat moss
_____ 3. Square foot
_____ 4. Loamy
_____ 5. Watermelon
_____ 6. Sweet pea
_____ 7. Bermuda grass

a. 12 × 12 inches
b. Fragrant flower
c. More than two growing seasons
d. Long-season vegetable
e. Persistent weed
f. Soil type
g. Form of organic matter

C. COMPLETION

1. A cold frame is a bottomless wooden box with a sloping _____.
2. You should store only the vegetables that are of good quality and at the proper stage of _____.
3. You should plant only the vegetables and flowers that are liked by _____.
4. Fertilizer and lime should be used to promote _____.
5. Watering should be done to a depth of _____.

UNIT 19

Vegetable Production

OBJECTIVE

To determine the opportunities in and identify the basic principles of vegetable production.

MATERIALS LIST

- seed catalogs
- scissors, index cards, and glue
- pen, paper, eraser, and ruler
- lima bean seeds, containers, and soil mix
- hand trowels
- tomato seeds, soil mix, and flats
- Internet access

COMPETENCIES TO BE DEVELOPED

After studying this unit, you should be able to:

- determine the benefits of vegetable production as a personal enterprise or career opportunity.
- identify vegetable crops.
- plan a vegetable production enterprise and prepare a site for planting.
- describe how to plant vegetable crops and use appropriate cultural practices.
- list appropriate procedures for harvesting and storing at least one commercial vegetable crop.

SUGGESTED CLASS ACTIVITIES

1. Invite a local fruit and vegetable wholesaler to visit the class, or take the class to the distribution center of a wholesale business. Discuss the different fruits and vegetables with respect to local demand, where they were produced, length of time they can be stored, proper storage conditions to maintain quality, and what precautions are taken to ensure they are free of dangerous pesticides. Discuss the amount of produce that the business markets in a week, a month, and a year.
2. Assign members of the class to choose a vegetable crop on which they will prepare an oral report. They should seek information about the vegetable from vegetable growers' organizations in major vegetable-producing states, such as the potato commission in Idaho, vegetable seed companies, state university extension office publications, and Internet sites. Help each class member organize

TERMS TO KNOW

olericulture
market gardening
truck cropping
olericulturist
angiosperm
monocotyledon (monocot)
dicotyledon (dicot)
aeration
green manure
transplant
arid
semiarid
pre-cooling
hydrocooling

his or her material in a predetermined format. Have students report their findings to the class.

3. As a class, create a business plan for a virtual vegetable production enterprise. The plan should include all estimated costs, such as for fuel, seed, hired help, and so on. The size of the property and the type of crop to be planted should be given. Assume that the business will have a fair yield and the crop will sell at the average market value. Find the estimated profit by subtracting the income from the costs. As a class, discuss the business plan.

INTERNET KEY WORDS: olericulture

A vegetable is the edible portion of an herbaceous plant (Fig ure 19-1). The study of vegetable production is **olericulture**. *Herbaceous* describes a plant that has a stem that withers away at the end of each growing season. The production of vegetables can be classified into three categories: home garden, market garden, or truck crop. Home gardening, as discussed in Unit 18, refers to the vegetable production for one family with most of the produce consumed at home. It usually does not involve any major selling of the crops. **Market gardening** refers to growing a wide variety of vegetables for local or roadside markets. **Truck cropping** refers to large-scale production of a few selected vegetable crops for wholesale markets.

FIGURE 19-1 Vegetables are important in the human diet because they supply basic nutrients. They also are good to eat and are good for you.

VEGETABLE PRODUCTION: HOME-BASED BUSINESS OR CAREER

Home Enterprise

Growing vegetables in a home garden is enjoyed by millions of people in the United States. It has become a part of the lifestyle of most families with access to a little bit of ground. The vegetables produced can be used for fresh table consumption or can be stored for later use. Vegetable gardening not only produces nutritious food but also provides outdoor exercise from spring until fall.

The gardener who enjoys this type of activity can plant enough to provide for the family and harvest vegetables for sale near home for extra income. Fresh, home-grown vegetables are usually superior in quality to those found in the supermarkets. In addition, the gardener can grow vegetables that may be expensive to buy.

Career Opportunities

The vegetable industry is a large and complex component of the horticultural industry today. Even though millions of homeowners raise gardens, the majority of vegetables consumed by the public are grown commercially. The commercial vegetable industry is fast moving, intensive, and competitive. It is a business that is continually changing as the demands for certain vegetables fluctuate with the tastes of consumers.

Numerous and varied career opportunities are available in the vegetable industry. These include being the owner of a small market gardening business, a member of a larger truck-crop business, a vegetable wholesaler or retailer, or a worker in a vegetable-processing plant. With adequate education and training, a person can become an **olericulturist**—someone who develops pest-resistant strains and new varieties of vegetables and does other specialized work. The opportunities are endless in the vegetable industry.

IDENTIFYING VEGETABLE CROPS

Vegetables can be identified in various ways: by their botanical classifications, according to their edible parts, or by the kind of growing season that is required by the plant.

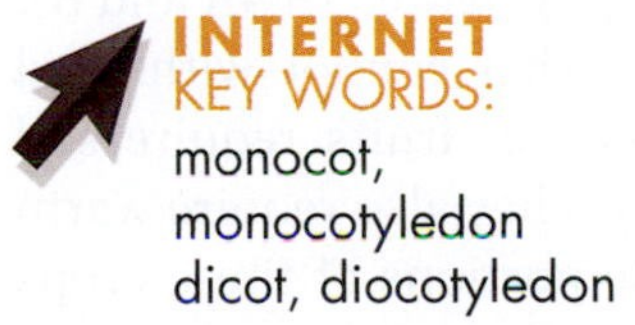

Botanical Classification

All vegetables belong to the division of plants known as **angiosperms**. These are plants with ovules and an ovary. From this division, the vegetables can be grouped into either Class I, **monocotyledons (monocots)** (having only one seed leaf), or Class II, **dicotyledons (dicots)** (having two seed leaves) (Fig ures 19-2A and B). A vegetable can be further grouped into a family, a genus, a species, and sometimes a variety. One of the more popular vegetable families is Cruciferae—the mustard family—which contains Brussels sprouts, cabbage, cauliflower, collards, cress, kale, turnips, mustard, watercress, and radish. Other families include Leguminosae—the pea family—which contains bush beans, lima beans, cow peas, kidney beans, peas, peanuts, soybeans, and scarlet runner beans; Cucurbitaceae—the gourd or melon family—includes pumpkins, cucumbers, cantaloupe, casaba melons, and watermelons; Solanaceae—the nightshade family—includes eggplants, ground cherries, peppers, and tomatoes.

A

B

FIGURE 19-2 Monocots have one seed leaf (A), whereas dicots have two (B).

Edible Parts

Vegetables can also be classified by the part of the vegetable that is eaten. Vegetables have three groups, according to their uses: (1) vegetables of which leaves, flower parts, or stems are used, (2) vegetables of which the underground parts are used, and (3) vegetables of which the fruits or seeds are used. (See Figure 19-3 for a list of these categories and the vegetables in each one.)

INTERNET KEY WORDS:
cool-season vegetable crops
warm-season vegetable crops

Growing Seasons

There are basically two growing seasons: warm and cool. Cool-season vegetable crops grow best in cool air and can withstand a frost or two. Some of these crops, such as asparagus and rhubarb, can even endure winter freezing. This group of crops is planted early in the spring and late in the season for fall and winter harvest. Cool-season crops include mostly leaf and root crops (Figures 19-4A and B).

Warm-season, or warm-weather, crops are those that cannot withstand cold temperatures, especially frosts. These vegetables and fruits require soil warmth to germinate and long days to grow to maturity. They also require warm temperatures to produce their edible parts. The edible portions of these crops are what can be picked off the standing plant or the fruit. Warm-season crops are listed in Figure 19-5A.

PLANNING A VEGETABLE-PRODUCTION ENTERPRISE

Before planting a vegetable garden or truck crop, the grower needs a plan. The gardener must decide where to plant, and when to plant as well as what to plant. Without some advance planning, the vegetable garden or truck farm is not likely to be successful.

Plants of which the Fruits or Seeds are Eaten		Plants of which the Leaves, Flower Parts, or Stems are Eaten		Plants of which the Underground Parts are Eaten	
Family	**Vegetable**	**Family**	**Vegetable**	**Family**	**Vegetable**
Grass, *Gramineae*	Sweet corn, *Zea mays*	Lily, *Liliaceae*	Asparagus, *Asparagus officinalis* var. *altilis* Chives, *Allium schoenoprasum*	Lily, *Liliaceae*	Garlic, *Allium satvium* Leek, *Allium porrum* Onion, *Allium cepa* Shallot, *Allium ascalonicum* Welsh onion, *Allium fistulosum*
Mallow, *Malvaceae*	Okra (gumbo), *Hibiscus esculentus*	Goosefoot, *Chenopodiaceae*	Beet, *Beta vulgaris* Chard, *Beta vulgaris* var. *cicla*	Yam, *Dioscoreaceae*	Yam (true), *Dioscorea batatas*
Pea, *Leguminosae*	Asparagus or Yardling bean, *Vigna sequipedalis* Broad bean, *Vicia faba* Bush bean, *Phaseolus vulgaris* Bush Lima bean, *Phaseolus limensis* Cowpea, *Vigna sinensis* Edible podded pea, *Pisum sativum* var. *macrocarpon* Kidney bean, *Phaseolus vulgaris* Lima bean, *Phaseolus limensis* Pea (English pea), *Pisum sativum* Peanut (underground fruits), *Arachis hypogaea* Scarlet runner bean, *Phaseolus coccineus* Sieva bean, *Phaseolus lunatus* Soybean, *Glycine max* White Dutch runner bean, *Phaseolus coccineus*	*Orach, Atriplex hortensis*	Spinach, *Spinacia oleracea*	Goosefoot, *Chenopodiaceae*	Beet, *Beta vulguris*
Parsley, *Umbelliferae*	Caraway, *Carum carvi* Dill, *Anethum graveolens*	Parsley, *Umbelliferae*	Celery, *Apium graveolens* Chervil, *Anthriscus cerefolium* Fennel, *Foeniculum vulgare* Parsley, *Petroselinum crispum*	Mustard, *Cruciferae*	Horseradish, *Armoracia rusticana* Radish, *Raphanus sativus* Rutabaga, *Brassica campestris* var. *napobrassica* Turnip, *Brassica rapa*
Martynia, *Martyniaceae*	Martynia, *Proboscidea louisiana*	Sunflower, *Compositae*	Artichoke, *Cynara scolymus* Cardoon, *Cynara cardunculus* Chicory, witloof, *Chichorium intybus* Dandelion, *Taraxacum officinale* Endive, *Chichorium endivia* Lettuce, *Lactuca sativa*	Morning Glory, *Convolvulaceae*	Sweet Potato, *Ipomoea batatas*
Nightshade, *Solanaceae*	Eggplant, *Solanum melongena* Groundcherry (husk tomato), *Physalis pubescens* Pepper (bell or sweet), *Capsicum frutescens* var. *grossum* Tomato, *Lycopersicon esculentum*	Mustard, *Cruciferae*	Brussels sprouts, *Brassica oleracea* var. *gemmifera* Cabbage, *Brassica oleracea* var. *capitata* Cauliflower, *Brassica oleracea* var. *botrytis* Collard, *Brassica oleracea* var. *viridis* Cress, *Lepidium sativum* Kale, Borecole, *Brassica oleracea* var. *viridis* Kholrabi, *Brassica oleraceae* var. *gongylodes* Mustard leaf, *Brassic juncea* Mustard, Southern Curled, *Brassica juncea* Pak-Choe, Chinese Cabbage, *Brassica chinensis* var. *crispifolia* Pe-tsai, Chinese cabbage, *Brassica pekinensis* Seakale, *Crambe maritima* Sprouting Broccoli, *Brassica oleracea* var. *italica* Turnip, Seven Top, *Brassica rapa* Upland cress, *Barbarea verna* Watercress, *Rorippa nasturtium-aquaticum*	Parsley, *Umbelliferae*	Carrot, *caucus carota* var. *sativa* Celeriac, *Apium graveolens* var. *rapaceum* Hamburg parsley, *Petroselinum crispum* var. *radicosum* Parsnip, *Pastinaca sativa*
Gourd or Melon, *Cucurbitaceae*	Chayote, *Sechium edule* Cucumber, *Cucumis sativus* Cushaw, *Cucurbita moschata* Gherkin, *Cucumis anguria* Cantalope (Muskmelon), *Cucumis melo* Pumpkin, *Cucurbita pepo* Summer squash (bush pumpkin), *Cucurbita pepo* Squash, *Cucurbita maxima* Watermelon, *Citrullus lunatus* Winter melon, *Cucumis melo* var. *inodorus*			Nightshade, *Solanaceae*	Potato, *Solanum tuberosum*
				Sunflower, *Compositae*	Black salsify, *Scorzonera hispanica* Chicory, *Chicorium intybus* Jerusalem artichoke, *Helianthus tuberosus* Salsify, *Tragopogon porrifolius* Spanish salsify, *Scolymus hispanicus*

FIGURE 19-3 Classification of vegetable crops by botanical family and crop use.

Cool-Season Crops		
• Asparagus	• Chinese cabbage	• Mustard
• Beet	• Chive	• Onion
• Broad bean	• Collard	• Parsley
• Broccoli	• Endive	• Parsnip
• Brussels sprouts	• Garlic	• Pea
• Cabbage	• Globe artichoke	• Potato
• Carrot	• Horseradish	• Radish
• Cauliflower	• Kale	• Rhubarb
• Celery	• Kohlrabi	• Salsify
• Chard	• Leek	• Spinach
• Chicory	• Lettuce	• Turnip

A

B

FIGURE 19-4 (A) Some cool-season crops. (B) Cool-season vegetable crops grow best in cool temperatures. They are even able to tolerate occasional freezing temperatures.

Selecting the Site

Choose a site that is convenient to a water supply. The site should also be exposed to the sun a minimum of 50 percent during the day. A minimum of 8 to 10 hours of direct sunlight is needed. Also consider the type of trees that are around the proposed site. Trees can provide excessive shade. They are also likely to compete for soil nutrients that are needed by the vegetable plants. Some kinds of trees are known to produce toxins that are harmful to specific vegetables. For example, the walnut tree is toxic to the tomato. Buildings and structures cast shade that can slow down or prevent maturation of a vegetable crop. Avoid trying to grow vegetables closer than 6 to 8 feet from the northern side of a one-story structure— farther for higher structures. Both the south and west sides of a building have access to good light and often radiate heat late in the day (Figure 19-6).

Warm-Season Crops	
• Cowpea	• Pumpkin
• Cucumber	• Snap bean
• Eggplant	• Soybean
• Lima bean	• Squash
• Muskmelon	• Sweet corn
• New Zealand spinach	• Sweet potato
• Okra	• Tomato
• Pepper, hot	• Watermelon
• Pepper, sweet	

A

B

FIGURE 19-5 (A) Some warm-season crops. (B) Warm-season vegetable crops respond best in summer conditions. They have little tolerance for temperatures that are near freezing.

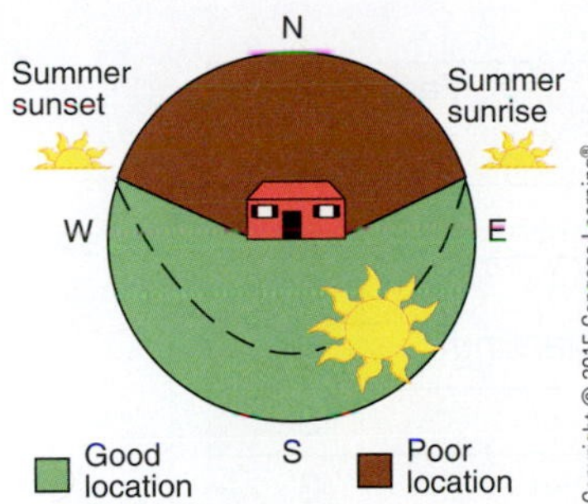

FIGURE 19-6 Avoid situating the garden in the shadow of a building or tree. The location of a building's shadow changes throughout the day and throughout the year.

The type of soil in the selected area is important as well. The majority of vegetables grow best in a well-drained, loamy soil. Avoid heavy clay soil that develops puddles after it rains. This is a sign of poor drainage. To test the drainage of a site, dig a trench 12 inches wide and 18 inches deep. Fill it with water and observe how long it takes for the water to drain away. If it takes 1 hour or less, the soil can be considered well drained. If the selected area has supported vegetation before, even if it has only been weeds, it will probably support a vegetable crop. If the selected site needs some alterations to its soil structure, adding organic matter can help. The organic matter can improve the drainage and allow air to move readily through the pores of the soil. If possible, the garden soil should be about 25 percent organic matter. To accomplish this, put a layer of organic matter 2 inches thick over the soil and work it in to a depth of at least 4 inches. If necessary, repeat this procedure until the final mix contains approximately 25 percent organic matter (Figure 19-7).

Scope of the Vegetable Enterprise

The size of the vegetable-growing area depends largely on the amount of ground that is available, the number of people served by the garden, and the use of the produce. A large garden that measures 50 × 100 feet will produce enough vegetables for the annual needs of a large family. Figure 19-8 shows a suggested planting plan for a 50 × 100 feet home garden.

If this much land is not available, the fresh vegetable needs for a medium-sized family may be met with a plot that is 40 × 50 feet. A planting guide for such a garden can be found in Figure 19-9. When deciding on the size, it would be wise to remember that the larger the vegetable production, the more care it will require. Vegetable production is more efficient with better crops if the area is a small well-managed garden than if you take on a bigger area that is neglected and full of weeds.

AGRI-PROFILE

CAREER AREAS: VEGETABLE PRODUCER/PROCESSOR/ DISTRIBUTOR/PRODUCE MANAGER

Good French fries require cooperative work of plant breeders, potato growers, food specialists, and vegetable processors, among others.

Career opportunities in vegetable production are similar to those in fruit production. Vegetables and fruits are frequently grown, processed, and marketed by the same people. In addition to the careers mentioned in Unit 20, many people find rewarding careers that pay well in the area of distributing and marketing produce. In this area, truckers, wholesalers, and retailers move the product from producer to consumer.

Jobs available in supermarkets include produce stockers and produce department managers. Products include fruits, vegetables, and nuts. Tasks may include inventory management, ordering, handling, stocking, and displaying produce to keep it fresh and attractive. In large cities as well as small towns, street vendors are frequently seen selling premium fruit and vegetables. Roadside stands provide opportunities for younger members of families to develop business skills while earning money for present and future needs.

Amounts of Organic Matter Needed to Cover 100 Square Feet at Various Depths	
Depth (inches)	To Cover 100 sq. ft.
6	2 cubic yards
4	35 cubic feet
3	1 cubic yard
2	18 cubic feet
1	9 cubic feet
1 cubic yard = 27 cubic feet	

FIGURE 19-7 Amounts of organic matter to apply per 100 square feet of garden soil.

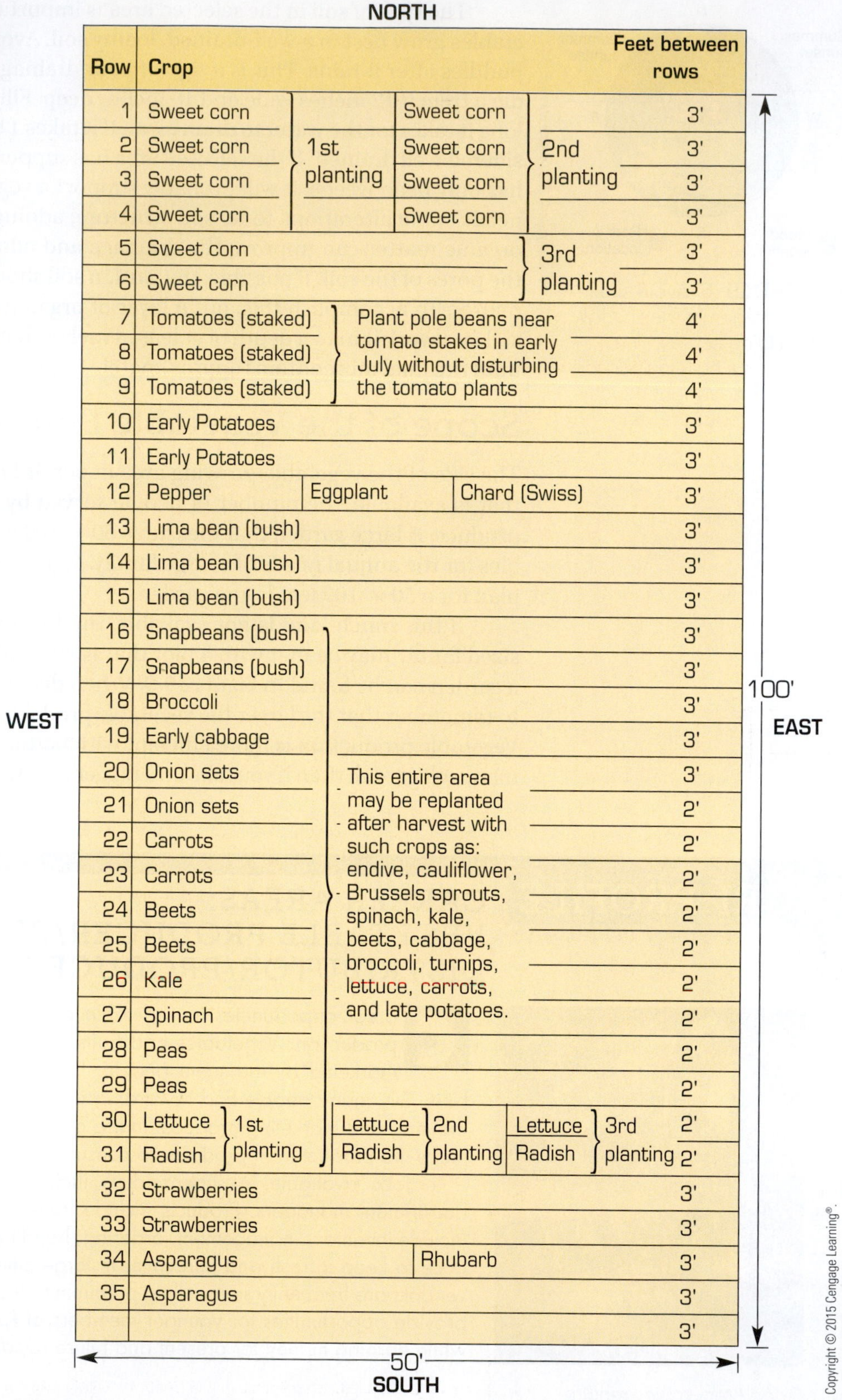

NORTH

Row	Crop			Feet between rows
1	Sweet corn	1st planting	Sweet corn — 2nd planting	3'
2	Sweet corn	1st planting	Sweet corn — 2nd planting	3'
3	Sweet corn	1st planting	Sweet corn — 2nd planting	3'
4	Sweet corn	1st planting	Sweet corn — 2nd planting	3'
5	Sweet corn		3rd planting	3'
6	Sweet corn		3rd planting	3'
7	Tomatoes (staked)	Plant pole beans near tomato stakes in early July without disturbing the tomato plants		4'
8	Tomatoes (staked)			4'
9	Tomatoes (staked)			4'
10	Early Potatoes			3'
11	Early Potatoes			3'
12	Pepper	Eggplant	Chard (Swiss)	3'
13	Lima bean (bush)			3'
14	Lima bean (bush)			3'
15	Lima bean (bush)			3'
16	Snapbeans (bush)	This entire area (rows 16–31) may be replanted after harvest with such crops as: endive, cauliflower, Brussels sprouts, spinach, kale, beets, cabbage, broccoli, turnips, lettuce, carrots, and late potatoes.		3'
17	Snapbeans (bush)			3'
18	Broccoli			3'
19	Early cabbage			3'
20	Onion sets			3'
21	Onion sets			2'
22	Carrots			2'
23	Carrots			2'
24	Beets			2'
25	Beets			2'
26	Kale			2'
27	Spinach			2'
28	Peas			2'
29	Peas			2'
30	Lettuce — 1st planting	Lettuce — 2nd planting	Lettuce — 3rd planting	2'
31	Radish — 1st planting	Radish — 2nd planting	Radish — 3rd planting	2'
32	Strawberries			3'
33	Strawberries			3'
34	Asparagus		Rhubarb	3'
35	Asparagus			3'
				3'

WEST — EAST — 100' — 50' — SOUTH

FIGURE 19-8 Example of a planting plan for a large home garden 50 × 100 feet.

Rows	Crop				Feet between Rows
					1½'
1	Early Cabbage	(Turnips)			2'
2	Early Potatoes	(Late Cabbage)			2'
3	Peas	(Beans)	Peas	(Kale)	2'
4	Kale	(Carrots)	Turnips	(Beans)	2'
5	Parsley		Parsnips		2'
6	Onions and Radishes (same row)		Carrots	(Spinach)	2'
7	Swiss Chard	(Spinach)	Spinach	(Beets)	2'
8	Lettuce	(Beans)	Beets	(Kale)	2'
9	Beans	(Spinach)	Beans	(Lettuce)	2'
10	Peppers		Eggplant		2'
11	Lima Beans, Bush				2'
12	Tomatoes, Late				2'
13	Tomatoes, Early				2½'
14	Squash		Cucumbers		2½'
15					2½'
16	Corn, Sweet interplanted with Pole Lima Beans				2½'
17	Corn, Sweet interplanted with Winter Squash				2½'
18					2'

40'

50'

NOTE: Items in parentheses are succession crops planted after the first crop is harvested or planted between the rows of mature plants to permit germination and early growth before the first crop is removed.

FIGURE 19-9 Example of a plan for a medium-sized garden 40 × 50 feet.

Deciding What to Plant

A potential commercial gardener or truck cropper must evaluate several considerations when deciding which vegetables to plant. First, consider the maintenance that some vegetables require. The easier they are to grow and maintain, the better.

A second factor to consider is the length of the harvest time for each vegetable. If the object is fresh consumption, then vegetables that can be harvested over a long period are best. For instance, carrots can be harvested for months, whereas cabbage should be picked at maturity. Another item to consider is the expected uses for which the vegetables are planted. If they are going to be for fresh use, plant only for that purpose. Surplus vegetables can be stored in a root cellar or freezer, but plantings should match the estimated needs.

The truck cropper has additional questions to consider. First, which vegetables are in high demand in the area? Which ones will produce the greatest return for the cost and labor involved? The truck-cropping enterprise is a business and must make a reasonable profit to be considered a good business. Second,

SCIENCE CONNECTION ENCOURAGING THE MASTER TILLERS

Earthworms are nature's master tillers. Always a favorite for fishing bait, the lowly earthworm is now regarded as an ally of homeowners, gardeners, farmers, landscapers, and conservationists. They enhance soil tilth and crop growth by consuming and digesting organic matter and mixing it with the mineral content of the soil. The new mixtures left behind by feeding worms are called worm casts and are rich in nutrients for growing plants. Worm holes in the soil start at the surface and go as deep as the worm needs to go to escape winter cold and summer heat. This may be 3 feet deep or more in the central area of the United States. This network of holes and tunnels collects rainwater as it falls, and the enriched soil absorbs and holds water for future plant use.

The North Appalachian Experimental Watershed at Coshocton, Ohio, has been the scene of no-till experiments for more than 30 years. In a major experiment there, night crawlers (*Lumbricus terrestris*), usually 4 to 8 inches in length and 3/8 inch in diameter, and other species were studied. It was found that earthworms may be encouraged by leaving crop residues on the surface or by incorporating them into the soil. However, if crop residues are removed from the soil, the earthworms do not have an adequate food source, and their populations do not expand as much.

Entomologist Edwin Berry of the National Soil Tilth Laboratory at Ames, Iowa, identified seven other species of earthworms. These all range in size from 2 to 6 inches in length and are about half the diameter of night crawlers. Most of their activity occurs in the first foot of topsoil. They make temporary burrows as they pass through the soil, consuming, digesting, and mixing soil particles and organic matter and excreting their rich casts.

Worm farming has become an important industry, and earthworms may be purchased for release for soil improvement. Worms grow and multiply the fastest where plenty of organic matter and moisture is available. Unfortunately, worms do not function as well in coarse, sandy soils as they do in soils where medium and fine particles are available. Gardeners, horticulturists, and farmers should consider practices that will encourage the work of the master tillers.

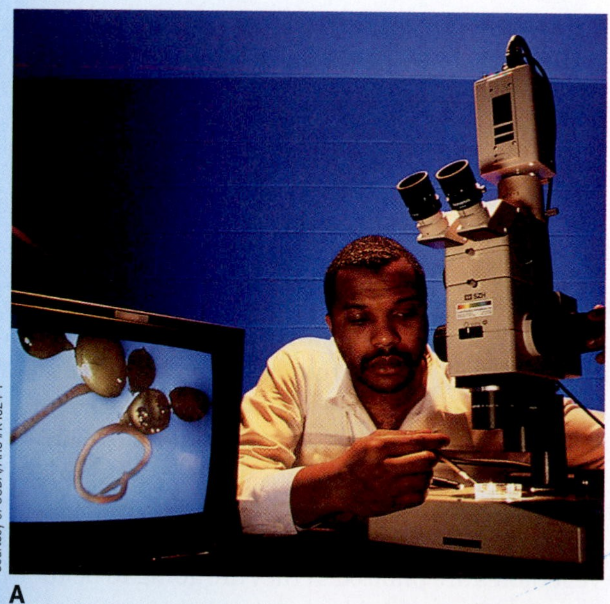

Courtesy of USDA/ARS #K4024-1

A

Courtesy of USDA/ARS #K3237-1

B

(A) Ken Ford at the Soil Tilth Research lab in Ames, Iowa, uses a video camera adapted for microscopic work to study earthworm cocoons (top of monitor), newly hatched earthworm larvae, and (B) adult earthworms—nature's master tillers.

what vegetables can be raised in the soil and climate? Truck cropping should be a high-income business, so high quality and quantity of produce are important. Third, what combination of vegetables can be grown to keep the land occupied and provide cash flow from sales throughout the season?

Other questions also should be explored: (1) Can the diseases and insect problems be managed? (2) Will continuous use of the land for the same crops create pest buildup? (3) Will the soil tolerate continuous cropping? (4) What perennials may be included to reduce the amount of labor?

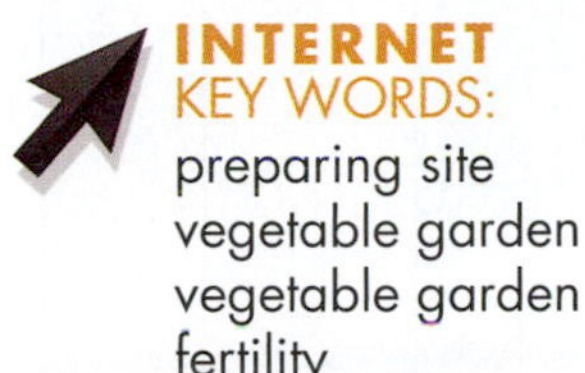

preparing site
vegetable garden
vegetable garden fertility

PREPARING A SITE FOR PLANTING

Preparing the Soil

Before planting vegetables, the soil must be properly prepared. This generally means adding organic matter, lime, and fertilizer. The land should be plowed if it is in sod and then left for 4 to 6 weeks for the sod to decay.

Plowing

Land should be plowed or spaded to a depth of 6 to 8 inches. This can be done either in the spring or the fall. Fall plowing has some advantages over spring plowing. First, the freezing and thawing action on the soil during the winter months improves the physical condition of the soil. Second, exposure of the soil to the weather can result in a reduced population of insects. Spring plowing should be done only a short time before planting. Take care not to plow the soil when it is too wet. Doing so tends to destroy the physical structure of the soil, resulting in hard clumps. The moisture content of loam or clay soil may be judged by pressing a handful of soil into a ball (Figure 19-10). If the ball crumbles easily, the soil is ready to plow. If the soil sticks together, it is too wet to plow.

Courtesy of DeVere Burton

FIGURE 19-10 Soil moisture in clay or loam may be judged by pressing a handful of soil into a ball. If the ball crumbles easily, the soil is ready to be worked.

Maintaining Organic Matter

Organic matter increases the water-holding and absorption capacity of the soil. It also helps prevent erosion and promotes aeration in the soil. **Aeration** refers to the movement of air within the soil profile. If it is available at a reasonable cost, animal manure is the best material for maintaining the organic content of the soil. It is also a good source of nutrients. Manure from animals other than sheep and poultry can be turned under at the rate of 15 to 20 tons per acre, or 20 to 30 bushels per 1000 square feet. Sheep and poultry manures can be used, but at half this rate. One note of caution: Animal manures are low in phosphorus; therefore, it may be necessary to use high-phosphorus fertilizer to balance the nutrients. For best results, apply fresh manure in the fall and well-rotted manure in the spring.

Another type of organic matter to use is green manure in the form of a cover crop, especially if it is a legume. **Green manure** is an active, growing crop that is then turned under to help build the soil. Green vegetation incorporated into the soil rots more quickly than dry material. A cover crop should be planted at the end of the growing season. A cover crop is a close-growing crop planted to prevent erosion.

Liming

The need for lime and fertilizer should be determined through the use of soil tests. A pH greater than 7.0 indicates alkalinity; a pH of 7.0 is neutral; and a pH less than 7.0 indicates acidity.

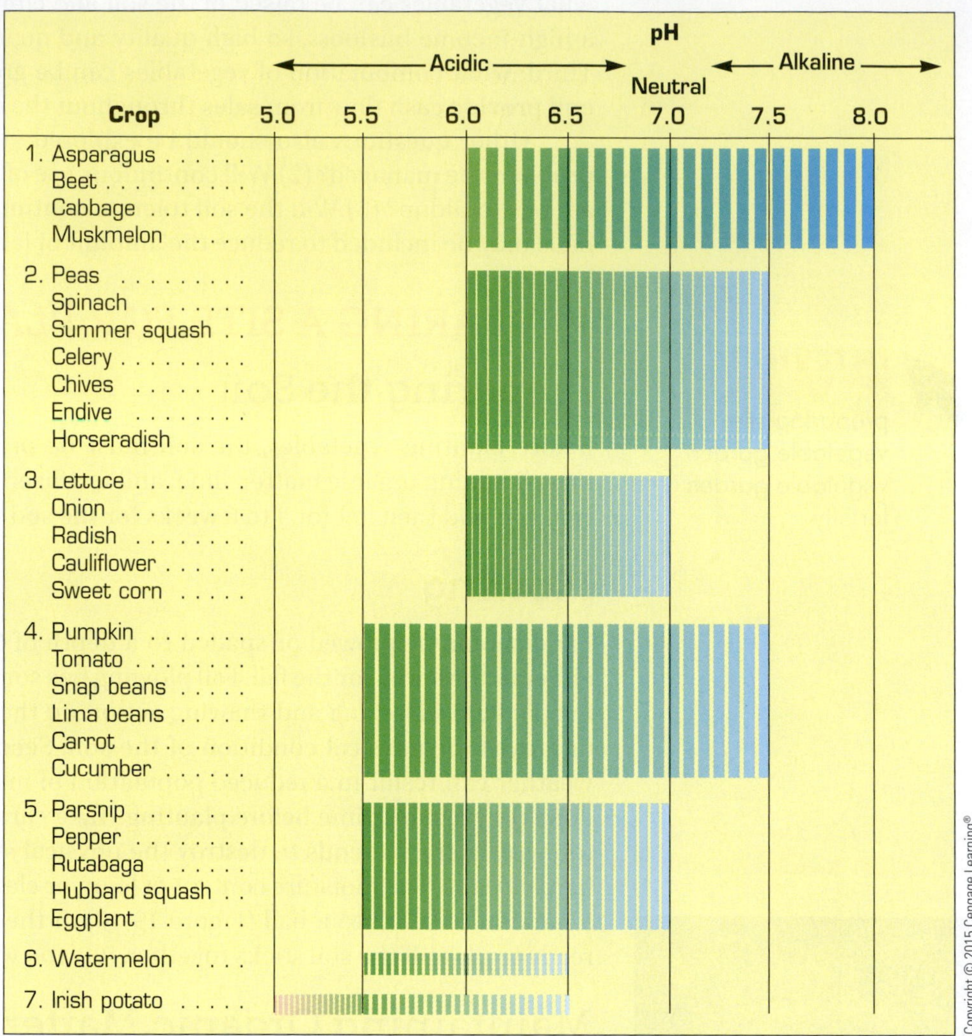

FIGURE 19-11 Optimum pH ranges for vegetable crops.

If the test indicates a pH of 6.0 or less, lime should be applied. The lime should be mixed into the top 3 to 4 inches of the soil. The amount of lime to be added depends on the type of soil, the desired pH, and the form of lime used. Some plants, such as blueberries, require acidic soils. For such plants, special materials are available to make the soil acidic if the pH is too high.

Vegetable crops grow best at specific pH levels. Figure 19-11 lists vegetables and their optimum pH ranges for maximum growth.

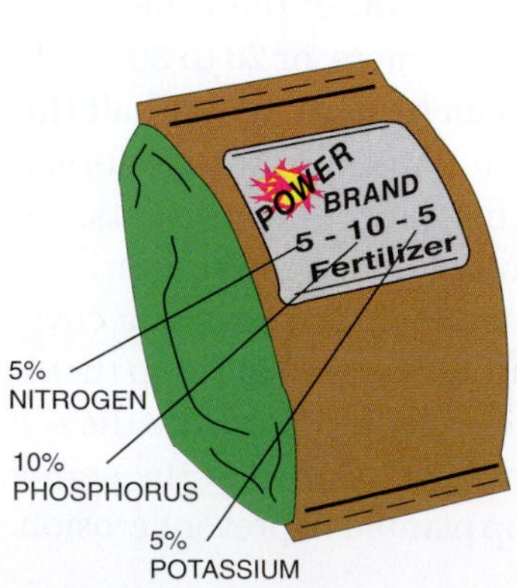

Fertilizing

The amount and type of fertilizer to add can best be determined by conducting a soil test. The test will help prevent application of too much or too little fertilizer and will determine the best timing. The application of commercial fertilizer is done to increase the availability of nutrients to the plants. Soils without additives are unlikely to provide the best combination of conditions for all vegetables.

Commercial fertilizers usually contain nitrogen, phosphoric acid, and potassium. Each vegetable crop needs varying amounts of each of these three elements. In general, fertilizer ratios used for home gardens are 5-10-10, 10-10-10, and 5-10-5 (Figure 19-12). The first number in a fertilizer ratio or grade represents the

POWER BRAND 5 - 10 - 5 Fertilizer
5% NITROGEN
10% PHOSPHORUS
5% POTASSIUM

FIGURE 19-12 The analysis on a fertilizer bag indicates the percentages of nitrogen, phosphoric acid, and potassium in the fertilizer.

percentage of nitrogen, the second represents the percentage of phosphorus, and the third indicates the percentage of potassium in the fertilizer. A rough guide for home gardeners is that most vegetables need about 3 to 4 lbs. of fertilizer per 100 square feet. Consult a county extension educator for exact fertilizer recommendations based on the soil types in the area. University extension educators are specialists on local and regional conditions that affect crops and gardens.

PLANTING VEGETABLE CROPS

Vegetable crops are initially grown from seeds. Some seeds can be sown directly in the location where the plants will grow. Others need to be started indoors with the seedlings transplanted to the garden at a later date. **Transplants** are plants grown from seeds in a special environment, such as a cold frame, hotbed, or greenhouse (Figure 19-13). The method used depends on the climatic requirements of the plant and the germinating characteristics of the seed.

Planting Seed

Depending on the size and scope of the garden and the quantity of seed to be planted, vegetable seeds are planted (1) by hand in hills or rows, (2) through broadcasting by hand or machine, (3) with one-row hand seeders, and (4) with single or multiple-row, tractor-drawn seeders.

Broadcasting is a planting method whereby seeds are scattered on the soil surface. Regardless of the type of planting method used, the seeds should be planted at the proper depth. The soil should also be left smooth and firm over the seed. For most vegetables, seed must be fresh. Seed more than 1 year old will probably not germinate at the rate necessary to produce good yields.

Courtesy of DeVere Burton

FIGURE 19-13 Some garden plants are grown in containers for later planting. This allows the gardener to extend the length of the growing season by starting frost-sensitive plants in the greenhouse or hotbed.

SCIENCE PROFILE WHAT IS GOOD ABOUT FUNGI?

Fungi are plentiful organisms that are found in the soil. Despite the species of fungi that are known to be harmful to crops, some fungi are useful to crops. Mycorrhizae are fungi that growers welcome in their fields. These microorganisms do not have chlorophyll, and they cannot use the process of photosynthesis to make their own food. To survive, they attach to the roots of plants to obtain carbohydrate nutrients. Even though these fungi attach to plant roots, they are considered beneficial.

Phosphorus is a nutrient that plants have a difficult time extracting from the soil. Mycorrhizae fungi produce an enzyme that frees nutrients like phosphorous from the soil so the plants can use them. They also help plants take in water and other nutrients. Both the plants and the fungi benefit from this symbiotic relationship. Scientists have estimated that 80 percent of all plants form this kind of relationship with fungi. Fields with adequate levels of these fungi require less water and less fertilizer. Mycorrhizae fungi are commercially available, and their benefits have been demonstrated to far outweigh their costs.

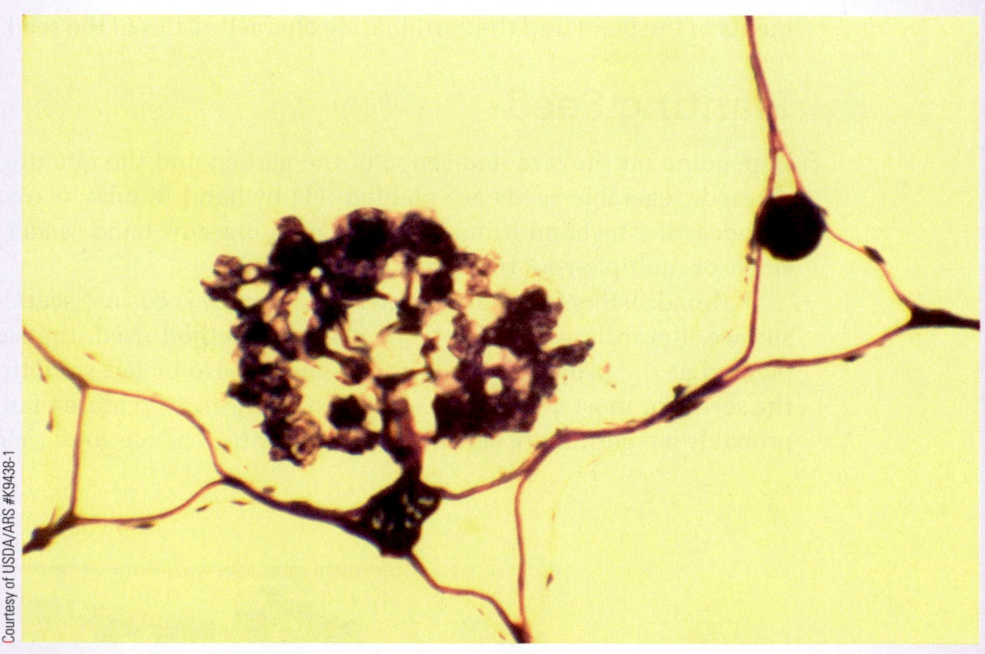

Courtesy of USDA/ARS #K9438-1

Mycorrhizae help host plants absorb nutrients from the soil and, in turn, they obtain carbohydrates from the plant.

Most vegetable seeds should be planted at a moderately shallow depth. It is best to plant in loamy-textured soil with adequate moisture after the danger of frost is over. The seed will germinate best when it is planted at a depth no more than four times the diameter of the seed. To *germinate* means to sprout, grow, and produce a plant from a seed. Regardless of how deep the seeds are planted, the soil surface should be level and firmly pressed after the seed is planted. This gives the seed close contact with moist soil particles, improving germination. It also helps prevent the seed from washing away or water from puddling above the seed after a rain.

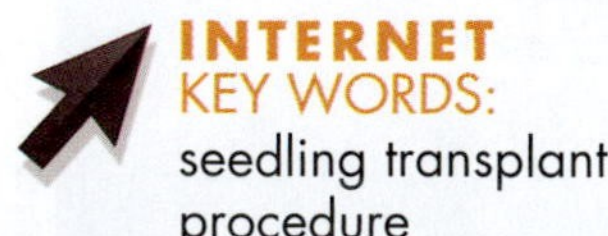

INTERNET KEY WORDS:
seedling transplant procedure

Transplanting Vegetable Seedlings

Three methods are used in transplanting, depending on the number of plants that are being transplanted: (1) hand setting, (2) hand-machine setting, and (3) riding-machine setting.

In home gardening situations, hand setting is most appropriate. Hand setting involves six steps: (1) dig a hole slightly bigger and deeper than the root ball of the plant being transplanted, (2) add some fresh soil to the hole, (3) place the plant in the soil a little deeper than it grew in its original container, trying to keep as much of the original soil as possible around the roots, (4) pull soil in and around the plant and firm it slightly, (5) add about a half pint of water and let it soak into the soil, and (6) pull in some dry soil to level off and cover the wet area (which helps prevent loss of moisture and baking of the soil).

Large commercial vegetable production practices rely on hand-machine setting or riding-machine setting methods for transplanting seedlings. Machines dig the planting furrows and pack the soil around the seedlings as workers place the seedlings. Soil conditions must be just right for the plants to survive. The soil must contain enough moisture to allow the seedling roots to become established. Stress on the seedlings can be greatly reduced by setting them out on a cloudy day or late in the afternoon. It is best to set plants just before or just after a rain.

CULTURAL PRACTICES

Cultivating

Cultivation, or intertillage of crops, is a proven agricultural practice. The benefits of cultivation are (1) weed control, (2) conservation of moisture, and (3) increased aeration. Cultivation increases the yield of most vegetable crops, mainly because the weeds are controlled.

All types of equipment are used to cultivate crops, from small hand tools to large tractors equipped with cultivating tools. The cultivation should be shallow and done only when there are weeds to be killed. Excessive cultivation is not beneficial and may even be damaging.

Controlling Weeds

Weeds are easier to control when they are small, using shallow cultivation. Weeds that are more established are more difficult to control. For them, deeper cultivation is required to rid the area of weeds. This can result in injury to the roots of vegetable crops (Figure 19-14). The key to weed control is to remove them as soon as they show up and as often as needed.

Using herbicides is appropriate for large home gardens and commercial vegetable production plots. They can greatly reduce the amount of effort that is required to remove weeds by mechanical methods. The use of any herbicide depends on the registration of the herbicide by federal and state environmental protection agencies (EPAs). Do not use any herbicide unless it is clearly stated on the label that it is intended for a particular crop. Always seek the latest information for use of herbicides on vegetables.

Courtesy of USDA/ARS # K5226-18

FIGURE 19-14 Mechanical weed control is applied regularly in the production of vegetable crops.

Irrigation

Nearly all commercial vegetable operations irrigate their crops. Irrigation is especially important in California and other **arid** and **semiarid**, or dry, regions of the West. Irrigation is important because rainfall is rarely uniform and adequate for high yields.

HOT TOPICS IN AGRISCIENCE ORGANIC VEGETABLE FARMING

A farmer who raises "organic" vegetables produces them without using the chemical pesticides and fertilizers on which most vegetable farmers depend. This does not mean that the vegetable crop is produced without insect controls or fertilizers. Fertilizers are obtained from animal manure and/or "green manure" crops such as legumes that are plowed under when they are 6 to 12 inches in height. It simply means that an organic farmer first tries to maintain a healthy environment for the plants. This is because a healthy plant located in an environment that supports its needs is less susceptible to pests and diseases than a plant that is not well adapted to its environment.

The organic farmer may also introduce the natural enemies of pests to the environment to control their populations. In addition, pepper and garlic sprays are common methods for discouraging insect pests. As a final resort, the organic farmer may apply sprays consisting of horticultural oils, isopropyl alcohol, soaps, ammonia, baking soda, and even bug juice (a spray made with water and insects that have been pulverized in a blender). Some chemicals such as rotenone and sulfur are acceptable to organic farmers because they break down easily and are degraded quickly to materials that are no longer harmful. Organic vegetables that are of high quality command a premium price among some consumers in comparison with the price of produce that was produced using standard farming methods that rely on chemical fertilizers and pesticides. Regulations are in place in most states that restrict the additives that may be used and the cultural practices that may be implemented to produce and market organic vegetables.

Water for irrigation can be obtained through established water rights from a stream, lake, well, spring, or stored stormwater. Definite laws and regulations guide the practice of irrigation in each state. Several types of irrigation systems are available, including sprinkler irrigation; drip irrigation, surface or furrow irrigation; and subirrigation. The selection of a system should be guided by local needs and proven practices in the region.

Mechanical sprinkler systems are versatile machines (Figure 19-15). They can be used in many situations, such as irrigating when soil moisture is in short supply. They also can deliver shallow and light irrigation to promote seed germination and can be used for the application of fertilizers. Drip irrigation uses a system of pipes and tubes to deliver the water to individual plants rather than wetting all of the soil (Figure 19-16). This method is gaining in popularity because of its water-conserving benefits.

Courtesy of DeVere Burton

FIGURE 19-15 Sprinkler irrigation is a proven method for providing water to vegetable crops. The water can be applied uniformly in the proper amounts.

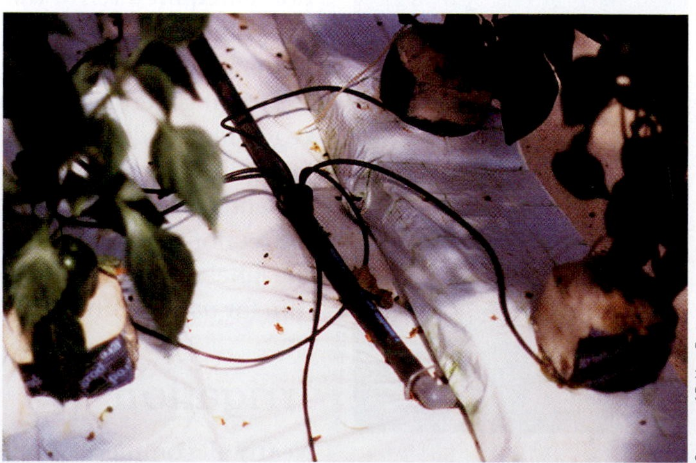
Courtesy of DeVere Burton

FIGURE 19-16 Drip or trickle irrigation delivers water to individual plants through small openings in pressurized tubes. It is the most efficient method of irrigation for vegetable crops.

Surface or furrow irrigation has the advantage of requiring a relatively low investment. The topography of the land is important for surface irrigation. The land must be gently sloping and uniform. This method is well suited for irrigating vast areas of land, but it generally requires land leveling to work well.

Subirrigation requires large amounts of water. With this system, the water is added to the soil so that it permeates the soil from below. This method is expensive and is suited only to commercial enterprises in locations where the depth of groundwater can be manipulated.

Drip or trickle irrigation is a form of subirrigation in which water is delivered under low pressure near the plants. The water seeps through porous hose under the rows. This system is expensive, but gives excellent results, with increased yields and less water consumed than with other methods.

Courtesy of DeVere Burton

FIGURE 19-17 Mulching is the practice of adding material such as straw, leaves, or plastic film to the surface of the soil to reduce or prevent the growth of weeds.

Mulching

Mulch is created whenever the surface of the soil is modified artificially by covering it with straw, leaves, grass clippings, paper, or polyethylene film (Figure 19-17). Mulching helps control weeds, regulates soil temperatures, conserves soil moisture, and provides clean, weed-free vegetables. The mulching material should be spread around the base of the plants and between the rows.

Plastic film is frequently used with cantaloupes, melons, cucumbers, eggplants, summer squash, and tomatoes. It is important that the plastic be laid over moist soil that has been freshly prepared. It is advisable to lay plastic mulch 1 to 2 weeks before planting the vegetables. This allows the soil to warm up underneath the mulch.

Pest Control

Good pest-control practices are essential for quality vegetables. A combination of cultural, mechanical, biological, and chemical methods can be used. Insecticides and fungicides can be effectively used by both home and commercial vegetable growers.

Insect and disease infestations bring about losses that can be devastating to vegetable crops (Figure 19-18). Types of losses include the following: (1) reduced yields, (2) decreased quality of produce, (3) increased costs of production and harvest, and (4) increased expenditures for materials and equipment. The use of chemical control of pests in vegetable production is sometimes restricted due to proximity of vegetables for which a particular chemical has not been approved. Government standards may apply when a crop will mature before sufficient time has passed to reduce the pesticide residues to acceptable levels.

Courtesy of USDA/ARS #K1291-2. Photo by Tim McCabe.

FIGURE 19-18 Insects such as the Colorado potato beetle (shown) and the Mexican bean beetle can strip the leaves from vegetable plants in a matter of hours.

Modern biological controls are expected to play a greater role in vegetable production in future years (Figure 19-19). Integrated pest management (IPM) practices call for a combination of control measures to be used to combat pests and diseases in plants. This approach puts priority on controlling pests with the least invasive methods; however, it does not rule out the limited use of chemical pesticides when other methods of control prove ineffective.

Many measures can be taken to help vegetable growers control insects and diseases. Some of these are the following:

- Dispose of old crop residues that serve to shelter pests, and thus enable them to overwinter or survive from one crop to another.
- Rotate the location of individual crops.

A

B

C

FIGURE 19-19 (A) Disease-resistant varieties of sweet potatoes. (B) Spraying with nontoxic Bt to spread insect-fighting bacteria. (C) A spined soldier beetle making a meal of a leaf-devouring Mexican bean beetle larva.

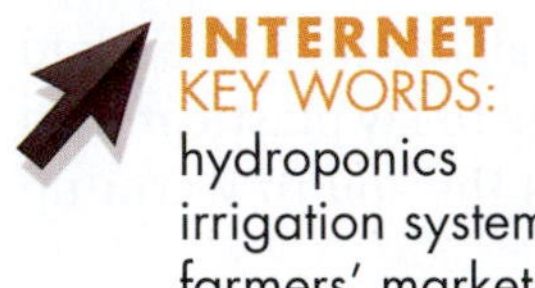

INTERNET KEY WORDS:
hydroponics
irrigation systems
farmers' markets

- Choose insect- and disease-resistant vegetable varieties.
- Treat seeds with fungicides and insecticides.
- Control weeds.
- Use adequate, but not excessive, amounts of fertilizer.
- Follow recommended planting dates.

Hydroponics

The ability to produce vegetables on a commercial scale without soil has contributed to the supply of tomatoes, cucumbers, and other fresh vegetable crops during the winter season (Figure 19-20). The plant most commonly produced using hydroponic methods is the tomato (see Unit 9). Although really a fruit, the tomato is used as a vegetable, and it is included in this unit for that reason. Most commercial greenhouses that produce tomatoes plan to deliver most of the crop during the winter season when tomato prices are highest. The plants are usually rooted in inert filler material through which water and nutrients flow, or they may be placed in containers to which the nutrients and water are delivered using drip irrigation methods. Large hydroponic operations harvest the crop daily or every other day during the production cycle. (See Unit 9 for more information on hydroponics.)

FIGURE 19-20 The use of hydroponic methods to produce vegetables has become a big business in recent years. Water and nutrients may be delivered through a channel in the floor or by drip irrigation tubes.

HARVESTING, MARKETING, AND STORING VEGETABLES

Harvesting

For the best possible quality, vegetables should be harvested at the peak of maturity. This is possible when the vegetables will be used for home consumption, sold at a local market, or processed. The best harvesting time varies with each vegetable crop. Some will hold their quality for only a few days, whereas others can maintain quality over a period of several weeks. Frequent and timely harvest of crops is needed if the vegetable grower wishes to supply the market or the kitchen table with high-quality produce over time. Specifications for harvesting each kind of vegetable crop should be consulted to achieve the best results.

Harvesting is done by hand and through the use of machines for a few crops (Figure 19-21). Mechanical harvesting is especially useful for the commercial grower. When harvesting a crop, care must be taken to prevent injury or bruising to the produce due to overly aggressive machine action or poor handling methods.

Marketing

The most important aspect of marketing fresh vegetables is to assure that the quality is high. This requires the vegetable plants to be protected from the stresses of disease and insects. Vegetables must be harvested when they are of optimum size and maturity. At this point in the process, it is important to market the produce as soon as possible.

Farmers' markets are ideal outlets for marketing fresh vegetables to local consumers. These markets are intended to bring several farmers together in a central location with vegetables and fruits of sufficient quality and quantities to attract local residents to the site. In some instances, these markets are established on private property. In other instances, they may be located on federal, state, or city property as long as permission is obtained from the proper government authorities. The U.S. Department of Agriculture has developed rules under which these markets can be legally established in local communities. It is important for a manager to assume the duties of running the business. Good business practices and high-quality fruit and vegetable products are keys to operating a successful farmers' market.

Storing

When storing vegetables, the proper stage of maturity is essential. There are ideal temperatures and humidity levels for storage of vegetable crops (Figure 19-22). Most crops need 90 to 95 percent humidity. Most homeowners find it difficult to

Courtesy of DeVere Burton

A

Courtesy of DeVere Burton

B

FIGURE 19-21 (A) Mechanical harvesting of vegetable crops such as potatoes requires special machinery that is designed to separate the crop from leaves, stems, and soil. (B) Metal parts are often rubber-coated to avoid bruising the tubers.

Vegetable Storage		
Crop	Temperature °F	Relative Humidity Percent
Asparagus	32	85–90
Beans, snap	45–50	85–90
Beans, lima	32	85–90
Beets	32	90–95
Broccoli	32	90–95
Brussels sprouts	32	90–95
Cabbage	32	90–95
Carrots	32	90–95
Cauliflower	32	85–90
Corn	31–32	85–90
Cucumbers	45–50	85–95
Eggplants	45–50	85–90
Lettuce	32	90–95
Cantaloupes	40–45	85–90
Onions	32	70–75
Parsnips	32	90–95
Peas, green	32	85–90
Potatoes	38–40	85–90

FIGURE 19-22 Ideal temperature and humidity levels for storing common vegetables.

reproduce the exact temperatures and humidity levels, but commercial growers can accomplish storage through specialized facilities (Figure 19-23).

Vegetables differ in the amount of time that they can be stored. Some may be stored for several months, others for only a few days. For the vegetables that store well, good storage conditions are essential to prolong the marketing period.

Fresh vegetables placed in storage should be free of skin breaks, bruises, decay, and disease. Such damage or disease will decrease storage life.

Refrigerated storage is recommended (Figure 19-24). This type of storage reduces respiration and other metabolic processes. It also slows ripening; slows moisture loss and wilting, reduces spoilage from bacteria, fungus, or yeast, and reduces undesirable growths, such as potato sprouts.

One practice that can increase the successful storage of vegetables is pre-cooling. **Pre-cooling** is the process of rapidly removing the heat from freshly harvested vegetables before storage or shipment. One effective method of pre-cooling is termed **hydrocooling**. This is a method wherein vegetables are immersed in cold water where they are immersed long enough to reduce the core temperature to the desired level.

Storage Conditions

Vegetable	Temperature (°F)	Relative Humidity (%)	Storage Life
Pea, English	32	95–98	1–2 weeks
Pea, southern	40–41	95	6–8 days
Pepper, chili (dry)	32–50	60–70	6 months
Pepper, sweet	45–55	90–95	2–3 weeks
Potato, early	___[1]	90–95	___[1]
Potato, late	___[2]	90–95	5–10 months
Pumpkin	50–55	50–70	2–3 months
Radish, spring	32	95–100	3–4 weeks
Radish, winter	32	95–100	2–4 months
Rhubarb	32	95–100	2–4 weeks
Rutabaga	32	98–100	4–6 months
Salsify	32	95–98	2–4 months
Spinach	32	95–100	10–14 days
Squash, summer	41–50	95	1–2 weeks
Squash, winter	50	50–70	___[4]
Strawberry	32	90–95	5–7 days
Sweet corn	32	95–98	5–8 days
Sweet potato	55–60[3]	85–90	4–7 months
Tamarillo	37–40	85–95	10 weeks
Taro	45–50	85–90	4–5 months
Tomato, mature green	55–70	90–95	1–3 weeks
Tomato, firm ripe	46–50	90–95	4–7 days
Turnip	32	95	4–5 months
Turnip greens	32	95–100	10–14 days
Water chestnut	32–36	98–100	1–2 months
Watercress	32	95–100	2–3 weeks
Yam	61	70–80	6–7 months

[1]Spring- or summer-harvested potatoes are usually not stored. However, they can be held 4–5 months at 40°F if cured 4 or more days at 60–70°F before storage. Potatoes for chips should be held at 70°F or conditioned for best chip quality.

[2]Fall-harvested potatoes should be cured at 50–60°F and high relative humidity for 10–14 days. Storage temperatures for table stock or seed should be lowered gradually to 38–40°F . Potatoes intended for processing should be stored at 50–55°F; those stored at lower temperatures or with a high reducing sugar content should be conditioned at 70°F for 1–4 weeks or until cooking tests are satisfactory.

[3]Sweet potatoes should be cured immediately after harvest by holding at 85° F and 90–95% relative humidity for 4–7 days.

[4]Winter squash varieties differ in storage life.

FIGURE 19-23 Ideal temperature, humidity level, and storage life for common vegetables.

Source: USDA, ARS. (Draft-Revised April 2004). The Commercial Storage of Fruits, Vegetables, and Florist and Nursery Stocks. Gross, K. C., Wang, C. Y., and Saltveit, M. (Eds.). Agriculture Handbook 66. Retrieved from http://www.ba.ars.usda.gov/hb66/contents.html

Courtesy of DeVere Burton

FIGURE 19-24 Climate-controlled vegetable storage units control humidity, temperature, and sometimes the level of oxygen in the atmosphere. This makes it possible to extend the storage life of vegetables and fruit.

STUDENT ACTIVITIES

1. Write the Terms to Know and their meanings in your notebook.
2. Using outdated seed catalogs, prepare flash cards of warm-season and cool-season vegetables.
3. Bring in an unusual vegetable to show to the class. Be prepared to tell the class which part is edible.
4. Select a site for planting a vegetable crop and prepare a planting plan to scale.
5. Take a soil sample from the site selected in Activity 4. Submit the sample to the local county extension service for analysis and lime and fertilizer recommendations for the vegetables planned.
6. Plant lima bean seeds in containers to observe the growth of the stem and leaves. Determine whether the plant is a monocotyledon or a dicotyledon.
7. Plant tomato seeds in a greenhouse for later transplanting.
8. Develop a chart with pictures illustrating the various insects that attack vegetable plants. Find pictures of insects in vegetable catalogs or extension service bulletins.
9. Store vegetables under different conditions (light/dark, dry/humid, warm/cool) in the classroom or at home. Record what happens under each set of conditions. Note which conditions are best for each vegetable.
10. During a grocery store visit, count the number of vegetables offered to consumers. Notice that a number of varieties of each vegetable are usually available. For example, how many varieties of lettuce are there? In your opinion, what are the advantages of producing more than one crop? What are the disadvantages?

SELF-EVALUATION

A. MULTIPLE CHOICE

1. The part of horticulture that has to do with the production of vegetables is called
   a. viticulture.
   b. olericulture.
   c. floriculture.
   d. pomology.
2. Vegetable gardening produces fresh, nutritious food and provides the gardener with
   a. fresh produce.
   b. weeds.
   c. pests.
   d. diseases.
3. The majority of vegetables grow best in
   a. well-drained, clay soil.
   b. well-drained, sandy soil.
   c. well-drained, silty soil.
   d. well-drained, loamy soil.
4. For vegetable gardening, the soil should be plowed to a depth of
   a. 1 to 2 inches.
   b. 3 to 5 inches.
   c. 6 to 8 inches.
   d. 10 to 12 inches.
5. What nutrient's percentage is reflected by the first number in the analysis of a fertilizer?
   a. nitrogen
   b. calcium
   c. phosphoric acid
   d. potash
6. The best time to transplant vegetable plants is on
   a. a bright, sunny day.
   b. a rainy day.
   c. a cloudy day.
   d. a snowy day.

B. MATCHING

________	1. Cultivation	a. Weed-control chemicals
________	2. Herbicides	b. Rapid removal of heat from vegetables
________	3. Polyethylene film	c. A person involved in the study of vegetable production
________	4. Olericulturist	d. Intertillage of crops
________	5. Monocotyledon	e. Having only one seed leaf
________	6. Pre-cooling	f. A type of mulching material

C. COMPLETION

1. The scientific name for the mustard family of vegetables is __________.
2. A site for the vegetable garden should be one that is exposed to the sun a minimum of __________ percent of the day.
3. To correct the acidity level in the soil, __________ is commonly added.
4. The type of irrigation that allows water to permeate the soil from below is called __________ irrigation.
5. Whenever the surface of the soil is artificially modified through the use of straw, leaves, plastic film, or paper, a __________ is created.
6. Tomatoes for use in the local market are best harvested in the __________ stage of maturity.

UNIT 20

Fruit and Nut Production

OBJECTIVE

To determine the opportunities in and identify the basic principles of fruit and nut production.

MATERIALS LIST

- paper, pencil, and eraser
- old seed catalogs
- pruning shears
- scissors, glue, and index cards
- Internet access

COMPETENCIES TO BE DEVELOPED

After studying this unit, you should be able to:

- determine the benefits of fruit and/or nut production as a personal enterprise or career opportunity.
- identify fruit and nut crops.
- plan a fruit or nut production enterprise and prepare a site for planting.
- describe how to plant fruit and nut trees and use appropriate cultural practices in fruit and nut production.
- list appropriate procedures for harvesting and storing at least one commercial fruit or nut crop.

SUGGESTED CLASS ACTIVITIES

1. Identify a farm in your area that produces fruit or nuts. Make arrangements with the owner/manager for the class to visit the farm or for the farm manager to visit the class. Prepare as a class to ask appropriate questions that will stimulate a discussion of the management practices that are involved in fruit or nut production. Focus some questions on how new varieties are chosen; how the crop is processed, packed, and marketed; and the storage requirements for the crop. Have him or her discuss the activities that must be performed at different times throughout the year to achieve the best possible harvest. Serve a piece of fresh fruit at the end of the activity.
2. Obtain grafting supplies, rootstock, and scions from apple trees to demonstrate grafting techniques. Demonstrate the most common grafts and have students practice each of the techniques. Provide an opportunity for each student to graft a

TERMS TO KNOW

pomologist
semidwarf
dwarf
pome
drupe
nursery
terminal
cane

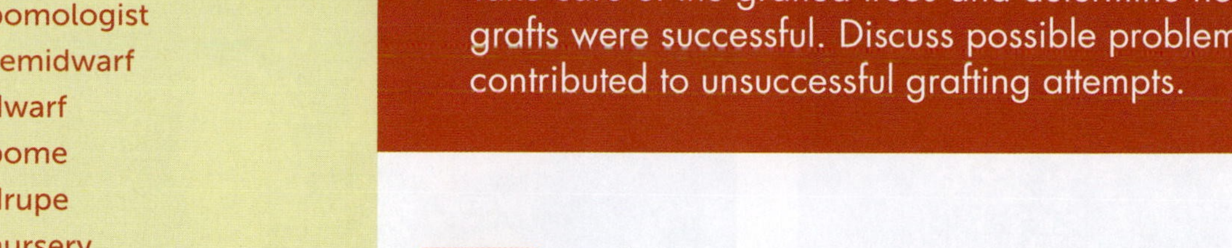

scion to a rootstock, and then plant it in a suitable container. Take care of the grafted trees and determine how many of the grafts were successful. Discuss possible problems that may have contributed to unsuccessful grafting attempts.

Home fruit operations can provide high-quality and bountiful varieties of fruit and nuts. Fruits grown at home can be enjoyed at the peak of ripeness. This quality is difficult to obtain from supermarket fruits. Home fruit and nut operations can be enjoyed as hobbies or turned into profitable commercial enterprises.

INTERNET KEY WORDS: pomologist career

CAREER OPPORTUNITIES IN FRUIT AND NUT PRODUCTION

Like other agricultural enterprises today, the production of fruit and nut crops is more a business than a way of life. The small-fruit grower has to be a good financial manager as well as be knowledgeable about science and farming. Producing fruit and nuts, however, can be an enjoyable and rewarding profession.

Career Descriptions

The production, harvest, and marketing of fruits is all part of the large fruit industry. The United States accounts for a sizable part of the combined world crops of apples, pears, peaches, plums, prunes, and oranges, grapefruit, limes, lemons, and other citrus fruits. A **pomologist** is a fruit grower or fruit scientist. There is currently a shortage of educated pomologists in the fruit industry.

Career opportunities exist at all levels of the industry. One can be an owner/grower, orchardist, plant manager, foreman, or technician. People who work in this field must be able to propagate fruit/nut trees and vines as well as plant, transplant, prune, thin, train, and fertilize them. They must also be able to provide protection from diseases and insects.

IDENTIFICATION OF FRUITS AND NUTS

Types of Fruits

There are three broad categories of fruit. Some fruits grow on trees, some grow on canes and small bushes, and others grow on vines. The shape and form of a plant, whether it is a tree, a bush, a cane, or a vine, is known as the growth habit of the plant. Fruit is produced on woody twigs and stems from plants that exhibit each of these growth habits.

Tree Fruits

Tree fruits are popular with home gardeners as well as large producers. One drawback, however, is that some types of trees can take several years to mature before producing the first harvest. Three types of fruit trees based on growth habit are in production in the fruit industry. They are standard, semidwarf, and dwarf. These

A B

FIGURE 20-1 (A) Seven-year-old semidwarf and (B) twelve-year-old dwarf trees.

three terms refer to the size of the tree and the length of time it takes from planting until the first crop is produced (Figure 20-1).

A standard tree is one that has its original rootstock and reaches normal size. A standard apple tree can grow to be 30 feet high. However, **semidwarf** and **dwarf** trees are smaller. They are standard varieties of trees, but they are grafted onto dwarfing rootstocks. Such rootstocks cause the tree to produce less annual growth. Thus, it remains smaller than a standard tree throughout its life. The semidwarf tree averages 10 to 15 feet in height, whereas dwarf trees average 4 to10 feet. Dwarf trees are better suited to the home garden because they bear earlier and take up less space than semidwarf or standard trees (Figure 20-2).

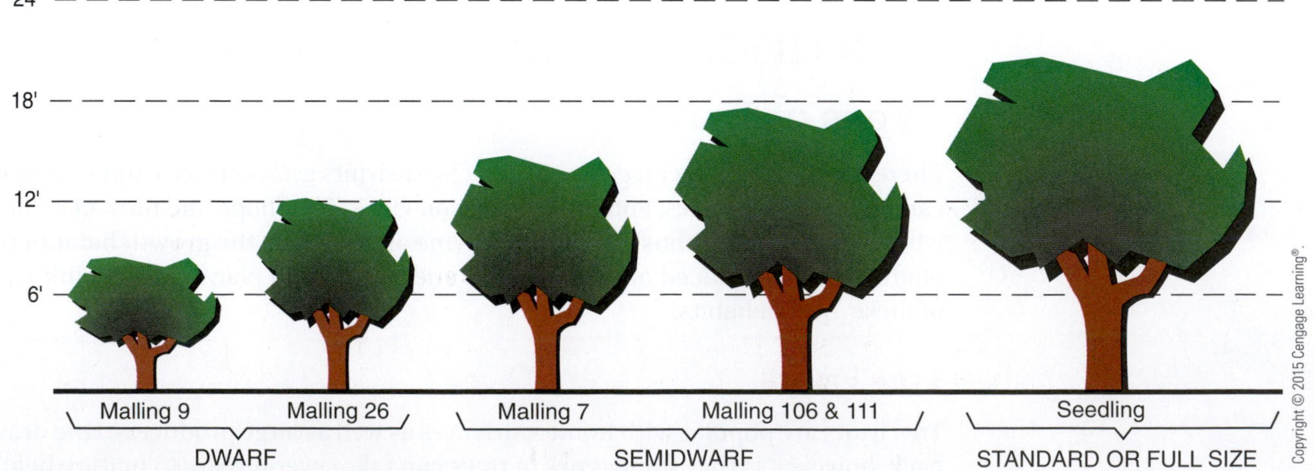

FIGURE 20-2 The different types of rootstocks can be depended on to produce trees of a certain size. Each of the codes refers to a particular rootstock with known "dwarfing" characteristics.

Courtesy of USDA/ARS #K3275-8

FIGURE 20-3 Peach-growing research at the Appalachian Fruit Research Station in Kearneysville, West Virginia.

Courtesy of USDA/ARS #K2949-6

FIGURE 20-4 Agricultural engineer David Peterson evaluates his spiked-drum shaker as it harvests a row of blackberry bushes.

Common fruits include apples, cherries, grapefruit, oranges, peaches, pears, plums, and quince. There are two types of tree fruits: pome and drupe. The common pome fruits are apples and pears. A **pome** is a fruit with a core and embedded seeds. A **drupe** is a fruit with a large, hard seed called a stone. Stone fruits include peaches, plums, quince, apricots, and cherries.

When growing tree fruits, particular attention must be paid to the climate of the area. Peaches, plums, and cherries are especially sensitive to climatic conditions (Figure 20-3).

Small-Bush Fruits

Small-bush and cane fruits that are popular with gardeners, and commercial operations include strawberries, blueberries, red raspberries, currants, gooseberries, and thornless blackberries (Figure 20-4). Small-bush fruits grow either low to the ground or only 3 to 4 feet high. They require less maintenance than tree fruits or vine fruits, and they tend to bear quickly—usually between 9 months and 1 year after planting. Some of these fruits will bear a harvest in the summer and again in the fall. A factor to consider in growing bush fruits is to buy plants from a reliable **nursery** that stands behind their products. It is vital that the new nursery stock required to establish a fruit enterprise is disease-free and healthy.

Courtesy of USDA/ARS #K3681-7

FIGURE 20-5 Grapes are popular as fresh fruit, dried fruit (raisins), juice, and wine.

Vine Fruits

The best-known vine fruit is the grape. Grapes are also the most prevalent vine fruit. They are easily grown and have a wide range of varieties and flavors. Grapes will occupy the land for many years, so site selection is important. The vines need to be trained and pruned for production and yield. Grapes require a growing season that has at least 140 frost-free days. A disadvantage of grape production is the 3- to 4-year wait for the vines to reach maturity before they produce fruit (Figure 20-5).

FIGURE 20-6 In addition to producing tasty nuts, pecan trees make attractive shade trees in landscapes.

FIGURE 20-7 The black walnut tree can be used for shade in the landscape. It also produces a good-quality edible nut.

TYPES OF NUTS

Nut trees are planted for numerous purposes. They can be useful for shade, landscaping, food, and wood. Wherever trees grow, a nut tree can also be grown. Popular nut trees grown in the United States are pecan, black walnut, filbert, hickory, almond, English walnut, chestnut, and macadamia (Figures 20-6 and 20-7).

PLANNING FRUIT AND NUT ENTERPRISES

Selecting Fruit or Nuts to Plant

Climatic Considerations

Certain varieties and cultivars of fruits or nuts are more suited to one type of climate than another. For example, citrus fruits can be grown only in southern-type climates, such as in Florida, Texas, and California. This information is a key factor when deciding which fruit to grow. Even though you may absolutely love nectarines, they may not grow successfully in your area. Recommendations for each fruit crop for a given area can be obtained from your local county agricultural extension educator.

Rootstock Selection

The variety of the root or rootstock of a grafted tree is important because many fruit crops are propagated by grafting. Select a rootstock that is true to its variety and is hardy enough to support the tree. Most nurseries will readily provide this information about the rootstock. The rootstock will determine whether the tree is a standard, semidwarf, or dwarf tree.

Frost Susceptibility

Locate crops in areas as protected from frost as possible. Also, select varieties that are late bloomers if your area is susceptible to late frosts. Planting areas on hillsides, near ponds, or near cities will remain warmer on clear, frosty spring nights. Such locations decrease the amount of damage caused by frosts. Apricots and sweet cherries are especially susceptible to late frost damage. The home gardener can protect some fruit crops, such as strawberries, from frost by re-mulching or irrigating and by placing cloth or plastic sheets over plants. Commercial growers use irrigation sprinklers, heaters, and fans to help reduce frost damage.

AGRI-PROFILE

CAREER AREAS: POMOLOGY/FRUIT GROWER/ NUT GROWER/PACKER

Courtesy of USDA/ARS #K3318-8

Development of new pecan tree varieties is monitored continuously during the growing season.

Pomology is the scientific study and cultivation of fruit, including nuts. Careers in pomology include work on farms and in nurseries, orchards, and groves. Fruit and nut specialists may be plant breeders, propagators, educators, food scientists, or consultants. Technical-level jobs are available for graders, packers, supervisors, and managers.

In the past, fruit production meant hard labor because of hand picking. However, the frequent shortages of labor for harvesting have created new opportunities for agricultural engineers, electronics technicians, plant breeders, systems managers, food scientists, and inventors to develop mechanical harvesting methods. Mechanical harvesters range from hydraulically controlled buckets, to tree shakers, to sensor-directed robots.

Fertility

The production of high-quality fruit is dependent on the fertility of the soil. On average, soil should have a pH of 5.5 to 6.0. However, for growing blueberries, the soil should be more acidic—a pH of 5.2 or less. Fruit crops should be fertilized annually after they are planted. However, too much fertilizer can damage roots and create other problems. One pound of 10-10-10 fertilizer per 100 square feet of surface soil should give good results.

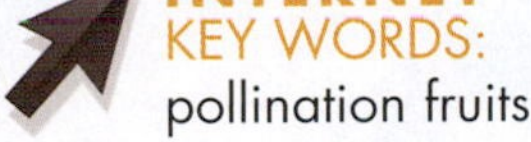

INTERNET KEY WORDS: pollination fruits

Planning for Pollination

Most fruit crops require pollination. Pollination is defined as the transfer of pollen from the male stamen to the stigma, or female part of the plant (Figure 20-8). Pollination is necessary for the fruit to set, or form. The gardener should choose crops that are known for having high-quality and abundant pollen. The pollen is

Courtesy of DeVere Burton

FIGURE 20-8 The honeybee is one of our most important insects because it is such a good pollinator. It transfers pollen from one plant to another as it gathers nectar.

distributed by wind or insects. Sometimes it is necessary to have several varieties that bloom at the same time for adequate cross-pollination. Similarly, it may be necessary to plant a tree just for its pollen quality and availability. Many fruit growers own or rent honeybees to help ensure good pollination (Figure 20-9).

When favorable conditions exist, just two varieties of a particular fruit crop are necessary for cross-pollination. Choose varieties that bloom at the same time or, at least, overlap for a period. Varieties of pears, plums, and cherries all tend to

© Tischenko Irina/Shutterstock.com

FIGURE 20-9 Honeybees are essential for good pollination of some fruit trees. The sale of honey can provide another source of income for the fruit grower.

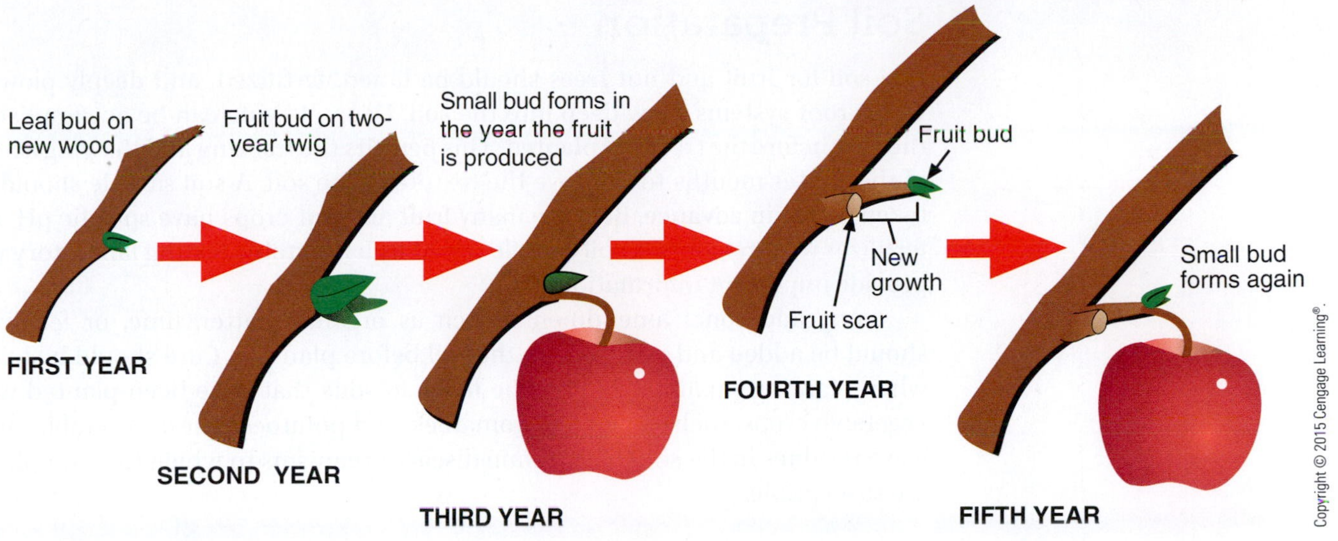

FIGURE 20-10 Growth of an apple-fruiting spur.

bloom at approximately the same time. However, apple varieties differ greatly in their times of bloom. Careful selection of apple varieties is important to ensure good cross-pollination.

Growing and Fruiting Habits

The various fruit and nut crops have different growth patterns and fruiting dates. This is especially true of fruit trees. Some varieties of apple trees require as long as 8 years before they bear fruit. Newly planted peach and sour cherry trees can bear fruit in as few as 3 years. It pays to understand what to expect from the variety you have chosen.

Most fruit trees go through two stages of growth: a vegetative period and a productive period. The vegetative period consists of rapid, vigorous growth of the tree. In younger trees, 4 to 5 feet of growth may occur in a single season. When the tree enters the productive period of fruit bud formation, a change is noticeable. Fruit buds are larger and plumper. On apple, pear, cherry, and plum trees, short, spur-like growths occur (Figure 20-10). On the peach tree, the peach buds become larger and occur in clusters, with two large ones surrounding a center one. The vigorous **terminal** growth slows down to only 18 to 20 inches in a season.

SOIL AND SITE PREPARATION

Selecting the Site

Because many fruits require full sunlight for maximum production and yield, choose a site that allows for maximum exposure to the sun. A southern exposure is best. To help prevent frost damage, select land with a slight slope and good air circulation. The soil for fruit crops should be well drained and of medium texture.

Soil Preparation

The soil for fruit and nut trees should be limed, fertilized, and deeply plowed. These root systems grow deep into the soil. Deep plowing can be accomplished the year before the trees are planted. This permits the freezing and thawing action of the winter months to improve the texture of the soil. A soil sample should be taken a year in advance. Because many fruit and nut crops have specific pH and fertilizer requirements, a soil sample tested by a reputable testing laboratory will provide important information.

Any additional amendments, such as organic matter, lime, or fertilizer, should be added and worked into the soil before planting. Care should be taken when planting small-bush and cane fruits in soils that have been planted with vegetable crops, such as peppers, tomatoes, and potatoes. These vegetables may leave residues in the soil that contain disease organisms to which the fruit plants are susceptible.

PLANTING ORCHARDS OR SMALL-FRUIT AND NUT GARDENS

When to Plant

Tree stock from reputable nurseries can be planted either in the spring or fall. There are advantages to planting in each season. Fall planting is advisable if the soil will be less than desirable or wet in the spring. Planting in the fall also allows the crop to get its root system established before rapid top growth occurs in the spring. Weather conditions may also be more stable in the fall than in the spring.

Planting in the spring has some advantages:

- The crop can make considerable top and root growth before the stress of the winter season.
- The threat of damage by rodents is reduced.
- The threat of winter kill is reduced.

Laying Out the Orchard

Each fruit or nut species and type will have its own space requirements. It is important to follow the specific spacing recommendations for each variety. When planting small trees, the distance between trees seems enormous. However, the trees will grow and fill the spaces in time. For aesthetic reasons, plant the trees in a straight line. A crooked row of trees will be obvious. Orchards are often organized in a grid pattern so that tractors and equipment may travel between rows and between the trees within the rows.

INTERNET KEY WORDS:
planting fruit and nut trees

Planting the Fruit or Nut Crop

Each fruit or nut variety has its own specifications for planting. To plant a tree, make the hole just wide enough to accommodate the roots without crowding them. The hole should be deep enough to allow the tree to be set 2 to 3 inches deeper than it was in the nursery. The two sides of the hole should be parallel, and the bottom should be as wide as the top. By making the hole to these specifications, the soil can be packed uniformly around the roots to avoid leaving

SCIENCE CONNECTION SAVING THE FRUIT

© iStockphoto/Juliann Itter

Ice on fruit blossoms can trap an insulating layer of water vapor inside the flower that prevents the fruit crop from being destroyed.

Fruit growers are at risk of losing money when their crops freeze. A number of techniques have been developed to reduce the risk of this happening. One successful method that is often used by fruit and vegetable growers appears to defy science. When air temperatures reach 32° F, the irrigation sprinklers are turned on. Water falls onto the fruit trees and becomes frozen. At this point, all a nervous farmer can do is wait to see if his or her efforts will save the harvest.

At first thought, this method makes no sense. Why would a person add water to trees that will surely become frozen? The explanation seems to be a trick of science. As the water is misted onto fruit blossoms, it begins to freeze. As this happens, heat trapped inside the blossom is released. The result is that a very thin layer of water vapor becomes trapped between the blossom and the newly formed ice crystals. The water vapor acts as insulation protecting the ovule of the flower from freezing.

air spaces. Air spaces permit the roots to dry out, reducing the likelihood that the plant will survive.

When planting trees, add some prepared soil to the hole. Then spread the roots, adding more soil until all roots are covered. Move the tree up and down several times, while firming the soil around the roots. This procedure allows all roots to be in contact with the soil and avoids air pockets.

Before the hole is completely filled, add about 2 gallons of water. After the water soaks in, add the final soil. Most trees need to be staked for good support. A metal pipe may be driven 8 inches from the tree and the tree tied to the pipe by using wire strung through a piece of rubber hose (Figure 20-11). Following planting, keep a 12-inch circular weed-free and grass-free area around the tree.

© Elena Elisseeva/Shutterstock.com

FIGURE 20-11 Provide support for the newly planted tree.

CULTURAL PRACTICES IN FRUIT AND NUT PRODUCTION

Fertilization

A complete fertilizer, such as a 5-10-10 or a 10-10-10, should be used annually on fruit and nut trees. After filling the hole of a newly planted tree, apply 1 lb. of the 5-10-10 or a half lb. of 10-10-10 fertilizer around each tree. Do not place the fertilizer closer than 8 to 10 inches from the trunk. Excessive fertilization can result in damage or death to young trees. After the first year, apply fertilizer at the rate of 1 lb. times the age of the tree. After the tree is 6 to 8 years old, do not increase the rate of application. As the tree becomes larger, fertilize all of the ground area under the outer dripline of the tree limbs.

These rates are useful for all tree fruits except pears. Pears should not be fertilized because increased susceptibility to fire blight more than offsets the benefit from fertilizing.

Fertilization requirements for the various bush and cane fruits vary considerably. Therefore, it is advisable to consult a grower's guide for each variety.

Pruning

The purpose of pruning and training young fruit trees is to establish a strong framework of branches that will support the fruit. This framework will resemble the shape best suited for the tree and for the fruit it will bear. Each fruit tree has a specific way that it should be pruned. For example, peach trees are pruned for an open-center, V-shape (Figure 20-12). However, apple trees should have a strong central leader and be pruned into a Christmas-tree scaffold (Figure 20-13).

Pruning a grapevine is based on its fruiting habit. Next year's grape clusters are on shoots produced during the current growing season. Therefore, this new growth should be carefully pruned. When the shoots have fruited and the leaves

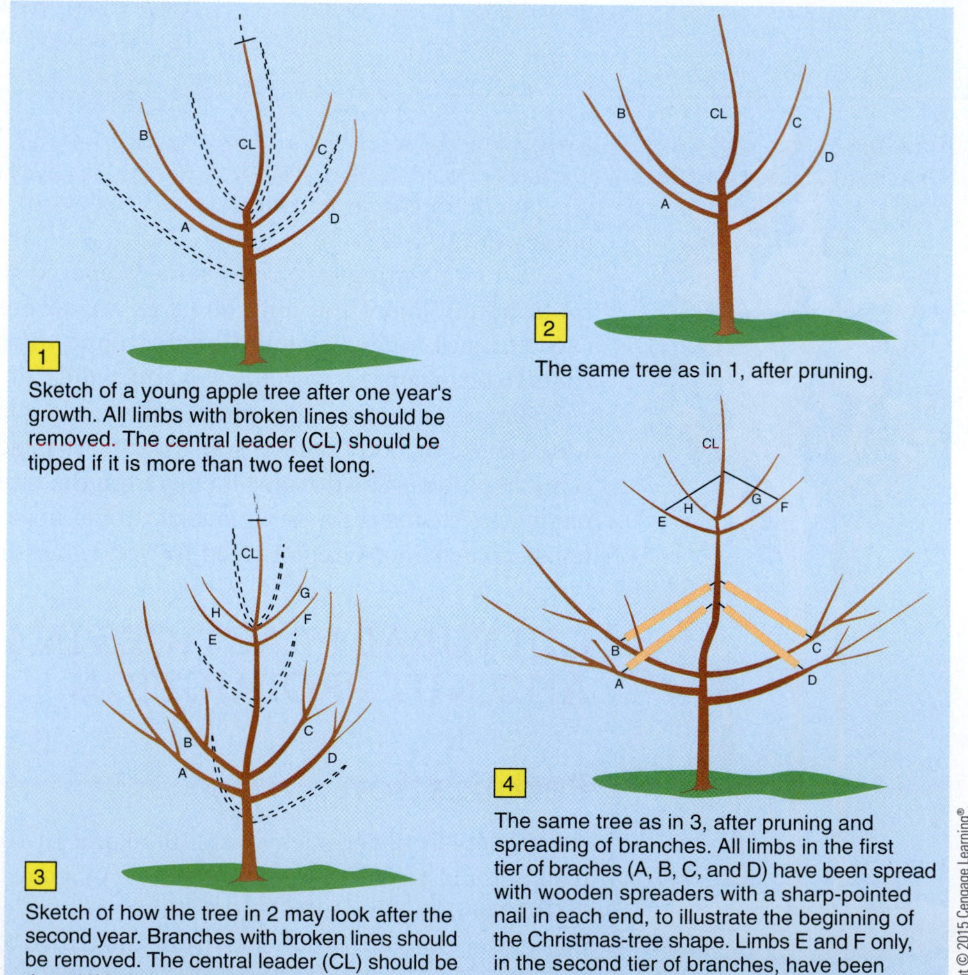

FIGURE 20-12 A peach tree should be pruned to open up a V-shaped pattern in the center of the tree.

FIGURE 20-13 A young apple tree should be pruned to maintain a strong central leader with strong, wide-angled main branches to support the weight of the fruit.

SCIENCE CONNECTION RIPENING GENES—KEY TO BETTER TASTE AND LONGER STORAGE

© iStockphoto/kokophoto

Tomatoes benefit from new gene sequences that produce firm, full-sized, flavorful fruits.

The discovery of gene sequences in tomatoes that control the ripening process is coming to *fruition* in the tomato industry, and perhaps the entire fruit industry will benefit. Scientists at the U.S. Department of Agriculture and the Boyce Thompson Institute for Plant Research, Inc., at Cornell University have discovered *ripening genes.* These genes are from a family of genes known as "MADS" and consist of over 100 gene sequences that are found in plants, animals, and microbes. They influence the developmental processes of living things, including the ripening process of fruits.

It is hoped that this discovery might soon make it possible to delay the ripening process until all fruits are fully grown and then make it possible to harvest full-sized, ripe tomatoes that are still firm enough to ship. Allowing the fruits to remain on the vine until they are ripe gives them the flavor of home garden tomatoes instead of the bland flavor of fruits that are shipped before ripening. These gene discoveries are expected to have similar ripening effects on several fruits and vegetables, including strawberries, bananas, melons and bell peppers. Imagine, perhaps some day, these commodities may be stored and shipped without spoiling and with little or no refrigeration!

have fallen, the shoots are called **canes**. The canes will produce the new shoots and next year's crop.

A common method of training grapevines is the Four-Arm Kniffin System. In this system, the vine is pruned to form a double-T on a two-wire trellis system. Two arms originating at the two wires are trained to grow in opposite directions from the trunk. The vines are pruned to remove all canes except those that are saved for fruiting and renewal growth (Figure 20-14).

Pruning and thinning small-bush and cane fruits should be accomplished according to the specific standards for each fruit. Pruning, thinning, and training these smaller fruits are essential practices for high-quality and high-volume yields. These practices also help prevent pests and diseases.

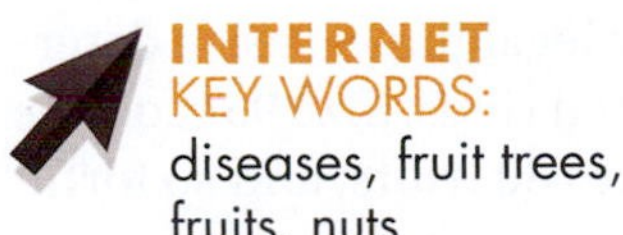
INTERNET KEY WORDS: diseases, fruit trees, fruits, nuts

Disease and Pest Control

A great deal of the time and money in fruit and nut production is spent controlling diseases, insects, and other pests. Control of diseases and insects is a major component of fruit and nut production (Figure 20-15). The type of pesticide needed and how often it is applied varies tremendously among all the species of fruits and nuts.

Each tree fruit, nut, and small-fruit variety must be treated individually. What may work on one variety may not be appropriate for another. Most orchardists will have a spray schedule worked out for each of their crops. Tree fruits require more pesticide than any other fruits. Before applying any pesticide, however, read the label thoroughly and follow it.

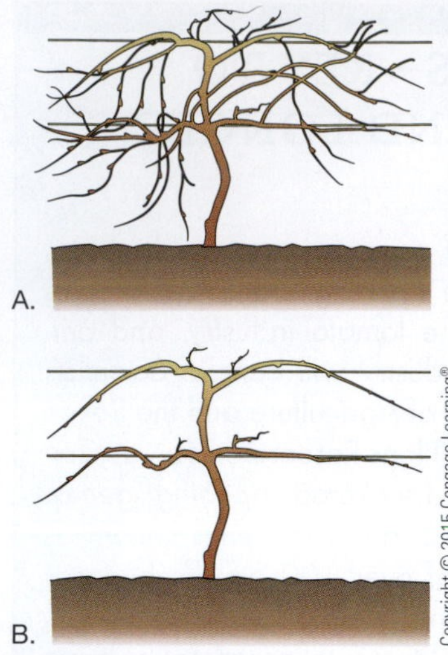

FIGURE 20-14 A grape vine trained to the Four-Arm Kniffin System (A) before and (B) after pruning.

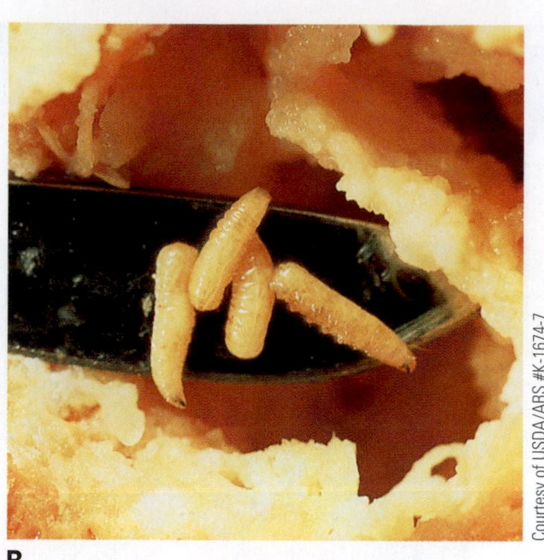

A B

FIGURE 20-15 (A) A Caribbean fruit fly laying eggs in the peel of a grapefruit. (B) Fruit fly larvae then hatch and devour the pulp.

HARVEST AND STORAGE

Harvesting

Harvesting fruits can be a daily practice. The advantage of homegrown fruit is that it can be picked at its peak of ripeness (Figures 20-16A, B, C, and D). However, fruits for commercial uses must be picked several days ahead so they will withstand shipping and handling.

For the best quality of fruit, harvest apples, pears, and quince when they begin to drop, soften, and become fully colored. Some varieties will ripen over a 2-week period, requiring picking each day. Other varieties will ripen all at once.

Peaches, plums, apricots, and cherries should be harvested when the green disappears from the surface skin of the fruit. A yellow under-color should be developed by this time. The fruit should be soft when pressed lightly in a cupped hand (Figure 20-17).

The nut varieties ripen from August to November. Harvest nuts immediately after they fall from the tree. Most of the nuts that do not fall can be knocked off the tree with poles or mechanical harvesters.

The nut varieties of pecan, hickory, chestnut, and Persian walnut will lose their husks when ripe. However, the husks of black walnuts and other types of walnuts need to be removed.

Citrus fruits are hand-picked, and the size of the fruits is the key in determining the time to harvest. Some citrus regulations set minimum size standards in order to market the fruit. For example, fruit size, for legal purposes, is determined by measuring individual fruits with standard-sized rings. Size 96 requires 96 fruits to fill a standard 1.6-bushel box; size 80 requires 80 fruits, and so forth.

Oranges are usually ring-picked only at the beginning of the harvest season. Smaller fruits are left on the tree, allowing them to grow to marketable sizes. As the season progresses, the increased size of oranges makes it more efficient to pick without using the rings, and fruits that are too small are discarded at the processing plant. The target size for most grapefruit harvests is 96, and ring picking is common practice.

Grapes should be harvested only when fully ripe. The best way to judge when the grape is at full maturity is to sample an occasional grape. Harvesting is

A

B

C

D

FIGURE 20-16 Harvest fruits at the peak of ripeness for home use. (A) Apple trees. (B) Orange trees. (C) Peach trees. (D) Plum trees.

FIGURE 20-17 Cherries ripen earlier in the season than most tree fruits, and they are a favorite fruit for eating fresh.

by hand or by using machinery. Hand harvesting is usually practiced with a small number of vines, but it may be employed with larger, high-value crops. Most large vineyards are harvested by machines. Clusters of grapes should be cut from the vine with part of the stem intact.

Small-bush and cane fruits, such as strawberries and raspberries, should be harvested when they are fully ripe. The best indicator is when the fruit is fully colored. When these fruits and berries reach maturity, they should be picked every day. When harvesting, pick when the fruit is dry to avoid mildew and mold damage.

Storage

Fruit storing is limited to those varieties that mature late in the fall or those that can be purchased at the market during the winter months. The length of time fruits can be stored depends on the variety, stage of maturity, and soundness of the fruit at harvest (Figure 20-18).

For long-term storage of apples, the temperature should be as close to 32° F as possible (Figure 20-19). Apples can be stored in many ways, but they should be protected from freezing. A cellar or other area below ground level that is cooled by night air is a good place for apple storage (Figure 20-20). The storage area should have moderate humidity. Pears have the same storage requirements as apples.

Commodity	Freezing Point	Storage Conditions		Length of Storage Period
		Temperature	Humidity	
	°F	°F		
Fruits:				
Apples	29°	32°	Moderate Moisture	Fall/Winter
Grapefruit	29.8°	32°	Moderate Moisture	4 to 6 weeks
Grapes	28.1°	32°	Moderate Moisture	1 to 2 months
Oranges	30.5°	32°	Moderate Moisture	4 to 6 weeks
Pears	29.2°	32°	Moderate Moisture	4 to 6 weeks

FIGURE 20-18 Fruit storage temperatures as adapted from the U.S. Department of Agriculture.

Source: USDA, ARS. (Draft-Revised April 2004). The Commercial Storage of Fruits, Vegetables, and Florist and Nursery Stocks. Gross, K C..Wang, C. Y., and Saltveit, M. (Eds.). Agriculture Handbook 66. Retrieved from http://www.ba.ars.usda.gov/hb66/contents.html.

Courtesy of USDA/ARS #K3853-5

FIGURE 20-19 Red Delicious apples ready for harvest in an orchard near Wenatchee, Washington.

Courtesy of DeVere Burton

FIGURE 20-20 A fruit storage facility in which temperature and humidity are controlled.

Nuts should be air dried before they are stored. Nuts, especially pecans, keep longer if they are left in the shell and refrigerated at 35° F. They can also be frozen if they are kept in their shells. Chestnuts have special requirements. They should be stored at 35° to 40° F and high humidity, shortly after harvest.

Small-bush and cane fruits, including grapes, are not suitable for storage. These fruits perish too quickly to keep very long and are best used as soon as they are harvested.

STUDENT ACTIVITIES

1. Write the Terms to Know and their meanings in your notebook.
2. Develop a bulletin board display explaining opportunities in fruit- and nut-related occupations. Include those that can be found on the local, state, and national levels.
3. Bring a specimen of some type of unusual fruit or nut to class. Discuss its origin and the techniques for growing the fruit or nut.
4. Prepare fruit and nut flash cards using pictures from seed catalogs.
5. Prepare a planting plan for fruits and nuts, drawn to scale. Indicate the types, quantity, and location of plants to be grown.
6. Many fruits and nuts consumed in our country are imported from other countries. How does the government ensure that the produce is safe? Research the process that is followed before imported crops are allowed into the United States. Why is this process necessary?

SELF-EVALUATION

A. MULTIPLE CHOICE

1. Most fruits and nuts require _______ to grow properly.
   a. full sunlight
   b. partial sunlight
   c. partial shade
   d. no sunlight
2. The stock for a fruit or nut crop can be planted in the _______.
   a. fall or spring
   b. spring only
   c. summer only
   d. fall only
3. When planting a fruit tree, you should make the hole deep enough to allow the tree to be set _______ deeper than in the nursery.
   a. 3 to 5 inches
   b. 2 to 3 inches
   c. 1 to 2 inches
   d. 6 to 8 inches
4. Nuts can be stored in the refrigerator at the optimum temperature of _______.
   a. 40° F
   b. 32° F
   c. 38° F
   d. 35° F
5. The pear is an example of a _______ fruit.
   a. drupe
   b. pome
   c. bramble
   d. aggregate

B. TRUE OR FALSE

1. A standard tree is one that has its original rootstock and grows to normal size.
2. A semidwarf tree averages a mature height of 30 to 40 feet.
3. Small-bush and cane fruits tend to bear quickly, usually between 9 months and 1 year after planting.
4. Rootstock selection is important to fruit and nut crops because most of these crops are not propagated by seed.
5. Frost susceptibility is not a factor to fruit crops.
6. Too much fertilizer can be a hazard to fruit trees.
7. Control of diseases and insects is a major component of fruit and nut production.
8. An apple is an example of a drupe fruit.

UNIT 21

Grain, Oil, and Specialty Field-Crop Production

OBJECTIVE

To determine the nature of and approved practices recommended for grain, oil, and specialty field-crop production.

MATERIALS LIST

- sample of field crops and products made from field crops
- bulletin board materials
- crop magazines
- reference materials on crops
- Internet access

COMPETENCIES TO BE DEVELOPED

After studying this unit, you should be able to:

- define important terms used in crop production.
- identify major crops grown for grain, oil, and special purposes.
- classify field crops according to use and thermal requirements.
- describe how to select field crops, varieties, and seed.
- prepare proper seedbeds for grain, oil, and specialty crops.
- plant field crops.
- describe current irrigation practices for field crops to meet their water needs.
- control pests in field crops.
- harvest and store field crops.

SUGGESTED CLASS ACTIVITIES

1. Most states have commissions to promote their major agricultural crops and products. For example, a Wheat Commission promotes and develops markets for wheat throughout the world. Obtain promotion and marketing materials from one of the agricultural commodity commissions within your state that is responsible for promoting and developing markets. Discuss the importance of marketing and promotion for the major crops that are produced in your state. Use the materials to demonstrate how this is done by the commodity commission that created the materials.
2. Obtain a variety of products such as cereals, vegetable oil, and pasta and use them as object lessons as you discuss the products that are produced in your area. Obtain statistics from your state Department of Agriculture about the ranking of your state in the

TERMS TO KNOW

field crop
malting
forage
oilseed crop
linen
linseed oil
ginning
cash crop
thermal requirement
cereal crop
seed legume crop
root crop
stimulant crop
conventional tillage
minimum tillage
sprinkler irrigation
surface irrigation
drip irrigation
mechanical pest control
genetic control

production of the various kinds of grains, oils, and specialty crops. Assign the students to make posters that show the state rankings in the production of these crops. Display the posters in the classroom. The rankings of states are available from the U.S. Department of Agriculture.

3. Bring field crop samples into class, including the roots and seeds. If fresh samples are not available, pressed samples or pictures may be used. Help students learn to identify each crop. Point out individual characteristics that are unique to each plant. This will help students with the identification process.

Anthropologists tell us that the cultivation of the land and the growing of crops began about 10,000 years ago in Africa. The need to produce food for the animals that humans had captured and begun to domesticate caused early humans to change from hunters to farmers. There were no guidelines to follow in selecting plants. Early agriculturists had to rely on observing what the animals were eating to decide which plants to grow. Trial and error and thousands of years of selection have led to the crops that are grown today. New types, varieties, and uses of plants continue to be developed in response to current needs and in anticipation of future demands for food and plant fiber by an ever-increasing world population.

INTERNET KEY WORDS:
field crops

In the United States, the production of grain, oil, and specialty crops occupies more than 450 million acres. These crops are called **field crops**. This acreage represents nearly 20 percent of the landmass in the United States. U.S. agriculturists are among the most efficient in the world, producing enough food for themselves and people in many other countries. As a result, less than 2 percent of U.S. workers are engaged in the production of food and fiber. The efficiency of the U.S. agriculturist allows the U.S. population to spend less of its income on food than citizens of most other nations. It also allows for sizable exports of food crops all over the world. This has helped maintain a favorable balance of trade for the United States in recent years. Today, we have so many products from agriculture that we no longer are aware of the plant and animal origins of many common nonfood items (Figure 21-1).

INTERNET KEY WORDS:
corn production

MAJOR FIELD CROPS IN THE UNITED STATES

Grain Crops

There are seven major grain crops in the United States. Grain crops are grasses that are grown for their edible seeds. These crops are corn, wheat, barley, oats, rye, rice, and grain sorghum. All information included in this section is based on the statistics and information available at the Web site of USDA's Economic Research Service, www.ers.usda.gov.

Courtesy of USDA/ARS #K-4796-20. Photo by Keith Weller.

FIGURE 21-1 New products from plants and animals through agriscience: (top row) plastics from vegetable oils and industrial uses for corn starch; (second row) cocoa butter substitute, improved cotton processing; (third row) tallow-based soap, detergents with improved surfactants, more uses for corn starch; and (fourth row) starch-based rubber products, plastics from vegetable oil, starch-based rubber.

Corn

Corn is the most important field crop grown in the United States. It is well adapted, high yielding, and is one of the most important crops in many states. About 35 to 40 percent of the corn produced annually in this country is grown in the Midwestern states, commonly called the Corn Belt. These states include Iowa, Illinois, Nebraska, Minnesota, Indiana, and Ohio, in descending order of production. The United States accounts for nearly 50 percent of the corn produced in the world (Figures 21-2A and 21-2B).

Originating in Central America, corn served as a major part of the diet of the native tribes of the area before the settlement of the New World by Europeans (Figure 21-3). Corn was unknown to the rest of the world until explorers found Native Americans growing corn.

Less than 10 percent of the corn grown in the United States is for human consumption. The rest is used for livestock feed, alcohol production, and hundreds of other products.

The major types or classifications of corn are dent corn, flint corn, popcorn, sweet corn, flour or soft corn, and pod corn. Most of the corn grown in the United States is dent corn.

A general discussion of cultural practices for crops appears later in this unit. However, the following approved practices are useful for growing corn:

- Select a hybrid that will mature during the growing season.
- Obtain a current soil test.
- Lime and fertilize according to the soil test and desired production.
- Select a field with deep, rich, well-drained soil.
- Select the proper tillage method.
- Prepare a firm seedbed or use a proper no-till planter.

© Zoran Karapancev/Shutterstock.com. © JIANG HONGYAN/Shutterstock.com.

FIGURE 21-2 Corn is a plant that produces high-quality grain for humans and animals and forage (silage) for livestock. (A) A corn field. (B) Corn up close.

Courtesy of DeVere Burton.

FIGURE 21-3 Corn is a Native American plant known as maize, or "Indian corn."

- Plant corn when soil temperature is 50° F or warmer.
- Calibrate the planter for the proper plant population.
- Match the plant population to soil-yield potential.
- Adjust the planter for proper depth of seed placement.
- Calibrate the sprayer for proper application of materials.

HOT TOPICS IN AGRISCIENCE CORN—NATIVE AMERICAN MAIZE

© argonaut/Shutterstock.com.

Maize, or corn, was an important food source for Native American cultures.

Long before the birth of the United States, and before corn became the number-one grain crop produced here, corn was known as maize. It was the most important grain crop for Native Americans. Maize was cultivated by native tribes as their principal food source for many years before European ships sailed to the New World. Plant breeders have modified the maize plant, also known as "Indian corn," as they have developed the plant for specific uses. For example, we now have popcorn, sweet corn, grain corn, and silage corn in many varieties. All these corn plants can be traced back to the Native American maize plants.

Corn is now used in many ways besides providing food for people and animals. It is used to make products that we use each day, such as cooking oil, biodegradable plastic bags, and ethanol (a component of gasohol). Large amounts of corn are exported throughout the world, which helps sustain the U.S. balance of trade. Corn also remains the most important feed grain for livestock in the United States.

- Use selected herbicides for controlling problem weeds.
- Cultivate to help control weeds.
- Use insecticides as necessary for proper insect control.
- Apply fungicides when needed.
- Make a yield check.
- Check moisture content during the harvest season.
- Harvest the crop as soon as moisture permits, to reduce losses.
- Take a forage analysis on ensiled corn.
- Store ensiled or high-moisture corn in sound structures.
- Track market trends and habits.
- Market the crop at an optimum time.
- Keep accurate enterprise records.
- Summarize and analyze records.

Wheat

Wheat is one of the most important grain crops in the world (Figures 21-4A and 21-4B). In the United States, wheat ranks second to corn in bushels produced. Leading states in the production of wheat are Kansas, Oklahoma, Washington, Texas, and Montana.

Wheat is used primarily for human consumption. Wheat is ground into flour, which is then made into products such as bread, cakes, cereal, crackers, macaroni, and noodles. Other uses of wheat include the manufacture of alcohol and livestock feed.

HOT TOPICS IN AGRISCIENCE BIOFUELS VERSUS FEED GRAINS

Is it possible to have the best of both worlds ... to have your cake and eat it, too? This dilemma is faced by the people of the world as corn and other grains are diverted in ever-increasing amounts to develop renewable energy sources. Ethanol and biodiesel are just two of the alternative energy products to come from feed grains and oilseed crops. Within the past decade, the monetary value of these crops has skyrocketed in comparison with the marginal grain prices that farmers historically received.

Now we face a new challenge: The price of food is rising, driven up by competition in the world markets. The *Law of Supply and Demand* is proving itself in dramatic fashion. Cereal grain foods and animal products cost more in the grocery stores than at any time in recent memory.

Livestock farmers operate at considerable risk as feed prices escalate, but they have also benefited from higher prices for their products. However, with higher prices come higher risks. Even a short delay in covering production costs through the sale of milk, meat, and eggs can drive a livestock producer out of business.

So what is the best result that can come from these apparently conflicting interests? Perhaps, in the final analysis, consumers will one day depend on renewable energy sources, and the value of agricultural products will be equitable with other sectors of the economy.

© Jim Parkin/Shutterstock.com.
A

© AZP Worldwide/Shutterstock.com.
B

Corn and other grains have found new markets. (A) Ethanol production plant in South Dakota. (B) Ethanol refinery setup.

Courtesy of USDA/ARS #K3597-18.
A

© Zeljko Radojko/Shutterstock.com.
B

FIGURE 21-4 Worldwide, wheat ranks second to rice as the most important grain crop.

Types of wheat grown in this country include common, durum, club, Poulard, Polish, Emmer, and spelt. Most of this wheat is the common type. Classes of common wheat include hard red spring, hard red winter, soft red winter, and white.

Barley

INTERNET KEY WORDS:
cultural practices, barley, oats, wheat, rye

The leading states in the production of barley in the United States include North Dakota, Montana, Minnesota, Idaho, South Dakota, and Washington (Figures 21-5A and 21-5B). Barley ranks fifth among the grain crops produced in the United States.

Most barley is used for livestock feed. It has slightly less food value than corn, but it can be grown in less favorable climates. The production of barley for malting is also important. **Malting** is the process of preparing grain for the production of beer and other alcoholic beverages.

Cultural practices for growing wheat, barley, oats, and rye are similar. The following list of approved practices for growing small grains should be useful for growing any of these crops:

- Select a variety adapted to your conditions.
- Plant certified seed.
- Obtain a current soil test.
- Add lime and fertilize according to soil test results and desired production.
- Select a field suitable for your grain choice.
- Select the proper tillage method.
- Prepare a firm and smooth seedbed.
- Calibrate the drill for the proper plant population.
- Match the plant population to soil-yield potential.
- Adjust the drill for proper depth of seed placement.
- Calibrate the sprayer for proper application of materials.

© Fedor A. Sidorov/Shutterstock.com.

A

© Madlen/Shutterstock.com.

B

FIGURE 21-5 (A) Most of the barley grown in the United States is used for livestock feed. (B) Barley is important as a feed grain, and it is also used in the production of malt.

- Use selected herbicides for controlling problem weeds.
- Use insecticides as necessary for adequate control of insects.
- Apply fungicides as needed.
- Make a yield check.
- Adjust the combine for proper harvesting adjustments.
- Monitor the grain moisture content during the harvest season.
- Harvest the crop as soon as the moisture level permits.
- Store grains in disinfected grain storage tanks.
- Bale straw as soon as moisture levels allow.
- Track market trends and habits.
- Market the crop at an optimum time.
- Keep accurate enterprise records.
- Summarize and analyze records.
- Make business decisions based on records.

Oats

Oats as a grain crop are fourth in acres produced in the United States (Figure 21-6). Major oat-producing states are South Dakota, Minnesota, Wisconsin, Iowa, and North Dakota.

KEY WORDS:
rice food staple rice production, culture, space oil crops, soybean production

The value of oats in adding bulk and protein to the diets of livestock is well documented. However, about 5 percent of the oats produced in the United States are made into oatmeal and cookies. Oats are also used in the production of plastics, pesticides, and preservatives. In addition, they are important in the paper and brewing industries.

Courtesy of DeVere Burton.

FIGURE 21-6 Oats provide bulk and protein to the diets of animals as well as food for humans.

Courtesy of DeVere Burton.

FIGURE 21-7 Rye is used as a livestock feed and for the production of specialty flour. Economically, it is the least important grain crop produced in the United States.

Rye

Although rye is grown in nearly every state in the United States, it is the least economically important grain crop. It does, however, have many uses (Figure 21-7). Most rye is grown in South Dakota, Georgia, Minnesota, North Dakota, and Nebraska.

About 25 to 35 percent of rye acreage is used for grain. The rest is used for forage, as a cover crop, or as a green manure crop. **Forage** is hay or grass grown for animal feed. Cover crops are planted to protect the soil from erosion. Green manure crops are grown to be plowed under to add nutrients and organic matter to the soil. The rye grown for grain is used for livestock feed, flour, whiskey, and alcohol production.

Rice

Rice is the most important grain crop grown for human food in the world. Most of the rice in the United States is grown in Arkansas, California, Louisiana, Texas, and Mississippi. Rice is the only commercially grown grain crop that can grow and thrive in standing water. The types of rice grown in the United States are short grain, medium grain, and long grain varieties (Figure 21-8).

The majority of the rice grown in the United States is used for human consumption. The excess that is produced is exported to other countries of the world.

Sorghum

Sorghum is grown in the United States primarily for livestock feed. It is about equal to corn in food value. Other uses of sorghum include forage, the manufacture of syrup or sugar, and the making of brooms (Figure 21-9). Leading states in the production of sorghum include Kansas, Texas, Nebraska, South Dakota, and Oklahoma. Based on acres harvested, sorghum is the third most important grain crop in the United States.

The five types of sorghum are grouped according to use. They are grain, forage, syrup, grass, and broomcorn.

Courtesy of USDA/ARS #K-3444-1.

A

© Mikus, Jo./Shutterstock.com.

B

FIGURE 21-8 Rice is the most important grain crop in the world. (A) Rice is the only commercially grown grain crop that will grow in standing water. (B) Most of the rice grown in the United States is for human consumption.

Courtesy of USDA/ARS #K9371-7. Photo by Peggy Greb.

FIGURE 21-9 Sorghum is a grain crop that is produced in a few Midwestern states. It is mostly used for livestock feed, either as grain or as forage.

Oilseed Crops

Crops that are grown for the production of oil from their seeds are called **oilseed crops**. These crops are increasing in importance each year as people in the United States rely more and more heavily on vegetable oils and less on animal fats in their diets. Important crops grown for the oil extracted from their seeds are soybeans, peanuts, corn, cottonseed, canola, safflower, flax, and sunflowers (Figure 21-10). Corn and cottonseed are discussed in other sections of this unit.

Soybeans

Approximately 75 million acres of soybeans are grown in the United States each year. With an average yield of close to 40 bushels per acre, the production of soybeans grosses more than $11 billion each year (Figures 21-11 and 21-12).

Oil and grain products are the major uses of soybeans. The meal resulting from the extraction of oil from soybeans is an important source of protein in livestock feeds. Soybeans are also used for hay, pasture, and other forage crops. Research has led to the development of hundreds of other uses for soybeans.

Large centers of production include the Midwestern states of the Corn Belt. Major soybean-producing states include Illinois, Iowa, Minnesota, Indiana, and Missouri. Internationally, Brazil has become a major producer of soybeans. Approved practices for raising soybeans are similar to those for raising corn, except that soybeans and other legume seeds should be treated with the proper inoculants (bacteria to ensure good nitrogen fixation).

Peanuts

The peanut is actually a pea rather than a nut, despite its nut-like taste and shell. It is grown primarily in the South, where warm temperatures and a long-growing season are keys to success. Leading peanut-producing states are Georgia, Texas, Alabama, North Carolina, Oklahoma, and Florida.

Courtesy of DeVere Burton.

FIGURE 21-10 The rape, or canola, plant is an old one that is getting new attention as an oil-producing crop. This plant belongs to the mustard family, as evidenced by the bright yellow flowers.

Courtesy of USDA/ARS #K5272-1. Photo by Keith Weller.

FIGURE 21-11 The late Edgar E. Hartwig devoted half a century to soybean research. Sometimes called the Soybean Doctor, this world-renowned agronomist developed productive plants with built-in resistance to insects, nematodes, and diseases.

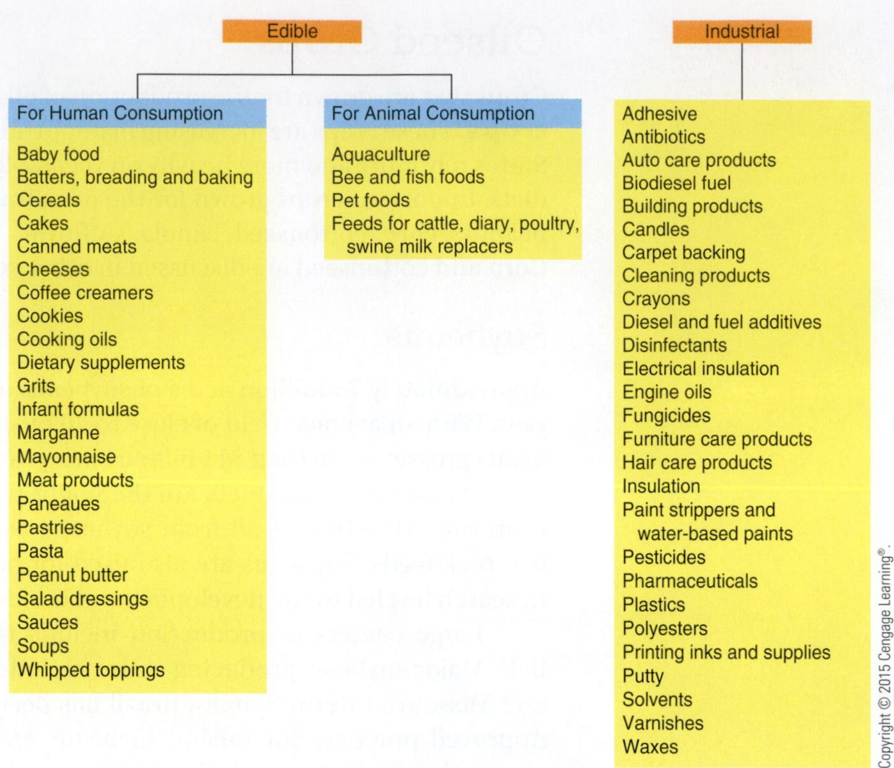

FIGURE 21-12 The soybean has become famous for its many products and is growing in importance in world markets.

One ton of peanuts in the shell will yield about 500 lbs. of peanut oil and 800 lbs. of peanut-oil meal. The remaining 700 lbs. are mostly shells. The oil meal is used for livestock feed and serves as a good protein source in human diets. Other food products produced from peanuts include peanut butter and dry-roasted peanuts (Figure 21-13).

FIGURE 21-13 George Washington Carver, the famous agricultural research pioneer, developed more than 300 products from peanuts.

Safflower

Production of safflower for oil occurs mainly in California. Safflower plants grow 2 to 5 feet in height and have flower heads that resemble Canadian thistles. The oil comes from the wedge-shaped seeds that the plant produces. The seeds contain from 20 to 35 percent oil.

Safflower oil is used in the production of paint and other industrial products. It is also used as cooking oil and in low-cholesterol diets.

Flax

Originally, the production of flax was mostly for fiber. Flax fibers from the stems of plants are used to produce **linen** (a cloth product).

The oil produced from the seed of the flax plant is called **linseed oil**. It is an important part of many types of paint and has hundreds of uses in industry. The linseed-oil meal that is left after extracting the oil is an excellent source of protein for animal feeds.

Most flax is grown in North Dakota, South Dakota, Minnesota, and Wisconsin.

© formiktopus/Shutterstock.com.

FIGURE 21-14 The sunflower is the state flower of Kansas, the "Sunflower State." A field of sunflowers is beautiful and it is the source of valuable oil-rich seed.

Courtesy of DeVere Burton.

FIGURE 21-15 Some vegetable oils, such as rapeseed oil and sunflower oil, can be used to produce fuels for diesel tractors. This kind of fuel is known as biodiesel.

INTERNET KEY WORDS:
sunflower production culture, vegetable oil, biodiesel

Sunflowers

Production of oil-type sunflower seed has been important in the United States in recent years (Figure 21-14). Most of the sunflower production is located in North Dakota, South Dakota, Kansas, Minnesota, and Texas.

Two types of sunflowers are grown commercially in the United States: oil-type and non–oil-type. About 90 percent of sunflower production is of the oil-type. Oil-type sunflower seeds contain 49 to 53 percent oil. The meal that remains after the oil has been removed is high in bulk and contains 14 to 19 percent protein. It is used for livestock feed. The oil is used for margarine and cooking oil. Sunflower oil can also be processed into biodiesel as a substitute for diesel fuel in tractors and trucks (Figure 21-15).

Specialty Crops

Fiber crops, sugar crops, and stimulant crops are grouped into the category called specialty crops. Specialty crops grown in the United States include cotton, sugar beets, sugarcane, and tobacco.

© Sherry Yates Sowell/Shutterstock.com.

FIGURE 21-16 Cotton has been an important crop since colonial times, but care must be taken to protect cotton plants from insects and diseases.

Cotton

Cotton originated in Central and South America. It has been an important crop in the South since colonial days. The cotton plant requires warm temperatures and a long growing season to reach maturity. The leading cotton-growing states include Texas, Mississippi, California, Louisiana, and Arkansas. Arizona's extensive irrigated acreages of cotton crops, producing multiple crops per year on the same land, also place it among the leading cotton-producing states.

More than 15 million bales of cotton are harvested per year in the United States. Because only about 9 million are needed by the textile industry, the rest is exported to other countries of the world (Figure 21-16).

FIGURE 21-17 Sugar cane and sugar beets are the sources of nearly all of our domestic sugar supply. Large factories and refineries process the raw product into syrup and then into the familiar sugar crystals we use to sweeten foods.

INTERNET KEY WORDS: sugar beets cane sugar production

Cotton seeds must be removed from cotton after it is harvested. This process is called **ginning**. The cotton seed is then processed to remove its oil, which is a major contributor to the vegetable-oil needs of the United States. Following the extraction of the oil, the seed is ground into a high-protein animal feed.

Sugar Beets

The production of sugar beets for sugar accounts for about 35 percent of the refined sugar produced in the United States (Figure 21-17). This crop is grown for its thick, fleshy storage root in which sugars are accumulated. Centers of production for sugar beets include the western states and the upper Midwest.

Sugarcane

Sugarcane production in the United States is concentrated in subtropical areas of Florida, the Gulf Coast states, and Hawaii. Sugarcane accounts for about 65 percent of the sugar refined in the United States.

This crop, which is a grass, is grown from sections of stalk called seed pieces, rather than from seed. It takes about 2 years for sugarcane to reach the harvesting stage in Hawaii. Compared with production in Hawaii, in the southern states, sugarcane is harvested about 7 months after planting, with a corresponding loss of yield. The same field of sugarcane can be harvested several times before it must be replanted because the plant regenerates its top growth (Figures 21-18A and 21-18B).

Tobacco

Tobacco is an original North American product that was used by American Indians in religious rites. It is produced as a cash crop in the southeastern states, predominantly in North Carolina, Kentucky, Virginia, South Carolina, Tennessee, and Georgia. A **cash crop** is a high-value crop. Production of tobacco declined in the United States in the 1980s as cigarette smoking and other uses of tobacco declined, but it increased again in the 1990s as export markets expanded.

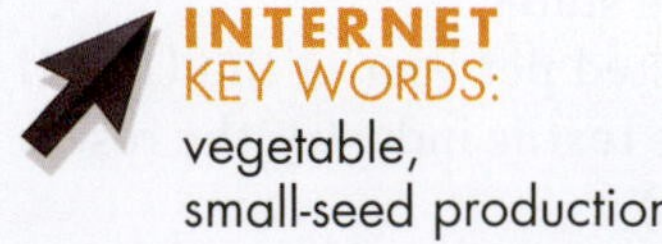

INTERNET KEY WORDS: vegetable, small-seed production

A

Copyright © 2015 Cengage Learning®./Courtesy of Elmer Cooper.

B

Courtesy of DeVere Burton.

FIGURE 21-18 (A) Sixty-five percent of U.S. sugar production comes from sugar cane produced in the southern states and in Hawaii. (B) Sugar beet production accounts for about 35 percent of the domestic sugar supply.

INTERNET KEY WORDS: certified foundation seed, AOSCA seed certification

Tobacco production requires large amounts of labor and is therefore best adapted to small farming operations. Warm temperatures and plenty of rainfall are required for optimum production of high-quality tobacco.

Seed Crops

The production of good-quality field crops depends on good seeds. An entire seed industry has emerged to supply the need for high-quality seeds for many varieties of field crops. Seeds must come from healthy plants that are free of diseases. Many of the vegetable crops that we raise are hybrid plants (see Unit 17) that come from different strains of parent stock. Seed crops are usually raised in areas that are

HOT TOPICS IN AGRISCIENCE BIODIESEL—A RENEWABLE RESOURCE

© Florian Augustin/Shutterstock.com.

Biodiesel can be manufactured from vegetable oils as well as animal fats and used cooking oil.

One of the great concerns about world dependency on fossil fuels has been that these fuels are nonrenewable resources. Once the oil reserves are used up, this important resource will be lost to the human race. Biodiesel is different because this product is produced by plants. It is obtained from oil seeds such as rapeseed and sunflower seeds.

Plant scientists and agricultural engineers have pooled their talents to develop ways that plant oils can be used to replace diesel fuel. Vegetable oils have shown great promise as fuels in diesel engines, but except when diesel prices are high, biodiesel has been expensive in comparison with the cost of petroleum-based diesel fuel. As the cost of crude oil rises, the cost to produce biodiesel also increases—mostly due to the increased cost of vegetable oils. Concurrently, consumers have had a tendency to blame the increasing cost of food on the use of grains and oil crops to produce ethanol and biodiesel.

Despite the cost, the consumers of the world will eventually run short on crude oil, and a renewable source of energy for fuel will become a necessity. A less expensive approach to producing biodiesel is to produce it from recycled vegetable cooking oils and from animal fat. These oils and fats form chemical compounds called esters when they are combined with methyl alcohol under alkaline conditions. This type of biodiesel is an improvement over earlier biodiesels because this fuel product works better in cold weather than the early biodiesel products.

isolated from related commercial crops. This helps maintain the purity of the genetic makeup of the seeds because stray pollen from undesirable sources is minimized.

Seeds for most field crops are produced in similar climates and growing conditions that will be encountered when the seeds are used tvo produce commercial crops. This assures that they are adapted to the conditions under which they are expected to grow and produce. Care must be taken, however, to ensure that the seed crops do not become infected by diseases and destructive organisms such as fungi. Such an infection of a seed crop could spread the infection wherever the seed is planted. In all cases, vseed crops must be subjected to inspection and testing to ensure that they are pure and that the quality of the seed is high.

Most vegetable seeds are produced in isolation from commercial crops. An arid climate is desirable in the production of most vegetable seed crops because such a climate is not favorable to fungi, bacteria, and many other organisms that cause damage to crops. As a result, most vegetable seeds are produced on irrigated farms at isolated locations in the western desert region of the United States (Figure 21-19). Seeds are high-value crops that must be protected from the stresses that are often found in the environment. The seeds must have favorable conditions that allow them to mature completely before they are harvested; otherwise, the germination rate will not be acceptable.

After the harvest, seeds are placed in storage under carefully controlled conditions to preserve their ability to germinate. Seed samples are routinely submitted to seed laboratories for certification. The seed laboratories test the germination rates of different seed samples, and they carefully inspect the seed samples for contamination with weed seeds and diseases. Seed crops must meet strict standards to be labeled as certified seed.

Classification of Field Crops

Field crops can be classified in a number of ways. Three classifications are by use, thermal requirements, and life span. **Thermal requirement** refers to the heat

FIGURE 21-19 Most vegetable seed crops are raised in arid climates at isolated locations, such as Melba, Idaho, to avoid contamination from the pollen of related field crops and to minimize exposure to fungal and bacterial diseases that thrive in damp environments.

requirements or length and characteristics of the growing season that are required for the crop to mature.

The classification of field crops according to use is as follows:

- **Cereal crops** are grasses grown for their edible seeds. They include corn, wheat, barley, oats, rice, rye, and sorghum.
- **Seed legume crops** are nitrogen-fixing crops that produce edible seeds. Included in this class are soybeans, peanuts, field peas, field beans, and cowpeas.
- **Root crops** are grown for their thick, fleshy storage roots. Beets, turnips, sweet potatoes, and rutabagas are root crops.
- Forage crops are grown for hay, silage, or pastures for livestock feed. Examples of forage crops include alfalfa, clover, timothy, orchard grass, and many other crops that are harvested for their stems and leaves.
- Sugar crops are grown for their ability to concentrate and store sugars in their stems or roots. They include sugarcane, sugar beets, and sorghum. Corn is also used to produce sugar.
- Oil crops are produced for the oil content of their seeds. Examples are soybeans, peanuts, cottonseed, flax, rapeseed, sunflower seed, safflower seed, and castor beans.
- Tuber crops are grown for their thickened, underground storage stems. Potatoes and Jerusalem artichokes are examples of tuber crops.
- **Stimulant crops** are grown for their ability to stimulate the senses of the user. Examples of stimulant crops are tobacco, coffee, and tea.

Crops can also be classified according to their thermal requirements. The two major thermal groups are warm-season and cool-season crops.

Warm-season crops must have warm temperatures to live and grow. They are adapted best to areas where freezing or frost seldom occurs. They also normally require longer growing seasons than cool-season crops. Examples of warm-season crops are cotton, tobacco, and citrus fruits.

Cool-season crops are normally grown in the northern half of the United States, where temperatures below freezing are normal. These crops often need a period of cool weather to attain maximum production. Most of the grains, tubers, and apples are cool-season crops.

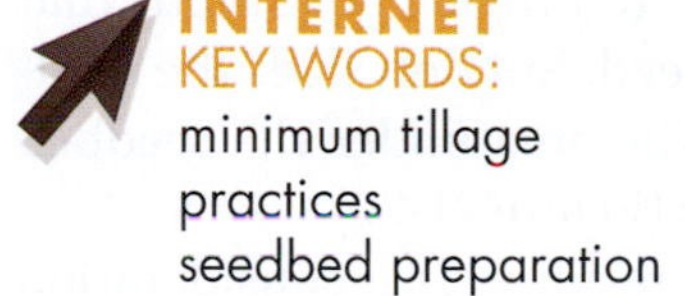

minimum tillage practices
seedbed preparation

Crops can also be classified according to their life spans. They may be annual, biennial, or perennial. An example of an annual crop is corn; red clover is a biennial crop; and alfalfa is a perennial crop.

Selection of Field Crops

A number of factors must be considered when selecting which field crops to grow. Some of these are the following:

- Select crops that will grow and produce the desired yields under the type of climate available. Be sure to consider length of growing season, average yearly rainfall, average daily temperature, humidity, and prevailing wind.
- Crops must be adapted to the type of soil available. Consider soil pH, soil type, soil depth, and soil response to fertilizers.
- Consider demand and availability of markets for the crop to be produced.
- Assess labor requirements and availability of labor for the crop.
- Identify machinery and equipment that will be needed to grow the crop.

AGRI-PROFILE

CAREER AREAS: BROKER/ELEVATOR MANAGER/GRAIN HANDLER/AGRONOMIST/PRODUCER

Courtesy of USDA/ARS #K3882-2.

Plant technicians and research specialists plan and conduct research projects leading to improved plant varieties, such as the pearl millet pictured here.

Corn, wheat and other small grains, sugarcane, soybeans, sugar beets, and other specialty crops are grown on large acreages in the United States, Canada, and many nations of the world. In the United States, grain, oil, and specialty crops account for large amounts of exports and do much to help maintain our balance of payments in foreign trade. Grain brokers, futures brokers, market reporters, grain elevator operators, crop forecasters, farmers, and others owe their jobs to these crop enterprises.

Custom combine operators, truck drivers, maintenance crews, cooks, and other service workers follow the grain harvest from Mexico to Canada. At season's end, the crews return south and prepare for the next season, when the cycle is repeated. At the same time, small farm owners and operators grow, harvest, and market millions of acres of field crops as part, or all, of the farm operations.

Along with crop enterprises, jobs are available in building and storage construction, systems engineering, machinery sales and service, welding/repair, irrigation, custom spraying, hardware sales, agricultural finance, chemical sales, and seed distribution.

- Consider the availability of enough land to justify production of the crop.
- Identify potential pest-control problems.
- Estimate yields.
- Anticipate production costs. Can a reasonable profit be expected?

Seedbed Preparation for Field Crops

The purpose of seedbed preparation for field crops is to provide conditions that are favorable for the germination and growth of the seed. Not only does the seedbed need to be prepared for seed germination but the area under the seedbed must also be prepared for the root growth of the crop (Figure 21-20).

Courtesy of DeVere Burton.

FIGURE 21-20 A good seedbed consists of crumbly, mellow soil that is free of clods, rocks, and debris.

Eliminating competition from weeds and crop residues is a consideration when preparing a seedbed for planting. Proper seedbed preparation can also increase the availability of soil nutrients to plants.

Seedbeds should not be overworked. The texture of the soil should be porous and allow for free movement of air and water. Small seeds require a seedbed with a finer texture than larger seeds require. The seedbed should contain enough fertility to encourage germination and growth until additional fertilizer can be applied.

Several methods can be used to properly prepare seedbeds for field crops. They can be divided into three general categories: conventional tillage; reduced, or minimum tillage; and no-tillage.

In **conventional tillage**, the land is plowed with a moldboard or disk plow, turning under all the residue from the previous crop. The soil is then worked with tillage machinery to smooth and further pulverize the soil for the seedbed (Figure 21-21).

FIGURE 21-21 With conventional tillage, the soil is turned using a moldboard plow so that all crop residues, livestock manure, lime, and fertilizer are mixed through the plow layer.

FIGURE 21-22 A chisel plow loosens the soil but leaves the crop residue on the surface.

Reduced or **minimum tillage** is a system of seedbed preparation that works the soil only enough so that the seed can make and maintain close contact with the soil and germinate. A chisel plow is frequently used (Figure 21-22). Minimum-tillage systems usually combine several operations into one pass across the field. This method reduces the amount of soil compaction, conserves soil moisture, and usually provides less opportunity for soil erosion (Figure 21-23).

No-till preparation of seedbeds involves planting seeds directly into the residue of the previous crop, without exposing the soil. Seed is usually planted in a narrow track opened by the seed planter. It is extremely important that good management practices be used when using the no-till method of seedbed preparation. In addition to coordinating planting with soil moisture conditions, practices such as controlling weeds, insects, and diseases should be implemented. Competition from previous crop residues must be managed (Figure 21-24). For example, as crop residue breaks down, it competes with a growing crop for soil nitrogen. Extra nitrogen may be needed to assure good crop yields.

FIGURE 21-23 A tillage tandem disc/chisel-point tiller cuts crop residue and loosens surface soil via the discs, loosens the deep soil via chisel points, and leaves a fine seedbed on the surface.

FIGURE 21-24 No-till planting retains crop residues on the surface for better erosion control.

SCIENCE CONNECTION DESIGNER FOODS

Golden rice has been genetically modified to address vitamin A deficiency, a leading cause of child blindness in developing nations of the world.

In the 1940s, food products were being fortified with vitamins and minerals. Before this time, many people experienced illness that was a result of vitamin and mineral deficiencies. Once the causes of such sicknesses were discovered, foods such as breads and cereals were supplemented with the needed nutrients. Currently, the back of any breakfast cereal box shows that fortification is still practiced. Biotechnologists are certain that there is another way the food we eat can improve our health. For example, a variety of bioengineered rice is available in the United States and Canada. Golden rice has been genetically altered to produce vitamin A.

Crops such as golden rice are created using a process called recombinant DNA technology. This process begins when a desired gene is found and cut out of the DNA of the cells from another plant. These cells are called donor cells. This is done using proteins called restriction enzymes. These enzymes cut DNA at a specific location. Next, the donor gene is put into a plasmid. A plasmid is a circular piece of DNA that can be found in some bacteria. The bacteria are then placed in a petri dish, where they begin to replicate. Many copies of the donor gene are made. Through bacterial action, the desired gene can be put into the cells of the target plant. The resulting plants reproduce themselves with the new donor gene in place.

In many developing countries, undernourished children experience childhood blindness. This condition is caused by vitamin A deficiencies in young children. Rice is often the only food source available to poor families in these countries. When golden rice is made available to the citizens of these nations, the number of children affected by childhood blindness can be demonstrated to decrease dramatically. Bioengineered foods have the potential to do for this generation what food fortification has done for generations in the recent past.

INTERNET KEY WORDS:
planting field crops

Planting Field Crops

The invention of the seed planter was one of the most important events for U.S. agriculture. Three general types of planters are used in planting field crops today. They are row crop planters, drill planters, and broadcast planters.

Row crop planters plant seeds in precise rows and with even spacing within the rows. Three types of row crop planters are drill planters, hill-drop planters, and checkrow planters. Hill-drop planters drop two or three seeds together in rows. Checkrow planters plant several seeds together in a checkered pattern in the field to permit cross-cultivation. Row crop planters are used to plant corn, beans, sugar beets, soybeans, sorghum, and cotton.

Drill planters plant seeds in narrow rows at high population rates. They drop seeds individually in a row at set distances apart (Figure 21-25). Drills are available in many row spacings and planter widths. Seeding rates are less accurate than with row crop planters. Among the field crops that are planted with drill planters are wheat, oats, barley, rye, forage legumes, and many grasses. Fertilizer and pesticides may be applied at the same time the seed is planted with drill planters.

Broadcast planters scatter the seeds in a random pattern on top of the seedbed. The seeding accuracy using this type of planter is the poorest of the planting methods. Broadcast seeders cover wide areas and usually plant seeds much faster than other methods. They are sometimes used when weather conditions make it difficult to get machinery into the fields. Airplanes are sometimes used in combination with broadcast planters, especially on difficult terrain following a fire. Knapsack seeders and spinners are other types of broadcast seeders.

FIGURE 21-25 Drill seeders are used to plant small seeds, such as small grains, grasses, alfalfa, and clover, at high population rates.

INTERNE KEY WORDS:
crop water requirements

One disadvantage of using broadcast seeders is that some kinds of seed need to be covered to germinate and to protect them from loss. This means that a second trip must often be made over the field to cover the seed. Small grains, grasses, and legumes such as clover and alfalfa are sometimes planted by broadcasting.

There are other considerations in planting field crops. These include the date to plant, germination rate of seeds, uniformity of seed, weather conditions, and insect- and disease-control problems.

Meeting Water Needs of Crops

The soil in which plants grow acts as a storage vat for the water needed by the plant. Under ideal soil conditions, approximately half of a plant's pore space is filled with water. About half of this water is available for use by plants. Unfortunately, this ideal condition seldom exists; often, much less water is available for plant use than is needed. Factors that affect water availability for crops include the type of soil, natural rainfall, water-table levels, and prevailing winds.

When conditions are such that sufficient water is not available for the crop, irrigation may be the answer to obtaining profitable yields. Irrigation is a means of providing adequate amounts of water to crops.

Irrigation of crops has been practiced for more than 5000 years. The Nile River was used by the Egyptians to irrigate crops grown in the fertile deserts of the area. The Chinese diverted the water from many rivers to irrigate their rice fields. American Indians of the American West used irrigation to ensure the production of corn in arid areas.

The major methods of supplying irrigation water to crops are sprinkler systems, surface irrigation, and drip irrigation.

Sprinkler irrigation is one of the most efficient uses of supplemental water because it places the water where it is needed at the time it is needed. The amount of water that is applied is easily controlled by the length of time that the

© Brenda Carson/Shutterstock.com.

FIGURE 21-26 Modern sprinkler irrigation systems are capable of delivering a uniform supply of water throughout the field.

Courtesy of DeVere Burton.

FIGURE 21-27 Cucumbers growing in a greenhouse are irrigated by a drip irrigation system that distributes water and nutrients.

INTERNET KEY WORDS: pest control, field crops

irrigation system is allowed to operate on the same setting. Unlike flood irrigation, which tends to over-irrigate part of the field while under-irrigating other areas, sprinkler irrigation delivers a uniform supply of water to all areas of the field (Figure 21-26).

Surface irrigation water is delivered to the crop by gravity, flowing over the surface of the soil or in ditches or furrows. It is an inexpensive means of providing water to crops. However, the cropland may need to be leveled before it can be irrigated.

Drip irrigation supplies water to the roots of crops in a uniform manner. Use of tubes located either above or beneath the soil to deliver the water to the crops makes a drip irrigation system expensive to set up. It has the advantage of low operating costs when in operation, however, and it permits the most efficient use of water (Figure 21-27).

Pest Control in Field Crops

Courtesy of USDA/ARS #K3965-17.

FIGURE 21-28 A few lesser Grain Borers left unchecked in stored grain soon reproduce to become a devastating horde of insects that is capable of ruining an entire crop of grain.

The control of pests in field crops is often the factor that determines whether a crop is profitable or not. Pests of field crops include diseases, weeds, insects, and animals. They may destroy the seed before it germinates, attack the growing crop, or render the harvested crop to be unusable and unfit for humans to eat. Economic losses from plant pests total billions of dollars each year (Figure 21-28).

There are three main categories of losses from plant pests. They are reduced yields, reduced quality, and storage losses.

Reduced yields occur when weeds germinate and grow more vigorously than the crop plants. Weeds compete successfully with the crop for moisture and nutrients, often causing the crop to be unhealthy. Parasitic plants draw nourishment from the host plants and may cause them to shrivel and die.

Damage from insects also causes the crop to yield less than expected. The damage usually occurs as adult insects feed or during the growth stages as immature insects feed on the crop. Reduction of yield may also occur as insects spread diseases from plant to plant.

Diseases can reduce yields by interfering with the ability of the plant to manufacture food. They can also cause other plant processes to be changed, affecting the health of the plant.

Courtesy of USDA/ARS #K3239-1.

FIGURE 21-29 The ridge tillage machine for cultivation is capable of reducing or eliminating pre-plant tillage under most soil conditions .

Courtesy of USDA/ARS #K5035-20.

FIGURE 21-30 Parasitic wasps provide biological control of some pests by laying their eggs within or on the larvae of specific pests.

INTERNET KEY WORDS:
biotech benefits
biotech debates

The quality of a crop is sometimes reduced from such things as weed seeds or rodent hairs and droppings intermixed with the crop. Foreign materials are usually unsanitary and may cause flavors that are objectionable to consumers. Foreign materials must be removed before the crop can be used.

Damage from insects and diseases can make the crop less desirable in appearance and can increase processing costs tremendously. Food crops may be deemed unsuitable for human consumption and result in a total loss if they are too severely infested with insects and diseases.

Spoilage of crops sometimes results when weeds hinder the drying process. Insects may also cause stored crops to overheat and mold, rendering them unfit for use.

Methods of controlling pests in field crops include mechanical control, cultural control, biological control, genetic control, and chemical control.

Mechanical pest control refers to anything that affects the environment of the pest or the pest itself. Cultivation is the normal mechanical control of weeds (Figure 21-29). Cultivation of the soil may also expose insects and soilborne diseases to the air. The effect of abrupt changes in temperatures as a result of exposure to the sun and air often proves fatal to some pests. Other types of mechanical control of pests include pulling or mowing weeds and the use of screens, barriers, traps, and electricity to exclude them from the crop.

Courtesy of DeVere Burton.

FIGURE 21-31 Plant scientists are able to transfer genetic-resistant genes from one species to another. Potatoes that are genetically resistant to the Colorado potato beetle no longer require pesticides to control this insect.

Cultural control refers to adapting or changing farming practices to control pests. Cultural controls include timing farming operations to eliminate pests, rotating crops, planting resistant varieties, and planting trap crops that are more attractive to insects than the primary crop.

Biological control of plant pests involves the use of natural predators or diseases as the control mechanisms. The release of sterile male insects and the use of baits and repellents are also examples of this type of pest control. When using insects or diseases to control crop pests, it is important that the control be specific to the intended pest (Figure 21-30).

The development of varieties of crops that are resistant to pests is called **genetic control** (Figure 21-31). This may involve making the crop less attractive to the pest because of taste, shape, or blooming time. Developing more rapidly

INTERNET KEY WORDS:
approved practices
grain storage
post-harvest
crop losses

growing crops that crowd out weeds is also an example of genetic pest control. Crops with resistance to diseases also fall into this category.

Chemical control of plant pests involves the use of pesticides to control pests of field crops. Excellent management practices must be exercised when using chemicals to control pests. Care should be taken to correctly identify the pest to be controlled and the chemical to be used. Dosage, runoff, and pesticide residues need to be carefully monitored (Figure 21-32).

Harvesting and Storing Field Crops

Harvesting field crops at the proper stage of maturity is a key to maximizing profits. The harvest of the crop is the culmination of a growing season of work and anticipation of the rewards for a job well done.

Development of mechanical harvesting equipment allows field-crop producers to harvest thousands of bushels of grain daily and with far less labor than previously required. This allows for tremendous increases in the amount of food available for people and animals and a greatly improved standard of living.

The primary harvesting machine for field crops is the combine (Figure 21-33). It performs the tasks of cutting the crop, threshing it, separating it from the straw,

SCIENCE CONNECTION NEW CROPS FOR NEW USES

Courtesy of USDA/ARS #K4692-2. Photo by Jack Dykinga.

Checking lesquerella for seed set. New crops, such as lesquerella and wormwood, promise advances in fabrics, plastics, lubricants, corrosion inhibitors, cosmetics, and medicines.

As the medical profession scrambles to find new drugs and treatments for new and old medical problems, agriscience is becoming an increasingly important tool. At the Southern Weed Science Laboratory in Stoneville, Mississippi, scientists are developing a weed-control system to ensure the supply of a medicine for patients with malaria. Annual wormwood has been used for 2000 years in China to treat malaria. Now a new, refined drug derived from the plant is in demand worldwide. The World Health Organization is seeking ways to increase the production of wormwood sufficiently to meet the growing need.

Annual wormwood grows throughout the United States as a weed. The woody-stemmed, cone-shaped plant reaches a height of 4 to 8 feet and could become a cash crop for farmers in the future. Wormwood is harvested with a sickle-type machine similar to that used to cut sugarcane. Cut plants must be protected from prolonged exposure to the sun because ultraviolet light breaks down artemisinin—the antimalarial substance.

Artemisinin is found in the leaves of growing tips, and it takes 2.2 lbs. of the leaves to process 1 gram of artemisinin. Authorities indicate that it would take thousands of acres of wormwood to meet the world demand for the drug. However, before commercial production of wormwood could occur, farming methods, including weed control, must be developed. According to the World Health Organization (2012), there were about 219 million cases of malaria in 2010.

Another newly emerging crop is lesquerella. Lesquerella has been developed from *Lesquerella fendleri*, a wild plant native to Arizona, New Mexico, Texas, and Oklahoma. The University of Arizona and two commercial firms joined with the U.S. Department of Agriculture to develop the crop. Machinery used for other grain and oil seed crops is suitable for lesquerella.

Cosmetic manufacturers and other companies are creating demand for the oils extracted from lesquerella seed. The oils can be used in resins, waxes, nylons, plastics, high-performance lubricants, corrosion inhibitors, coatings, and cosmetics such as hand soap and lipstick. The oil meal that remains after the oils are extracted is suitable as a high-protein livestock feed.

Courtesy of USDA/ARS #K5287-4. Photo by Keith Weller.

FIGURE 21-32 Environmentally friendly, ultra-low-volume herbicide application methods can significantly reduce the amount of agricultural chemicals used.

Courtesy of DeVere Burton.

FIGURE 21-33 Late corn harvest using a combine near College Park, Maryland.

and cleaning it. Threshing refers to the separation of grain from the rest of the plant materials. Many types of combines have been adapted to the harvest of specific crops.

The proper storage of crops after harvesting is important. Threats to the quality of stored crops include heat, moisture, fungi, insects, and rodents (Figure 21-34).

Drying grain to reduce moisture and heat is important for successful long-term storage. Much grain is harvested with a moisture content that is much too high. If stored without drying, the grain may heat up and encourage the growth of

© Denton Rumsey/Shutterstock.com.

FIGURE 21-34 Modern weather- and rodent-proof grain storage facilities provide optimum storage and conditions to minimize insect and mold damage in stored grain.

fungi, causing the grain to spoil. Foreign materials, such as high-moisture weed seeds, may also cause stored crops to spoil.

Stored crops must be protected from insects and rodents if quality is to be maintained. Rodent droppings, hair, and urine, as well as insect parts, render crops unfit for human consumption. Reduction of food value and spoilage are other hazards of stored crops when insects and rodents are not controlled.

The production of field crops generates more income for U.S. agriculturists than any other production enterprise. More than 20 percent of the land in the United States is currently being used for growing crops, and we are fortunate to harvest much more than we need each year. The excess production contributes to a more favorable balance of trade. Crop production will surely remain our most vital national product.

STUDENT ACTIVITIES

1. Write the Terms to Know and their meanings in your notebook.
2. Compile a list of the field crops grown in your area.
3. Write a report on a field crop of interest to you.
4. Select a field crop and determine as many uses for it as possible.
5. Visit a local crop farm and talk to the operator about the advantages and disadvantages of growing a particular crop.
6. Prepare an advertisement to promote a field crop.
7. Construct a bulletin board about field crops and products made from them.
8. Visit a farm-machinery dealer. Make a list of all the equipment sold there that is used in the production and harvesting of field crops.
9. Make a collection of as many different field crops as you can.
10. Do a germination test on the seeds of several types of field crops. Observe the number of days required for germination to take place and the percentage of the seeds that germinate. Also conduct germination tests under a variety of environmental conditions, such as warm and cool, wet and dry, or with and without light. Compare the results to determine optimum conditions for the germination of crop seeds.
11. Make sketches of each of the field crops (and their seeds) provided by your teacher for identification. Label each drawing with its name. Write any notes that will help you identify the plant in the future. Review this activity from time to time to ensure that you remember how to identify each field crop.

SELF-EVALUATION

A. MULTIPLE CHOICE

1. Flax is an example of a/an _______ crop.
   a. oilseed
   b. grain
   c. sugar
   d. fiber
2. The most important grain crop in the world is
   a. corn.
   b. wheat.
   c. rice.
   d. barley.

3. The moldboard plow is the primary tillage machine for the _______ tillage system.
   a. no-till
   b. minimum
   c. conventional
   d. none of the above
4. About 65 percent of the refined sugar produced in the United States comes from
   a. sugarcane.
   b. sugar beets.
   c. corn.
   d. sorghum.
5. The use of airplanes is an example of _______ seeding.
   a. row crop
   b. drill
   c. aerial
   d. broadcast
6. Grasses grown for their edible seeds are _______ crops.
   a. grain
   b. legume
   c. oil
   d. sugar
7. Ginning is the process used to
   a. prepare a seedbed for planting.
   b. prepare crops for storage.
   c. remove seeds from cotton.
   d. prepare grain for use as alcohol.
8. An example of genetic control of pests is
   a. planting a crop when insects are not present.
   b. releasing sterile male insects.
   c. genetically altering a plant so that insects that eat it will be killed.
   d. planting a variety of a crop that grows more rapidly than do weeds.
9. Tobacco falls into the category of crops called
   a. fiber.
   b. biennial.
   c. thermo.
   d. stimulant.
10. Soybeans are grown for
   a. oil.
   b. hay.
   c. rain.
   d. all of the above.

B. MATCHING

_______ 1. Most important U.S. crop
_______ 2. Provides raw material for breads and pasta
_______ 3. Irrigation
_______ 4. Preparing barley for alcohol production
_______ 5. Primary harvesting machine
_______ 6. Most important world grain crop
_______ 7. Oil crop
_______ 8. Cloth made from flax
_______ 9. Warm-season crop
_______ 10. Separation of grain from plant

a. Malting
b. Linen
c. Combine
d. Sugarcane
e. Corn
f. Sprinkler
g. Peanut
h. Rice
i. Threshing
j. Wheat

C. COMPLETION

1. Heat, moisture, fungi, rodents, and insects are all problems to be dealt with in the _______ of crops.
2. _______ crops protect the soil from erosion.
3. The primary use of wheat is for _______.
4. Cotton is grown for fiber and _______.
5. Mowing is a means of _______ control for weeds.

UNIT 22

Forage and Pasture Management

OBJECTIVE

To determine the nature of and approved practices recommended for forage and pasture production and management.

MATERIALS LIST

- several specimens of forage and pasture crops
- quart jars for Student Activity
- bulletin board materials
- samples of lactic and butyric acids
- agriscience magazines
- Internet access

COMPETENCIES TO BE DEVELOPED

After studying this unit, you should be able to:

- define important terms used in forage and pasture production and management.
- identify major crops grown for forage and pasture.
- select varieties for forage and pasture.
- prepare proper seedbeds for forage and pasture crops.
- plant forage crops and renovate pastures.
- control pests in forage crops and pastures.
- harvest and store forage crops.

SUGGESTED CLASS ACTIVITIES

1. Invite a producer of forage crops to discuss the production of high-quality forages with the class. Prepare the class with specific questions to ask. Have each class member prepare written answers to each of these questions as a follow-up to the presentation.
2. Have class members identify each of the important forage crops by studying laminated specimens of each of the crops. Prepare these materials in advance of the class.
3. Invite a representative from a local or state cattlemans' or sheep producers' organization to come and discuss how range animals (ruminants) obtain adequate nutrition from grazing without being fed exact amounts of proteins, fats, and carbohydrates. Have the representative discuss the advantages and disadvantages of grazing animals on rangelands.

TERMS TO KNOW

hay
silage
pasture
nurse crop
overseeding
carrying capacity
tiller
haylage
silo

Forages are crop plants that are produced for their vegetative growth. Forage and pasture crops rank first in total acres among crops grown in the United States. There are more than 475 million acres of pasture and rangeland. Another 61 million acres are used to produce hay. The importance of forages is further emphasized when you consider that half of the pasture and rangeland in the United States is not suited for the production of cultivated field crops.

Forage production is divided into three general categories: hay, silage, and pasture. **Hay** is forage that has been cut and dried until its moisture content is reduced to safe storage levels. **Silage** is green, chopped forage that has been allowed to ferment in the absence of air. **Pasture** is forage that is harvested by livestock as they graze. Forages are generally planted or maintained with one of these uses in mind.

FORAGE AND PASTURE CROPS

Legumes

Alfalfa

The most important forage crop in the United States is alfalfa (Figure 22-1). It is often called the "queen of the forages." Alfalfa is a legume that adds nitrogen to the soil. It is high in protein and other nutrients and is productive in fertile soils (Figure 22-2). It is also one of the oldest cultivated forage crops; it was mentioned in the Bible and in other early writings.

When properly harvested and stored, alfalfa is the most economical source of nutrients for ruminant animals. With production of alfalfa and alfalfa mixtures accounting for nearly 40 percent of the hay produced in the United States, annual alfalfa production ranks third to corn and soybeans in dollar value.

The North Central states account for about two-thirds of the yearly alfalfa production in the United States. However, alfalfa can be grown in nearly every state. California ranks number one in alfalfa production, followed by Idaho, Wisconsin, Iowa, Minnesota, Nebraska, Kansas, and Montana.

Courtesy of DeVere Burton.

FIGURE 22-1 Alfalfa is a forage crop that is high in protein and other nutrients that are needed by livestock. It is the most important forage crop in the United States.

Courtesy of USDA/ARS.

FIGURE 22-2 Bacteria in the nodules on alfalfa roots are capable of converting nitrogen gas from the atmosphere into nitrates, a required plant nutrient.

Courtesy of Sharon Rounds.

FIGURE 22-3 Clover is an important hay and pasture crop in the United States.

True Clovers

True clovers include about 300 species; however, only about 25 have agricultural importance. Clovers of greatest economic importance in the United States include red clover, white clover, crimson clover, ladino clover, and alsike clover (Figure 22-3).

Clover ranks second among hay crops grown in this country. It is a legume with the ability to add nitrogen to the soil. Clover has the disadvantage of smaller yields than alfalfa under similar growing conditions. It is popular as hay, pasture, and silage and grows well in combination with many forage grasses. The Northeast and North Central states produce most of the clover grown in the United States.

Sweet Clover

Sweet clover is used most often in areas that are hot, drought stricken, or both, where its ability to survive and produce a crop is unsurpassed. It is used as hay, pasture, and sometimes as a green manure crop. Sweet clover is also used in Texas in rotation with cotton to help control a cotton-root disease. It is also an excellent source of nectar, which honeybees make into honey.

Three species of sweet clover are grown in the United States: biennial yellow, biennial white, and annual. Biennial white sweet clover yields more than the other species, although biennial yellow sweet clover is usually of better quality. Sweet clover will grow in all areas of the United States. Areas of large production include the Dakotas, Minnesota, Wisconsin, Michigan, and the Central states to Texas.

Courtesy of USDA/ARS #K2610-10.

FIGURE 22-4 Bird's-foot trefoil is a legume containing tannin, which reduces the danger of life-threatening bloat to grazing cattle.

Bird's-foot Trefoil

Bird's-foot trefoil is a comparatively new crop in the United States. It originated in Europe, where it has been a forage crop for 300 years. Approximately 2 million acres are in production in the United States. Bird's-foot trefoil is used as pasture in most cases, though some is also grown for hay (Figure 22-4).

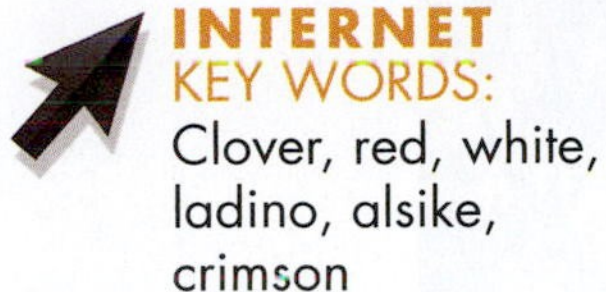

INTERNET KEY WORDS:
Clover, red, white, ladino, alsike, crimson

The food value of bird's-foot trefoil is about equal to that of alfalfa. Because of smaller yields, bird's-foot trefoil is unlikely to seriously challenge alfalfa in importance. It does have the advantage over true clovers in being longer-lived. States that report large acreages of bird's-foot trefoil include California, Ohio, Iowa, New York, and Pennsylvania.

Lespedeza

This legume is grown primarily in the South, where 1 to 2 million acres are harvested as pasture and hay each year. It can grow and thrive in soils that are low in fertility. Because lespedeza has a lower nutritive value than true clovers and alfalfa, it is recommended as feed for beef cattle but not for dairy cattle.

Types of lespedeza include annual and perennial varieties. Most of the lespedeza grown in the United States is of the annual type (Figure 22-5).

Peanut Hay

Peanut hay is a by-product of peanut production and is used in the southeastern region of the United States. Although it lacks the protein content of forage legumes, the addition of nitrogen compounds such as urea improves its quality. The extra nitrogen in the hay can be converted to protein by ruminant animals such as cattle and sheep.

Grasses

Bromegrass

Bromegrass is an important forage grass throughout the northern half of the United States. It is extremely hardy and grows to a height of 2 to 3 feet (Figure 22-6). Because bromegrass produces many rhizomes, thin stands rapidly

Courtesy of DeVere Burton.

FIGURE 22-5 Lespedeza is a forage plant that provides pasture and hay for cattle. Most of the roughage from lespedeza is fed to beef cattle.

Courtesy of DeVere Burton.

FIGURE 22-6 Bromegrass is grown throughout the northern half of the United States.

Courtesy of DeVere Burton.

FIGURE 22-7 Orchard grass is known for its rapid, early-spring growth.

thicken with age. Rhizomes are horizontal, underground stems from which new plants arise. Fertile soil is required for the production of bromegrass.

Orchard Grass

Orchard grass is known for its rapid germination and early-spring growth (Figure 22-7). It also recovers quickly after being harvested. Timing of the harvest of orchard grass is important because quality decreases rapidly after the plant reaches maturity. Orchard grass makes excellent pasture and is often used for hay in combination with legumes.

Timothy

Timothy is a cool-season grass that grows best when temperatures are between 65° and 72° F (Figure 22-8). The use of timothy as a forage grass has declined in recent years because bromegrass and orchard grass out-yield it in those areas where all three species are adapted. Most timothy is grown in the northeastern part of the United States.

INTERNET KEY WORDS:
timothy grass
orchard grass
reed canary grass
Kentucky bluegrass
fescue grass

Reed Canary Grass

In areas that are wet or poorly drained, reed canary grass is often the only answer to producing a forage crop (Figure 22-9). It can produce more than 4 tons of forage per acre in the cool, damp areas where it thrives. This grass grows as tall as 7 feet and can produce high-quality feed if harvested before reaching maturity. Most reed canary grass is grown in Washington, Oregon, California, Iowa, and Minnesota.

FIGURE 22-8 Timothy grows best when temperatures are between 65° and 72° F.

FIGURE 22-9 Reed canary grass grows in areas that are too wet for the production of other grasses.

INTERNET KEY WORDS:
Bermuda grass
Dallis grass
Johnson grass
corn silage

Kentucky Bluegrass

Kentucky bluegrass is raised over much of the United States, even though it is best adapted to the Northeast. It is the major grass of many pastures, lawns, and even golf courses in areas where summers remain moderately cool. Kentucky bluegrass is a good native, permanent pasture grass, but it is out-yielded by other grasses, and it is not practical to use for hay due to low yields. It also goes dormant during hot weather, making it necessary to use bluegrass in combination with other grasses such as ryegrass to have forages available all summer (Figure 22-10).

A

B

FIGURE 22-10 Kentucky Bluegrass is a cool-season grass. Its primary uses are (A) lawn and golf course turf when mowed regularly and (B) pasture and forage crops for livestock.

HOT TOPICS IN AGRISCIENCE RANGELAND FORAGE

Courtesy of DeVere Burton.

Range forage is a valued resource to owners of cattle and sheep. These ruminant animals are able to convert the forage to valuable meat for human consumption.

In many parts of the country, ranchers use rangelands to graze their animals. Some of these lands are privately owned, while the government owns others. Some people are opposed to grazing on public land, citing the fear that the animals will destroy the environment. However, most public rangeland is managed under the multiple-use system, allowing shared use of the resource. This includes grazing, mining, and recreation.

On rangelands, the forage is generally provided by nature, in the form of grass or edible wild forage. In native areas, forage management is much different compared with forage production in pasture settings. Animals must be carefully monitored to avoid overgrazing. Overgrazing occurs when animals are allowed to feed in an area for too long. The result is damage to the soil and loss of valuable native plants. When this occurs, the area may not produce enough forage in subsequent years. Ranchers who rely on rangelands as feed for their animals must do all that they can to maintain the health of rangelands.

Some of the common techniques that are used to prevent overgrazing are fencing, rest rotation, deferred rotation, supplementation, and herding. Fencing off areas of the rangeland helps keep the animals where they are supposed to be. Nutrient supplements such as salt are also used to entice animals to stay where they are wanted. Fencing also plays a part in both rest rotation and deferred rotation. Rangelands can be divided into pastures; for example, a rancher may divide his or her range into four different pastures. Rest rotation is when one pasture is not grazed for a given period. This gives plants a chance to complete at least one life cycle. Ideally, this should be done every other year. Deferred rotation occurs when the normal rotation between pastures is changed.

Range riders and herders are people who take care of range animals. They move them from place to place, making sure that the animals have access to sufficient forage and that the land and plant cover are preserved. When ranchers properly apply these methods, both the animals and the environment benefit.

Fescue

Tall fescue and meadow fescue are perennial grasses that are used for pasture and hay, usually in combination with other grasses and legumes. Fescue is best adapted to the Southeast, where 10 to 20 million acres are grown each year. The quality and palatability of the fescues are less than most other forage grasses.

Bermuda Grass

Bermuda grass is a warm-season grass that is dormant during cool weather. It is adapted to pastures and lawns because it grows only 6 to 12 inches tall. Common Bermuda grass is considered to be a weed in some areas because it tends to crowd out other grasses. However, improved varieties of Bermuda grass make good pastures in the Southeast (Figure 22-11).

Dallis Grass

Dallis grass grows 2 to 4 feet tall in the warm areas of the South. It cannot stand continuous use as pasture because it needs to be able to recover from close grazing. However, it is productive earlier in the spring than other warm-season grasses.

FIGURE 22-11 Improved varieties of Bermuda grass make good pastures in the Southeast.

Corn and Oats as Forages

Alfalfa and clovers are often planted with a nurse crop of oats. A **nurse crop** consists of plants that are planted with other plants that require protection during early stages of development. The oats protect the tiny legume plants during the early stages of growth. They are then cut for silage while the leaves are still nutritious and the grain is immature. Oats make excellent forage when they are harvested at this stage of maturity. The new alfalfa or clover crop responds to the reduction in competition, often producing a hay crop in the first growing season.

Corn is produced as a forage crop on many farms, where it is used for making silage. Corn silage is a highly nutritious feed for beef and dairy cattle, and high yields are possible when adequate water and fertilizer are provided. The corn plant is chopped and ensiled when the grain is in the full dent stage of maturity. The acids that are produced as the silage cures are highly digestible, and the nutritional value of the forage is maintained at a high level. It is a high-moisture feed; therefore, it is not recommended as a large part of the diet for young cattle. They are unable to eat enough silage to satisfy their nutritional needs. Dairy cows require dry hay in combination with silage to maintain high levels of milk production.

Selection of Forages

Many considerations must be made when selecting forages for hay, silage, and pasture, including the following:

- Intended use of the forage
- Expected yield

- Nutrient value of the crop
- Climatic conditions under which the forage will be grown—warm season versus cool season, summer and winter temperatures, humidity, soil type, anticipated rainfall/access to irrigation water, nutrient level, length of growing season
- Pest-control measures
- Methods of establishment
- Compatibility with other forages when grown in mixtures
- Expected life span of the crop
- Care and maintenance
- Equipment and labor necessary for growing, harvesting, and storing the crop

Forage crops that are adapted to the production of hay include alfalfa, clover, bromegrass, orchard grass, timothy, and fescue. Forages used for pastures are clover, lespedeza, Kentucky bluegrass, and Bermuda grass. Almost any legume or grass crop can be used for silage. In addition, corn and most small-grain crops make excellent silage when harvested when the grain is nearing maturity and before the leaves and stalks lose their nutritional value.

AGRI-PROFILE

CAREER AREAS: AGRONOMIST/PLANT PHYSIOLOGIST/ FORAGE MANAGER/RANGE MANAGER

Courtesy of USDA/ARS #K4264-8.

Summer employee Aimee Crago stains an alfalfa cotyledon for microscopic examination.

Forage and pasture provide a variety of career options, ranging from farm and ranch duties to plant breeding and physiology. Grains such as silage corn provide enormous volumes of feed for dairy and beef cattle. Crops harvested for silage, as well as those cut for hay, are numbered among the forages. Similarly, pasture and range plants are forages. Those who specialize in the science of forage growth and use are called agronomists.

Forage specialists are hired by universities and agricultural research centers as well as by large farms and businesses. In certain parts of the United States, hay businesses are on the increase as more people engage in part-time farming, especially those with pleasure horses. In addition, hay and straw are needed by large commercial dairies, beef feedlots, and racetracks, creating strong markets for these forages in many areas.

Hay business managers, truckers, and dealers also conduct thriving businesses distributing hay and straw. Large concentrations of dairy cattle and pleasure horses have created markets for cross-country shipping of hay, a practice once regarded as too expensive to be worthwhile. Large amounts of hay are also shipped during periods of drought.

Seedbed Preparation

In general, preparing a seedbed for forages is much the same as it is for the production of field crops. Residues from previous crops must be incorporated in the soil. The soil must be prepared for the planting of seed or vegetative pieces used in the propagation of some grasses. The soil also must be amended so that its pH and fertility are suitable for the germination and growth of the intended forage crop. With some crops, the moisture level of the seedbed may need to be regulated to ensure germination.

One difference between preparing a seedbed for most forages and one for field crops is that the texture of the soil in the seedbed must be finer and the seedbed somewhat firmer for forage crops than is necessary for most field crops. This is because most forage plants have very small seeds that have difficulty making firm contact with the soil in seedbeds unless the soil is finely textured.

Conventional tillage methods are used when preparing seedbeds for starting grass and legume crops for forage. The residues from previous crops are plowed down or shredded. The soil is then pulverized using modern tillage machinery, followed by smoothing and leveling. For very small seeds, further preparation may be necessary. Final preparation of the seedbed should occur immediately before planting the seed. New seeding equipment has permitted grass seeding into established sod. This greatly reduces soil losses and provides protective vegetation while the seedlings become established (Figure 22-12).

Planting Forages and Renovating Pastures

Forage crops are usually planted by drilling or by broadcasting. The same grain drills that are used for planting small grains are often equipped with metering devices for small seeds, allowing them to be used to plant forages. A separate seedbox on the drill is needed for this purpose. A nurse crop is used to protect another crop while it becomes established. Forage seed is often planted along with a grain crop. Many grasses are planted in this way. The grain crop germinates faster than the forage and acts as a nurse crop until the forage plants can become established.

Many forage crops are also planted separately, using either drills or broadcast planters. With some forages, the seed must be covered after it is broadcast onto the soil by broadcast planters. This is done to protect the seed from pests, such as birds, and to reduce the drying effects of sun and wind. Covering the seed

FIGURE 22-12 The no-till grassland drill places seed, fertilizer, and pesticides in one pass across the field.

Courtesy of DeVere Burton.

FIGURE 22-13 The packing wheels of this modern small seed drill planter create a firm seedbed, bringing the seeds in close contact with the soil particles.

may be accomplished by a light harrowing of the soil surface or by the use of the corrugated wheels of a cultipacker, which presses the seed into the seedbed (Figure 22-13). Some types of forage seeds are broadcast into growing crops. They must be able to germinate fairly easily because contact with the seedbed is often minimal. Red clover may be overseeded into stands of forage grasses to make mixed hay. **Overseeding** is the practice of seeding a second crop into one that is already growing. This is usually done during late winter and early spring, when freezing and thawing of the soil help provide contact between the seedbed and the seeds.

Forages may be planted with no-till planters through crop residues. The no-till planter opens a narrow furrow in which the seed is planted. Advantages include fewer trips across the field with heavy tractors and tillage equipment. This means less compaction of the soil and little exposure of the seedbed to erosion. Care must be taken to control pests in no-till planting. Another concern is to minimize competition for water and nutrients from other plants growing in the seedbed.

INTERNET KEY WORDS:
forage pest control

There are several methods of renovating pastures in the United States. The existing pasture can be killed with an herbicide; the soil worked down to prepare a fine seedbed; and the pasture then reseeded with the desired types of grasses or legumes. The pasture is usually treated for insects and diseases at this time if necessary. It is also an ideal time to apply fertilizers and lime to the soil.

Another method of renovating a pasture involves using selective herbicides to kill unwanted species of plants. The pasture is then tilled to break up the existing sod. This allows easier entry of moisture and nutrients into the soil. The pasture may or may not be overseeded with desirable species to improve yield. It is also fertilized and limed at this time, according to soil test recommendations.

No-till grassland seeders may be used to place new seed in existing sod, which permits the thickening of the existing plant population. These seeders can also introduce new, improved, and aggressive species or varieties of forage crops.

FIGURE 22-14
Grasshoppers are serious pests in forage crops because they eat the leaves and growing tips of the plants. This grasshopper represents only one of more than 600 species of grasshoppers in the United States.

Pest Control in Forage Crops

The control of weeds, insects, diseases, and rodents in forage crops helps result in optimum yields. Because many forage crops grow for more than one growing season, pest control is usually an ongoing part of forage crop management (Figure 22-14).

The proper identification of the pest or pests affecting a crop is important. To that end, personal experience and the advice of trained professionals are often necessary.

The actual methods of pest control are many, and they are discussed in previous units. Chemicals, cultural practices, biological control, genetic control, and the timing of crops are all tools to be used in controlling the pests of forage crops. Not to be overlooked in pest control management is the control of pests in stored forages. When chemicals are used to control pests, care must be taken to ensure that the pesticide is properly applied at the recommended rates. Timing the application of the pesticide for maximum effect is also essential. Care must also be taken to ensure that overspray and runoff of pesticide materials do not adversely affect organisms other than the intended pest. Measures to ensure that pesticide residues do not end up in food sources are of utmost importance.

Harvesting and Storage of Forage Crops

The proper harvesting and storage of forage crops is extremely important in the management of the forage enterprise. For maximum profits, special care must be taken to control moisture levels in the harvested crop. It is also important for the forage to be removed from the field as soon as possible. Once it is in storage, forage quality is much easier to maintain.

Too much moisture in ensiled crops causes nutrients to be leached away in the liquids that drain from the storage area. Too much moisture in hay causes it to become moldy, sometimes resulting in fires ignited by spontaneous combustion due to a buildup of heat.

Pastures

The harvesting of pastures involves several factors. One of the most important is that the **carrying capacity** of the pasture must be determined to figure out how many animals can be fed by the pasture that is available. The carrying capacity of a pasture equals the number of animals the pasture is capable of feeding.

Pastures also need time to recover from the ravages of the animals that graze them. The rotation of the animals using the pastures is important if the pastures are to meet their potential yields. Pasture rotation also helps control parasites of the animals grazing on them by breaking the pests' life cycles.

Hay

Harvesting hay involves several operations that must be accurately timed if high-quality hay is to be produced. The hay crop must be cut at the optimum time, with an eye on the weather forecast for the next several days. Hay is normally dried in the field by the sun. Nothing ruins hay faster than rain on the crop after

HOT TOPICS IN AGRISCIENCE PRODUCING HIGH-PROTEIN ALFALFA HAY

The alfalfa plant is a great forage plant because it is a succulent with a lot of leaves. Just as it begins to bloom, it reaches the peak of quality. If it is harvested any sooner, valuable tonnage is lost. Later harvesting results in a smaller percentage of protein. At just about the same time, young **tillers**, or shoots, begin to grow at the base of the plant. These tillers are the new stems for the next hay crop. If the harvest is delayed beyond this stage of maturity, the plant goes into seed production. The stems become thickened, and the proportion of stems to leaves will increase. This reduces the percentage of protein in the hay.

High-quality alfalfa hay can contain more than 20 percent protein when it is harvested at the proper time. Hay of high quality commands a premium price as dairy hay. Cows that are fed top-quality, high-protein hay have the potential to maintain a high level of milk production. Only the best-quality alfalfa hay should be fed to dairy cows.

Another key element in producing high-quality hay is to process the hay immediately and move it into a covered stack to protect it against rain. Hay that is rained on between cutting and baling never is high-quality hay. This is because nutrients and color are leached out, and leaves are usually lost during the extra days that are required to get it dry enough to bale.

Most of the forages grown on U.S. farms and ranches are fed to livestock maintained on the same farm. Forages are usually the least expensive sources of nutrients available for cattle, horses, and sheep.

Type of Hay	When to Harvest
Alfalfa	Pre-bud to 1/10 bloom stage
Clover	1/4 to 1/2 bloom stage
Bird's-foot trefoil	1/4 bloom stage
Sweet clover	Start of the bloom stage
Brome grass	Medium head stage
Timothy	Boot stage to early bloom stage
Lespedeza	Early bloom stage
Orchard grass	Full head but before blooming
Reed canary grass	When the first head appears

FIGURE 22-15 Maturity levels at which selected forages should be harvested for hay.

it starts to dry and cure. The maturity of the forage being harvested for hay is also critical (Figure 22-15). The more mature the forage becomes, the lower will be the quality and nutritional value of the hay.

Terms that describe forage crop maturity include boot stage (immature seed development), pre-bud (early stage of flower formation), and 1/10 bloom (1 in 10 flower buds have opened).

Most hay is cut with either a sickle-bar mower or a rotary mower. The sickle-bar mower cuts with the same action as a pair of scissors. These types of mowers are best adapted for forages that are standing upright.

The rotary mower cuts with blades that move in a rapid circular motion parallel to the ground. This type of mower will cut any type of forage, although greater amounts of horsepower are required when the stands of forage are extremely thick and heavy.

A swather, also known as a mower/conditioner, is the most commonly used machine for cutting hay. It performs two processes in a single pass through the field. It cuts the forage plant off just above the ground, and it places the cut forage in a windrow for convenience in the baling or chopping operations.

Crushing and mashing of the stems is also performed by conditioners that are built into most swathers to reduce the length of the drying time. In areas where rain is frequent, the hay may have to be cut with a mowing machine and allowed to lie in the swath where it falls. This spreads the hay sufficiently for the sun to dry it quickly. In areas of high humidity, chemicals to speed the drying of the forage are sometimes sprayed on the forage as it is cut.

Hay is often raked at least once before it is baled to allow it to dry more evenly and more rapidly. Legume-type hay should be raked during the part of the day when humidity is highest, to keep the loss of leaves to a minimum. Too much raking reduces hay quality because the high-protein leaves fall off. Raking hay also puts it into windrows so that it can be baled.

© iStockphoto/Cameron Pashak.

FIGURE 22-16 Hay is harvested by mowing, conditioning, air drying, baling, and hauling to storage areas.

When hay has dried to the desired moisture level, it is baled or chopped and removed from the field. In some cases, it may be cubed or pelleted, particularly if it is being exported or transported long distances. Most hay is baled and hauled with automated hay wagons that pick up the bales from the fields and transport them to the storage area, where they are stacked in place.

Most hay is harvested by forming it into square or rectangular bales or into round bales of various sizes (Figure 22-16). The baler gathers the dried hay from the windrow, compresses and shapes it into a bale, and ties twine or wire around the bale to hold it together, allowing for handling. The bale is then expelled from the baler. It is dropped on the ground for later transfer to storage. A recent trend has been to bale hay with balers that make large bales of approximately 1 ton each. These bales are preferred for shipping long distances (Figure 22-17).

A number of machines are designed to move baled hay from the field to the storage site. These include trucks, wagons, bale loaders, bale handlers, and bale stackers (Figure 22-18).

Hay may also be harvested with machines called stack wagons. These machines gather forages from the field and form dense stacks or loaves of weather-resistant hay. With special handling equipment, stacks can be moved with few problems.

Hay cubes were developed to allow for the full mechanization of the feeding of hay. The cubing machine gathers the dry forage and compresses it into cubes about 1.25 × 1.25 × 2 to 3 inches. Normally, only legumes are cubed because of difficulty in getting grass cubes to bind together. Artificial binders are sometimes used.

Hay may be stored in buildings or in the open field. Regardless of the method of storage, care must be taken to ensure that quality is maintained. Hay that is stored in buildings must be kept free from pests, particularly rodents. Care must also be taken to ensure that the moisture level of stored hay is low enough that spoilage does not occur.

When hay is stored outdoors, it should be placed on land that is well drained and sloped to drain water away from the hay. Stacks of baled hay stored

Courtesy of DeVere Burton.

FIGURE 22-17 Large, rectangular bales are preferred for shipping hay by truck because they can be loaded by machine, they are easier to secure on the truck, and the load is not easily shifted.

Courtesy of USDA.

FIGURE 22-18 Bale stackers have done much to replace manual labor with machine labor. One person operating a bale wagon can haul more hay in a day than a crew of four using flatbed wagons.

FIGURE 22-19 Protective covers prevent baled hay from spoiling in the stack due to moisture entering the baled hay.

outside should be covered with plastic or other materials so that little moisture can penetrate them and to protect the hay from adverse weather conditions (Figure 22-19).

Silage

INTERNET KEY WORDS:
alfalfa hay harvest
forage, silage

Harvesting forages for silage is easier than making hay. Less equipment and labor are required, and the entire harvesting operation is usually completed in minimal passes over the field. A forage harvester cuts the green forage, chops it into small pieces, and deposits it into a wagon or truck to be hauled to the storage facility. In instances where the crop is unusually succulent, the forage may need to be cut and allowed to dry for a few hours in the field to reduce the moisture content before it is ensiled. Forages used for **haylage** are cut and allowed to dry to a moisture content of 40 to 55 percent before they are placed in a silo. A **silo** is an airtight storage facility for silage or haylage. Haylage has the advantage of being lower in moisture than silage. This results in a product with enough dry matter to allow it to be fed to young cattle, and it requires less storage space than silage.

Silage may be stored in silos or in trenches or bunkers. It may also be stored in piles on top of the ground and sealed with plastic or other materials to make it airtight. The production of silage is dependent on sealed storage in the absence of air or oxygen. The fermentation that occurs in the absence of air preserves the silage. Spoilage occurs when green forages are stored in the presence of air (Figure 22-20). The chopped forage must be tightly packed to remove the air.

The Fermentation Process

Silage-making is a fermentation process whereby bacteria break down sugars and starches to produce the acids that preserve the forage. Silage-making and silage-keeping require that several conditions be met and maintained. First,

SCIENCE CONNECTION KENAF: SUPER FORAGE?

Kenaf, a bamboo-like plant with high-protein forage, shows promise to become a super foliage crop of the future.

Kenaf was viewed for several years as a possible new source of fiber for newsprint, rope, and other fiber-based products. However, it appears that its best use may be as a roughage and protein source for sheep and cattle. Research at the Agricultural Research Service Forage and Livestock unit at El Reno, Oklahoma, indicates that animals will eat the leaves and stems, and the digestibility of certain parts of the plant is about the same as that of alfalfa. This combination of high fiber and high protein has attracted interest in the plant as cubes or pellets for cattle. The leaves are of growing interest as a small-animal feed.

Kenaf grows rapidly, and a 30-day harvest rotation seems to work best. Kenaf leaves contain 20 to 29 percent crude protein when harvested at this maturity stage. Stalks must be cut at least eight inches above the ground, however, to ensure good survival. When cut, two or more shoots spring from each stalk to produce new plants. The potential is to grow multiple forage crops in a single season. Being an annual, the crop is easy to establish and has few or no insect and disease pests. Production in the United States is increasing as research continues on kenaf as a forage crop.

the crop must be chopped into short pieces and have sufficient moisture to pack tightly enough to exclude most of the air. Second, the crop must contain a sufficient amount of sugars and starches for the fermentation process to occur. Third, the silo or silage container must be airtight so air cannot reenter the silage after it has fermented.

INTERNET KEY WORDS: forage kenaf

FIGURE 22-20 Forage silos must restrict the movement of air to preserve and store silage.

What happens in the silo to change the taste, texture, and chemical content of the forage? If the ensiled forage contains grain, such as corn, sorghum, wheat, barley, oats, rye, or other seeds, it will ferment when packed in an airtight container or silo with about 60 percent or more moisture. If the mixture does not have sufficient sugars and starches, or if the moisture content is not within the acceptable range, then grains or preservatives must be added to the forage to make good silage.

When the moisture content is less than 60 percent and the silo is sufficiently tight, a low-moisture silage known as haylage may be made with the addition of preservatives.

Immediately after suitable forage has been chopped and packed in an airtight space, bacteria that live in the presence of oxygen in the air that fills the pores of the forage start to multiply. They feed on the sugars and starches in the forage and consume the oxygen in the process. They feed and multiply until the oxygen is consumed, so long as there are sufficient sugars and starches in the forage to keep them feeding. The bacteria make lactic and acetic acids, which smell good and give the forage a pleasant taste for livestock.

The frantic burst of bacterial activity in freshly packed forage causes the temperature of the forage to increase within hours after the process starts. This continues until the oxygen is consumed and the bacteria die. The resulting level of acid in the forage, now called silage, prevents any other bacterial action so long as oxygen-laden air does not reenter the silage. Without bacterial action, the silage cools down and is preserved until air enters the silage again. In practice, air may get into silage and cause spoilage around poorly fitted silo doors and holes left in the plastic film used to seal silos. Air movement through poorly packed or excessively dry silage also leads to spoilage. Spoiled silage can be detected by the presence of mold or the foul odor of butyric acid, which forms when undesirable fermentation takes place.

Approved Practices for Forage Crops

Many crops are usable as forages. Forages may be in the form of grazing of young plants by livestock, cut and dried as hay, or harvested and stored in silos as medium- or high-moisture haylage or silage. Some approved practices for the production of forages are as follows:

- Select a perennial species for long-term stands.
- Select a perennial or annual species for short-term stands.
- Plant certified seed of the selected variety.
- Obtain a current soil test.
- Apply lime and fertilizer according to the soil test results and desired production.
- Select a field with soil suitable for your forage choice.
- Select the proper tillage method to prepare a firm and smooth seedbed.
- Calibrate the drill for a proper plant population.
- Match the plant population to the soil-yield potential.
- Adjust the drill for proper depth of seed placement.
- Decide if forage may best be grown with a companion crop.
- Calibrate the sprayer for proper application of materials.

- Use selected herbicides for controlling problem weeds.
- Use insecticides as necessary.
- Make sure chemicals have been approved on forage crops.
- Decide the stage of maturity to harvest the crop.
- Make a yield check.
- Keep harvesting equipment in good repair to save downtime.
- Check the moisture content of harvested forage.
- Harvest the crop as soon as the moisture level permits.
- Store hay in a clean, dry structure.
- Store silage in an airtight structure.
- Obtain a forage analysis.
- Track market trends and habits.
- Market the crop at the optimum time.
- Keep accurate enterprise records.
- Summarize and analyze records.

The production of forages in the United States accounts for about half of all the land in agricultural use. Much of this land is unsuitable for production of other crops. With the production of hay ranking third only to corn and soybeans, forages are extremely important to U.S. agriculture. Because forages are the least expensive sources of nutrients for cattle, sheep, and horses, they are likely to remain essential for years to come.

STUDENT ACTIVITIES

1. Write the Terms to Know and their meanings in your notebook.
2. Put fresh-cut forage in each of two quart jars. Pack the forage tightly into the jars. Seal one jar and leave the other open to the air. Compare the contents of the two jars after 10 days.
3. Make a collection of as many different forage seeds as you can.
4. Write a report on a forage crop of economic importance in your area.
5. Make a collection of local forages.
6. Visit a machinery dealership to learn about the types of seedbed-preparation, planting, and harvesting equipment for forages.
7. Make a bulletin board about forages.
8. Obtain samples of hay and silage, and compare them for firmness of leaves and stems, odor, and color.
9. Ask your instructor to obtain samples of lactic acid (found in good silage) and butyric acid (found in spoiled silage) from the science laboratory, and compare the odors produced by these acids. Do not smell the vapors of these acids directly from the container.
10. A cow eats about 90 lbs. of feed each day and drinks 35 to 50 gallons of water. Hay makes up about 30 lbs. of its total daily feed intake. Determine how many pounds of hay a cow will eat in one year in a confined feedlot with no supplemental grazing.
11. Determine the cost for 1 ton of hay in your area by calling an area hay farmer. Determine how much it will cost for hay in the ration of one cow for one year, using the answer you calculated in Activity 10.

SELF-EVALUATION

A. MULTIPLE CHOICE

1. The type of mower that cuts with a scissors-type action is a _______ mower.
   a. rotary
   b. flail
   c. sickle-bar
   d. vibrating
2. A silo is used to store
   a. silage.
   b. hay.
   c. pasture.
   d. cubes.
3. The most important legume hay in the United States is
   a. clover.
   b. lespedeza.
   c. bird's-foot trefoil.
   d. alfalfa.
4. A forage that grows in wet or poorly drained soil is
   a. reed canary grass.
   b. timothy.
   c. fescue.
   d. bird's-foot trefoil.
5. _______ is suitable only for pasture and lawns because it grows only 6 to 12 inches high.
   a. Dallis grass
   b. Sweet clover
   c. Kentucky bluegrass
   d. Orchard grass
6. Forage that is cut and allowed to dry to 40 to 55 percent moisture before storing is called
   a. silage.
   b. haylage.
   c. hay.
   d. cubes.
7. The harvesting of _______ requires the most labor of any of the forages.
   a. pasture
   b. hay
   c. silage
   d. haylage
8. The production of forages in the United States equals _______ percent of the land used for crops.
   a. 25
   b. 35
   c. 50
   d. 65
9. A warm-season grass is
   a. sweet clover.
   b. bromegrass.
   c. reed canary grass.
   d. Bermuda grass.
10. Hay is raked to
    a. allow even drying.
    b. allow the hay to ferment.
    c. reduce its maturity.
    d. remove leaves and weeds.

B. MATCHING

________	1. Timothy	a. Number one grass hay crop
________	2. Alfalfa	b. Excellent source of nectar
________	3. Bromegrass	c. Grows very early in the spring
________	4. Sweet clover	d. Number one forage crop
________	5. Orchard grass	e. Popular in mixes with clover
________	6. Lactic acid	f. Indicates silage is spoiled
________	7. Butyric acid	g. Gives a good smell to silage

C. COMPLETION

1. A disadvantage of _______ mowers is that they tend to chop up the forage.
2. Sweet clover types include annual and _______.
3. Seedbeds for forages should be _______ than those for most other crops.
4. _______ is forage harvested by livestock.
5. Horizontal underground stems are called _______.
6. A _______ weed is prohibited or banned by state law.
7. Crops that are harvested for their vegetative growth are called _______.
8. Two methods of planting forages are _______ and broadcasting.
9. A _______ gathers dry forage, compresses it, and ties it with twine and wire.
10. The more mature the forage, the _______ its quality.
11. _______ converts sugars and starches to acids in the fermentation process.
12. Silage will spoil if _______ is permitted to reenter.

SECTION 7

ORNAMENTAL PLANTS—AN INDICATOR OF ECONOMIC HEALTH

When one considers the fastest growing segment of U.S. agriculture, the image of great fields of golden grain, corn "as high as an elephant's eye," or feedlots of cattle as far as the eye can see come to mind. Many would not envision acres of greenhouses filled with flowers, golf courses with gorgeous turf, or suburban communities of beautifully landscaped homes. But flowers, potted plants, and landscape trees and shrubs are some of the hottest agricultural products during periods of economic prosperity. They not only beautify our surroundings but they also help purify the air.

A conservative estimate puts the worth of the floral and landscape industries in the United States at about $17 billion at the grower level, according to the U.S. Department of Agriculture (USDA) Economic Research Service. The value of production has grown at 5 percent per year nearly every year since 1989. In 1992, flowers and house and nursery plants advanced into the sixth largest commodity group in the United States. In 21 states, this segment placed among the top five agricultural commodities.

People tend to take our decorative plants for granted and assume they are simply gifts from nature. This is far from reality, however. Years of painstaking research and development have gone into most ornamental plants, and research must continue on an ongoing basis to stay ahead of pests and the public appetite for new decorative plants. For the most part, we seldom see plants from the wild—nearly all the plants in common use are products of plant-improvement efforts.

Consider the poinsettia, for example. Research and genetic improvements contributed greatly to this plant's phenomenal growth in commercial value during the last two decades. The poinsettia industry grew from a wholesale value of $38 million in 1976 to an economic value of approximately $1 billion in 2011. The poinsettia has become the number-one potted plant, based on numbers sold in the United States, even though its annual sales period is only 6 weeks. About 118 million pots are marketed annually.

Most of the USDA's work on the poinsettia and other ornamentals is centered at the National Arboretum in Washington, DC, and the Agricultural Research Service's Florist and Nursery Crops Laboratory in nearby Beltsville, Maryland. Scientists there have developed many new ornamentals and are constantly improving others for commercial use. An ongoing project investigates the biochemical and physical bases for color in plants and our perception of it. Such information may help make development of new kinds of flowers easier and may provide a scientific basis for variety identification.

Applied research answers immediate needs of growers, such as solving pressing problems occurring in today's commercial ornamental enterprises. However, the long-range research goals are to stay well ahead of the current technology. From all indications, the ornamental use of plants will play an increasing role in the health and well-being of our citizens as we all pursue the great American Dream.

Courtesy of DeVere Burton

UNIT 23

Indoor Plants

OBJECTIVE

To use plants indoors for beautification, air quality, and pollution control.

MATERIALS LIST

- indoor plant reference books
- seed catalogs
- pencil, grid paper, and eraser
- indoor plants
- pots of various sizes
- Internet access

COMPETENCIES TO BE DEVELOPED

After studying this unit, you should be able to:

- identify plants that grow well indoors.
- select plants for various indoor uses.
- grow indoor foliage plants.
- grow indoor flowering plants.
- describe elements of design for indoor plantscapes.
- describe career opportunities in indoor plantscaping.

SUGGESTED CLASS ACTIVITIES

1. Go as a class on a field trip to a local nursery, floral shop, or the garden section of a large retail store. Observe the variety of plants that are available, and then ask the manager which plant is the most important and profitable product. Observe the tools and materials that are available for working with ornamental plants. Have each class member write a summary of the important things that were observed and discussed during the field trip.
2. Conduct an Internet search for such key words as floriculture, research, statistics, decorative plants, and ornamental plants. The class may suggest other key words that will identify sites for learning about the floral industry. Print and discuss key articles that are found during the Internet search.
3. Many indoor plants stop growing or begin to look unhealthy when they are kept in a container that is too small. Demonstrate the proper way to repot a plant. If the plant is a variety that can be divided easily into three or four individual plants, demonstrate this technique as well.

TERMS TO KNOW

floriculture
succulent
variegated
herb
foot-candle
phototropism
rootbound
plantscaping
basic color
accent color
texture
accent
sequence
balance
formal
symmetrical
informal
asymmetrical
scale

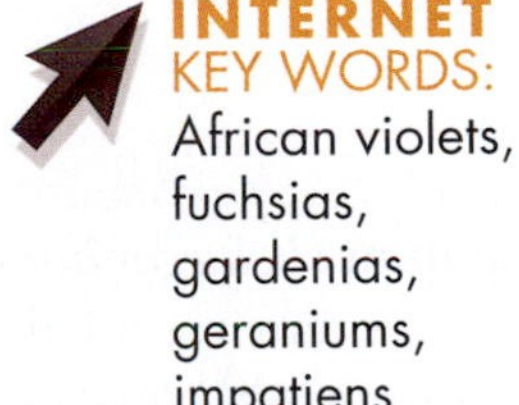

INTERNET KEY WORDS:
African violets, fuchsias, gardenias, geraniums, impatiens

The world of indoor plants is a fascinating one. There are plants that have magnificent blooms, unusual shapes, fancy foliage, and fragrant smells. If selected and cared for properly, they can last for years and may even be passed from generation to generation. The industry of indoor plants is the floriculture part of ornamental horticulture. **Floriculture** involves the production and distribution of cut flowers, potted plants, greenery, and flowering herbaceous plants. Indoor plants present a challenge to the homeowner who wishes to add the living color of plants to the rooms of his or her home. They also challenge those who decorate public areas (Figure 23-1).

PLANTS THAT GROW INDOORS

Almost all plants can be grown indoors. There are plants, however, that favor indoor conditions. Many of the trees and shrubs do better when grown outdoors. Small succulent plants are best for indoor use. **Succulent** means having thick, fleshy leaves or stems that store moisture. A wide variety of shapes and sizes of indoor plants are available. These plants can be divided into two major groups—those that flower and those that are grown only for their foliage. Foliage consists of stems and leaves.

Popular and Common Indoor Flowering Plants

African Violets

One of the most popular and common indoor flowering plants is the African violet (*Saintpaulia ionantha*). This plant can be recognized by its small size and hairy leaves. The leaves are oval-shaped, dark green, and covered with soft, short hairs. The flowers contain four to five petals arranged in a clover pattern. They vary in color from deep purple to brilliant white (Figure 23-2).

FIGURE 23-1 Indoor plants add beauty and atmosphere to homes and public areas.

FIGURE 23-2 African violet (*Saintpaulia ionantha*).

FIGURE 23-3 Fuchsia (*Fuchsia triphylla*).

FIGURE 23-4 Gardenia (*Gardenia jasminoides*).

Fuchsias

Fuchsias (*Fuchsia triphylla*) are plants with colorful flowers that cascade from the plant. Most have flowers that are two-tone pinks and reds. The foliage is dark green, and the leaves tend to be long and oval in shape, with a bronzy hint of color (Figure 23-3).

Gardenias

Gardenias (*Gardenia jasminoides*) are particularly fragrant, flowering, indoor plants that have deep-green shiny foliage and pure white flowers. The leaves are in clusters of three and are pointed (Figure 23-4).

Geraniums

One of the most versatile flowering indoor plants is the geranium (*Pelargonium zonale*). These indoor plants are among the oldest. The leaves are rounded, yellowish green with scalloped edges. The flower is borne on a stem and consists of many petals in a cluster shaped like a ball. The flower color ranges from the most popular red, to white and pink (Figure 23-5).

FIGURE 23-5 Geranium (*Pelargonium zonale*).

Courtesy of H. Edward Reiley.

FIGURE 23-6 Impatiens (*Impatiens wallerana*).

INTERNET KEY WORDS:
asparagus fern
emerald fern
foxtail fern

Impatiens

If an indoor plant with many blooms is desired, the impatiens (*Impatiens wallerana*, or *Impatiens sultanii*) is a good choice. The flowers are small and rounded, with five petals. One petal is shaped like a tube that protrudes from the underside of the flower. Flower colors include white, pink, salmon, coral, lavender, purple, and red. The leaves are lance-shaped and have succulent stems (Figure 23-6).

AGRI-PROFILE

NUMBER ONE POTTED PLANT: POINSETTIA

© Scott Prokop/Shutterstock.com.

The sales leader for potted plants in the United States is the poinsettia.

The poinsettia (*Euphorbia pulcherrima)*, which is native to Mexico, was introduced to the United States in 1825. Nearly two centuries later, it is the most popular potted plant sold in this country. This festive decorative plant was reported by North Carolina State University to have a $1 billion annual economic impact. The poinsettia far outsells other potted plants and is typically only sold in November and December. The brilliant red leaves are often confused for flowers. However, the red petal-like foliage is actually a form of specialized leaves called bracts. Its true flowers are small and unassuming.

Additional Indoor Flowering Plants

Many other varieties of plants can be grown indoors for their flowers. Some flowering plants are both indoor and outdoor plants, such as wax begonias, ageratums, verbenas, and petunias. Any other flowering plant that is capable of withstanding the rigors of an indoor environment may be considered here.

Popular and Common Indoor Foliage Plants

Indoor foliage plants can be divided into five groups for easy identification. These groups are ferns, indoor trees, vines, cacti/succulents, and specimen plants. A foliage plant is grown for the appearance of the leaves and stems (Figure 23-7).

FIGURE 23-7 Business and public areas with foliage plants are attractive and appealing.

Ferns

Ferns come in a variety of types. Some of the most popular indoor ferns are Boston fern (*Nephrolepis exaltata*) (Figure 23-8), asparagus fern (*Asparagus sprengeri*; Figure 23-9), maidenhair fern (*Adiantum capillusveneris*), sword fern (*Nephrolepsis cordifolia*), rabbit's foot fern (*Davallia canariensis*), and staghorn fern (*Platycerium biforcatum*). Ferns are categorized by their long and often multicut leaves. Most ferns are feathery in appearance. Some ferns are used extensively as greens in floral arrangements.

Indoor Trees

An indoor tree can be an excellent accent to a room or an attractive addition to a patio or hallway. Trees can grow to be 6 to 7 feet tall or higher indoors. Some

FIGURE 23-8 Boston fern (*Nephrolepis exaltata*).

FIGURE 23-9 Asparagus fern (*Asparagus sprengeri*).

of the more popular indoor trees are Norfolk Island pine (*Araucaria excelsa*; Figure 23-10), fiddleleaf fig (*Ficus lyrata*), umbrella plant (*Schefflera actinophylla*; Figure 23-11), rubber plant (*Ficus elastica*; Figure 23-12), fragrant dracaena (*Dracaena fragrans*), weeping fig (*Ficus benjamina*; Figure 23-13), and croton (*Codiaeum variegatum*). Indoor trees vary in the type and size of their foliage, but all have woody-type stems.

Vines

Vines are characterized by their habit of climbing or draping from the sides of the pot. Some of the more widely recognized vine-type indoor plants are philodendrons (*Philodendron* sp.; Figure 23-14), wandering Jew (*Tradescantia fluminensis* "Variegata"; Figure 23-15), grape ivy (*Rhoicissus rhomboidea*), and English ivies

FIGURE 23-10 Norfolk Island pine (*Araucaria excelsa*).

FIGURE 23-11 Umbrella (*Schefflera actinophylla*).

FIGURE 23-12 Rubber plant (*Ficus elastica*).

Courtesy of H. Edward Reiley.

FIGURE 23-13 Weeping fig (*Ficus benjamina*).

Courtesy of H. Edward Reiley.

FIGURE 23-14 Philodendron (*Philodendron sp.*).

© Stephen VanHorn/Shutterstock.com.

FIGURE 23-15 Wandering Jew (*Tradescantia fluminensis*).

(*Hedera* sp.). These plants can be trained to climb up a piece of wood, around a planter box, or up a wall.

Cacti and Succulents

Cacti and succulents are a unique group of indoor plants that originated in desert areas. They are among the easiest plants to grow indoors. These plants tend to hold water within their stems and leaves. Cacti and succulents survive dry heat, low humidity, and varying temperatures. Cacti usually have some type of prickly needles, whereas succulents do not. One of the most popular succulents is the jade plant (*Crassula arborescens*; Figure 23-16).

Specimen Plants

There are numerous other types of indoor plants that do not fit into the categories of ferns, indoor trees, vines, and cacti and succulents. These indoor plants

Courtesy of H. Edward Reiley.

FIGURE 23-16 Jade plant (*Crassula arborescens*).

FIGURE 23-17 Peperomia (*Peperomia caperata*).

FIGURE 23-18 Passion plant (*Gynura aurantica*).

FIGURE 23-19 Snake plant (*Sansevieria trifasciata*).

tend to have special characteristics or features that people like to display; thus, they are called specimen plants. A few examples of more popular specimens are the spider plant (*Chlorophytum elatum vittatum*), which shoots off baby plants or "spiders"; peperomia (*Peperomia caperata*), which has pale pink to red stems with deeply grooved heart-shaped leaves (Figure 23-17); purple passion plant (*Gynura aurantica*; Figure 23-18), which has rich royal purple leaves that are covered with velvet-type hair; bromeliads, one variety of which has a deep pink color at the center; and snake plant (*Sansevieria trifasciata*), which has long, spike-like, thick leaves that are variegated with gold (Figure 23-19). **Variegated** means having streaks, marks, or patches of color.

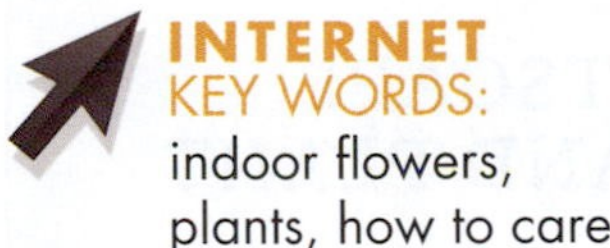

INTERNET KEY WORDS:
indoor flowers, plants, how to care

SELECTING PLANTS FOR INDOOR USE

Two rules of thumb should be followed when selecting plants for use indoors. The first is to be selective when you purchase the plant. The second is to choose the right plant for the growing conditions available in the location you want the plant to live. Careful attention to these two rules greatly increases the chance for success with indoor plants.

Purchasing an Indoor Plant

When buying an indoor plant, look for one that appears to be healthy. Look closely at the plant for insects, being careful to check the undersides of the leaves where many insects hide or lay their eggs. Select plants with even green color, no yellowing leaves, and without any spots or blotches on the leaves. Avoid plants that have spindly growth or appear wilted. If possible, purchase the plant during its growing season. Finally, look for new growth, such as leaf or flower buds.

Choosing the Right Plant

Before selecting an indoor plant, decide where the plant will be placed. Note particularly the light intensity and duration, as well as the temperature of the location. Each species of indoor plant has specific conditions for optimum growth. The plant should match the location for best results.

Courtesy of H. Edward Reiley.

FIGURE 23-20 Chinese evergreen (*Aglaonema modestum*).

Light

The amount of available light for a plant is a factor that is difficult to control. A shadow test can help determine this. To conduct this test, hold a piece of paper up to the light (window or lamp) and note the shadow it makes. A sharp shadow means you have bright or good light. However, if there is barely a shadow visible, the light is dim or poor. It is important to know how much light the plant needs. If a plant needs direct or full sun, then exposure of the sun is needed for at least half of the daylight hours. When a plant needs indirect or partial sun, the light should be filtered through a curtain or slats. Plants that fit into this category are Chinese evergreen (*Aglaonema modestum*; Figure 23-20) and dracaena (*Dracaena fragrans*; Figure 23-21). Even for plants preferring no direct sunlight, the room should be bright and well lit. Plants that need shade should be kept in a well-shaded part of the room. In summary, too much or too little light can greatly affect the health of the plant.

Courtesy of H. Edward Reiley.

FIGURE 23-21 Dracaena (*Dracaena fragans*).

Temperature

Plants are adapted to specific temperature ranges, and they do not do well when these temperatures are exceeded. Indoor plants are grouped into three temperature categories: cool, moderate, and warm. The cool temperature range is 50° to 60° F, with temperatures not falling to less than 45° F. The moderate temperature range is from 60° to 70° F, with a minimum of 50° F. The warm temperature range is from 70° to 80° F, with a minimum limit of 60° F.

AGRI-PROFILE

CAREER AREAS: PLANTSCAPE DESIGNER/PLANTSCAPE CONTRACTOR/GREENHOUSE AND PLANT TECHNICIAN/FLORIST

Courtesy of DeVere Burton.

The florist industry provides cut flowers and floral arrangements for all occasions. Interior plants are grown in greenhouses, nurseries, and fields and become part of floral and interior arrangements and plantscapes.

Foliage plants, including small trees, have become popular decor for interior public areas such as shopping malls, office buildings, institutions, and housing complexes. Flowering and foliage plants have long been part of interiors of attractive homes and have provided excellent hobbies for many people. Their extensive use in commercial settings is a relatively recent development that has stimulated many new job opportunities in ornamental horticulture.

Youth in suburban settings have new career and employment opportunities caring for plants, installing interior plantscapes, rotating plants between growing areas and display areas, and contracting to maintain interior plantscapes. Early experiences frequently lead to life-long careers.

With more experience and education, one may become a plantscape designer, business owner–operator, grower, wholesaler, plant doctor, or extension specialist. Ornamental horticulture specialists are hired by universities, research institutes, and business firms to develop improved plants and horticultural practices. The florist industry provides cut flowers and floral arrangements for all occasions and provides jobs in design, arranging, care, and delivery.

USES OF INDOOR PLANTS

Uses of indoor plants are varied. Plants can be used as room dividers or to brighten up dull spots in the kitchen or bathroom. Plants can be used to divide a room into specific living areas. They may be placed in containers with special watering devices, or several plants may be grouped together to form a natural barrier. Containers are available in several sizes and shapes, some with castors on the bottom so they can be moved easily. Indoor trees and climbing plants are best used as living-area dividers (Figure 23-22).

Specimen or showy plants, such as the Boston fern or date palm (*Phoenix dactylifera*), are good for brightening up dull areas (Figure 23-23). Plants may be placed on ornate plant stands or in decorative pots. A series of wall shelves may also be used to display interior plants. Plants on shelves should be compact, and perhaps trailing, for a more dramatic effect.

Hanging baskets filled with indoor plants are useful when space is at a premium. Because baskets can be easily suspended from the ceiling, they can utilize the available space above head level. Care should be used when selecting plant containers so that water does not drip on the furniture or the floor. Select baskets that are designed with drip trays attached to the bases.

Some interesting areas for hanging baskets are stairwells, offices, and hallways. An important element that is unique to hanging baskets is the hook securing the basket to the ceiling. It should be secured firmly enough to support the basket when the soil is wet and when the plant is being cared for (Figure 23-24).

Bathrooms are excellent places for indoor plants. These rooms are humid and warm, which provides a good atmosphere for plants. They can be placed on windowsills or around the tub. Shelves can also be installed in a bathroom to display indoor plants. Plants will add a touch of color to a bathroom, especially today because some bathrooms are expanding to hold a hot tub or Jacuzzi (Figure 23-25).

The kitchen is another room where plants can add a beautiful touch. Refrigerators, other appliances, and kitchen windowsills all make good places for plants that

FIGURE 23-22 Decorative plants make excellent area dividers.

FIGURE 23-23 Date palm (*Phoenix dactylifera*).

FIGURE 23-24 Hanging baskets are versatile and attractive.

FIGURE 23-25 The high humidity of a bathroom provides an excellent environment for plants when light is adequate.

FIGURE 23-26 Aloe (*Aloe variegate*).

require cooler temperatures. Most plants in the kitchen should be relatively small because space is usually at a premium. Several plants suitable for the kitchen are aloes (*Aloe variegata*; Figure 23-26), maidenhair ferns, peperomias, and even a variety of herbs. **Herbs** are plants kept for aroma, medicinal purposes, or seasoning.

GROWING INDOOR PLANTS

Growing plants indoors is much like growing them outdoors. Consideration needs to be given to the environment in each instance. Aspects of the plant environment that must be considered are light, temperature, water, drainage, and nutrients. Although indoor plants are not harvested, they need to be maintained. Some aspects of maintenance that need to be addressed are grooming, repotting, and propagation.

Light

Light is one of the most crucial factors to consider when growing indoor plants. Light is measured in foot-candles. A **foot-candle** is the amount of light found one foot from a burning standard candle, known as a candela. Two aspects of light that need to be determined are intensity and duration. The intensity and duration of indoor light vary during different seasons, at different times of the day, and through different windows. A plant near a window with southern exposure receives more intense sun, and for a longer period, than a plant in a window with northern exposure. Pulling shades across a window can reduce the intensity of light, and a tree outside a window can reduce both the intensity and duration of light that enters the window. Summer sun is much more intense than winter sun because the angle of the sun to the Earth is near perpendicular during the summer.

Too little light will cause some indoor plants to grow tall and spindly and lose leaves because the long, thin, weak stems can no longer support the plants because of limited photosynthesis. Conversely, too much light causes indoor

FIGURE 23-27 Plants generally must be moved or rotated periodically to provide correct lighting to all parts of the plant.

INTERNET KEY WORDS:
plants, phototropism

plants to wilt and lose their vibrant, green colors. The youngest leaves on plants are affected first by unfavorable conditions. Another item concerning light is the tendency of plants to grow toward the strongest source of light. If you notice that a plant is leaning toward the light, you may need to rotate the plant container periodically so that it maintains a balanced shape (Figure 23-27).

There are five basic light categories for indoor plants. These are full sun, some direct sun, bright indirect light, partial shade, and shade. Full sun is a location that receives at least 5 hours of direct sun a day. This amount of sun can be found in areas that have southern exposure. Some direct sun occurs in areas that are brightly lit but receive less than 5 hours of direct sun a day. This usually occurs in windows facing east and west.

Bright indirect light describes areas that receive a considerable amount of light but no direct sun. An example might be an area 5 feet away from a window that receives full sunlight. Partial shade refers to areas that receive indirect light of various intensities and durations. Areas 5 to 8 feet away from windows that receive direct sun are partial-shade areas. Shade refers to poorly lit sections away from windows that receive direct sun. Few plants can survive low-intensity light.

Temperature

Plants survive best at constant temperatures. Temperatures that fluctuate up and down are not ideal for plant growth. Temperature interacts with light, humidity, and air circulation. It is best to maintain temperatures in a range from 60° to 68° F for optimum indoor plant growth. As temperatures increase beyond this range, the air tends to become warmer, and the available moisture in the air decreases. Although the thermostat in a house reads one temperature, each room usually varies as much as 5 degrees in one direction or the other. Because plants respond differently to various temperature ranges, it is advisable to place plants in rooms that match their specific temperature requirements.

Water

Water is an essential ingredient for growth of any living organism. The amount of water needed by indoor plants is usually not as much as one would think.

SCIENCE PROFILE PHOTOTROPISM IN PLANTS

Plant leaves respond to light by turning to face the light source. Sometimes an entire plant will grow toward light, requiring a potted plant to be rotated regularly.

The tendency of plant parts to move or turn toward or away from a light source is known as **phototropism**. This phenomenon occurs most often when the leaves of a plant turn toward the sun to maximize the sun's intensity on the leaf surface. It is a survival mechanism that allows photosynthesis to occur in the leaf at the maximum rate. It is also observed in flowers that turn to face the sun each morning, following the track of the sun across the sky until it sets in the evening. Sometimes the petals of flowers close at night and open the next morning when the sun comes up. Each of these adaptations by plants to the sun or to other light sources is an example of phototropism.

FIGURE 23-28 Plants need to receive water on a regular basis, but too much water causes most plants to do poorly or even die.

The most likely problem with unhealthy indoor plants is too much water. More plants die from overwatering than from any other cause. As with other environmental factors such as light and temperature, each plant varies in its requirements for water. Unless you know the particular needs of the plant, it is best to water when the soil around the plant is a little dry. Most plants do best if allowed to dry out between applications of water. When indoor plants are watered properly, the roots remain more active than if the soil becomes waterlogged or excessively wet.

There are numerous ways to water plants. The basic rule about watering is to use water that is neither hot nor cold. Water should be tepid, or a moderate temperature. One way to water is to soak the pot in a bucket of warm water for half an hour, remove the pot, and drain it. A second way is to pour the water on top of the soil slowly, filling the pot to the top with water. Allow it to absorb the water until the excess drains from the hole in the bottom of the pot. Do not water again until the soil becomes dry to the touch (Figure 23-28).

Besides water in the media around their roots, plants also need moisture in the air. The moisture content in the air is referred to as humidity. It is expressed as relative humidity (a relationship between the amount of water vapor in the air compared with the maximum moisture the air will hold at a given temperature). Almost all indoor plants prefer 50 percent relative humidity. The simplest method for humidifying the air around plants is to set pots in trays filled with gravel and add water to just cover the gravel. Misting is a good way to add humidity, but it should be done several times each day to be effective. The humidity around plants can also be increased by grouping them together. However, plants should have enough space around them to allow for adequate air circulation.

Drainage

Good drainage is achieved by using pots with porous materials in the bottom and drainage holes in the pot. Good drainage is essential for indoor plants. Adding coarse material such as sand or perlite to the soil will improve soil drainage. The addition of gravel or bits of broken clay pot material to the bottom of a pot before adding the soil is a substantial aid to good drainage.

Fertilizing

Plants need nutrients on a regular basis for good health. A balanced fertilizer, such as a 5-5-5, should be applied at regular intervals. Fertilizers with a greater proportion of nitrogen than of phosphorus and potash are often used to keep foliage plants green and healthy looking. Plants should be fertilized at 2- to 6-week intervals, depending on the type of fertilizing material. Slow-release–type fertilizers dissolve slowly and release nutrients evenly over weeks or months. In contrast, liquid fertilizers suitable for foliar application are used by the plant within a few days. They may also be applied at weekly or biweekly intervals (Figure 23-29).

Plants should be fertilized while they are actively growing. Fertilization should be discontinued when the plant is dormant, or in a resting stage. Flowering plants need more fertilizer. Addition of fertilizer once every 2 weeks is recommended from the time flower buds first appear until the plants stop blooming. Avoid fertilizing plants when the soil is excessively dry. Under such conditions, the fertilizer solution is likely to be too highly concentrated and may cause burned leaves or roots.

FIGURE 23-29 Plants need to be fertilized with carefully selected materials that correspond to the needs of the plants.

Courtesy of DeVere Burton.

Grooming

Grooming plants is important even with the best combination of light, temperature, humidity, water, and drainage. Grooming should be done weekly. This task includes removing wilted or withering leaves, flowers, and stems using sharp scissors or shears (Figure 23-30). The plant should be observed to determine if it is

FIGURE 23-30 Plants need to be examined and groomed weekly.

© auremar/Shutterstock.com.

© Christina Richards/Shutterstock.com.

FIGURE 23-31 Repotting at appropriate times prevents the plant from becoming rootbound and promotes good plant health.

getting spindly or thin. If so, pinch out new growth to force the plant to branch out. To avoid crooked stems, stake plants when they are young. Climbing or trailing plants need a stake made of bark to enable them to climb and cling to the stake.

Once a week, the plant should be dusted. Dust accumulates on the leaves and blocks the stoma so that the plant cannot breathe or transpire as well as it should. The soil around the base of the plant should be loosened with a fork or small spade to allow air to enter the soil and water to percolate through. To help control insects, mist the infested plants with a diluted solution of mild dishwashing soap and water.

Repotting

In time, plants develop root systems that are restricted by the pot or container. When this happens, the plant is said to be **rootbound**. The roots have no place to continue to grow, so repotting is necessary for plant health. In general, repotting is done in the spring or the fall. A good rule of thumb for determining pot size is to use one with a diameter at the top of the rim equal to one-third to one-half the height of the plant. This rule does not apply to plants whose growth habit is tall and slender, as opposed to a balanced top growth. When repotting, it is not desirable to move an established plant to a new pot that is more than 2 inches wider than the original pot. Excessively large pots result in wasted soil, water, and nutrients (Figure 23-31).

Flowering plants are best repotted after the flowers have faded. During repotting, check the roots for insects and root damage. Remove any roots that look or feel unhealthy.

Propagation

Propagating most indoor plants is relatively easy. The method chosen depends on the type of plant (whether it is herbaceous or woody, flowering or foliar). Both sexual and asexual methods of propagation are used.

In sexual propagation, seeds may be started in containers with a good potting soil, plenty of moisture, and adequate air circulation. Keeping pots in warmer areas of the house increases the speed of seed germination. Covering the pots with glass or plastic held up by stakes will also aid in the germination process.

Some popular methods of asexual propagation of indoor plants are leaf and stem cuttings, removal of plantlets from parent plants, and air layering. Each of these procedures results in the multiplication of the plant.

Flowering Plants

Indoor flowering plants may take some extra, special care to ensure blossoming. These plants are more sensitive to the availability of light, so some artificial lighting may be needed. They are also more sensitive to temperature changes. Flowering plants have particular seasons in which they flower. It is important to know when to expect the plants to bloom. Some examples of plants with specific bloom times are the Christmas cactus (*Schlumbergera x buckleyi*), which is expected to bloom between October and late January, and the Easter cactus (*Rhipsalidopsis gaertneri*), which blooms in April or May. These two plants are similar in their leaf types and flowers, but they bloom in opposite seasons of the year. Indoor plants may flower in the winter, spring, summer, or year-round.

HOT TOPICS IN AGRISCIENCE GENETICALLY ENGINEERED FLOWER: BLUE ROSE

© Veniamin Kraskov/Shutterstock.com.

The world's first blue rose was developed by splicing a pansy gene into the DNA of a rose.

The world's first blue rose was genetically engineered by splicing a pansy gene into a rose plant. Suntory Flowers spent 20 years of research developing the new variety that was released in Tokyo for the first time in 2009. The new rose is capable of producing a blue pigment named *delphinidin*, which is not found naturally in any rose. The first "blue" rose was not true blue, but somewhat violet in color. More recent releases are available with a beautiful blue hue.

Roses have been cultivated for more than 5000 years, and they are considered to be among the most revered flowers on the planet. Their common colors are red, pink, yellow and white, but the blue rose sells for much more than any of them. It has been one of the most popular floral developments in a very long time.

Foliage Plants

Foliage plants are nonflowering plants grown and sold for their attractive leaves. Many shopping malls, offices, and other public buildings prefer foliage plants over flowering or fruiting plants because they are very easy to maintain and add a nice decorative element to interiors. In the early 1970s, wholesale customers purchased an average of $29 million worth of foliage plants in the United States. In 2005, that number soared to $721 million. This increase was not only spurred by an increase in popularity, it also was driven by a technique called *new tissue culture technology*. This new method speeds the rate of propagation, allowing growers to significantly increase production.

A wide variety of foliage plants is grown in homes and offices, and specific instructions for their care compose entire books. However, the correct management of light is probably the greatest single factor for growing foliage plants. Light is the source of energy for the process of photosynthesis, whereby the leaves produce sugars and starches to feed all parts of the plant. Different foliage plants have different light requirements. It is recommended that a good reference book on indoor plants be used to determine the particular requirements for any given plant. Regardless of the amount and duration of light, it is desirable to rotate plants so that each side receives the same amount of light over time.

INTERIOR LANDSCAPING OR PLANTSCAPING

Plantscaping is the design and arrangement of plants and structures in indoor areas. This design and arrangement is an art. Interior plantscaping is an activity that is fun. However, it should be approached seriously, just as the arrangement of furniture or other interior decorations must be done carefully. The indoor plants should be used to complement people-oriented spaces. The more creative you are in designing, the more distinctive indoor plantscapes will become. There is no right or wrong way to design with indoor plants, but there are elements that enhance the interior plantscaping technique.

Design

Before beginning a plantscape design, the designer must know the purpose or intent of the plants. Are the plants to be used as a space divider? Are they being used to accent existing furnishings? Will they fill empty space? Answers to these questions will result in an organized design, rather than a happenstance.

Another question to consider is the function or functions of the plants. Are they to create a specific shape, emphasize a specific area, or support a specific architectural feature of the room? If the function of the plantscape is not considered, the end result is not likely to be successful.

INTERNET KEY WORDS:
interior plantscape design principles

The physical characteristics of color, form, and texture should be considered when selecting plants for a plantscape. These are determined in concert with the perceptual characteristics of accent, sequence, balance, and scale. These characteristics are the basic tools a designer uses to create an interior plantscape (Figure 23-32).

Color

Color is the most important physical characteristic of a plantscape. It can influence emotions, create specific feelings, and add beauty and harmony to the environment. Two types of color must be considered by the designer in creating an interior plantscape. First is the basic, or background, color. **Basic color** is the color of the walls, ceiling, and floor. These colors should influence the selection of the flowering characteristics and foliage of plant material. Accent is the second type of color. **Accent color** is the color of the plants or other attention-getting objects.

Form

Form refers to shape. There are different forms that plants possess naturally. The most common shapes are round; oval; weeping (or drooping); upright; spiky; and spreading or horizontal. These shapes can be used individually or grouped to form an artificially sculptured shape and form. Each plant shape or cluster shape adds its own particular feature to the design.

SCIENCE CONNECTION

LIKE THE SEASON'S FIRST SNOWFALL

A serious threat to ornamentals and nonornamentals alike, whiteflies settle on plants like the season's first snow.

With adults not much larger than the head of a pin, whiteflies suck the life from plants, including fruits, vegetables, flowers, shrubs, and field crops. The costs associated with control of this pest and the damage it causes run an annual tab in the millions of dollars. Greenhouse growers, nurseries, and retail outlets wage a continuous battle against whiteflies in an effort to save crops from damage and infestations that would prevent the sale of these products. Similarly, homeowners and interior plantscapers must be on a constant vigil, lest unobserved eggs hatch into an infestation of damaging proportions. Like the season's first snowfall, the eggs can suddenly appear as white specks all over the plant. When the plant is moved or the insects disturbed, the masses of adults can take flight and create the appearance of a snowstorm in the air!

FIGURE 23-32 Interior plantscaping is an important element in shopping malls and other public areas.

Texture

Plant **texture** refers to the visual or surface quality of the plant or plants. Texture is influenced by the arrangement and size of leaves, stems, and branches. It is described in terms of coarseness or fineness, roughness or smoothness, heaviness or lightness, and thickness or thinness. Coarse-textured plants, such as fiddleleaf fig or prickly pear cactus, should be used in large spaces, whereas a maidenhair fern should be placed in small spaces, such as on shelves.

Accent and Sequence

Accent means a distinctive feature or quality. An accent captures the attention of the viewer. It has a dramatic effect on the visual appearance of the room or part

First observed in Florida in 1986, an especially viral strain of the whitefly spread rapidly from Florida to California. It is challenging scientists and growers alike as it spreads nearly unchecked by either pesticides or natural enemies. Around the country, researchers are showing the textile industry how to wash off the sticky material the insects leave on cotton fiber. They are also uncovering ways to deal with plant viruses left in the wake of the troublesome, sucking insects. They are testing legions of environmentally friendly controls such as insect parasites, predators, fungi, plant extracts, and dish detergent/vegetable oil mixtures.

Scientists have identified more than 30 predators and 25 parasites. So far, the big-eyed beetle is the most promising. It secretes a sticky substance from its mouth to stick the whitefly to the leaf. Then it leisurely eats the fly. Scientists have released some big-eyed beetles on a pilot basis. Another tiny but promising insect that has been identified devours an estimated 10,000 whitefly eggs or 700 nymphs in its 6- to 9-week lifetime. However, not enough is yet known about this insect to release it. Similarly, a number of parasitic wasps are under examination. One of the more promising spray materials is a biosoap, which provides some control over whiteflies as well as other damaging insects.

of the room. You can create an accent with the use of color, form, or texture. It can also be created through the use of a sequence. **Sequence** refers to a related or continuous series. Sequence is created with plant material by repeating the same color, texture, or form. The overuse of accents, however, will detract from their function to capture the attention of the viewer.

Balance and Scale

Balance is the state of equality and calm that is created between items in a design. In a design, two types of balance are used: formal, or symmetrical; and informal, or asymmetrical. **Formal**, or **symmetrical**, balance occurs when the items are equal in number, size, or texture on both sides of the center of the design. **Informal** or **asymmetrical** balance occurs when the items are not equal in number, size, or texture on the two sides of the center. **Scale** refers to the size of items. A large patio with African violets as accents would be out of scale. Plant materials need to complement the size and, therefore, the scale of the room. A better selection for a patio would be Norfolk Island pines and dracaenas.

A Design Process

There are various types of design processes. A good one to use has three phases. The three-phase process emphasizes the how and why of using plants for the particular space.

Phase One

The first phase of the process is preplanning, which includes three steps. The first step is to develop the design objectives by determining the purpose of the plantscape. The second step is to determine the space capacities, which include the habitat and the circulation of people. Circulation is how people tend to move in the area. The amount of light, temperature, and humidity in the room is referred to as habitat. The third step is to determine the development limitations. Development limitations include the amount of money available to implement the design; room and space characteristics; and habitat limitations created by light, temperature, people, pets, and other factors.

Phase Two

The second phase is to develop the plan. The plan will include the basic design and arrangement of plants and materials. The physical and perceptual characteristics of the plants are important in this phase. Tentative selection of plants with visual placement in the room is part of this step. The planning is done on paper, and alternatives are examined. The details of the final plan include the selection of all the plants that will be used.

Phase Three

Implementation is the third phase of the interior plantscaping process; this is where the design comes to life. This phase has three steps. The first step is the preparation of the documents. If any construction is involved, such as platforms, decks, or planter boxes, drawings for these objects must be completed. Drawings and specifications for the installation and maintenance of water lines and fixtures, electrical devices, and plants should also be finalized at this point. This

© R. Gino Santa Maria/Shutterstock.com.

FIGURE 23-33 Greenhouses, nurseries, florists, and suppliers are needed to provide a constant supply of plant materials.

step is important, especially if you are designing for someone other than yourself. The second step in this phase is the installation of all physical modifications. Selection of plant containers and the planting of the indoor plants take place in this phase. The plants are put in place. The final step is evaluating the project. A thorough look at the plants, their containers, and their position in the room or area must be done. If everything looks balanced, no changes need to be made. If not, minor adjustments may be made. The interior plantscape is now complete.

Plantscape Maintenance

Maintenance of the plantscape with attention to light, moisture, humidity, and grooming is important. These factors are discussed in preceding parts of this unit. The nursery and greenhouse industries often provide critical expertise in developing and maintaining plantscapes (Figure 23-33).

CAREERS IN INDOOR PLANTSCAPING

Interior plantscaping is a career field within the large industry of horticulture. Horticulture is the study of plants, especially garden crops, both indoor and outdoor, for human consumption, for aesthetic purposes, or for medicinal purposes. Interior plantscaping is a part of floriculture, which is a division of horticulture. Plantscaping is both a science and an art requiring practical skills and understanding. It is a career area that allows for creativity and use of scientific knowledge and technology.

The opportunities for careers in indoor plantscaping are numerous. An individual may be involved in growing indoor plants, designing interior plantscapes, installing plantscapes, maintaining plantscapes, or selling or servicing plantscape materials. The plantscapes can be small-scale within a residence or home, or in large-scale public areas such as shopping malls or office buildings. This area of horticulture offers many individuals the chance to be entrepreneurs with relatively low investments.

An interesting occupation in the field of plantscaping is the plant rental business. Many office and building managers want the benefits of plantscaping but cannot spend the time required for maintenance. Plant rentals, which include regular maintenance, give building managers the best of both worlds.

The education required for a career in indoor plantscaping varies. Persons with high school technical experience in horticulture, two-year college degrees, or four-year college degrees can be successful in this career.

STUDENT ACTIVITIES

1. Write the Terms to Know and their meanings in your notebook.
2. Take an inventory of the species, type, and number of indoor plants in your home.
3. Select indoor plants that are suitable for your home from reference books and seed or plant catalogs.
4. Develop a plan for an interior plantscape of a room in your home, the school office, or another location.
5. Repot an indoor plant.
6. Propagate indoor plants using one or more methods.

7. Perform a shadow test at several windows in your home or classroom to determine the intensity of the light available.
8. Prepare hanging baskets for use in the home or office.
9. Check with a local greenhouse or flower shop for any discarded plants you could try to save. Try to rehabilitate the plants using the knowledge you gained from this unit about fertilizer, watering, and light requirements. Donate any nice-looking plants to a nursing home, school, or community organization.

SELF-EVALUATION

A. MULTIPLE CHOICE

1. One of the most popular flowering indoor plants is the
   a. spider plant.
   b. petunia.
   c. African violet.
   d. Norfolk Island pine.
2. An example of an indoor tree is a
   a. philodendron.
   b. cactus.
   c. fiddleleaf fig.
   d. zebra plant.
3. When purchasing an indoor plant, check the underside of the leaves for
   a. powdery mildew.
   b. insects.
   c. price tags.
   d. brown spots.
4. The moderate temperature range for indoor plants is
   a. 60° to 70° F.
   b. 45° to 55° F.
   c. 90° to 100° F.
   d. 35° to 45° F.
5. If an indoor plant needs direct or full sun, then the plant will need sun for at least
   a. the entire day.
   b. 1 hour.
   c. a fourth of the daylight hours.
   d. half of the daylight hours.
6. The place in a home not usually considered an ideal location for an indoor plant is the
   a. bathroom.
   b. basement.
   c. kitchen.
   d. living room.
7. Which of the following is not considered an environmental factor when growing indoor plants?
   a. weather
   b. light
   c. temperature
   d. humidity
8. When watering plants, the ideal temperature for the water is
   a. ice cold.
   b. hot.
   c. tepid.
   d. cold.
9. Which of the following is not a physical characteristic of an indoor plant?
   a. texture
   b. form
   c. color
   d. balance
10. Interior plantscaping is a career area in which area of horticulture?
   a. agronomy
   b. floriculture
   c. forestry
   d. arboriculture

B. MATCHING

________	1. *Ficus lyrata*	a. A fern-type foliage plant
________	2. Gardenia	b. Has unusually colorful leaves
________	3. Croton	c. A succulent-type foliage plant
________	4. Wandering Jew	d. A fragrant, flowering indoor plant
________	5. Jade plant	e. A tree-type indoor plant
________	6. *Aspargus sprengeri*	f. A vine-type foliage plant

C. COMPLETION

1. The two environmental factors that are most important to the survival of indoor plants are ______ and ______.
2. An indoor plant that is a tree can be identified by its usually ______ stem.
3. The bathroom is a great place for indoor plants because of the ______ usually found in a bathroom.
4. Indoor plants survive best at a ______ temperature.
5. More indoor plants die of ______ than any other cause.
6. Adding a coarse material such as sand to the soil will improve an indoor plant's ______.
7. A 5-5-5 fertilizer is an example of a______ fertilizer.
8. When repotting an indoor plant, do not move a plant to a pot that is more than ______ wider than its original pot.

UNIT 24

Turfgrass Use and Management

OBJECTIVE

To understand growth and development of turfgrasses and the establishment and cultural practices involved in managing these plants.

MATERIALS LIST

- writing materials
- samples of grass seed
- various types of turf samples
- brochures from lawn care services
- lawn equipment catalogs
- Internet access

COMPETENCIES TO BE DEVELOPED

After studying this unit, you should be able to:

- identify and describe careers available in the turfgrass industry.
- identify turfgrass plant parts.
- select turfgrass species for various purposes and locations.
- state the basic cultural practices for turfgrass production and maintenance.
- list the basic steps for turfgrass establishment.

SUGGESTED CLASS ACTIVITIES

1. Establish some turfgrass demonstration plots at your school. Demonstrate the effects of different management practices, such as height of mowing and differences in fertilizer applications, on the various plots. This is a long-term project that can be extended for several years.
2. Visit a turf farm or invite a local turfgrass farmer to speak to the class about the turfgrass business. Include these topics of discussion: (1) preparing the soil for planting, (2) planting methods, (3) establishing a good stand of grass, (4) weed and pest control, (5) harvesting practices, and (6) marketing the turfgrass product.
3. Find a small area in the school yard or a nearby area that either has no grass or has grass in poor condition. Remove any unhealthy grass that may be present. Next, break up the soil a little with a rake and evenly spread grass seed. Cover the seed with straw mulch. Protect the plot with a tape barrier or fence. Take care to water the plot daily with a light application of water until new turf is established.

TERMS TO KNOW

turfgrass
seminal root
crown
extravaginal growth
vegetative reproduction
intravaginal growth
seed culm
inflorescence
induction
sheath
blade
ligule
collar
recuperative potential
fungal endophyte
thatch
syringing
seed blend
seed mixture
sodding
turf
sprigging
stolonizing
plugging

Who would have expected 50 years ago that farmers would one day raise fields of grass for a purpose other than feeding livestock? Yet, a huge industry has risen for the purpose of growing "instant lawns" that are transported from the fields to the front yards where green grass suddenly appears and thrives. It is a classic case of the instant gratification we have come to expect in modern society.

THE TURFGRASS INDUSTRY

It has been estimated that the total economic impact of the turfgrass industry in the United States is approximately $100 billion. This industry is large, diverse, and still growing. Career opportunities in this field expanded rapidly in the last decade and are projected to continue growing in the years ahead. **Turfgrass** includes all of the grasses that are mowed frequently to maintain a short and even appearance.

Golf course superintendents and other athletic-field managers maintain turfgrass at a certain level of playability. Playability level means suitability for the intended use. It is determined by the type of sporting or recreational event. For example, putting greens are areas used for playing golf, and the turfgrass is very short. These areas are maintained to provide a surface for consistent, yet adequate, putting speeds (Figure 24-1). A football field may be managed to offer secure footing and sufficient turfgrass resiliency for player safety.

Lawn care services offer many job positions and are among the largest employers within the industry. Job opportunities include lawn care specialist, branch manager, owner-operator, and others.

The maintenance of turfgrasses at large government, apartment, university, commercial, and private complexes requires personnel trained in turfgrass management.

Sod-production farms and landscaping businesses are involved in the establishment and installation of turfgrass for lawns (Figure 24-2). Turfgrass specialists are also needed for these segments of the industry (Figure 24-3).

The turfgrass industry supports a substantial sales force. Companies that produce or distribute seed, fertilizer, pesticides, and turfgrass equipment require an extensive support and sales staff (Figure 24-4).

Federal and state governments and private companies hire turf specialists and scientists with advanced degrees. Career opportunities in these areas offer challenging positions in research, teaching, and extension.

TURFGRASS GROWTH AND DEVELOPMENT

Turfgrasses are plants grouped into the Poaceae family. These grasses differ from other grass plants because they can withstand mowing at low heights. They can also tolerate vehicle and foot traffic. Turfgrasses are frequently used to hold the soil and as ornamental plants. These traits have made turfgrasses the most widely used ornamental crop in the United States. To properly maintain turfgrasses, an understanding of their growth and development is required.

THE TURFGRASS PLANT

The grass plant can be divided into two broad areas known as the root and shoot systems. The root system consists of adventitious and seminal roots. The shoot system includes the stem and leaves of the plant.

A

B

FIGURE 24-1 (A) Sports turf requires maintenance practices that provide an acceptable level of playability. (B) Soils near golf greens tend to become compacted due to heavy use. Aeration involves removing small plugs of soil to restore infiltration of air and water in the root zone.

Root System

The **seminal roots** develop from the seed during seed germination. They initially anchor the seed into the soil. The seminal root system will be active for 6 to 8 weeks. The adventitious roots develop from the nodes of stem tissue. They usually compose the entire root system of a mature turfgrass stand.

Turfgrass roots are multibranching and fibrous. Their function is nutrient and water absorption. They also prevent soil erosion by effectively stabilizing and anchoring the soil particles in place.

FIGURE 24-2 Turfgrass is widely used to establish new lawns that create beautiful and pleasing environments around homes, businesses, and public buildings.

FIGURE 24-3 Turfgrass is grown in large fields from which it is harvested for the purpose of creating "instant lawns."

© iStockphoto/BanksPhotos.

FIGURE 24-4 Turfgrass requires regular maintenance such as fertilizing and mowing.

Seasonal changes in root growth are dependent on soil temperature and moisture. Active root growth for warm-season turfgrasses occurs in the summer. A warm-season turfgrass is one of a group of grasses adapted to the southern region of the United States. A cool-season turfgrass is a plant adapted to the northern region of the United States. These turfgrasses have active root growth in the fall and early spring. Optimum growth occurs at temperatures from 60° to 75° F.

AGRI-PROFILE CAREER AREAS: TURFGRASS GROWER/GROUNDSKEEPER/ LANDSCAPE MAINTENANCE TECHNICIAN

© cappi thompson/Shutterstock.com.

Turfgrass production, establishment, and maintenance provide many good jobs and create attractive surrounding for homes, business, and recreation.

In the last three decades, turfgrass production and management have become big business in the United States. In one eastern state, turfgrass recently became the number one crop, based on total acres in production. Starting in the 1950s, growth and development of golf courses stimulated the turfgrass industry with high-paying salaries for golf course superintendents and other turfgrass specialists.

Currently, career opportunities in turfgrass production, management, service, supervision, research, and consultation are extensive. Turfgrass technicians and specialists generally work in attractive and appealing surroundings. Many work outdoors in sunny weather and indoors when the weather is bad. In many localities, salaries have become quite attractive, even for laborers.

Educational programs in turfgrass science and management are available in high school technical programs, technical schools, colleges, and universities. With the movement toward urbanization, interest in open spaces, concern for the environment, and increasing population, the outlook for careers in turfgrass production and management is excellent.

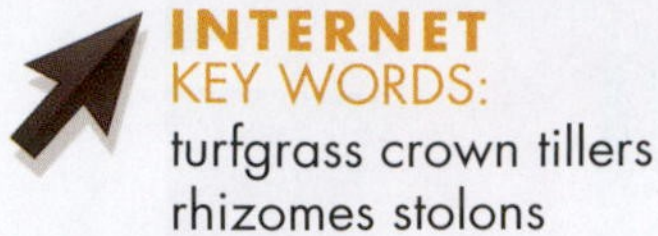
INTERNET KEY WORDS:
turfgrass crown tillers rhizomes stolons

Rooting depth is affected by plant species, soil factors, and cultural or maintenance practices. Average rooting depth for turfgrasses is 6 to 12 inches. The warm-season grasses have deeper root systems than the cool-season grasses. Well-drained, sandy-loam soils with neutral soil pH are best adapted for turfgrass roots.

Cultural practices that influence rooting depth and growth include mowing, fertilization, and irrigation. Frequent mowing and low mowing heights reduce rooting depth. In addition, fertilization programs that emphasize only shoot growth are likely to impair root growth. Light, frequent irrigation results in shallow-root grasses. Heavy but infrequent irrigation is more conducive to deep-root penetration and growth.

Turfgrass maintenance practices should attempt to optimize the rooting potential of turfgrass plants. An extensive root system allows a plant to recover from drought and other stress conditions more rapidly.

Shoot System

The shoot system consists of stems, leaves, and seed head, or inflorescence. These plant parts are involved in capturing solar energy through photosynthesis and storing it in forms that the plant can use for growth and seed production.

Stems

Turfgrass stems include the crown, tillers, rhizomes, stolons, and seed culms. The crown is the meristem tissue of the grass plant from which new growth occurs. The **crown** is a stem with the nodes stacked on top of each other (Figure 24-5). All root, leaf, and other shoot growth originates from this area. The crown is located at the base of the grass plant in the soil surface area.

INTERNET KEY WORDS:
grass, axillary bud, crown

Rhizomes and stolons are horizontal stems. A rhizome is a creeping underground stem, whereas a stolon is an aboveground stem (Figure 24-6). They both originate from an axillary bud on the crown and will penetrate through the

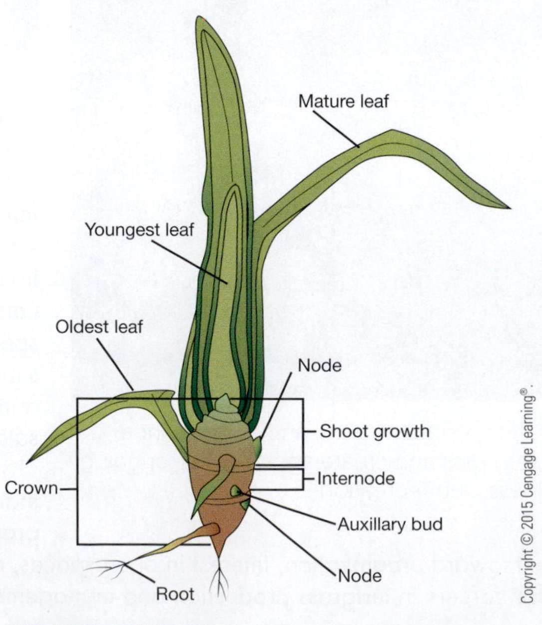

FIGURE 24-5 The crown of a turfgrass plant.

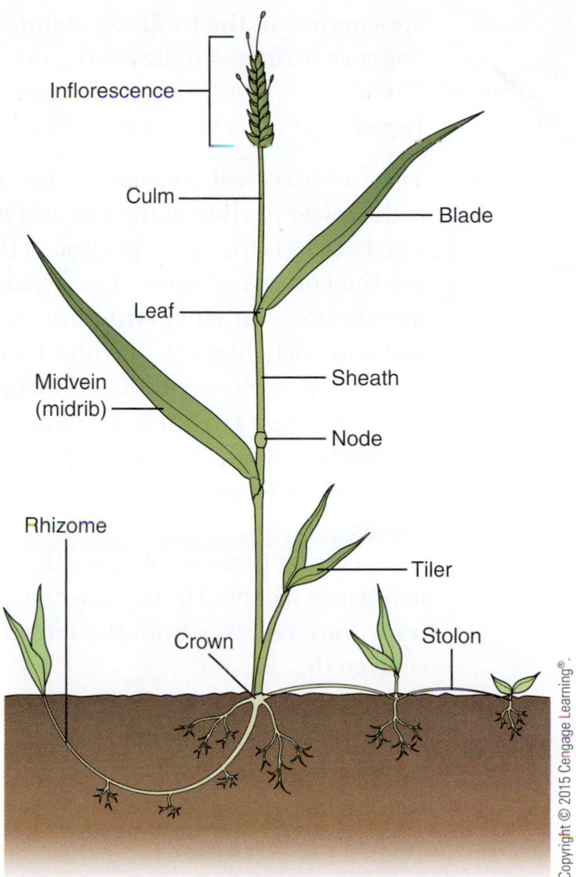

FIGURE 24-6 The major parts of a grass plant.

lower leaf sheath. This type of growth is referred to as **extravaginal growth**. Rhizome and stolon growth allows for vegetative spreading of turfgrasses. **Vegetative reproduction** is reproduction from plant parts other than seeds.

Tiller

Tillers are new shoots of a grass plant that develop at the axillary bud of the crown. They form within the lower leaf sheath of the plant. This type of growth is referred to as **intravaginal growth**. Increased tillering will enhance turfgrass density. All turfgrasses produce tillers. Optimum tillering for the cool-season grasses occurs in the spring and fall months. Warm-season grasses have optimum tillering during the summer. Under low-moisture and high-temperature stress conditions, tiller, rhizome, and stolon development is reduced.

Seed Culm and Inflorescence

The **seed culm**, or seed stem, supports the inflorescence of the plant. The seed culm originates at the top of the crown. **Inflorescence** is the arrangement of the flowering parts of a grass plant. Cool-season grasses produce their inflorescence in the spring. Warm-season turfgrasses produce their inflorescence in the late summer.

Flower **induction** or initiation is caused by several environmental conditions. Temperature and photoperiod are the two induction processes for grasses. When seed heads form, they will cause a decrease in playability and

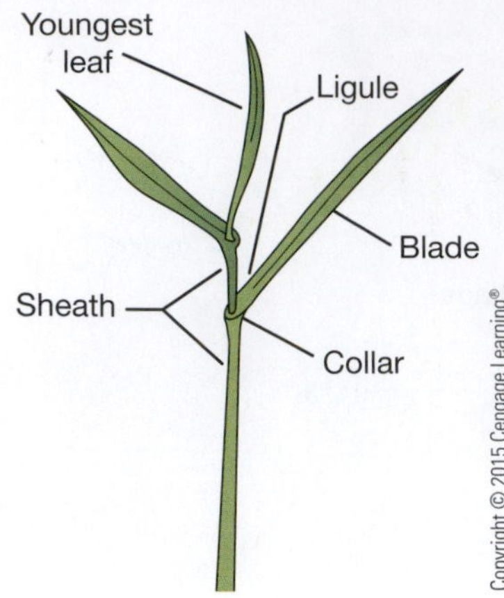

FIGURE 24-7 The leaf blade and leaf sheath of a grass plant.

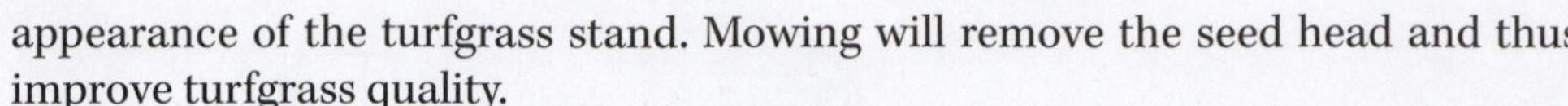

appearance of the turfgrass stand. Mowing will remove the seed head and thus improve turfgrass quality.

Leaf

The turfgrass leaf consists of the sheath and blade (Figure 24-7). The **sheath** is the lower portion of the leaf and may be rolled or folded over the shoot system. The **blade** is the upper portion of the leaf. At the junction of the blade and sheath are the collar and ligule. The **ligule** is located on the inside of the leaf and is a membranous or hairy structure. The **collar** can be found on the outside of the leaf and is a light green or white-banded area (Figure 24-8). These two features are important vegetative traits for turfgrass identification.

Turfgrass growth is dependent on the production and use of carbohydrates. The turfgrass leaf is responsible for photosynthesis and, ultimately, carbohydrate production. Reserve carbohydrates will be stored in crown, rhizome, and stolon tissue.

During unfavorable growth conditions, the plant will go into dormancy. All leaf tissue will die. However, when favorable environmental conditions recur, carbohydrate reserves from the crown and other stem tissue will be used for plant re-growth.

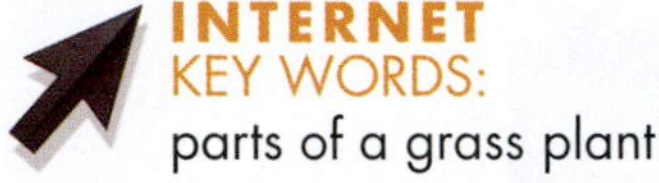

INTERNET KEY WORDS:
parts of a grass plant

Turfgrass maintenance programs attempt to optimize root growth and carbohydrate accumulation. Greater recuperative potential and plant persistency occur when these two basic concepts of growth are understood and managed. **Recuperative potential** is the ability of a plant to recover from drought or damage.

TURFGRASS VARIETIES

INTERNET KEY WORDS:
turfgrass varieties, cool-season turfgrass

Approximately 7500 plants are classified as grasses. Only a few dozen are considered useful for turfgrass. Turfgrasses are divided into two major groups based on climatic adaptation. The two groups are cool-season and warm-season

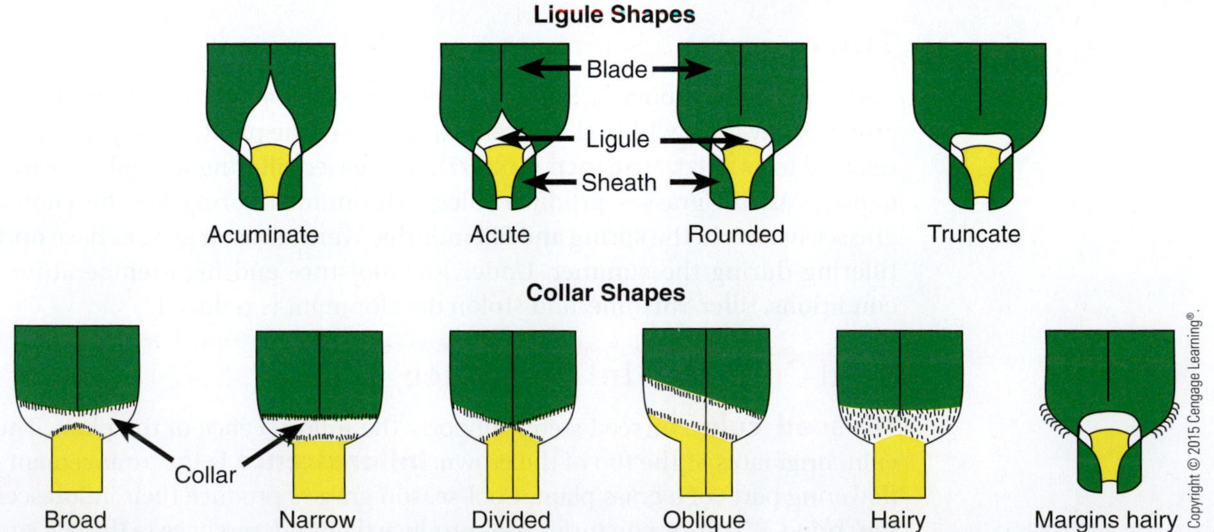

FIGURE 24-8 Grasses can be identified by using keys that eliminate incorrect identities until only the correct identity remains. The shapes of ligules and collars are used to describe plant characteristics that are included in the identification key for grasses.

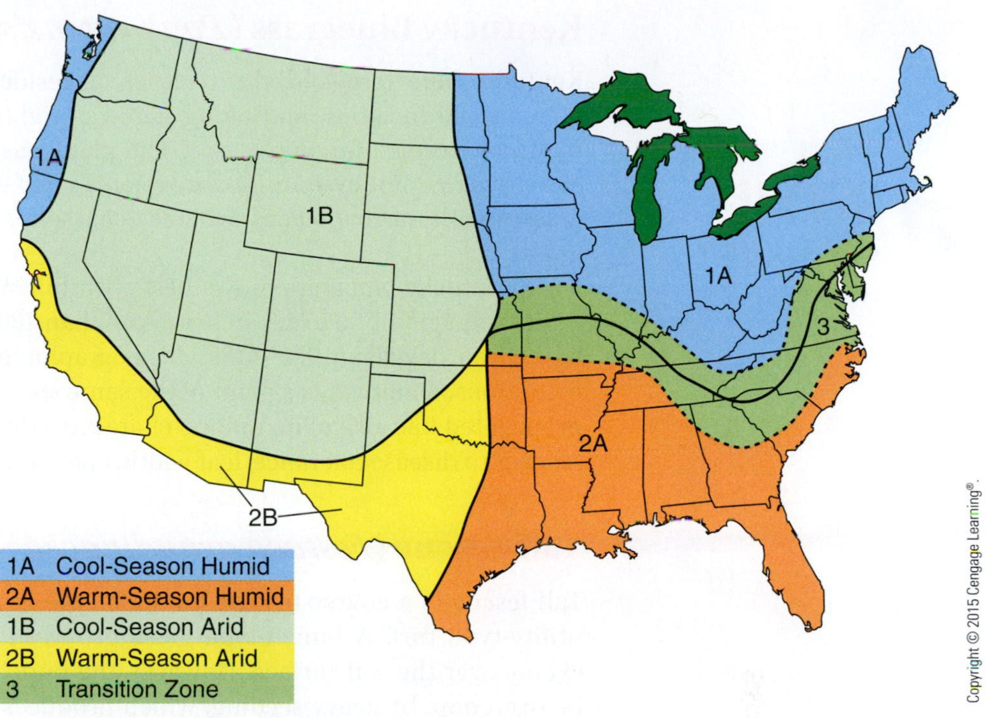

FIGURE 24-9 The major turfgrass adaptation zones.

grasses. The United States may be divided into a number of climatic zones based on temperature and moisture conditions. It is helpful to be aware of these zones when selecting turfgrass (Figure 24-9).

Cool-Season Turfgrasses

Cool-season turfgrasses originated in Europe and Asia. They have optimum growth at temperatures from 60° to 75° F. They predominate in the northern and central regions of the United States. Species adaptation within the cool-season group is determined by rainfall, soil fertility, and turf use. Major cool-season turfgrasses in the United States include Kentucky bluegrass, tall fescue, red fescue, perennial rye grass, creeping bentgrass, and crested wheatgrass (Figure 24-10).

Cool-Season	Warm-Season
Colonial Bentgrass (*Agrostis tenuis*)	Bermuda grass (*Cynodon dactylon* L)
Creeping Bentgrass (*Agrostis palustris*)	Zoysia grass (*Zoysia japonica* Steud.)
Kentucky Bluegrass (*Poa pratensis*)	Buffalo grass (*Buchloe dactyloides*)
Tall Fescue (*Festuca arundinacea*)	St. Augustine grass (*Stenotaphrum secundatum*)
Fine Fescue (various *Festuca* species)	
Red Fescue (*Festuca rubra* L)	
Perennial Ryegrass (*Lolium perenne*)	
Crested Wheatgrass (*Agropyron cristatum*)	

FIGURE 24-10 Some major warm- and cool-season turfgrasses in the United States.

Kentucky Bluegrass (*Poa pratensis* L)

Kentucky bluegrass is used extensively in residential and commercial lawns, in recreational facilities, and along highway rights-of-way. It performs well with moderate levels of maintenance. The plant has a medium leaf texture and an extensive rhizome system. Texture refers to leaf width. Fine-texture turf contains grasses with narrow blades, whereas coarse-texture turf consists of wide-blade grasses.

Kentucky bluegrass grows best with full sun, moist and fertile soil, and a mowing height of 1.5 to 2.5 inches. More than 100 cultivars of Kentucky bluegrass have been developed for different geographic areas and specific maintenance conditions. A cultivar is a plant of the same species that has been discovered and propagated because of its unique characteristics. Cultivar differences exist with respect to disease tolerance, leaf width, color, and other traits.

Tall Fescue (*Festuca arundinacea* Screb)

Tall fescue is a coarse-textured, bunch-type grass used in home lawns or as a utility-type turf. A bunch-type grass grows in clumps rather than spreading evenly over the soil surface. However, the bunching tendency of tall fescue can be overcome by heavy seeding, which produces thick stands of grass. A utility-type grass refers to a turfgrass adapted to low maintenance levels. Tall fescues have extensive root systems and are among the most drought-tolerant, cool-season species. However, they are prone to winter injury in the northern range of the cool-season zone.

Recent genetic developments have introduced many new and improved cultivars of tall fescue. These new cultivars have medium leaf texture with more aggressive rhizome development. This turfgrass will not tolerate low mowing heights and should be mowed at 2.5 to 3 inches. Insect resistance and plant persistency are excellent for tall fescues infected with fungal endophytes. A **fungal endophyte** is a microscopic plant growing within a plant. They tend to improve turfgrasses.

Red Fescue (*Festuca rubra* L)

Red fescue is a fine-textured turfgrass well adapted to shady, dry locations. It has excellent drought tolerance and can persist on rather infertile soils and under acidic soil conditions (pH 5.5–6.0).

Red fescue is often seeded in mixtures with Kentucky bluegrass for lawn turf. It is not used on athletic fields as a permanent turf because it has poor recuperative potential. However, in the South, it may be overseeded into dormant Bermuda grass greens to provide winter play endurance and color.

Red fescue functions satisfactorily if it is mowed at 1.5 to 2.5 inches in height and is provided with minimal levels of nitrogen fertilizer and water. The major pest problem of red fescue is leaf spot disease.

Perennial Ryegrass (*Lolium perenne* L)

Perennial ryegrass has a medium leaf texture and is most often used in seed mixes with other turfgrasses. It has a rapid germination and excellent seedling vigor. It is often used in seed mixes to provide soil stabilization during the establishment period.

The perennial ryegrasses are used extensively for recreational turf. They have good wear tolerance and can be rapidly established if the turfgrass stand is damaged. They are also used in winter overseeding programs with warm-season turfgrasses.

Perennial ryegrasses require moderate levels of maintenance to form attractive turf. As a group, they have poor disease resistance. Ryegrasses express improved plant persistence and insect resistance in turf plantings that have been inoculated with fungal endophyte. A symbiotic relationship exists between ryegrass and fungal endophyte.

Creeping Bentgrass (*Agrostis palustris* Huds)

Used in close-cut, high-maintenance areas, this is an extremely attractive turfgrass and fits well when used on putting greens or bowling greens mowed at five-sixths of an inch. It can be maintained at higher mowing heights (0.5–0.75 inch) for use on football fields, golf course fairways and tees, lawns, and formal gardens.

The plant has extensive stolon growth and is a fine-textured turfgrass. It is best adapted to slightly acidic soil with a pH of 5.5 to 6.0 that has good internal drainage and is not prone to compaction. Creeping bentgrass has excellent cold tolerance but poor heat tolerance.

Creeping bentgrass requires a high level of maintenance to produce a quality turf. It requires proper irrigation, disease control, mowing, and cultivation practices when grown as a sports turf. Because of these high-maintenance requirements, creeping bentgrass is not recommended as a lawn turf.

Crested Wheatgrass (*Agropyron cristatum*)

Crested wheatgrass is also known as fairway crested wheatgrass and is used on nonirrigated lawns and fairways in dry, cold regions. It is a coarse-textured, non-creeping, bunch grass. Crested wheatgrass is not considered a high-quality turf, but it is durable in semiarid, northern areas of the Great Plains. Its extensive, deep root system enables the plant to survive lengthy drought conditions without irrigation.

Recommended mowing height for crested wheatgrass is 1.5 to 2.5 inches, and low-to-moderate fertility is required. Heavy watering will stress the plant; therefore, irrigation should be minimal.

Warm-Season Turfgrasses

Warm-season turfgrasses originated in Africa, North and South America, and Southeast Asia, but they are adapted to the southern United States. They make optimum growth at temperatures of 80° to 95° F. These grasses go into winter dormancy as temperatures decrease to lower than 50° F. Fourteen species-of-interest of warm-season turfgrasses are found throughout the world.

Some major warm-season turfgrasses in the United States include Bermuda grass, zoysia grass, St. Augustine grass, and buffalo grass.

Bermuda grass (*Cynodon dactylon* L)

Bermuda grass is considered the most important and widely used warm-season turfgrass in the United States. It is principally used as a lawn turf and sports turf. Improved breeding lines have provided fine-textured Bermuda grasses capable of being used on putting greens and fairways. The common type of Bermuda grass is a medium-textured turfgrass used for airport runways, rights-of-way, and other low-maintenance areas.

SCIENCE CONNECTION WASHINGTON, DC, CAPITOL MALL: TOUGH TURF

Courtesy of the D.C. Committee to Promote Washington.

Turf created by a mixture of zoysia grass and tall fescue.

Each summer, millions of people visit the museums, monuments, and Capitol Mall in Washington, DC. The daily foot traffic of Washington's employed, as well as the continuous flow of tourists, means many thousands of feet pounding the turfgrass sections throughout the summer months. Furthermore, the seasonal use of the Mall for national rallies, concerts, inaugurations, and demonstrations pounds and punishes the turfgrass and soil like no other place. The end of summer generally leaves the turfgrass thin and vulnerable at best and bare, muddy, and void at worst. The best of traditional turfgrasses have not been up to the task, and the most heavily traveled areas have been converted to gravel and concrete.

Turfgrass is still the preferred surface for large areas from the standpoint of both beauty and function. The unique needs of the Mall have stimulated the development of a turfgrass mixture that fills the need for toughness, sustainability, and year-round green color. This is a large order in the Washington, DC, climate, which may swing from less than 0° F in the winter to more than 100° F in summer.

Enter U.S. Department of Agriculture (USDA) agronomists with an idea for a solution. Why not use one high-quality turfgrass with exceptional durability in hot weather in a mixture with one having exceptional performance in cool-to-cold weather, resulting in a high-quality, tough, year-round turf?

For example, zoysia grass forms a dense, tough, fine-textured, attractive, and highly competitive turf from May to October. Unfortunately, it turns brown with October's first frost and stays brown until late April or early May. It is usually propagated with plugs.

In contrast, the new, improved, tall fescue turfgrasses shine in the cooler weather of fall and spring and generally stay green through Washington's winter months. They are propagated with seed. In fact, the tall fescue seed germinates with exceptional ease, and the seedlings become established and grow in competitive environments.

However, both grasses are tough competitors and tend to crowd out other plants. Could new strains, varieties, or cultivars of each species be found or bred that would survive together and combat the common enemies—excessive heat, cold, wet, dry, and the constant pounding of foot traffic in all kinds of weather?

The agronomists screened hundreds of varieties of grass in search of a fine-bladed fescue that could hold its own against zoysia grass. Eventually, they were able to bring together a mixture of fescue and zoysia grass that was worth field trials. Some 200,000 square feet, or the area of four football fields, were eventually sodded with zoysia and overseeded with tall fescue on the Washington Mall. This turfgrass appears to be enduring the test of time.

Meanwhile, a zoysia grass that will produce seed has been developed, and the turfgrass industry has picked up on the new products. Similarly, agronomists of the National Turfgrass Evaluation Program have established a testing program for zoysia grass and mixtures at more than 25 locations throughout the country. The greatest promise for the zoysia–fescue mixture is in the transition zone that stretches from New Jersey to Georgia and as far west as Kansas. Perhaps one plus one can equal one: One warm-season turfgrass plus one cool-season turfgrass equals one exceptional year-round turfgrass!

The plant spreads by both stolon and rhizome growth. It has excellent wear resistance, recuperative potential, and drought tolerance. Bermuda grass can persist on a wide range of soil types and soil pH. However, it has poor shade tolerance and low winter hardiness.

The improved types of Bermuda grass require a high level of maintenance and must be vegetatively established. Recommended mowing heights for these grasses range from 0.25 to 1 inch. The common type of Bermuda grass is established by seed, requires a greater mowing height than the improved cultivars, and in some situations is considered a weed.

Zoysia grass (*Zoysia japonica* Steud.)

Zoysia grass has excellent low-temperature hardiness and is found in home lawns as far north as New Jersey. It can also be found on golf course fairways and tees within the transition zone. The transition zone is a geographic area of the United States where the warm-season and cool-season adaptation zones overlap.

Zoysia grass has good drought and shade tolerance. It can survive in a wide range of soil types. However, this plant will not perform well in poorly drained soils that remain waterlogged. Though zoysia grass has excellent wear resistance, it has a low recuperative potential. It is not as aggressive as Bermuda grass.

zoysia grass
buffalo grass

Zoysia grass requires a low-to-moderate level of maintenance. It is vegetatively established because seed germination is poor; however, new strains show promise for seed propagation. This grass should be mowed at a height between 1 and 2 inches. Two major pest problems of zoysia are nematodes and billbugs.

Buffalo grass (*Buchloe dactyloides*)

Buffalo grass is adapted to the dry, semiarid Great Plains region of the United States. It is native to Oklahoma, Texas, Arizona, Kansas, Colorado, North Dakota, South Dakota, and Montana. Regarded as the most drought-tolerant turfgrass in the United States, it survives both high and low temperature extremes and persists through droughts without irrigation. However, its shade tolerance is poor.

Buffalo grass makes a gray–green turf with fine texture. Its vertical growth is slow, but it spreads by stolons, creating a continuous turf. It tolerates alkaline soil and prefers soils of the heavier, fine-textured type. It is used on nonirrigated lawns and golf fairways with cutting heights ranging from 0.5 to 2 inches. Buffalo grass is also used along roads and other low-maintenance areas for erosion control. Little or no fertilizer is necessary for survival, but turf quality may be improved by light applications of fertilizer. The grass is propagated by seed or plugs. It is virtually pest-free if not overfertilized or overirrigated.

St. Augustine grass (*Stenotaphrum secundatum*)

St. Augustine grass is a major lawn grass in the Deep South and is used in the warmest area of the subtropical zone. It is adaptable to many soil conditions but does best on moist, well-drained sandy soils. Drought resistance is only fair, so irrigation is required in dry weather. The grass has thick stolons and produces a coarse turf of good color, medium density, and excellent shade tolerance.

St. Augustine grass is propagated with sprigs or sod and has a good establishment rate. The grass has a vigorous growth rate with moderate maintenance requirements. Medium fertility is required, and acceptable cutting heights range from 0.5 to 3 inches. Thatch buildup, chinch bugs, and the St. Augustine decline virus are some of the problems encountered with this grass.

TURFGRASS CULTURAL PRACTICES

Mowing, fertilization, and irrigation are the most common and most important cultural practices performed to maintain turfgrass stands. These practices have tremendous effects on turfgrass quality and persistency. Improper mowing, fertilization, and irrigation are major causes of poor lawns (Figure 24-11). Once a choice of turfgrass has been made, it is important to understand and address the unique requirements of the variety or mixture.

Mowing

Mowing will influence the functional use, persistency, and aesthetic value of a turfgrass. Grasses used for recreational purposes must be playable. The playability of an athletic-field turf is principally determined by its mowing height. A uniform turf surface, fine-leaf texture, and freedom from weed encroachment require proper mowing height and frequency.

Turfgrasses are capable of being mowed because the crown, their growing point, is located just below or at the soil surface. However, mowing does have several adverse effects on the grass plant. Reduced rooting depth and decreased carbohydrate reserves occur in mowed turfs. Improper mowing practices only accentuate these and many other adverse effects, causing a decline in turfgrass quality (Figure 24-12).

Causes of Poor-Quality Lawns

- Using the wrong turfgrass species or cultivars
- Using poor-quality seed
- Mowing the lawn too closely
- Permitting excessive growth between mowings
- Using too little or too much lime or fertilizer
- Improper watering
- Too much shade
- Droughty or poorly drained soils
- Too much traffic
- Damage by insects or disease
- Improper use of chemicals

FIGURE 24-11 Major causes of poor-quality lawns.

FIGURE 24-12 The cutting deck of a home lawnmower should be set for the proper depth of cut to ensure a healthy lawn.

Recommended Mowing Heights	
Species	Mowing height range in inches
Bahia grass	2–4
Bermuda grass	
Common	0.5–1.5
Hybrids	0.25–1.0
Carpet grass	1–2
Centipede grass	1–2
St. Augustine grass	1.5–3.0
Zoysia grass	0.5–2.0
Creeping bent grass	0.2–0.5
Colonial bent grass	0.5–1.0
Fine fescue	1.5–2.5
Kentucky bluegrass	1.5–2.5
Perennial ryegrass	1.5–2.5
Tall fescue	1.5–3.0
Crested wheatgrass	1.5–2.5
Buffalo grass	0.7–2.0
Blue grama	2.0–2.5

FIGURE 24-13 Recommended mowing heights for different turfgrasses.

Mowing Height

The recommended ranges for mowing height are listed in Figure 24-13. If one mows below or above these ranges, various problems will occur. Mowing below the desired range will reduce photosynthesis, thus preventing carbohydrate production within the plant. This causes a reduction in rooting and decreases the recuperative potential of the plant.

Mowing above the recommended height will increase thatch buildup and leaf texture. It will also decrease turf density and appearance. **Thatch** is the buildup of organic matter on the soil around the turfgrass plants. Excessive thatch causes poor water infiltration, increased disease activity, and decreased rooting depth.

Mowing Frequency

Frequency of mowing is determined by the mowing height and the growth rate of the plant. No more than one-third of the top growth should be removed per mowing (Figure 24-14). This allows sufficient leaf area for photosynthesis after mowing. The lower the mowing height, the more frequent the mowing if the one-third rule is observed. For example, creeping bentgrass mowed at a quarter inch will be mowed five to six times per week. However, tall fescue cut at 3 inches requires only one mowing per week. Under ideal growing conditions, grasses require more frequent mowing than they do under poor growing conditions.

Fertilization

Fertility programs attempt to supply adequate levels of plant nutrients to allow for favorable plant growth (Figure 24-15). Proper fertilization should increase

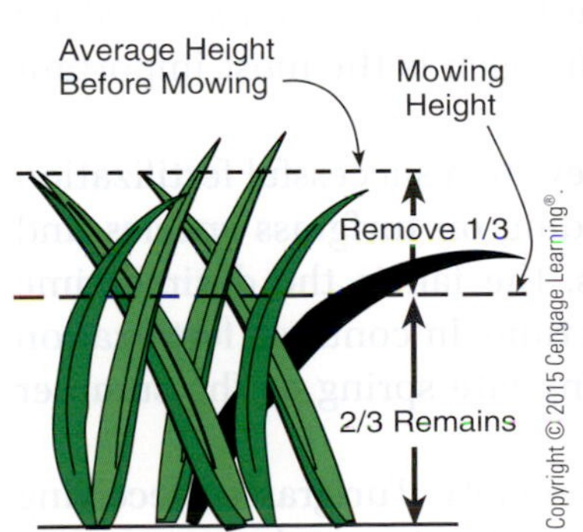

FIGURE 24-14 Mowing more than one-third of the top growth of turfgrasses is likely to result in reduced rooting and less recuperative potential in the plant.

FIGURE 24-15 Healthy lawns require fertilization with the proper nutrients to maintain vigor and color.

mist to a turfgrass, particularly on golf putting greens. It may be used to reduce plant temperatures or to remove dew and frost from the turfgrass leaf.

The amount of irrigation needed to maintain optimum plant growth is dependent on many factors. Some of these factors are turfgrass species, geographic location, soil type, weather conditions, and turfgrass use. General recommendations are to apply 1 inch of water per week during the summer months. It takes approximately 620 gallons of water per 1000 square feet to provide 1 inch of water. This amount of water will produce a green and actively growing turfgrass stand.

Heavy but infrequent irrigation will force root growth deep into the soil. Light and frequent irrigation will keep the surface soil moist, thus encouraging shallow rooting.

The best time to irrigate is at night, when evaporation and wind are low. However, an increase in disease activity will occur at this time. Therefore, early morning watering is often selected because the effects of evaporation, wind, and disease activity are at a minimum.

TURFGRASS ESTABLISHMENT

Turfgrasses may be established by seeding or vegetative propagation. Regardless of the method used, proper establishment practices, including site preparation, should be performed. Correct establishment practices will ensure adequate turfgrass quality and persistency.

Turfgrass Selection

Selection of the correct turfgrass species is one of the most important decisions in the establishment process. Selecting a turfgrass seed blend or seed mixture should be based on the intended use and performance data of the turfgrasses. A **seed blend** is a combination of different cultivars of the same species. A **seed mixture** is a combination of two or more species. Many state land-grant universities evaluate turfgrass species and new cultivars and provide information on their performances (Figure 24-19).

Site Preparation

Proper site preparation may include any or all of the following activities:

- Debris removal
- Nonselective weed control
- Installation of a subsurface drainage or irrigation system, or both
- Soil modification, tillage, and grading

The removal of woody vegetation, leftover construction debris, and any large stones will reduce future maintenance problems. If woody debris is buried on the site rather than removed, the soil will settle and diseases such as fairy ring may result. Excessive amounts of stone and rock present in the seedbed will cause localized dry spots and interfere with future cultivation practices.

Nonselective weed control may be necessary if perennial grassy weeds or difficult-to-control weeds are present. A nonselective herbicide or a soil fumigant should be applied before seeding. The herbicide glyphosate (Roundup) is often used to provide nonselective weed control. A soil fumigant, such as metham, may also be used.

INTERNET KEY WORDS:
grass seeding practices

Kentucky Bluegrass Cultivar	Quality Rating*					
	April 1	May 4	June 8	July 15	Sept. 2	Oct. 17
Adelphi	4.0*	5.6	5.4	4.5	5.9	5.7
Baron	3.8	5.1	4.8	4.4	5.4	5.5
Bensun (A-34)	3.6	5.3	6.0	5.9	6.1	5.4
Cheri	3.2	5.2	4.7	4.9	5.8	6.0
Emmundi	4.0	4.9	4.6	4.5	5.4	5.7
Glade	4.2	5.1	5.4	5.5	6.3	6.3
Majestic	3.9	6.1	6.0	5.8	6.0	5.8
Newport	3.7	3.6	3.8	4.0	4.2	4.9
Parade	4.5	6.8	5.9	4.4	5.7	5.8
Ram 1	4.3	6.9	5.9	6.0	6.2	6.4
Sydsport	3.8	6.2	6.0	5.9	6.4	6.1
Touchdown	4.1	6.3	5.9	5.7	6.2	6.5

*It is important to note that quality ratings can vary significantly from one region or location to another. A rating of 1 = no live turf, 9 = ideal turf, >5 = acceptable quality.

FIGURE 24-19 Quality ratings for Kentucky bluegrass cultivars.

If the site is to be used as a sports turf, the installation of drainage and irrigation systems may be necessary. This should be done before final grading.

Soil modifications, followed by tillage and grading, are the last steps in site preparation (Figure 24-20). Soil amendments, such as organic matter and sand, may be incorporated to improve nutrient retention and/or drainage. Topsoil may also be used to provide a favorable growth medium for turfgrasses. The soil should be worked to a depth of 6 inches and then lightly tilled to provide a uniform seedbed and adequate surface drainage (Figure 24-21). Incorporation of lime and

FIGURE 24-20 The site must be prepared properly by tilling and grading the soil before grass seed is planted.

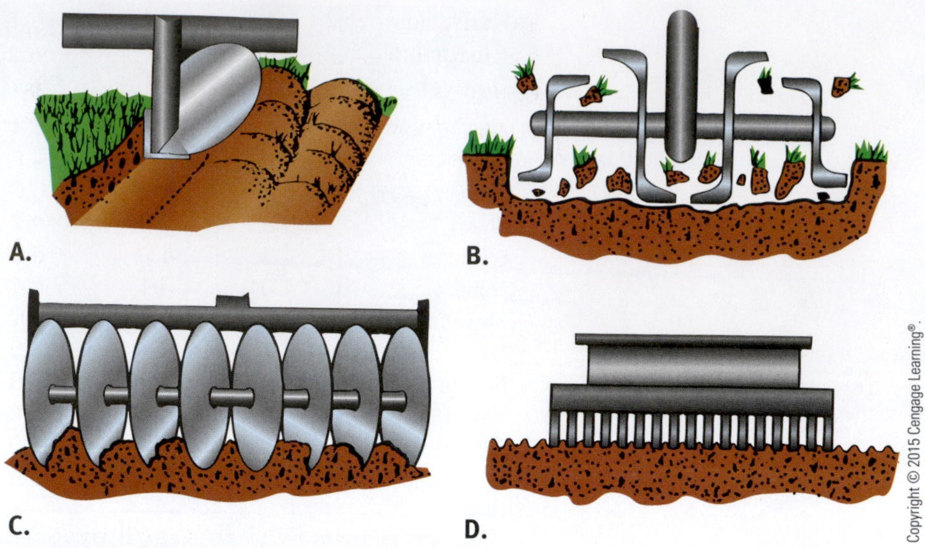

FIGURE 24-21 Tillage equipment frequently used for turfgrass establishment: (A) plow, (B) rototiller, (C) disc cultivator, and (D) harrow.

fertilizer should be done during this time. Excessive tillage will destroy soil tilth. A seedbed with soil aggregates of 0.25 to 1 inch in diameter is ideal.

Additional preparation is necessary on sites where bunch grasses are selected as turfgrass crops. The lack of a strong fibrous root system makes it difficult to hold the harvested turf plugs together while they are shipped and transplanted. Netting materials are used to compensate for the lack of an extensive root system. Netting materials must be placed on the soil surface before the grass is planted. As the turfgrass matures, it becomes bound into the netting material. This allows it to be harvested and handled while causing minimal damage to the turf (Figure 24-22).

FIGURE 24-22 Some grass species that lack extensive root systems require netting to be applied to the soil surface to hold the turf together when it is harvested.

Planting

Turfgrasses can be established by seeding, sodding, stolonization, sprigging, and plugging. The last four methods are vegetative means of planting. Vegetative establishment is usually practiced when the selected turfgrasses produce infertile seed or have low seed yields. Also, if quick establishment is required, sod may be installed.

Seeding

Seeding is the principal method for turfgrass establishment because it is the least expensive. Figure 24-23 provides the seeding rates for some popular grasses. These rates will vary depending on seed size and percentage of seed germination.

The seed label informs the buyer or consumer of the quality and ingredients of a seed blend or mixture. Information on purity, germination, weed seed content, inert matter, and test date is present on the label (Figure 24-24). Certification programs are available for turfgrass seed. Seed certification programs are administered by state governments, and they ensure that seed is true to type. They also require the seed to meet other minimum quality standards concerning percentage of germination and weed seed content.

The best time to plant turfgrass depends on the type of turfgrass that is to be established. The optimum planting time for cool-season grasses is in the fall.

Seeding Rates	
Species	Pounds of Seed per 1000 ft^2
Bahia grass	3–8
Bent grass	
Colonial	0.5–1.5
Creeping	0.5–1.5
Bermuda grass	
(hulled)	1–2
Bluegrass	
Kentucky	1–2
Buffalo grass	3–7*
Carpet grass	1.5–5
Centipede grass	0.25–2*
Fescue	
Fine	3–5
Tall	5–9
Grama, blue	1.5–2.5
Ryegrass	
Annual	5–9
Perennial	5–9
Wheatgrass,	
crested	3–6
Zoysia grass	1–3

*The higher rates are best, but lower rates are commonly used because the seed is expensive.

FIGURE 24-23 Seeding rates for the major turfgrass species.

Ideal weather conditions and less annual weed competition are the principal reasons for fall establishment. The plant will also have sufficient time to develop an adequate root system before summer stress conditions occur. Spring seeding is less desirable because there is insufficient time to develop a mature stand capable of competing with undesirable grasses.

Warm-season turfgrasses are established during the spring and summer months. The best planting time is in late spring. Late-spring planting gives the new grass the longest period for optimum growth and development after establishment.

Different types of equipment may be used to apply grass seed. Fertilizer drop spreaders, overseeders, hydroseeders, and cultipacker seeders are some types of equipment used for seeding. Shallow seed placement (a quarter inch) is important for proper germination. This can be done by hand raking for small areas or the use of a drag mat for larger areas. Specialized seeding equipment, such as a hydroseeder or cultipacker, will place the seed at the right soil depth (Figure 24-25).

After seeding, the use of mulch provides a favorable environment for seed germination. Mulch conserves soil moisture and prevents soil erosion. This is extremely important because adequate surface moisture must be present for seed germination. Straw mulch is preferred and is usually applied at the rate of 2 bales per 1000 square feet. Other mulches are wood and paper by-products, net or fabric, and peat moss.

Sodding

Sodding refers to removing a rectangular piece of grass and a shallow layer of the soil beneath it and moving it to another location. The grass and the soil immediately beneath are referred to as **turf**. Sodding offers the ability to establish turf at any time of the year. It also provides an instant cover. However, the cost is comparatively high, and the sod must be installed soon after harvesting. Specialized equipment for harvesting sod has been developed (Figure 24-26). Sod is cut

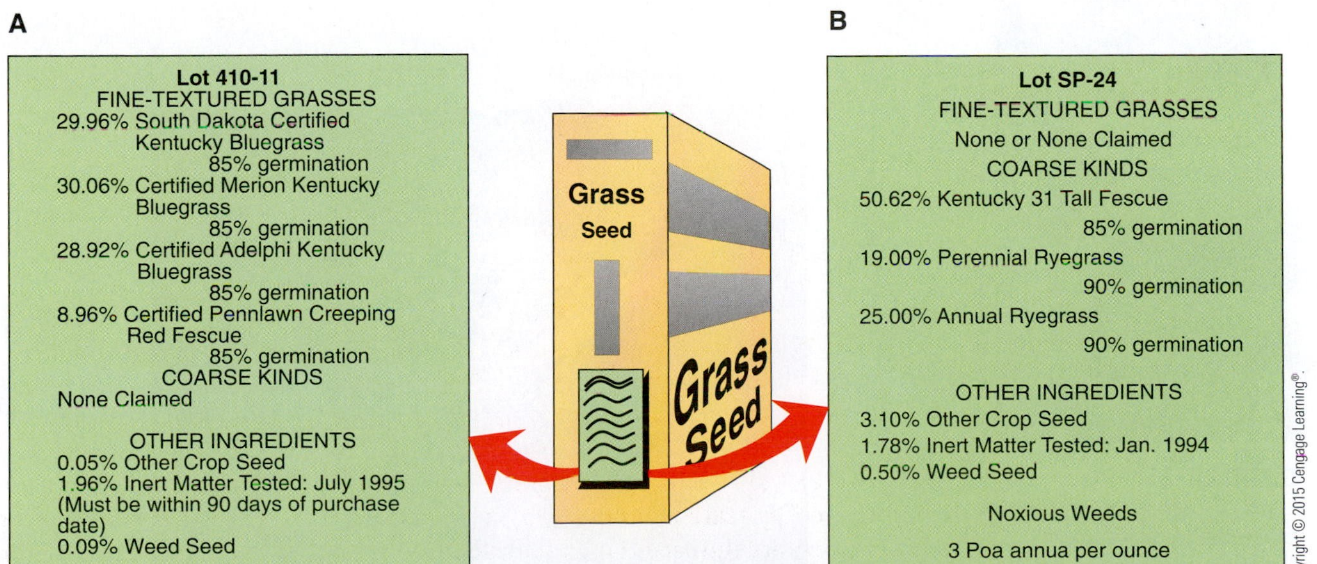

FIGURE 24-24 Sample labels for lawn and turf seed. (A) Label describing a recommended seed mixture of cool-season grasses. (B) Label describing a poor-quality seed mixture of cool-season grasses.

AGRI-PROFILE CAREER AREA: GOLF COURSE MAINTENANCE SUPERVISOR

Golf course fairways and greens require regular early morning care, especially during periods of heavy use.

The game of golf is one of the most popular leisure-time activities in which people participate. Many new golf courses are designed and constructed every year, and all of the existing courses must be maintained. This kind of work requires a person who knows how to establish turfgrass plantings and keep them healthy. Whether the golf course is new or well-established turf, it requires constant maintenance and care. Among the tasks that are performed as part of a golf course maintenance plan are controlling weeds, insects, and rodents; reseeding damaged turf; irrigation management; soil testing; fertilization; mowing; caring for trees; and establishing and maintaining other plantings. All of these tasks require expert knowledge.

A person who is interested in a career in golf course maintenance can obtain the required training by enrolling in a 2-year technical college program in turfgrass management or by obtaining a Bachelor's degree in plant science or a related field of study. A summer internship on a golf course would be a valuable component of either of these education options. A person who becomes a good manager of golf course resources can expect to have good employment opportunities

into 12- to 24–inch widths and lengths of up to 3 feet or in carpet-like pieces. The depth of cut will vary from 0.3 to 0.5 inch. The cut sod is then rolled, folded, or stacked on pallets for transport and later placed on a carefully prepared seedbed. The new turf is then rolled to ensure good contact with the soil. After the sod is in place, it must be watered thoroughly.

FIGURE 24-25 One method of seeding turfgrass is to use the hydroseeding technique, which mixes seed, mulch, fertilizer, and water and sprays it on the planting surface.

FIGURE 24-26 Cutting sod for placement in a new location as turfgrass.

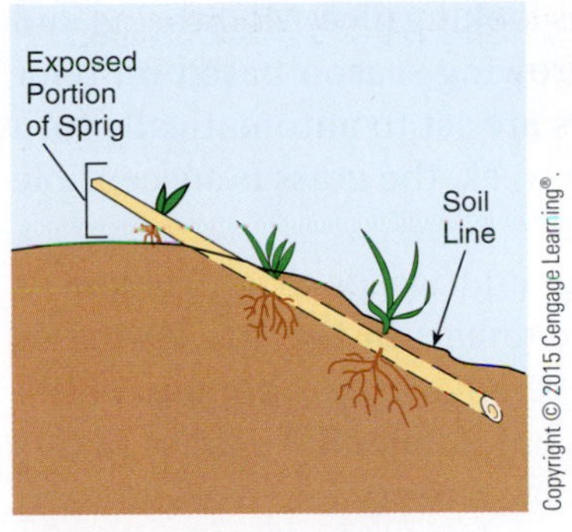

FIGURE 24-27 When sprigging, two or more nodes with shoots should be placed in the soil and watered well.

Sprigging

Sprigging is the planting of a section of a rhizome or stolon, referred to as a sprig. A sprig may be up to 6 to 8 inches in length. For successful establishment, a section of the sprig with several nodes must be properly placed into the soil (Figure 24-27). After sprigging, irrigation must be applied to prevent drying out. Equipment for sprigging has been developed and will plant sprigs in rows spaced 6 to 18 inches apart.

Stolonizing

Stolonizing is similar to sprigging, in that 6- to 8-inch sections of rhizomes or stolons are used. However, in stolonizing, sprigs are broadcast onto the soil surface. These sprigs may be lightly top-dressed with soil and rolled before irrigation. Survival rates are lower than with sprigging. Therefore, stolonizing requires greater quantities of sprigs.

Plugging

Plugging is the establishment of a turfgrass stand by using plugs, or small pieces of existing turf. Plug sizes vary, as noted in Figure 24-28. Spacing of plugs can be on 6- to 18-inch centers, depending on how quickly the turfgrass needs to be established. Plugging can be done by hand or with specialized equipment. It requires a longer time to cover the soil than using the other establishment techniques.

In summary, the turfgrass industry is expanding. As more space is used for non-farm purposes, the use of turfgrass for conservation, recreation, and beautification is expected to continue to grow. Excellent career opportunities exist, and suitable educational programs are available to prepare new workers for this increasingly popular area of agriscience.

Maintenance

After investing a lot of time and money establishing a grassy area, it would be unfortunate if a lack of maintenance ruined the lawn. Homeowners and grounds

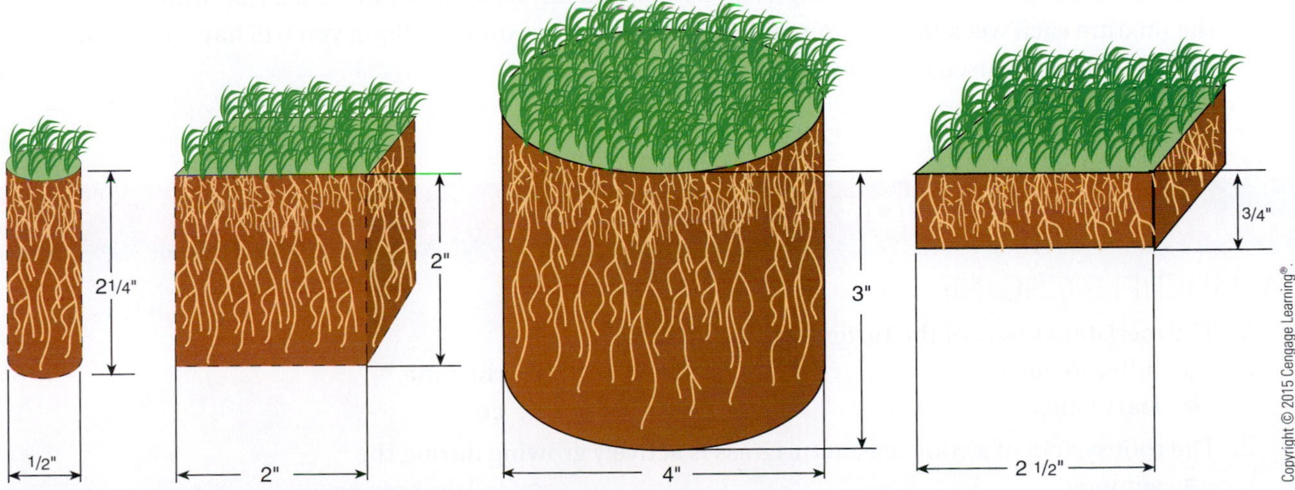

FIGURE 24-28 Turfgrass plugs may be used to plant lawns and playing fields. As the plugs grow, they expand to fill in the space between plants.

keepers need to do a number of things to keep lawns looking nice. Monitoring and adjusting water should be done throughout the growing season based on temperature and rainfall. For example, when sprinklers are set to automatically turn on every other day for 20 minutes and it rains for a week, the grass is susceptible to rot and fungus infestations.

Grassy areas need to be monitored for pests and weeds. If a problem is discovered, it should be treated quickly to avoid damage. In the late spring or early summer, older lawns should be thatched to remove excess organic build-up. Organic matter is good in moderate amounts, but it can choke grass plants at excessive levels. It is also important to reseed a lawn every few years in early spring. Reseeding is important to keep desired grasses strong and to keep undesirable grasses such as crab and quack grasses at bay.

STUDENT ACTIVITIES

1. Write the Terms to Know and their meanings in your notebook.
2. Determine whether your community is located in a warm-season, transitional, or cool-season zone.
3. Visit or call your county cooperative extension office and obtain copies of the various publications available on turfgrass or lawn production and maintenance.
4. Determine the recommended cultural practices for establishing and maintaining a major turfgrass species or mixture for your locality. Report your findings to the class.
5. Arrange to visit a golf course and discuss with the superintendent the duties of turfgrass workers on the golf course.
6. Obtain a map of your county and attach it to the bulletin board. Place pins at the locations of all businesses, schools, government buildings, and institutions that have extensive lawn or turfgrass areas around them. Ask your classmates to help identify them. Use different colored pins to identify different institutions.
7. Obtain a grass identification key from your teacher or the cooperative extension service office. Collect 10 specimens of different turfgrasses and identify them by using the identification key.
8. Make your own grass clipping mulch. Inside a compost bin or in a plastic garbage container with ventilation holes, mix grass clippings, leaves, soil, and any other plant material. Add water and stir the mixture each week to add moisture and oxygen to the compost. Soon you will have useful nutrient mulch that can be used in a flower bed or a gardenplot.

SELF-EVALUATION

A. MULTIPLE CHOICE

1. The meristem tissue of the turfgrass plant is the
   a. inflorescence.
   b. leaf blade.
   c. rhizome.
   d. crown.
2. The root system of a cool-season turfgrass is actively growing during the
   a. summer.
   b. fall and winter.
   c. fall and spring.
   d. late spring.

3. The estimated value of the turfgrass industry in the United States is
   a. $147 billion.
   b. $50 billion.
   c. $100 billion.
   d. $200 billion.
4. Which irrigation practice will develop a deeper, more extensive root system?
   a. frequent and light applications
   b. infrequent and heavy applications
   c. frequent and heavy applications
   d. infrequent and light applications
5. Turfgrasses are in the _______ family.
   a. Compositae
   b. Cyperacese
   c. Cruciferue
   d. Poacea
6. _______ is a warm-season turfgrass.
   a. Kentucky bluegrass
   b. Zoysia grass
   c. Red fescue
   d. Perennial ryegrass
7. Which cultural practice will have an adverse effect on rooting depth?
   a. low mowing
   b. heavy, infrequent irrigation
   c. moderate fertilizer applications during the time of root growth
   d. high mowing
8. A tiller is an example of
   a. rhizome growth.
   b. extravaginal growth.
   c. stolon growth.
   d. intravaginal growth.
9. Which of the following is not a turfgrass stem?
   a. rhizome
   b. stolon
   c. leaf sheath
   d. crown
10. Which cool-season turfgrass is adapted to shady, dry locations?
   a. red fescue
   b. Kentucky bluegrass
   c. tall fescue
   d. creeping bent grass

B. MATCHING

_________	1. Rhizome	a. An induction process for inflorescence development
_________	2. Stolon	b. A turfgrass stem that grows horizontally aboveground
_________	3. Creeping bent grass	c. A cool-season turfgrass that is very drought tolerant
_________	4. Sprigging	d. A wide-leaf blade
_________	5. Photoperiod	e. A cool-season turfgrass used on putting greens
_________	6. Tall fescue	f. A turfgrass stem that grows horizontally below ground
_________	7. Coarse texture	g. A buildup of organic matter on the soil around turfgrass plants
_________	8. Syringing	h. A complete fertilizer
_________	9. 10-10-10	i. Light application of water to a turfgrass
_________	10. Thatch	j. Vegetative establishment

C. COMPLETION

1. _______ is a warm-season turfgrass having good winter hardiness.
2. _______ roots are present at the time of seed germination.
3. _______ is an example of intravaginal growth.
4. A seed _______ consists of different cultivars of the same species.
5. _______ is a turfgrass species with rapid seed germination.

6. As mowing heights decrease, mowing frequency will _______ .
7. _______ is considered the most important element in turfgrass fertilization.
8. A planting depth of _______ is recommended for turfgrasses.
9. _______ is an establishment practice of applying the sprig to the soil surface.
10. A _______ mulch is preferred over other types of mulch for seeding turfgrasses.

UNIT 25

Trees and Shrubs

OBJECTIVE

To use trees and shrubs for beautification, wood products, improvement of air quality, and pollution control.

MATERIALS LIST

- nursery catalog
- pictures of plants commonly grown in an area
- list of insects and diseases of trees and shrubs that are problems in the area
- Internet access

COMPETENCIES TO BE DEVELOPED

After studying this unit, you should be able to:

- identify ornamental trees and shrubs.
- select trees and shrubs for appropriate landscape use.
- classify trees and shrubs according to growth habit, growth habitat needs, and other requirements.
- identify trees and shrubs using proper nomenclature.
- purchase plant material for installation in a landscape.
- plant and maintain plant material.

SUGGESTED CLASS ACTIVITIES

1. Conduct an inventory of the trees in a park or along the streets of a small community. Identify the species and determine which species are most often found. A map of tree species may be available from the park or forestry department in some cities and towns. Teach students how to use a key to identify different species of trees.
2. Identify a potential location at the school or other community property where a landscape project could be conducted by the class. Contact the appropriate officials and obtain permission to develop a landscaping plan. Work with community groups and local government to obtain financing for the project. Install the trees, shrubs, and ornamental plants in the landscape according to the landscape design.
3. Create a hypothesis based on this question: What does a tree's age have to do with the circumference of its trunk? Take a field trip to a native tree stand in your area. Measure the circumference, at breast

TERMS TO KNOW

- specimen plant
- border planting
- group planting
- hardy
- nomenclature
- genus
- bare-root
- heel in
- balled and burlapped (B & B)
- root pruning
- containerized plant
- stake
- drip line
- pruning
- canopy

height, for 15 trees of the same species. Core each tree to determine its age. Calculate the average circumference and age. Write a short paper discussing how your results compare with your hypothesis.

Trees and shrubs are major components of the environment. They are important in providing natural beauty and in providing oxygen to humans and animals. As a result, the need for trees and shrubs has risen in recent years. This need has created a demand for horticulture technicians who understand the production and care of landscape plants.

TREES AND SHRUBS FOR LANDSCAPES

A well-designed landscape increases the value of a property (Figure 25-1). Residential properties are much more attractive when trees and shrubs are used to enhance their beauty and balance.

Trees and shrubs provide more than natural beauty. An acre of good, healthy trees or shrubs will produce enough oxygen to keep 16 to 20 people alive each year. These plants are also valuable in helping keep our air clean by using the carbon dioxide produced by people, automobiles, and factories.

Trees and shrubs cut noise pollution by acting as barriers to sound. They can deflect sound as well as absorb it. When used properly in the landscape, trees and shrubs can provide shade and act as insulators to keep the house cooler in the summer and warmer in the winter.

Many urban areas depend on trees and shrubs to soften the concrete, blacktop, and steel environment. In fact, many of these areas now employ urban foresters to help design, install, and maintain trees and shrubs in large cities (Figure 25-2).

FIGURE 25-1 Trees and shrubs add character and value to a property.

FIGURE 25-2 Trees and shrubs add beauty, provide sound insulation, purify the air, and provide privacy.

FOREST RESOURCES

Vast regions of North America produce forest products such as wood, cardboard, paper, solvents, medicines, fuels, and many other products. An entire industry depends on trees for the raw materials that are used to process these important products. Thirty percent of the land area in the world is forest land, and forest products are important to the economies of developed countries. Some of the trees that are used for this purpose are produced on private land. Vast tracts of public land are also devoted to timber production.

Forests are important resources in many ways besides production of wood and forest products. A forest also functions as a biological filter system that cleans the environment by removing impurities from air and water. Forest plants restore oxygen to the atmosphere and improve watersheds that contribute to consistent supplies of fresh, pure water. Forests furnish habitat to many kinds of wild animals and birds. They also provide recreation opportunities for people who enjoy outdoor activities such as camping, going on picnics, hiking, boating, fishing, and hunting. Forests provide many kinds of resources that are important to people.

PLANT SELECTION

Trees and shrubs are distinct groups of plant materials. Trees are woody plants that produce a main trunk and grow to a height of 15 feet or more. Shrubs are woody plants with a low growth habit, produce many stems or shoots from the base, and do not reach more than 15 feet in height (Figure 25-3).

USE

Ornamental trees and shrubs in the landscape are both beautiful and functional. They are used as specimen plants, border plantings, or groupings. A **specimen plant** is used as a single plant to highlight or provide some other special feature

FIGURE 25-3 A tree has a single stem and a height of 15 feet or greater. A shrub has multiple stems and does not grow to a height greater than 15 feet.

to the landscape. A **border planting** is used to separate some part of the landscape from another or to serve as a fence or a windbreak (Figure 25-4). A **group planting** consists of a number of trees or shrubs that are planted together so they point out a special feature, provide privacy, or create a small garden area.

The location of plant material in the landscape will play an important part in the selection process. Color of the leaves, texture of the plant, and color of the flowers are just some of the factors that must be considered before purchasing or planting trees or shrubs.

INTERNET KEY WORDS:
plant hardiness zone map

FIGURE 25-4 A border planting is used to separate one landscape feature from another.

Geographic Location

The geographic region in which a plant will be used is an important factor in the selection of a tree or shrub. Some plants are not **hardy**—that is, they may not be able to survive or even grow properly in an area for which they are not adapted. The hardiness of a plant is affected by the intensity and duration of sunlight, length of the growing season, minimum winter temperatures, annual precipitation, summer droughts, and humidity. The U.S. Department of Agriculture (USDA) has issued a revised plant hardiness zone map. (See Unit 18.) There are 13 zones and 26 subzones in the United States, including Alaska and Hawaii. Each zone represents an area of winter hardiness that is based on the average annual minimum winter temperatures. It is possible that local climates may vary from the general zone map. They may be colder or warmer than is indicated on the map. Local nursery personnel are generally willing to help in the selection of the best plant material. Most nursery catalogs list the plant for the coldest zone in which it can reasonably be expected to grow. Two examples of such catalog listings are the following:

1. *Cornus florida* (flowering dogwood)—Zones 5–9: A low-branched, flat-topped tree that has a horizontal branching habit. It will grow 20 to 30 feet

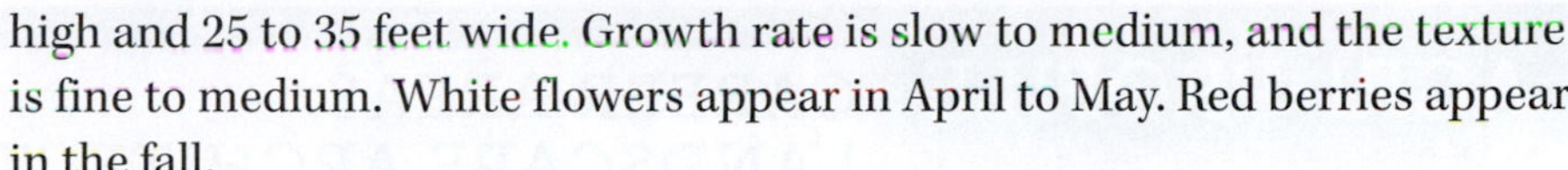

high and 25 to 35 feet wide. Growth rate is slow to medium, and the texture is fine to medium. White flowers appear in April to May. Red berries appear in the fall.

2. Pyrus calleryana "Bradford" (Bradford pear)—Zone 5: A dense, pyramidal tree that becomes brittle with age. It will grow 30 to 50 feet high and 30 to 35 feet wide. The growth rate is medium, and the texture is medium to fine. Its use as an all-purpose tree, shade, or street tree has declined in recent years due to the tendency for the trunk to split or limbs to break when it is exposed to ice, heavy snowfall, or strong winds. Profuse white flowers one-third-inch in diameter appear in late April or early May in clusters up to a diameter of 3 inches.

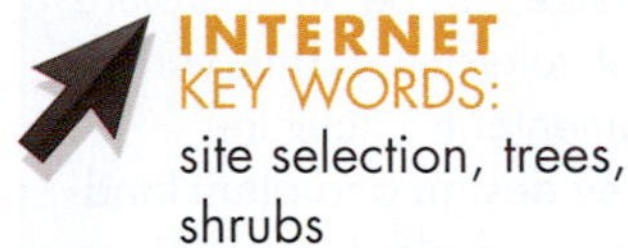

site selection, trees, shrubs

A plant can also be expected to live in a warmer zone than indicated if rainfall, soil, and summer conditions are comparable. In some cases, it may be necessary to adjust these conditions through irrigation, correction of soil conditions, wind protection, and alteration of shade or sun exposure. It is also possible to grow some plants in areas north of the indicated zone. Such plants may need special attention to protect them from wind or cold. Without such protection, they may not perform normally and are likely to suffer winter injury.

Site Location

When planning the location for planting trees and shrubs, many of the following factors should be given consideration:

- Flower color: Is it compatible with the house, fence, patio, and any other plants in the area?
- Fruit size and type: Many trees, such as some crabapples or cherries, drop messy fruit. Therefore, they should not be placed near an area that will be walked on or heavily used by people. Such areas include driveways, walkways, patios, and swimming pools.
- Other structures: Do not plant trees directly in front of doors; near wells, cesspools, or field drains; or under utility lines, where interference is likely when the plant matures.
- Ornamental characteristics of the plant: Flowering time, shape, foliage texture, fall color, pest resistance, landscape suitability, and mature size are all part of any consideration when selecting a plant.

Type and Growth Habit

Plants have many different types of growth habits. The type of growth habit is an important consideration when selecting plants (Figure 25-5). A good landscape planner or designer will be aware of the type of growth habit for each plant used in a landscape design.

Plant size in relation to the structure around which the landscape is constructed is important. A common mistake, made by homeowners and landscape workers, is planting a tree or shrub without considering what the plant will look like in 20, 50 or even 100 years. Some trees, like the cottonwood (*Populus deltoids*), cover lawns with a thick cotton-like layer of seeds. Many of these seeds

AGRI-PROFILE

CAREER AREAS: LANDSCAPE ARCHITECT/LANDSCAPE TECHNICIAN/LANDSCAPE CONTRACTOR/ PLANT SPECIALIST

Plant care and management require many skills, including the diagnosis of disease and insect problems.

Career opportunities related to trees and shrubs span both the arts and the sciences. Ornamental trees and shrubs vary greatly in size, shape, temperature preference, light preference, fertility needs, and pest tolerance. This variety stimulates diversity in jobs and specialties in the area of ornamental horticulture.

Landscape architects practice the art of design in that they design and plant landscapes pleasing to their clients. They must know plant species and plant materials to develop plans that are attractive yet functional in the environment. Landscape architects generally have bachelor's degrees in landscape design and considerable experience in ornamental horticulture.

Horticulture specialists and technicians have careers centering on nursery or greenhouse management, landscape contracting, groundskeeping, wholesaling, retailing, public information, writing, and consulting. They may specialize in just trees, just shrubs, or both. Some specialize in pruning, tree planting, pest control, or landscape maintenance. Ornamental horticulture seems to provide an endless array of career opportunities in many localities.

germinate and result in undesirable woody shoots. The willow (*Salix babylonica*) tree is a beautiful addition to a landscape. However, its roots grow shallow and wide. Often, when planted too close to structures, willow tree roots grow into foundations of buildings. This can cause expensive structural damage. This tree is better suited farther away from structures.

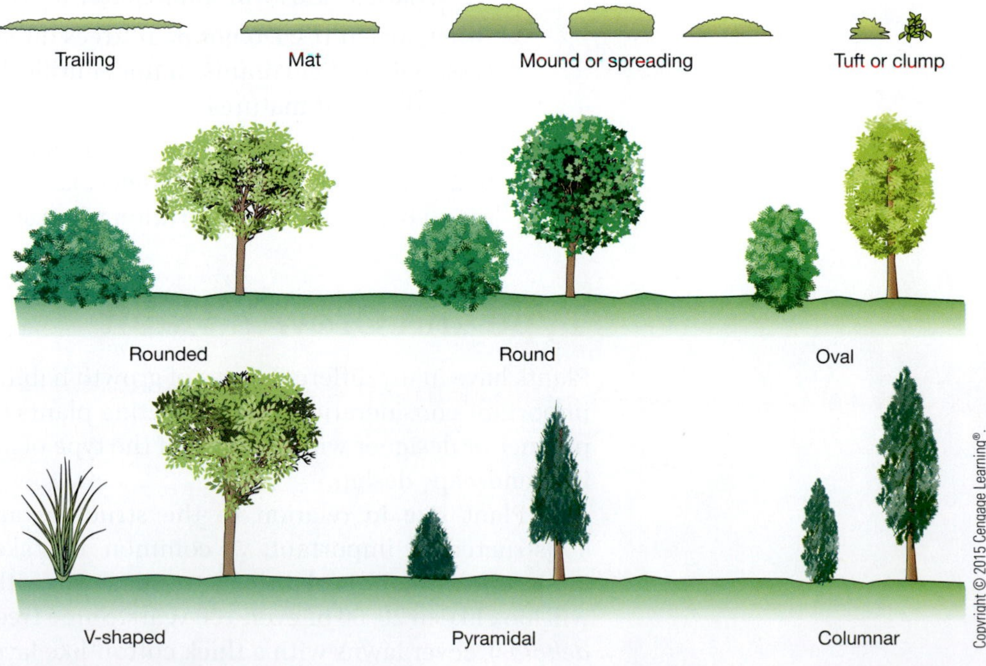

FIGURE 25-5 Types of growth habit.

INTERNET KEY WORDS:
plant taxonomy
nomenclature

Maintenance and frequent pruning of plants is expensive. Pruning of shrubs is frequently done without climbing, whereas most trees require the use of a hydraulic lift or they must be climbed to prune them. The use of a tree that is not in proportion to nearby structures can result in an expensive and frequent pruning program. Figure 25-6 illustrates some common trees and their mature sizes. A tree that is too tall will require extra work to keep it in proper relationship to adjacent structures. Such a design error would cancel out the goal of property enhancement.

FIGURE 25-6 Tree sizes in relation to a two-story house.

FIGURE 25-7 Typical shapes of landscape trees.

Shape of the Plant

The shape of a plant is an important factor in the selection of plant material. Some common undesirable effects can be avoided by asking some basic questions. Does the plant grow straight up and give little shade? Does the plant have a trunk that divides and spreads to cast unwanted shade? Is the plant a low-growing tree that gives little shade? Consider which shapes of trees and shrubs are needed in the landscape. Then select the appropriate plant to achieve the objective (Figure 25-7). Before plants are selected, research is necessary to better understand the types of plants to be used in a particular area. To do this, study the types of plants that are sold and used in the locality, and note the various characteristics of each plant. Catalogs are available from nurseries that sell plants in the locality. Also, a visit to local garden centers and nurseries will pay dividends in gathering information on plant selection.

PLANT NAMES

Trees and shrubs must be ordered by their proper names. Common names such as flowering dogwood and upright juniper are not governed by any formal code of nomenclature. **Nomenclature** is a systematic method of naming plants or animals. The botanical or scientific name is recognized internationally. Scientific designations are always written in Latin and consist of two names. The first name is the genus, and the second is the species. The genus name always begins with a capital letter and is a noun. The species name is usually written in all lowercase letters and is an adjective.

The **genus** (plural is genera) is defined as a group of closely related and definable plants composed of one or more species. The common, definable characteristics are fruit, flower, leaf type, and arrangement. The species (plural is also species) is the basic unit in the classification system whose members have similar structures and common ancestors, and that maintain their characteristics.

Variety (var.) is a subdivision of species. A variety has various heritable characteristics of form and structure that are perpetuated through both sexual and asexual propagation. The term for variety is written in lowercase letters and underlined or *italicized.* Often in catalogs and on plant labels, the abbreviation var. is used. An example of how this is used might be *Cornus florida rubra* or *Cornus florida* var. *rubra.*

A cultivar (cv.) is a group of plants within a particular species that has been cultivated and is distinguished by one or more characteristics and that, through sexual or asexual propagation, will keep these characteristics. The term is written inside single quotation marks. An example is *Pyrus calleryana* 'Bradford' or *Pyrus calleryana* cv. 'Bradford.'

It is important to become familiar with this plant-naming system because common names vary from area to area. There are no standards for creating common names. Professional horticulturists order plants using the method of naming just described. Landscape architects place scientific names on landscape drawings, and nurseries use the scientific names in their catalogs to avoid confusion by anyone who orders plant material.

OBTAINING TREES AND SHRUBS

After determining the specific plants needed in a landscape, the plants must be purchased. Plants are normally dug and shipped bare-root, balled and burlapped (B & B), or containerized (Figure 25-8).

Bare-Root Plants

A **bare-root** plant has been dug, and the soil has been shaken or washed from the roots. Normally, only deciduous trees, shrubs, and trees with taproots are shipped this way. They are dug while the trees are dormant. Plants ordered from nursery catalogs that can survive as bare-root stock are shipped this way because of lower transportation costs and easier handling. It is not economical to ship soil with the roots because it is so heavy.

Bare-root plants are best planted while they are dormant. In all instances, the roots must be protected to keep them from drying out. If the plant cannot be planted when it is received, the roots must be protected by putting them in a container of water, wrapping them in burlap after a good watering, or placing

Photo courtesy of Boise National Forest.
A

© J. Bicking/Shutterstock.com.
B

© Vaidas Bucys/Shutterstock.com.
C

FIGURE 25-8 Methods of preparing plants for shipping and handling.

HEELING IN

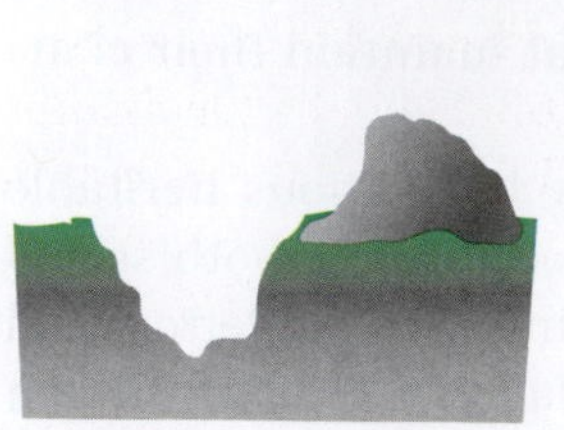
1. Dig V-shaped trench in moist, shady place.

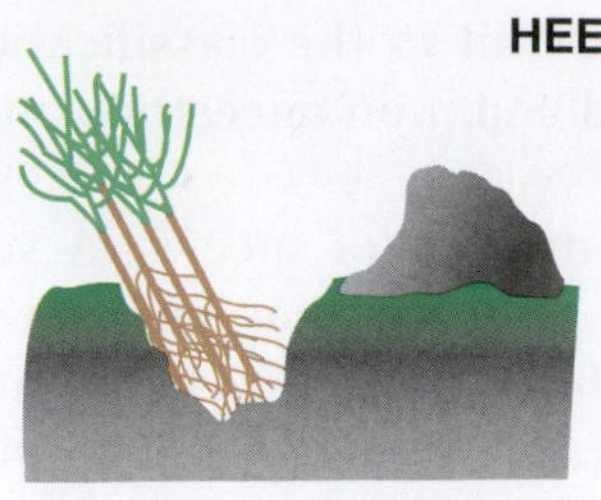
2. Break bundles and spread out evenly.

3. Fill in loose soil, and water well.

4. Complete filling in soil and firm with feet.

HANDLING SEEDLINGS IN FIELD

CORRECT
In bucket with sufficient wet moss to cover roots.

INCORRECT
Incorrect handling promotes drying of the root.

CORRECT AND INCORRECT DEPTHS

CORRECT
At same depth or 1/2" deeper than seedling grew in nursery.

INCORRECT
Too deep and roots bent.

INCORRECT
Too shallow, roots exposed.

FIGURE 25-9 Handling and planting seedling trees.

wet newspaper around them. If plants cannot be planted for many days, the plants may be heeled in. To **heel in** a plant, a trench is dug in the soil deep enough to hold the roots of the plant, the plant roots are placed in the trench, and then they are covered with soil (Figure 25-9). Then the soil around the roots is compacted with the heel of a shoe or boot. It is important to wet the soil well after heeling in. Frequently, bare-root plants will need additional pruning at planting time.

Balled and Burlapped Plants

Balled and burlapped (B & B) plants are dug with a ball of soil remaining with the roots. This is wrapped with burlap and laced with twine. Normally, these plants have been root pruned. **Root pruning** is a process whereby roots are cut close to the trunk so that a good root system develops close to the trunk before the plant is dug. The result is that when the ball is dug, the tree or shrub will have a compact root system in the ball. This gives the plant a good chance to reestablish itself in a new environment.

It is important that transplanted trees and shrubs have every opportunity to grow after transplanting. Plants typically balled and burlapped for transplanting are deciduous trees with branching root systems. Other plants that are sold as B & B are conifers, azaleas, rhododendrons, and other plants that have fibrous root systems.

As is true with bare-root plants, it is important that B & B plants' root systems not dry out. However, because there is soil around the roots, the drying-out

process is more gradual. B & B plants should be avoided if they do not have a sound ball of soil surrounding the roots. Handling tends to break the ball up and strip the hair roots from the plants. Therefore, a good root ball will help substantially to avoid damage when handling. Plant material that is shipped as B & B may be planted anytime, as long as the soil can be worked.

The mechanical tree spade is becoming more popular in the industry today. A tree spade is an expensive piece of equipment that will dig a tree in a matter of a few minutes with a very specific-sized ball. In addition, it is used to dig holes for trees and shrubs quickly and efficiently. The tree spade saves many hours in digging and planting trees. Frequently, B & B trees and shrubs are shipped in burlap with wire cages, specially prepared baskets, or other containers. The tree spade has helped make this an efficient way to ball and burlap.

These types of plants require little pruning at planting time. This makes B & B a popular way of handling plants by the mechanized professional horticulturist.

FIGURE 25-10 Nurseries package and ship many of their trees and shrubs to customers in disposable containers.

Containerized Plants

The use of container-grown plants is increasing in the nursery industry. A **containerized plant** is grown and shipped in a pot or can (Figure 25-10). Normally, the smaller types of plants are handled in this manner. They are grown for a reasonable period in the containers in which they are shipped and purchased.

The disadvantage of using containerized plants is that they need to be planted carefully. The roots of the plant have developed in a limited space and are generally rootbound. The roots also may have grown back around the trunk. To prevent the plant from strangling itself and also to encourage the roots to grow out of the confined area, these plants must have the "container ball" broken. This is done by placing a sharp shovel through the root mass or by breaking and spreading the root mass as it is planted in the new hole.

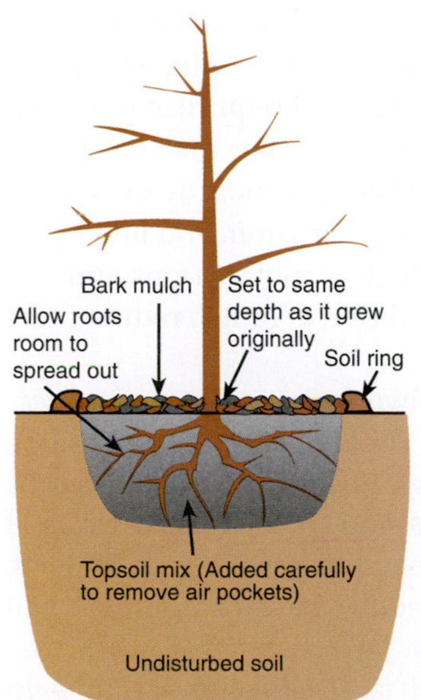

FIGURE 25-11 It is important to follow carefully all of the instructions for planting trees and shrubs.

PLANTING TREES AND SHRUBS

Planting may be done in spring, summer, or fall, as long as soil can be worked. A good practice for preparing the hole is to dig it about one-third wider than the ball or container. If the soil is hard, compacted, or of poor quality, it is advisable to dig the hole 4 to 5 inches deeper than the ball. This will allow peat moss or another soil conditioner to be added to the soil and placed under and around the root ball. Fill the hole with enough good soil to allow the ball to be the proper depth when it is put in the hole. The ball or container should be about 2 to 5 inches above the top of the hole. After cutting the twine that holds the ball together, peel back and remove the burlap. Backfill the hole around the plant, tamping the soil lightly to ensure removal of all air pockets. Continue backfilling the hole until it is level with the surrounding soil. Place a ring of soil about 4 inches high around the backfill to retain water.

In the case of a container plant, remove the plant from the container and slice the root zone with a knife or shovel. Then place the plant in the hole and backfill carefully.

After backfilling, fill the ring with several inches of water to saturate the soil to the bottom of the backfill. This will settle the soil and provide moisture for the plant (Figure 25-11).

SCIENCE CONNECTION BIOCONTROL—TAMER OF THE EUONYMUS SCALE?

"Scale insects that attack euonymus plants rank among the most insidious enemies of trees and bushes in the United States," said entomologist John J. Drea of the U.S. Department of Agriculture (USDA) Insect Biocontrol Laboratory in Beltsville, Maryland (Bryan and Drea, 1992). A hitchhiker and intruder from Asia, *Unaspis euonymi* is commonly known as the euonymus scale. It has become established and has multiplied in the United States without the normal hassle of natural enemies.

Euonymus plants are said to rank twelfth among the top 20 plants used by the multibillion-dollar landscaping industry. They are used as ground covers, hedges, vines, shrubs, and trees. Yet scale insects have become such a problem that many nurseries have stopped selling them. Euonymus scale attacks many other trees and shrubs, such as pachysandra, hibiscus, and camellia, but the euonymus plants are quite vulnerable. In one survey of suburban properties, nearly 70 percent of the euonymus plants had problems, due mostly to the scale.

The female euonymus scale is brownish in color, about one-sixteenth inch in length, and shaped like an oyster shell. It is possible for the scale to have three generations in 1 year, so the buildup can occur at a rapid rate. The female forms an armor-like covering of wax, inserts her mouth parts, and settles into uninterrupted feeding and egg laying until she dies. Pesticide sprays cannot get to her. After the eggs hatch, the flying males live about 1 day—sufficient time to mate. The fertile females then form their waxy shelters, feed, lay eggs, and start another generation. Controls based on sprays and dusts are not useful because of the brief window of insect exposure during the life cycle. Also, the use of pesticides on plants infested with scale insects tends to kill any natural enemies of the scale that might exist. Therefore, biocontrol such as the use of natural enemies is about the only likely solution to the problem.

The search for the scale's natural enemies in Asia started in the early 1980s. Finding those that could be introduced to the United States without becoming pests themselves and testing such organisms for effectiveness and safety took more than 10 years. Producing and distributing most natural or biocontrol agents adds years to such a project.

MULCHING

Trees and shrubs will benefit from the addition of mulch around the planted area. Mulch in the form of shredded hardwood or pine straw will reduce evaporation and help hold in the moisture. A newly planted tree will require about 15 to 20 gallons of water twice a week in a hot, dry environment. The conservation of water is important. Other advantages of mulch are that it helps provide a more constant soil temperature, helps control weeds, prevents erosion, and helps prevent soil compaction. Some species of plants have shallow root systems and will compete with any ground cover for nutrients. The addition of mulch reduces this competition.

The use of mulch will prevent damage from lawn mowers or *weed whackers* by keeping grass and ground covers away from the trunk. Mulch should be applied 3 to 4 inches deep. For estimating the mulch needs for a bed of plants, use the rule of thumb that 1 cubic yard of mulch can be spread over an area of 100 square feet. Mulch will last for 1 to 3 years, depending on the type applied. Pine bark must be replaced annually, whereas shredded hardwood mulch will last about 3 years.

As an additional bonus, consider the use of landscape fabrics to help control weeds. Such materials are placed under the mulch. The fabric is made of fiberglass and lasts many years. It is better than sheet plastic because it allows water and fertilizer to move through it, whereas the plastic sheet blocks water and fertilizer.

Two species of beetles were found and tested, and they are in the process of being multiplied, distributed, and observed. They are the red-spotted, black Asian lady beetle, *Chilocorus kuwanae*, and the one-twenty-fifth-inch nitidulid, *Cybocephalus prob.nipponicus*. Both lay their eggs under the body of the euonymus scale and in cracks of the bark and other protected places on the host plant. When the larvae hatch, they feed voraciously on the scale insects and their offspring. The powerful jaws of Asian lady beetles enable them to chew through the scale's armor, burrow under the protective cover, and devour the scale insects.

The USDA and the nursery and landscaping industries have successfully released and tested these and other predators in their efforts to bring the euonymus scale under control. Predatory wasps prey on scale insects during their larval stage. Other natural enemies are also being evaluated. It is expected that these and other biological controls will help solve the problems with scale insects.

A

B

(A) A red-spotted ladybug devours euonymus scale insects and (B) deposits an egg to start a new generation of predators.

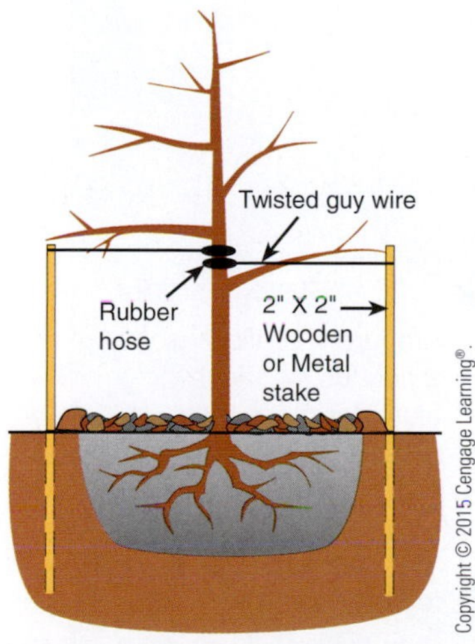

FIGURE 25-12 A newly planted tree should be anchored in position with guy wires.

STAKING AND GUYING

Newly planted trees and shrubs need to be staked or guyed. This is to avoid loosening of the soil and disfiguration of the plant by the wind. To **stake** a tree, drive a wooden pole or metal post into the ground near the plant and tie the upper part of the plant to it with rope or wire. Guying is a form of staking, and it is accomplished by tying a tree to two to four stakes with wire or rope (Figure 25-12). Stakes and guys should be left in place for at least one growing season. When using wires to stake or guy a tree, make sure the wires do not come in contact with the trunk. Using old water hose with the wire running through is a good practice to protect the trunk. Another good practice is to tighten the wires at the stake and not at the trunk. The best method is to use a double strand of wire twisted together for tightening. Guy wires should be checked for tightness several times during the growing season.

FERTILIZING

Plants need nutrients to maintain their vigor and to make healthy new growth. If the soil is fertile, it is not necessary to add fertilizer at planting time. Trees and shrubs should not be fertilized during the first year of growth. This practice is followed to prevent the plant from developing too much top growth in relation to the root growth. Plants should be fertilized every 3 to 5 years, starting with the growing season after the first year. Generally speaking, the best time is in the early spring. It is not a good practice to fertilize a plant after middle to late July. This practice would force new growth that will not mature enough to escape damage by the winter cold.

Unless the plants are in a bed, it is not a good practice to place fertilizer on the surface of the soil. It will wash away in the rains or will not penetrate into the root-zone area. To fertilize a tree properly, it is necessary to place the fertilizer into a hole 1 inch in diameter and 18 inches deep. Care should be taken to avoid underground utility wires and pipes, such as electric lines, telephone wires, gas pipes, and water pipes.

Holes for fertilizer are made using a 1–inch steel stake placed in concentric circles starting about halfway out from the trunk and the drip line. The **drip line** is the outer edge of the tree where the branches stop. Holes should be about 24 inches apart, and the circles should be about 24 inches apart (Figure 25-13). Each hole receives the appropriate amount of fertilizer and is sealed with soil or by closing the hole with a heel of the foot. The amount of fertilizer needed by a tree is determined by measuring the trunk about 4.5 feet above the ground. If the trunk of the tree is less than 8 inches, multiply the number of inches of diameter by 3. If the trunk is greater than 8 inches, multiply by 6. This will give the number of pounds of fertilizer the tree needs. Divide the number of pounds of fertilizer by the number of holes to determine the amount of fertilizer to put in each hole. For example, if the tree measures 6 inches at 4.5 feet, multiply 6 × 3. The answer is 18 lbs. of fertilizer. If the tree measures 12 inches at 4.5 feet, then add 12 × 6 = 72 lbs. of fertilizer. If the tree has more than one trunk, combine the diameters of all the trunks and multiply by the appropriate factor.

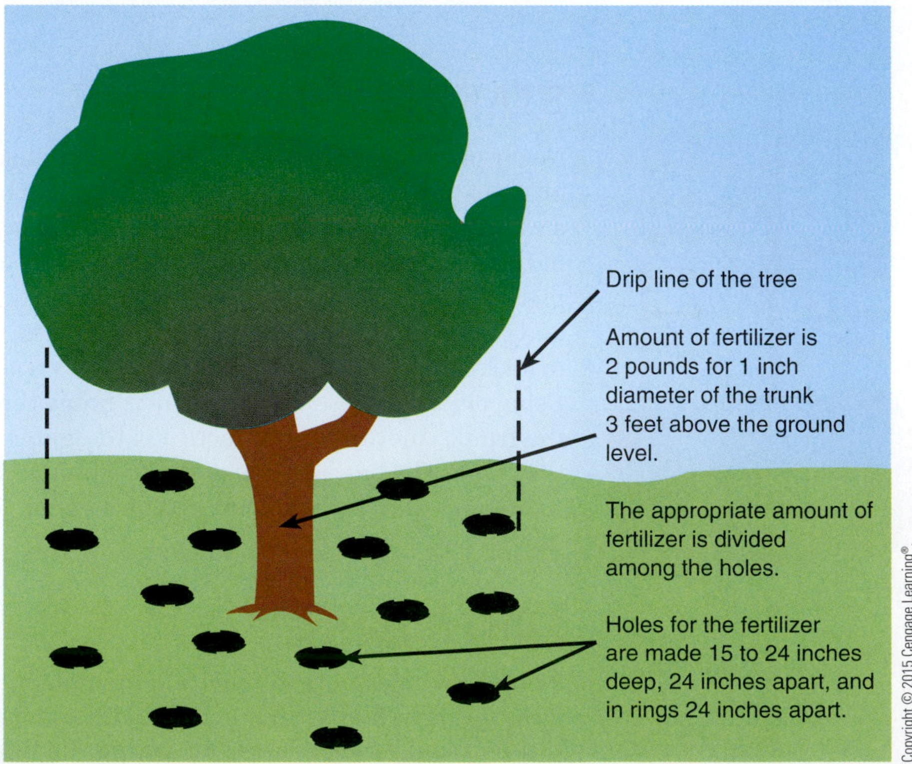

FIGURE 25-13 Holes for fertilizer applications should be placed within the drip line of the tree.

SCIENCE PROFILE TREE GALL INFESTATION

Courtesy of DeVere Burton.

Gall consists of growth of extra tissue formed by a tree in response to injury or infestation by parasytic organisms.

A tree can be a host to many different kinds of organisms that do not cause any problems. Other guests are less cordial and take advantage of their host. Some organisms use trees and other plants to provide them with a place to stay and food to eat. This happens when a particular organism attacks a tree either by eating it or by laying its eggs in it. When this happens, the tree will form extra tissue called a gall. The parasitic organism will move into its newly formed home where it can live, reproduce, and eat. Galls are believed to form in response to a plant-growth-regulating chemical produced by the attacking organism. The growths can be caused by many different kinds of organisms, including fungi, viruses, insects, mites, and nematodes. They can be found on leaves, branches, trunks, and roots. Galls usually do not cause serious damage to a well-established tree and are more of a bother than a serious threat.

Proper pruning techniques and timing are important in growing good ornamental trees and shrubs. Incorrect pruning can leave a plant in worse condition than before pruning. A plant in weak condition is susceptible to insects and diseases. Poor pruning can cause the loss of a season of flowers or fruit. In general, flowering plants should be pruned just after they bloom, or produce flowers. This is done so that the next season's flowering wood is not cut off.

INTERNET KEY WORDS:
prune trees, shrubs

A fertilizer with an analysis of 10-6-4 or 10-10-10 is generally acceptable. Never apply more than 100 lbs. of fertilizer to any given tree in a year. If the rate to be applied is more than 100 lbs., make the application over a 2-year period. Trees should be fertilized every 3 to 5 years.

PRUNING

Pruning is the process of removing dead or undesirable limbs from a tree or shrub. Removing dead, broken, diseased, and insect-infested wood helps protect the plant from additional damage. Trees may have waterspouts, bad-angle crotches, branches that interfere with other branches, or branches that form an asymmetrical habit. These need to be taken out or corrected. Sometimes it is necessary to reduce the top growth of the plant to match the root ball on a newly transplanted plant (Figure 25-14). After a tree is transplanted, the size of the **canopy**, or branches, should be reduced to more closely match the size of the root ball.

It is easier to prune deciduous trees and shrubs in the late fall or winter, after the leaves drop. The framework is bare and easier to see. When removing large limbs, care should be taken to avoid splitting and tearing of the limb (Figure 25-15). Specific plants have specific pruning requirements. It is best to determine how each plant should be pruned before the pruning operation is started. Local plant specialists are usually willing to help or give suggestions for proper pruning. Good reference books are available with accurate information on pruning particular plants.

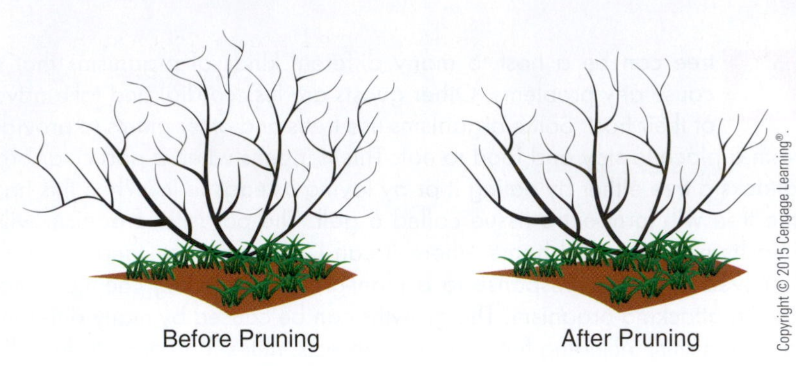

FIGURE 25-14 Careful selection of branches for removal is important when pruning deciduous shrubs.

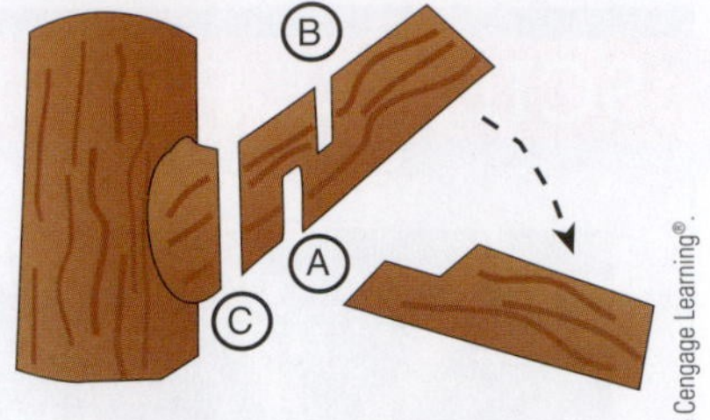

FIGURE 25-15 It is important to follow the recommended order of cuts when removing a large tree branch. The wrong cutting order is likely to damage the tree.

INSECTS AND DISEASES

Most plants are subject to damage by insects and diseases. It is easier to prevent such damage than to control it, and several preventive measures are available. Pest-resistant varieties should be used in areas where pest populations are well established. Trees should be carefully managed to remain healthy and vigorous. Weakened trees and shrubs are easily overcome and damaged by insects and diseases (Figure 25-16).

It is important that plants that are selected for the environment in which they are to be planted. Plants cannot tolerate stressful conditions continuously. They may have some ability to adapt to various conditions, but some plants adapt better than others. Some plants can withstand air pollution,

HOT TOPICS IN AGRISCIENCE IMPROVING FOREST HEALTH

One of the great difficulties forest managers experience is court intervention by individuals who have political agendas. In recent years, many management decisions of forest managers have been challenged in courts of law. Regardless of the outcomes in these cases, extra costs are incurred in the management of forest resources, and the resulting time delay is sometimes catastrophic. For example, an infestation of pine beetles can often be contained to a small area if it is treated with appropriate chemicals in the early stages. A challenge in court will prevent or delay treatment until a major outbreak of the beetles has occurred, damaging an entire forest.

In some instances, the harvest of trees that have been killed by fire has been prevented or delayed until the trees have become completely dehydrated. When this occurs, large cracks develop in the wood, destroying its value, and a useful resource has been wasted.

A second issue related to forest health is that some forests are simply not managed for healthy growth. Trees must be thinned to allow sufficient access to sunlight and nutrients. Forest practices must be consistent with good management practices for soil and water. New trees must be planted after timber harvests and wildfires. Pest management must be practiced, and political issues must be dealt with in a fair and timely manner. All of these practices and issues have impacts on forest health. Healthy forests can sustain our needs for wood and forest products. Unhealthy forests cannot fill these needs.

Courtesy of USDA/ARS #K2849-1.

FIGURE 25-16 Insect infestations can build up fast on susceptible plants.

drought, heat, or even wet soils. Other types of plants can tolerate infertile conditions, but not hot, dry weather. Still others may not survive in heavy shade or in full sun.

In its lifetime, any plant can be expected to experience stress from insects and diseases. When an infestation is suspected, it is important to act quickly. Often, caretakers do not know what is causing damage to a tree. State agricultural universities have county extension offices with plant experts on staff who can identify issues, offer suggestions, and help save the plant. It is wise to become aware of the plants in your area and to consult cooperative extension professionals for more specifics on the types of insects and diseases that can be expected to infect various plants in a given locality.

STUDENT ACTIVITIES

1. Write the Terms to Know and their meanings in your notebook.
2. Contact a local nursery and obtain a nursery catalog.
3. Visit a garden center or nursery and compile a list of trees and shrubs that are available and recommended for your community.
4. Prepare a chart of popular ornamental trees and shrubs for your community. For each plant, list the scientific name, common name, mature size of the plant, time of flowering, color of flowers, spring leaf color, and fall leaf color.
5. Survey your home or other assigned area and make a list of the trees and shrubs that are present. Use books, nursery catalogs, and other resources to help identify the plants.
6. For a given area, determine the diameter of each tree 4.5 feet above the ground. Work up a recommended fertilizer program for the trees, including the amount of fertilizer per tree, local fertilizer prices, cost of fertilizer for each tree, time of year to fertilize, and the total cost of fertilizer for the area.
7. Prepare a pruning schedule for plants around your home or other area based on local recommendations. Include in the schedule the common and scientific names of each plant, time of pruning, and special requirements for pruning.
8. Prepare a sketch or a scale drawing of your home property or other assigned area. Locate the existing trees and shrubs on the sketch, using a circle with the initials of the scientific name to represent each plant. Add new plants that you believe will enhance the property you have surveyed.
9. Volunteer at a local park to help plant new trees.
10. Create a game, song, or activity that will help you remember tree names by their shapes shown in Figure 25-6.

SELF-EVALUATION

A. MULTIPLE CHOICE

1. An acre of plant material can produce enough oxygen to keep
   a. 5 to 10 people alive each year.
   b. 10 to 16 people alive each year.
   c. 16 to 20 people alive each year.
   d. nobody alive. It does not produce enough oxygen to be of any value.
2. A border planting
   a. is used as a single plant to highlight a fence or some other special feature of the landscape.
   b. is used to separate some part of the landscape from another.
   c. is a number of trees or shrubs planted together as a point of interest.
   d. is a collection of plants that are placed in the landscape as needed.
3. When planning the location for planting trees and shrubs, which is not a major consideration?
   a. fruit size and type
   b. flower color
   c. other structures
   d. bare-root plants
4. The term *Cornus florida rubra* is a
   a. common name.
   b. scientific name.
   c. name that was developed in Florida.
   d. type of annual deciduous plant.
5. Planting of trees and shrubs may be done
   a. in spring, summer, or fall.
   b. in spring only.
   c. in fall only.
   d. in spring and summer.
6. Mulch should be applied
   a. 1 to 2 inches deep.
   b. 2 to 2.5 inches deep.
   c. 3 to 4 inches deep.
   d. 6 inches deep to keep out the weeds.

B. MATCHING

________	1. Urban foresters	a. The removal of dead, broken, unwanted, diseased, and insect-infested wood
________	2. Shrubs	b. Used as a single plant to highlight it or some other special feature of the landscape
________	3. Specimen plant	c. A systematic method of naming plants
________	4. Nomenclature	d. Woody plants that normally grow low and produce many stems or shoots from the base
________	5. Pruning	e. The top of the plant; has the framework and leaves
________	6. Canopy	f. Help install and maintain trees in large cities

C. COMPLETION

1. Plant material can cut noise pollution by ________.
2. A rule of thumb is that 1 cubic yard of mulch can be spread over a(n) ________ square-feet area.
3. Newly planted trees may have to be ________ to prevent wind damage.
4. Plants should be fertilized every ________ to ________ years, starting with the growing season after the first year.

5. Proper pruning techniques and timing are important in growing good ornamental trees and __________.
6. __________ is the systematic cutting of the roots by hand or machine to encourage the roots to develop close to the trunk.

D. TRUE OR FALSE

1. Trees and plants are valuable only for beauty.
2. Trees can act as insulators, keeping a house cool in the summer and warm in the winter.
3. A well-designed landscape does not increase the value of a property.
4. All plants have the same type of growth habit.
5. It is not a good practice to fertilize trees after middle to late July.
6. Most plants are subject to damage from insects, diseases, or both at some time.

SECTION 8

ANIMALS OF THE FUTURE

MAKING BETTER SHEEP

Fifty years ago, desert range sheep averaged a 70 percent lamb crop—that is, each 100 female sheep averaged a total of 70 live lambs per year. Today, that figure is 85 percent, and scientists at the U.S. Sheep Experiment Station at Dubois, Idaho, believe the figure will be 150 percent in 50 years. They also project that under intensive confinement production, it is reasonable to expect a national average lamb crop of 400 percent. How will we do this? The Booroola Merino sheep, which originated in Australia, has some individual female sheep that have five or six lambs instead of the typical one or two. There is hope that the gene responsible for the multiple lambs can be isolated and bred into the genetic makeup of our most productive breeds. That would solve the numbers game, but how will the sheep and their managers keep so many newborn lambs alive and growing?

MORE MILK FROM FEWER COWS

Fifty years ago, the United States had 25 million dairy cows, and the national average milk production per cow was 4600 lbs. In 2010, the figures were 9.197 million cows averaging 21,335 lbs. of milk. Some believe that in 50 years we will have herds averaging 40,000 lbs. of milk per cow per year. Will production increases in the future require larger cows? Dairy scientists at the U.S. Department of Agriculture (USDA) Dairy Forage Research Center in Madison, Wisconsin, do not believe so. Better feed could make conversion of feed to milk a more efficient process. Currently, only 30 to 40 percent of the material in plant cell walls is available to the animal. The rest simply passes through animals. If plant cell walls could be modified to make them more digestible without diminishing plant health and performance, then the animal could extract more nutrients from a given amount of feed.

Discovery of the hormone bovine somatotropin (BST) is affecting the way a cow's body divides calories obtained from feed between milk production and other body functions. The use of BST causes the cow to shift more of her calories into milk, thus increasing her conversion rate. The widespread use of BST results in an immediate increase in milk production by 15 percent or more. Research will continue to identify ways to influence production increases in this and other livestock enterprises.

PREVENTING POULTRY DISEASES

Devastating diseases, such as Avian Influenza and Newcastle disease, still threaten to eliminate entire flocks in the poultry industry. Both viruses have strains with disease-producing capability that range from those that do not kill any birds to those that kill 100 percent of the birds. So, when only a few birds are diagnosed

with the disease, the entire flock is destroyed. The movement of people and poultry in the locality, and even within the state or region, may be restricted. The economic loss is staggering. Scientists struggle to keep ahead of such diseases by developing new vaccines and practicing strict quarantines.

Courtesy USDA/ARS #K-4134-6.

Many of the improvements leading to greater animal productivity are the result of genetic engineering techniques and other advances in laboratory procedures.

Veterinary researchers at the USDA Southeast Poultry Research Laboratory at Athens, Georgia, hope to find better ways to manage poultry diseases. With genetic markers, scientists are able to more easily identify specific disease pathogens. They report progress in understanding the fundamental genetic traits of the viruses associated with poultry diseases. Perhaps research efforts will soon enable prediction of specific virus behavior in poultry that will be dependable in determining if a given virus in a bird or in sample of birds is mild or lethal. The decision about destroying birds can then be made more intelligently. DNA probes for examining bird tissue promise to be a useful diagnostic tool to aid in the process.

OCCUPYING THE PARKING SPOT

Eradicating the African swine fever is a dream of the swine industry. So far, it has been an elusive dream. However, a genetic-engineering technique using what is referred to as the *nonsense gene* has some promise. Certain viruses, such as the African swine fever virus, have the ability to incorporate themselves into the genetic material of an animal and be carried into the animal's offspring. However, the virus has to establish itself in a particular spot on the genetic material. If a modified version of the virus that cannot cause the disease can be engineered, then animals can be infected with the safe virus. The safe virus would become "nonsense genes," occupying the "parking spot" on the chromosome and preventing the disease-causing virus from establishing itself.

In summary, animals of the future will continue to improve in productivity and function. Much research on farm animals focuses on increasing productivity to help feed the world's expanding population. In addition, research focuses on the feeding and management of pets and other animals for recreational uses. Some research deals with learning the needs and behavior of animals in the wild and providing improved habitat according to the unique needs of wildlife.

UNIT 26

Animal Anatomy, Physiology, and Nutrition

OBJECTIVE

To determine the nutritional requirements of animals and learn how to satisfy those requirements.

MATERIALS LIST

- commercial feed tags from various feeds
- charts of various animal digestive systems
- Internet access

COMPETENCIES TO BE DEVELOPED

After studying this unit, you should be able to:

- compare animal digestive systems.
- understand the basics of animal physiology.
- understand how nutrients are used by animals.
- identify classes and sources of nutrients.
- identify symptoms of nutrient deficiencies.
- explain the role of feed additives in livestock nutrition.
- compare the composition of various feedstuffs.

SUGGESTED CLASS ACTIVITIES

1. Conduct a feeding trial with baby chickens using different-quality rations. Weigh the chicks in each group at the beginning of the trial and at the end of the trial. Record any observations that you make as the trial is conducted. Calculate the differences in total weight gain after 2 weeks. Place all of the groups on the ration that resulted in the greatest weight gain and proceed with the trial for 2 more weeks. Record the results and offer an explanation for the results.
2. Examine the anatomy of a farm animal by dissecting a fetal pig or by butchering a lamb, hog, or beef animal. Observe each of the systems of the animal and explain the physiology or function of each system as you examine the organs that make up each system.
3. Divide the class into small groups. Each group will research the daily nutrient requirements of a different domestic animal of their choosing. The groups may use the Internet, library resources, or other reliable source. After compiling their information, the groups will share their findings with the class.

TERMS TO KNOW

nutrition
ration
deficiency disease
vitamin
mineral
anatomy
skeletal system
muscular system
voluntary muscle
involuntary muscle
protein
circulatory system
carbohydrate
respiratory system
central nervous system
peripheral nervous system
urinary system
endocrine or hormone system
hormone
digestive system
ruminant
rumen
roughage
monogastric
concentrate
fructose
galactose
sucrose
maltose
lactose
starch
cellulose
fat
feed additive
antibiotic
dry matter
TDN

Feed is animal food. It represents the largest single-cost item in the production of livestock. Therefore, it is important to understand the complex nature of animal nutrition and how animals use the feed they eat. **Nutrition** is the process by which animals eat food and use its nutrients to live, grow, and reproduce (Figure 26-1).

"You are what you eat" is an axiom that is true to a considerable extent, especially as it relates to good health of both humans and animals. This unit explores the relationship between good nutrition and good health.

NUTRITION IN HUMAN AND ANIMAL HEALTH

The relationship between proper nutrition and health has long been recognized. Early sailors stocked their sailing vessels with limes when going to sea for long periods. This was to prevent the dreaded disease of scurvy. Scurvy is a disease of the gums and skin caused by a deficiency of vitamin C in the diet. Even today, the effects of poor nutrition such as in cases of anorexia and obesity are widely observed in the human population. In simple terms, anorexia is a result of too little nutrition, and obesity is the result of too much or improper kinds of food.

In animals, proper nutrition is just as important as it is in humans. Feed efficiency, rate of gain, and days-to-market weight are all uppermost in the minds of people who raise livestock for meat. Proper nutrition is just as important for animals that produce milk, wool, or fur. Slow growth, poor reproduction, reduced production, and poor health are generally the results of less-than-adequate animal rations. The amount and content of food eaten by an animal in one day is referred to as the animal's **ration**. When the amount of feed consumed by an animal in 24 hours contains all of the needed nutrients in the proper proportions and amounts, the ration is a balanced ration.

Numerous diseases may result from improper amounts or balance of vitamins and minerals. Such diseases are called **deficiency diseases**. These diseases usually occur because of inadequate diets or digestive disorders. **Vitamins** are organic substances that are required in small amounts for normal

Courtesy of DeVere Burton.

FIGURE 26-1 Animal health, growth, and reproduction all are directly related to nutrition.

metabolism. **Minerals** are elements found in nature that are essential for normal body functioning of humans and animals alike. A shortage of either vitamins or minerals in the diet can lead to a deficiency disease. It should be noted, however, that not all types of animals require the same vitamins and minerals to maintain good health.

ANIMAL ANATOMY AND PHYSIOLOGY

The internal functions and vital processes of animals and their organs are referred to as animal physiology. The various body systems, such as the skeletal, muscular, circulatory, respiratory, nervous, urinary, endocrine, digestive, and reproductive, must each be properly fed and working together for the animal to be healthy and productive. To this end, proper nutrition is a must. The various organs and parts of the body are collectively known as **anatomy**.

Skeletal System

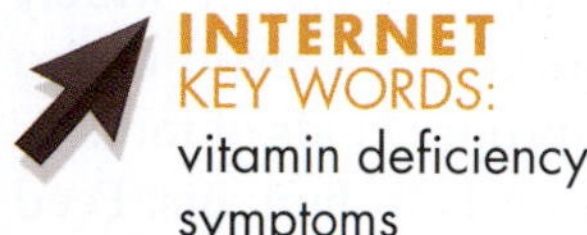
INTERNET KEY WORDS:
vitamin deficiency symptoms

The **skeletal system** (Figure 26-2) is made up of bones joined together by cartilage and ligaments. The purpose of the skeletal system is to provide support for the body and protection for the brain and other soft organs of the body.

Bone is the main component of the skeletal system. It is composed of about 26 percent minerals. This mineral material is mostly calcium, phosphate, and calcium carbonate. Another 50 percent of bone is water, 20 percent is protein, and 4 percent is fat.

AGRI-PROFILE

CAREER AREAS: ANIMAL NUTRITIONIST/ FEED FORMULATOR/ANIMAL MANAGER/ PHYSIOLOGIST/VETERINARIAN

© iStockphoto/Alexandru Nika.

Science has helped us determine the animal rations that are most healthy for farm animals. Healthy dairy cows, such as these, are capable of producing more and higher-quality milk than cows with inadequate nutrition.

Animal nutrition is an interesting field of study that can lead to several types of career opportunities. Some careers are involved in basic research in which scientists investigate how an animal uses its food supply to grow or reproduce. They also determine the nutrient values of feeds, including how digestible they are. This career field extends to production, processing, and sale of feeds. Some people in this career field raise farm animals. Raising and managing farm animals requires an understanding of animal nutrition.

Elements of nutrition are basic, such as the composition of feed grains, animal by-products, and the basic nutrients needed by animals. However, the nutritive content of forages and other feedstuffs varies considerably according to stage of growth, condition, and quality. The digestive capabilities of animals vary considerably from species to species. Nutritional needs of animals vary with age, stage of development, production, and pregnancy.

Careers in animal nutrition may be predominantly in the basic sciences or may be in the application of science to the nutrient requirements of an animal species. They may focus on fish, small animals, pets, horses, poultry, livestock, dairy, or wild animals.

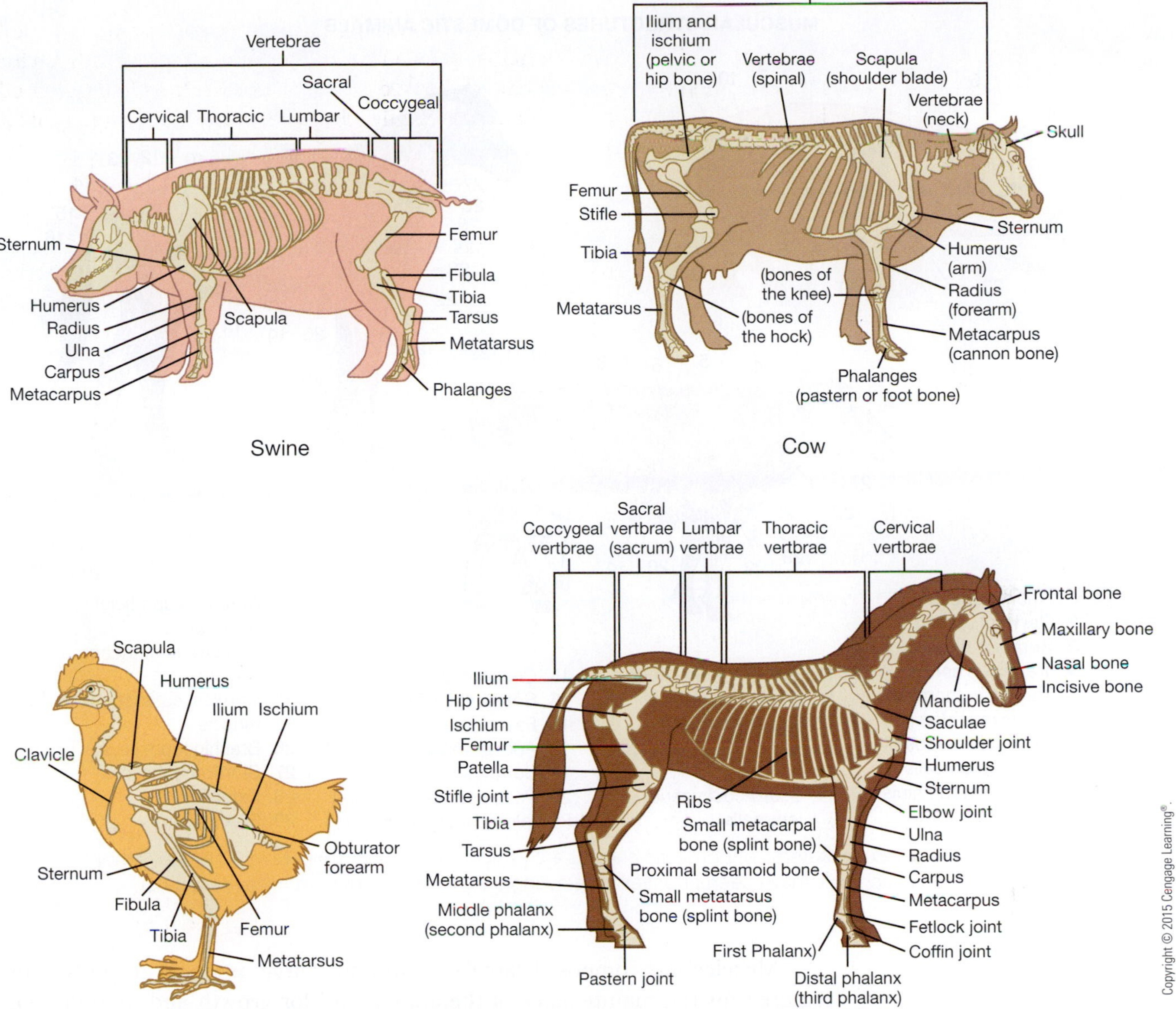

FIGURE 26-2 The skeletal system provides support for the body and protection for the soft organs.

The material inside bones is called bone marrow, and it produces the body's blood cells. The growth and strength of bones are greatly affected by the minerals and vitamins in animal rations.

Muscular System

The **muscular system** (Figure 26-3) is the lean meat of the animal. This is the part of the animal used for our steaks, chops, and roasts. Muscles provide for body movement in tandem with the skeletal system, and they support life (as in the heart muscle and the diaphragm).

Muscles may be voluntary or involuntary, depending on whether they can be physically controlled by the animal. **Voluntary muscles** can be controlled to do such things as walk and eat food. **Involuntary muscles** operate in the body without control by the will of the animal, and they function even during sleep. Examples of involuntary muscles are heart and diaphragm muscles.

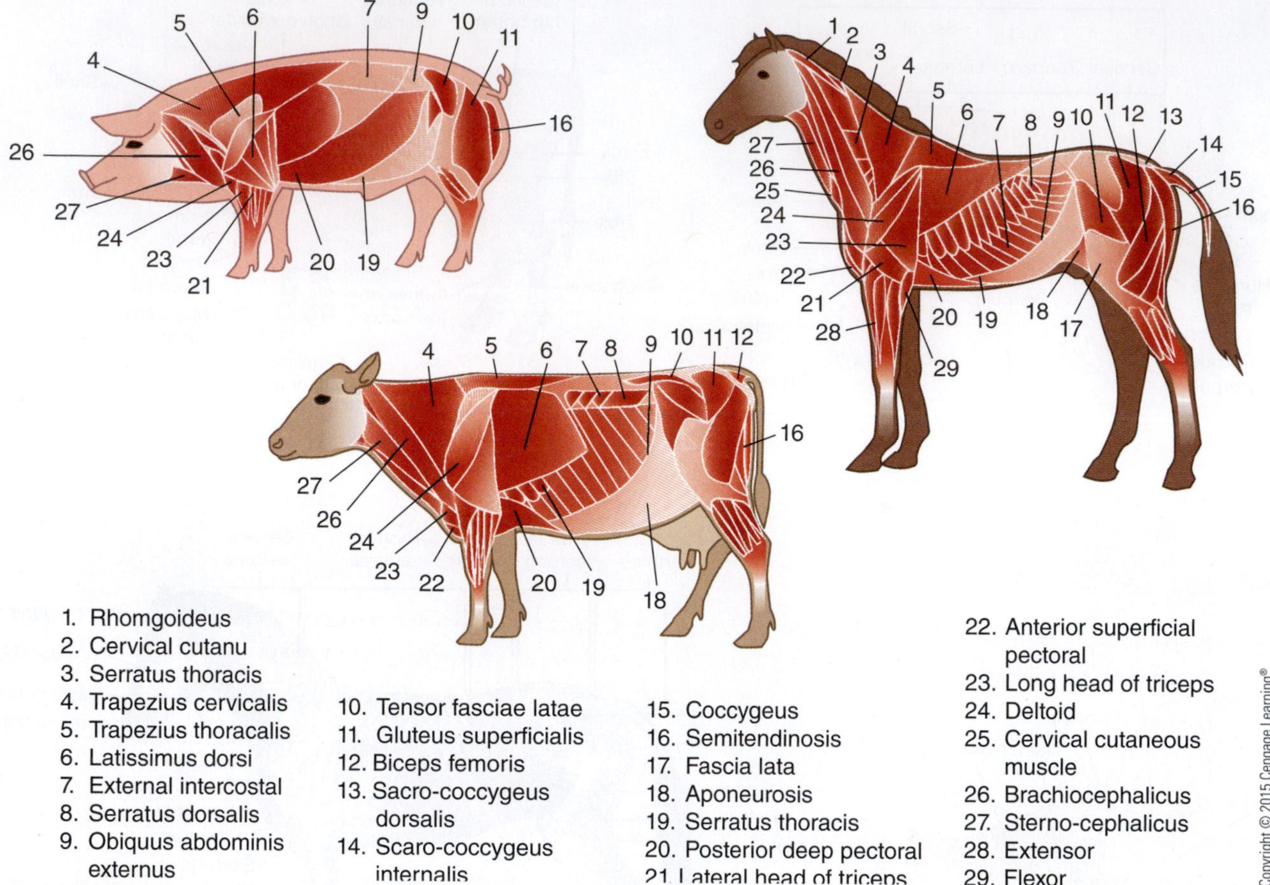

FIGURE 26-3 The muscular systems of domestic animals are similar in many ways; however, they vary among species depending on which traits have been priorities in selecting breeding stock over many generations.

Muscles are composed largely of protein. Large amounts of protein are required for the maintenance of the animal and for growth and reproduction. **Proteins** are nutrients made up of amino acids, the building blocks of muscles.

Circulatory System

The heart, veins, arteries, capillaries, and lymph system compose the **circulatory system** (Figure 26-4). This system transports blood that contains nutrients and oxygen to the cells of the body and it filters waste materials from the body. Lymph glands secrete disease-fighting materials into the body. They are part of the lymphatic circulatory system that operates within the blood circulatory system. A key function of this system is to transport excess water from the cells of the body.

Vitamins, minerals, proteins, and carbohydrates are all essential for the smooth function of the circulatory system. **Carbohydrates** provide sugars and starches that supply energy to the animal.

Respiratory System

The **respiratory system** provides oxygen to the blood of the animal and removes waste gases such as carbon dioxide from the blood. The respiratory system is composed of the nostrils, nasal cavity, pharynx, larynx, trachea, and lungs

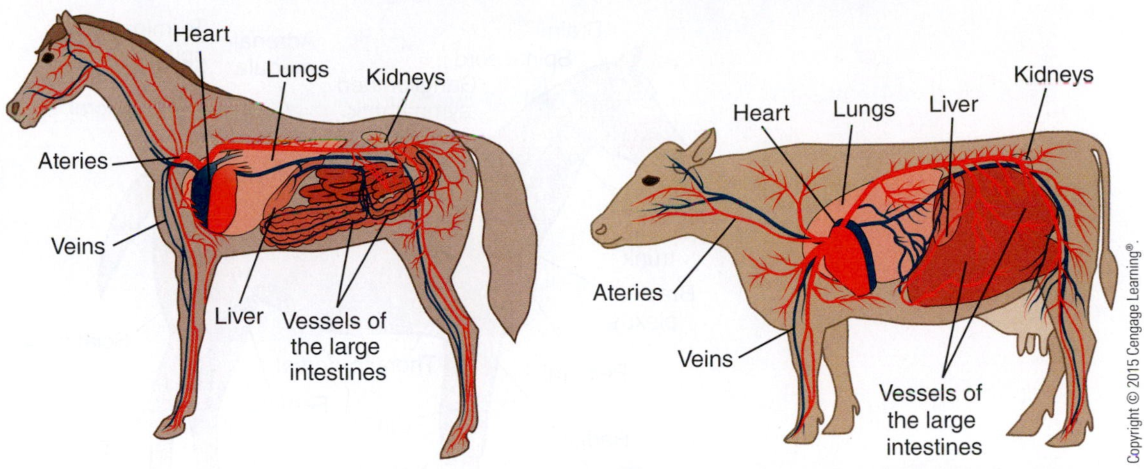

FIGURE 26-4 The circulatory system distributes food and oxygen that is dissolved in the blood to the cells of the body.

(Figure 26-5). This system controls breathing and uses the muscular and skeletal systems to draw air in and out of the lungs. Oxygen passes from the lungs to the blood.

Nervous System

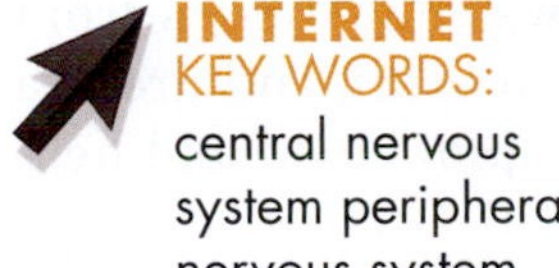

INTERNET KEY WORDS:
central nervous system peripheral nervous system

The nervous system of an animal is composed of the central nervous system and the peripheral nervous system.

The **central nervous system** includes the brain and the spinal cord. It is responsible for coordinating the movements of animals and also responds to all of the senses. The senses are hearing, sight, smell, touch, and taste.

The **peripheral nervous system** controls the functions of the body tissues, including the organs. The nerves transmit messages to the brain from the outer parts of the body (Figure 26-6).

Because the nervous system is composed primarily of soft tissues, proteins are particularly important in maintaining its health.

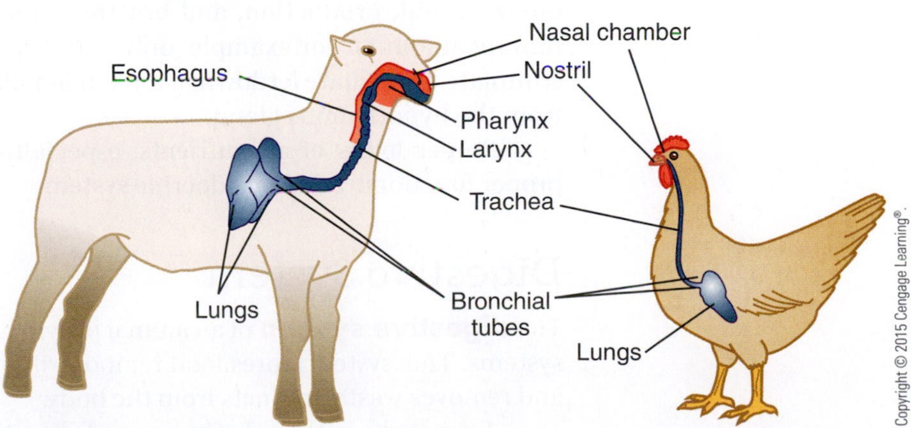

FIGURE 26-5 The respiratory system provides oxygen to the cells of the body.

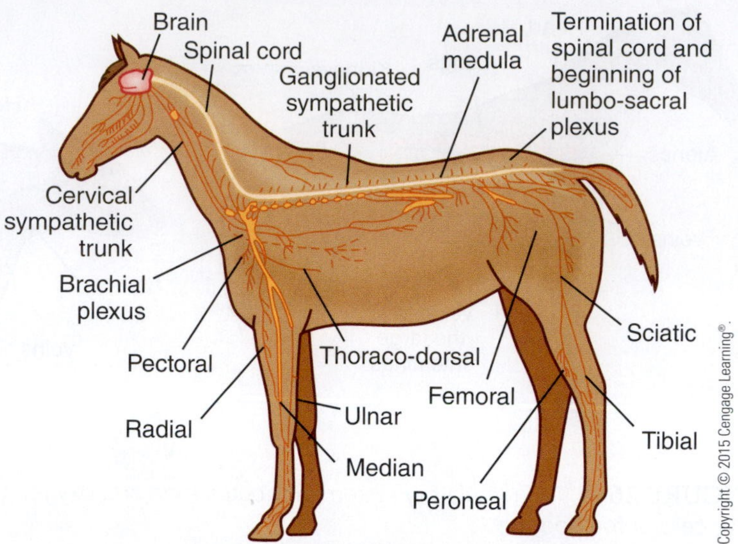

FIGURE 26-6 The nervous system consists of the brain, spinal cord, and the nerves that are distributed throughout the body. This system coordinates all of the other body systems.

Urinary System

The function of the **urinary system** is to remove waste materials from the blood. The primary parts are the kidneys, bladder, ureters, and urethra (Figure 26-7). The kidneys also help regulate the makeup of blood and help maintain other internal systems.

Abnormal levels of proteins fed to animals have been known to cause stress to the urinary system, which rids the body of excess protein. Greater-than-recommended levels of minerals may also cause kidney problems.

Endocrine System

The **endocrine or hormone system** is a group of ductless glands that release hormones into the body. **Hormones** are chemicals that regulate many of the activities of the body. Some of these body functions are growth, reproduction, milk production, and breathing rate. Hormones are needed in only minute amounts. For example, only 1/100,000,000 g. of oxytocin hormone will stimulate immediate letdown of milk in female animals. Oxytocin is a hormone from the hypothalmus gland.

Proper levels of all nutrients, especially minerals, are important for the proper functioning of the endocrine system.

Digestive System

The **digestive system** of an animal provides food for the body and for all of its systems. This system stores food temporarily, prepares food for use by the body, and removes waste products from the body.

Animals have three basic types of digestive systems: polygastric, or ruminant; monogastric; and poultry.

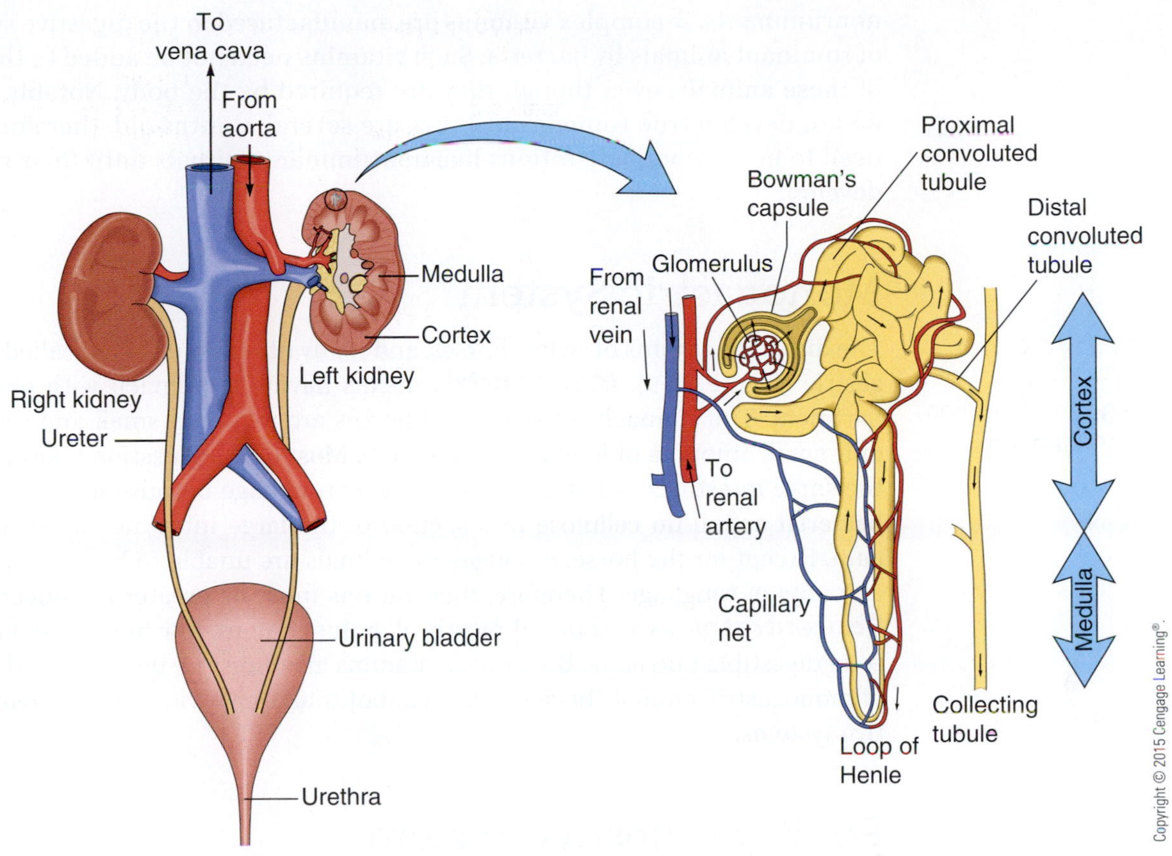

FIGURE 26-7 The urinary system removes waste materials from the blood.

Polygastric, or Ruminant, System

Ruminants are a class of animals that have stomachs with more than one compartment (Figure 26-8). Cattle and sheep are ruminants and have multicompartment stomachs. The largest compartment is called the **rumen**. The rumen can store large amounts of roughage. Examples of **roughage** include grass, hay, silage, or other high-fiber feed. Ruminants have the ability to break down plant fibers and to use them for food far more efficiently than

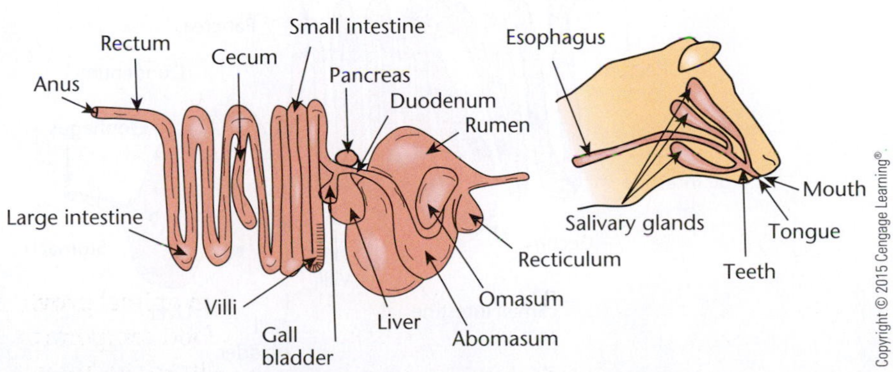

FIGURE 26-8 The ruminant digestive system processes large amounts of roughage.

nonruminants. B-complex vitamins are manufactured in the digestive systems of ruminant animals by bacteria. Such vitamins need not be added to the diets of these animals, even though they are required by the body. Notably, calves do not develop true rumens until they are several months old; therefore, they need to be fed complete rations like nonruminant animals until their rumens develop.

Monogastric System

INTERNET KEY WORDS:
monogastric digestion
poultry digestion

The digestive systems of swine, horses, and many other animals are called monogastric (Figure 26-9). **Monogastric** means having a stomach with one compartment. The stomachs of swine and horses are relatively small and can store only small amounts of food at any one time. Most of the digestion takes place in the small intestines. A horse is able to digest roughage because it benefits from bacterial action on cellulose in a section of the large intestine called the caecum. Except for the horse, monogastric animals are unable to break down large amounts of roughage. Therefore, their rations must be greater in concentrates. **Concentrates** are composed mostly of grains that are low in fiber and high in total digestible nutrients. B-complex vitamins also must be included in the diets of monogastric animals because they cannot make such vitamins in their digestive systems.

Poultry Digestive System

Although poultry have monogastric digestive systems, their digestive systems are different enough to discuss separately (Figure 26-10). Chickens have no teeth and must swallow their food whole. The food is stored in the crop and passed on to the gizzard, which grinds it up. The gizzard contains pebbles and other hard objects (eaten by the bird) that crush seeds and other large food particles as the muscular gizzard contracts and agitates its contents. Food particles then pass on to the small intestine where digestion occurs. Poultry rations must be high in food value because birds have no true stomachs and little room for storage of food they have eaten.

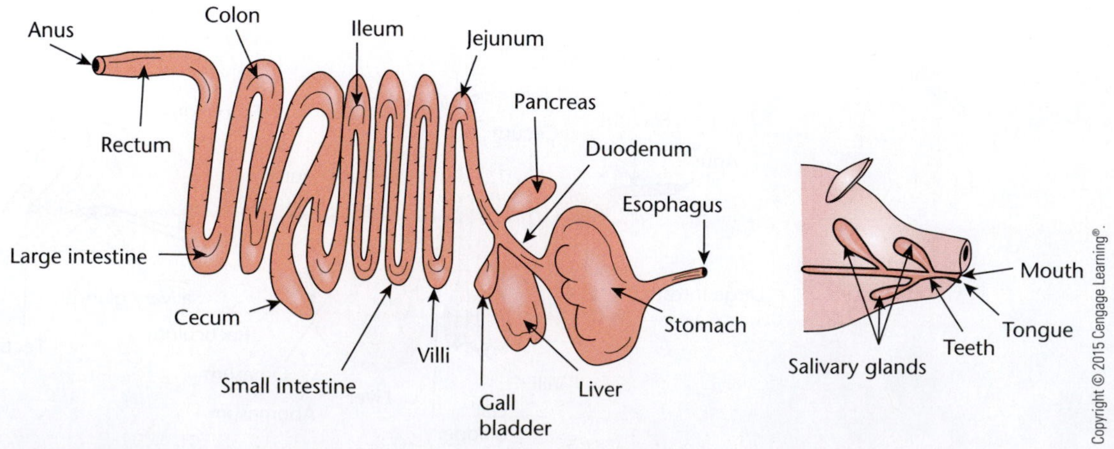

FIGURE 26-9 The monogastric digestive system processes food in a simple stomach.

SCIENCE PROFILE A FOUR-CHAMBERED STOMACH

Ruminant animals have distinct advantages over their nonruminant counterparts. They can digest roughage. The main source of energy in roughage is cellulose. Cellulose is a complex, nutrient-rich plant material that is unusable to nonruminant animals because they cannot digest it. The ruminant stomach is host to specific strains of bacteria that produce enzymes with the capability of unlocking the complex structure of cellulose and digesting it. Not only do these microbes aid in breaking down cellulose but they become a valuable source of protein for the animal as well.

The ruminant animal has four compartments that make up its stomach: the rumen, reticulum, omasum, and abomasum. The compartments work together to digest food that is high in fiber. Food can flow easily between the rumen and the reticulum. Most of the roughage is broken down there. Next, food enters the omasum, where the food particles become smaller and water is reabsorbed into the body. Next, food enters the abomasum or true stomach. Here, digestive enzymes are secreted that break down the food into basic nutrients. The nutrients then move into the small intestines where they are absorbed. Unused materials move into the large intestine where more water is absorbed, and the remaining material leaves the body as feces.

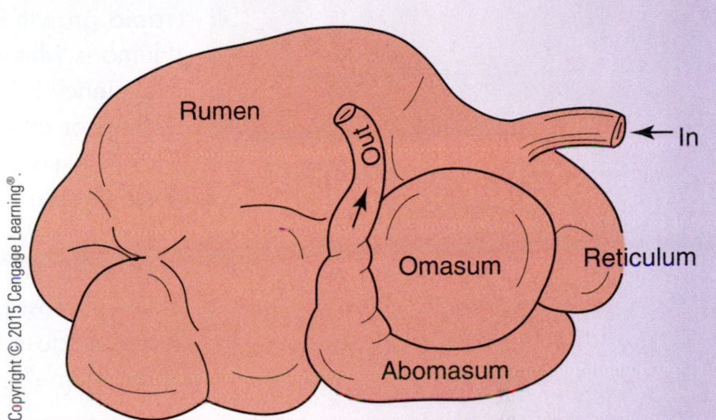

The stomach chambers of a ruminant animal include the rumen, reticulum, omasum, and abomasum.

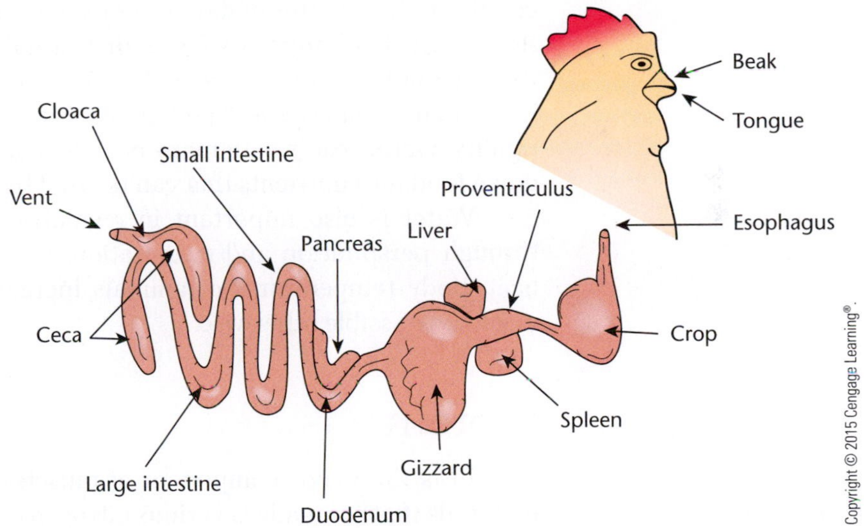

FIGURE 26-10 The poultry digestive system has no true stomach, but it does have an organ called the crop, which stores small amounts of feed. It also has a gizzard, which is the organ that grinds the seeds and other materials eaten by birds.

MAJOR CLASSES OF NUTRIENTS

Water

Water is the largest component of nearly all living things. Growing plants are usually 70 to 80 percent water. Similarly, the muscles and internal organs of animals contain 75 percent or more water.

HOT TOPICS IN AGRISCIENCE THE AMINO ACID "LYSINE"

© iStockphoto/Carol Gering.

Much of the nation's corn crop is fed to pigs. However, the amino acid, lysine, is deficient in most corn varieties. The development of high-lysine corn has greatly benefited the health of pigs and the efficiency of pork production.

Animals require nutrients of the right kinds and in the right amounts to produce the proteins that make up muscles and other body tissues. When an amino acid is unavailable, a protein that requires that amino acid can no longer be manufactured by the body.

The amino acid lysine is too low in some corn varieties to support rapid growth in hogs. Lysine must be added to corn to avoid this problem. Humans who rely on diets that are high in corn are also at risk for a lysine deficiency in the diet. However, people can supplement their diets with fish, turkey, or other high-lysine foods to correct the deficiency.

Researchers have developed new corn varieties known as "high-lysine" corn that have been shown to correct the nutrient deficiency. However, the new variety is lower in grain yield than most dent-corn varieties, and fields of "high-lysine" corn must be separated from other corn fields to prevent cross-pollination. Cross-pollinated corn yields the usual "low-lysine" corn because the gene for "low-lysine" corn is dominant over the gene for "high-lysine" corn.

Water is the least expensive nutrient for animals. However, most animals can live only a matter of days if they do not have access to it. Water is the solution in which all nutrients for animals are dissolved or suspended for transport throughout the body (Figure 26-11). Water provides rigidity to the body, allowing it to maintain its shape. The liquid solution in each cell is responsible for this rigidity. Water reacts with many chemical compounds in the body to help break down food into nutrients that can be used by the body.

Water is also important in regulating the body temperature of animals through perspiration and evaporation. Because water absorbs and transports heat, body temperatures of animals increase and decrease more slowly than would be possible otherwise.

Protein

INTERNET KEY WORDS:
monogastric protein requirement
animal nutrition, amino acids
animal nutrition, carbohydrates
animal nutrition, minerals
animal nutrition, vitamins
animal nutrition, fat

Protein is the major component of muscles and tissues. Proteins are complex materials that are made of various nitrogen compounds called amino acids. Some amino acids are essential for animals and some are not. Therefore, the quality of proteins fed to animals must be considered.

Monogastric animals need specific amino acids, so it is important that they receive high-quality proteins containing the appropriate amino acids. However, in ruminant animals, quantity of protein is more important than quality. Ruminants can convert amino acids in their rumens to different amino acids to meet their needs.

Protein is used continuously by animals to maintain the body because cells are continually dying and being replaced. In young animals, large amounts of protein are used for body growth. Protein is also important for healthful reproduction.

Carbohydrates

Carbohydrates make up a class of nutrients composed of sugars and starches. They provide energy and heat to animals. Carbohydrates are composed primarily of the elements carbon, hydrogen, and oxygen.

The energy obtained from carbohydrates is used for growth, maintenance, work, maintaining body heat, reproduction, and lactation (milk production). Carbohydrates come in several forms, with the sugars being the simplest. Examples of simple sugars used in animal feeds are glucose, **fructose**, and **galactose**. Compound sugars include **sucrose**, **maltose**, and **lactose**. Complex forms of carbohydrates include **starch** and **cellulose**.

Carbohydrates make up about 75 percent of most animal rations, but there is little carbohydrate in the body at any one time. Carbohydrates in the diet that are not used quickly are converted to fat and stored in the body. **Fat** is a tissue that stores energy in a concentrated form; it contains 2.25 times as much energy per gram as carbohydrates.

Minerals

Minerals have many functions in the animal. The skeleton is composed mostly of minerals. Minerals are important parts of soft tissues and fluids in the body. The endocrine system is heavily dependent on various minerals, as are the circulatory, urinary, and nervous systems.

Fifteen minerals have been identified as being essential to the health of animals. These are calcium, phosphorus, sodium, chlorine, potassium, sulfur, iron, iodine, cobalt, copper, fluorine, manganese, molybdenum, selenium, and zinc. In the past, most of these minerals were provided naturally by feeds grown on fertile soils and by contact with the soil itself. Today, it is increasingly important to provide additional mineral matter to the diet of animals. Mineral supplements are especially important for animals that spend their lives in confinement. A mineral fed as a separate feed is called a supplement.

Vitamins

Vitamins are acquired by animals in several different ways. Some are available in roughages and concentrates, some are available in feeds containing animal by-products, and some are made by the body itself.

Vitamins are required in only minute quantities in animals. They act mostly as catalysts for other body processes. There are large variations in the necessity for vitamins in various species of animals that are important to agriscience.

Some of the specific ways that vitamins are used in animals include clotting of blood, forming bones, reproducing, keeping membranes healthy, producing milk, and preventing certain nervous system disorders.

Fat

Only small amounts of fat are required in most animal diets. The addition of fat to the diets of animals improves the palatability, flavor, texture, and energy levels of feed. The addition of small amounts of fat to the diet has also been shown to increase milk production and to aid in the fattening of meat animals. Fats are also necessary in the body as carriers of fat-soluble vitamins.

SCIENCE PROFILE ENERGY CONTENT

Like humans, livestock need food for energy. But it is important to know how much energy an animal will receive from the food it eats. The calorie content of a feed source can be calculated. A *calorie* is a term used to describe available energy. One calorie has enough energy to raise one gram of water one degree Celsius. Animals need a specific number of calories to maintain good health and be productive. It is a balancing act. Too much feed and an animal will gain an inappropriate amount of weight and not be as productive as it could be. In contrast, if an animal does not receive an adequate amount of calories, it will lose weight and will not reach its production potential.

Macronutrient Calorie Content

Macronutrient	Calories per Gram	Livestock Feed Example
Carbohydrate	4 calories per gram	Grain
Protein	4 calories per gram	Alfalfa
Alcohol	7 calories per gram	Silage
Fat	9 calories per gram	Plant oil additives

In the preceding table, each macronutrient is listed with the calorie content available in one gram of each. An example is provided, showing a feed source for each nutrient.

*Note: The feed examples listed may contain other nutrients. For example, alfalfa is also a source of carbohydrate.

The table on macronutrient calorie content shows the amount of energy (in calories) available for each macronutrient. This is useful information for individuals who need to provide proper nutrition to livestock and receive the highest levels of production possible. For example, if a grain contains 50 grams of carbohydrates, 20 grams of protein, and 10 grams of fat in a cup of feed, how many calories will each cup of feed contain?

50 g. carbohydrates × 4 calorie/gram = 200 calories from carbohydrates

20 g. protein × 4 calorie/gram = 80 calories from protein

10 g. fat × 9 calorie/gram = 90 calories from fat

200 + 80 + 90 = 370 calories per cup of feed

*With this information, an exact measurement of feed can be calculated based on an animal's weight and energy requirements.

SOURCES OF NUTRIENTS

The sources of nutrients for animals are many and extremely varied. Important animal feed components include roughages, concentrates, animal by-products, minerals from mineral deposits, and chemically made nutrients called synthetic nutrients.

Proteins

The major sources of protein for animals include oil seeds such as soybeans, peanuts, cottonseed, and linseed. These seeds are processed by cooking and other procedures to remove the bulk of the oil from them. The remainder of the seed content is then dried and ground up for feed. Feed consisting of ground oil seeds with the oil removed is called oil meal.

Cereal grains provide lesser amounts of protein than oil meal, but they are also important protein sources. Good-quality legume hay, such as alfalfa or clover, is another good plant source of protein for ruminant animals.

Animal protein is generally of greater quality than plant protein. More specifically, animal protein usually contains more of the essential amino acids than plant protein. Sources of animal protein include meat and bone meal, meat meal, fish meal, blood meal, skim milk, whey, feather meal, and meat products.

It should be noted that most meat by-products obtained from mammals cannot be fed to ruminant animals. Some exceptions include products made from blood obtained from slaughter facilities and milk by-products.

Nonprotein nitrogen in the form of urea can be used as a substitute for some of the protein required by ruminant animals. Urea is a synthetic source of nitrogen made from atmospheric nitrogen, water, and carbon. The rumen bacteria found in ruminant animals such as cattle and sheep can convert urea and other nitrogen sources to amino acids and proteins. Rumen bacteria are also a source of protein because they are digested together with the feed. The feeding of urea should be limited to not more than 1 percent of the total dry matter in the ration. Young ruminant animals and all nonruminants are unable to digest urea.

Carbohydrates

Carbohydrates are found in all plant materials. The major sources of carbohydrates for animal feed are the cereal grains. Corn is the most important of these grains in the United States, followed by wheat, barley, oats, and rye. Other sources of carbohydrates include nonlegume hays such as orchard grass, timothy, other grasses, and molasses. Animal rations generally contain adequate levels of carbohydrates.

Fats

Because fats are needed in fairly small amounts in the diets of animals, it is seldom necessary to identify specific sources of dietary fat. Carbohydrates are converted to fats when a surplus exits in the diet, and most proteins are also sources of fat. This is especially true for the oil seeds and animal by-products.

Vitamins and Minerals

Vitamins and minerals are part of all the normal feeds for animals. Ruminants manufacture B-complex vitamins in their rumens. Exposure to sunlight allows the body to manufacture vitamin D. Contact with the soil, coupled with other feeds grown on fertile land, provides most of the mineral requirements for animals that have access to pasture and high-quality feeds. However, it is sometimes necessary to supplement natural sources of vitamins and minerals. Commercial vitamin and

HOT TOPICS IN AGRISCIENCE MAD COW DISEASE

Mad cow disease, also known as bovine spongiform encephalopathy (BSE) or chronic wasting disease, was discovered in 2003 in the United States near Yakima, Washington, and a 2005 case was subsequently confirmed in Texas. Within hours of the first case, Japan, South Korea, and Taiwan halted imports of U.S. beef, and other countries threatened to do so. An earlier case of the disease in Canada closed the border between the United States and Canada for beef imports.

The disease is known to be spread by feeding contaminated animal products such as processed meat scraps to cattle. A ban has been placed on the feeding of animal meat products to cattle. An extensive search was also implemented that traced the origin of the diseased animal and all known animals that were herd mates. Offspring of the affected cow were also located and destroyed. Government agencies went to great lengths to isolate the disease and eradicate it. Mad cow disease in the United States has been assigned the international status of *Controlled Risk* for beef exports. Control of the disease in Europe resulted in thousands of animals being killed and burned in efforts to control and eradicate it.

mineral supplements are formulated for specific classes of animals and their special needs. Such supplements are available in the developed countries of the world.

SYMPTOMS OF NUTRIENT DEFICIENCIES

Animals must be fed appropriate types and amounts of feed regularly to remain healthy and to produce milk, meat, wool, eggs, fur, work, or healthy offspring. Shortages, or deficiencies, of various nutrients will generally produce observable effects in animals. Some common symptoms of nutrient deficiencies are described in Figure 26-12.

FEED ADDITIVES

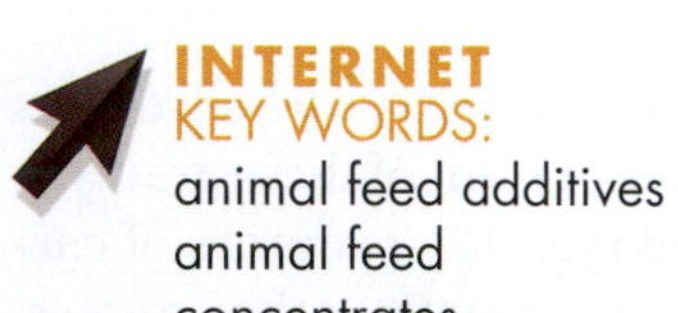

animal feed additives
animal feed concentrates

A **feed additive** is a nonnutritive substance that is added to feed to promote more rapid growth, to increase feed efficiency, or to maintain or improve health. Feed additives fall into two major groups: growth regulators (mostly hormones) and antibiotics. **Antibiotics** are substances used to help prevent or control diseases.

Some common growth regulators include hormones such as progesterone, estrogen, and testosterone. They are known to increase growth rates and feed efficiency by as much as 5 percent. Use of growth regulators is controlled by government meat inspectors, who test for the presence of these substances in meat.

A wide range of antibiotics are added in low levels to the diets of animals such as swine and poultry. Antibiotics keep certain low-grade infections at bay. Antibiotics added to feed allow growing animals to gain weight at their greatest potential rate.

In past years, a good deal of controversy has arisen over including antibiotics and growth hormones in the feed of animals. Of major concern is the possibility of these substances remaining in the meat of animals slaughtered for human consumption. To reduce this possibility, the substances must be removed from the feed well before the animals are marketed.

COMPOSITION OF FEEDS

All feeds are composed of water and dry matter. The material left after all water has been removed from feed is **dry matter**. Water makes up 70 to 80 percent of most living things. However, dry feeds generally contain only 10 to 20 percent water.

Dry matter is made up of organic matter and ash or minerals. The organic-matter portion of animal feed consists of protein; carbohydrates, such as starch and sugar; fat; and some vitamins. The proportion of these materials varies widely among different feeds.

ROLE OF WATER IN LIVING ORGANISMS
• Dissolve or suspend nutrients • Break down foods to forms the organism can use • Create rigidity to maintain the shape of the organism • Regulate the body temperature of animals

FIGURE 26-11 Water is one of the most important requirements of living organisms. Without it, most animals can't survive for more than a few days.

Disorders Caused by:

Vitamin Deficiency

Vitamin A
1. Night blindness
2. Loss of young
3. Poor growth
4. Nasal discharge
5. Diarrhea

Vitamin C
1. Scurvy
2. Gum inflammation
3. Hemorrhages
4. Slow healing of wounds

Vitamin D
1. Bone weakness and deformities
2. *Rickets* and *osteomalacia* are bone diseases in young and old animals, respectively
3. Thin egg shells
4. Weak, deformed young
5. Lowered milk production

Vitamin E
1. Reproductive failures
2. Degeneration of certain muscles
3. *Stiff lamb disease*, or muscle degeneration in lambs
4. *White muscle disease*, or muscle degeneration in young calves
5. Poor egg hatchability

Vitamin K
1. Poor blood clotting
2. Internal hemorrhages

Thiamine
1. Poor appetite
2. Slow growth
3. Weakness
4. Nervousness

Riboflavin
1. Slow growth
2. *Dermatitis*, or skin disorder
3. Eye abnormalities
4. Diarrhea
5. Weak legs in pigs

Niacin
1. Dermatitis
2. Retarded growth
3. Digestive troubles

Pyridoxine
1. *Anemia*, or low red-blood-cell count
2. Poor growth
3. Convulsions in pigs

Pantothenic Acid
1. "Goose-stepping" in pigs
2. Unhealthy appearance
3. Digestive problems

Biotin
1. Dermatitis
2. Loss of hair
3. Retarded growth

Choline
1. Poor coordination
2. Poor health
3. Fatty liver
4. Poor reproduction in swine

Folic acid
1. Blood disorders
2. Poor growth

Vitamin B_{12}
1. Slow growth
2. Poor coordination
3. Poor reproduction

Mineral Deficiency

Calcium
1. Rickets
2. Poor growth
3. Deformed bones
4. Milk fever

Phosphorus
1. Lameness
2. Stiff joints
3. Rickets
4. Poor milk production

Sodium Chloride
1. Lack of appetite
2. Unhealthy appearance
3. Slow growth

Potassium
1. Slow growth
2. Joint stiffness
3. Poor feed efficiency

Sulfur
1. General unthriftiness (lack of strong growth)
2. Poor growth

Iron
1. Anemia
2. Labored breathing
3. Edema (swelling) of the head and shoulders
4. Flabby, wrinkled skin

Iodine
1. *Goiter*, or enlarged thyroid gland in the neck
2. Weak or dead offspring at birth
3. Hairlessness
4. Infected navels, especially in foals

Cobalt
1. Delayed sexual development
2. Poor appetite
3. Slow growth
4. Decreased milk and wool production

Copper
1. Abnormal wool growth
2. Poor muscular coordination
3. Anemia
4. Weakness at birth

Fluorine

Poor teeth

Manganese
1. Poor fertility
2. Deformed young
3. Poor growth

Molybdenum

Poor growth rate

Selenium
1. Muscular degeneration
2. Heart failure
3. Paralysis
4. Poor growth

Zinc
1. Poor growth
2. Unhealthy wool or hair
3. Slow healing of wounds
4. *Parakeratosis*, or a skin disease similar to mange

FIGURE 26-12 Some disorders caused by vitamin and mineral deficiencies in animals.

4. Night blindness may be caused by a deficiency of
   a. zinc. c. protein.
   b. vitamin A. d. biotin.
5. Improved feed efficiency can be accomplished by including __________ in animal feeds.
   a. fats c. additives
   b. minerals d. roughages
6. Grain, oil meal, molasses, and meat by-products are all forms of
   a. roughages. c. carbohydrates.
   b. concentrates. d. dry matter.
7. Dry matter is made up of organic matter and
   a. minerals. c. fat.
   b. water. d. concentrate.

B. MATCHING (GROUP I)

__________	1. Goiter	a. Iron
__________	2. Parakeratosis	b. Vitamin D
__________	3. Rickets	c. Vitamin E
__________	4. White muscle disease	d. Zinc
__________	5. Anemia	e. Iodine
__________	6. Dermatitis	f. Niacin

C. MATCHING (GROUP II)

__________	1. Ration	a. Feed high in TDN, low in fiber
__________	2. Ruminant	b. Disease-control substance
__________	3. Carbohydrate	c. Essential element
__________	4. Mineral	d. Amount of feed fed in one day
__________	5. Antibiotic	e. Fermented green roughage
__________	6. Silage	f. Sugars and starches
__________	7. Concentrate	g. Multicompartment stomach

UNIT 27

Animal Health

OBJECTIVE

To determine the most effective strategies to maintain animal health.

MATERIALS LIST

- different gauges of needles and syringes
- various containers or labels from containers of animal drugs
- fresh discarded ears from a meat-processing plant
- Internet access

COMPETENCIES TO BE DEVELOPED

After studying this unit, you should be able to:

- identify physical signs of good and poor animal health.
- identify symptoms of animal diseases and parasites.
- understand how to prevent animal health problems.
- explain various methods of treating animal health problems.

SUGGESTED CLASS ACTIVITIES

1. Compile a list of animal diseases and identify some sources of information about each of the diseases. Assign class members to work separately or in pairs to prepare a class report on the disease they have chosen to study. (Do not assign more than one group to study the same disease.) Encourage the students to use Internet sources as well as more traditional sources such as textbooks and extension service bulletins. Have each study group make a class presentation of their material.
2. Visit a veterinary clinic as a class, or ask a veterinarian to visit your class as a guest speaker. Ask the veterinarian to discuss the animal health problems that he or she deals with most often. Also include discussion of the career opportunities that are available in the animal health field.
3. Divide the class into six groups. Each group will have an orange, a syringe and needle (a plastic needle that cannot penetrate skin is recommended), and a cup of water. Demonstrate to each group the technique for giving intramuscular and subcutaneous injections, using the fruit. Each group member will practice this process on the orange. Administrative approval should be sought before beginning this activity.

TERMS TO KNOW

disinfectant
host animal
contagious
noncontagious
abortion
roundworm
fluke
protozoa
secondary host
mange
balling gun
drench
intravenous
intramuscular
subcutaneous
intradermal
intraruminal
intraperitoneal
infusion
cannula
immune

Maintaining animal health is the key to a profitable and satisfying animal enterprise. Several considerations need to be addressed in dealing with the health of animals. These include being able to recognize signs of good and poor health, maintaining a healthy environment, being able to identify animal diseases and parasites, and knowing how to treat health problems when they occur. These topics are explored in this unit. Diseases are infective agents that result in declining health of living things. Parasites are organisms that derive their food from live organisms by ingesting body fluids such as blood or mucus.

SIGNS OF GOOD AND POOR ANIMAL HEALTH

Having the ability to recognize the signs of good health or the symptoms of health problems is the single most important key to being efficient in maintaining good animal health. A keen sense of observation is important as well as the innate ability to know when something is not right with an animal (Figure 27-1).

Signs of Good Health

One of the best signs of good health is a contented animal. Of course, a good deal of experience in dealing with animals is necessary to recognize contentment. Alertness and the chewing of the cud in ruminant animals is a good sign. A shiny hair coat, bright eyes, and pink membranes are other signs that an animal is healthy. Normal body discharges of urine and feces are further evidence that animals are not experiencing serious health problems. On the technical side, a healthy animal should have a normal body temperature, pulse rate, and respiration, or breathing, rate (Figure 27-2).

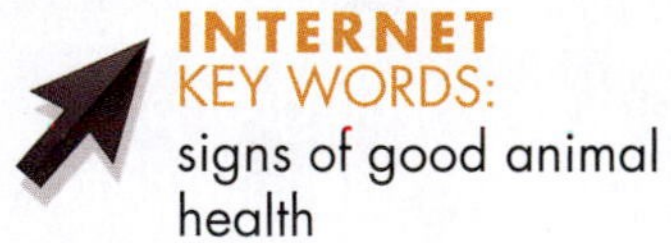
INTERNET KEY WORDS: signs of good animal health

FIGURE 27-1 The good animal manager develops a keen sense of observation and learns to recognize the symptoms of an animal under stress.

HOT TOPICS IN AGRISCIENCE MODERN WAYS TO MONITOR HERD HEALTH

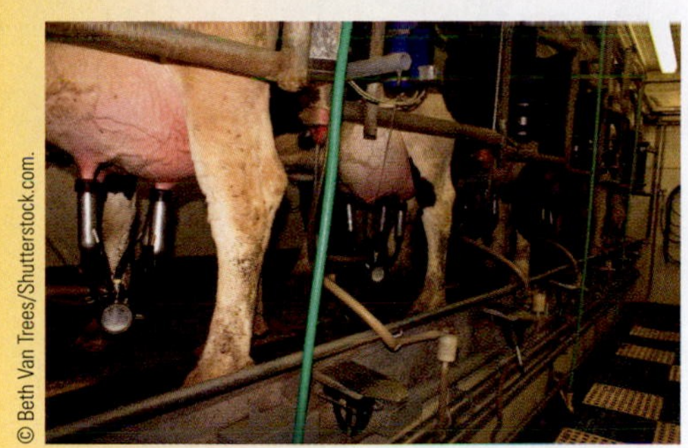

Abnormal milk temperature is a good indicator of an unhealthy cow. An electronic temperature sensor, placed in the milk pipeline at each milking station, helps identify cows that need medical attention.

The earlier that a herd health problem is detected and treated, the more likely it is that a sick animal will recover. Modern electronic devices make it possible to detect health problems in animals during the early stages of infection. For example, electronic sensors are available that can take the body temperature of an animal as it walks by such sensors. These devices are placed on alleyway fences that lead to water or from milk parlors. When an animal with an abnormal body temperature approaches the sensor, a hydraulic gate sorts the animal into a holding pen to be checked by a veterinarian.

A different type of sensor is available to monitor the health of dairy cows. This sensor is placed in the milk line that carries milk from each cow to the main pipeline. The sensor measures the temperature of the milk as it comes from the cow. The sensor is useful on those dairy farms where each cow is identified by a computer chip as she enters the milking area. The sensor enters the milk temperature in the computer record for each cow, and a sorting gate is activated that sorts each cow with an abnormal milk temperature into a holding pen as she leaves the milking area. These types of devices make it possible to identify sick animals in the early stages of infection. Early detection and treatment contribute to early recovery from infections and diseases.

Class of Livestock or Poultry	Degree F Average	Degree F Range
Cattle	101.5	100.4–102.8
Sheep	102.3	100.9–103.8
Goats	103.8	101.7–105.3
Swine	102.6	102.0–103.6
Horses	100.5	99.9–100.8
Poultry	106.0	105.0–107.0

FIGURE 27-2 Normal body temperatures for animals.

SIGNS OF POOR HEALTH

Often, it is easier to tell when an animal is sick than to tell when it is healthy. A rough hair coat and dull, glassy eyes are often the first signs that an animal is not well. Sick animals often isolate themselves from other animals and stay alone with their heads down. Such animals may be drawn up and walk slowly when forced to walk. Abnormal feces, either too hard or too soft, as well as discolored urine also are indicators that an animal is experiencing a health problem. Loss of production, especially in dairy cattle, is often the first sign that the animal is not well. Abnormal body temperature, labored breathing, and rapid pulse rate are other indications of poor health in an animal.

HEALTHFUL ENVIRONMENTS FOR ANIMALS

Maintaining a healthy environment for animals is a key factor in a complete animal-health program. It is often much less expensive to maintain a healthy environment for animals than it is to treat animals that are unhealthy due to a poor living environment (Figure 27-3).

Sanitation

Good sanitation is important to good health. Factors related to good sanitation include keeping animal facilities clean. Sanitation also requires the use of clean equipment when dealing with animals. This includes feed containers, milking equipment, artificial-breeding equipment, needles and syringes, and surgical equipment. A syringe is an instrument used to give injections of medicine or to draw blood and other body fluids from animals (Figure 27-4). Simple, on-farm

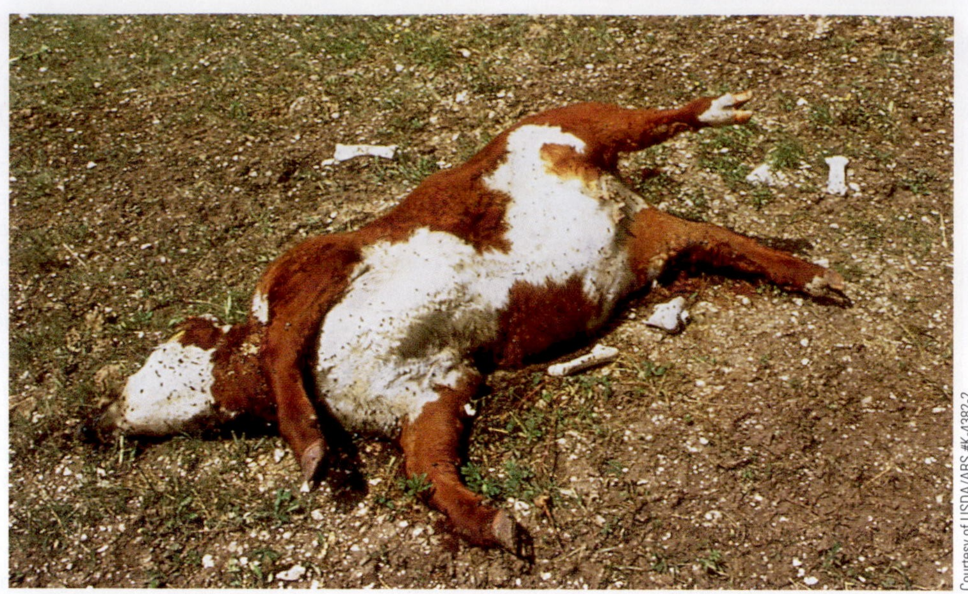
Courtesy of USDA/ARS #K-4382-2.

FIGURE 27-3 A dead animal is a grim reminder that untreated diseases or parasites, poisonous plants, or predators can quickly take the life of a valuable animal.

INTERNET KEY WORDS: livestock, facilities, housing

surgical procedures should always be performed with the strictest sanitation possible. The liberal use of disinfectants when dealing with animals is important. A **disinfectant** is a material that kills disease organisms.

Housing

Maintenance of proper housing is an important consideration in maintaining good animal health. Housing should be clean and free from cold drafts. However,

AGRI-PROFILE CAREER AREAS: VETERINARIAN/VETERINARY TECHNICIAN

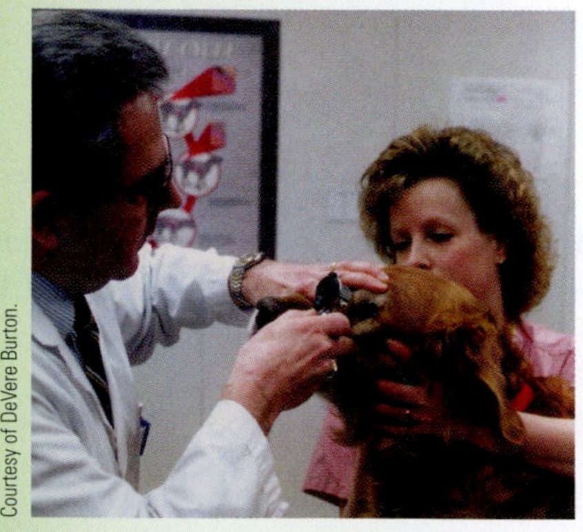
Courtesy of DeVere Burton.

Veterinarians and their associates work constantly to treat animals having disorders and to improve animal health.

Animal pathologists, animal behaviorists, physiologists, biologists, zoologists, microbiologists, geneticists, nutritionists, and others must work together to understand the complexities of animals. Animals exist as pets, production animals, work animals, pleasure animals, fish, fowl, birds, wild animals, and specimen animals in zoos. The need for health services for animals varies with the species and its primary use.

The desire to become a veterinarian has been high among youths in recent years. This interest has permitted the numbers of veterinarians to increase, despite the rigor of the college curriculum and tough competition to be accepted into veterinary schools. Colleges of animal sciences offer curricula in other career areas in animal sciences, such as nutrition, breeding, education, and production.

In urban and suburban areas, veterinary clinics for pets offer many opportunities for licensed veterinarians. Animal shelters, hospitals, kennels, and pet stores provide many career opportunities for those interested in animal health at the technician level. In rural areas, large-animal veterinarian, veterinary assistant, laboratory veterinarian, and laboratory technician are typical positions in the animal-health industry.

FIGURE 27-4 Syringes are instruments that are used to inject medications and vaccines into animals. They are also used to withdraw blood and other body fluids.

good air circulation throughout the shelter is important to help decrease high temperatures in the summer and reduce humidity in the cold of winter. Proper circulation also helps reduce the spread of germs through the air. Extremely dry and dusty conditions are to be avoided where possible. Proper maintenance of animal housing also is important. Loose boards, roofing materials, and nails often cause injuries and pose other problems in poorly maintained facilities.

Handling Manure

Piles of manure, dirty pens, and dirty feedlots are often sources of serious health problems in animals. It is important that manure not be allowed to accumulate in areas frequented by animals (Figure 27-5). Manure piles sometimes harbor

FIGURE 27-5 Manure can be treated to eliminate offensive odors by creating favorable conditions for organisms that break it down. Manure that has been converted to compost is a valuable product for use in fields, nurseries, and gardens.

diseases and parasites. They also attract flies, which may spread diseases. Cages and pens soiled continually with animal waste products may also decrease the quality of the air the animals breathe. Wet, poorly drained, manure-soiled feedlots usually reduce the rate of weight gain in beef cattle and swine. Feedlots are areas in which large numbers of animals are grown for human consumption. Foot and leg problems can often be traced to poorly maintained feedlots.

Consumers are becoming more concerned with where their food is coming from. Farmers, ranchers, and feedlot managers need to consider how the grocery store shopper might perceive the sanitation of his or her livestock operation, and then think about how it may affect profits. History provides a classic example. In the early 1900s, a book written by Upton Sinclair titled *The Jungle* drew mass consumer attention to the poor meat-handling and butchering practices of the time. Consumer reaction resulted in major losses of revenue from the farmer to the grocer. The public placed so much pressure on the government and the industry that the Meat Inspection Act and the Pure Food and Drug Act of 1906 were passed. Today's producers should be aware that they are once again under the scrutinous eyes of consumers and government and environmental groups, and they should strive always to maintain animal environments that are as clean as conditions allow.

Controlling Pests

The control of pests and parasites is an important consideration in the maintenance of animal health and welfare. Regular use of disinfectants to control parasites such as lice and flies is necessary in a good disease-prevention program. Regular, close observation of animals also is necessary to determine when outbreaks of parasites occur. Prevention of parasites is preferable to controlling outbreaks. To that end, the development of a good prevention program is a wise decision (Figure 27-6).

The control of other pests, such as birds and wild animals, is also part of a good animal-health program. Many birds carry parasites on their bodies and in their droppings. When the parasites move from infected animals to healthy

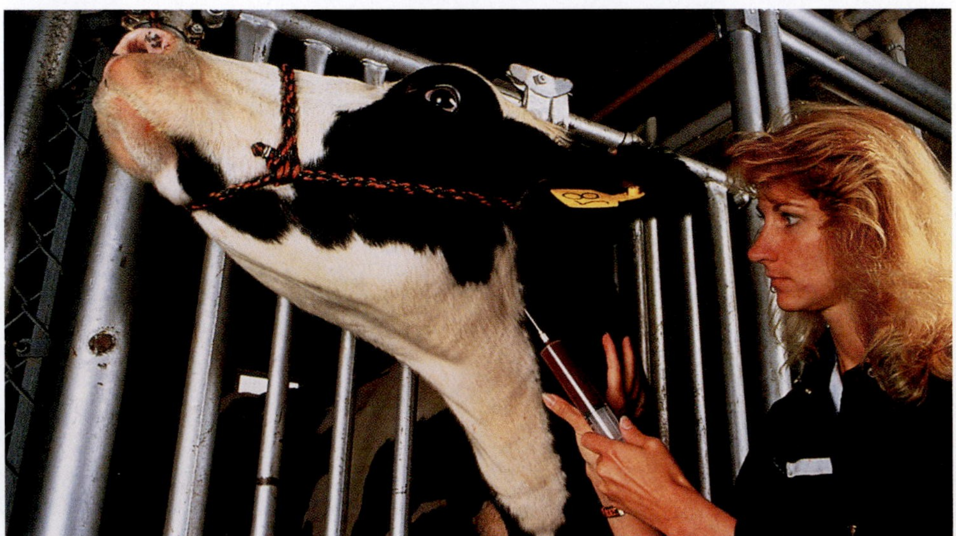
Courtesy of USDA/ARS #K-4135-11.

FIGURE 27-6 The drawing and analyzing of blood samples is a technique for diagnosing and treating many diseases and disorders in animals.

ones, they sometimes carry diseases. Wild animals and pets may also cause serious health problems when allowed to roam freely around farm animals. Dogs and coyotes will sometimes chase animals and cause injuries. Bites from these animals may cause infection, other health problems, and death. Just the presence of pets around farm animals can cause farm animals to be nervous, affecting their production.

Isolation

The isolation of animals that are new to a herd is an important part of any good preventive health program. New animals may be harboring diseases or parasites that are not readily apparent. It is wise to keep them isolated from other animals for a time, usually a minimum of 30 days. This gives the new owner time to observe the isolated animals closely for health problems (Figure 27-7).

Similarly, isolation of diseased animals is important. Animals with contagious diseases that can be spread by contact should never be allowed to remain in contact with healthy animals. It is difficult to treat unhealthy animals when they are living with large groups of animals. Healthy animals tend to become aggressive with unhealthy ones, making it especially difficult for such animals to recover their health.

Pasture Rotation

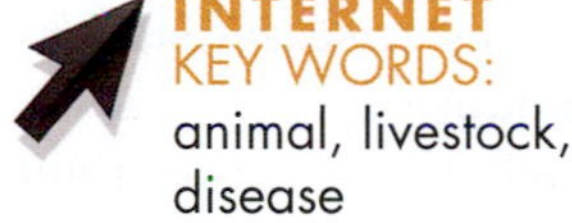

INTERNET KEY WORDS: animal, livestock, disease

The rotation of pastures is a consideration in maintaining a healthy environment for animals and in preventing health problems. Many disease organisms that infect animals are harbored in the soil. They are killed only by not being able to come into contact with host animals for extended periods. A **host animal** is an animal in or on which disease organisms and/or harmful parasites live. Rotating pastures for grazing animals on a regular basis breaks the life cycle of most parasites, thus contributing to control of parasite populations (Figure 27-8).

Courtesy of USDA/ARS #K-4702-6.

FIGURE 27-7 It is important to isolate animals that are being added to a herd for up to 30 days. This allows time for disease symptoms to develop if they are present, and it protects the rest of the herd from exposure to any disease that does develop.

Courtesy of USDA/ARS #K-3714-2.

FIGURE 27-8 Pasture rotation is necessary to break the life cycles of diseases, insects, and internal parasites, and it pays additional dividends in greater pasture production.

ANIMAL DISEASES AND PARASITES

Diseases

Animal diseases can be divided into two major classes: contagious and noncontagious. **Contagious** diseases are those that can be passed on to other animals by contact. **Noncontagious** diseases cannot be spread through contact with other animals.

The handling of these two classes of diseases varies somewhat. It is important that animals with contagious diseases be isolated from other animals in the herd as soon as the disease is identified. Because some contagious animal diseases can be transmitted to humans, care must be taken when handling infected animals. Similarly, humans who handle animals should become familiar with the proper techniques, vaccinations, and precautions to avoid human disease and parasitic infections from animals.

Noncontagious diseases pose no threat to humans or other animals, except to the animals afflicted with the diseases. Therefore, there is more leeway in dealing with these animals. It is still a good idea to isolate them from the herd for their own good, however.

Causes

Contagious diseases are mostly caused by bacteria and viruses. They can be spread by direct contact with infected animals, from shared housing, or from contaminated feed or water. In some cases, the spread of infectious diseases takes place through intermediary hosts, such as birds, rodents, or insects. Diseases such as sore mouth and brucellosis have the potential to infect humans.

Noncontagious diseases may be caused by nutrient deficiencies or nutrient excesses. Eating poisonous plants or foreign material and any open wounds that become infected may cause or lead to noncontagious disease.

SCIENCE CONNECTION THE UNSEEN HARVESTERS

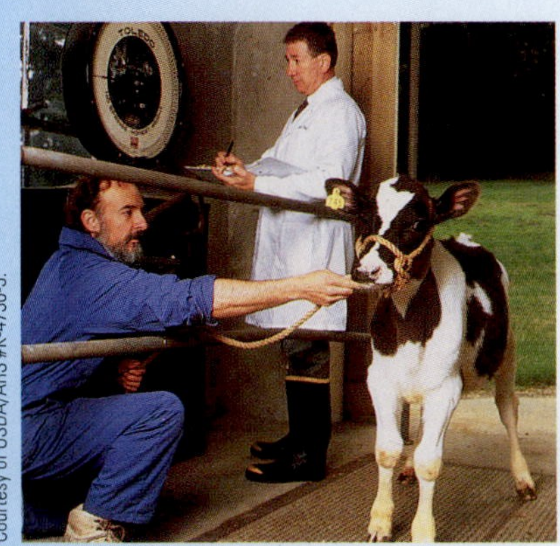
Courtesy of USDA/ARS #K-4736-3.
Researchers weigh a parasite-infected calf that has been treated.

The control of internal parasites in production animals and pets is probably the most persistent animal-health problem. Generally out of sight and frequently microscopic in size, parasites are small organisms that live in the flesh or internal organs of larger animals, where they draw nutrients from the bodies of their hosts. Animals can become infected with parasites, and they are helpless to do anything about it when symptoms begin to appear. Fortunately, the body has some ways of keeping such intruders in check. If otherwise healthy and free of reinfection, the hosts can sometimes function fairly well even though they are infected with parasites. Sometimes they can even rid their systems of the parasites. Generally, however, the host needs outside help.

Farmers, ranchers, veterinarians, and scientists have long known that animals with parasites grow poorly and seldom reach their full potential. No matter how much they eat, they are undersized and underachievers. Researchers confirm that internal parasites do more than just consume nutrients intended for the host.

INTERNET
KEY WORDS:
parasite roundworm

Symptoms

General symptoms of disease are extremely varied and may include the following:

- Poor growth, reduced production, or both
- Reduced intake of feed
- Rough, dry hair coat
- Discharge from the nose or eyes
- Coughing or gasping for breath
- Trembling, shaking, or shivering
- Unusual discharges, such as diarrhea or blood in feces or urine
- Open sores or wounds
- Unusual swelling of the body, including lumps and knots
- **Abortion**, or the loss of a fetus before it is fully developed
- Peculiar gait, or walking pattern, or other odd movements

Some diseases may have little or no external symptoms and may even progress so rapidly that death of the animal occurs before symptoms are noticed.

FIGURE 27-9 Roundworm parasites come in many sizes. They infect the digestive tracts of animals and humans and are common in most parts of the world. The worms are round, as the name implies, and tapered on both ends.

Parasites

Parasites may also be grouped into two general classifications. These are internal—inside the animal—and external—living on the outside of the animal.

The most important internal parasites that infest animals are the **roundworms** (slender worms that are tapered on both ends) (Figure 27-9). Other types of internal parasites include flukes and protozoa. **Flukes** are very small flat worms, and **protozoa** are microscopic, one-celled animals. In nearly every instance, an internal parasite spends at least a part of its life cycle outside

The animal's immune system responds to invading organisms by producing chemical signals that modify the host's metabolism. These immune-response signals are small proteins called cytokines that manipulate the hormones regulating feed intake, nutrient use, and, ultimately, growth of the animals. Dairy and beef calves infected with a protozoan parasite called Sarcocystis run fevers, lose their appetites, and become emaciated. Even after an animal has been treated and the parasites have been brought under control, some calves are incapable of growing normally.

Research was conducted on calves at Beltsville, MD, to find some explanations. Before, during, and after acute infection with the parasite, the concentration of growth-regulating hormones in the blood was measured. It was found that after acute infection, the concentration of a hormone essential for growth decreased and the concentration of another hormone that blocks growth-hormone secretion increased. These hormone changes persisted in the infected calves even after the symptoms of infection were gone.

It is thought that the body implements a survival strategy in response to the invasion of the parasites. The strategy is one of growth restriction so more energy can be available to fight off the intruders. It is therefore essential that every effort be made to keep animals in surroundings that are free from parasites and to use precautionary measures whenever applicable. When symptoms appear, it is important to get a professional diagnosis and promptly administer the proper treatment.

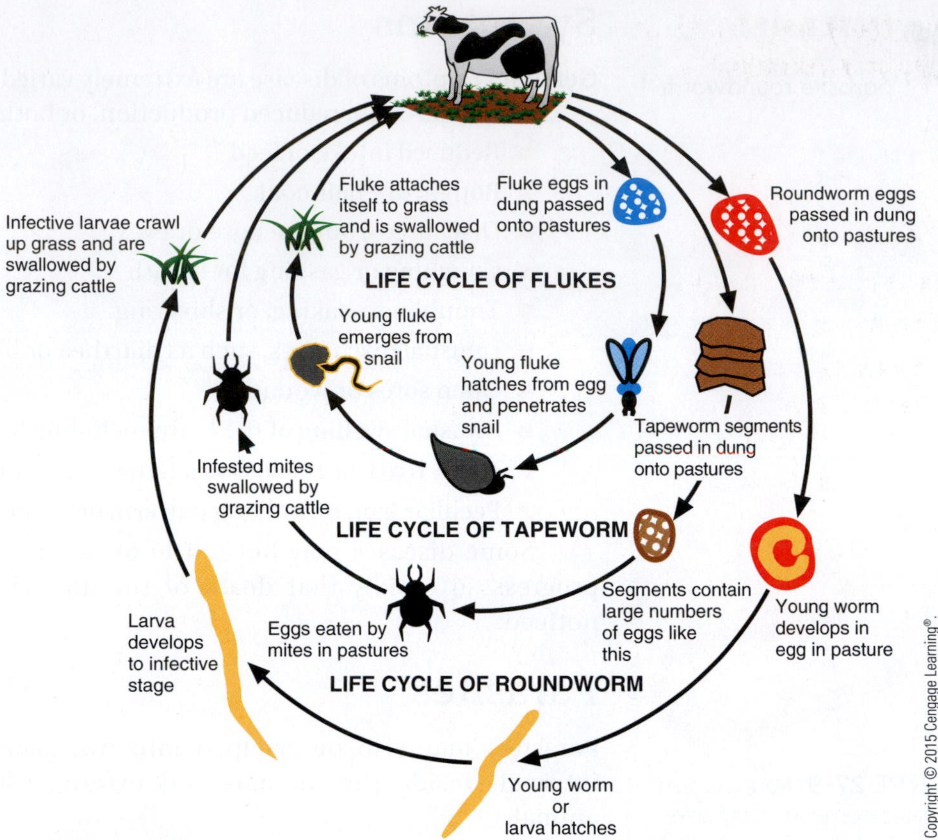

FIGURE 27-10 Life cycles of some common internal parasites.

the host animal (Figure 27-10). It is during this period that the parasite is most easily spread to other animals. Contact with discharges from infested animals, contaminated feed, water, housing, or secondary hosts may result in the spread of internal parasites. A **secondary host** is a plant or animal that carries a disease or parasite during part of its life cycle. Some internal parasites are also spread by insects such as flies and mosquitoes.

External parasites include flies, ticks, lice, mites, and fleas. They are spread from animal to animal through physical contact.

Symptoms of parasite infestation may include the following:

- Poor growth
- Weight loss
- Constant coughing and gagging
- Anemia
- Reduced production and reproduction
- Diarrhea or bloody feces
- Worms in the feces
- Swelling under the neck
- Poor stamina
- Loss of hair
- The occurrence of **mange**
- Visible evidence of the parasite itself

PREVENTING AND TREATING ANIMAL HEALTH PROBLEMS

A number of activities and procedures are used to prevent and treat health problems found in animals. Some of these include administering drugs, dipping with insecticides, and restraining animals. The roles of feed additives and vaccination will also be explored in the next sections.

Practical Veterinary Skills

All owners of animals should have a few basic veterinary skills. Farmers and ranchers would spend far more on animal medical costs if they had a veterinarian administer all of the routine vaccinations required by their animals. Some procedures that are commonly performed by animal owners are birthing assistance, castration, vaccination, treating for lice or mites, minor cut repair, mastitis treatment, and many more. Vaccination is the injection of a modified disease organism into an animal to stimulate development of immunity to prevent disease. Although many tasks can be done by animal caregivers, many more-serious procedures still must be completed by a licensed veterinarian.

Many common animal illnesses have distinct symptoms and can be treated initially in the field. Medicines for these illnesses can be purchased at a farm supply store or from a veterinarian. Some knowledge on how to provide medical care to animals can be learned through observation and experience. Other skills are learned through organized education classes and by enrolling in animal technician programs at a college or university. Agricultural extension workshops, clinics, and veterinary handbooks also teach basic veterinary skills that are needed on the farm or ranch. In addition, a lot can be learned about preventing and treating routine animal health conditions by consulting with a local veterinarian.

Administering Drugs

Several factors must be considered before administering drugs to an animal. These include determining the amount or dosage to be administered, type of drug to use, purpose of the drug, site of administration of the drug, and type of animal to be treated. Most of this information can be found on the drug container. It is important that the drug manufacturer's recommendations be followed closely. Another factor that needs to be considered is the amount of time that is required for the drug to be inactivated in the animal's body. This is important when determining how long the drug will contaminate milk from a dairy animal or meat from a meat animal. Contaminated milk must be discarded to make sure it is not in the human food supply. Also, it must be determined how long to wait before a treated animal can be slaughtered for meat.

Drugs are manufactured and sold as pills, powders, pastes, or liquids (Figure 27-11).

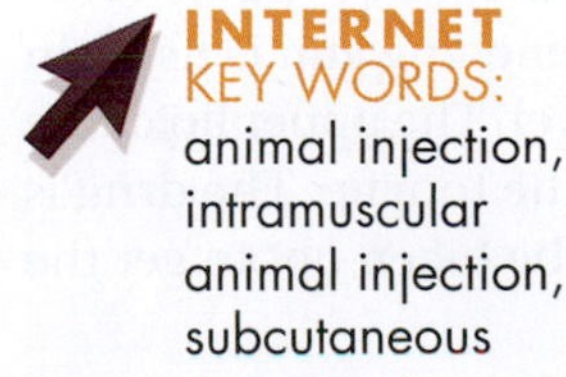

Pills

The procedure for giving a pill to an animal is to restrain the animal and lift its head so that the mouth opens. Gently place the pill as far back on the tongue as possible, using a **balling gun** (a device used to place a pill in an animal's throat). Massage the animal's throat until it swallows the pill.

FIGURE 27-11 Pills, liquids, powders, and pastes are methods of giving medicine to animals. It is important that the drug manufacturer's recommendations be followed closely.

Powders

Powder drugs are normally mixed in the feed or water of the animal. Powders are also mixed with water and administered as a drench. Powders mixed in the water or feed are effective only when the animal is well enough to consume it.

Paste

Paste is normally used for treating horses for worms. The preparation is placed on the back of the horse's tongue with a caulking gun, and the horse is forced to swallow. Pastes are used for horses because it is often nearly impossible to treat them for worms by any other method.

Liquids

Liquid drugs administered orally (by mouth) can be placed directly in the animal's stomach by drenching. To **drench** an animal, a fairly large amount of liquid medication is force-fed to an animal by mouth. A syringe or drenching gun is used. In the process, the animal is restrained, with the head held level. The upper lip of the animal is lifted, and the tube is inserted along the side of the tongue. The drug is released, and the animal is allowed to swallow. Care must be taken not to get the drug into the animal's lungs.

The injection of drugs into animals takes many forms, based on the location of the injection. Injection is the process of administering drugs by needle and syringe. Some of the injection sites include **intravenous** (in a vein), **intramuscular** (in a muscle), **subcutaneous** (under the skin), **intradermal** (between layers of skin), **intraruminal** (in the rumen), and **intraperitoneal** (in the abdominal cavity). One factor in determining where injections are made is how fast the drug needs to work. A drug injected into the blood is available faster than one injected under the skin. Often, it is desirable for drugs to be released slowly over a long period. Growth hormones are generally administered in this way.

To give an injection, use the following procedure:

1. Fill the syringe, making sure that all air is removed.
2. Restrain the animal.
3. Select the location for the injection.
4. Disinfect the area to be injected.
5. If the injection is to be made intradermally, clip the hair from the area to be injected.
6. Pop the needle into the desired area without the syringe attached (this prevents the loss of the drug if the animal jumps).
7. Attach the syringe to the needle and inject the liquid.

Infusion

Infusion is another method of getting drugs to the site of the infection. It is used most often to treat dairy animals with udder and teat problems. The udder is the milk-secreting gland of an animal. Teats are the appendages of an udder through which milk is obtained by mechanical means or by suckling offspring. To perform an infusion, a sterile **cannula** (blunt needle) is inserted into the opening of the teat, and the drug is forced into the teat canal from a syringe (Figure 27-12).

Dipping

Dipping is a process for treating animals, mostly cattle and sheep, for external parasites. It involves filling a vat with medicated water and forcing the animal to

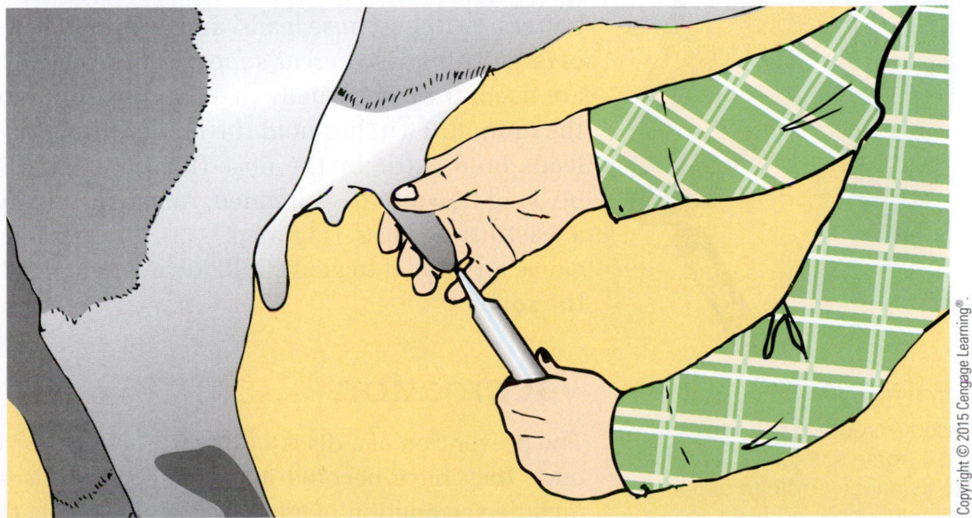

FIGURE 27-12 Infusion is a method of treating cows and other large animals for infections in the udder.

walk or swim through it. This process is also used to treat dogs for ticks and fleas. Dipping is used in cases where large numbers of animals must be completely covered with the medication.

Taking Temperatures

Taking an animal's temperature is basically the same as taking a human baby's temperature. It is usually taken in the rectum. The rectum is the last organ in the digestive tract. Animal thermometers are normally longer and heavier than those used in human medicine. An animal thermometer also has an eye at one end and should have a string attached to it to prevent loss of the thermometer in the body cavity of the animal.

To use an animal thermometer, first shake down the column of mercury. Coat the thermometer with sterile jelly to make insertion easier. Do not force the thermometer into the rectum. If there is resistance, injury may result. Correct the conditions that are causing the resistance, and then reinsert the thermometer. A digital thermometer gives an immediate readout, but those containing mercury take a little longer. Leave the thermometer in place for 2–3 minutes before removing it and reading the temperature on the scale or digital screen.

Determining Pulse and Respiration Rates

The pulse rate for a large animal can be taken by holding your ear against the animal's chest or using a stethoscope and listening to the heartbeat. The number of heartbeats in 1 minute is the pulse rate.

The respiration rate of an animal can be determined by watching its rib cage move. Counting the number of breaths that the animal takes in 1 minute will indicate the rate of respiration. Readings are compared to normal temperature and respiration rates for healthy animals of the same species.

FIGURE 27-13 A twitch is used to redirect the attention of a horse so that it is attentive to its handler instead of the person who is administering a diagnosis or medical treatment.

Restraining Animals

A number of ways can be used to restrain animals for observation and treatment of diseases and parasites. These include head gates, squeeze chutes, halters, twitches, nose leads, and casting harnesses. Head gates trap the heads of large animals, whereas squeeze chutes hold the whole animal. When halters are used, they are usually tied to a post or something else substantial to hold the animal. Twitches hold the tender upper lip of a horse (Figure 27-13). Nose leads hold cattle by the nose (Figure 27-14). Sometimes, large animals must be lying down to be examined. An effective way to accomplish this is by using a casting harness (Figure 27-15). Properly applied, a casting harness will cause an animal to gently fall down when steady pressure is applied by pulling the rope.

Vaccination

The prevention of a disease is nearly always less expensive than treating animals once they have become infected. A good disease-prevention program should include vaccination of animals. Vaccination is the injection of an agent, such as a modified disease organism, into an animal to immunize it against a specific

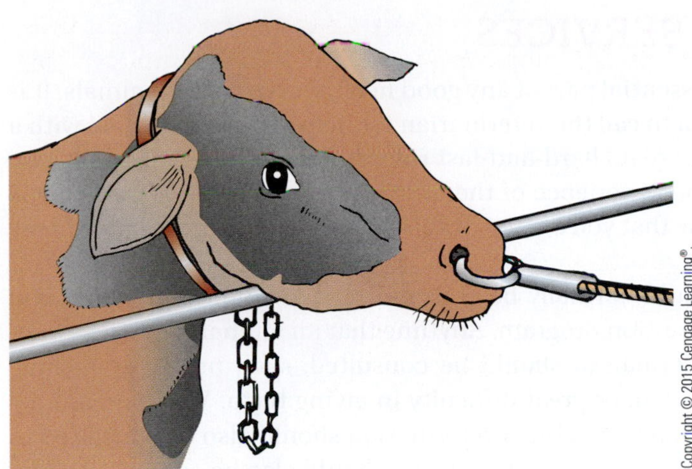

FIGURE 27-14 Cattle can be restrained during examination or treatment by using a nose lead to hold them in position. The nose lead distracts the animal's attention from the procedure by applying moderate pressure on the tissue between the nostrils when the animal struggles.

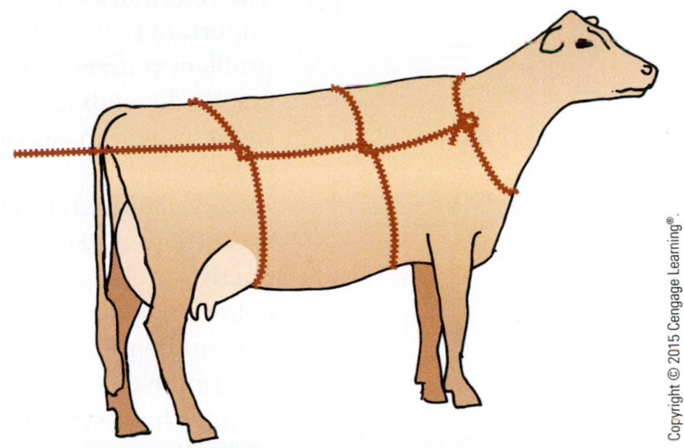

FIGURE 27-15 A properly used casting harness causes a large animal to lie down on its side while it is examined and treated.

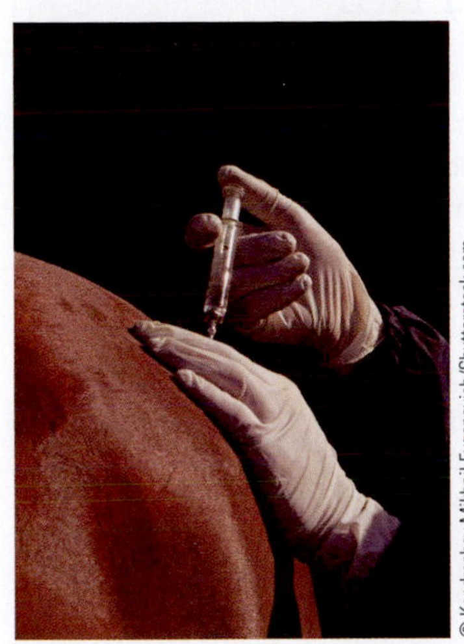

FIGURE 27-16 Most animal diseases can be prevented by a timely injection of a vaccine that targets a specific disease organism.

disease (Figure 27-16). The agent causes the animal to develop immunity to the disease. An animal that is **immune** to a disease should no longer be affected by the disease organism. Vaccination programs are usually part of the services of a veterinarian (an animal doctor). Vaccination programs vary widely with the type of animal, the disease organism involved, and the area of the country in which the program is being implemented.

Feed Additives

Feed additives are used primarily to control the incidence of low-level infections in growing animals. The materials used are primarily antibiotics that help increase feed efficiency and rate of gain as well as control of infections and diseases. Feed additives are sometimes used to control internal parasites. Caution must be taken to always follow the manufacturer's recommendations concerning the use of these materials. Failure to do so may lead to illegal contamination of animal products that are used for human consumption.

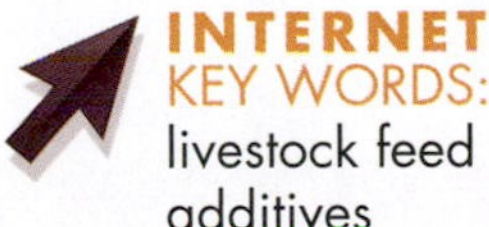

INTERNET KEY WORDS:
livestock feed additives

Medication Withdrawal

Drugs that are used to medicate food animals such as dairy cows, poultry, hogs, beef, and sheep must be withdrawn before the animal products can be used for human consumption. It often requires several days for drugs to leave an animal's system. If the withdrawal period has not been observed before slaughtering a meat animal or consuming eggs or milk from treated livestock, the medicine is likely to end up in human food. People who are allergic to antibiotics sometimes experience serious allergic reactions to these drug residues. Frequent use of foods that are contaminated with drugs can also build up a tolerance for antibiotics among disease organisms that infect humans. Such tolerances for antibiotics make medications less effective when they are needed for the treatment of animal and/or human infections.

VETERINARY SERVICES

The veterinarian is an essential part of any good health program for animals. It is important to know when to call the veterinarian for help and when to deal with a problem yourself. There are no hard-and-fast rules in this regard, and it will vary greatly depending on the experience of the individual. In general, it pays to call the veterinarian any time that you are not absolutely sure of the problem and how to handle it.

A veterinarian should normally be consulted when you are planning and executing a disease-prevention program. Anytime that an animal is having reproductive problems, a veterinarian should be consulted; such problems include failure to conceive, abortion, or great difficulty in giving birth. When an animal dies suddenly for no apparent reason, a veterinarian should also be consulted to determine the cause of the death. A veterinarian should also be contacted when animals have symptoms of a contagious disease. This will help minimize spread of the disease.

SCIENCE PROFILE FIGHTING "FIRE" WITH "FIRE"

The number one cause of death in newborn pigs is an infection caused by the *Escherichia coli* bacteria. As the result of many deaths, the swine industry loses millions of dollars every year. Up to this point, antibiotics have been used to treat the sick animals. Unfortunately, current techniques are becoming less and less effective because of the tendency of bacteria to develop resistances to the antibiotics. When a bacterium has developed a resistance, it means that an antibiotic no longer kills it. The scientist Roger Harvey heads a research project committed to developing an alternative to antibiotics.

A promising alternative is a mixed culture of "good" bacteria from the digestive systems of healthy pigs. These bacteria, known as "RPCF" (recombined porcine continuous-flow), are then introduced in the digestive tracts of newborn piglets. When this is done, the good bacteria attach themselves to the limited number of attachment sites in the intestines, thereby preventing *E. coli* from colonizing and causing an intestinal infection. Using "good" bacteria is a viable alternative to prevent infection by *E. coli*, the deadly bacteria that are responsible for so many infant pig deaths.

Newborn pigs frequently die from infection caused by *E. coli* bacteria. Treating them with a mixed culture of good bacteria in the digestive tract is proving to be a good alternative to antibiotics to which harmful bacteria are becoming resistant.

STUDENT ACTIVITIES

1. Write the Terms to Know and their meanings in your notebook.
2. Compare the methods of administration, dosage rates, and times of withdrawal of several drugs for animals.
3. Practice giving injections of water using discarded animal ears from a meat-processing plant.

4. Develop a complete disease-prevention program for a livestock operation.
5. Check the housing of production animals for animal safety and proper sanitation at home, on someone else's farm, or at someone's business.
6. Visit a farm supply store or a pet center and record the names of animal medicines, animal pest controls, and disinfectants that are for sale. Also list the use(s) of each item.
7. Interview the manager of a small animal or livestock operation concerning the disease-prevention and health-maintenance practices that are used. Report your findings to the class.
8. At your home, a friend's home, or a willing farmer's home, practice some restraint techniques on an animal. Determine what works best for you and for the animal.
9. Contact a local veterinarian and ask to spend a few hours observing him or her as he or she works.

SELF-EVALUATION

A. MULTIPLE CHOICE

1. Subcutaneous injections are made
   a. in a vein.
   b. under the skin.
   c. between layers of skin.
   d. in the body cavity.
2. An animal that lives and feeds on other animals is a
   a. parasite.
   b. disease.
   c. vaccination.
   d. host.
3. When taking the temperature of an animal, use a/an
   a. catheter.
   b. oral thermometer.
   c. rectal thermometer.
   d. syringe.
4. Which of the following is not a means of administering drugs orally?
   a. drench
   b. balling gun
   c. pills
   d. infusion
5. The purpose of vaccination is to
   a. prevent parasites.
   b. prevent diseases.
   c. control parasites.
   d. treat diseases.
6. Which of the following is not a sign of good health in animals?
   a. smooth hair coat
   b. increased pulse rate
   c. alertness
   d. contentment
7. The purpose of isolating sick animals is to
   a. prevent spread of contagious diseases.
   b. keep other animals from hurting the sick animal.
   c. allow for easier treatment of the problem.
   d. all of the above.
8. Which of the following is not an internal parasite?
   a. fluke
   b. mite
   c. stomach worm
   d. protozoan

9. Call a veterinarian when the animal
   a. has bright eyes.
   b. chews its cud.
   c. aborts.
   d. starts to lose hair.
10. A twitch is used to
   a. restrain horses.
   b. treat internal parasites.
   c. give injections.
   d. isolate infected animals.

B. COMPLETION

1. ______ diseases cannot be passed from one animal to another.
2. An______ injection is one that is made in the vein of an animal.
3. Parasites that are very small, flat worms are called ______.
4. ______ refers to the number of heartbeats in 1 minute.
5. An agent that prevents the growth of a disease organism is called an ______.
6. Protozoa are a form of______ parasites.
7. Oral administration of medicine refers to administering the drug through the animal's ______.
8. Vaccination is a form of disease ______.
9. An animal doctor is called a______.
10. A nose lead is used to ______ animals.

UNIT 28

Genetics, Breeding, and Reproduction

OBJECTIVE

To determine the current roles of genetics, breeding, and reproduction in animal science.

MATERIALS LIST

- illustrations of animal reproductive systems
- animal magazines
- bulletin board materials
- Internet access

COMPETENCIES TO BE DEVELOPED

After studying this unit, you should be able to:

- define terms associated with genetics and reproduction.
- understand the principles of genetics.
- identify systems of breeding.
- identify parts of reproductive systems.
- understand new technologies in animal reproduction.
- evaluate animals for type and production.
- understand types of testing programs.

SUGGESTED CLASS ACTIVITIES

1. Obtain a video that describes the processes by which artificial insemination (AI) and/or embryo transfer are accomplished in livestock. View the video as a class. Take time to discuss various aspects of the process by stopping the video at strategic points. Discuss key questions such as the following: What are the advantages of AI and embryo-transfer procedures as opposed to maintaining a herd sire on the farm? Why are embryos collected only from the most productive female animals?
2. Obtain female reproductive tracts from a biological supply house or other reliable source and use them to teach students how AI is performed. Obtain some disposable gloves and plastic pipettes. Teach the students to thread the pipette through the cervix of a cow to simulate the AI process.
3. In groups of two or three, search the Internet for information involving animal genetics. Choose information based on current research from a reputable source. Prepare a 4- to 5-minute class presentation of your findings.

TERMS TO KNOW

ova
sperm
zygote
gene
chromosome
gamete
dominant
recessive
cell
protoplasm
nucleus
cell membrane
nuclear membrane
lipid
mitochondrion
ribosome
endoplasmic reticulum
mitosis
meiosis
deoxyribonucleic acid (DNA)
homozygous
heterozygous
incomplete dominance
genotype
phenotype
sex-linked
mutation
heritability
testosterone
estrogen
estrus
progesterone
ovulation
fetus
gestation
inbreeding
semen

Perhaps the fastest-growing area of technology in agriscience is in genetics and reproduction. Artificial insemination has allowed for more improvement in milk production in the last 50 years than had occurred in the previous 200 years. Artificial insemination (AI) is the process of placing sperm in a female animal in close proximity with female reproductive cells, called eggs or **ova**, by a method other than natural mating. **Sperm** consists of male reproductive cells. AI allows for use of a superior male animal to sire many times more offspring than would be possible naturally.

Embryo transfers have made it possible for superior female animals to produce far more offspring than would be possible otherwise. It is a process that removes fertilized eggs, or **zygotes**, from a female animal and places them in another female animal to be carried and nurtured until birth (Figure 28-1). Embryo transfers are even being used to reproduce endangered species, such as ferrets, faster than would normally be possible.

Genetic engineering has made it possible to increase resistance to diseases, improve production, and improve efficiency of animals. Gene splicing, recombinant DNA, and biotechnology are common terms used by geneticists today. A geneticist is a scientist who studies genetics, or heredity. Heredity is the passing on of traits or characteristics from parents to offspring.

This unit explores some new technologies as well as the basics of animal breeding. Also notable are current directions of research in animal breeding.

ROLE OF BREEDING AND SELECTION IN ANIMAL IMPROVEMENT

The English geneticist Robert Bakewell is generally credited with being the father of animal husbandry. His work in the selection of Merino sheep for fine wool production and quality encouraged other farmers of his era to try

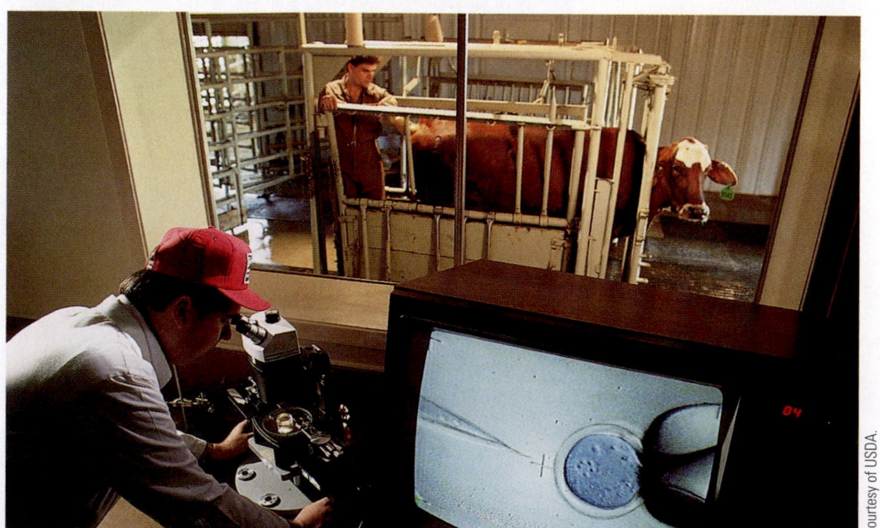
Courtesy of USDA.

FIGURE 28-1 Highly trained veterinarians and technicians backed up by sophisticated laboratory equipment have developed processes for producing more offspring from our best animals.

FIGURE 28-2 Mating the most desirable female animals with the best male animals has been the method used to produce modern livestock breeds. Most offspring of such matings prove to be as good as or superior to their parents.

INTERNET KEY WORDS:
livestock embryo transfer

to improve their livestock. Bakewell and others took great pains to always mate the most desirable female animals with the best male animals, with the expectation that the offspring would be as good as or superior to their parents. These practices have continued through the years and have resulted in major advances in animal agriscience (Figure 28-2).

By continually selecting animals for a specific type or characteristic, the resulting generations of animals tend to conform to the characteristics for which they were selected. For example, 200 years ago, cattle were not identified as dairy

AGRI-PROFILE

CAREER AREAS: GENETICIST/GENETIC ENGINEER/ ANIMAL PHYSIOLOGIST/VETERINARIAN/ INSEMINATION TECHNICIAN

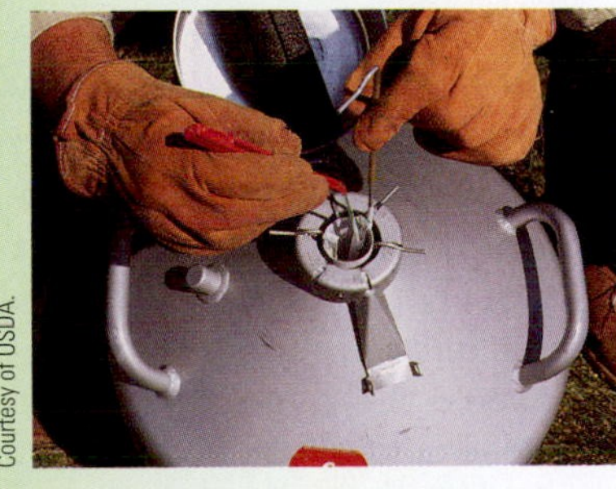

A technician removes frozen semen from liquid nitrogen for artificial insemination of a farm animal.

The greatest improvements in animal production and performance have come about as a result of selective breeding. Recent advances in the field of genetics and the ability to stimulate multiple ovulations and enabling embryo transplants to surrogate mothers have greatly increased the genetic influence of superior animals within the breeds. It is now possible to freeze embryos for future incubation and output of superior individuals. These new technologies have made possible the sale of extra embryos to introduce genetic superiority in livestock herds far removed from the female donors.

However, genetic engineering, in which the genetic coding of cells can be manipulated, holds forth the greatest opportunities for changing animal characteristics and capabilities in the future. These changes are products of biotechnology. In the foreseeable future, the greatest number of new positions in agriscience is expected to be for scientists, engineers, and related specialists. The animal industry will provide many of these careers.

INTERNET KEY WORDS:
livestock breed associations
horse breed associations
sheep breed associations
cattle breed associations
swine breed associations

and beef types. Through careful selection of those animals with superior milk production and those types with excellent meat production, two distinct types of animals emerged from the same ancestors. Many distinct breeds of domestic animals have been developed in the same way. A breed is a population of animals having common ancestors and similar physical characteristics that are passed on to their offspring. It should also be noted that selection is an extremely important part of animal agriscience today. This is especially true as consumer demands for animal products change.

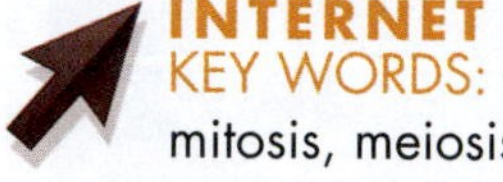

INTERNET KEY WORDS:
mitosis, meiosis

PRINCIPLES OF GENETICS

Gregor Mendel, an Austrian monk, is generally given credit for having discovered the basic principles of genetics. He did this through keen observation as he raised peas in his garden. These principles have become the foundation of modern genetics. They are summarized as follows:

- In every living organism, there is a set of paired genes in every cell that determines every trait in that individual. A **gene** is a unit of hereditary material located on a chromosome. A **chromosome** is the rod-like structure that functions as a carrier for genes.
- Individuals receive one gene for each trait from each parent.
- Genes are transmitted from parent to offspring as unchanging units.
- A **gamete** is a reproductive cell. In the production of gametes, gene pairs separate; only one gene for each trait is contained in each gamete.
- When different genes are present for a single genetic trait, in most instances, only one trait will be expressed. The trait that is expressed is known as the **dominant** trait, whereas the trait that is masked is called the **recessive** trait.

SCIENCE PROFILE HUMAN GENOME PROJECT

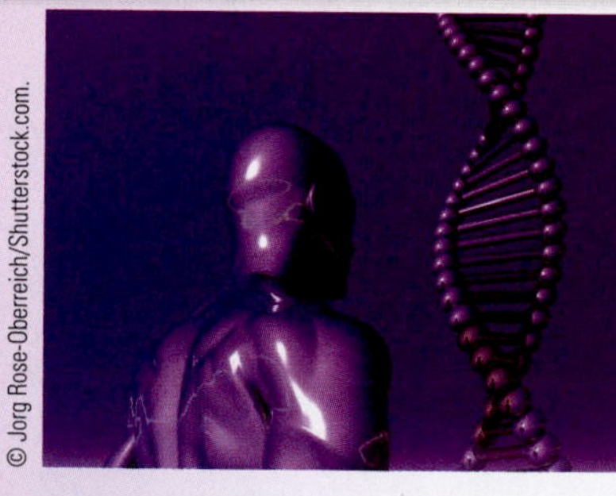

In 1990, a team of international scientists started the Human Genome Project. Their goal was to identify the sequence of genes on each chromosome, including the entire DNA in a human. Ten years later, in June of 2000, these scientists announced that their work was essentially complete. Before mapping a human's genome, the scientists sequenced the genomes of other organisms to perfect their techniques. They started with bacteria, which have much less DNA than a human. Researchers also worked on the cattle genome. Once the human genome was finished, comparisons could be made between organisms of different species. The results of one comparison involved the DNA of cattle and a human. This comparison showed that the two species share many of the same genes. They also found that four whole chromosomes shared identical genes. Scientists can identify a gene in one organism and apply that understanding to another organism. The application of this kind of information is immeasurable. Could the gene responsible for some cancers be identified and eliminated? Could the lactation gene be manipulated in such a way that a single cow could produce as much milk as 20 cows on less feed? The future of animal genetics is compelling, and only time will tell where these types of discoveries will lead agriscientists.

The Cell and Its Parts

Cells are the basis of all genetic activity. A **cell** is a single unit of living material; it is the basic structure or building block of all living things. Inside the cell are distinct structures with specific functions. The cell contents are collectively called **protoplasm** (Figure 28-3). Cells are microscopic in size, requiring a microscope to see them.

All plant and animal life begins as a single cell. The **nucleus** of the cell contains pairs of chromosomes on which genes are located at specific locations. Each gene controls a genetic trait. The gene for a specific trait is always located in the same place on the same pair of chromosomes in a given plant or animal species.

The structures within a cell are called organelles. The nucleus is typically the largest organelle in a cell, containing all of the instructions needed to direct the cell's growth and activities. Cells in an organ or a tissue are separated from each other by a **cell membrane**. It forms a protective barrier around the cell contents and controls which substances enter or exit the cell. The **nuclear membrane** works similarly to guard the nucleus. The cytoplasm is a gel-like substance that gives structure to a cell and cushions the other organelles. Often, in an animal cell, **lipid** droplets can be seen through a microscope. These are stored fats, which will be used by the cell as needed.

Many processes take place within cells that are vital to the entire animal's existence. For example, the **mitochondrion** plays a major part in converting an animal's food into usable energy. A **ribosome** is responsible for making proteins, which are continually needed by an organism. Some ribosomes float freely within the cell, some cluster together into groups, while others are attached to the endoplasmic reticulum. The **endoplasmic reticulum** stores proteins and facilitates their movement to other parts of the cell as needed. Similarly, the golgi body apparatus stores and packages lipids and proteins from the endoplasmic reticulum and distributes them where needed. The last organelles discussed here are the centrosomes. These structures form microtubules that function like strings to pull cells apart during cell division.

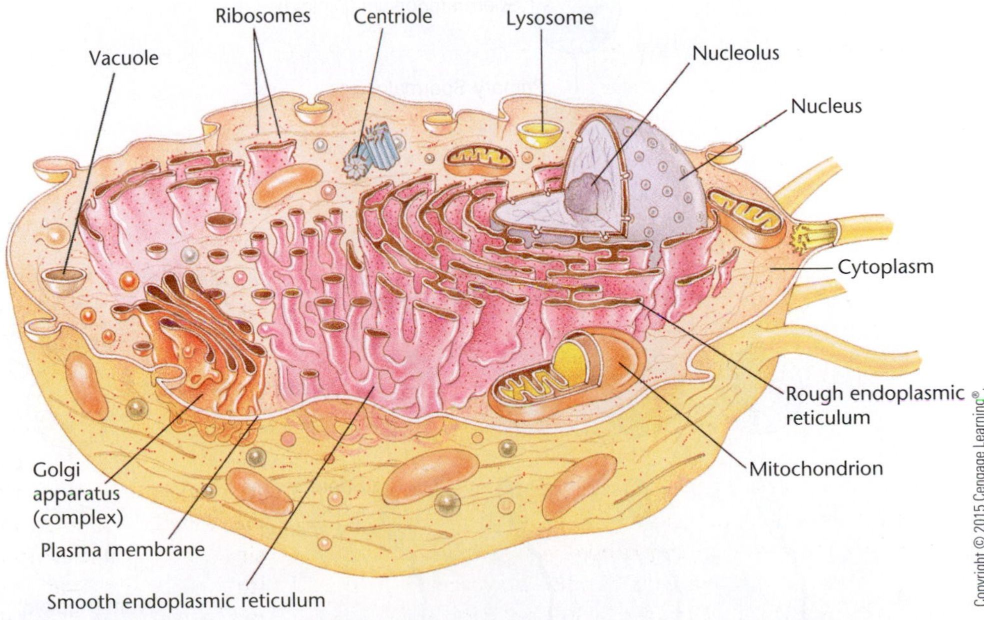

FIGURE 28-3 Each animal cell consists of distinct structures that carry out the functions of the cell.

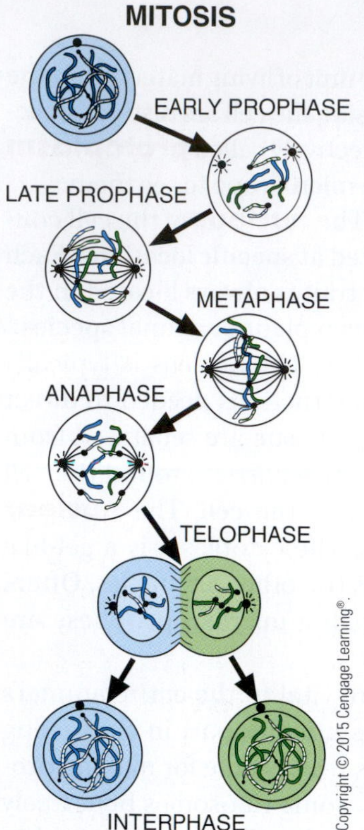

FIGURE 28-4 Mitosis is the process of division or duplication of a typical cell.

Cell Division

Animal growth and reproduction take place through cell division. Simple cell division for growth is called **mitosis**. The process begins as each chromosome first divides into two parts. The wall of the nucleus disappears, and the chromosomes move to opposite sides of the cell. A new nucleus wall forms around each of the groups of chromosomes. The walls on opposite sides of the original cell then move toward each other until they divide the cell into two new cells complete with nuclei and pairs of chromosomes (Figure 28-4).

The cell division that occurs during reproduction resulting in the formation of gametes is called **meiosis** (Figure 28-5). It differs from mitosis primarily in that instead of the chromosomes dividing and moving in pairs to the opposite sides of the cell, they separate and move individually to the cell walls. When the new cells are formed, each cell contains only one of each chromosome pair rather than both chromosomes. In animals, meiosis occurs only in the reproductive organs. When an egg is fertilized, the sperm contributes one-half of each chromosome pair; the egg contributes the other.

Genes

Genes are the units of genetic material that are responsible for all of the traits or characteristics that an animal inherits from its parents. Genes occur at specific locations on structures called chromosomes. Chromosomes control the production of specific enzymes and proteins that influence the physical traits in animals. The chromosomes are themselves composed of a protein covering surrounding two chains of **deoxyribonucleic acid (DNA)**. This structure serves as the coding mechanism for heredity.

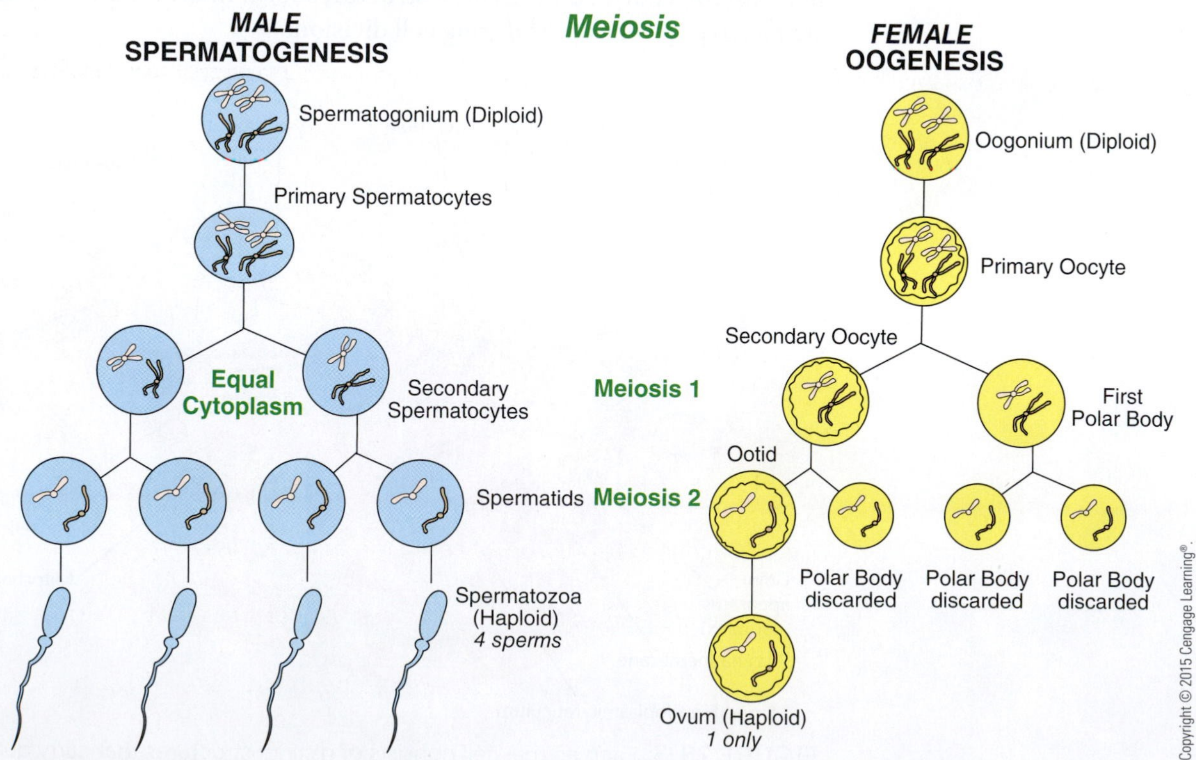

FIGURE 28-5 Meiosis is the division or duplication of egg and sperm cells.

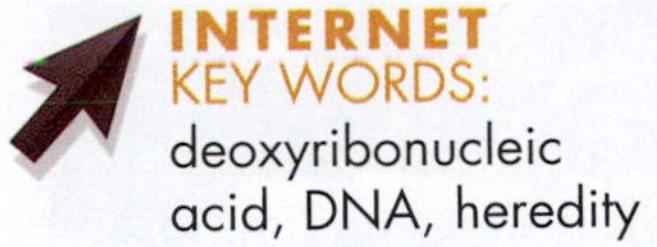

In pairs of genes on matching chromosomes, the genes may be either alike or different. Pairs of genes that are alike are said to be **homozygous**, whereas those pairs that are different are called **heterozygous**. When the two genes in a pair are different, one gene usually expresses itself, and the expression of the other remains hidden. The gene that expresses itself is referred to as dominant. The gene that remains hidden and expresses itself only in the absence of a dominant gene is called recessive. Sometimes, neither gene of a pair expresses itself to the exclusion of the other. When this happens, the gene pair is referred to as expressing partial dominance, or **incomplete dominance**. The actual configuration of genes in an animal is called the **genotype**, whereas the physical appearance of the animal is referred to as **phenotype**. All of this is important when exploring the basics of genetics and the use of genetics in animal breeding.

Some traits are controlled by genes that are located on the chromosomes that determine the sex of the animal. These are called **sex-linked** traits. The chromosomes that determine an animal's gender are a pair: one chromosome from the mother and one from the father. The chromosome from the female that determines gender is shaped like an X. The gender-determining chromosome from the male is either an X chromosome (resulting in female offspring) or a Y chromosome (resulting in male offspring). For example, when a male passes on a Y-shaped chromosome, the offspring will be male. If the male passes on an X-shaped chromosome, the offspring will be female. Other genes are also found on these X and Y chromosomes. When a mutation occurs on one of these gender-determining chromosomes, it results in a sex-linked trait. Both males and females can express this type of mutation, but the majority of sex-linked traits occur in male offspring. Color blindness is an example of a sex-linked trait.

Genes normally duplicate themselves accurately. However, sometimes accidents or changes occur. These genetic accidents or changes in genes are called **mutations**. Sometimes these mutations result in desirable changes in animals. One such example is the polled characteristic in breeds of cattle that are normally horned. Polled means naturally or genetically hornless. In other cases, the mutation results in a lethal characteristic, which causes an animal to be born dead or to die shortly after birth.

Once an egg has been fertilized, cell divisions occur and the embryo begins to grow. Mitotic cell division takes over as cells divide, forming the organs and tissues and increasing the size of the developing fetus. Occasionally, a fertilized egg divides into two identical eggs, and identical twins result (Figure 28-6). It has become common practice with superior cattle to mechanically divide the growing embryo during the stage of growth when it consists of 16–32 cells. Each cell mass develops as before, and two or more embryos develop, having identical genetic make-up. They are clones, manipulated through technology to become multiple births.

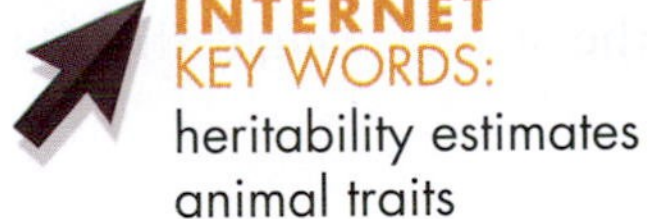

Genetics in the Improvement of Animals

The improvement of animals through genetics can be either natural or planned. In natural selection, the "survival of the fittest" occurs. In other words, as changes in genes occur naturally, only those animals experiencing gene changes that make them better adapted to their environment will survive. Popular examples include protective colorations, ability to digest certain feeds, and ability to survive in extreme heat or cold.

In planned or artificial selection, people decide which traits they want in animals. They then use the animals with the desirable traits in the breeding program. Over a period of time, the offspring that result from such selection practices show more and more of the desired traits.

Courtesy of USDA/ARS #K-4323-18.

FIGURE 28-6 These two polled calves are identical twins that resulted when a fertilized egg was divided into two identical cells or masses of cells in the early stages of embryonic development.

Heritability is the capacity of a trait to be passed down from a parent to offspring. Many of the traits for which people are selecting animals are the result of a combination of more than a single pair of genes. Because of this, few traits are 100 percent heritable from parents. For example, the extent of heritability for loin-eye size in pigs is 50 percent. A boar with a 9-inch loin-eye is crossed with a sow that has a 7-inch loin-eye. The expected average loin-eye size for the resulting offspring would be 8 inches if loin-eye size were 100 percent heritable. However, because loin-eye size is only 50 percent heritable, the offspring can only be expected to have 7.5-inch loin-eyes. The heritability rates of other traits can be found in Figure 28-7. These rates should be used as a guide when attempting to improve animal traits through genetic selection.

Environmental factors often play a part in the expression of genetic traits, masking to some extent the true potential of the animal. For example, an animal that is improperly fed or cared for may never reach the size or weight that its genetic potential would permit.

Genetic Engineering

Genetic engineering is the process of transferring genes from one individual to another by inserting them in the chromosomes of the new individual or organism. Geneticists have been able to link specific genes to specific traits. They have also developed procedures for removing the genes from the cells of one animal and inserting them into the cells of another animal. Genetic engineering has much potential for improving animals for the use of humans.

The potential for change in animals using genetic engineering is tremendous. For example, if an animal species is genetically resistant to a certain disease, genes that make that animal resistant can be inserted into cells of an animal species that is not resistant. Because genes are passed on to offspring from parents, resulting generations of animals are also resistant to that disease.

Estimated Percent Heritability						
Trait	Cattle	Sheep	Swine	Poultry	Rabbits	Horses
Fertility	0–10	0–15	0–15	0–15	–	Low
Number of young weaned	10–15	10–15	10–15	–	3	–
Weight of young at weaning	15–30	15–20	15–20	–	35	–
Postweaning rate of gain	50–55	50–60	25–30	–	60	–
Postweaning gain efficiency	40–57	20–30	30–35	–	–	–
Fat thickness over loin	40–50	–	40–50	–	–	–
Loin-eye area	50–70	–	45–50	–	60	–
Percent lean cuts	40–50	–	30–40	–	60	–
Milk production (lb.)	25–30	–	–	–	–	–
Milk fat (lb.)	25–30	–	–	–	–	–
Milk solids, nonfat (lb.)	30–35	–	–	–	–	–
Total milk solids (lb.)	30–35	–	–	–	–	–
Body weight	–	–	–	35–45	40	–
Feed efficiency	–	–	–	20–25	–	–
Total egg production	–	–	–	20–30	–	–
Age at sexual maturity	–	–	–	30–40	–	–
Viability	–	–	–	5–10	–	–
Speed	–	–	–	–	–	25–50
Wither height	–	–	–	–	–	25–60
Body length	–	–	–	–	–	25
Heart girth circumference	–	–	–	–	–	34
Cannon bone circumference	–	–	–	–	–	19
Points for movement	–	–	–	–	–	40
Temperament	–	–	–	–	–	25

FIGURE 28-7 The rates of heritability for certain traits of domestic animals.

Some of the areas being explored by geneticists working with genetic engineering include disease resistance, cancer research, vaccines, increased growth and production, and immunology.

REPRODUCTIVE SYSTEMS OF ANIMALS

Male Reproductive System

The male reproductive system functions to produce, store, and deposit sperm cells. Secondary functions include production of male sex hormones and elimination of urine from the body (Figure 28-8).

The actual structural makeup of the male reproductive system varies widely with different species of animals. The testes are the organs that produce sperm cells. They also produce **testosterone**, the male sex hormone. The testes are attached to the body by the spermatic cord and are protected by the scrotum. A coiled tube called the epididymis stores and transports the sperm cells. The vas deferens carries the sperm cells to the urethra, which extends through the penis. Other glands of the male reproductive system include the seminal vesicles and the prostate and Cowper's glands. Seminal vesicles secrete seminal fluid.

The prostate gland provides nutrition to the sperm cells, and the Cowper's gland prepares the urethra for the passage of the sperm cells. The penis serves to deposit the sperm cells into the female reproductive tract.

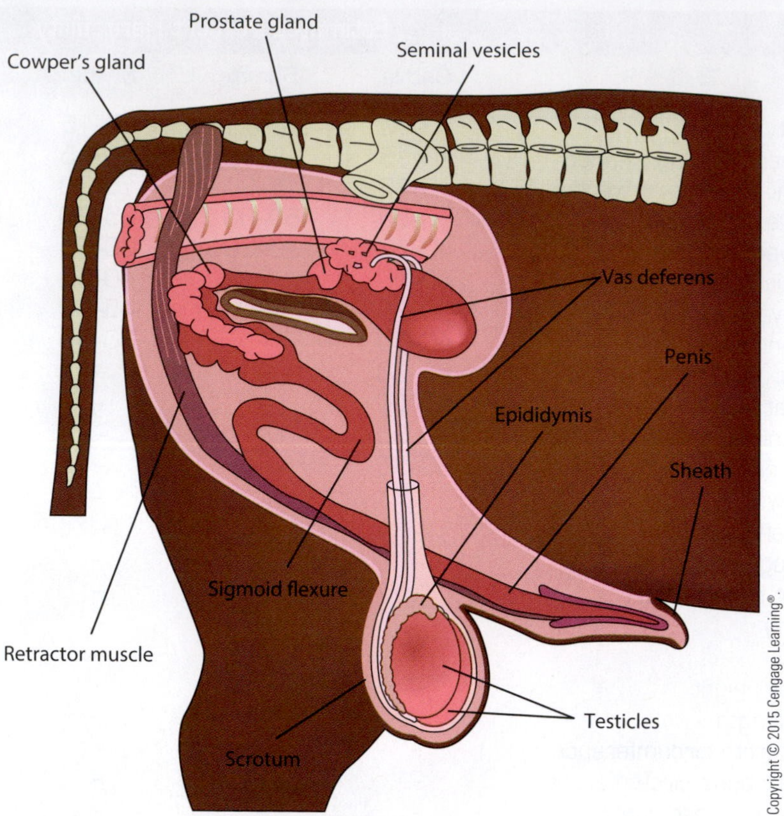

FIGURE 28-8 Reproductive system of a bull.

Female Reproductive System

The female reproductive system produces the ova, or egg, together with the female sex hormones estrogen and progesterone (Figure 28-9). **Estrogen** regulates the heat period, **estrus**, whereas **progesterone** prevents estrus during pregnancy and causes development of the mammary system. The mammary gland produces milk, the food of young mammals.

The parts of the female reproductive system include two ovaries, which produce ova, also referred to as eggs. The infundibulum is a funnel-shaped structure that catches the eggs during **ovulation** (the process of releasing a mature egg from the ovaries). The egg then passes to the fallopian tube, which is also called the oviduct. A fallopian tube is associated with each ovary, and they are the sites where fertilization of the eggs takes place. The fertilized egg, or embryo, then moves to the uterine horn, where it attaches to the wall of the uterus and remains until birth (also known as parturition). When the embryo attaches itself to the uterine wall, it becomes known as a **fetus**. The time period between fertilization of the egg by a sperm cell and birth is called the **gestation** period. The vagina, which is separated from the uterine horn by the cervix, serves as the passageway for the sperm cells. The external opening, or vulva, protects the rest of the female reproductive system from outside infection.

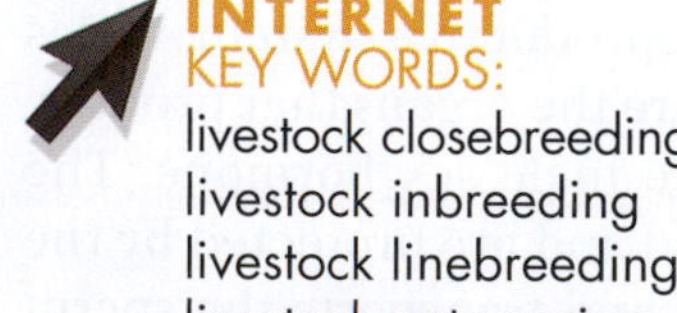

Reproductive Problems

Sterility is the inability of an animal to reproduce. A number of conditions may result in sterility or reproductive failure. Some of these problems are physical and others are genetic. Examples of genetic reproductive problems include sterile female cattle (also some female sheep) that are born as a twin or triplet with male

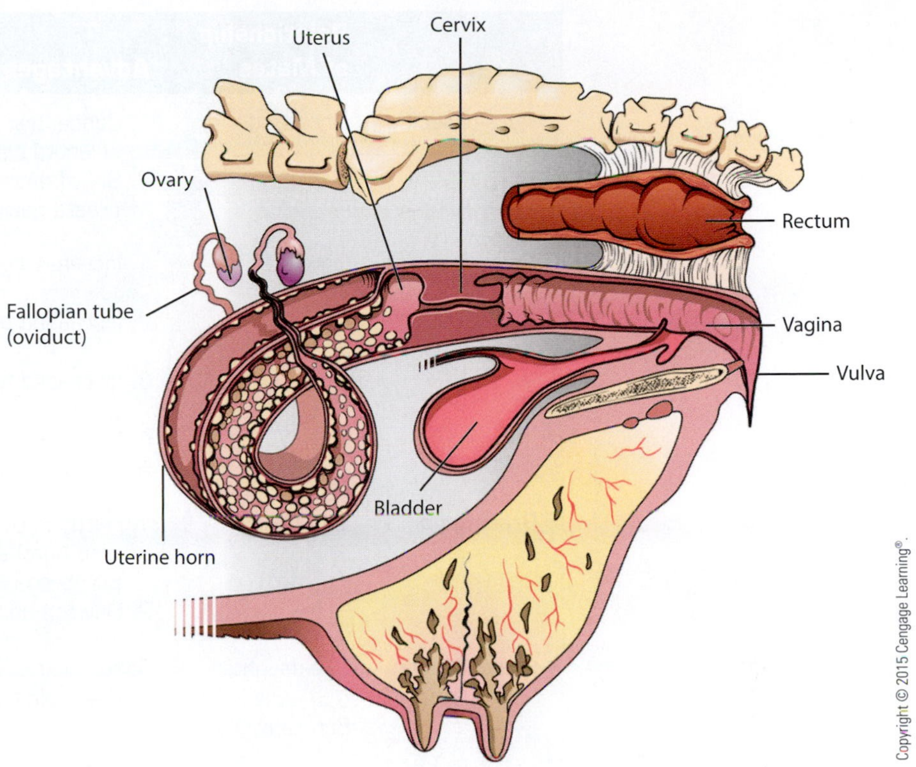

FIGURE 28-9 Reproductive system of a cow.

offspring. These female animals are often exposed to male hormones before birth that interfere with the development of their female organs. A female animal that exhibits this problem is called a freemartin. Another problem of a genetic nature includes the scrotal hernia. This is a muscle tear that is usually caused by a hereditary weakness in the muscle. Undeveloped or missing ovaries are often thought to be the result of genetic or developmental accidents.

Infections and diseases are important physical causes of sterility. Physical damage to the reproductive system and nutritional deficiencies also are known to contribute to reproductive failures.

SYSTEMS OF BREEDING

Several breeding systems are important to animal science. Which system or systems to use depends on several factors. Some of the considerations include type of operation, available markets, available resources, climatic conditions, size of operation, goals of the breeder, and personal preference.

Commonly recognized systems of breeding include pure-breeding, crossbreeding, grading up, inbreeding, and outcrossing (Figure 28-10).

Pure-breeding

Pure-breeding occurs when a purebred animal is bred to another purebred animal. A purebred animal is one of a recognized breed and one whose ancestors can all be traced to the foundation stock of the breed. A registered animal is a purebred that can qualify for breed registry. Registration papers are records of ancestry. Although there are no guarantees, purebred animals are usually considered to be

	Relationship of Mates	Advantages	Disadvantages
Pure-breeding	Unrelated	1. Concentration of selected traits 2. Breed assn. to create demand	1. May result in less-desirable traits 2. Loss of hybrid vigor
Crossbreeding	Unrelated	1. Increased growth 2. Increased prod. 3. Increased hybrid vigor 4. Higher fertility 5. Disease resistance	1. Less uniformity of offspring 2. Not eligible for registry
Grading up	Unrelated	1. Herd improvement w/o purchasing purebreds 2. Develop uniformity	Slow process of improvement
Closebreeding	Sire-daughter Son-dam Brother-sister	Concentrates desirable traits	1. Concentrates undesirable traits 2. Expression of abnormal traits
Linebreeding	Not closer than half-brother to half-sister	Concentrates desirable traits of one individual	1. Concentrates desirable traits 2. May result in expression of abnormal traits
Outcrossing	Unrelated	Produces hybrid vigor within a breed	

FIGURE 28-10 Systems of breeding for animal improvement.

superior in some ways to animals that are not purebreds. They are used as show animals and are important parts of the crossbreeding and grading-up systems of breeding.

People who elect to use the pure-breeding system need to have ample resources and a good knowledge of genetics. They should also be good salespersons capable of marketing the purebred animals that reflect the superior qualities of the animals at premium prices.

Crossbreeding

Crossbreeding is the breeding of one recognized breed of animal to another recognized breed. The resulting offspring are called hybrids. Hybrid animals have a number of advantages. They tend to grow faster, be stronger, and be capable of greater production as a result of the combination of desirable traits from the two breeds. This is called hybrid vigor. They also tend to be more fertile and more disease resistant. Crossbreeding is generally used by commercial producers who are more interested in offspring that are efficient producers than in maintaining a specific breed of animals.

Grading Up

Livestock producers who are raising animals of mixed breeding sometimes use the grading-up breeding system to improve their herds. When a non-purebred female animal, called a *grade*, is mated with a purebred male animal, the process is called *grading up*. The idea is that the purebred male animal should be superior to the grade female animal and the resulting offspring should be superior to their mother. Succeeding generations of female animals are also mated to superior-quality purebred males of the same breed. Over several generations, the breeding herd takes on the best qualities of the breed from which the sires were selected. The purposes of grading up include improvement of quality and production in the progeny or offspring. The development of uniformity in the herd is also a reason for grading up a breeding herd.

Inbreeding

Inbreeding is the mating of animals that are genetically related. The purpose of inbreeding is to intensify the desirable characteristics of a particular animal or family of animals. Unfortunately, inbreeding also intensifies the undesirable and abnormal characteristics.

There are various degrees of inbreeding, based on how closely related the individual animals are. When a father, or sire, is mated with his daughter, a son is mated with his mother, or dam, or a brother is mated to his sister, the term *close-breeding* is used.

Linebreeding is the mating of less closely related individuals that can be traced back to one common ancestor within a few generations. Normally, the most closely related cross made in linebreeding is half-brother to half-sister.

Inbreeding must be used carefully because inbred animals tend to exhibit more undesirable characteristics than animals produced by other systems. Unless a breeder is willing to be highly selective among the progeny resulting from inbreeding, this system should be avoided. A few highly superior individuals may come at the cost of discarding even greater numbers of inferior animals that are the result of concentrating undesirable traits.

Outcrossing

Outcrossing is the mating of unrelated animal families within the same breed. This is probably the most popular system of breeding used in purebred breeding herds. It also has many of the advantages of crossbreeding, including increased production and improved type.

METHODS OF BREEDING

Three general methods are used for breeding animals: natural service, AI, and in vitro. Natural service occurs when the male animal is allowed to mate directly with the female animal. Several factors impact the selection of the breeding method to be used. These factors include the type of operation you now operate or wish to have, the amount of labor available, the number of animals in the herd, the location of the herd during the breeding season, requirements of breed associations, and personal preference.

Pasture mating is a system of natural service in which the male animal is allowed to roam freely with the female animals in the herd. The male animal detects the heat period of each female in the herd and mates with her at the appropriate time. There are some disadvantages to this system. One male can

BIO-TECH CONNECTION EMBRYO CO-CULTURE

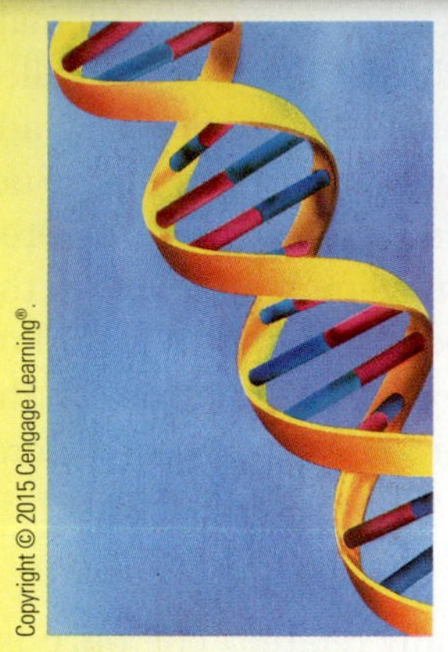

Various genetic-engineering techniques are being used for embryo technology.

Sheep and cattle embryos can be kept alive and growing outside the mother's womb for up to 6 days when tissue-cultured oviduct cells are used for nourishment. This technique was first developed for sheep and is known as co-culture. The technique permits embryos to be held for observation and checked for success of genetic engineering before being implanted into surrogate mothers.

Single-celled embryos that have had a gene inserted are placed in cultures of cells from the oviduct. Scientists believe that certain nutrients from the cultured cells keep the embryos alive and growing. Microscopic examination of the cultured embryos is used to determine if they are developing properly. An embryo cultured for 3 days should have more than eight cells. If there are eight or fewer cells, it signals that development is not progressing normally and implanting should not be performed.

Another technique that can be used is to place gene-implanted embryos into rabbits for 5 days and then remove them for implanting. Greater labor requirements and other disadvantages related to the use of rabbits make the use of oviduct cells the more promising of the techniques, however.

mate with only a limited number of females. Also, if more than one female is in heat at the same time, all may not get bred. In addition, breeding records are difficult to keep. When this system is used, the possibility exists that a sterile male may not be detected in time, and the females may not be bred at all. Commercial beef ranchers usually modify this system by placing several bulls with a large herd of cows to make sure each of the cows has access to a fertile and healthy bull.

Hand mating is accomplished by bringing the female animal to the male animal for mating. More labor and better management are required because someone must determine when the female is in heat and get the female to the male animal so mating can take place. Advantages of hand mating include being able to keep more accurate breeding records and being able to mate more females to a single male.

Mating animals through AI has numerous advantages. There are a few disadvantages as well. AI involves collecting semen from a male animal and placing it in the reproductive tract of the female. **Semen** is composed of the sperm cells and accompanying fluids. The technique of AI has been responsible for tremendous increases in animal productivity in recent years.

Advantages of AI are that semen, collected from the male, can be used fresh or stored frozen in liquid nitrogen for later use. One ejaculate, or the amount of semen produced at one time by a male animal, can be diluted and used to breed many females. Because semen can be frozen and stored for long periods, the use of an outstanding male can be greatly extended. AI also greatly reduces the need to keep male animals and reduces the danger of having dangerous male animals around. There is less chance of injury to people and their animals, reduced spread of reproductive diseases, and improved recordkeeping.

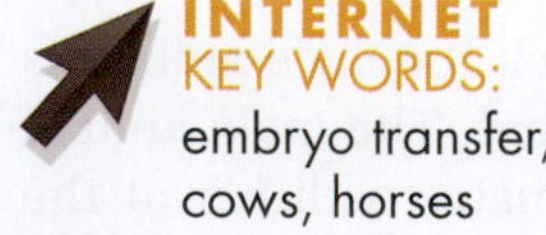

INTERNET KEY WORDS: embryo transfer, cows, horses

There are some disadvantages to AI as a method of mating animals. A person trained in AI must be available when the animal is in heat. Semen collection from the male animal can be a dangerous activity requiring the services of trained technicians, special equipment and facilities, and excellent management. Finally, genetic defects of the male must be identified early on to prevent defects from spreading faster and more broadly.

In vitro mating occurs outside the animal's body. Mature eggs are flushed from the female animal and fertilized by sperm cells collected from a male.

The fertilized eggs are then placed in host females for development into offspring. Although this process is very exacting in its requirements and facilities, at times this is the only way that a viable fetus can be obtained. This means of mating is often used as part of the new technology of genetic engineering.

SELECTION OF ANIMALS

There are several methods by which animals may be selected. These methods fall into two major types: selection based on physical appearance and selection based on performance or production of either the individual or its progeny.

Selection based on physical appearance is generally used when choosing purebred animals. Often, the sole criterion for selection of an animal is how well he or she performs in the show ring. This is an acceptable means of selection when animal breeders are raising animals for show and to sell to others for the same purpose. However, this method often leaves much to be desired when the animal so chosen is expected to produce a product or perform a desired activity. Animals fitted to perform or look their best in the show ring often fall sadly short of expectations in the milking parlor or in other performance settings.

The use of comparative judging helps individuals develop skills in evaluating animal appearance. In livestock-judging events, one animal is compared with another and frequently judged in groups of four (Figure 28-11).

Selection based on production or performance is usually a more reliable means of choosing animals. If you are selecting dairy cows to produce milk, it makes more sense to select cows with high-production records and high-producing relatives. In meat animals, progeny testing, or the testing of performance of the offspring, is often the only way to predict the breeding value of the parents. Other measures of production on which various animals may be selected include

HOT TOPICS IN AGRISCIENCE

EMBRYO CLONING—MORE OFFSPRING FROM OUTSTANDING FEMALE ANIMALS

Outstanding cows are routinely given hormone treatments to increase the number of ova they produce. This allows them to produce more than one calf per year when "extra" embryos are placed in other cows of lesser quality. In addition, the number of embryos can be expanded by splitting the cells that make up each embryo mass. In this instance, each of the cell masses obtained from the procedure is capable of producing identical individuals, or clones. An embryo that is in the 16- to 32-cell stage of development can be divided into as many as four to eight different clones.

When these cloned embryos mature in the uterus of a cow that is not the genetic mother, they retain the genetic makeup that they inherited from the cow and the sire that produced the embryos. Each of these calves may be nourished by its surrogate mother, but it is the genetic offspring of the cow from which the embryos were collected. In this manner, numerous offspring can be obtained from a single outstanding cow.

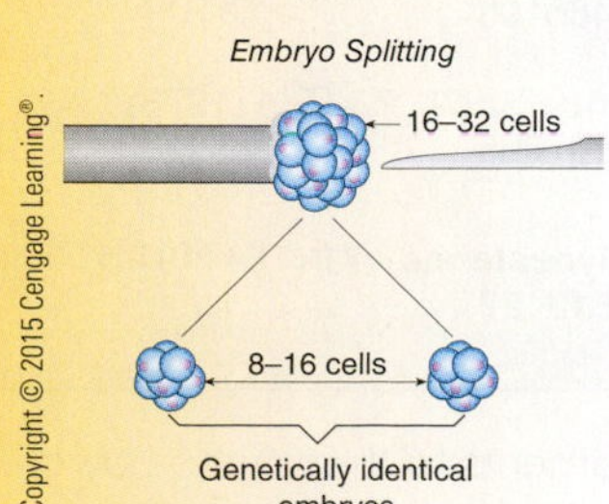

Embryos in the 16- to 32-cell stage can be successfully split into genetically identical "clones." These embryos are placed in surrogate mothers, but they retain the genetic characteristics of the "mother" that produced the original ovum.

FIGURE 28-11 Comparative judging is a good technique for studying the details of the animal, such as feet, legs, udder, head, body, and overall appearance.

rate of gain, feed efficiency, butterfat production, back-fat thickness, loin-eye area, yearly egg production, and pounds of wool produced.

Sometimes, selection of an animal is based on its pedigree, a record of an animal's ancestry, and is included on registration papers for purebred animals (Figure 28-12). Although the consideration of pedigree can be important in the

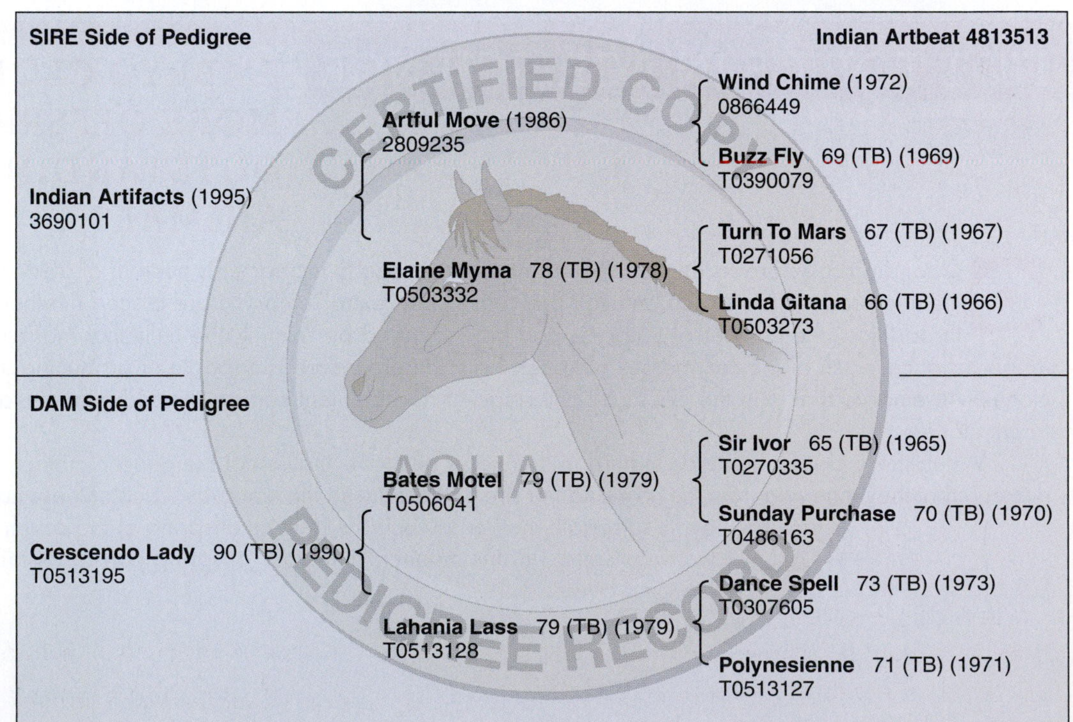

FIGURE 28-12 Pedigree and registration papers are valuable documents to those who buy or sell purebred animals because they document the purity of the bloodlines.

selection process, it should always be used in combination with other methods of selecting animals.

In summary, tremendous gains have occurred in the productivity of domestic animals in the last 100 years. Fewer and fewer animals are providing for the needs of an ever-growing human population. Genetics and gains in animal breeding have been responsible for much of the increase in productivity. This trend is expected to continue as technology in the field of animal agriscience continues to advance.

STUDENT ACTIVITIES

1. Write the Terms to Know and their meanings in your notebook.
2. Label the parts of the male and female reproductive systems.
3. Sketch and label the various stages in mitosis and meiosis.
4. Suppose you mate a Hereford bull that has horns to six female animals that were born without horns (polled). If the female animals were homozygous for the polled characteristic, how many would probably bear calves having no horns? Why?
5. Select a species of animal and determine the origin of the popular breeds in that species.
6. Clip pictures of animals from magazines and newspapers and make a collage of popular breeds for the bulletin board.
7. Develop a bulletin board showing the popular breeds of sheep, cattle, horses, swine, dogs, rabbits, and cats in your community.
8. Suppose that you mated a black male rabbit with a white female rabbit and that the female gave birth to eight bunnies. If white is dominant to black and the female is heterozygous for hair color, how many of the bunnies are likely to be black?
9. If a red pig with floppy ears was crossed with a white pig with erect ears and all possible characteristics were homozygous, what is the probability that the offspring will be red with erect ears? In pigs, white hair color and erect ears are dominant.
10. Dissect the reproductive organs of a female animal and identify the major parts.
11. Visit a farm or other animal facility and observe AI or other reproductive techniques being conducted.

SELF-EVALUATION

A. MULTIPLE CHOICE

1. The male sex hormone is called
   a. estrogen.
   b. testosterone.
   c. progesterone.
   d oxytocin.
2. The gestation period is the
   a. length of pregnancy.
   b. time during which an animal is in heat.
   c. period when an animal is fertile.
   d. time it takes the egg to mature.
3. Single-cell division is called
   a. meiosis.
   b. parturition.
   c. mitosis.
   d. estrus.

4. The Golgi apparatus
   a. makes energy for the cell.
   b. makes microtubules.
   c. stores fat.
   d. stores and packages protein.
5. The mitochondrion is responsible for
   a. producing energy.
   b. assembling proteins.
   c. protecting the nucleus.
   d. transporting protein.
6. A mutation that is associated with an animal's gender is called a
   a. gender-linked trait.
   b. bipolar anomoly.
   c sex-linked trait.
   d. gender-linked recessive.
7. A chromosome
   a. is composed of DNA.
   b. is a carrier of genes.
   c. occurs in pairs.
   d. is all of the above.
8. A pair of genes with characteristics that are alike is said to be
   a. heterozygous.
   b. homozygous.
   c. genotype.
   d. recessive.
9. Mating of an animal of one breed to an animal of another breed is
   a. pure-breeding.
   b. grading up.
   c. crossbreeding.
   d. outcrossing.
10. The phenotype of an individual is
   a. what the genes look like.
   b. the physical appearance.
   c. the type of animal.
   d. the expected production.
11. When genes are transferred from one individual to another other than through mating, it is referred to as
   a. artificial insemination.
   b. in vitro fertilization.
   c. genetic engineering.
   d. hybrid vigor.
12. The inability to reproduce is
   a. ovulation.
   b. meiosis.
   c. biotechnology.
   d. sterility.
13. Mating a sire to his daughter is
   a. closebreeding.
   b. crossbreeding.
   c. pure-breeding.
   d. grading up.

B. MATCHING

__________	1. Gamete	a. Developing embryo
__________	2. Genotype	b. Mother
__________	3. Fetus	c. Sex cell
__________	4. Progeny	d. Offspring
__________	5. Dam	e. Configuration of genes

C. COMPLETION

1. The __________ manufacture male sex cells.
2. A __________ studies genetics.
3. The passing of traits from parents to offspring is __________.
4. The crossing of half-brother to half-sister is __________.
5. __________ is the coding mechanism for heredity.

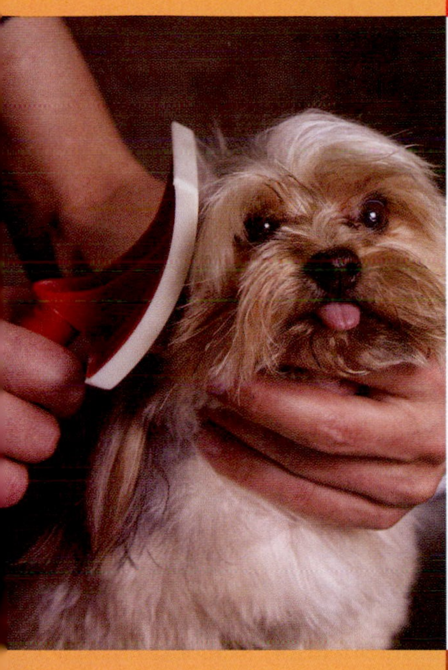

UNIT 29

Small Animal Care and Management

OBJECTIVE

To determine the types, uses, care, and management of small animals.

MATERIALS LIST

- an observation beehive
- bulletin board materials
- magazines and other materials with pictures of poultry and rabbits
- labels from poultry and rabbit feed bags
- Internet access

COMPETENCIES TO BE DEVELOPED

After studying this unit, you should be able to:

- describe the domestication and history of small animals.
- determine the economic importance of the various classes of small animals.
- list the types and uses of the various classes of small animals.
- describe the approved practices in feeding and caring for small animals.

SUGGESTED CLASS ACTIVITIES

1. Obtain a merit badge booklet for small animal care from the local center for the Boy Scouts of America. Carefully review the requirements for this merit badge, and use the booklet to organize the class discussion on small animal care and management.
2. Invite a person who raises rabbits to bring some animals to class to discuss the care and management of rabbits. Prepare class members to ask key questions of their guest during the presentation. Discuss rabbits as a source of meat that can be produced with minimal space in the backyard. Identify appropriate breeds of rabbits that might be used for this purpose. Ask the guest lecturer to address these topics in his or her presentation.
3. Invite class members who own small animals to bring one of them to class. Have these students describe the care needed to keep their animal healthy. Allow class members to ask questions at the end of each presentation. Be sure to get approval for this activity with the school administration first so that there is a clear understanding about where the animals will be kept while students are attending other classes.

TERMS TO KNOW

domestic
jungle fowl
waterfowl
broiler
layer
chick
cockerel
rooster
pullet
hen
tom
poult
drake
duckling
gander
goose
gosling
squab
angora
buck
doe
apiculture
queen
worker
drone
apiary

FIGURE 29-1 Pet ownership and the animal care industry have grown extensively in the last two decades.

As our world population continues to expand and less and less room is left for humans and large animals to coexist, the importance of small animals continues to increase. They are more efficient than large animals in converting the feed they eat into food and other products that humans use. They are less intrusive on the lives of people, and more people are keeping animals as pets (Figure 29-1). There are many important species of small animals. This unit explores poultry, rabbits, pets, and honeybees and their contributions to our daily lives.

POULTRY

History and Domestication

A **domestic** animal is one that has been tamed and cared for by humans who use such animals for labor and food. The domestication of chickens occurred about 4000 BC in Southeast Asia. The **jungle fowl** is an ancestor of our modern chickens. Its association with humans benefited both the jungle fowl and its human neighbors. Humans made small clearings in the jungle that attracted insects and other food for the jungle fowl. The jungle fowl provided some eggs and meat for humans. This association over centuries gradually led to the domesticated chicken of today. Chickens came to the New World with the earliest settlers, and the people of the Jamestown settlement had pens of chickens.

Turkeys are the only domesticated animals of agricultural importance to have originated and been domesticated in the New World. When early explorers arrived here, they found that the natives of Central America were raising domesticated turkeys as food for themselves and their animals. Although modern domestic turkeys are direct descendants of the wild turkeys of the Americas, their physical makeup has changed to such a degree that they are totally dependent on humans and cannot survive in the wild.

The various types and breeds of ducks and geese have originated from places all over the world. Ducks and geese are also known as **waterfowl**.

ECONOMIC IMPORTANCE

The consumption of red meat decreased by 40 lbs. per person in the period between 1970 and 2009. Negative publicity regarding fat and cholesterol was probably responsible for the declining consumption. In contrast, consumption of poultry and poultry products has been increasing. Poultry is a group name given to all domesticated birds. During the same period mentioned earlier, U.S. consumption of poultry meat increased by 61 lbs. per person and fish consumption rose by 4 lbs. People in the United States also eat about 250 eggs per person each year. As a result, the poultry industry is and will continue to be an important part of the U.S. agricultural industry. Currently, poultry production ranks first, ahead of beef and swine production in pounds of meat consumed. Some of the largest farms in the United States are poultry operations.

Broilers are young chickens grown for meat. In 2009, the United States produced 45 billion lbs. of broilers. The top producers by state were Georgia, Arkansas, Alabama, Mississippi, North Carolina, and Texas. During that same year, Americans consumed 92 lbs. of broiler

SCIENCE PROFILE TAXONOMY OF FARM POULTRY

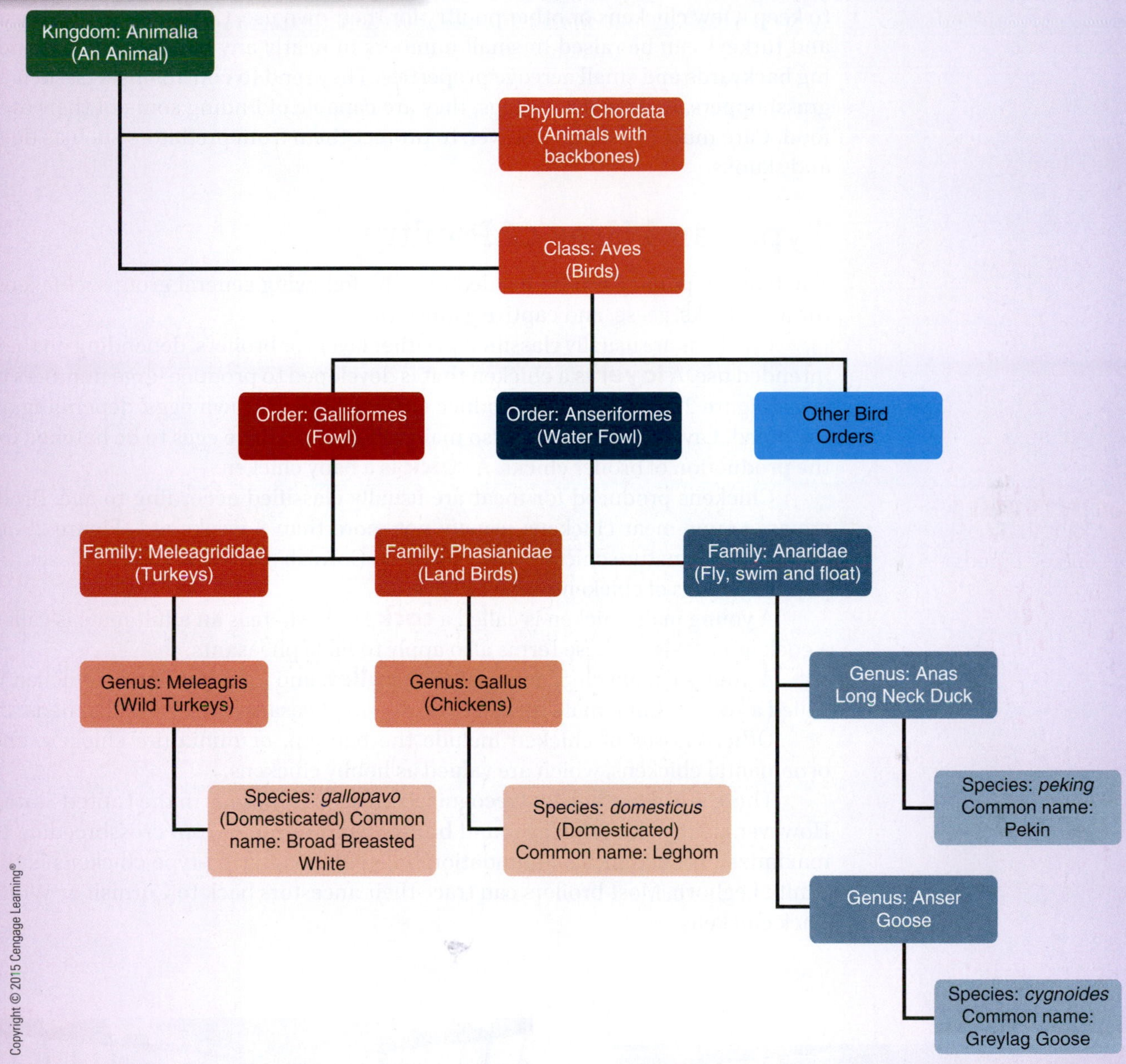

Kingdom: Animalia (An Animal)

Phylum: Chordata (Animals with backbones)

Class: Aves (Birds)

Order: Galliformes (Fowl)

Order: Anseriformes (Water Fowl)

Other Bird Orders

Family: Meleagrididae (Turkeys)

Family: Phasianidae (Land Birds)

Family: Anaridae (Fly, swim and float)

Genus: Meleagris (Wild Turkeys)

Genus: Gallus (Chickens)

Genus: Anas Long Neck Duck

Species: *gallopavo* (Domesticated) Common name: Broad Breasted White

Species: *domesticus* (Domesticated) Common name: Leghom

Species: *peking* Common name: Pekin

Genus: Anser Goose

Species: *cygnoides* Common name: Greylag Goose

Scientists organize and classify organisms in levels based on commonness. This chart shows the levels of classification for common farm birds.

chickens per person. Iowa, Ohio, and Pennsylvania are the leading producers of eggs in the United States, followed by Indiana and California.

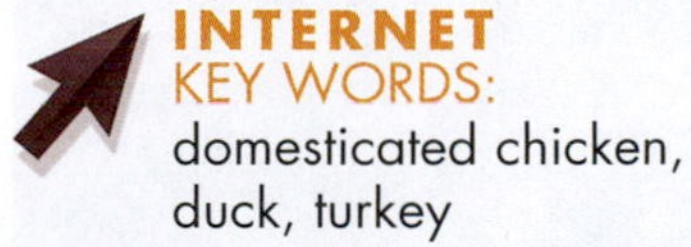

INTERNET KEY WORDS:
domesticated chicken, duck, turkey

The production of turkeys is spread over a wide area, with Minnesota, North Carolina, Arkansas, Virginia, and Missouri being leading states. In 2007, consumption of turkey meat was 13.7 lbs. per person, and approximately 256 million turkeys were produced. Nearly 60 percent of the more than 10 million ducks produced in the United States each year comes from Long Island, New York. New York, Missouri, Iowa, South Dakota, and Minnesota are major producers of geese.

Most poultry farms are large, with thousands of birds in production. This is because of the greater potential to earn a profit when fixed costs such as buildings are divided among more animals. Despite this, it is still attractive to many people to keep a few chickens or other poultry for their own use. Chickens, ducks, geese, and turkeys can be raised in small numbers in nearly any environment, including backyards and small acreage properties. They tend to control insects, such as grasshoppers. In outdoor settings, they are capable of finding some of their own food. Care must be taken, however, to protect them from predators such as dogs and skunks.

Types and Uses of Poultry

The types of poultry can be divided into the following general groups: chickens, turkeys, ducks, geese, and captive game birds.

Chickens are usually classified as either layers or broilers, depending on their intended use. A **layer** is a chicken that is developed to produce large numbers of eggs (Figure 29-2). They may produce either white or brown eggs, depending on the breed. Laying chickens are also maintained to produce eggs to be hatched for the production of broiler chicks. A **chick** is a baby chicken.

INTERNET KEY WORDS:
turkey breeds
duck breeds

Chickens produced for meat are usually classified according to age. Broilers are young-meat chickens usually not more than 8 weeks old (Figure 29-3). A roaster is a mature chicken used for meat. Cornish and White Rock crosses are popular breeds of chickens raised for meat.

A young male chicken is called a **cockerel**, whereas an adult male is called a cock or **rooster**. These terms also apply to male pheasants.

A young female chicken is called a **pullet**, and an adult female chicken is called a **hen**. Adult female turkeys, ducks, and pheasants are also called hens.

Other classes of chicken include the bantam, or miniature chicken, and ornamental chickens, which are valued as hobby chickens.

There are more than 200 recognized breeds of chickens in the United States. However, nearly all of the layer and broiler types resulted from crossbreeding to maximize production. The foundation breed of most laying-type chickens is the White Leghorn. Most broilers can trace their ancestors back to Cornish or White Rock chickens.

Courtesy of Bill Muir, Purdue University.

FIGURE 29-2 Laying hens are bred to convert feed into eggs rather than body flesh.

Courtesy of USDA.

FIGURE 29-3 Broilers are chickens that are bred for meat production. They are produced in large numbers, especially in the southern states.

Turkeys are of a single breed but of eight varieties, including Beltsville Small White, Black, Bourbon Red, Bronze, Narragansett, Royal Palm, Slate, and White Holland. Of these, only the White Holland, which includes the Broad-Breasted White (Figure 29-4), and the Bronze, which includes the Broad-Breasted Bronze, are used extensively for commercial production. The broad-breasted birds have been intensively bred for superior meat qualities. They are produced by artificial breeding because they are no longer capable of breeding naturally. However, these two varieties account for nearly all of the turkey meat that is produced in the United States each year. A male turkey is called a **tom**, a female turkey is called a hen, and a young turkey is called a **poult**.

Ducks can be classified as meat producers or egg producers. The primary meat breed is the Pekin (Figure 29-5). Ducks reach a market weight of about 7 lbs. in 8 weeks. This makes them faster growing than broilers, which reach 4 lbs. in the same period. Other breeds of duck used for meat production are Aylesbury, Muscovy, Rouen, and Call.

Egg-laying ducks are generally either Khaki Campbell or Indian Runner. The Khaki Campbell is the champion egg layer of the bird world, often averaging more than 350 eggs per year (Figure 29-6). This compares to an average of about 250 eggs laid per year for laying chickens. A male duck is called a **drake**, and a young duck is a **duckling**.

Geese are raised primarily for meat. There is also a limited market for geese used for weeding certain crops. The Chinese breed is popular for this use. Other breeds of goose are Toulouse, Emden, Pilgrim, and African. A male goose is called a **gander**, a female is a **goose**, and a young goose is a **gosling**.

Captive game birds include pheasants, quail, chukar partridge, and pigeons (Figure 29-7). The uses of game birds include meat and eggs. Some game birds are also raised to release to the wild or on game preserves for hunting.

Pigeons are often raised for sport (racing) or as a hobby, but some breeds are raised for meat purposes. Many people maintain pigeon lofts for the sheer pleasure of watching these birds. They require little space, and they can be raised almost anywhere. The adult male pigeon is a cock; an adult female is a hen. A young pigeon that has not left the nest is called a **squab**. When a pigeon is used for meat, it is processed before it learns to fly, and the meat is called *squab*. It is usually considered a *delicacy* meat served in restaurants.

Courtesy of USDA.

FIGURE 29-4 The most popular commercial variety of turkey in the United States is the Broad-Breasted White.

Courtesy of Jurgielewicz Duck Farm.

FIGURE 29-5 Pekin ducks are the most popular breed of duck used for meat.

SCIENCE PROFILE GOT ENERGY?

Hunger leaves an individual feeling tired and weak. This is because the brain has ways of letting the body know that it needs food. Food provides the basic building blocks for the energy supply of the body. Inside cells, the mitochondria (the cell's power plant) works tirelessly producing energy for the cell. The process whereby food is transformed into chemical energy is known as the Kreb cycle. Before the Kreb cycle can begin, another process breaks a sugar molecule called glucose in half, creating two energy molecules called adenosine triphosphate (ATP). ATP is the molecule that cells use for energy. Next, the two sugars are sent into the Kreb cycle, where they go through several chemical reactions that result in 34 more ATPs.

Animals require a lot of energy. From the beating of a heart to running, every function of the body takes energy. Which animal do you think needs more energy, a horse or a rabbit? The answer may surprise you: It is the rabbit. Small animals need more energy per ounce than larger animals. In general terms, the bigger an animal is, the less energy it needs. Eighty to 90 percent of the energy of the body is spent regulating the body's temperature. Small animals lose heat much faster than large animals, so small animals use more energy or ATP to maintain their bodies at the right temperature.

© Alexander Jache/Shutterstock.com.

A

© Makarova Viktoria (Vikarus)/Shutterstock.com.

B

Small animals (A) require more energy per unit of body weight than large animals (B).

Courtesy of John Metzer, Metzer Farms.

FIGURE 29-6 Khaki Campbell ducks are the champion egg layers of the bird world, producing as many as 350 eggs per year, compared with an average of approximately 250 eggs per year for chickens.

Some Approved Practices for Poultry Production

Up-to-date information on poultry and other small animals is available from suppliers of breeding stock, equipment, medicine, and building materials. Recommended or approved practices are available from your state university. Approved practices for the production of poultry include the following:

- Purchase young poultry with a specific use in mind.
- Purchase young poultry or eggs for hatching only from reputable hatcheries or breeders. A hatchery is a business that hatches young poultry from eggs.
- Purchase chicks, pullets, or poults that are immunized and free from disease.
- Purchase young poultry at the proper time for meeting a target market date. Broilers should be 7 to 8 weeks old before marketing them; ducks, 7 to 8 weeks; turkeys, 12 to 14 weeks; and geese, 12 to 14 weeks old. Layer chicks should be purchased 20 to 22 weeks before you expect them to produce eggs.

A

B

FIGURE 29-7 Two types of game birds are pictured here: (A) Chinese Ring-necked Pheasant and (B) Bobwhite Quail.

- Ensure that proper housing is available for the type and number of poultry you are planning to raise. Housing considerations include size, ventilation, ease of cleaning, lighting, heating and cooling, feed storage, and maintenance requirements.
- Secure and maintain the proper equipment for the type of poultry operation planned. Consider feeder and waterer space and brooder size.
- Feed a balanced ration designed for the type of poultry being grown.
- Develop and implement a plan to control external parasites.
- Plan and follow a flock health program.
- Plan for marketing at the optimum time.
- Properly clean and disinfect facilities before introducing a new flock of poultry.

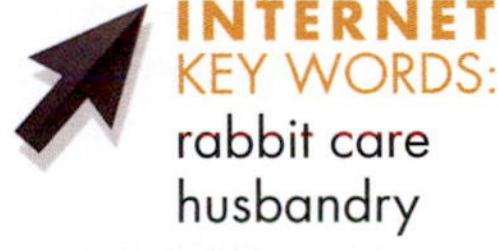
INTERNET KEY WORDS:
rabbit care
husbandry

RABBITS

History and Domestication

Much of the early history of the rabbit is obscure. It is believed that the Phoenicians brought rabbits to Spain about 1100 BC. They are also given credit for having introduced rabbits to most of the world that was known at that time.

Romans kept rabbits in special enclosures. Roman women were known to have eaten large quantities of rabbit meat. They believed it enhanced their beauty.

Early monasteries produced large amounts of rabbit meat and fur. These religious institutions are given credit for having domesticated the rabbit. It is known that great pride was taken in producing high-quality rabbits and that much rabbit trading existed between monasteries.

Rabbit meat has long been an important component of the diets of people in densely populated countries of Europe. Rabbits are efficient converters of feed to meat. They take up relatively little space and reproduce rapidly.

Rabbits have been raised in the United States since the time of the early settlers, but serious rabbit production did not begin until the turn of the twentieth century. An intense advertising campaign was conducted for "Belgian Hare" at that time to promote commercial production of rabbits. Rabbit production also got a boost in the United States during the two world wars. At a time when shortages and rationing of food products occurred, rabbits became an inexpensive source of lean, red meat.

Economic Importance

Rabbit production is an important agricultural enterprise in the United States. Each year, 6 to 8 million rabbits are raised. The U.S. population consumes 25 to 30 million lbs. of rabbit meat each year. Another 600,000 rabbits are used each year for biomedical teaching and research.

Rabbit production is an ideal enterprise for a young person because it can be started with limited capital (Figure 29-8). With only a small investment in housing and equipment, a person with one pair of rabbits can produce 50 to 60 rabbits each year to eat or to sell. Because they are small and generally accepted by people, rabbits are better adapted for production in suburban/urban areas than most other types of animals. Rabbit meat is low in fat (4 percent), sodium, and cholesterol and is high in protein (25 percent).

The outlook for rabbit production in the future is variable with a need for good market analysis, good market development, and careful production management. As the competition between humans and animals for grain products increases, rabbits are capable of playing an important role in meeting the protein needs of humans in the future.

Types and Uses

Rabbits come from two families and three genera, with distinct differences. These include rabbits, cottontails, and hares. Rabbits bear their young in underground burrows in the wild. The young are born blind, hairless, and completely helpless. In contrast, cottontails and hares usually give birth in nests above the

Courtesy of FFA.

FIGURE 29-8 Rabbit production can be done on a small basis with very little capital and space.

ground. The young are born with their eyes open and with hair. They are able to fend for themselves shortly after birth. Hares also have larger hind legs and longer ears.

Hares include the jackrabbit, arctic hare, and snowshoe rabbit. Because they belong to different genera, cottontails, hares, and rabbits cannot interbreed.

Domestic rabbits can be divided into a number of groups based on use. These groups include meat, fur, pets, show, and laboratory use. Many breeds fall into several of these use groups.

The primary use of rabbits in the United States is for the production of meat, with pelts being a by-product. A pelt is an animal skin with the hair attached. Almost 100 million rabbit pelts are used in the United States each year. Most of these are imported from other countries because, in the United States, rabbits are slaughtered at too young an age to have desirable pelts.

Although all the breeds of rabbit will produce meat, some breeds are far more efficient in producing desirable-quality meat. The New Zealand White is the most popular breed of rabbit in the United States for meat production. This rabbit occurs in three colors: white, red, and black. It is of medium size. It may be grown into a 4-lb. rabbit at 8 weeks of age, using about 4 lbs. of feed for each pound of rabbit produced. The Californian and Champagne D'Argent are also popular breeds used for meat production (Figure 29-9).

Some breeds of rabbit are grown for their lustrous fur for use in the manufacture of fur coats and many other rabbit-fur products. The Satin, Rex, and Havana are examples of rabbit breeds grown for their fur (Figure 29-10).

Rabbits have long been important for use in laboratory work. They are used for research in the development of drugs for treating a wide range of diseases. They are also important in nutritional studies and in genetic research. Commonly used breeds of rabbit for laboratory work include New Zealand White, Dutch, and Florida White. Producers who breed rabbits for laboratory work must be aware that many laboratories will use only white rabbits of medium size (Figure 29-11).

The Angora rabbit is used strictly for the production of wool called **angora**. The wool from Angora rabbits is sheared or pulled from the rabbit about every 10 to 12 weeks. Mature Angora bucks may produce 1 to 1.5 lbs. of wool each year. A **buck** is a male rabbit. A female rabbit is a **doe**. There are two breeds of Angora rabbit: French and English (Figure 29-12). All of the 40 breeds of rabbit recognized by the American Rabbit Breeders Association can be used for pets, for show, or for both. They range in size from the Flemish Giant, which can weigh nearly 20 lbs., to the Netherland Dwarf, which seldom weighs more than 2 lbs. and makes a popular pet (Figure 29-13).

Courtesy of American Rabbit Breeders Association.
A

Courtesy of American Rabbit Breeders Association.
B

Courtesy of American Rabbit Breeders Association.
C

FIGURE 29-9 Important meat-producing breeds of rabbit include the (A) New Zealand, (B) Californian, and (C) Champagne D'Argent.

A

B

C

FIGURE 29-10 Important fur-producing breeds of rabbit include the (A) Satin, (B) Rex, and (C) Havana.

A

B

C

FIGURE 29-11 The (A) Dutch and (B) Florida White join the (C) New Zealand White as important breeds in biomedical research and education.

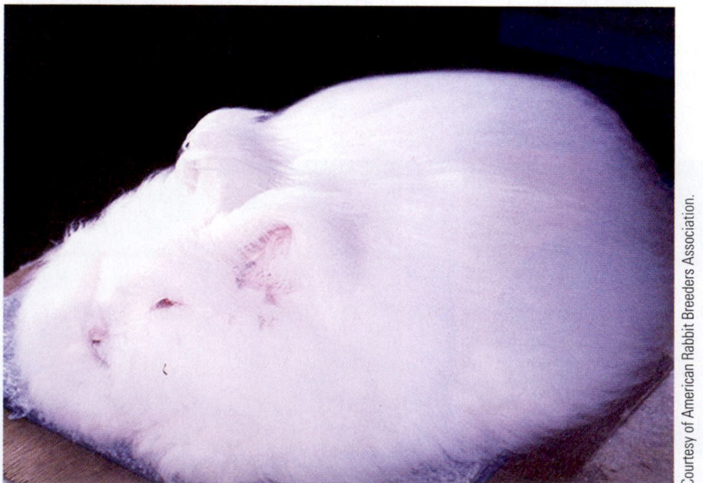

FIGURE 29-12 Angora rabbits, such as the Giant Angora (pictured here), are raised for the production of angora wool.

FIGURE 29-13 Netherland Dwarf rabbits, generally weighing in at less than 2 lb., are popular pets.

SCIENCE CONNECTION TAXONOMY OF RABBITS

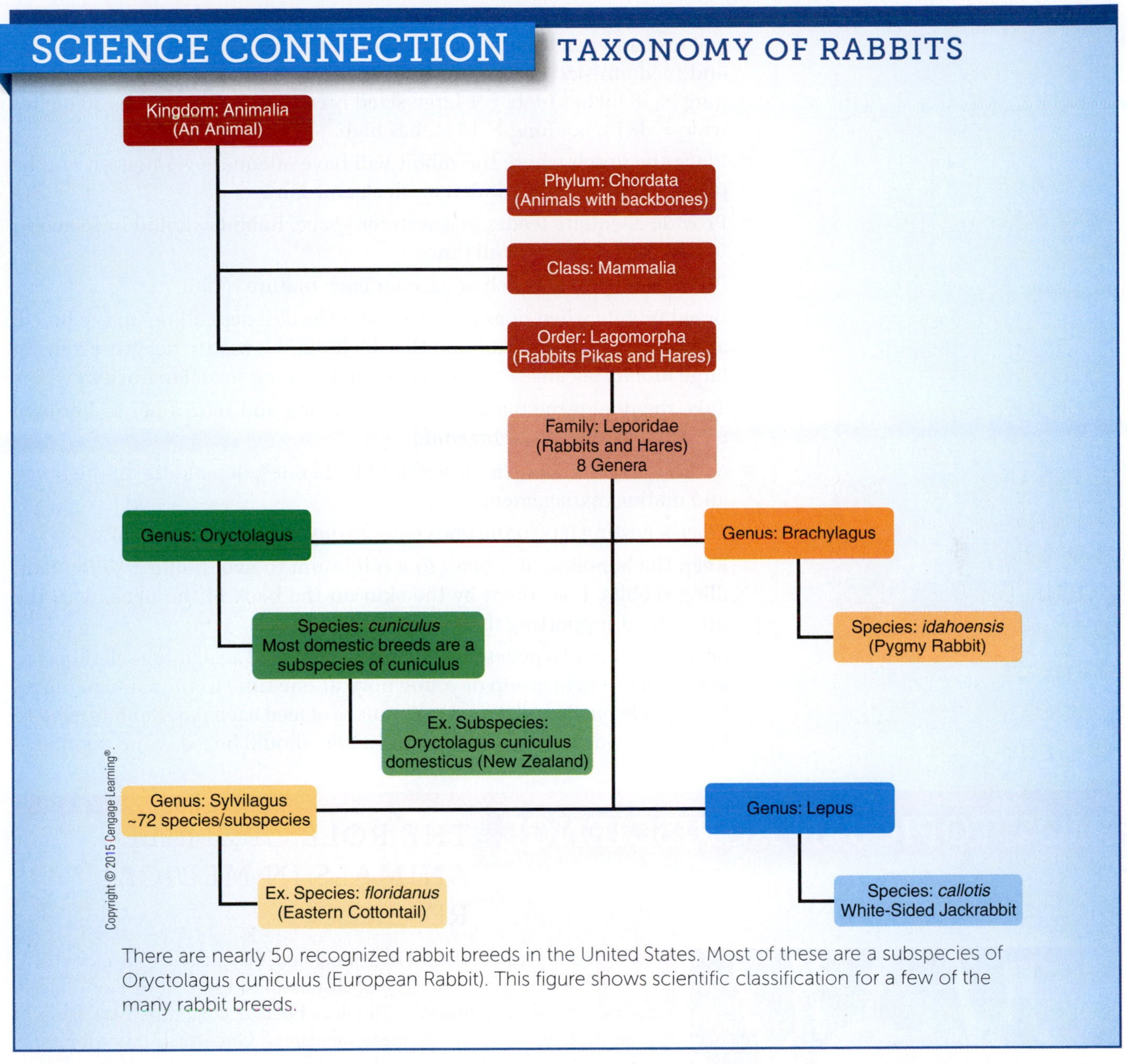

There are nearly 50 recognized rabbit breeds in the United States. Most of these are a subspecies of Oryctolagus cuniculus (European Rabbit). This figure shows scientific classification for a few of the many rabbit breeds.

Personal preference and the availability of breeding stock usually determine what breed or breeds of rabbit to raise for pets or show.

Approved Practices for Rabbit Production

Approved practices for the production of rabbits include the following:

- Select the correct breed for the intended use.
- Use purebred stock if you plan to sell breeding stock and to maintain uniformity in your herd.
- Purchase breeding stock only from reputable breeders with accurate records.

- Build or choose a hutch of the proper size for the breed of rabbit that you are growing. A hutch is a cage or house for a rabbit. For small-sized and medium-sized breeds, provide hutches 30 inches wide × 36 inches long × 18 inches high. For large-sized breeds, provide hutches 30 inches wide × 48 inches long × 18 inches high.
- Place the hutch where the rabbit will have adequate ventilation and be protected from heat, wind, rain, sleet, and snow.
- Provide adequate feeder and waterer space. Rabbits should have access to fresh, clean water at all times.
- Provide a separate hutch or cage for each mature rabbit.
- Breed rabbits when does are 5 to 8 months old, depending on the breed, and when bucks are 6 to 7 months old. It may be wise to delay breeding of large rabbits because they are slower in reaching sexual maturity.
- Take the doe to the buck's cage for breeding and return her to her own cage immediately after breeding.
- Maintain 1 mature buck for every 10 to 25 does, depending on the breed and mating management.
- Place a nesting box in the doe's cage 25 days after mating occurs.
- Keep the handling of rabbits to a minimum to avoid injury. When handling rabbits, hold them by the skin on the back of the neck, with the other hand supporting the weight of the rabbit.
- Feed a commercial pelleted-feed free choice (feed available at all times) to does and litters (a group of young born at one time to the same parents). Feed single bucks and does 3 to 6 ounces of feed each day. Rabbits need to be fed only once a day, and preferably, they should be fed in the evening.

HOT TOPICS IN AGRISCIENCE

THE ROLE OF SMALL ANIMALS IN MEDICAL RESEARCH

© Sebastian Duda/Shutterstock.com.

Small animals such as rats and rabbits are used in medical research to test new vaccines and other medications before they are used to treat humans.

Medical researchers have identified the causes and many of the cures for a variety of diseases and infections that afflict people and animals. This has been accomplished by studying the organisms that cause these infections. Once such an organism has been identified, it is cultured, modified, weakened, and tested in animals in an effort to create a vaccine that will prevent the infection from becoming established. Fertile chicken eggs are often used to culture such organisms in live tissue. These eggs are supplied to medical research facilities from farms that specialize in producing the quality of eggs that is needed for this purpose.

Rabbits and other small animals are used in medical research to test the effects of potential vaccines and medications on animal life. Much of the progress that medical science has made in controlling diseases and infections of various kinds has depended on the use of animals for research purposes. Laws prevent researchers from using human subjects until medicines and vaccines have been proved to be effective in laboratory animals. Without research animals, we would not have the ability to control many of the diseases and infections that are known to afflict humans. The average life span of humans would be much less in the absence of research animals.

- Maintain accurate breeding, production, and health records for all rabbits.
- Tattoo all breeding rabbits for identification. A tattoo is a means of marking rabbits and other animals for identification. Rabbits are tattooed in the ear.
- Plan for and maintain a strict herd health program.
- Dispose of sick and dead rabbits promptly.
- Market rabbits as soon as they reach market size or weight.

honeybee
honeybee pollination

HONEYBEES AND APICULTURE

History and Domestication

Honeybees (*Apis mellifera*) are an important part of human history. Cavemen drew pictures of bees and honey collection on cave walls. Honey is a thick, sweet substance made by bees from the nectar of flowers. In Egypt, mummies were embalmed and stored in a liquid based on honey. Jars of honey have been found in many Egyptian tombs.

The Bible makes many references to honey and the use of honey for food. During biblical times, honey was not produced in nice, neat combs as it is today. Rather, in most cases, the hive was destroyed in the process of removing the honey and comb. A hive is a home for honeybees. The comb is the wax foundation in which bees store honey.

Greeks and Romans were very familiar with honeybees and honey. Pompey used poisoned honey to defeat his enemies in at least one engagement. Aristotle wrote in great detail about bees and their production of honey.

Most early civilizations considered honey to be the food of gods. Athletes, competing in Olympic games, often ate honey before their events to gain extra strength and endurance.

Early beekeepers kept their bees in hollow logs, straw hives, or even in crude clay cylinders. All of these containers had to be destroyed to remove the honey.

With the invention in the 1850s of movable combs with wax foundations to encourage bees to make neat, straight honeycombs, the whole beekeeping industry changed. Honey was finally a commodity to be enjoyed by nearly everyone (Figure 29-14). Soon after the development of movable comb, the discovery that honey could be whirled out of the comb led to the invention of the honey extractor. It was no longer necessary to destroy the comb to get to the honey. The comb, after being emptied of honey, now can be placed back in the hive to be refilled by the bees.

Today, the production of honey in the United States is a large and profitable business. Far more important than the production of honey is the work that honeybees do in the pollination of important agricultural crops. Modern beekeeping is known as **apiculture**.

Courtesy of USDA/ARS #K-5064-2.

FIGURE 29-14 Biological technician Gary Delatte and graduate student Lilia de Guzman examine a honeycomb and check the health of the honeybees.

Economic Importance

It is difficult to accurately gauge the true economic importance of honeybees. They are responsible for about 80 percent of insect pollination of plants. Without honeybees, many of the crops important to agriculture would simply disappear from the Earth.

Pollination of orchard crops such as citrus, peaches, and apples by honeybees is so important that many beekeepers rent their bees to orchardists when these trees are in bloom. Many commercial beekeepers make more money from

SCIENCE CONNECTION BEES IN BIOCONTROL BUSINESS

Courtesy of Paul C. Pecknold.

Fire blight is caused by a bacterium that enters apple and pear trees by colonizing the stigma of the fruit blossom. Honeybees, carrying a fire blight–fighting bacteria, are used to colonize the stigma with beneficial bacteria as well as pollinate the fruit blossoms.

An often-expressed adage is, "If you want to get a job done, ask a busy person to do it." Noting another saying, "busy as a bee," entomologists have found a way to have busy honeybees spread biocontrol agents to plants as they pollinate billions of dollars worth of crops each year. Scientists are using unsuspecting honeybees to deliver biocontrol agents right where the agents are needed on plants. Two serious pests are targeted for control by the process. One is the fire blight bacterium, which infects pear and apple trees; and the other is the corn earworm, one of the worst insect pests of corn and cotton.

Fire blight is caused by *Erwinia amylovora*, a bacterium that first colonizes a flower's stigma, the part that receives pollen grains during pollination. The bacteria then multiply and spread quickly. The disease causes cankers on twigs and branches and causes the leaves to have a burnt appearance, hence, the name "fire blight." The disorder weakens or kills the trees. On the

bee rental than from honey. Such beekeepers use flatbed trailers to move their hives and operate from Florida to Maine and from Texas to Washington State (Figure 29-15).

There are about 300,000 beekeepers in the United States, of which about 99 percent are hobby or part-time beekeepers. These 300,000 people care for about 6 million hives of bees. In a normal year, a hive of bees will produce 100 to 150 lbs. of honey in excess of the amount needed by the bees to live (about 150 lbs.). The production of honey is big business.

Approved Practices for Beekeeping

The following is a list of approved practices to be used in the keeping of bees:

- Check local regulations before starting a beekeeping operation.
- Locate bees out of direct contact with people and neighbors' yards and gardens.
- Place hives facing away from prevailing winds. They should also be protected from hot summer sun.
- Thoroughly clean and disinfect hives before allowing new groups of bees to use them.
- Purchase bees from reputable sources. It is usually far more profitable to purchase a 3-lb. package of bees with a purebred queen than to rely on a swarm to populate a new beehive. A swarm is a group of bees complete with a queen that leaves an overcrowded hive to find a new home.

positive side, scientists have found that spraying blossoms with beneficial bacteria helps prevent fire blight disease. The beneficial bacteria apparently compete with the harmful bacteria by consuming the nutrient-rich stigma, and the pathogen cannot get a foothold to invade the tree.

Rather than spraying with the beneficial bacteria, entomologists and plant pathologists at Utah State University and Oregon State University are doing research to identify the honeybees that deliver the beneficial bacteria to the point-of-entry for the fire blight pathogen—the stigma. Such a strategy maximizes the use of the biocontrol material. The procedure is being field-tested, wherein numerous beehives are fitted with devices called pollen inserts. The pollen insert is a slotted passageway for bees that automatically dusts the bees with the beneficial bacterium, which is the biocontrol in this case, each time they leave the hive. A hive contains from 10,000 to 100,000 bees, each visiting perhaps 100 blossoms per hour. The beneficial bacteria do no apparent harm to the bees.

Meanwhile, entomologists in Tifton, Georgia, use honeybees to carry a natural virus, Heliothis, that destroys the larval stage of the corn earworm. The device used to "load" the bees is one in which the bees use a different entrance and exit to the hive. The exit device includes a metal tray that contains the virus material. The exiting bees carry the material on their feet, legs, and undersides and deposit it on their rounds. The technique reduced earworm infestations in clover, and it should be effective on other flowering plants.

- Replace queens every 2 years. A **queen** bee is the only fertile, egg-laying female bee in each hive (Figure 29-16).
- Have your bees inspected for contagious diseases by a federal bee inspector each year.

Courtesy of USDA/ARS #K-4715-1.

FIGURE 29-15 Bees are responsible for up to 80 percent of all insect pollination of plants, and they are absolutely necessary for pollination of fruit crops such as citrus, peaches, apples, and other orchard crops.

Courtesy of USDA/ARS #K-5069-22.

FIGURE 29-16 Worker bees tend the queen (marked), the hive, and the brood, as well as making and storing honey for the future.

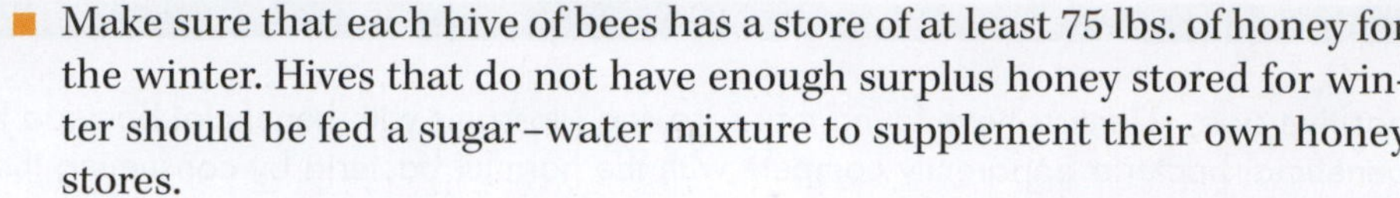

- Make sure that each hive of bees has a store of at least 75 lbs. of honey for the winter. Hives that do not have enough surplus honey stored for winter should be fed a sugar–water mixture to supplement their own honey stores.
- Always be sure that bees have ample room to store the honey they produce.
- Remove surplus honey as soon as the bees have capped it over with wax.
- Remove honey in the evening or at night when nearly all the bees are in the hive. Supers containing surplus honey can be freed from bees by blowing cool smoke over the bees and brushing them off the comb with a bee brush. A super is a box filled with a movable wax foundation that is used by the bees to store honey. You can also use a bee excluder between the honey to be removed and the hive body.
- After moving a hive, put a deflector in the entrance of the hive so the bees will notice that they have been moved. Hives must be moved at least 5 miles to prevent bees from returning to the former site of their hive.
- Inspect beehives at least monthly to determine the strength of the hive and the queen. Be sure to observe the number of eggs being laid by the queen. Also note whether the worker bees are building drone or queen cells in the hive. Drone and queen cells look like peanuts. Such cells should be destroyed. **Worker** bees are undeveloped female bees and constitute all of the working force of the hive. A **drone** is a male bee whose only purpose in life is to fertilize the queen once in his life span.
- Reduce or prevent swarming of bees by providing ample hive space for the bees and eliminating queen cells as they are found. Overcrowding often causes bees to develop a second queen. A new queen will attract a group of worker bees and leave the hive to start a new colony. This process is called swarming. Bees will not swarm without a queen because she is their only hope of survival.
- Be aware of pesticides that are used in the area that could kill bees or be stored in the honey.
- Secure the proper equipment before starting an apiary. An **apiary** is an area for keeping beehives.
- Keep honey that has been removed from bees in an area that bees cannot get to. Otherwise, they will steal all of the honey in a short time.
- Extract honey from the comb as soon as possible after harvesting it. Honey stored for long periods in the comb may granulate, which makes it impossible to extract.
- Develop a market for your honey.

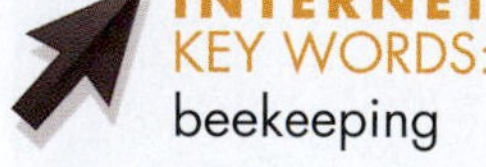

INTERNET KEY WORDS:
beekeeping

PET CARE AND MANAGEMENT

Animals make excellent companions and can provide specialized functions for children and adults alike. For many animals, the line between pet, companion animal, and work animal is a fine one. For example, guide dogs are essential "eyes" for the blind and must make intelligent choices for their masters when crossing streets (Figure 29-17). Similarly, dogs and other pets provide companionship for the elderly and warn of intruders. Good dogs may ward off attackers and provide other protection when appropriately trained. Guard dogs are essential for herding and controlling sheep and protecting sheep from predators.

Courtesy of Guide Dog Foundation for the Blind, Inc.

FIGURE 29-17 Guide dogs serve dual purposes. When at work, they act as "eyes" for the blind. When at rest or play, they provide companionship.

Many kinds of caged birds and reptiles are used for pets. Their needs are rather specialized, and advice for their care is usually obtained from pet shops and other suppliers. Similarly, a large assortment of fish and other aquatic creatures makes attractive and challenging aquarium projects for all ages (Figure 29-18). Even exotic animals such as llamas and kangaroos find their way onto farms and into the hearts of those in the United States (Figure 29-19).

Dogs and cats are probably the most prevalent companion animals in the United States. Although they have different moods and behave in different ways, they have similar needs (Figure 29-20). The following is a list of approved practices for the care of dogs and cats (Source: Pennsylvania State University):

- Select animals that are alert and healthy.
- Vaccinate for rabies at age-determined intervals.
- Select a breed adapted to your situation.
- Prepare a clean, draft-free living area.
- Provide an adequate number of feed dishes.
- Provide a diet of high-quality food for the breed and age.
- Provide toys, treats, or both.
- Clean the food and water dishes daily.
- Keep clean, fresh bedding in the sleeping area.
- Provide plenty of clean, fresh water.
- Clean the pen or box and exercise area daily.
- Develop and implement a sound health plan.
- Consult with your veterinarian to prevent and control internal and external parasites.
- Vaccinate animals routinely at proper times.
- Properly bathe and groom the animal.
- Use a proper carrier for transport.
- Use proper restraint procedures.

Courtesy of FFA/Photos by Michael Wilson.

FIGURE 29-18 Aquatic animal and plant projects create interest and provide opportunities to accept responsibility for living creatures that depend on them for nutrients and safe environments.

Courtesy of FFA.

FIGURE 29-19 The care of exotic animals creates new challenges for experienced pet handlers.

AGRI-PROFILE

CAREER AREAS: ANIMAL TECHNICIAN/GROWER/ BEEKEEPER/MANAGER

Career opportunities for small animal care and management are available in many areas. The extensive use of laboratory animals for research, small animals for pets, fish for home aquariums and garden pools, small animals for fur, and animals for zoological parks ensures attractive jobs in the future.

The operation of animal hospitals, kennels, grooming services, pet stores, training programs, boarding facilities, public aquariums, animal shelters, and humane societies provides opportunities in a number of fields. These include animal nutrition, facilities construction and maintenance, feeds, health services, care and management, production, breeding, and marketing.

Curricula in small animal care and management have been added to many high school agriscience programs. Similarly, programs are available in many technical schools, community colleges, and universities.

© iStockphoto/Joe Brandt.
A

© Danny E. Hooks/Shutterstock.com.
B

(A) Animal trainers teach animals to obey commands and to perform tricks for audiences. (B) Domesticated animals require human care to provide for basic needs such as food, shelter, and health care.

- Maintain accurate breeding and production records.
- Select a male animal to mate with your female animal.
- Use a superior proven male animal.
- Take extra precautions for female animals in heat.
- Prepare a clean area for whelping (giving birth).
- Train properly.
- Market and sell young animals.
- Develop a private market for the animals.
- Complete a registration application if animals are purebred.
- Neuter animals that are not intended for breeding.
- Summarize and analyze records.

INTERNET KEY WORDS:
care for animals

Courtesy of FFA.

FIGURE 29-20 Cats make great pets and provide companionship to people. They also have needs, such as food and veterinary care, that must be provided by their owners.

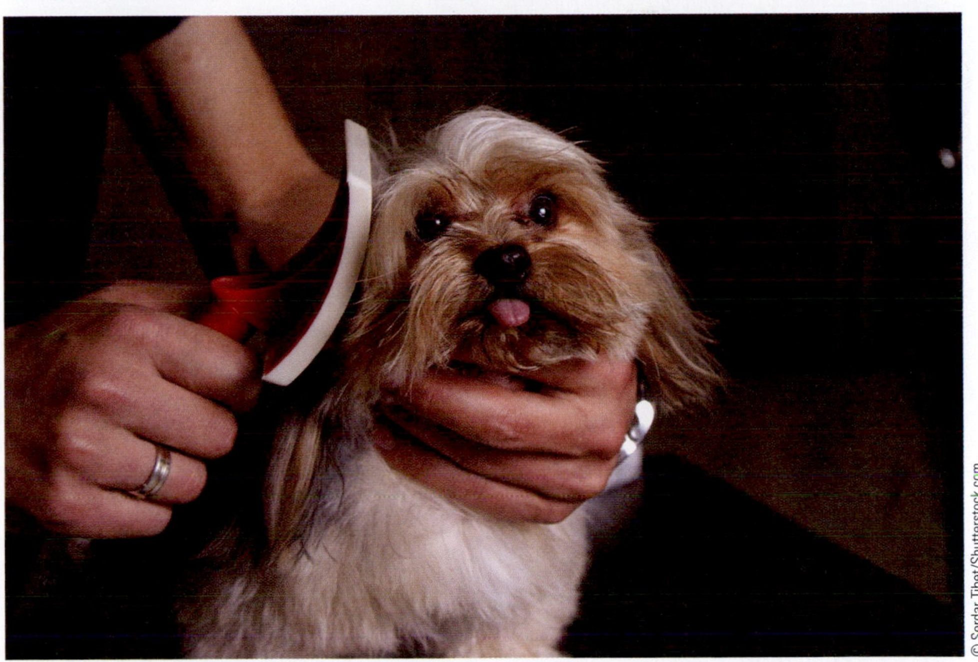

© Serdar Tibet/Shutterstock.com.

FIGURE 29-21 Businesses are needed to supply specialized care, food, health supplies, and equipment for pets and small-animal projects.

What is one person's pet always becomes someone else's business in providing replacement animals, feed, supplies, and equipment for that pet (Figure 29-21).

Raising small animals provides the opportunity for individuals with limited capital and facilities to get a start in animal agriculture. Most small animals are better adapted to production in urban and suburban areas than larger animals. The same experiences in planning for, caring for, managing, and marketing can be learned with small-animal enterprises without the large outlay of cash needed for the production of large animals.

STUDENT ACTIVITIES

1. Write the Terms to Know and their meanings in your notebook.
2. Make a bulletin board display of breeds and types of poultry and rabbits.
3. Attend a fair or show and record the names of the breeds of poultry and rabbits shown there.
4. Interview a local beekeeper about beekeeping practices in your area.
5. Compare the label from a bag of poultry feed with one from a bag of rabbit feed. Determine the differences in ingredients, percentage of protein, additives, and fiber.
6. Set up an observation beehive in the school.
7. Make a list of local crops that are of importance to agriscience and that bees pollinate.
8. Develop a crossword puzzle or word search using the Terms to Know.
9. Take your small animal to the veterinarian for a checkup. While there, have all of the scheduled vaccinations given.
10. Ask your veterinarian to discuss with you the proper care that your animal needs. Be sure to talk about how much food and water it should be getting as well as any special care it may require.
11. Select a rabbit breed. Using the Internet, identify its full scientific name from kingdom to subspecies.

SELF-EVALUATION

A. MULTIPLE CHOICE

1. A hive of honeybees needs about __________ lbs. of honey stored to live during the winter.
   a. 25
   b. 50
   c. 75
   d. 100
2. A gosling is a baby
   a. chicken.
   b. duck.
   c. pigeon.
   d. goose.
3. A pair of rabbits can produce __________ lbs. of meat per year.
   a. 10 to 30
   b. 30 to 50
   c. 50 to 80
   d. 200 or more
4. A male duck is a
   a. drake.
   b. capon.
   c. cockerel.
   d. rooster.
5. The only bee in a hive capable of laying eggs is the
   a. king.
   b. queen.
   c. worker.
   d. drone.
6. Honeybees account for about __________ percent of all insect pollination.
   a. 20
   b. 40
   c. 60
   d. 80
7. The __________ breed of chicken is the foundation of nearly all types of laying hens.
   a. Cornish
   b. Pekin
   c. Leghorn
   d. game

8. A type of rabbit that produces wool is
   a. pelt.
   b. fur.
   c. Angora.
   d. mohair.
9. The champion egg-laying breed of bird is the
   a. Leghorn.
   b. quail.
   c. Khaki Campbell.
   d. Toulouse.
10. More than __________ lbs. of rabbit meat are produced in the United States each year.
   a. 1 million
   b. 10 million
   c. 25 million
   d. 50 million

B. MATCHING

__________ 1. Drone
__________ 2. Cockerel
__________ 3. Gander
__________ 4. Doe
__________ 5. Drake
__________ 6. Chick
__________ 7. Buck
__________ 8. Worker
__________ 9. Hen
__________ 10. Poult

a. Male goose
b. Young chicken
c. Male bee
d. Female bee
e. Female pheasant
f. Young turkey
g. Male rabbit
h. Female rabbit
i. Male chicken
j. Male duck

C. COMPLETION

1. More than 90 percent of the turkeys produced in the United States are __________.
2. Ducks raised for meat reach a weight of 7 lbs. in about __________ weeks.
3. The most popular breed of rabbit for meat production is the __________.
4. A __________ is a home for honeybees.
5. The average laying hen produces about __________ eggs each year.
6. Another word for beekeeping is __________.
7. __________ and __________ are probably the most common companion animals in the United States.
8. Whelping means __________.

UNIT 30

Dairy and Livestock Management

OBJECTIVE

To determine the history, types, uses, care, and management of dairy and livestock.

MATERIALS LIST

- copies of various livestock breed magazines and bulletin board materials
- paper, glue, scissors
- paper for notebook, covers
- examples of various animal products or things made from animal products
- Internet access

COMPETENCIES TO BE DEVELOPED

After studying this unit, you should be able to:

- describe the history and economic importance of dairy and livestock.
- recognize major types and classes of livestock.
- list major uses of livestock.
- understand basic approved practices in the care and management of dairy and livestock.

SUGGESTED CLASS ACTIVITIES

1. Take a class field trip to a livestock farm located near your community. Ask the farm owner or manager to talk to the class about the critical management decisions that he or she makes that affect the profitability of the farm. Be prepared to ask leading questions about the nutrition, health, breeding, and marketing strategies that are used on the farm. Prepare individual written student reports that address these and other important issues that were discussed during the visit.
2. Study the breeds of livestock by assigning each class member to write to a breed association to obtain information about the breed. Have students prepare an oral presentation and report their information to the class. Provide class members with a worksheet on which they record key points about each livestock breed as the reports are given.
3. Invite a dairy, sheep, swine, or goat judge to come to the class and give a presentation outlining the physical characteristics that

TERMS TO KNOW

mammal
veal
calf
colostrum
heifer
cow
draft
bull
steer
implant
pork
sow
gilt
boar
farrow
wool
lamb
mutton
ewe
ram
mohair
chevon
lactation
kid

make a good show animal, as well as a good producer. You may also want him or her to provide a presentation and/or a demonstration on showmanship. Students should be ready to ask questions and participate in the discussion.

Large animals, including dairy, beef, sheep, goats, and swine, are the backbone of animal agriculture. Keeping wild animals made it necessary for early humans to provide feed for them. It also changed their lives, in that it forced humans to tend the animals. The capturing and domestication of animals changed humans from hunters to farmers. Today, much of agriscience is centered on the production of animals and animal products and the production of feed for those animals.

DAIRY CATTLE

Origin and History

A **mammal** is an animal that produces milk, a highly nutritious white or yellowish liquid secreted by the mammary glands of animals for the purpose of feeding their young. In the wild, mammals normally produce only enough milk to feed their offspring. When early humans realized that milk was good to drink and that some types of animals produced more milk than others, the domestication of milk animals began. Although the cow, buffalo, camel, goat, ewe, and mare have been and are currently being used for the production of milk in various parts of the world, this unit focuses primarily on cattle with a minor emphasis on goats in discussing dairy production.

Early historical records note the use of cattle to produce milk, and the Bible also has a number of references to milk and the production of milk. Hippocrates, the father of modern medicine, recommended milk as a medicine in his writings around 400 BC.

The New World had no dairy cows until Columbus brought them with him on his second voyage in 1493. Cattle were also transported to the New World in 1611 with the Jamestown colonists. The production of milk was limited to a few cows per family during the colonial period. It was not until the late 1800s that dairy farming became an important agricultural industry in the United States. Since then, intensive efforts have been directed toward breeding and developing the dairy type of cattle (Figure 30-1).

Economic Importance

The production of milk is the second most important animal enterprise in the United States, if sales dollars are the criterion for importance. The consumption of milk and other dairy products has remained nearly steady during the past several years, after several decades of steady decline (Figure 30-2). The average person in the United States uses more than 580 lbs. of milk and dairy products each year. About 9.197 million cows in the United States produce about 170.3 billion lbs. of milk annually. Average production in a recent report by the United States Department of Agriculture was 20,267 lbs. of milk per cow per year. Dairy is truly big business.

The production of milk is not the only income-generating part of dairy production. Bull calves enter the meat production product stream as **veal**

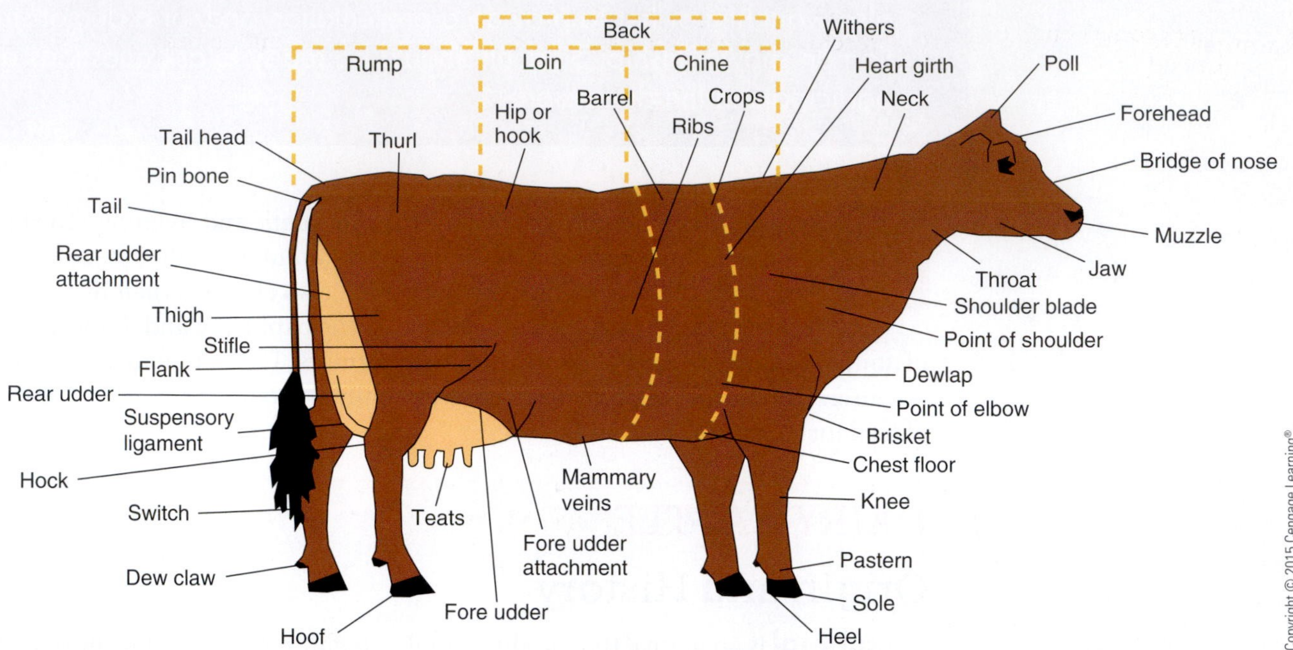

FIGURE 30-1 Anatomy of a dairy cow.

(the meat of young calves), or they are grown to market size for beef (meat from grown cattle). Similarly, cows that are no longer profitable producers of milk are sold for beef. Female calves become tomorrow's cows (Figure 30-3).

Dairy-cattle operations are generally divided into two types: Class A and Class B. The class refers to the intended use for the milk that is produced. Class A milk is produced under strict standards and is intended for consumption as fluid

FIGURE 30-2 Consumption of milk products has remained relatively steady for the past several years.

FIGURE 30-3 Today's female calves become tomorrow's dairy cows, so good calf care and management are top priorities on dairy farms.

INTERNET KEY WORDS:
milk production, United States

milk. Fluid milk includes whole milk, reduced-fat milk, and cream. Class B milk, which can be produced under less strict standards, is intended to be used to make butter, cheese, ice cream, nonfat dry milk, and other manufactured dairy products.

Types and Breeds

More than 90 percent of all dairy cattle in the United States are of the Holstein breed. These familiar black and white cattle are the greatest average producers of milk of any breed in the country. Because of the large numbers of cattle involved, the breed also has been able to make the most genetic improvement in recent years.

The second most popular breed of dairy cattle is the Jersey. They are the smallest in size of the dairy breeds, but they rank number one in the amount of butterfat in their milk. Butterfat is the fat in milk from which butter is made. Their rich milk is also high in milk solids such as protein, from which cheese is made. Another popular breed of dairy cattle is the Guernsey. This breed is known for the yellowish tint to the color of its milk. Ayrshire and Brown Swiss round out the top five breeds of dairy cattle in the United States (Figure 30-4).

Courtesy of the Holstein Association USA, Inc.

A

Courtesy of the American Jersey Cattle Club.

B

Courtesy of the American Guernsey Association.

C

© smereka/Shutterstock.com

D

continues

continued

E

Courtesy of the Brown Swiss Cattle Breeder's Association.

FIGURE 30-4 Major breeds of dairy cattle in the United States include (A) Holstein, (B) Jersey, (C) Guernsey, (D) Ayshire, and (E) Brown Swiss.

APPROVED PRACTICES

Raising Calves

General approved practices for raising dairy calves include the following:

- Make sure the newborn **calf** receives colostrum as its first food as soon after birth as possible, and continue to feed colostrum for at least the first 36 hours of its life. **Colostrum** is the milk that a cow produces for a short time after calving. It contains antibodies that protect the newborn animal from diseases until it can build up its own natural defenses.
- Feed milk or milk replacer daily at 8 to 10 percent of the calf's weight until the calf is 4 weeks old. Milk replacers are dry dairy or vegetable products that are mixed with warm water and fed to young calves instead of milk. Such products are less expensive than whole milk.
- Start feeding calf starter, a grain mixture, free choice after about 10 days. Free choice means making feed available at all times.
- Wean calves from milk when they are eating 1.5 lbs. of calf starter per day.
- Feed calves green, leafy hay and water free choice at 2 to 9 months of age. Up to 4 lbs. of grain can be fed daily.
- Make up the bulk of the ration with forages fed free choice after 9 months of age.
- Remove horns at an early age, preferably as soon as the horns begin to develop.
- Remove extra teats at an early age.
- Identify calves with ear tags or tattoos as soon as possible after birth.
- Prevent calves from sucking the ears of other calves or licking each other.
- Keep hooves properly trimmed.
- Vaccinate for cattle diseases at the recommended times.

AGRI-PROFILE CAREER AREAS: FARM MANAGER/HERD MANAGER/FARMER/ RANCHER/BREED ASSOCIATION REPRESENTATIVE

Courtesy of USDA.

Farming, ranching, and feedlot management have become big business, and they are based on scientific research.

Science and production can no longer be separated. Individuals and teams of scientists visit farms, ranches, and feedlots frequently to gather data, analyze problems, and seek solutions. Farmers, ranchers, and feedlot managers are usually graduates of agricultural colleges and may have advanced degrees in business, management, or animal sciences.

In addition to scientists and managers, technicians find meaningful careers in the animal production businesses of the nation. Livestock enterprises include beef, dairy, sheep, and swine. There are many non-farm career opportunities as field representatives for breed associations, feed companies, marketing cooperatives, supply companies, animal-health products, herd improvement associations, and financial-management firms.

Many individuals use their farming and ranching backgrounds in agriscience writing, publishing, and telecasting careers. Farm and ranch experience provides an excellent background for careers in agriscience teaching and extension work.

INTERNET KEY WORDS:
Holstein cattle
Jersey cattle
Guernsey cattle
Brown Swiss cattle
Ayrshire cattle

- Maintain the calf in clean and sanitary conditions.
- Plan for and maintain a disease- and parasite-prevention and control program.
- Breed heifers to calve (give birth) at 20 to 24 months of age. A **heifer** is a female that has not given birth to a calf.
- Maintain heifers and calves in uniform groups according to size and weight.

Dairy Cows

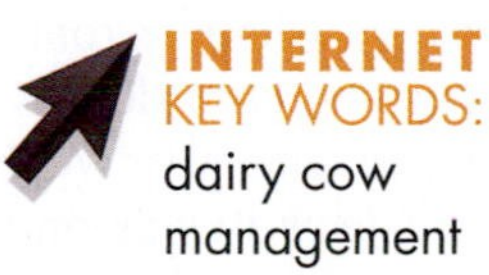

Some approved practices for dairy cows include the following:

- A **cow** is a female animal of the cattle family that has given birth. Cows should be bred to calve once every 12 months. Rebreed cows 45 to 60 days after calving.
- Observe cows for evidence of heat period twice daily, mornings and evenings (Figure 30-5).
- Check cows to determine whether they are pregnant. This should be done 45 to 60 days after breeding them.
- Provide a dry period of about 60 days before calving to allow the cow to rebuild her body. The dry period refers to the time when a cow is not producing milk.
- Feed dairy cows according to their levels of production and stages of pregnancy.
- Maintain complete health, breeding, and production records for every cow in the herd (Figure 30-6).

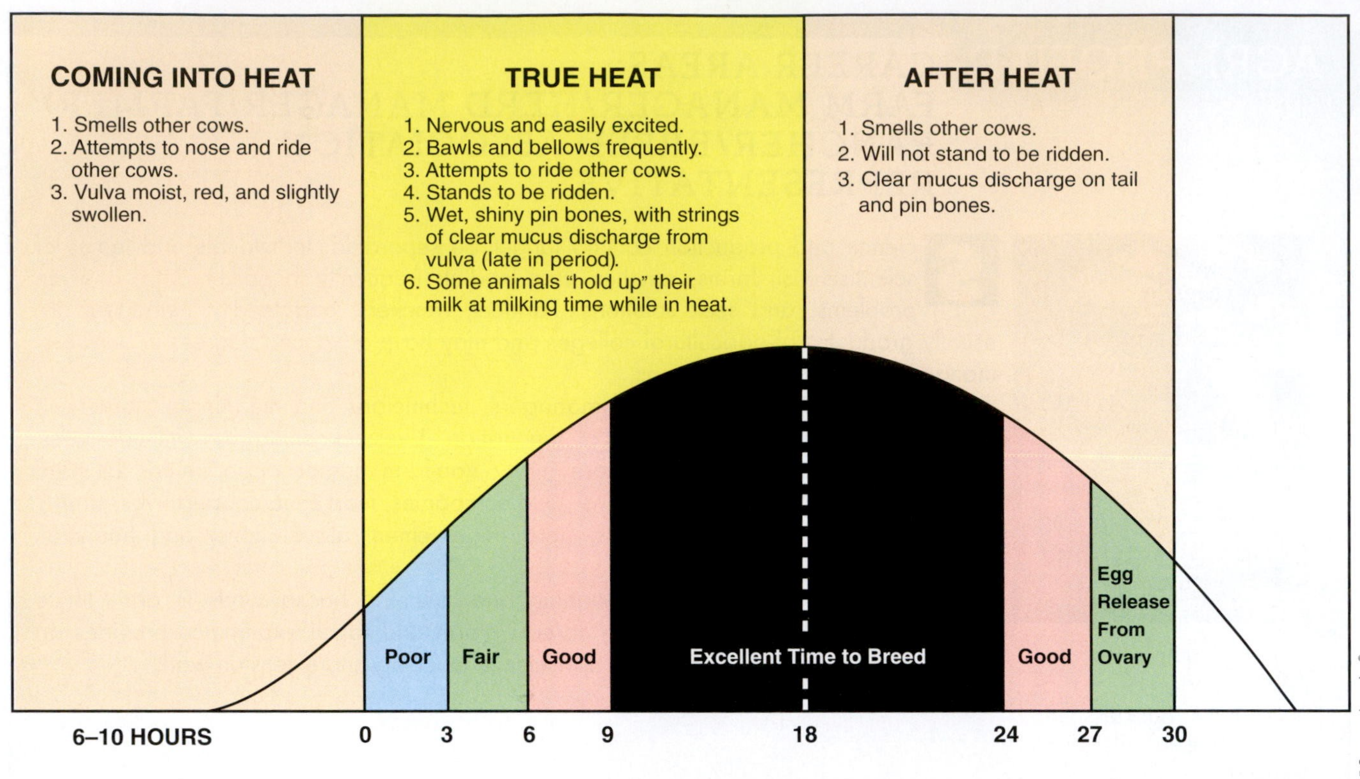

FIGURE 30-5 Breeding at the proper stage of the heat period is essential to maximize pregnancy rates.

SCIENCE CONNECTION OF COWS AND CARS

Transponders and other sensing units permit better research and management. The unit being applied here will record motion and other data so scientists can determine how much time the animal spends grazing.

Cows and cars are sometimes the recipients of the same technology. Transponders are small devices that can be attached to the ears of cows and other livestock to identify them. As animals come into the electronic fields of certain other devices, their identities can be noted and recorded by computer. The computer can be programmed to simply note and remember what the animal does, or it can direct other devices to provide selected kinds and amounts of feed, water, medicine, pesticide, or other treatment. This technology is becoming more common in modern milking parlors and feeding systems.

At the U.S. Department of Agriculture (USDA) Fort Keogh Livestock and Range Laboratory in southeastern Montana, scientists found transponders to be the answer to a weighty problem! To conduct feeding, grazing, pasture-improvement, and pasture management experiments, they had to weigh their research subjects, range cattle, at frequent intervals. Animal response to treatment

Factors Determining Efficiency			
Factor	Results	Goals	State Average
______ 1. Number of cows in herd			
______ 2. Total lb. of milk produced			
______ 3. Total lb. of butterfat produced			
______ 4. Average annual milk production/cow (lb.)			
______ 5. Average annual butterfat production/cow (lb.)			
______ 6. Average annual percent of butterfat for student's herd			
______ 7. Feed cost/lb. of butterfat produced			
______ 8. Average feed cost/cow			
______ 9. Percent of calves sold or kept until 3 months of age			
______10. Profit or loss			
______11. Lbs. of milk sold/hour of total labor			
______12. Dollar returns/hour of self labor			
______13. Production cost/lb. of milk			

FIGURE 30-6 Efficiency factors are good indicators of how the dairy business is operating.

- Establish and maintain a disease- and parasite-prevention and control program.
- Contact a veterinarian anytime that you are unsure of how to treat a dairy health problem.

generally shows up in the form of weight gain or loss. At the same time, any disturbance or change in an animal's routine, such as cattle roundup, forced weighing, or catching in a head gate, will influence an animal's weight. Good research required a system that could weigh the animals at frequent intervals without disturbing them—that is, they needed to weigh themselves!

The answer was found in a system using a scale, a computer, and a transponder. Cattle need water. A continuous supply of fresh water must be readily available if body functions are to be optimal. Therefore, the researchers set up a scale and guide railing so the animal had to cross over the scale to drink. An ear-tag transponder on the animal identifies the animal as it enters the range of the electronic pickup device and steps on the scale. The electronic scale sends continuous weight readings to the computer, indicating the animal's weight before, during, and after drinking. The computer records the weight data, together with time, temperature, weather, and other information monitored at the site.

As drivers zip across California's Golden Gate Bridge and blissfully bypass the toll booth, they are probably legal motorists simply using the same transponder device that was pioneered by the USDA's Agricultural Research Service. If so, the car-mounted transponder in their car gave the computer the auto's account number, and the computer will have noted the time, date, and auto number for future billing to the motorist. The system is being used by many toll facilities across the country.

- Milk dairy cows at regular intervals each day. Two milkings each day, approximately 12 hours apart, is a normal routine. However, milking three times per day results in greater milk production, and is recommended in high-producing herds.
- Maintain a regular routine in handling dairy cattle to maintain maximum production.
- Cull unprofitable dairy cows. To cull means to remove the least productive animals from the herd.
- Properly maintain dairy housing and milking equipment.

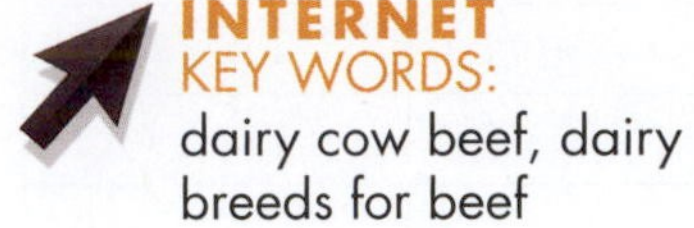
INTERNET KEY WORDS:
dairy cow beef, dairy breeds for beef

BEEF CATTLE

Origin and History

It is likely that cattle were domesticated in Europe and Asia at some time during the New Stone Age. Domesticated cattle of today are probably all descended from one of two wild species, *Bos taurus* or *Bos indicus* (Figure 30-7). Some of these wild cattle stood as tall as 7 feet at the shoulders. They were domesticated for meat, milk, and draft. **Draft** is a term for an animal that is used for work. Oxen are draft animals.

Owning cattle was a symbol of wealth in early times. They were worshipped in some cultures and used in others. Several early civilizations had "gods" fashioned to look like cattle. The use of cattle in sports such as bullfighting also developed during early times.

Cattle came to the New World with the earliest settlers, who were more interested in animals that could do heavy work than in those that could produce meat. As the European explorers came to the New World, they also brought cattle with them. Spanish settlers brought longhorn-type cattle as a source of food for Christian missions in the Southwest. As demand increased for beef, the cattle industry developed on the frontier, where grass and the large open spaces required for cattle were abundant. Great cattle drives originated in these areas

FIGURE 30-7 Bos taurus was one of the wild ancestors of today's cattle.

because cattlemen drove their cattle to transportation centers to market them. Today, the beef cattle industry is concentrated in the Midwest and South, where plenty of feed is available and production costs are less than in some areas of the country.

A discussion of beef production includes several important terms. A **bull** is a male animal of the cattle family; a mature female is a cow. A calf is a young animal, and a heifer is an immature female. A **steer** is a castrated male. A group of several animals is a herd. Meat from a calf is veal, and meat from older animals is beef.

Economic Importance

INTERNET KEY WORDS: beef cattle breeds, exotic foreign beef breeds

The beef cattle industry is the number one red-meat production industry in the United States. The U.S. population eats about 65 lbs. of beef per person per year. This is part of a total consumption of about 200.4 lbs. of meat, poultry, and seafood. Other products besides meat are obtained from cattle. They convert inedible grasses into food for people. Cattle manure provides fertilizer for crops, and meat by-products are made into many nonfood products that we use every day (Figure 30-8).

Types and Breeds

The general types of beef-cattle operations include purebred breeders, cow–calf operations, and slaughter-cattle, or feedlot operations. In a purebred operation, only cattle of a single, pure breed are raised. Operations are geared to produce purebred bulls for cow–calf operations and to produce animals to be sold to other purebred producers. Breeders of purebred cattle have been responsible for much of the genetic improvement in beef cattle in recent years.

Cow–calf operations produce feeder calves that are placed in feedlots owned and managed by producers of slaughter cattle. Cow–calf operations are located

By-products from the Production of Meat

- Bone for bone china.
- Horn and bone handles for carving sets.
- Hides and skins for leather goods.
- Rennet for cheese making.
- Gelatin for marshmallows, photographic film, printers' rollers.
- Stearin for making chewing gum and candies.
- Glycerin for explosives used in mining and blasting.
- Lanolin for cosmetics.
- Chemicals for tires that run cooler.
- Binders for asphalt paving.
- Medicines such as various hormones and glandular extracts, insulin, pepsin, epinephrine, ACTH, cortisone, and surgical sutures.
- Drumheads and violin strings.
- Animal fats for soap and feed.
- Wool for clothing.
- Camel's hair (actually from cattle ears) for artists' brushes.
- Cutting oils and other special industrial lubricants.
- Bone charcoal for high-grade steel, such as ball bearings.
- Special glues for marine plywoods, paper, matches, window shades.
- Curled hair for upholstery. Leather for covering fine furniture.
- High-protein livestock feeds.

FIGURE 30-8 Many important substances are obtained from the meat-processing industry. They range from raw materials for cosmetics to medicines and many other products that are used for a variety of purposes.

mostly in the upper Great Plains states and in the western range states where grass is in abundance and much of the land is unsuited to produce other crops. Calves are usually born in the spring, remain with their mothers during the summer, and are weaned in the fall. They are then sold to feedlot operations. Most cow–calf producers use purebred bulls to breed the grade cows in the herd, producing superior calves.

The feedlot operator buys calves from cow–calf operators and feeds them until they reach slaughter weight. These operations are concentrated in the Midwest, where there is an abundance of corn and other grain for feed. Large feedlot operations are located wherever an abundance of cheap feed exists.

Until about 40 years ago, there were basically three breeds of beef cattle in the United States: Hereford, Angus, and Shorthorn. Although these three breeds are still important today, they have been joined by more than 50 other breeds from all over the world (Figure 30-9). These breeds can be divided into many classifications. However, for our purposes, they are designated as English, exotic, and American.

English breeds originated in the British Isles and include Hereford, Angus, Shorthorn, Galloway, Devon, and Red Poll. In general, English breeds of cattle are of medium size and are noted for the excellent quality of meat they produce.

Exotic breeds of cattle were first imported to the United States from all over the world when consumers began demanding leaner beef. Beef producers also became aware that calves from these breeds of cattle grew faster and much more efficiently than most of the English breeds. Exotic-breed bulls are often used in grade cow–calf operations to increase the weight of calves being produced. Examples of exotic breeds of cattle include Charolais, Limousin, Simmental, Blond D'Aquitaine, and Maine Anjou.

U.S. breeds of beef cattle were developed from necessity. To grow beef cattle in the South and Southwest, heat tolerance and disease- and parasite-resistance were needed. To develop cattle that met these requirements and still had desirable meat quality, breeders crossed Brahman cattle from India with the English breeds. Examples of U.S. breeds of beef cattle are Brangus, Beefmaster, Santa Gertrudis, and Barzona. Many commercial herds are of crossbred, or mixed, breeds (Figure 30-10).

INTERNET KEY WORDS:
growth implants, beef
beef management practices

An understanding of beef production includes knowing and using the terms that are used to describe beef cattle (Figure 30-11).

Approved Practices for Beef Production

The following are some of the considerations in the production of beef cattle:

- Select breeds and individual beef cattle according to intended use, area of the country, and personal preference.
- Buy cattle only from reputable breeders.
- Select purebred cattle according to physical appearance, pedigree, and available records.
- Isolate new animals from the herd for at least 30 days to observe for possible diseases and parasites.
- Provide enough human contact so that beef cattle can be handled when necessary.
- Break calves to lead as soon as possible if they are going to be exhibited at fairs and shows.
- Plan a complete herd-health program and follow through with it.
- Use the services of a veterinarian for serious or unknown health problems.
- Vaccinate to prevent diseases of local concern.

Courtesy of the American Hereford Association.

A

Courtesy of the American Angus Association.

B

Courtesy of the American Simmental Association.

C

Courtesy of the North American Limousin Foundation.

D

Courtesy of the Beefmaster Breeders United.

E

Courtesy of the American-International Charolais Association.

F

FIGURE 30-9 continued

continued

G

H

FIGURE 30-9 Today's predominant beef breeds in the United States are (A) Hereford, (B) Angus, (C) Simmental, (D) Limousin, (E) Beefmaster, and (F) Charolais. More of today's predominant beef breeds in the United States are (G) Brahman and (H) Shorthorn.

FIGURE 30-10 Many commercial producers have crossed and mixed breeds of cattle, such as the Brangus bull pictured here.

- Castrate, dehorn, and permanently identify calves at an early age.
- Breed heifers to calve at 2 years of age.
- Implant steers and heifers grown for slaughter with approved growth hormones. An **implant** is a substance that is placed under the skin. It is released slowly over a long time period to stimulate efficiency and growth.
- Wean calves at 205 days of age and at 450 to 500 lbs.
- Provide supplemental nutrition to cattle when natural forages are in short supply.
- In a pasture-breeding system, allow one mature bull for every 25 to 30 cows. With pen mating, one bull can be used to service 30 to 50 cows.
- Restrict the length of the breeding season so that all calves are born in a 30- to 60-day period.
- Assure that cows calve in clean stalls or pastures.

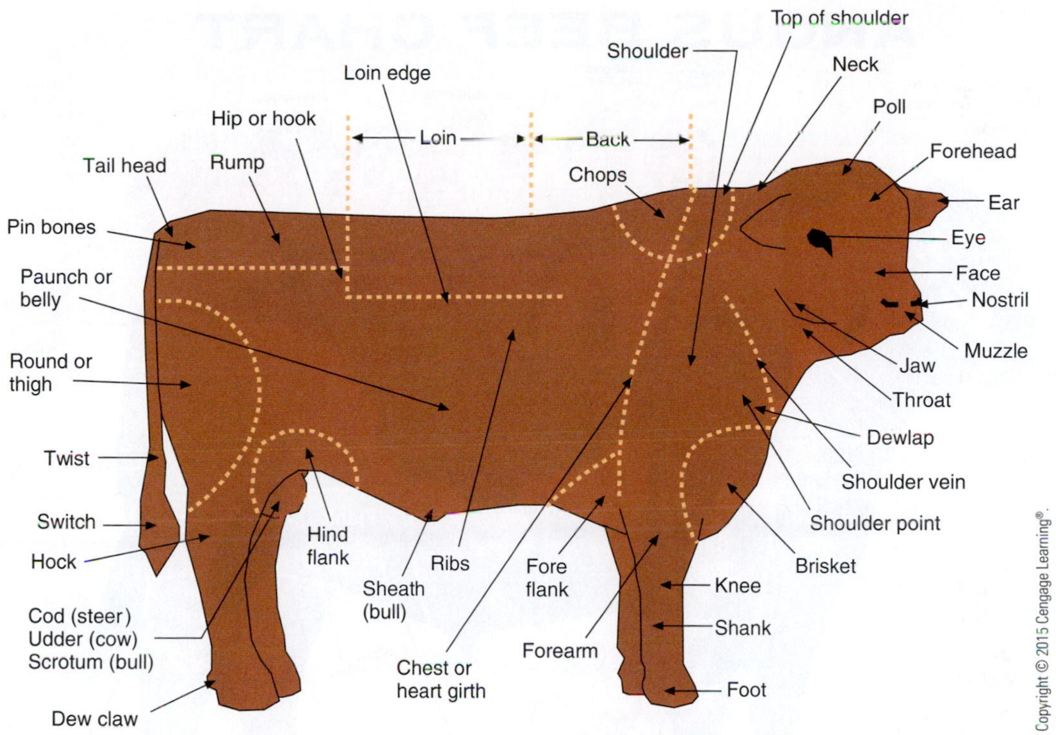

FIGURE 30-11 Learning the proper names of cattle parts is necessary for discussing and evaluating animals.

INTERNET KEY WORDS:
National Pork Producers
annual pork consumption

- Group the cattle being fed for slaughter according to size and sex.
- Provide shelter from inclement weather.
- Provide access to clean, fresh water at all times.
- Feed slaughter cattle to reach market weight at 15 to 24 months of age.
- Have proper facilities and equipment available for the type of operation.
- Market animals at the optimum time to maximize profits (Figure 30-12).
- Maintain complete and accurate records.

HOT TOPICS IN AGRISCIENCE

THE "GENE FOR LEANNESS"

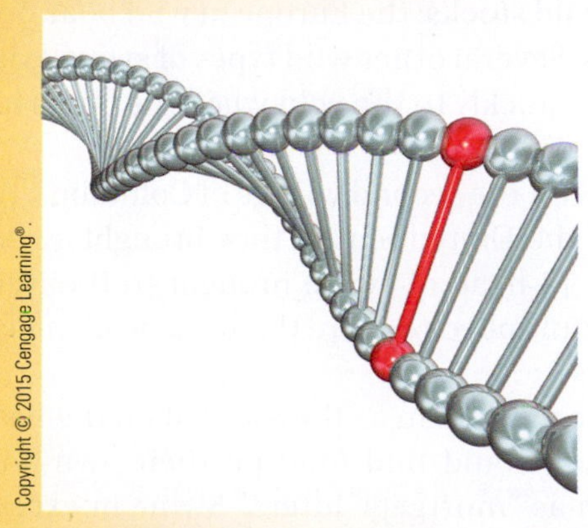

Genetic engineering techniques make it possible to control and lock in desirable hereditary traits in food animals.

Lean, tender beef is the product that is in demand from cattle producers. These qualities have been pursued in the past by attempts to breed cattle that will develop marbling in the meat without putting on too much fat on the outside of the carcass. Now science has identified a new approach to producing lean, tender beef. A gene has been isolated that codes for a protein called myostatin. In its active form, this gene controls the number of muscle fibers that develop in the fetus of a calf. Another form of the gene produces inactive myostatin. When inactive myostatin genes are paired in an animal, a condition known as double-muscling occurs.

Double-muscling is considered undesirable because the extra muscle mass tends to produce stress in animals, and it has been shown to cause difficult births. When this gene is paired with another form of the gene, more extensive muscling occurs without the related stress. The animal produces a carcass with approximately 7 percent more beef and 14 percent less overall carcass fat than cattle that have active myostatin.

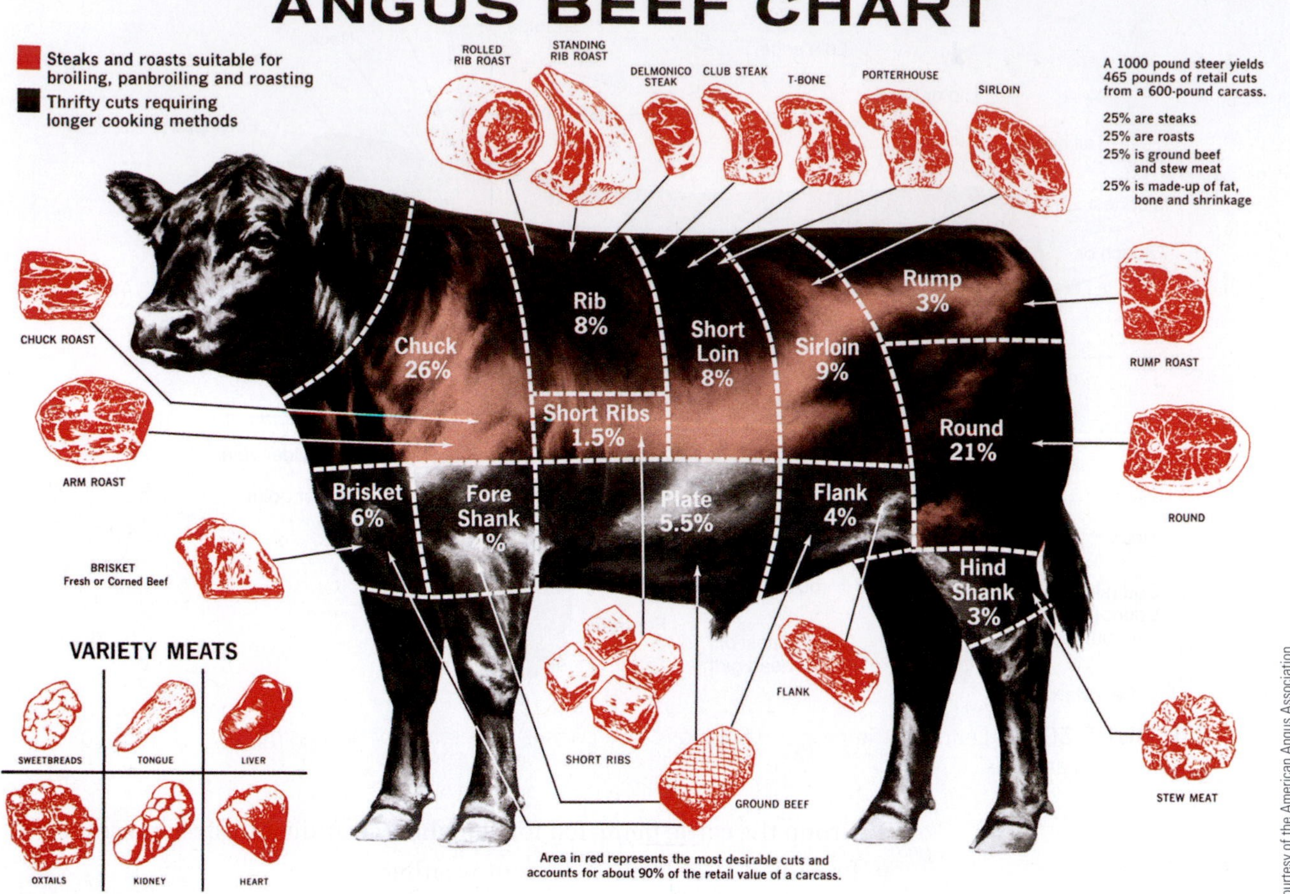

FIGURE 30-12 Many different cuts of meat are obtained from beef animals.

SWINE

Origin and History

Swine were apparently domesticated in China during the Neolithic Age. The first written record of keeping swine appears in 4900 BC. Mention of swine occurs in the Bible as early as 1500 BC.

Domestic swine originate from two wild stocks: the European wild boar, *Sus scrofa*, and the East Indian pig, *Sus vittatus.* Several other wild types of swine exist today. Even domesticated swine can revert quickly to the wild when the opportunity or need arises.

Swine came to the New World in 1493 on the second voyage of Columbus. As European explorers came to what is now the United States, they brought swine with them as a food supply. The original 13 head of swine brought to the New World by the explorer Hernando de Soto multiplied to more than 700 head in just 3 years and provided food for the explorers.

European settlers also brought swine with them as they settled on the East Coast of the United States. Because swine could find food on their own and reproduce rapidly, they were soon known as "mortgage lifters." Swine in excess of local needs were exported as pork and lard. **Pork** is meat from swine. Lard is pork fat that has been obtained by cooking and pressing the oil from fatty tissue. This process is called *rendering.*

SCIENCE CONNECTION OH, THE SWEET SMELL OF SUCCESS

The suburbs are moving farther and farther away from the cities. Farms and subdivisions find themselves becoming neighbors. As a result, conflicts are becoming more frequent because of the smells produced by animal wastes. Fortunately, researchers at the University of Florida have designed a piece of equipment called a fixed-film digester that may be a solution that will keep everyone breathing deeply. A system of plastic pipes lines a 100,000-gallon tank. Bacteria inside the pipes transform smelly waste into compounds that do not have an odor problem. The system can reduce offensive smells by 90 percent. Besides clearing the air, the digester also produces a free fuel source for the farmer, biogas (methane). Another benefit is that water can be cleaned and recycled. The system currently has a high price tag, but as it becomes commercially available, the cost is expected to come down and make the system a reasonable farm expense that will keep both farmers and their neighbors breathing the sweet smell of success.

As westward expansion occurred, swine production followed. In time, the center of swine production settled in the Corn Belt. This area provided large amounts of corn and other feed grains necessary for large-scale pork production.

ECONOMIC IMPORTANCE

Pork production is the number-two red-meat industry in the United States, and it is also the number-two red-meat industry in the world. With a population of 65.93 million hogs, the United States ranks second in the world in swine production, behind China. About 780 million head of swine are in the world today, with more than 50 percent of them in China alone. China produced 52 million metric tons of pork in 2006, while U.S. production was 9.39 million metric tons. U.S. production of pork makes up approximately 11.9 percent of the world supply.

Since 2002, the consumption of pork in the United States has declined each year. In 2010, U.S. consumption was about 48 lbs. of pork per person annually. Consumption of red meat has also been slowly decreasing.

The modern hog of today is vastly different from its ancestors. It is lean and trim compared with the lard-type hog in demand 50 years ago and even the meat-type hog of 20 years ago. Hogs are the most efficient converters of feed into meat among the large red-meat animals. It takes about 3.5 lbs. of feed to gain 1 lb.

Leading states in the United States in the production of swine include Iowa, North Carolina, Minnesota, Illinois, and Nebraska. More than 25 percent of the swine production in the country takes place in Iowa alone.

Swine Operations

The basic types of swine operations in the United States are feeder-pig producers and market-hog producers. Operations that produce feeder pigs usually maintain large herds of sows that produce approximately 2.1 litters of piglets each year. A **sow** is a female animal of the swine family that has given birth; a **gilt** is a young female that has not given birth. A **boar** is a male member of the swine family. A piglet is a baby pig. Once they are weaned, young pigs are called feeders, and they are often sold to other producers who feed them until they reach market weight. In many operations, specific hybrid sow lines are bred to compatible hybrid boar lines to produce offspring with superior hybrid vigor. These superior hog types have predictable meat characteristics that will command top market prices (Figure 30-13).

Courtesy of Cole Swine Farms, Inc.

FIGURE 30-13 Feeder-pig producers keep brood sows and boars that provide young stock for the swine industry.

Market-hog operations normally purchase pigs at 5 to 8 weeks of age from feeder-pig producers. They feed the pigs until they reach a market weight of about 240 to 260 lbs. They are then marketed and sent to slaughter plants for processing.

Another type of swine operation is the purebred producer. This kind of swine operation is responsible for producing high-quality boars and gilts for feeder-pig operations and purebred stock for other purebred farms. They contribute much to the genetic improvement of swine in general.

Types and Breeds

Purebred swine in the United States consist of meat-type hogs. With the decreased demand for lard and the demand for lean pork nearly devoid of fat, the lard-type hog has been bred out of existence.

Popular breeds of swine that were refined in the United States include Duroc, Hampshire, Chester White, Poland China, and Spotted Hog. The Berkshire, Yorkshire, and Tamworth breeds were developed in England, and the Landrace originated in Denmark. Duroc, Hampshire, and Yorkshire are the most popular breeds of swine in the United States today (Figure 30-14).

INTERNET KEY WORDS:
swine management

Approved Practices in Swine Production

Approved practices in swine production include the following:

- Buy pigs only from reputable producers or at certified feeder-pig sales.
- Observe newly purchased animals for signs of disease and parasites.
- Group pigs according to size in groups of not more than 20 to 25 animals.
- Feed a complete, balanced ration based on the age and weight of the animal being fed.
- Ensure that access to an unlimited supply of fresh water is available at all times.

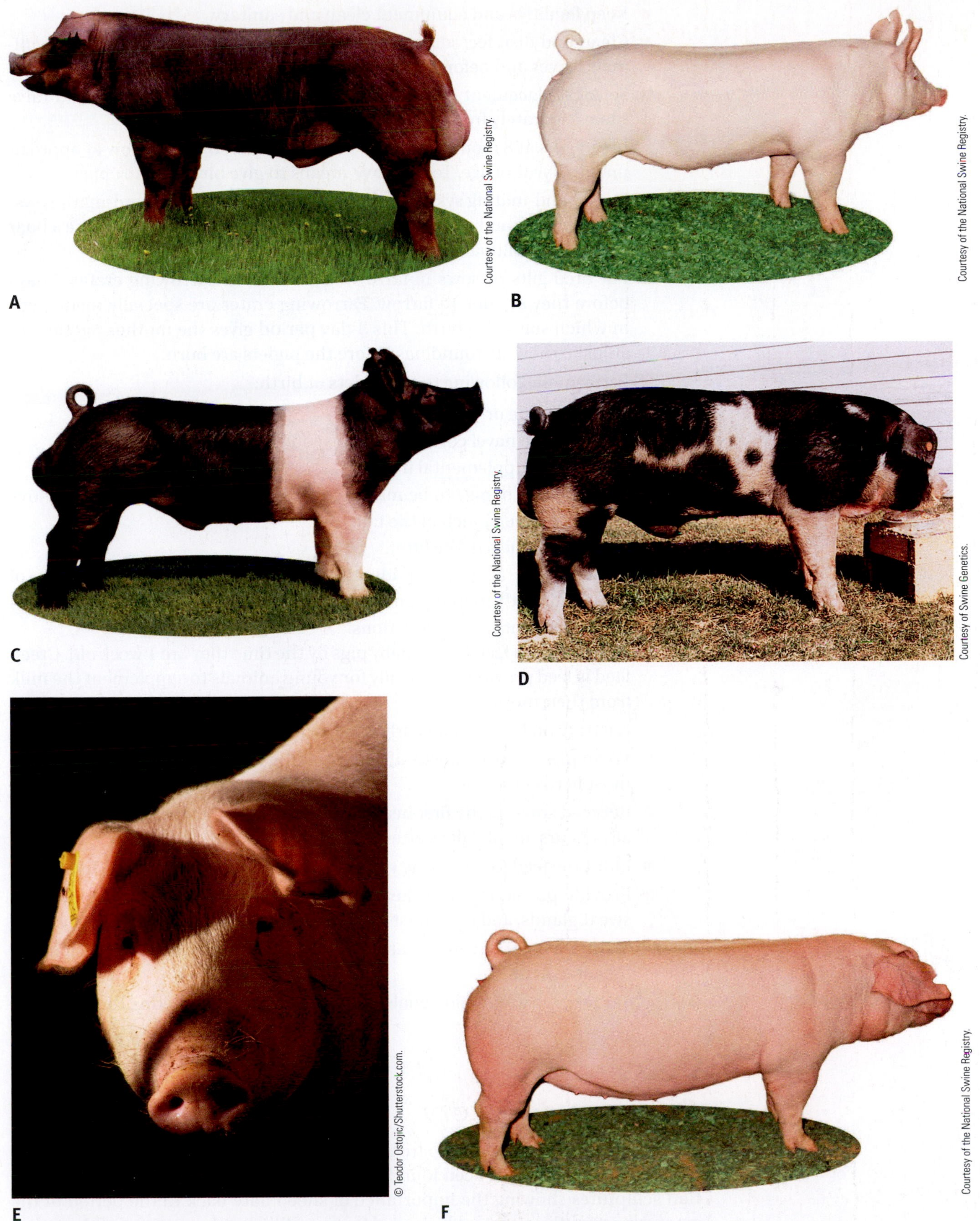

FIGURE 30-14 Major breeds of swine in the United States include (A) Duroc, (B) Yorkshire, (C) Hampshire, (D) Spotted, (E) Chester White, and (F) Landrace.

- Keep facilities and equipment clean and sanitary.
- Clean and disinfect all facilities and equipment after each group of animals leaves and before the next group arrives.
- Select replacement gilts for the breeding herd at an early age and raise them separately from market hogs.
- Breed gilts at 8 months of age or 250 to 300 lbs. so they farrow at approximately 1 year of age. To **farrow** means to give birth to baby pigs.
- Use a hand-mating system to breed gilts and sows. In a hand-mating system, the boar and sow are kept separate except during mating. Use a boar to check for animals in heat.
- Put bred gilts or sows in farrowing facilities or farrowing crates 3 days before they are due to farrow. Farrowing crates are specially made pens in which sows give birth. This 3-day period gives the mother pig time to adjust to new surroundings before the piglets are born.
- Perform the following to the piglets at birth:
  1. Clip needle or wolf teeth.
  2. Clip or tie navel cord and dip the end in iodine.
  3. Provide supplemental iron.
  4. Dock tails of pigs to be marketed for meat. To dock means to remove all but about 1 inch of the tail.
  5. Weigh all pigs in the litter.
  6. Ear notch all pigs for identification. Ear notching is a system of permanently marking animals for identification by cutting notches in their ears at specific locations.
- Provide creep feed for the baby pigs by the time they are 1 week old. Creep feed is feed provided especially for young animals to supplement the milk from their mothers.
- Castrate male pigs at an early age.
- Wean pigs at 4 to 6 weeks of age. Weaning at about 6 weeks is normal in most herds of swine.
- Rebreed sows on the first heat period after weaning the piglets. This usually occurs about 3 days after the pigs are weaned.
- Limit the feed for gestating sows to prevent them from getting too fat.
- Provide protection from heat and cold, especially heat. Swine have no sweat glands, and care must be taken to keep them cool in hot weather.
- Maintain complete health and production records for each animal in the breeding herd.
- Set realistic production goals and cull animals not meeting the goals.

SHEEP

Origin and History

The domestication of sheep apparently occurred before the time of recorded history. Fibers of wool have been found in ruins of the earliest Swiss villages. Egyptian sculptures showing the importance of sheep date back to the period of the great dynasties. Even ancient historical texts are filled with mentions of sheep and shepherds.

Sheep were probably domesticated from wild types in Europe and Asia by early humans to use for meat, wool, pelts, and milk. **Wool** is a modified hair with superior insulating qualities. It is usually obtained from sheep, but it is also obtained from some other animals such as llamas. Wool has been an important product of international trade for many centuries, and it is still one of the most important fibers known to humans.

As civilization advanced, the production of wool became a priority in sheep production. As a result, specific wool-producing breeds were developed in Europe. Most of today's breeds can be traced back to these breeds developed 500 to 1000 years ago.

Columbus had sheep with him on his second voyage to the West Indies in 1493. Cortez brought sheep when he explored Mexico in 1519. Spanish missionaries also kept sheep and taught American Indians of the Southwest how to weave wool into cloth. English settlers on the East Coast also raised sheep for the production of wool. Lamb and mutton were of secondary concern. The term **lamb** refers to a young sheep as well as the meat of young sheep. **Mutton** is meat from mature sheep.

Courtesy of USDA/ARS #K-4166-5.

FIGURE 30-15 Sheep can use low-quality forage, graze land unsuited for other purposes, survive severe weather conditions, and produce meat and wool.

Centers of sheep population gradually moved from the Northeast to the West as populations expanded. Areas of open spaces and abundant grasses such as western rangelands have proved to be ideal for the production of sheep.

Economic Importance

Production of sheep in the United States has declined in the last 50 years, while dairy, beef, and swine production has expanded. The U.S. population eats less than a pound (0.8 lb.) of lamb and mutton per person each year. This amount is rising slowly but is not expected to seriously challenge beef and pork producers. Each person in the United States also uses about 0.8 lb. of wool each year.

There were approximately 5.35 million sheep on U.S. farms in 2012. The production of wool yields about 32.5 million lbs. Leading states in the production of lamb include Texas, California, and Wyoming, followed by Colorado, South Dakota, and Utah.

Because sheep have the ability to survive in areas of limited feed and harsh climates, they are of more economic importance than would be expected. In many such areas, only sheep can survive, and they constitute a major animal enterprise (Figure 30-15).

Types and Breeds

Sheep operations can be divided into two basic types: farm flocks and range operations. *Flock* is a term describing a group of sheep. Farm-flock operations are generally small and are often part of diversified agricultural operations. They may raise either purebred or grade sheep. These flocks usually average fewer than 150 animals. They are responsible for about one-third of the sheep and wool produced in the United States.

The other two-thirds of the approximately 6 million sheep in the United States are produced on range operations. Many flocks contain 1000 to 1500 head. They are nearly 100 percent grade sheep. Range production is concentrated in the 12 western states. In the United States, there are five basic classifications of sheep according to wool type. They are fine wool, medium wool, long wool, crossbred wool, and fur sheep.

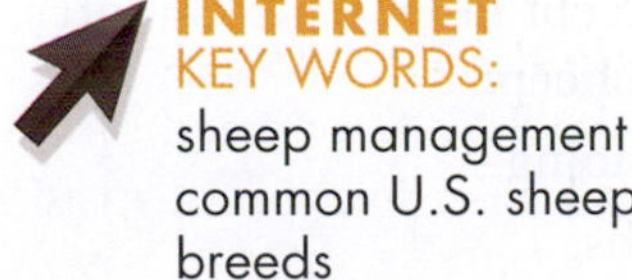
INTERNET KEY WORDS:
sheep management
common U.S. sheep breeds

Fine-wool breeds of sheep produce wool that is fine in texture with a moderate staple length. It has a wavy texture, is very dense, and is used to make fine-quality garments. Fine-wool sheep sometimes produce up to 20 lbs. of wool per sheep per year. Breeds of fine-wool sheep all originated from the Spanish Merino breed. Fine-wool breeds in the United States include the American Merino, Delaine Merino, Debouillet, and Rambouillet (Figure 30-16).

Medium-wool breeds of sheep were developed for meat, and little emphasis was placed on the production of wool. Popular medium-wool breeds include Suffolk, Shropshire, Dorset, Hampshire, and Southdown (Figure 30-17). The long-wool breeds of sheep were developed in England. They tend to be larger than most of the other breeds. Their wool tends to be long and coarse in texture. Long-wool breeds in the United States include Leicester, Lincoln, Romney, and Cotswold.

Crossbred-wool breeds are the result of crossing fine-wool breeds of sheep with long-wool breeds. They were developed to combine good-quality wool with good-quality meat. Because they tend to stay together as a group better than other breeds of sheep, they are said to have a *herding instinct.* They are popular in the western range states. Crossbred-wool breeds include Corriedale, Columbia, Panama, and Targhee (Figure 30-18).

There is only one breed of fur sheep in the United States. The Karakul is grown for the pelts of its lambs. Young lambs are killed shortly after birth, and the pelts are made into expensive Persian lamb coats. The production of wool and meat is of little importance in this breed.

Approved Practices in Sheep Production

Some of the approved practices in the production of sheep and wool include the following:

- Select lambs that are growing rapidly and are large for their age.
- Select purebred stock based on physical appearance, production, and pedigree.
- Select breeding stock with a history of multiple births.
- Provide shelter from severe weather conditions during the lambing season.
- Provide good-quality forages and unlimited fresh, clean water.
- Vaccinate for disease problems of local concern.
- Treat regularly for internal and external parasites.
- Breed ewes to lamb at no more than 2 years of age. A **ewe** is a female animal of the sheep family.
- Use marking harnesses on rams to tell when ewes have been bred. A **ram** is a male animal of the sheep family.
- Use a system of identification to distinguish ewes from one another.
- Do not unnecessarily disturb ewes during lambing. With sheep, lambing means giving birth.
- Shear ewes at least 1 month before lambing except in extreme climates. Shearing is the process of removing wool from sheep.
- Provide a clean, warm, dry environment for lambing.
- Make sure that ewes accept their newborn lambs.
- Dock (remove) lambs' tails at 7 to 10 days of age.

Photo courtesy of Dr. Richard Coffey (University of Kentucky).

A

Courtesy of the American Delaine and Merino Record Association

B

FIGURE 30-16 Fine-wool breeds of sheep include (A) Rambouillet and (B) Delaine Merino.

Courtesy of the American Hampshire Sheep Association.

A

B

FIGURE 30-17 Medium-wool breeds of sheep include (A) Hampshire and (B) Suffolk.

Courtesy of the Columbia Sheep Breeders Association of America.

A

Courtesy of the American Corriedale Association.

B

FIGURE 30-18 Crossbred wool breeds of sheep include (A) Columbia and (B) Corriedale.

- Castrate ram lambs that will be marketed for meat.
- Keep the hooves of farm sheep properly trimmed.
- Maintain a complete flock disease- and parasite-prevention and control program.
- Cull ewes that do not lamb or those that have health problems.
- Maintain a complete and accurate recordkeeping system.

GOATS

Origin and History

The domestication of goats probably took place in Western Asia during the Neolithic Age, between 7000 and 3000 BC. Remains of goats have been found in Swiss lake villages of that period. Mention of the use of mohair from goats is made in the Bible. **Mohair** is hair from Angora goats that is used to make a shiny, heavy, wooly fabric.

Goats were imported to the United States from Switzerland for milk production early in the colonial period. Angora goats, to be used for mohair production, were also imported from Turkey. Nearly 95 percent of the mohair-producing goats in the United States are located in Texas. Milk or dairy-type goats can be found throughout the United States; however, there are concentrations on the East and West coasts.

Economic Importance

Goats are of relatively low economic importance in the United States. Most dairy goats are raised in small numbers by suburbanites and small farmers to produce milk and meat for their own families. There are few large herds of milk goats. Finding a processor to bottle goats' milk is often difficult.

INTERNET KEY WORDS: goat production, management, common U.S. goat breeds

The United States produces nearly 60 percent of the mohair in the world. Texas provides more than 95 percent of the 17 million lbs. produced in the United States annually.

Goats provide little competition for food with cattle and sheep. They prefer to eat twigs and leaves from woody plants rather than grass.

Types and Breeds

Three types of goats are raised in the United States: hair producing, meat producing, and milk producing.

The only hair-producing goat is the Angora (Figure 30-19). It produces 6 to 7 lbs. of mohair per year. Mohair ranges in length from 6 to12 inches. Angora goats are best adapted to a dry climate with moderate temperatures. In addition to producing mohair, Angora goats are used for meat and to help control weeds and brush. The current interest in Boer, Kiko, and Spanish meat goat breeds has resulted in a market for meat-producing goats (Figure 30-20). Goat meat is called **chevon**.

FIGURE 30-19 The Angora goat is the only commercial breed of hair-producing goat in the United States.

Milk-producing dairy goats are found in every state in the United States. Normal production (per goat) averages 3 to 4 quarts of milk per day during a 10-month lactation period. **Lactation** is the period during which a female mammal produces milk. The common breeds of dairy goat are LaMancha, Nubian, Alpine, Saanen, and Toggenburg (Figure 30-21).

A

B

FIGURE 30-20 The (A) Boer and (B) New Zealand Kiko goats are the most widely used commercial breeds of meat-producing goats in the United States.

A

B

FIGURE 30-21 The (A) LaMancha and (B) Nubian goats are widely used commercial breeds of milk-producing goats in the United States.

Approved Practices in Goat Production

Some approved practices in the production of goats and goat products include the following:

- Select goats according to intended use.
- Use physical appearance, pedigree, and records as a basis of selection.
- Purchase replacement animals from reputable breeders.
- Provide adequate feed for hair goats on the range in winter.
- Feed dairy goats supplemental grains based on the amount of milk production.

Courtesy of the American Dairy Goat Association.

FIGURE 30-22 Dairy goats are highly adaptable, thriving in mountain pasture settings to confinement operations.

- Breed goats in the fall to have **kids** (young goats) in the spring.
- Breed does (female goats) for the first time at 10 to 18 months of age.
- Use one male goat (buck or billy) for every 20 to 50 does.
- Shear or clip hair goats twice each year.
- Maintain clean and sanitary conditions for the production of milk.
- Castrate bucks that are not to be used for breeding at an early age.
- Maintain a herd-health program.
- Milk dairy goats twice daily.
- Dehorn dairy-type goats at an early age.
- Maintain complete and accurate records of reproduction, production, and health.

Nearly every facet of life is affected in one way or another by animals and animal products. There are types of livestock production that are adapted to almost every locality and situation. The production of dairy and livestock in the United States is big business. When one considers the value of products and the income from jobs that are created by the animal sector of the United States economy, the total economic impact is approximately $289 billion annually. Although food from animals is more expensive than food from crops, animal products add variety and quality to the human diet. Similarly, animals are sources of high-quality fabrics, leather, and many other products. Therefore, animals and the production of animals will be important enterprises far into the future.

STUDENT ACTIVITIES

1. Write the Terms to Know and their meanings in your notebook.
2. Develop a word search or other puzzle using the Terms to Know. Trade your puzzle with someone else in class and solve his or her puzzle.
3. Make a chart showing when and where the various types of animals were domesticated.
4. Participate in a class discussion on how and why certain animals were domesticated and others were not.
5. Take a class survey of all of the breeds of animals owned by students and their families.
6. Invite a breeder of purebred livestock and a breeder of commercial, or grade, livestock to class to discuss advantages and disadvantages of each type of operation.
7. Conduct a survey of your local school district to determine the types of livestock being raised, the breeds, and the numbers of each.
8. Make a bulletin board showing the various types of livestock in your area and the importance of each.
9. Visit a livestock operation to determine the types of jobs that need to be performed there and what training would be necessary to perform the tasks.
10. Survey various purebred livestock magazines to determine prices of animals being sold at various purebred sales.
11. Make a notebook-cover collage showing as many breeds as possible of the particular type of livestock in which you are interested.
12. Make a bulletin board showing some items made from animal products.

SELF-EVALUATION

A. MULTIPLE CHOICE

1. The number one livestock industry in the United States is __________ production.
   a. dairy
   b. beef
   c. swine
   d. sheep
2. Milk for the purpose of making cheese is produced in __________ operations.
   a. Grade A
   b. Grade B
   c. Class A
   d. Class B
3. Merino sheep belong to the __________ -wool type.
   a. fine
   b. medium
   c. long
   d. crossbred
4. The average Angora goat produces about __________ lbs. of mohair each year.
   a. 2 to 3
   b. 3 to 5
   c. 6 to 7
   d. 10 to 15
5. The most popular breed of dairy cattle in the United States is
   a. Jersey.
   b. Ayrshire.
   c. Brown Swiss.
   d. Holstein.

6. The leading state in swine production is
   a. California.
   b. Texas.
   c. Iowa.
   d. Pennsylvania.
7. Docking refers to
   a. fees charged for selling animals.
   b. removal of tails.
   c. backing a cattle trailer to a loading ramp.
   d. none of the above.
8. Sows normally have __________ litter(s) of piglets per year.
   a. one
   b. two
   c. three
   d. four
9. The average person in the United States eats about __________ lbs. of beef each year.
   a. 12.1
   b. 55.5
   c. 62.5
   d. 76
10. Dairy heifers should be bred to calve when they are __________ year(s) old.
   a. 1
   b. 2
   c. 3
   d. 4

B. MATCHING

__________ 1. Buck
__________ 2. Mutton
__________ 3. Milk
__________ 4. Calve
__________ 5. Veal
__________ 6. Chevon
__________ 7. Farrow
__________ 8. Shepherd
__________ 9. Shear
__________ 10. Butterfat

a. Give birth to cattle
b. Remove wool from sheep
c. Calf meat
d. Male goat
e. One who takes care of sheep
f. Meat from mature sheep
g. Fat in milk
h. Give birth to piglets
i. Liquid from mammary glands
j. Goat meat

C. COMPLETION

1. __________ were domesticated in China.
2. Goats are of two types: hair and __________.
3. __________ is the most common system of swine identification.
4. __________ is rendered pork fat.
5. Milk-producing animals are called __________.
6. The __________ period is the time when cows are not producing milk.
7. Three classifications of beef-cattle breeds are English, exotic, and __________.
8. Consumption of red meat in the United States has been __________ in recent years.
9. __________ feed is feed provided for young animals to supplement milk from their mothers.
10. Breeds of sheep grown especially for meat belong to the __________ wool class.

UNIT 31

Horse Management

OBJECTIVE

To determine the role of horses in our society and to learn how to care for and manage them.

MATERIALS LIST

- bulletin board materials
- horse breed magazines
- examples of horse tack
- Internet access

COMPETENCIES TO BE DEVELOPED

After studying this unit, you should be able to:

- understand the origin and history of the horse.
- determine the economic importance of the horse in the United States.
- recognize the various types and breeds of horse.
- understand approved practices for the care and management of horses.
- understand the basics of English and western riding.
- list rules of safety for handling horses.
- understand the vocabulary associated with horses.

SUGGESTED CLASS ACTIVITIES

1. Attend a horse show or other event that involves horses. Observe the horses and record evidence of training that the horses have received to perform as they are expected. Discuss some of the methods that are used to teach horses to perform on farms and ranches as well as in horse shows.
2. Assign class members to research the history of the horse and to prepare a written paper reporting what they have learned. This assignment could be modified to include the history of a particular breed of horse. If breeds of horses are researched, give the assignment in time for class members to request and receive printed information from the individual horse breed associations.
3. Invite a horse judge to come and talk to the class about the physical characteristics that make a good show animal. You may also want him or her to provide a demonstration on showmanship. Students should be ready to ask questions and participate in the discussion.

TERMS TO KNOW

equine
light horse
draft horse
donkey
horse
pony
hand
color breed
mule
jack
mare
hinny
stallion
jennet
tack
gait
equitation
filly
foal
colt
gelding
farrier
blemish
unsoundness
vice

FIGURE 31-1 Horses are friends, working companions, and pets to youths and adults.

© Atumer/Shutterstock.com.

Horses have been closely associated with humans for much of recorded history. One important way they have served humans is by providing transportation for people and materials. Horses and mules have filled important roles in military operations, making it possible to move soldiers and supplies. For close to 5,000 years, they provided power for farming, and teamsters used them to haul goods and materials, much as our modern trucks move the products of commerce today.

Horses continue to be used extensively on working cattle and sheep ranches, particularly in areas where other forms of transport are difficult, impossible, or illegal (such as roadless areas). In addition, they are the foundation of a significant racehorse industry that generates millions of dollars every year. Horses are used in competitive events such as rodeo, jumping, polo, and competitive riding events. But, most important, most horses are owned by people who take pleasure in training, riding, and caring for them (Figure 31-1). The term **equine** means horse or horse-like. From this term come the names of activities associated with horses, such as equestrian events that feature horses and riders performing skills for which the horses are highly trained to respond to their riders.

ORIGIN AND HISTORY

Fossil remains of the horse family have been found in the United States dating back nearly 58 million years. However, by the time of the discovery of America, no horses remained in the New World. It should be noted that the earliest horses were only about the size of a collie. The descendants of these horses are the horses of today that humans have domesticated. Even though horses became extinct in the New World, it is a common belief that these animals migrated to Europe and Asia across the Bering Strait when Alaska and Siberia were connected.

The horse was probably one of the last animals to be domesticated. This occurred first in Central Asia or Persia about 5,000 years ago. Horses came to Egypt in about 1680 BC from Asia. From there, the Egyptians introduced horses over the known world.

The Arab horse is the ancestor of most modern light breeds of horse. **Light horses** are breeds used for riding. However, horses were not used to any great extent until AD 500 to 600.

Draft horses probably originated from the heavy Flanders horse of Europe. **Draft horses** are used for work. At the time of domestication, they were often used in religious rites for sacrifice or for food.

Donkeys were domesticated in Egypt some time before 3400 BC, when they appeared on slates of the First Dynasty. A **donkey** is a member of the horse family and has long ears and a short, erect mane. Mention of using donkeys appears in many places in the Bible. They were generally used as saddle animals or beasts of burden.

Horses were imported to the New World by Columbus in 1493. When Cortez came to Mexico in 1519, he brought nearly 1,000 horses with him. The death of the explorer Hernando de Soto in the upper Mississippi region led to the abandonment of many of his horses. These horses, coupled with some of those lost or stolen from Spanish missionaries, probably formed the nucleus of the horses used by the American Indians of the Plains. Herds of wild horses roam the desert regions of the American West even today.

SCIENCE PROFILE THE ORIGIN OF THE HORSE

Courtesy of the Illinois Racing News.

The fossil record of the horse is probably the most complete record science has, documenting the physical changes that horses have undergone. Since the time horses appeared on Earth, they have been successful. Because of their survival success, many fossils of animals that appear to be ancestors of the modern horse have been found. The story of the horse begins about 60 million years ago with an animal that has been called both *Hyracotherium* and *Eohippus*.

The first ancestor is a small, dog-sized animal that is thought to be from the horse family because it had teeth like the unique ridges seen in molars of the modern horse. Over time, many different horse ancestors appeared. Probably the most significant change that took place was the adaptation of the foot. Early horses had between three and four toes. As mutations occurred and the most successful species were left to reproduce, gradual changes took place. The modern horse has only one toe with a thick toenail or hoof. Studying the fossil record left behind by the ancient relatives of the horse can provide an interesting glimpse of what our modern horse once might have been.

The introduction of the horse to the American Indians of the Plains completely changed their culture. Horses permitted easier hunting of buffalo in wider areas. This led to competition between various tribes of Plains Indians and a nearly constant state of war in an area where peace had previously prevailed.

The European settlers on the East Coast were far less dependent on the horse than were the explorers. It was of little advantage to have a riding horse with no place to ride. As draft animals, oxen or cattle were more popular because they were stronger than horses, and they produced meat and milk as well as work.

It was only as Americans prospered that horses became an important part of their lifestyles. Horses have had an illustrious past, and a bright future looms ahead.

ECONOMIC IMPORTANCE

Approximately 3.6 million horses and ponies are in the United States today. Many horses are owned by people other than farmers and ranchers. They are used primarily for pleasure, racing, breeding, and companionship.

Horses have a direct impact of approximately $39 billion on the U.S. economy each year. Horse racing is one of the most popular spectator sports in the United States. Individual racing events may offer as much as $1 million or more in purses.

INTERNET KEY WORDS: horse origin, development

Horse shows are also popular events, especially for young people. Recently, one breed association alone sponsored more than 4,000 shows in 1 year.

Horses provide numerous benefits besides economic ones. They help develop a sense of responsibility in young people. They provide physical activity for the young and old alike. They provide opportunities for families to participate in outdoor activities together, and they provide companions when needed (Figure 31-2). Horseback riding is also therapeutic for the healing of certain injuries and disabilities (Figure 31-3). Horses are still essential for managing and caring for cattle and sheep on the vast western rangelands.

AGRI-PROFILE WILD HORSES IN THE UNITED STATES

Courtesy of DeVere Burton.

Wild horse roundups and government wild horse sales make wild horses available to anyone who has the facilities and the interest to own one.

As of 2011, scattered across 43 million acres of wild desert rangelands in the western United States were approximately 33,000 wild horses and 5,500 burros. Most of these animals live on federal lands where the ranges are shared with livestock and other wildlife. The wild horse population is capable of doubling every four years under good conditions. Agencies managing the herds have been hard-pressed to maintain the population at a consistent level that can be supported by the available supply of forage. Legislation mandates that the herds may not be reduced by lethal means; therefore, horses are routinely rounded up, and excess animals are offered to the public for adoption.

It is generally believed that the wild herds are the remnants of horses that escaped from the Spanish conquistadors in the years after America was settled by Europeans. The wild horses are smaller than most domestic light horse breeds. Wild stallions weigh about 1,000 lbs., whereas stallions of domestic breeds generally weigh close to 1,200 lbs. Controversy has surrounded these wild animals in recent times because of the tendency of the horses to overrun the ranges as their numbers expand. Adoptions are useful in controlling the expansion of wild herds, but they are also expensive. Adopted horses are usually penned for several months, requiring feed, immunizations, and transportation.

TYPES AND BREEDS

For the purpose of this discussion, members of the equine family are divided into horses, ponies, donkeys, and mules. Horses are further divided into light horses, coach horses, and draft horses. Coach horses were developed to pull heavy coaches and freight wagons.

Members of the horse family that are 14.2 hands or taller are called **horses**. Animals of the same family that are less than 14.2 hands tall are **ponies**. A **hand** measures 4 inches and is used as a unit of measure for members of the horse family to determine the size of horses, mules, donkeys, and ponies.

© tepic/Shutterstock.com.

FIGURE 31-2 Horses provide opportunities for families to participate in outdoor activities together, and they provide companionship.

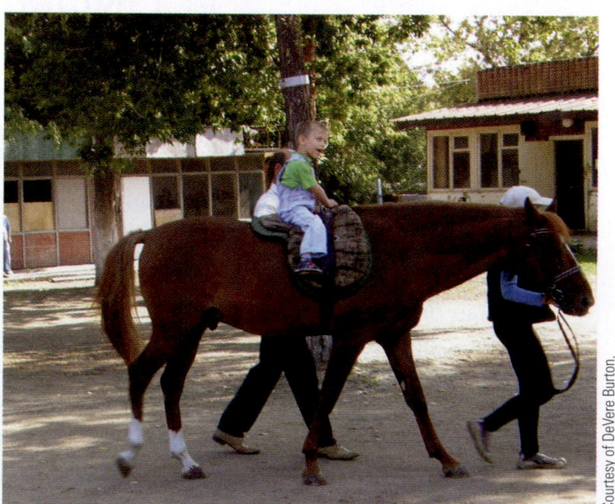
Courtesy of DeVere Burton.

FIGURE 31-3 Horseback riding is good therapy for some people and promotes healing of some injuries and medical conditions.

The light breeds of horses comprise by far the most popular horses in the United States. Light breeds are further divided into true breeds, or purebreds, and color breeds. True breeds have registration papers and parents of the same breed. **Color breeds** need only be specific colors or color patterns, although other characteristics may be considered for registration. They need not have purebred parents.

Examples of the nearly 50 true breeds of horse in the United States include Arabian, Morgan, Thoroughbred, Quarter Horse, Standardbred, Tennessee Walking Horse, and American Saddle Horse (Figure 31-4). The most popular of the light breeds of horse in the United States is the Quarter Horse. Quarter Horses are used for riding, hunting, racing, sports, and ranch work with cattle and sheep.

AGRI-PROFILE

CAREER AREA: HORSE BREEDER/TRAINER/RANCHER/MANAGER/JOCKEY/FARRIER

Courtesy of Barbara Lee Jensen, After Hours Farm.

Maintaining working horses in peak condition is vital to the daily operations of many ranch and recreation enterprises.

Careers in horse management are available in a variety of settings. Urban police forces and park services have rediscovered the advantages of patrolling on horseback, and horse-drawn carriages have captured the fancy of sightseers in many areas. In addition, many families have adopted the horse for teenage animal-production projects in semirural, as well as rural, settings. These frequently involve the entire family in fairs, shows, or rodeos. Horses are used extensively for herding and handling cattle on ranches and in cattle feedlot operations. Pack mules and riding horses are used extensively for wilderness travel, hunting, and other recreational pursuits. The horse-racing industry is huge in the United States, involving many people and extensive career possibilities.

Although many people raise horses as an avocation, rich and varied career opportunities are available in owning, managing, or caring for horses. Veterinarians, trainers, managers, attendants, jockeys, and farriers are common around racetracks and breeding farms. Tack, feed, and equipment supply centers provide business opportunities for those wishing to interface with the horse industry but preferring not to handle horses directly.

Horse health, soundness, breeding, behavior, and performance command the attention of those in research, veterinarian services, and education. Horse-related jobs may be indoors or outdoors and run the spectrum from laborer to scientist.

INTERNET KEY WORDS: horse breeds

Among the first things to learn about horses are the names of their body parts (Figure 31-5). Differences in body parts define the different breeds. Knowing these terms also helps a person understand what *horse people* are talking about when they discuss the merits of a particular horse.

Some of the color breeds of horses include the Buckskin, Palomino, Appaloosa, and Pinto or Paint. Color breeds may or may not breed true to color. The uses of the color breeds are the same as those of the true breeds of light horse.

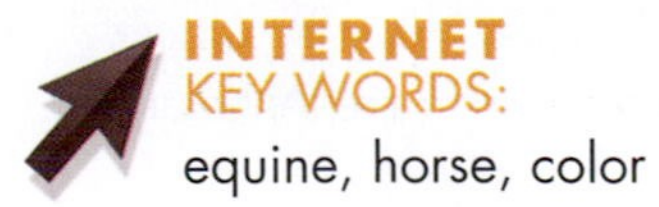

INTERNET KEY WORDS: equine, horse, color

When coaches disappeared from the U.S. scene, coach horses became rare in the United States because there was little need for them. While they were used, though, they combined the qualities of both the light horse and draft breeds. They were fast (like the light horse) and strong enough (like a draft horse) to pull the heavy stagecoaches of their time. The only example of the old coach-style horse in the United States today is the Cleveland Bay.

FIGURE 31-4 The most popular breeds of light horses in the United States, based on number registered, are (A) Quarter Horse, (B) Thoroughbred, (C) Arabian, (D) Standardbred, (E) Paint Horse, and (F) Appaloosa.

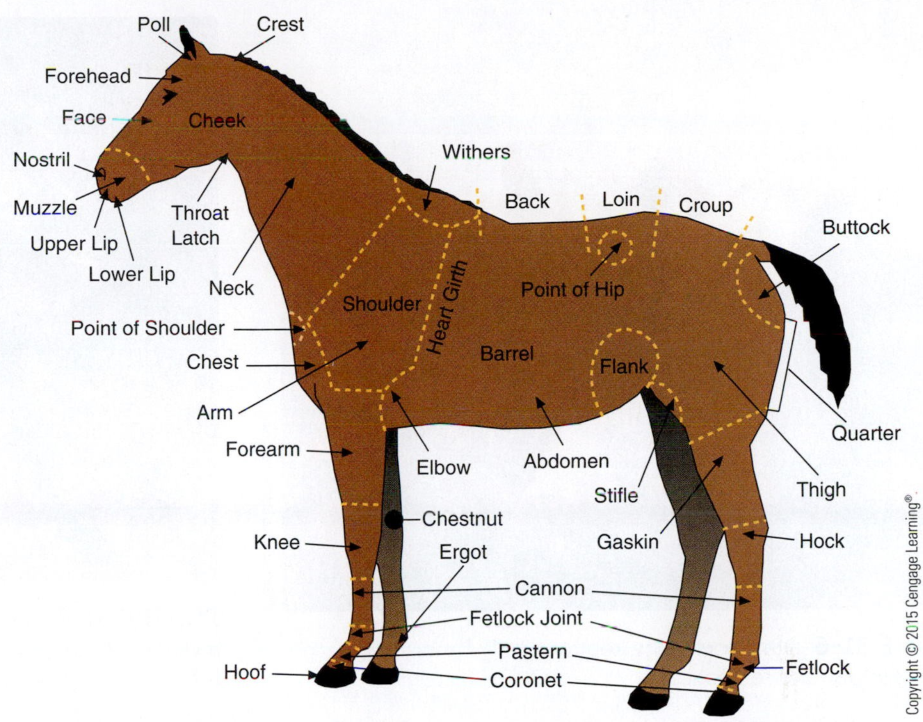

FIGURE 31-5 Learning the proper names of the parts of a horse aids in discussing, evaluating, and caring for horses.

INTERNET KEY WORDS: mule, hinny

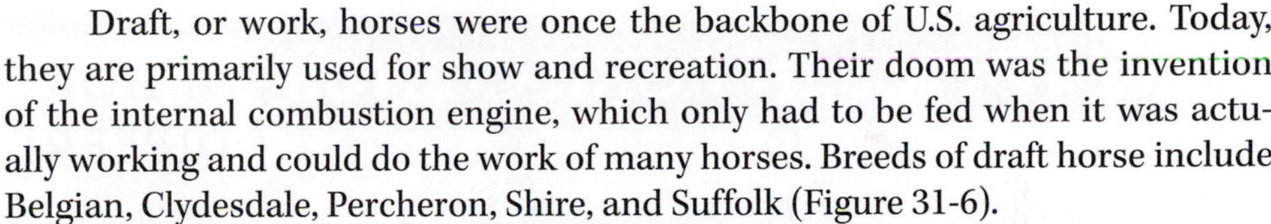

Draft, or work, horses were once the backbone of U.S. agriculture. Today, they are primarily used for show and recreation. Their doom was the invention of the internal combustion engine, which only had to be fed when it was actually working and could do the work of many horses. Breeds of draft horse include Belgian, Clydesdale, Percheron, Shire, and Suffolk (Figure 31-6).

Ponies are smaller versions of the horse. They are used for riding and driving and as pets. Some breeds were originally developed to work in coal mines and in other pursuits, where small size is important. The common breeds of pony include Shetland, Welsh, Gotland, Appaloosa, Connemara, and Pony of the Americas. The Shetland is the most popular breed of pony in the United States (Figure 31-7).

Donkeys are used in the United States for work and as pack animals. Miniature donkeys also make excellent pets. The male donkey, called a jack, is used as the male parent of mules. A **mule** is a cross between a jack and a mare. A **jack** is a male donkey or mule. **Mares** are mature female horses or ponies. The opposite cross, a stallion with a jennet, results in a **hinny**. A **stallion** is an adult male horse or pony. A **jennet** is a female donkey or mule (Figure 31-8).

Mules are thought to be more intelligent than horses and can do more work than comparable-sized horses. There are several types and sizes of mule, depending on the breed or type of horse that serves as the female parent. Mules may be used for work, sports, or pleasure.

RIDING HORSES

The styles of riding horses can be divided into two general classifications: English and western. The style of riding and the **tack** (horse equipment) vary greatly between the two classifications. Even the **gaits** (the ways an animal moves) of the horses involved have different names.

© Becky Swora/Shutterstock.com.

FIGURE 31-6 Draft horses are kept primarily for historic interest, show, and recreation.

Courtesy of Multi-World Champion Shetland Harness Pony.

FIGURE 31-7 The Shetland pony can be trained for a variety of uses, ranging from riding to show pony. It is the most popular breed of pony in the United States.

HOT TOPICS IN AGRISCIENCE

"A HORSE OF A DIFFERENT COLOR"

© Alexia Khruscheva/Shutterstock.com.

The dominant gene for color in horses results in bay coloring or one of its shades.

The color of a horse is determined by the combination of several genes and the presence or absence of two important pigments in the hair and skin. Two primary pigments are found in the coats of mammals. These pigments are eumelanin, a black or brown pigment, and pheomelanin, which is red. All of the horse colors are combinations of these pigments. In many instances, these pigments are diluted by certain gene combinations, resulting in different degrees or shades of color. Spots of white indicate the absence of the pigments. Genes control the distribution of the pigments in the skin and hair. This explains the white and colored spots of the Appaloosa and Paint breeds of horses.

Overriding all other genes for color is the dominant gene A, which results in the coloring of bay or one of its shades. Every time the A gene is present, the horse will be bay in color. Another interesting color combination is "roan." In this instance, hair of different colors is mixed together. A "sooty" or "smutty" color occurs when black is mixed with hair of other colors.

Many different colors are possible in horses. However, once the offspring of a stallion has been observed for color, it is possible to work out the genetics that are involved. When the color genetics of both the stallion and the mare are known, the possible color combinations in the foal can be predicted with reasonable accuracy. However, it is beyond the scope of this book to expand on the specific genetic combinations and the colors they produce.

SCIENCE CONNECTION AN INTERNATIONAL SOLUTION

Most U.S. horses are vulnerable to equine babesiosis, a disease that is often contracted from horses entering the country from other parts of the world. Imported horses must be held in quarantine until it has been demonstrated that they are free of disease.

U.S.veterinarians and other animal health officials are constantly on the alert for a lurking menace that could threaten the horse industry if allowed to get out of hand. A malaria-like disease of horses—equine babesiosis—infects the horses of the world in varying degrees. Horses of Brazil, Argentina, Russia, and Poland have lived with the parasite so consistently that they have developed a level of immunity that protects them from serious consequences of the disease.

Most U.S. horses have not been exposed to the parasite; therefore, they have developed little or no resistance to it. Such horses are vulnerable, and a bout with equine babesiosis can be fatal. Babesiosis resembles malaria in humans because, in both diseases, insects transmit the infectious agent, which then attacks and destroys red blood cells. Sick animals become feverish and lethargic and refuse to eat. Microscopic, single-celled parasites, called *Babesia equii* or *Babesia caballi*, enter the horse through infected blood-sucking ticks. The parasites then multiply in the bloodstream and form tiny pear-shaped or ring-shaped bodies called merozoites. The merozoites then invade the red blood cells. The horse's immune system reacts by forming antibodies in an attempt to fend off the invaders.

The United States tries to protect its horse population by keeping the parasite out of the country. However, horses coming into the United States from any other part of the world may be carriers of the parasite, even though they themselves appear healthy. Several field tests have been developed to detect the presence of this parasitic disease. Each test involves taking a blood sample from a horse and testing it for specific antibodies that are produced when a horse becomes infected. It is now possible to perform a single test for both *B. equi* and *B. caballi*.

The National Veterinary Services Laboratory in Ames, Iowa, screens all blood samples from animals attempting entry to the United States. However, it is expensive to ship a horse to an international border only to be required to ship it back home because of rejection for health reasons. As diagnostic tests become more economical and reliable, courtesy tests can be run on blood samples taken before the animal is shipped. If no problems are indicated, the animal can be shipped, and then retested at the point of entry. This procedure is expected to make it easier to address the old international dilemma of how to move horses across borders economically without the danger of spreading the dreaded equine babesiosis.

FIGURE 31-8 The mule is a hybrid cross between a jack donkey and a horse mare.

English Equitation

Equitation is the art of riding on horseback. A saddle is the padded leather seat for the rider of a horse. Stirrups are foot-supports hung from each side of the saddle. In English equitation, the saddle is smaller and lighter and has shorter stirrups than the western saddle (Figure 31-9). English riding clothing includes closefitting breeches, jodhpurs, jacket, and hard hat or derby (Figure 31-10).

The gaits of the English-equitation horse are walk, trot, canter, gallop, rack, and pace. The walk is a slow, four-beat gait. The trot is a fast, two-beat diagonal gait. The canter is a slow, three-beat gait, whereas the gallop is a fast, three-beat gait in which all four feet come off the ground at the same time. The rack is a fast, four-beat gait, and the pace is a side-to-side, two-beat gait (Figure 31-11).

INTERNET KEY WORDS:
Western horsemanship
horse safety

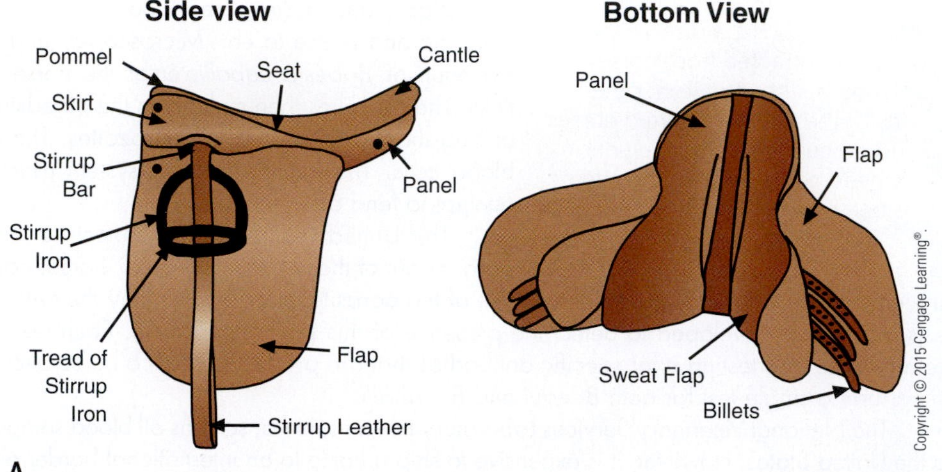

B

FIGURE 31-9 The English saddle is lighter and is structured differently from the western saddle.

A

B

C

D

FIGURE 31-10 Proper riding attire for riders includes (A) western, (B) gaited, (C) dressage, and (D) jumping.

The English-equitation horse is controlled by reins that may be held either in one or both hands. Reins are leather or rope lines attached to the bit. The bit is a metal mouthpiece used to control the horse.

Western Horsemanship

Horsemanship is the art of riding and knowing the needs of the horse. Western horsemanship differs from English equitation in many ways. The western saddle is heavy and has a horn that is used to secure the rope when roping livestock. This saddle has longer stirrups than the English saddle and may be plain or have fancy

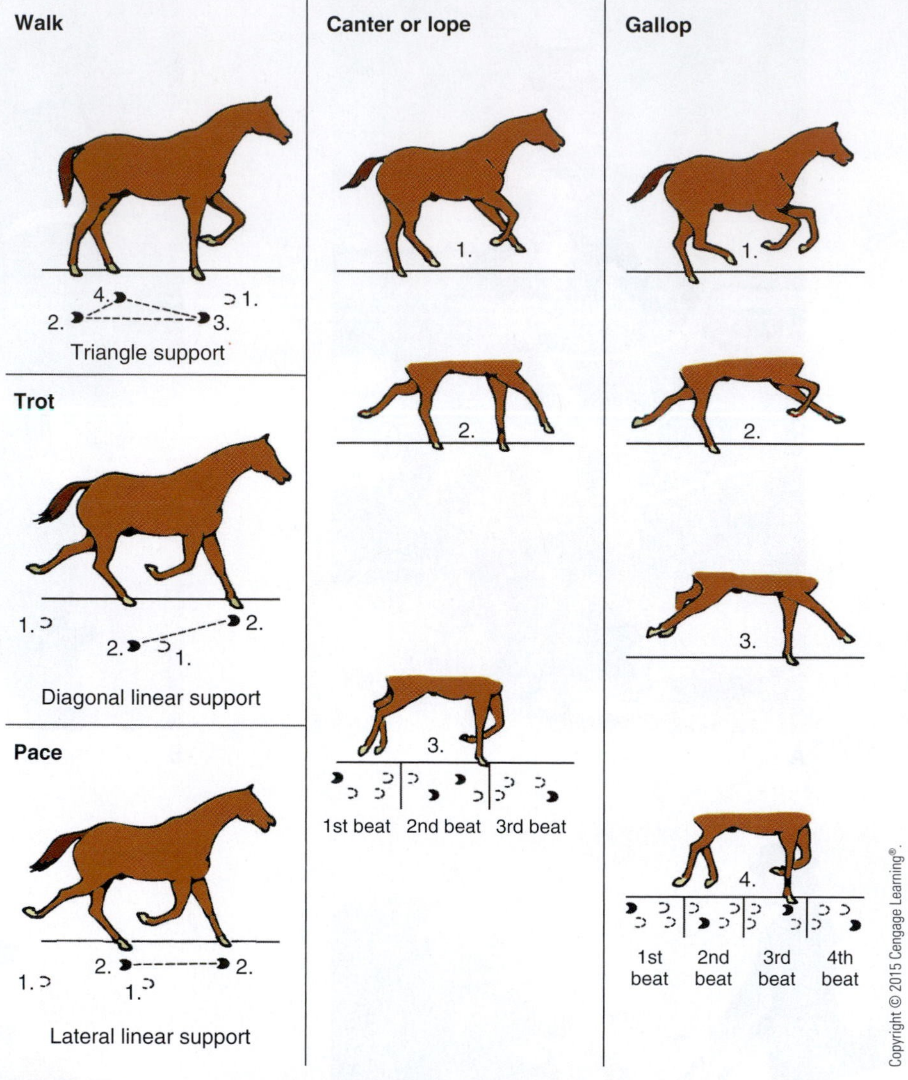

FIGURE 31-11 Basic gaits of English equitation include walk, trot, pace, canter, and gallop.

ornamentation. The western saddle was designed for the comfort of cowboys, who spend much of their time in a saddle (Figure 31-12).

Western riding attire was also designed for comfort. It consists of jeans and a western shirt and hat. Cowboy boots and chaps complete the typical western outfit. The gaits of the western horse are walk, jog, lope, and gallop. The jog is a slow, smooth, two-beat diagonal gait. The lope is a slow canter. The western rider controls the horse with the reins held in one hand (Figure 31-13). The hand holding the reins cannot be changed during competition riding.

Working horses are still found on farms and ranches (Figure 31-14). They are particularly useful in working with livestock that graze in areas that do not have road access. Range riders depend on their horses to get them into the canyons and sometimes-difficult terrain where cattle and sheep graze. Horses are also used extensively to check cattle in large feedlots where it is necessary to separate individual animals from the herd for treatment of illnesses or injuries. A well-trained horse is a great asset for this task.

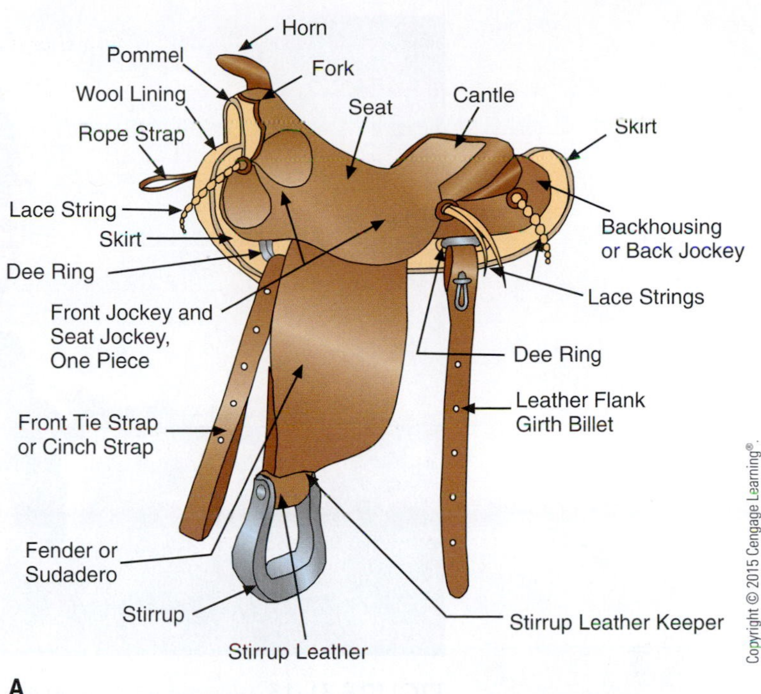

FIGURE 31-12 Western saddles were originally developed for cowboys who worked with cattle. However, present-day designs of these saddles vary from basic and functional to highly ornamental.

HORSE SAFETY RULES

Some safety rules to observe when riding and caring for members of the horse family include the following:

- Always approach a horse from the front-left side.
- Never do anything to startle or scare a horse. It may kick or rear up.
- Be sure that the horse always knows exactly what you intend to do to or with it.
- Pet the horse by putting your hands on its shoulder, not on its nose.
- Tie horses that are strangers to each other far enough apart that they cannot fight.
- Walk beside the horse when leading it, not in front of or behind it.

FIGURE 31-13 Western riding requires control of the horse with one hand on the reins and the other hand free to handle a rope or open a gate.

FIGURE 31-14 Horse and rider must coordinate their thinking and their movements to control herds of livestock. Training of both horse and rider is the key to success.

- Never wrap the lead rope around your hand when leading a horse.
- Use care in adjusting the saddle on the horse. Be sure that it is tight enough so that it will not slip or slide.
- Always mount the horse from the left side (Figure 31-15).
- Always keep the horse under control.
- Do not allow the horse to misbehave without disciplining it.

SCIENCE PROFILE TAXONOMY OF HORSES

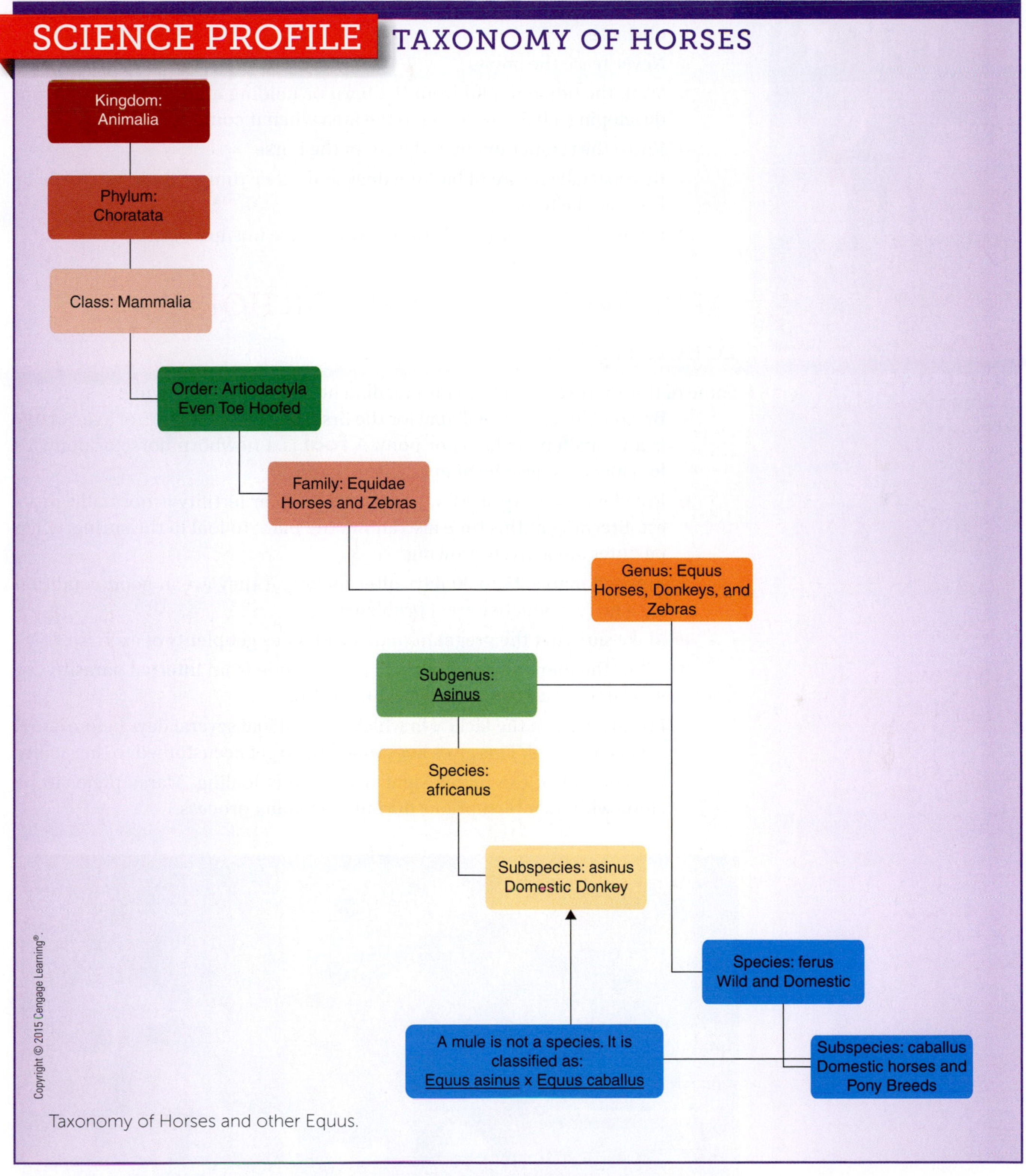

Taxonomy of Horses and other Equus.

- Walk the horse up and down steep slopes, on rough ground, and across paved roads.
- Reduce speed when riding on rough terrain or in wooded areas.
- In groups, ride in single file and on the right side of the road.
- Be calm and gentle when dealing with your horse.

© DDCora/Shutterstock.com.

FIGURE 31-15 Always mount a horse from the left side.

- Always wear appropriate riding attire, including hard hats when jumping obstacles.
- Never tease the horse.
- Walk the horse to and from the barn or holding area to prevent it from developing a habit of riding to the area when it comes into sight.
- Know the temperament and vices of the horse.
- Be especially aware of barking dogs and other things that may startle or frighten the horse.
- Do not allow other people to ride your horse unsupervised.

APPROVED PRACTICES FOR HORSES

Breeding Horses

Some of the approved practices for breeding horses are the following:

- Breed fillies so they will foal for the first time at 3 to 4 years of age. A **filly** is a young female horse or pony. A **foal** is a newborn horse or pony. To foal means to give birth in the horse family.
- Breed mares in April, May, or June, when their fertility is normally greatest. Breeding at this time also allows the mare to foal in the spring, when pastures are actively growing.
- Rebreed mares 25 to 30 days after foaling, if they are in good condition with no reproductive-tract problems.
- Make sure that the pregnant mares and fillies get plenty of exercise.
- Allow the mare to foal in a clean pasture, free from internal parasites, or in a large, clean pen or stall (Figure 31-16).
- Put the mare in the facility in which she will foal several days before she is expected to foal. This gives her some time to get accustomed to the facility.
- Remain out of the mare's sight when she is foaling. Mares prefer to be alone with no interruptions during the foaling process.

© Karel Gallas/Shutterstock.com.

FIGURE 31-16 Mares and foals should be kept in clean, parasite-free surroundings.

- Contract the services of a veterinarian if the mare shows signs of difficulty in foaling or if the position of the foal is abnormal for delivery.
- At birth, the following actions should be performed:
  1. Make sure that the foal is breathing. Tickling the foal's nose with a piece of straw often will stimulate it to start breathing after it is born. More drastic measures, such as artificial respiration, may be needed in serious situations.
  2. Remove mucus from the nose of the foal, and dry the foal off.
  3. Dip the end of the navel cord in iodine after tying it off. This prevents infection from entering the foal through the open navel cord.
  4. Make sure that the foal nurses as soon as it can stand. It should be allowed to stand on its own. Normally, nursing first takes place within one-half hour after birth.
  5. Check to see that the foal has a bowel movement within the first 12 hours of birth. The foal may need an enema if the bowel movement does not occur naturally.
  6. If the foal shows signs of diarrhea, reduce the amount of milk that it is allowed to drink.
- Provide creep feed for the nursing foal when it reaches 10 days to 2 weeks of age.
- Begin training the foal as soon as possible. A foal that is trained from an early age seldom needs to be broken.
- Do not mistreat the mare or foal at any time.
- Wean the foal at 4 to 6 months of age. The foal and the mare should not be allowed to see each other for several weeks after weaning to break the bond between mother and offspring.

equine, horse management

- Castrate colts that are not to be used for breeding purposes. A **colt** is a young male horse or pony. A castrated male horse is called a **gelding**. Geldings are usually much more docile and easier to handle than stallions.

Care and Management of All Horses

Approved practices for the care and management of all horses and ponies include the following:

- Groom the horse daily or weekly at a minimum (Figure 31-17).
- Always use soft brushes when grooming a horse. Its skin is sensitive and easily damaged by rough treatment.
- Shorten the mane and tail when needed by pulling the hairs from the underside.
- Be especially careful not to get water in the horse's ears when washing. Be sure to dry off the horse quickly so that it does not catch a cold in cool weather.
- Inspect and clean the horse's hooves daily and always before and after riding.
- Trim the horse's hooves every 4 to 6 weeks if the horse is not shod. Shod means wearing shoes.
- Replace the shoes of a shod horse every 4 to 6 weeks. A **farrier** (a person who shoes horses and cares for their feet) should be engaged to perform this task.
- Shoe a horse for the first time when it is 2 years old or when it starts being worked.

A

B

FIGURE 31-17 (A) Grooming a horse helps it develop trust for people. It is also important to keep the horse clean. While grooming, take that opportunity to make a complete inspection of the horse. (B) Frequent inspection and care of a horse's feet are important because a lame horse has no value. It is much easier to prevent lameness that it is to cure it.

- Cool down the horse thoroughly after every exercise period and after riding. This should be done before the horse is allowed to drink large quantities of water.
- Do not overfeed a horse. The total daily consumption of concentrates and roughages should not total more than 2 to 2.5 percent of the horse's weight.
- Feed a horse at a regular time each day. The number of times that the horse is fed per day makes little difference, as long as it is the same every day.
- Do not abruptly change the ration. The stomach of the horse is temperamental and slowly adjusts to changes in feeding practices.
- Never feed moldy feed to a horse.
- Provide unlimited access to fresh, clean water at all times.
- Maintain a strict health-care program for all horses.
- Maintain clean and sanitary facilities at all times.
- When buying a horse, be sure to do the following:
  1. Note blemishes and unsoundness characteristics of the horse. A **blemish** is an abnormality that does not affect the use of the horse. A condition of **unsoundness**, such as chronic lameness, is an abnormality that does affect the use of the horse.
  2. Check the age of the horse by examining its teeth (Figure 31-18). Examining the teeth can also indicate how long a horse might be useful.
  3. Determine that the horse is not blind or having other problems that may affect its value and use.
  4. Insist on evidence of good health.
  5. Try to determine if the horse has vices. A **vice** is a bad habit.
  6. Try to determine the personality and spirit of the horse.
  7. Consider the price of the horse and whether you can afford the cost of owning one.
  8. Check the pedigree of the horse, if it is a purebred.

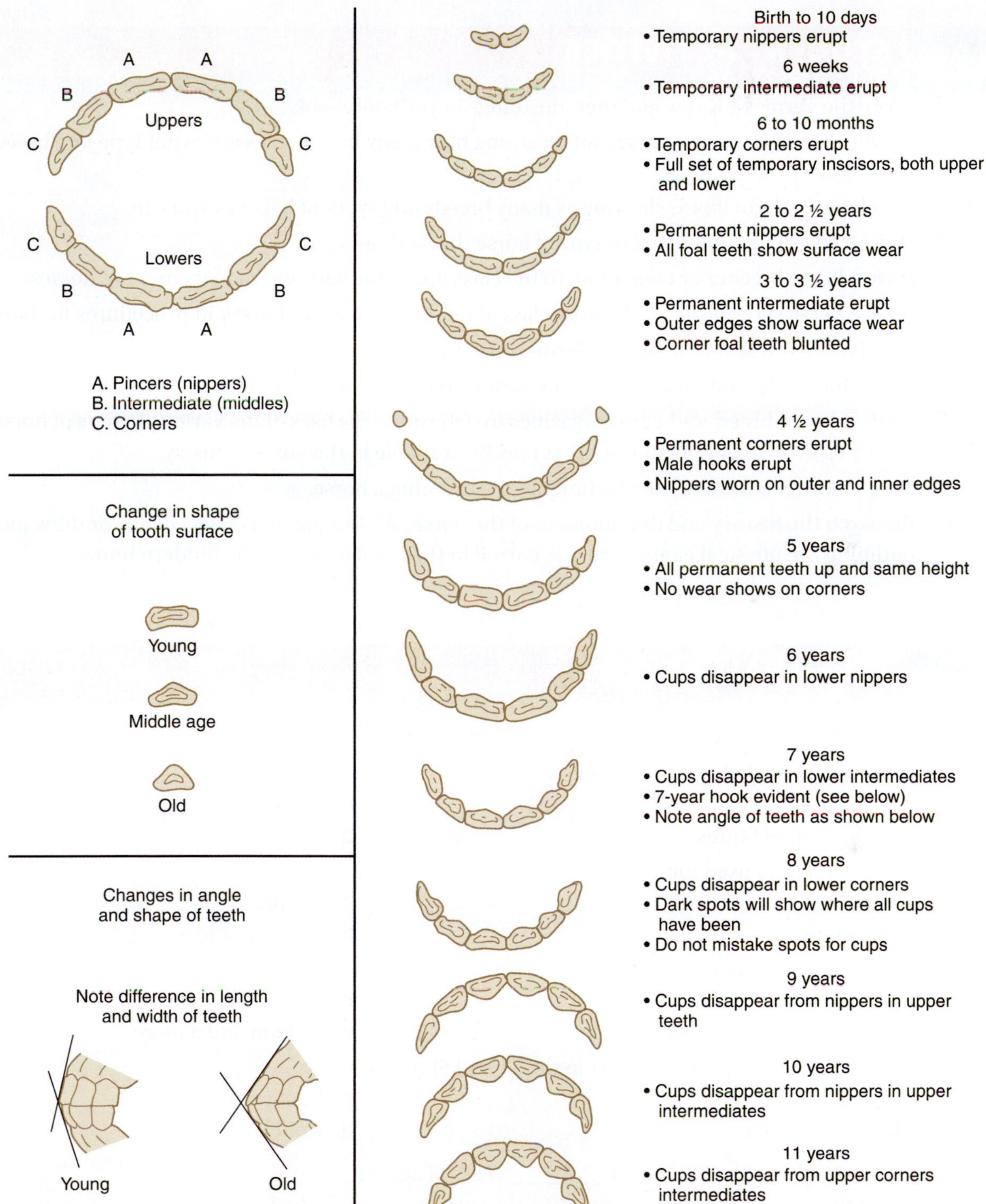

FIGURE 31-18 The age of a horse can be determined by the number and shape of its teeth.

Owning and caring for horses and ponies is an excellent way to get outdoor exercise. Horses have given many young people companionship during the trials of youth. Competitors find horses and horse events to be excellent outlets for their excess energy. Cattle producers find horses to be an essential part of a team when working with cattle. The horse was, is, and will continue to be an important part of the world of many people in the United States.

STUDENT ACTIVITIES

1. Write the Terms to Know and their meanings in your notebook.
2. Survey the students in your school regarding how many have horses and what type and breed of horse they own.
3. Develop a bulletin board showing as many breeds and types of horse as you can.
4. Write a report about a breed or type of horse that interests you.
5. Have a horse breeder or owner talk to the class about the care and management of horses.
6. Have a horse veterinarian talk to the class about health-care and first-aid procedures for horses.
7. Develop a word search using the Terms to Know.
8. Visit a tack shop and make a list of the various types of horse equipment sold there.
9. Look at horse-breed-and-care magazines to determine the uses of the various breeds of horse. Also look for the types of job opportunities that may be available in the horse industry.
10. Demonstrate to the class the techniques of grooming a horse.
11. Research the history and development of the horse. Write a paper, create a chart, or draw pictures outlining the physical changes that occurred in the fossil record of the modern horse.

SELF-EVALUATION

A. MULTIPLE CHOICE

1. Horses were probably domesticated in
   a. Russia.
   b. the United States.
   c. Persia.
   d. China.
2. Light horses are used for
   a. riding.
   b. rodeos.
   c. racing.
   d. all of the above.
3. A mule is a cross between a
   a. jack and a mare.
   b. stallion and a jennet.
   c. jack and a jennet.
   d. stallion and a mare.
4. The number-one spectator sport in the United States is
   a. rodeos.
   b. horse racing.
   c. horse shows.
   d. riding.
5. Fillies should be bred to foal at ________ years of age.
   a. 1 to 2
   b. 2 to 3
   c. 3 to 4
   d. 4 to 5
6. Always approach a horse from the
   a. rear.
   b. right side.
   c. left side.
   d. none of the above.
7. Horses should be re-shod every ________ weeks.
   a. 2 to 4
   b. 4 to 6
   c. 6 to 8
   d. 8 to 10

8. The most popular breed of light horse in the United States is the
   a. Thoroughbred.
   b. Appaloosa.
   c. Morgan.
   d. Quarter Horse.
9. A fast, three-beat gait is the
   a. gallop.
   b. canter.
   c. pace.
   d. trot.
10. There are approximately ________ horses in the United States.
   a. 100,000
   b. 1 million
   c. 3.6 million
   d. 100 million

B. MATCHING

__________	1. Stallion	a. Way of moving
__________	2. Jennet	b. Young female horse
__________	3. Gelding	c. Female donkey
__________	4. Mare	d. Newborn horse
__________	5. Gait	e. Adult male horse
__________	6. Filly	f. Adult female pony
__________	7. Tack	g. Male donkey
__________	8. Jack	h. Horse equipment
__________	9. Foal	i. Young male horse
__________	10. Colt	j. Castrated male horse

C. COMPLETION

1. A __________ is a person who shoes horses.
2. Horses originated in what is now the __________.
3. __________horses were developed for work.
4. __________breeds of horse need only to be certain colors or color patterns to be registered.
5. The art of riding on horseback is called __________ .
6. An abnormality that affects the use of a horse is called a(an) __________ .
7. Bad habits in horses are called __________ .
8. The result of the cross between a stallion and a jennet is a __________ .
9. The most popular breed of pony is the __________.
10. The unit of measurement for determining the size of a horse is the __________.

SECTION 9

FOOD SCIENCE AND TECHNOLOGY

Protein-rich, edible food coatings, use of bacterial endospores to take up oxygen from within food-grade bottles, biodegradable food packaging, and even in vitro meat production are topics of research now and a wave of the future. Soybean film types promise a variety of new products. These films may have uses ranging from films on citrus and other fruits to keep them fresh to edible fast-food wrappers. Plastics from soybeans have been around since the 1940s, but plastic-like digestible and biodegradable films are new and still developing.

The process involves the separation and isolation of proteins of the soybean by freeze-drying protein to remove the water and grinding the protein into fine powder. One product is soy concentrate containing 70 percent protein; another is soy isolate containing 90 percent protein. The proteins can then be mixed with various ingredients and additives before being cast into films for coatings of food products. By using enzymes and other treatments, the protein can be modified for films and coatings with specific uses. Researchers have developed films and coatings from corn and wheat starches, but soy-protein films promise some additional adaptability.

Proteins are compatible with oils, and this quality is important in processing products with moisture-resistant characteristics. Soy protein can be mixed with starch and gums to form films that keep moisture in but keep oxygen out. This creates an ideal barrier for maintaining food quality. By adjusting other factors, films can be tailored for effective food packaging and preservation. For instance, fat-containing foods, such as meats, can develop off-flavors after being cooked. This creates problems in institutional cooking, such as in restaurants and school lunch programs. Coating such foods with the appropriate protein film could solve the problem of off-flavors. The use of protein and other edible films must be approved by the Food and Drug Administration before being put into commercial practice. The development and use of "agriplastics" will surely continue because they present the advantages over petroleum-based films of being edible, nutritionally valuable, and environmentally safe.

Wave of the Future

Courtesy of USDA/ARS #K-4472-8.

A

Courtesy of USDA/ARS #K-3517-5.

B

Courtesy of USDA/ARS #K-3183-10.

C

(A) Chemist Fred Shih analyzes the permeability of various soybean film types, (B) fruits are treated with a protective film, and (C) entomologist Jennifer Sharp evaluates the effectiveness of heat treatment of shrink-wrapped grapefruit to prevent insect damage.

UNIT 32

The Food Industry

OBJECTIVE

To explore elements, trends, and career opportunities in the food industry.

MATERIALS LIST

- bulletin board materials
- State Department of Agriculture reports on commodities grown and foods processed in particular states
- Internet access

COMPETENCIES TO BE DEVELOPED

After studying this unit, you should be able to:

- explain what is meant by the term *food industry.*
- determine the importance of the food industry to the consumer.
- describe the economic scope of the food industry.
- identify government requirements and other assurances of food quality and sanitation.
- compare the major crop and animal commodity production areas in this nation and in the world.
- discuss the major food commodity groups and their predominant origins.
- explain the major operations that occur in the food industry.
- describe career opportunities in food science.
- discuss future developments predicted for the food industry.

SUGGESTED CLASS ACTIVITIES

1. Invite a representative of a food-processing plant or business in your area to speak to the class about the food-processing industry. Key topics for discussion should include food safety laws and regulations, food science careers, quality-control methods, sources of raw products, transportation, and marketing.
2. Create a map that shows the major food production areas in your state or region. Obtain statistics from your state department of agriculture or the Economic Research Service of the U.S. Department of Agriculture that indicate the ranking of your state or region in the production of the foods that you identified. Create a chart with the state rankings for each of these food items.

TERMS TO KNOW

food industry
retailer
wholesaler
distributor
processor
grader
packer
trucker
harvester
producer
grade
climatic condition
harvesting
maturity
underripe
overripe
microorganisms
processing
bran
endosperm
germ
edible

3. Follow one locally grown product on its trip from the field to the shelf. For this task, you will need to contact producers and harvesters, processing plants, distributors, wholesalers, and grocery store retailers. When interviewing individuals, plan the questions that you will ask before starting the interview. End the project with a class discussion and project summary.

Food is all around us—in the school cafeteria, on the dinner table at home, at the fast-food chains that dot our nation's cities, towns, and highways, and in our nation's supermarkets. Less visible are the gourmet and specialty food stores and restaurants that satisfy special dietary needs and tastes. *Gourmet* means sensitive and discriminating taste in food preferences. Learning about other cultures is frequently accomplished by tasting their foods and learning how those foods are prepared.

The **food industry** is involved in the production, processing, storage, preparation, and distribution of food for consumption by living things. Pet and animal food as well as human food require a chain of people, places, equipment, regulations, and resources to change farm products into edible foods (Figure 32-1).

This unit explores the many operations in the food-processing industry. Careers are plentiful in this industry. As the different operations are explored, remember that new people are needed to maintain and expand the vital functions that keep the abundant food before us.

ECONOMIC SCOPE OF THE FOOD INDUSTRY

When you purchase groceries in the food store or a hamburger at the local fast-food restaurant, does most of your food dollar go directly to the farmer who grew the beef? How about the lettuce, tomato, pickle, bun, and sesame seeds that adorn your hamburger? Many businesses and individuals join the farmer in dividing your food dollar. The economic chain reaction that begins with your food purchase sends signals to the retailer, wholesaler, distributor, processor, grader,

FIGURE 32-1 Most modern food products are prepared for the consumer, as seen in ready-to-heat-and-eat grocery items.

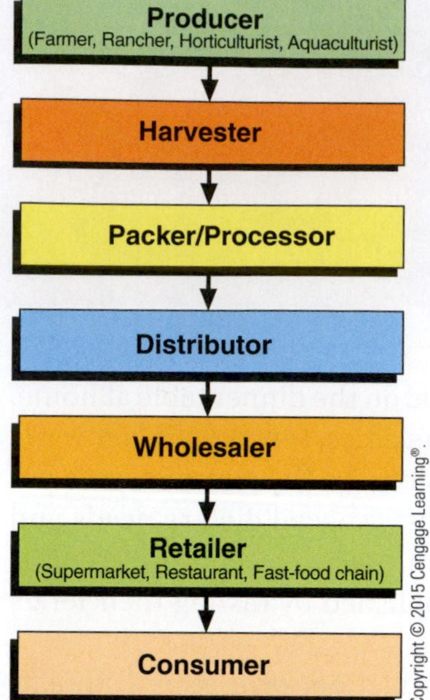

FIGURE 32-2 Producers, such as farmers, ranchers, and aquaculturists who actually produce food, are joined by many others before most food items reach the consumer.

packer, trucker, harvester, producer, and others to replace that food for your next purchase (Figure 32-2).

A **retailer** is the person or store that sells directly to the consumer. The retailer is the end of the marketing chain, whereas a **wholesaler** is a person who sells to the retailer, having purchased fresh or processed food in large quantities. A **distributor** stores the food until a request is received to transport the food to a regional market. A **processor** is anyone involved in cleaning, separating, handling, and preparing a food product before it is ready to be sold to the distributor. A **grader** is the person who inspects the food for freshness, size, and quality and determines under what criteria it will be sold and consumed. A **packer** is the person or firm responsible for putting the food into containers, such as boxes, crates, bags, or bins, for shipment to the processing plant. A **trucker** is the person responsible for transporting the product anywhere along the way from farm to consumer. A **harvester** is the person who removes the edible portions from plants in the field. Finally, the **producer** grows the crop and determines its readiness for harvest. It seems like everyone gets a part of your food dollar (Figure 32-3)! Moreover, the dollars spent on food and fiber in the United States provide jobs for approximately 20 percent of the country's working population (Figure 32-4).

Where you spend your food dollar also has an influence on who gets how much of your dollar. A meal purchased in a restaurant costs considerably more than a meal prepared from raw food products at home. In what ways have our change in lifestyles and the shift to families with two or more people employed outside the home influenced how and what we eat? Many more meals are eaten outside the home than was the case a generation ago. Convenience foods for use at home are more in demand today. In the United States in 2010, the expense for food eaten away from home was

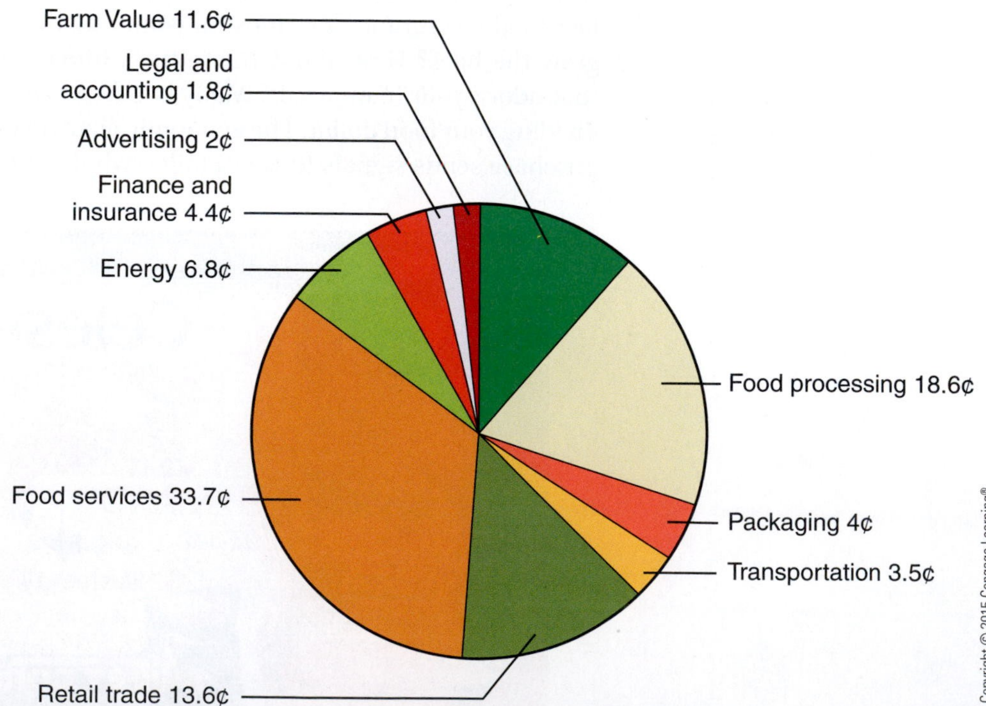

FIGURE 32-3 The farmer's share of the food dollar was reported by ERS/USDA to be 11.6 cents in 2011. Numerous other services and value-added inputs to a product account for the remaining amount of each dollar spent by consumers for food.

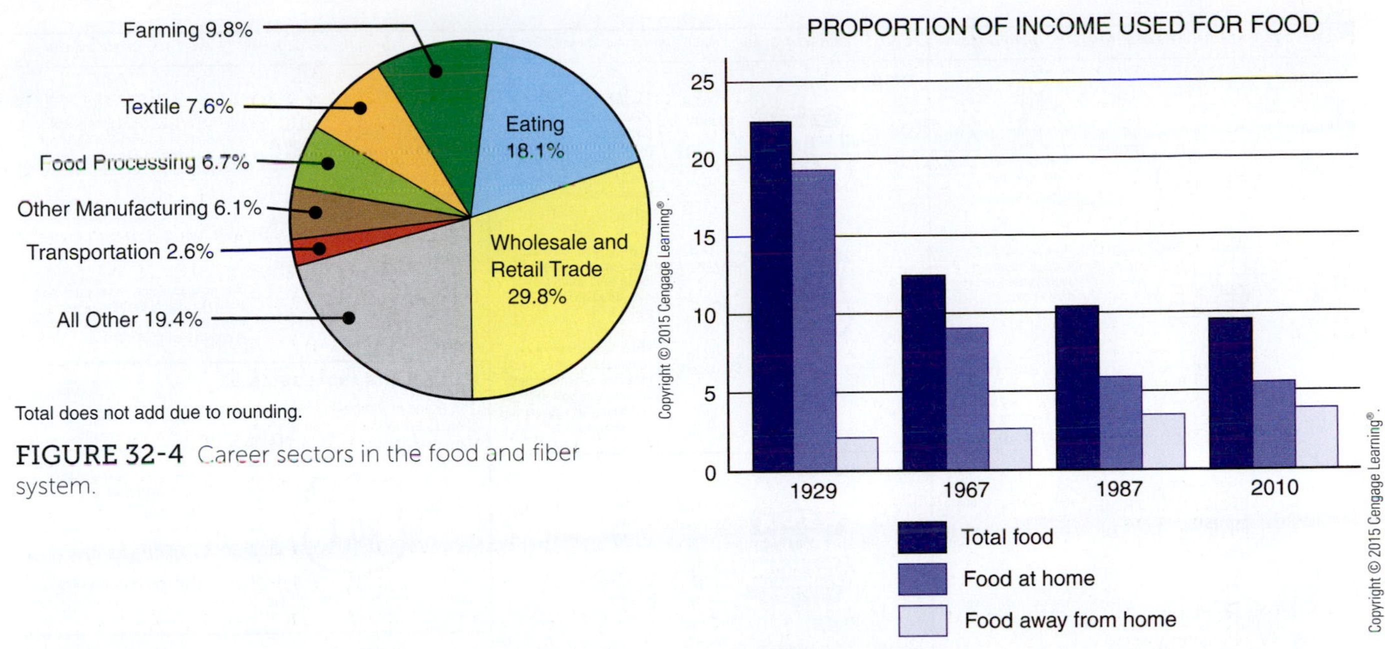

FIGURE 32-4 Career sectors in the food and fiber system.

FIGURE 32-5 Proportion of disposable income used for food in the United States.

41 percent of all income spent for food. In the same year, the percentage of disposable income needed for food was 9.4 percent in the United States (Figure 32-5).

QUALITY ASSURANCE

Grading and Inspecting

In the United States, we have become accustomed to high-quality food in every state and every store. The grading system established by the U.S. Department of Agriculture (USDA) has provided a uniform set of trading terms known as grades. **Grades** are based on quality standards. These improve acceptability of products by the consumer.

Grade standards are established for the following commodities: meat, cattle, wool, poultry, eggs, and dairy products; fresh, frozen, canned, and dried fruits and vegetables; cotton, tobacco, and spirits of turpentine; and rosin. Grades indicate freshness, potential flavor, texture, and uniformity in size and weight, depending on the commodity (Figure 32-6).

Sanitation

Additional quality-assurance programs administered by the USDA include inspection of slaughterhouses and processing plants and oversight of processing operations (Figure 32-7). The USDA oversees food labeling and enforces regulations regarding representation on such labels (Figure 32-8). The National Shellfish Sanitation Program, the U.S. Public Health Service, and the U.S. Food and Drug Administration work with the USDA to ensure the safety of food and food products. States, counties, and municipalities also have inspectors. They regulate local conditions to ensure sanitation and safe food handling, especially in restaurants and food-preparation areas.

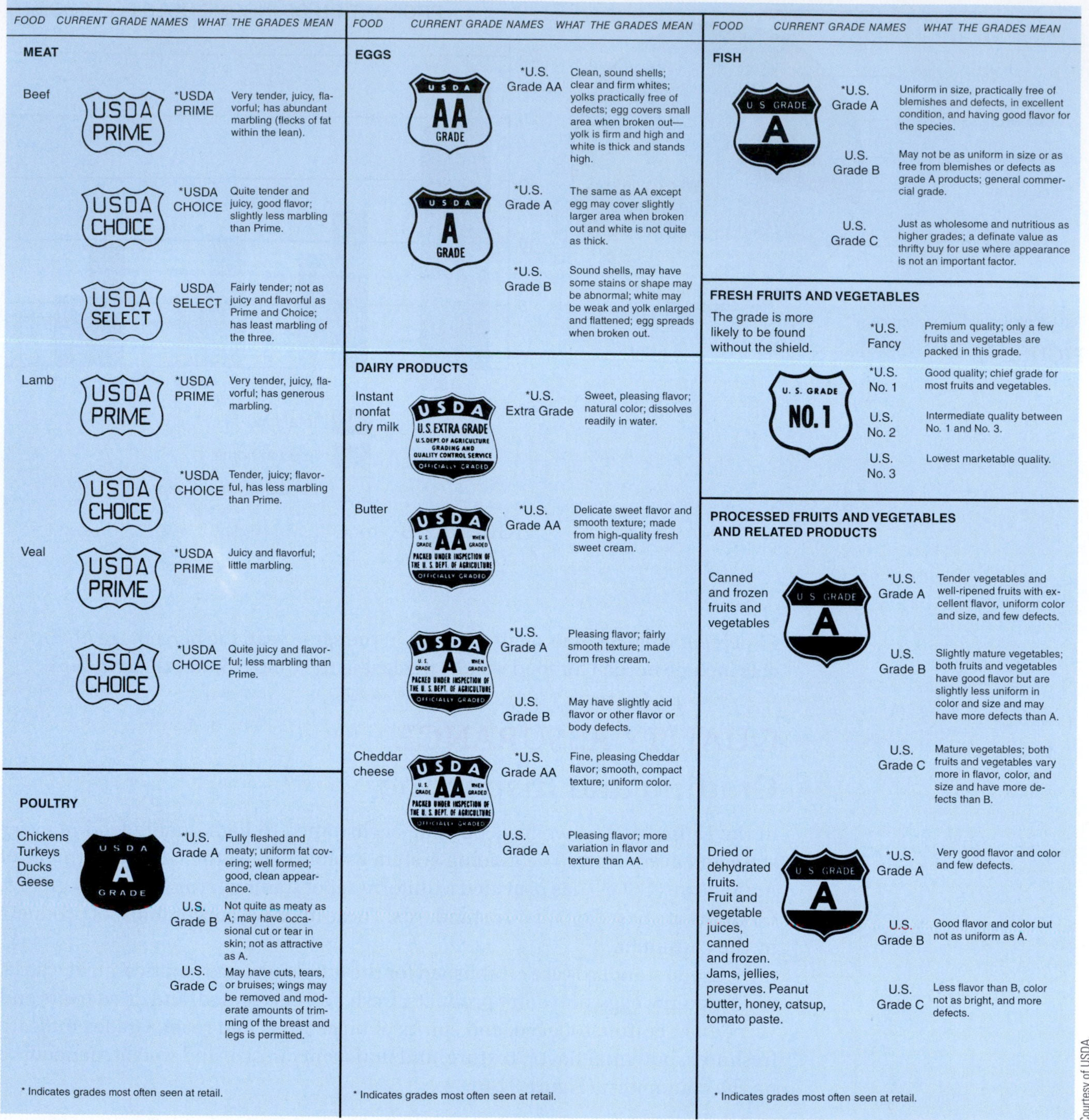

FOOD	MARK	CURRENT GRADE NAMES	WHAT THE GRADES MEAN
MEAT			
Beef	USDA PRIME	*USDA PRIME	Very tender, juicy, flavorful; has abundant marbling (flecks of fat within the lean).
	USDA CHOICE	*USDA CHOICE	Quite tender and juicy, good flavor; slightly less marbling than Prime.
	USDA SELECT	USDA SELECT	Fairly tender; not as juicy and flavorful as Prime and Choice; has least marbling of the three.
Lamb	USDA PRIME	*USDA PRIME	Very tender, juicy, flavorful; has generous marbling.
	USDA CHOICE	*USDA CHOICE	Tender, juicy; flavorful, has less marbling than Prime.
Veal	USDA PRIME	*USDA PRIME	Juicy and flavorful; little marbling.
	USDA CHOICE	*USDA CHOICE	Quite juicy and flavorful; less marbling than Prime.
POULTRY			
Chickens Turkeys Ducks Geese	USDA A GRADE	*U.S. Grade A	Fully fleshed and meaty; uniform fat covering; well formed; good, clean appearance.
		U.S. Grade B	Not quite as meaty as A; may have occasional cut or tear in skin; not as attractive as A.
		U.S. Grade C	May have cuts, tears, or bruises; wings may be removed and moderate amounts of trimming of the breast and legs is permitted.

* Indicates grades most often seen at retail.

FOOD	MARK	CURRENT GRADE NAMES	WHAT THE GRADES MEAN
EGGS			
	USDA AA GRADE	*U.S. Grade AA	Clean, sound shells; clear and firm whites; yolks practically free of defects; egg covers small area when broken out—yolk is firm and high and white is thick and stands high.
	USDA A GRADE	*U.S. Grade A	The same as AA except egg may cover slightly larger area when broken out and white is not quite as thick.
		*U.S. Grade B	Sound shells, may have some stains or shape may be abnormal; white may be weak and yolk enlarged and flattened; egg spreads when broken out.
DAIRY PRODUCTS			
Instant nonfat dry milk	USDA U.S. EXTRA GRADE U.S. DEPT. OF AGRICULTURE GRADING AND QUALITY CONTROL SERVICE OFFICIALLY GRADED	*U.S. Extra Grade	Sweet, pleasing flavor; natural color; dissolves readily in water.
Butter	USDA U.S. GRADE AA WHEN GRADED PACKED UNDER INSPECTION OF THE U.S. DEPT. OF AGRICULTURE OFFICIALLY GRADED	*U.S. Grade AA	Delicate sweet flavor and smooth texture; made from high-quality fresh sweet cream.
	USDA U.S. GRADE A WHEN GRADED PACKED UNDER INSPECTION OF THE U.S. DEPT. OF AGRICULTURE OFFICIALLY GRADED	*U.S. Grade A	Pleasing flavor; fairly smooth texture; made from fresh cream.
		U.S. Grade B	May have slightly acid flavor or other flavor or body defects.
Cheddar cheese	USDA U.S. GRADE AA WHEN GRADED PACKED UNDER INSPECTION OF THE U.S. DEPT. OF AGRICULTURE OFFICIALLY GRADED	*U.S. Grade AA	Fine, pleasing Cheddar flavor; smooth, compact texture; uniform color.
		U.S. Grade A	Pleasing flavor; more variation in flavor and texture than AA.

* Indicates grades most often seen at retail.

FOOD	MARK	CURRENT GRADE NAMES	WHAT THE GRADES MEAN
FISH			
	U S GRADE A	*U.S. Grade A	Uniform in size, practically free of blemishes and defects, in excellent condition, and having good flavor for the species.
		U.S. Grade B	May not be as uniform in size or as free from blemishes or defects as grade A products; general commercial grade.
		U.S. Grade C	Just as wholesome and nutritious as higher grades; a definate value as thrifty buy for use where appearance is not an important factor.
FRESH FRUITS AND VEGETABLES			
The grade is more likely to be found without the shield.		*U.S. Fancy	Premium quality; only a few fruits and vegetables are packed in this grade.
	U. S. GRADE NO. 1	*U.S. No. 1	Good quality; chief grade for most fruits and vegetables.
		U.S. No. 2	Intermediate quality between No. 1 and No. 3.
		U.S. No. 3	Lowest marketable quality.
PROCESSED FRUITS AND VEGETABLES AND RELATED PRODUCTS			
Canned and frozen fruits and vegetables	U S GRADE A	*U.S. Grade A	Tender vegetables and well-ripened fruits with excellent flavor, uniform color and size, and few defects.
		U.S. Grade B	Slightly mature vegetables; both fruits and vegetables have good flavor but are slightly less uniform in color and size and may have more defects than A.
		U.S. Grade C	Mature vegetables; both fruits and vegetables vary more in flavor, color, and size and have more defects than B.
Dried or dehydrated fruits. Fruit and vegetable juices, canned and frozen. Jams, jellies, preserves. Peanut butter, honey, catsup, tomato paste.	U S GRADE A	*U.S. Grade A	Very good flavor and color and few defects.
		U.S. Grade B	Good flavor and color but not as uniform as A.
		U.S. Grade C	Less flavor than B, color not as bright, and more defects.

* Indicates grades most often seen at retail.

Courtesy of USDA.

FIGURE 32-6 Grades of food commodities, as established by the U.S. Department of Agriculture (USDA).

COMMODITY GROUPS AND THEIR ORIGINS

What Foods Are Grown Where

Food is grown all over the world. Climatic conditions and available technology result in some foods growing better and in greater abundance in certain areas of the world. **Climatic conditions** refer to average temperature, number of days with a certain temperature range, length of growing season, and amount

INTERNET KEY WORDS:
grain crops
oil crops
fruit crops
vegetable crops

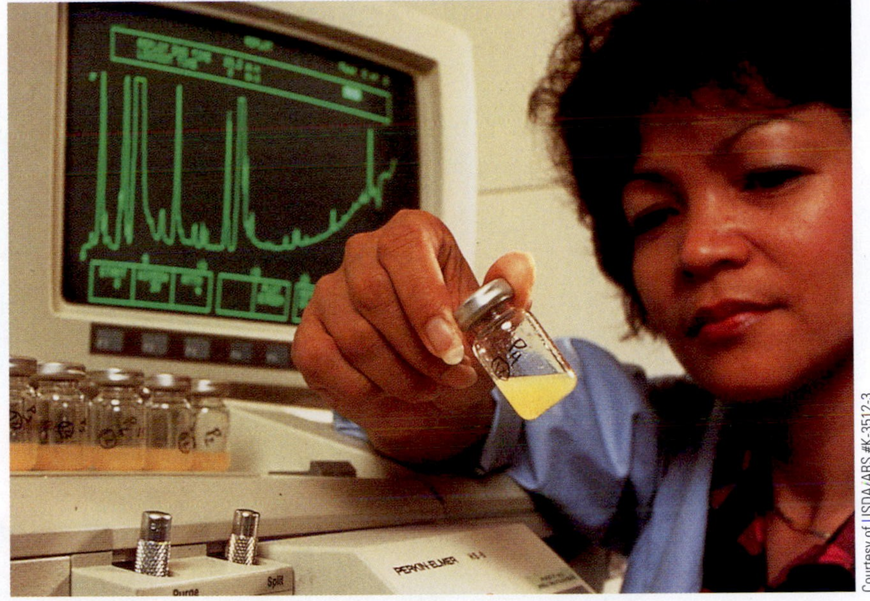

FIGURE 32-7 Quality-control personnel as well as government inspectors use USDA standards to monitor foods for cleanliness, wholesomeness, and quality.

Nutrition Facts
Serving Size 1 Cup (55 g/2.0 oz)
Servings per Container 13

Amount Per Serving	Cereal	Cereal with 1/2 Cup Vitamins A & D Skim Milk
Calories	170	210
Fat Calories	10	10
	% Daily Values**	
Total Fat 1.0 g*	**2%**	**2%**
Sat. Fat 0 g	**0%**	**0%**
Cholesterol 0 mg	**0%**	**0%**
Sodium 300 mg	**13%**	**15%**
Potassium 340 mg	**10%**	**16%**
Total Carbohydrate 43 g	**14%**	**16%**
Dietary Fiber 7 g	**28%**	**28%**
Sugars 17 g		
Other Carbohydrate 19 g		
Protein 4 g		
Vitamin A	15%	20%
Vitamin C	0%	2%
Calcium	2%	15%
Iron	45%	45%
Vitamin D	10%	25%
Thiamin	25%	30%
Riboflavin	25%	35%
Niacin	25%	25%
Vitamin B_6	25%	25%
Folate	25%	25%
Vitamin B_{12}	25%	35%
Phosphorus	20%	30%
Magnesium	20%	25%
Zinc	25%	25%
Copper	15%	15%

*Amount in cereal. One-half cup skim milk contributes an additional 40 calories, 65 mg sodium, 6 g total carbohydrate (6 g sugars), and 4 g protein.
**Percent Daily Values are based on a 2,000-calorie diet. Your daily values may be higher or lower depending on your calorie needs:

		Calories	2,000	2,500
Total Fat		Less than	65 g	80 g
Sat. Fat		Less than	20 g	25 g
Cholesterol		Less than	300 mg	300 mg
Sodium		Less than	2,400 mg	2,400 mg
Potassium			3,500 mg	3,500 mg
Total Carbohydrate			300 g	375 g
Dietary Fiber			25 g	

Calories per gram:
Fat 9 • Carbohydrate 4 • Protein 4

FIGURE 32-8 The commodity label is the consumer's best assurance of food quality and value.

of precipitation for a given geographic area. Technology refers to the equipment and scientific expertise available to cultivate, store, process, and transport the crop for consumption in a variety of forms after harvest. Food production in the United States has always been influenced by geography and climate (Figure 32-9).

Since early times, when humans traveled and traded, foods have been introduced outside the areas where they are grown naturally. The origin of the soybean, for example, can be traced back 3000 years to China, where it is still produced and consumed. However, major growing areas for the soybean today include the United States, Brazil, and Western Europe.

Modern technology has allowed producers to raise crops somewhat artificially with irrigation and in greenhouses where the conditions of temperature and moisture are controlled (Figure 32-10). Different varieties of food have been developed to grow under different climatic conditions, such as extreme heat or cold. Similarly, in aquaculture, seafood and fish are produced under controlled conditions (Figure 32-11).

In the United States, we are accustomed to having almost every food available fresh at any time of the year; but all foods are not grown in all parts of the country. Citrus fruit, including oranges and grapefruit, require warm climates, such as those found in Texas, California, and Florida. More than 70 percent of the fresh vegetables that are consumed in the United States are grown in California, Florida, and Arizona. Every day, however, people in North Dakota and Maine enjoy the nutrients and good taste of fresh or processed citrus and vegetable products, such as orange juice. Our country is not only "America the Beautiful," it is also "America the Bountiful."

The many operations of the food industry that are discussed later in this unit explain how we can enjoy foods that are not grown naturally in our particular region of the country and world. Realizing where some of our food products originate also makes you appreciate the size and scope of the food industry.

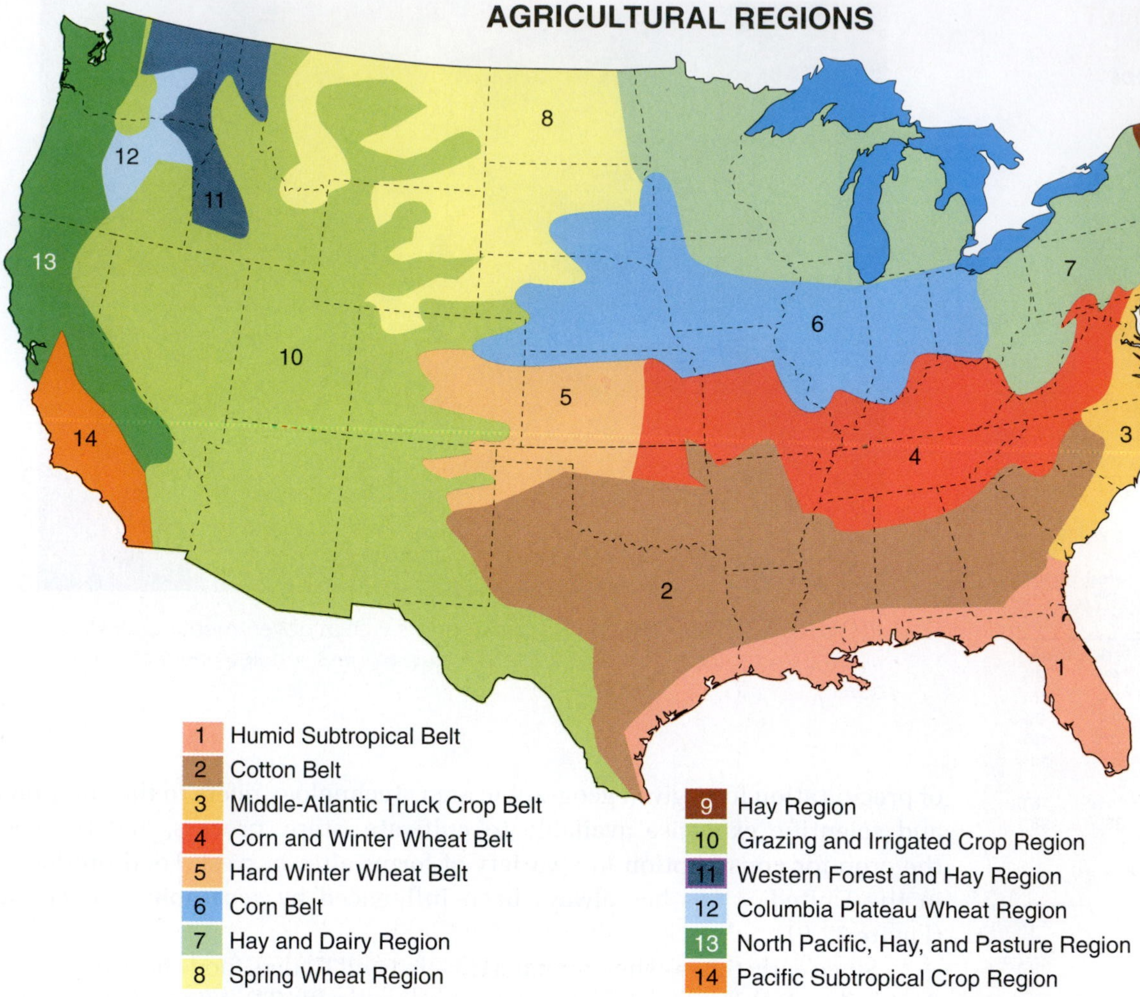

AGRICULTURAL REGIONS

1 Humid Subtropical Belt
2 Cotton Belt
3 Middle-Atlantic Truck Crop Belt
4 Corn and Winter Wheat Belt
5 Hard Winter Wheat Belt
6 Corn Belt
7 Hay and Dairy Region
8 Spring Wheat Region
9 Hay Region
10 Grazing and Irrigated Crop Region
11 Western Forest and Hay Region
12 Columbia Plateau Wheat Region
13 North Pacific, Hay, and Pasture Region
14 Pacific Subtropical Crop Region

Courtesy of USDA.

FIGURE 32-9 Major agricultural regions of the United States.

Courtesy of DeVere Burton.

FIGURE 32-10 Technologies such as modern irrigation systems have made it possible to produce crops in arid environments such as deserts.

Courtesy of DeVere Burton.

FIGURE 32-11 Controlled living environments such as fish runs make it possible to produce fish under controlled conditions.

© Zeljko Radojko/Shutterstock.com.

FIGURE 32-12 Corn is adapted to a wide range of growing conditions and is grown in every state.

Crop Commodities

Grains

Various grains have different growing requirements. Therefore, they are produced in different parts of the United States and the world. Wheat, which originated in Asia, is grown in the cooler climates of the United States. Different varieties have been developed to accommodate different growing seasons and climatic conditions around the world. Corn is a warm-weather crop, but the many varieties and types permit its growth in every state in the United States (Figure 32-12). Rice, however, has special moisture requirements; therefore, its production is limited to specific areas of the country.

Oil Crops

Oil crops are sometimes thought of as the invisible food product. Soybeans, corn, cotton, flax, sunflowers, coconut, peppermint, and spearmint are all significant oil crops in the United States. The United States is a world leader in the production of soybeans, corn, cotton, and peanut oils (Figure 32-13). Soybean products are referred to in ancient Chinese literature, and the origin of the peanut may be traced to Brazil and Paraguay. Sunflowers are native to the United States and are growing in importance. Sunflowers are also grown in Spain, China, Russia, Bulgaria, and Mediterranean areas. Safflower oil is a relatively minor oil in terms of proportion to the total oil crop worldwide. It originated in northern India, North Africa, and the Middle East. Its origin is indicative of its drought tolerance.

Sugar Crops

Sugar beets and sugarcane are the principal sugar crops in the United States. Corn is a secondary source of sugar. Sugar beets are grown in temperate areas, with most of the production in the states of Minnesota, North Dakota, Idaho, and

© iStockphoto/Valentyn Volkov.

FIGURE 32-13 Edible oils are obtained from the seeds of many different kinds of plants, making the United States a world leader in the production of vegetable oils.

Michigan (Figure 32-14). Sugarcane is grown in tropical and subtropical locations around the world. Florida, Hawaii, and Louisiana are the three largest producers of cane sugar in the United States.

Citrus

Oranges, limes, lemons, and grapefruit all require warm temperatures to survive. They cannot tolerate freezing conditions. Consequently, the southern states with warmer climates, such as Florida, Texas, Arizona, and California, are the major producers of citrus. The industry provides consumers with both fresh citrus and frozen concentrate products.

Tree Fruits

The many varieties of fruit that grow on trees require specific weather conditions. Therefore, various fruits are adapted to different parts of the country. Apples and pears require cooler temperatures and do well in mountainous areas (Figure 32-15). Washington is particularly well adapted for these fruits. Bananas require very warm conditions and grow best in tropical areas. The United States imports most of the bananas we eat. Hawaii is the only state with significant commercial production.

Vegetables and Berries

Vegetables and berries are consumed shortly after harvest, or they are processed by canning, drying, or freezing for future consumption. Some vegetables require warm climates and some require cool climates. Vegetables that require cooler climates include cabbage, broccoli, potatoes, and cauliflower. Vegetables requiring warmer environments include beans, tomatoes, and sweet corn. Vegetable and berry production occurs in most regions of the country.

© iStockphoto/Paul Jackson.

A

© iStockphoto/Jason Lugo.

B

FIGURE 32-14 Sugar beets are grown in a temperate climate and are a significant source of processed sugar.

© iStockphoto/Marek Mnich.

FIGURE 32-15 Apples are produced in areas with a relatively cool season. Apples are one of our favorite fresh fruits, and large volumes of apples are consumed every year.

Meat Commodities

Animals, like crops, are typically raised in locations with some regard to climatic conditions. Artificial cooling or heating of livestock is practiced in some regions, but it is expensive. Where fewer artificial conditions are introduced, the cost of production is minimized.

The type, cost, and availability of livestock feed are important factors influencing where animals are raised in the United States and around the world. Large amounts of water must be available for livestock. This can be a limiting factor for animal production in some areas.

Beef

Most beef is raised near corn, the main feed source for feedlot cattle. More than half of the corn in the United States is grown in Nebraska, Iowa, Illinois, and Minnesota. Therefore, beef is raised extensively in the Midwest. The open ranges in the western part of the United States provide other important areas where beef is raised.

INTERNET KEY WORDS:
meat, beef, pork, lamb products

Pork

Corn is also the primary food of hogs. Similar to beef, the primary area where hogs are raised is the Midwestern part of the United States. The mid-Atlantic and southern states are also important hog-production areas.

Lamb

Sheep are animals that require large amounts of grazing area. Therefore, they are raised extensively in the range states of the far West. However, as is true with beef and pork, lamb products are produced in other states.

AGRI-PROFILE

CAREER AREAS: SCIENTIST/INSPECTOR/QUALITY CONTROLLER/BUYER/SELLER/PROCESSOR/TRUCKER/WHOLESALER/RETAILER

© iStockphoto/Eliza Snow.

Processors convert raw farm produce to products that are convenient for consumers to use and for which there is a ready market.

The food industry is massive and includes both plant and animal products. It includes the producers, processors, distributors, wholesalers, retailers, fast-food establishments, restaurants, and the home kitchens where food is prepared.

Career opportunities include many that have been discussed previously and some new ones, too. The meat processing-and-packaging industry is massive and employs a large number of individuals in the United States. Field supervisors and coordinators direct the work of crews to harvest crops at the peak of their quality and transport them to processing or packing plants. Similarly, crews take fish, oysters, clams, lobsters, and other seafood from production habitat to processing centers. In many cases, huge packing and/or processing machines are used in the field, orchard, or boat. Quality-control personnel collect food specimens, label them, test them, and maintain records to ensure quality control on each batch of food coming out of the plant.

Careers in food science, store management, produce management, meat cutting, laboratory testing, field supervision, research, diet and nutrition, health and fitness, and promotion are all possibilities in the food industry.

FIGURE 32-16 The U.S. dairy industry provides a wide range of milk products to consumers.

Dairy Products

Wisconsin has long been called the "Dairy State." Indeed, many of our dairy products are produced in Wisconsin and other northern-tier states. California also has a large dairy industry with many cooperatives and processing plants. These provide dairy products to consumers nationwide (Figure 32-16). Dairy animals prefer moderate environments, so the industry is extensive in the northern part of the United States. During recent years, however, the industry has expanded significantly in Idaho, New Mexico, and other western states where high-quality forages are available.

Game

Each state has native game. Whether it is harvested as an agricultural product depends on the demand for the product. Venison is the tasty and popular meat of deer. In some states, deer are raised in captivity to help meet the market demand for venison.

Seafood

States that border the Atlantic and Pacific Oceans and the Gulf of Mexico are considered the primary suppliers of seafood in the United States. However, the science of aquaculture permits the production of fish in interior states.

Poultry

Poultry can be raised in a variety of settings. Typically, chickens and turkeys are raised indoors, where ventilation and temperature are carefully controlled. Most poultry is raised in the mid-Atlantic and southern states; however, important poultry-producing areas are also found in California and other states.

OPERATIONS WITHIN THE FOOD INDUSTRY

The food industry begins with the process of photosynthesis in plants. It progresses through plant and animal growth and on to processing and distribution of commodities. The discussion in this unit focuses primarily on the food industry after the crops and animals have been grown.

Harvesting

Harvesting means taking a product from the plant where it was grown or produced. This may involve taking potatoes out of the ground, picking oranges off a tree, removing bean pods from bean plants, or removing and threshing grain from stalks (Figure 32-17). It is most important that the crop be harvested in a timely and careful fashion. The plant should be harvested at the correct stage of maturity. **Maturity** means the state or quality of being fully ripe or mature. Harvesting a crop when it is of proper maturity means that it is not underripe, overripe, or spoiled. **Underripe** means that it has not reached maturity. **Overripe** means that the plant is past the optimum maturity; stalks or limbs can easily break or shatter, or fruit can drop. Spoiled means that chemical changes have taken place in the food or food product that either reduce its nutritional value or render it unfit to eat.

Spoilage is usually caused by microorganisms. **Microorganisms** that contribute to food spoilage are bacteria, fungi, and nematodes. Bacteria consist

Courtesy of DeVere Burton.

FIGURE 32-17 Harvesting is the timely removal of crops from the field. Most crops are harvested mechanically using specialized machines, such as this potato-harvesting equipment.

of a group of single-celled plants. Fungi are plants that lack chlorophyll and obtain their nourishment from other plants, thus causing rot, mold, and plant diseases. Nematodes are small worms that feed on or in plants or animals. They live in moist soil, water, or decaying matter. They pierce the cells of plants and feed on the juices.

The moisture content of a product determines how well it can undergo processing. It also dictates the types of handling procedures and storage facilities that are required for certain foods. Some fruits, such as bananas and tomatoes, continue to ripen after they have been picked from the tree and vine. Other foods, such as beans and oranges, do not continue to ripen once picked, and they must be handled accordingly. Knowledge of the complete growth process and handling requirements of each commodity is essential for the producer and harvester.

Harvesting involves the use of equipment and labor. Because timing is critical, migratory labor is often hired at harvest time. Migratory labor is provided by workers who move from place to place, following the harvest as it occurs throughout the year. As crops ripen, laborers migrate to new geographic locations where the harvest season is just beginning. Usually, workers move from the southern tier of states to central and northern regions as crops mature. Many of these workers return to the same farms year after year, and they become efficient and skilled. For example, picking fruit is usually done by hand, and the crop must be harvested within a very short time period. A dependable crew is needed, but only for a short season. Local laborers are not always available, nor are they as skilled as workers who pick fruit for several months each year. When dependable workers are not available, fruit and vegetable crops are at risk of being left unharvested.

Some food crops are adapted to mechanical harvest methods. Engineers and technicians continue to develop and market machines to perform many harvesting operations. Such machines make harvesting much more efficient than it was in the past (Figure 32-18). However, machines may not always be as gentle

Courtesy of USDA/ARS #K-4416-14.

FIGURE 32-18 Mechanical harvesters replace the backbreaking and difficult work of hand harvesting. This machine is separating cranberries from other plant materials.

with the fruits and vegetables as human hands. Consequently, plant breeders have developed new crop varieties that lend themselves better to mechanical harvesting. Such varieties may not be as appealing to the human touch or taste, though. For instance, tomato varieties having skins tough enough for mechanical harvesting are harder to slice and may not be as juicy or flavorful as varieties suitable for the home garden.

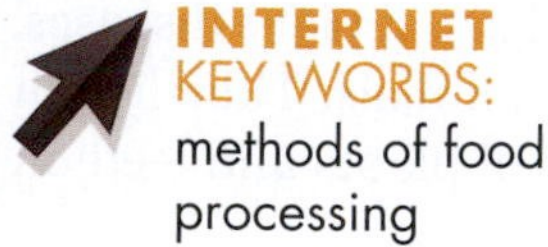

INTERNET KEY WORDS: methods of food processing

Processing and Handling

The steps involved in turning the raw agricultural product into an attractive and consumable food are collectively known as **processing**. Processing factories, or plants, clean, dry, weigh, refrigerate, preserve, store, and turn the commodity into a variety of other products (Figure 32-19).

Wheat is cleaned, dried, weighed, and graded for quality. It is then ground into flour. However, before this occurs, it may be separated into bran and germ. The skin or covering of a wheat kernel is known as **bran**. Inside the bran is the **endosperm**, which will become flour, and the **germ**, which is a new wheat plant inside the kernel. Wheat flour is used for breads, cereals, cakes, and pasta. Other grains have similar parts and are used to make similar products.

Processing of tomatoes results in a variety of products (Figure 32-20). Some people claim that the fresh tomato defines summer. The peak of the North American growing season occurs when the tomato is ripe. To the gardener, this means picking tomatoes directly from the vine in a backyard garden and consuming them immediately. However, tomatoes are harvested year-round somewhere in the world, and the food industry can get them to us in edible form. **Edible** means fit and safe to eat. Processing preserves tomatoes for future use.

INTERNET KEY WORDS: importance of food transport

After tomatoes are cleaned and separated for size and quality, they may be canned whole; chopped, cooked, and strained for juice; or made into other products. Such products include salsa, spaghetti and hamburger sauces, paste, relish, catsup, and many other foods.

Courtesy of FFA.

FIGURE 32-19 Processing is one of many intermediate steps between the producer and the consumer.

Courtesy of W. Altee Burpee Company.

FIGURE 32-20 Tomatoes can be eaten fresh or processed into a variety of products.

Transporting

Trucks, planes, boats, cars, trains, carts, and bicycles are vehicles used by the food industry in various parts of the world (Figure 32-21). The transporting of fresh and processed food products composes 5.5 percent of the marketing cost within the food industry in the United States. Timing and the distance foods must travel contribute to the ultimate cost of the foods. The efficiency of transportation can influence food quality in terms of freshness and spoilage. Insulated and refrigerated trucks enable food products to move in fresh form to most parts of the country year-round (Figure 32-22). This luxury is not available to most people of the world.

Approximately 90 percent of our perishable food is shipped by truck. Much of the less-perishable foods, such as wheat, potatoes, and beets, are shipped by rail. Air transportation allows us to enjoy perishable foods from distant regions

SCIENCE CONNECTION

DO NOT CHUCK ROTTEN TOMATOES

© Emin Ozkan/Shutterstock.com.

A valuable antioxidant, lycopene, can be recovered from spoiled tomatoes and marketed as a health product.

The vitamin and mineral industry may spell big money for tomato growers. A valuable antioxidant, lycopene, is found in tomatoes. It is the substance responsible for the red color of the tomato. Lycopene helps protect cells from harm that can be caused by the oxidation process. It has been found to reduce the risk for prostate cancer and other diseases. Until now, the substance has been expensive, costing $2500 a kilo in its pure form.

Agriscientists at the University of Florida have created an inexpensive way to extract lycopene from tomatoes. This new discovery may earn the tomato industry millions of dollars. Every year, tomato growers lose money because some of their tomatoes are unfit for market because of blemishes that are unappealing to consumers. With the new extraction technique, these blemished tomatoes can be marketed to the health industry. The technique will provide consumers with a less expensive product, and it will provide growers with a profit, instead of a costly loss.

Courtesy of FFA, Photo by Bill Stagg.

FIGURE 32-23 Wholesale terminals provide the facilities for trucks and trains to bring food commodities together, allowing buyers to obtain commodities for their retail outlets.

Courtesy of FFA.

FIGURE 32-24 Superstores may stock 15,000 or more items.

Courtesy of USDA.

FIGURE 32-25 Farmers' markets provide fruits, vegetables, honey, and other farm products that are fresh from the fields to consumers. They have become popular marketing outlets in many areas.

Consumers can purchase their food items from many types of retail stores. Such marketing sources meet the needs of consumers in different locations and situations. Superstores carrying 15,000 or more items, conventional supermarkets, limited-assortment and box stores, convenience stores, unconventional food stores, small stores, corner stores, food cooperatives, farmers markets, roadside stands, pick-your-own businesses, and other farm outlets are the most common places consumers purchase their food items (Figures 32-24 and 32-25). The major differences among the various types of stores are the number of items stocked and the physical size of the facilities.

CAREER OPPORTUNITIES IN THE FOOD INDUSTRY

Each of the areas discussed in this unit and in Unit 33 require people who manage, operate, and carry out the many and varied elements of the food industry. As with all careers in agriscience, those in the food industry present many challenges and rewards. Career opportunities await individuals at the local, county, state, national, and international levels. Careers in the food industry can be divided into eight, often overlapping, categories. Career opportunities in each area are numerous (Figure 32-26).

THE FOOD INDUSTRY OF THE FUTURE

The food industry is ever-changing, with new developments occurring each day. Some areas that may attract the food researcher include new food products, new processing and preserving techniques, and new equipment for harvesting labor-intensive crops.

Aquaculture is meeting the increasing demand for fish and will continue to supplement the catches of commercial fishermen. The use of extreme heat and cold in processing has contributed to the development of many food items that meet the demand for convenience foods. The convenience food store is expected

Some Careers in Food Science and the Food Industry

Business	Processing	Retailing/Food Service
Accountant	Butcher	Baker
Buyer	Efficiency expert	Cook/Pizza maker
Distributor	Engineer	Counter salesperson
Financial analyst	Plant line worker	Deli operator
Loan officer	Plant supervisor	Meat cutter
Marketing specialist	Refrigeration specialist	Nutritionist
Salesperson	Safety expert	Produce specialist
Statistician		Restaurant owner/operator
	Quality Assurance	Waiter/waitress
Communications	Food analyst	
Advertising specialist	Grader	**Transportation**
Broadcaster	Inspector	Dispatcher
Media specialist	Lab technician	Trucker
TV Producer/Demonstrator	Quality-control supervisor	Rail operator
Writer	Quarantine officer	Merchant marine
Education	**Research and Development**	
College professor	Distribution analyst	
Extension specialist	Biochemist	
Industry educator	Microbiologist	
Dietician	Packaging specialist	
Teacher	Process engineer	

FIGURE 32-26 Career opportunities in food science and the food industry are many and varied.

HOT TOPICS IN AGRISCIENCE IRRADIATION OF FOODS

© kurt_G/Shutterstock.com.

Whether food items are imported or exported, damaging insects must not be transported from regions that are infested with them to regions that are not. Irradiation is a proven method for controlling insects, such as fruit flies, on fruits and vegetables.

Harmful insects, such as the Mediterranean fruit fly, enter the United States through imported fruits and vegetables. One method of reducing the threat of these and other invading insects is to subject incoming fruits and vegetables to radiation. In high doses, this treatment is capable of killing many insects. At lower exposure levels, many insects become sterile or unable to reproduce. Irradiated insect eggs do not hatch; thus, irradiated produce is essentially free of dangerous insects even though some of them may survive the radiation treatment. The process works by breaking the chromosomes in the cells of the insect. Irradiation is expected to play a major role is controlling insects in the future.

Countries such as Japan require fumigation treatments of cherries and other produce from the United States before they are allowed into the country. Irradiation is less damaging to the produce than fumigation, and it is hoped that wider acceptance of irradiated fruit and vegetables will lead to acceptance of the irradiation process on U.S. exports of fruits and vegetables. U.S. military installations continue to use irradiated foods, and it is hoped that consumers can be educated to accept widespread use of this technology for protecting fruits and vegetables.

to continue to play a larger role in the food chain. Economic efficiency in convenience foods and convenience stores is under constant review. In addition, the USDA and other agencies continue their vigilance regarding safety and nutritional standards at all steps of the food chain.

Improved harvesting equipment for products such as grapes is being tested to reduce the labor costs of such crops. Fuel alternatives for cost-effective transportation and refrigeration with carbon dioxide snow instead of conventional diesel-powered mechanical refrigeration are some of the many developments under constant review in the food industry. This industry must continue to meet the demands for high-quality food in the United States and the world through effective research and qualified employees.

STUDENT ACTIVITIES

1. Write the Terms to Know and their meanings in your notebook.
2. Keep a food-dollar diary to document where your food dollars are spent. Record the cost of meals and snacks eaten in and outside the home.
3. Do a cost comparison of meals prepared at home and similar meals consumed at fast-food places and restaurants.
4. Trace the activities that occur in transforming wheat in the field to a hamburger roll consumed in your home.
5. Draw a diagram tracing the individual food components of a deluxe hamburger back to the places where the components were produced. Label each component, process, and commodity along the way.
6. Ask your instructor to arrange a field trip to a butcher shop or supermarket to observe demonstrations on meat cutting and packaging.
7. Make a collage illustrating the various activities of the food industry.
8. Prepare a report on some of the safety regulations that are in place in the United States to prevent illness caused by food contamination.
9. Refer to Figure 32-22, which lists a variety of career opportunities in the food industry. Choose a career that interests you. Find out what steps you will have to take to get the job in which you are most interested. Be sure to include all training and education that will be required.

SELF-EVALUATION

A. MULTIPLE CHOICE

1. Approximately what percentage of the U.S. food dollar is spent on meals away from home?
   a. 15 percent
   b. 25 percent
   c. 35 percent
   d. 45 percent
2. Which of the following products is native to North America?
   a. soybeans
   b. wheat
   c. sunflowers
   d. peanuts

3. In the United States, more than one-half of the fresh fruits and vegetables are grown in which states?
   a. Montana, Oregon, and Washington
   b. California, Florida, and Texas
   c. New Jersey, North Carolina, and Georgia
   d. Arizona, Nebraska, and Ohio
4. Migratory workers would harvest wheat last in which state?
   a. Arizona
   b. Nebraska
   c. Montana
   d. Ohio
5. When you spend one dollar for food, approximately how much goes into the labor required to harvest and then process that food after it leaves the farm?
   a. $0.94
   b. $0.64
   c. $0.34
   d. $0.04
6. Approximately what percentage of all the jobs in the food and fiber system is related to wholesale and retail sales?
   a. 60 percent
   b. 50 percent
   c. 40 percent
   d. 30 percent
7. Which product is consumed away from home the most?
   a. fruits
   b. beverages other than milk
   c. vegetables
   d. meat
8. Superstores are likely to carry approximately how many items?
   a. 15,000
   b. 1500
   c. 150
   d. 15

B. MATCHING

________	1. Harvester	a. Purchases food in large quantities
________	2. Grader	b. Stores food until requested
________	3. Retailer	c. Follows crop harvesting geographically
________	4. Wholesaler	d. Is involved in the transportation of food
________	5. Migrant worker	e. Takes the crop from the field
________	6. Trucker	f. Is involved in cleaning, sorting, and preparing a product
________	7. Processor	g. Inspects food and determines how it will be sold
________	8. Distributor	h. Sells directly to the public

C. COMPLETION

1. Three careers that you could pursue in the quality-assurance area of the food industry include ________, ________, and ________.
2. Differences in how grocery stores are categorized are primarily determined by ________ and ________.
3. When buying food at a pick-your-own farm, the producer is also the ________.
4. A beef grade of ________ would indicate very tender, juicy, and flavorful, with abundant marbling.
5. Grocers purchase their supplies through ________.

UNIT 33
Food Science

OBJECTIVE

To explore the nutrient requirements for human health and the processes used in food science to ensure an adequate and wholesome food supply.

MATERIALS LIST

- bulletin board materials
- Internet access

COMPETENCIES TO BE DEVELOPED

After studying this unit, you should be able to:

- discuss nutritional needs of humans and the food groups that meet these needs.
- categorize foods in the U.S. Department of Agriculture *MyPlate* nutrition initiative.
- discuss food customs of major world populations.
- relate methods used in processing and preserving foods.
- list the major steps used in slaughtering meat animals.
- list the major cuts of red-meat animals.
- identify methods of processing fish.
- describe techniques used to enhance retail sales of food commodities.

SUGGESTED CLASS ACTIVITIES

1. Conduct a class competition to identify the greatest number of processed food products. Divide the class into teams of four to five students. Assign each team to make a collage of photos or graphics that depict food items, using magazines and other printed materials. Each food item must be a completely different product, not just different brand names of the same thing. Ten points are scored when the product is represented in the collage by a graphic from its package. Five points are earned when the product is represented in the collage by a picture from an advertisement. One point is scored when the product name is printed. Only the highest category is scored for each

TERMS TO KNOW

nutrient
fermentation
blanching
canning
dehydration
freeze-drying
oxidative deterioration
dehydrofrozen product
retortable pouch
irradiation
dry-heat cooking
moist-heat cooking
convection oven
dehydrator
smoker
casein
vacuum pan
shackle
hide
viscera
carcass
split carcass
shroud
age (ripen)
block beef
disassembly process
fabrication and boxing
giblet
kosher
dressing percentage
sweetbreads
tripe
tankage
collagen

item. An additional 50 points is awarded for neatness and imagination. Use the collages for decorating bulletin boards.

2. Have a class discussion about possible new food products that you have recently heard about or even some new ideas for foods that do not now appear on the market. Encourage students to use their creativity to imagine some new way to prepare a food or to combine foods to make new products. For example, consider the fruit flavors and colors for single-serving cartons of milk. Create a list of the ideas that come from the class.
3. Have the students keep a food journal for up to 1 week before beginning this unit. Introduce them to the *MyPlate* nutrition initiative. Be sure the students understand how much one serving is for each of the food groups. Next, have the students reflect on their food choices for the last week. Students should then write a short essay highlighting some positive food choices they made and areas in which they could improve.

You are what you eat. Have you ever stopped to consider what that statement means? *Food* is the material needed by the body to sustain life. It consists of carbohydrates, fats, proteins, and supplementary substances, such as minerals and vitamins. Food is used to sustain growth, repair cells, sustain vital processes and furnish energy to the body. This unit explores the foods humans need to maintain health and sustain growth (Figure 33-1). In addition, it explores how those foods reach our tables from their beginning as raw products.

Courtesy of USDA.

FIGURE 33-1 The future well-being of children is largely dependent on the nutrition they receive. Good nutrition coupled with good health habits are the best practices to keep a body productive and vibrant for a lifetime.

NUTRITIONAL NEEDS

The body is a complex system that has many nutritional demands. **Nutrients** are substances that are necessary for the functioning of an organism. More than 50 specific nutrients are required for bodily functions. Nutrition involves the combination of processes by which all body parts receive and use materials necessary for function, growth, and renewal. Nutrition includes the release of energy, the building up of body tissues (both hard and soft), and the regulating of body processes. After food is digested, basic nutrients enter the bloodstream, which then transports them to the cells of the body. Nutrients are classified into six major groups, each supporting different functions in the body (Figure 33-2).

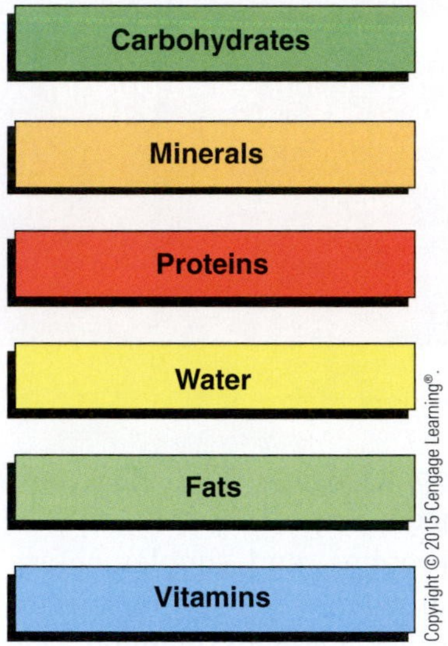

FIGURE 33-2 Nutrients are classified into six categories. Each category of nutrients supports different functions in the body.

Carbohydrates

Carbohydrates serve as the main source of energy for the body. Carbohydrates contain four calories per gram. There are three different types of carbohydrates: sugars, starches, and fiber (Figure 33-3). Sugars are simple carbohydrates and are found naturally in many foods, such as fruit, milk, and peas. Refined sugar, or the sweet substance added to many household foods, comes from sugar beets and sugarcane. Starch is a complex carbohydrate that is found in foods such as bread, potatoes, rice, and vegetables. Starches and sugars are converted to glucose in the body and serve as the major body fuel. Some of the fuel that is generated is stored by the body for later use. This occurs when the glucose is not fully used by the body and it is converted to fat. Fiber is also a complex carbohydrate and is found in the walls of plant cells. Humans are unable to digest fiber, yet it plays an important role in moving food through the body and expelling waste after digestion.

Fats

Fats are another source of energy for the body. Fats contain nine calories per gram. They are considered to be a compact source of energy because they have 2.25 times the number of calories as the other two energy sources—carbohydrates

FIGURE 33-3 Fruits, vegetables, and grains are healthful sources of carbohydrates.

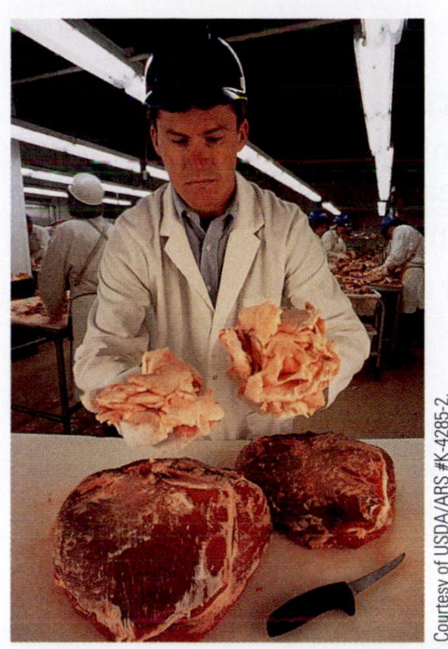
Courtesy of USDA/ARS #K-4285-2.

FIGURE 33-4 Animal products provide high-quality protein, but their fatty parts, such as skin and fatty layers, are best removed to avoid excess fat in the diet.

and proteins. Some vitamins are retained in the body by becoming dissolved in fats. They require fats to carry them to the parts of the body where they are needed. These vitamins are classed as fat-soluble vitamins. Although fat is necessary in the body, too much fat results in obesity and serious diseases, such as heart problems, diabetes, and high blood pressure. Fats are present in differing amounts in most foods. Foods that are known to be high in fat content include cheese, meat, poultry skin, avocados, and numerous others. Some foods we are accustomed to eating require using fats as part of their preparation. Baked goods, such as cakes and cookies; salad dressing; and fried foods acquire fats through preparation (Figure 33-4).

Proteins

The body needs food with proteins to build and rebuild its cells. Proteins have four calories per gram. Hair, skin, teeth, and bones are all parts of your body that require protein. Proteins are in a continuous cycle of building up and breaking down. Approximately 3 to 5 percent of your body's protein is rebuilt each day. Beans, peanut butter, meats, eggs, and cheese are high-protein foods.

Vitamins

Vitamins are also essential to the functions of the body. Some vitamins are dissolved in body fat and are stored in the body. Fat-soluble vitamins are not required in the diet each day because they can be stored in the body. These include vitamins A, D, E, and K. Nine other vitamins—vitamin C and eight B vitamins—are water soluble and must be replenished daily. Specific vitamins have specific jobs in the body, and some foods are known to be rich in specific vitamins (Figure 33-5).

VITAMINS

Functions and Sources					
A	**B**	**C**	**D**	**E**	**K**
Functions					
vision bones skin healing wounds	using protein, carbohydrates, and fats to keep eyes, skin, and mouth healthy brain nervous system	wound healing blood vessels bones teeth other tissues works with minerals	needed for using calcium and phosphorus bones teeth	preserve cell tissue	blood clotting
Sources					
yellow, orange, and green vegetables	whole-grain and enriched cereals, breads, meats, beans	citrus fruits, melons, berries, leafy green vegetables, broccoli, cabbage, spinach	fatty fish, liver, eggs, butter, added to most milk	vegetable oils, whole-grain cereals	leafy green vegetables, peas, cauliflower, whole grains

FIGURE 33-5 Vitamins have specific functions, and it is important to choose foods that supply adequate vitamin content.

Minerals

More than 20 minerals are needed by the body. The amounts needed may be small, but they are required nonetheless. The 20 minerals are divided into four major groups. Some minerals are required for healthy bones, others regulate bodily functions, some are needed to make special materials for cells, and others trigger chemical reactions in the body (Figure 33-6).

Water

The human body is more than 50 percent water. Water carries nutrients to cells, removes waste, and maintains the body's proper temperature. Fluid foods, such as milk and juice, obviously help supply the body with water. However, foods such as meat and bread also provide water.

Food Groups That Meet Needs

Each food is different in the types of nutrients it contains and ultimately provides to the body. The new *MyPlate* nutrition initiative replaces the food pyramid. This new program was introduced by the U.S. Department of Agriculture in 2012, and it divides foods into five major food groups, which represent the nutritional needs of the body. The five food groups are the following:

- Fruits—Any fruit or 100 percent fruit juice
- Vegetables—Any vegetable or 100 percent vegetable juice
- Grains—Foods made from cereal grains: wheat, rice, oats, corn, barley, or others
- Protein foods—Seafood, meat, poultry, beans, peas, eggs, processed soy products
- Dairy—Fluid milk and many foods made from milk that retain their calcium content. Calcium-fortified soymilk belongs to this food group; milk products with little or no calcium, such as cream, cream cheese, or butter do not belong. Choices should be fat-free or low-fat.

Functions of Minerals			
Bone Development	**Fluid Regulation**	**Materials for Cells**	**Trigger other Reactions**
Sources of Minerals			
calcium milk products *magnesium* nuts, seeds, dark-green vegetables, whole-grain products *phosphorus* no specific food group *fluorine* some seafood, some plants, may be added to drinking water	*sodium* salt *potassium* bananas *chlorine* salt	*iron* meats, liver, beans, leafy green vegetables, grains *works with vitamin C* *iodine* iodized salt added to salt	*zinc* whole-grain breads and cereals, beans, meats, shellfish, eggs *copper* fish, meats, nuts, raisins, oils, grains

FIGURE 33-6 The careful selection of foods can ensure a correct balance of minerals in our diets.

Essential oils are not one of the food groups, but essential nutrients are provided by them. Fats that are liquid at room temperature are oils. Oils from plants and fish are included in this group.

Other categories included in *MyPlate* include physical activity and discretionary calories. The recommendations vary for physical activity according to size and other characteristics of the individual.

The U.S. Department of Agriculture (USDA) represents these groups in the format of a dinner plate to help us remember the relative proportions that each group should make up in a daily diet (Figure 33-7). You generally should need not seek foods in the essential oils because plenty of these are used to prepare foods, and they show up in adequate amounts in most people's diets in the United States. The names of the groups suggest some of the typical foods that they include. However, the quality of diet can be increased by selecting the most nutritious items from each group (Figure 33-8).

Eating from each of the major food groups daily will ensure a well-balanced diet and provide the essential nutrients needed for growth and development. The number of portions consumed per day differs with each group. At different stages in life, requirements within each group may vary to some degree, but no food group should dominate or be eliminated from the diet. More information on nutrition is provided in Unit 26.

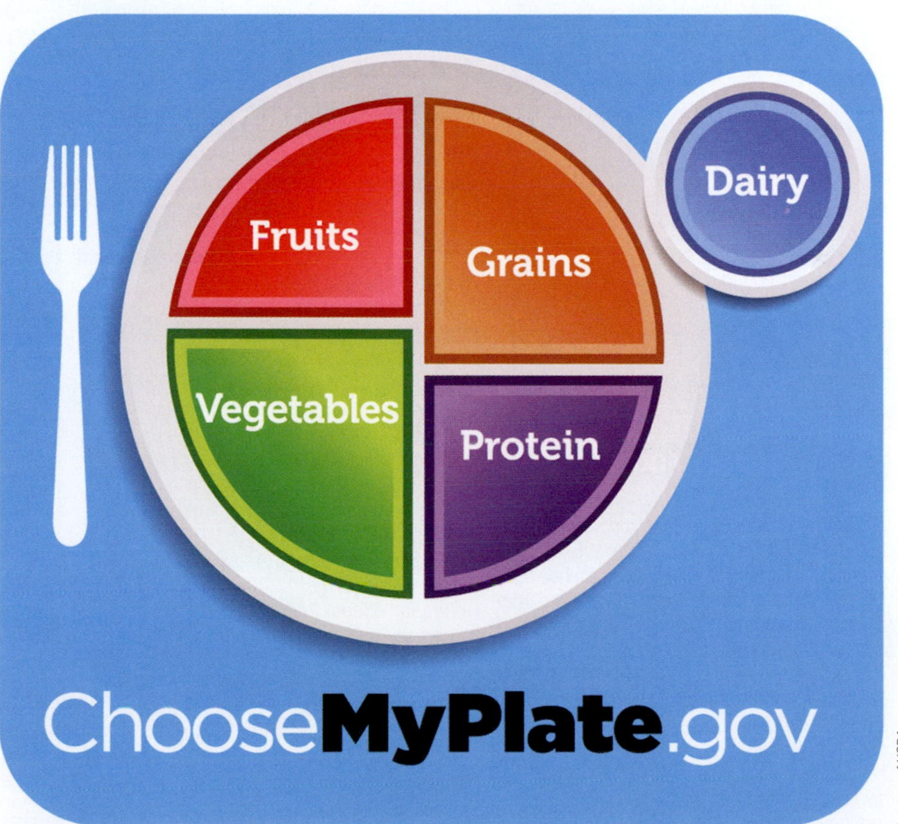

Courtesy of USDA.

FIGURE 33-7 The nutrition icon provides guidelines for the human diet based on research findings. Food groups are organized differently today from those of the food pyramid with which we are most familiar. "ChooseMyPlate" is becoming a familiar way to organize our foods. The following interactive Web site is available for further information: www.choosemyplate.gov.

POULTRY, FISH, MEAT, AND EGGS

Lower-scoring foods tend to be high in calories, cholesterol (eggs), fat (red meat), or sodium (processed meats). The foods near the top are relatively low in fat. Most of the foods are rich in protein and iron. (All servings are 4 ounces broiled, baked, or roasted, unless noted otherwise.)

	NUTRITION SCORE
clams, steamed	19
turkey breast, skinless	10
tuna canned in water (3 oz.)	6
cod	1
egg white (1 large)	1
salmon, canned (3 oz.)	1
scallops, steamed	1
flounder	−1
lobster meat, boiled	−4
salmon fillet	−5
blue crab meat, steamed	−6
chicken breast, skinless	−8
turkey breast luncheon meat, 3 slices (2 oz.)	−13
tuna canned in oil (3 oz.)	−18
shrimp, steamed	−21
chicken breast with skin	−23
Canadian bacon, fried, 2 slices (2 oz.)	−24
veal cutlet, breaded, pan-fried	−25
round steak, trimmed (5 oz.)	−29
ham, luncheon meat, 2 slices (2 oz.)	−32
pork chops	−48
bacon, fried, 4 1/2 slices (1 oz.)	−54
egg (1 large)	−59
shrimp, fried	−63
bologna, 2 slices (2 oz.)	−70
leg of lamb	−76
salami, luncheon meat (2 oz.)	−80
hamburger, lean	−81
sirloin steak (5 oz.)	−86
hamburger, regular	−92
round steak, untrimmed (5 oz.)	−97
chicken thighs, fried, home recipe (2)	−103
sausage links, 2 (2 oz.)	−112
pot roast	−114

GRAIN FOODS

Contrary to myth, starchy grain foods are not fattening. Most people would do well to eat more grain foods in place of meat. Grains, especially whole grains, are a nicely balanced, low-fat source of carbohydrate, vitamins, minerals, and protein. (All serving sizes are 1 cup cooked, unless noted otherwise.)

	NUTRITION SCORE
bulgur (cracked wheat)	69
wheat germ (1/4 cup)	61
pearled barley	60
brown rice	45
spaghetti or macaroni	45
oatmeal	38
hominy grits	35
whole-wheat bread (2 slices)	31
hamburger or hotdog roll (1)	18
corn muffin (1)	1

FRUITS

Fruits can give you, naturally, all the sweetness you want, plus fiber, vitamins A and C, and other nutrients. Go easy on the dried fruits! Their sugars are sticky and promote tooth decay. (All servings are one medium piece, unless noted otherwise.)

	NUTRITION SCORE
papaya (1/2 medium)	74
cantaloupe (1/4 medium)	67
strawberries (1 cup)	65
orange	62
prunes, uncooked (5)	51
dried apricots (5)	49
tangerine	41
watermelon (2 cups cubed)	36
apple	36
pear	36
blueberries (1 cup)	36
pink grapefruit (1/2)	35
pineapple, fresh (1 cup)	34
banana	32
cherries (1 cup)	32
honeydew melon (1/10 melon)	31
raisins (1 oz.)	28
plums (2)	27
applesauce, unsweetened (1/2 cup)	18
peach	17
grapes (30)	16
peaches in heavy syrup (1/2 cup)	2

DAIRY

While most dairy foods are rich in protein and calcium, the lower-scoring foods are high in saturated fat, cholesterol, and sodium.

	NUTRITION SCORE
yogurt, nonfat (1 cup)	58
milk, skim (8 oz.)	40
yogurt, plain lowfat (1 cup)	36
milk, 1% lowfat (8 oz.)	28
milk, 2% lowfat (8 oz.)	16
yogurt, fruit-flavored lowfat (1 cup)	13
chocolate milk, 2% lowfat (8 oz.)	6
cottage cheese, 1% fat (1/2 cup)	3
sour cream, lowfat (2 Tbsp.)	−2
ricotta cheese, part skim (1 oz.)	−3
yogurt, plain (1 cup)	−5
mozzarella cheese, part skim (1 oz.)	−5
milk, whole (8 oz.)	−7
cheddar cheese, reduced fat (1 oz.)	−7
nondairy powder coffee creamer (2 tsp.)	−12
half and half cream (2 Tbsp.)	−15
Swiss cheese (1 oz.)	−15
cottage cheese, 4% fat (1/2 cup)	−16
mozzarella cheese (1 oz.)	−19
sour cream (2 Tbsp.)	−26
cheddar cheese (1 oz.)	−32
American cheese (1 oz.)	−34
whipped cream (2 Tbsp.)	−59

VEGETABLES

Most vegetables are great sources of vitamins—especially A and C—and minerals. Try a new vegetable today! (All serving sizes are 1/2 cup cooked, unless noted otherwise.)

	NUTRITION SCORE
sweet potato, baked (1 medium)	184
potato, baked (1 medium)	83
spinach	76
kale	55
mixed vegetables, frozen	52
broccoli	52
winter squash (acorn, butternut), baked	44
Brussels sprouts	37
cabbage, chopped, raw (1 cup)	34
green peas	33
carrot (1)	30
okra	30
corn on the cob (1 ear)	27
tomato (1 medium)	27
green pepper (1/2)	26
cauliflower, raw	25
artichoke (1/2)	24
romaine lettuce, raw (1 cup)	24
collard greens	23
asparagus	22
celery (four 5" pieces)	19
green beans	18
turnips	16
sauerkraut	15
summer squash (zucchini)	12
green beans, canned	10
iceberg lettuce, raw (1 cup)	8
bean sprouts (1/4 cup)	7
onion, chopped, raw (1/4 cup)	7
eggplant	6
cucumber slices, raw	4
mushrooms, raw (1/4 cup)	2
dill pickle (1/2 large)	−3
avocado (1/2 medium)	−25

LEGUMES

Beans are excellent sources of dietary fiber, protein, vitamins, and minerals. They are also very low in fat. (All serving sizes are 3/4 cup cooked, unless noted otherwise.)

	NUTRITION SCORE
kidney beans	91
navy beans	82
black beans, black-eyed peas, or lima beans	78
lentils	74
chickpeas	68
split peas	56
tofu/bean curd (4 oz.)	33

DESSERTS

Most desserts are high in fat, sugar, and calories. Next time, try fresh fruit or nonfat frozen yogurt for a change. (Serving sizes are 1 cup, unless noted otherwise.)

	NUTRITION SCORE
angelfood cake (2 oz.)	1
chocolate pudding (1/2 cup)	−2
Jell-O (1/2 cup)	−7
brownie with nuts (1 3/4" square)	−23
sherbet	−37
vanilla ice cream	−73
cheesecake (4 1/2 oz.)	−161

FIGURE 33-8 Estimated relative nutritional values of selected food items.

FOOD CUSTOMS OF MAJOR WORLD POPULATIONS

What people eat from each of the major food groups varies around the world. Food habits reflect what is most readily available.

In hot and wet climates such as in Southeast Asia, a lot of rice is consumed. In areas of the world where corn grows well, many food items contain corn in some form. For example, cornmeal may be used to make tortillas or pancakes and may be mixed as a cereal in countries such as Mexico. Availability of food and technology to prepare food have dictated eating habits over the years. For instance, introducing dairy products in countries where dairy cows are not raised presents educational as well as transportation and processing challenges.

METHODS OF PROCESSING, PRESERVING, AND STORING FOODS

One of the oldest ways to preserve food for delayed use is fermenting and pickling. **Fermentation** is a chemical change that involves foaming as gas is released. Long ago, it was determined that some foods did not spoil when allowed to ferment naturally or when fermented liquids were added to the foods. Controlled fermentation is now used to produce cheeses, wines, beers, vinegars, pickles, and sauerkraut (Figure 33-9).

Courtesy of USDA.

FIGURE 33-9 Fermentation is a food-preservation process whereby bacteria convert sugars to acids or alcohol that protects the food from spoilage.

Today, the primary objective of processing and preserving is to change raw commodities into stable forms. With refrigeration and various processing techniques, we now expect that almost all foods should be available at any time during the year and in any part of the world.

Slowing deterioration is the primary goal in food preservation. Tomatoes and cucumbers that will be sold raw are waxed to retard shriveling while they are in the grocery store. Apples may be treated with a decay inhibitor. Table grapes are fumigated with sulfur dioxide to control mold. Similarly, silos where grains are stored are purged with 60 percent carbon dioxide to control insects.

Carbon dioxide inhibits the growth of bacteria. Controlled atmosphere (CA) is the process whereby oxygen and carbon dioxide are adjusted to preserve or enhance particular foods. An example of food preservation by CA is the transporting of cut lettuce in CA to prevent the edges from turning brown.

There are many other processing and preservation techniques. These techniques slow deterioration and allow one to enjoy foods in a variety of forms throughout the year and around the globe.

Courtesy of USDA.

FIGURE 33-10 Food storage and preservation through ice, refrigeration, and freezing are the cornerstones of milk, meat, fish, fruit, and vegetable handling today.

Refrigeration is an important key to many processing and preservation techniques. Refrigeration is the process of chilling or keeping cool. Low temperatures reduce or stop processes that contribute to the deterioration of products. Refrigeration retards respiration, aging, ripening, textural and color change, moisture loss and shriveling, insect activity, and spoilage from bacteria, fungi, and yeasts. Refrigeration is effective, but it is expensive and is used widely only in well-developed countries.

When crops are harvested in the field, their temperatures are between 70° and 80° F. The goal of refrigeration is to quickly reduce that temperature to near 32° F. How quickly the food is cooled varies depending on the type of cooling technique used. Some vegetables are precooled in the field by cold air blasts, hydrocooling, or vacuum. Hydrocooling means cooling with water. Milk is quickly cooled in refrigerated lines and tanks on dairy farms (Figure 33-10).

SCIENCE CONNECTION NEW SOLUTION TO AN OLD PROBLEM

Courtesy of USDA.

Irradiation of fruits and vegetables has proven to be a safe and effective way to control the spread of insects that infect these foods.

After a lengthy period of inquiry and research, the USDA approved the irradiation of raw packaged poultry as a safe and effective procedure for protection against food-borne illness. Irradiation of food is being done in more than 35 countries.

Proper cooking is still the last stage in food handling to eliminate microorganisms that can cause food poisoning or other illness to those who consume the food. The use of water for washing and cooling fruits, vegetables, and meats during and after processing may spread harmful microorganisms throughout the batch and leave every item with low concentrations of the microorganisms—that is, all of the food will have low-level contamination, and the organisms will multiply and increase their hazards if conditions are right to do so. Therefore, handling raw meat that has pathogens on its surface carries the threat of contamination of hands, countertops, cooking areas, and, ultimately, cooked food.

The logical step after a product is cooled is to continue cooling it until it is frozen (Figure 33-11). When foods are kept at 0° F or lower, little deterioration occurs. However, even frozen foods have storage limits. Fruits and vegetables should be consumed within 1 year after freezing. Meats should be consumed within 3 to 6 months. Vegetables that are to be frozen often require blanching before freezing. **Blanching** is the scalding of food for a brief time before freezing it. This process inactivates enzymes that cause undesirable changes when plant cells are frozen.

Canning is a popular preservation technique. **Canning** involves putting food in airtight containers and sterilizing the food to kill all living microorganisms that could cause spoilage. Temperatures of 212° to 250° F are required to successfully can food products. Metal cans are coated to reduce chemical reactions between the metal can and its contents. A 2-year shelf life for canned food is considered normal.

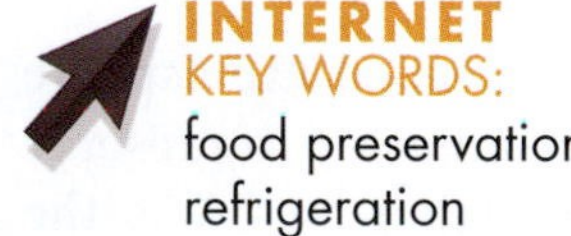
INTERNET KEY WORDS:
food preservation
refrigeration

Another popular way to process and preserve food is dehydration. **Dehydration** occurs as the moisture content is reduced to inhibit growth of microorganisms. Moisture can be removed by the sun, by indoor tunnel or cabinet dehydrators, or by freeze-drying. **Freeze-drying** is the newest method of dehydration. It involves the removal of moisture by rapid freezing at very low temperatures. When foods are dehydrated, they have a moisture content of 2 to 10 percent. The shelf life of dehydrated foods is as long as 2 years. **Oxidative deterioration** is loss of quality because of a reaction with oxygen. This can occur when air reaches dried foods. Glass or metal containers are more airtight than plastic ones. Dried soup mixes, packaged salad dressing, spices, and dried fruits are foods that have been dehydrated. Dehydrated foods are lighter in weight and lower in volume than the same whole foods.

Another processing technique is dehydrofreezing. A **dehydrofrozen product** is processed by precooking, evaporating water, and freezing. Potatoes are being used in the development of this technique. They are precooked as cubes or slices, and water is evaporated to reduce the weight by 50 percent. The potatoes are then frozen.

Another factor to consider in the processing and handling of foods is humidity during storage and transportation. Humidity is the amount of moisture in the air. A humidity level of 90 to 95 percent is required for high-moisture products

Science has long sought a way to eliminate food-borne pathogens without altering or damaging the product or leaving poisonous chemicals on the food. Irradiation seems to be the best answer. Irradiation provides the same benefits as processing food by heat, refrigeration, or freezing. It also replaces treatments with chemicals—to destroy insects, fungi, or bacteria that cause food to spoil or cause human disease. Irradiated foods are wholesome and nutritious. The treatment process involves passing food through an irradiation field, but the food itself never contacts a radioactive substance. The process has been subjected to extensive study and declared safe by special scientific committees in Denmark, Sweden, the United Kingdom, and Canada, and it has received official endorsement from the World Health Organization and the International Atomic Energy Agency. No known risks exist from consuming food that is irradiated using a legal procedure.

Notably, the approved level of irradiation does not kill all microorganisms—in other words, it does not sterilize food. Therefore, other food-preserving measures must be followed when storing irradiated food and preparing it for consumption. Irradiation does not alter the food in any perceptible way for the consumer. The Cordex Alimentarius, an international committee on food safety, has developed a green irradiation logo. Irradiated foods from the United States bear this green logo along with the words "Treated with Radiation" or "Treated by Irradiation."

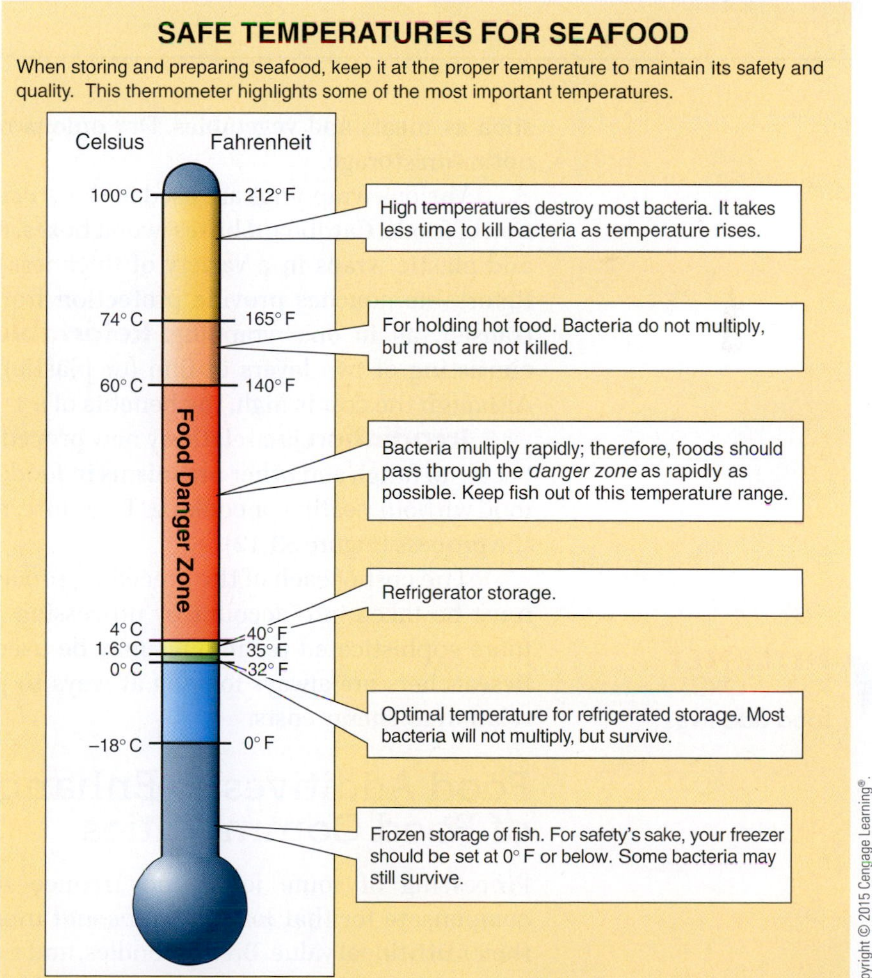

FIGURE 33-11 It is important to choose the appropriate heat intensity and duration for the intended purpose.

SCIENCE CONNECTION MILK STAYS COOL UNDER PRESSURE

© Larisa Lofitskaya/Shutterstock.com.

A new process for homogenization of milk uses extreme pressure to disperse the milk fat throughout the milk. The final product is more stable with a longer shelf life.

Alexander Lin, a graduate student at Texas A&M University, has improved an old technique for processing milk. The result is a longer shelf life for milk products. During processing, milk is homogenized. Homogenization is performed by pumping milk into a chamber that applies between 2000 and 3000 lbs. of pressure per square inch (psi). The pressure causes the fat in the milk to break into tiny random-shaped particles that stay suspended in the milk instead of rising to the surface like fat does in nonhomogenized milk. Lin decided to find out what would happen if the pressure were increased to 14,000 psi. The result of the greater pressure was a smaller, more uniform fat globule disbursed throughout the milk. This made the final product more stable with an increased shelf life. This research may change the way dairy products are processed and may help keep them fresher longer. The improved high-pressure homogenization will benefit milk processors, grocers, and consumers.

such as meats and vegetables. Dry onions require only 75 percent humidity for optimum storage.

Various wrappings are used in the processing, preservation, and transportation of foods. Cardboard boxes, wood boxes, molded pulp trays to reduce bruising, and plastic wraps in a variety of thicknesses or plies all meet different needs. Retortable pouches provide protection from light, heat, moisture, and oxygen transfer, all in one wrapping. **Retortable pouches** are flexible packages consisting of two layers of film (or plastic) with a layer of foil between them. Although the cost is high, the benefits of a 1- to 2-year shelf life are appealing.

Irradiation is a relatively new procedure using gamma rays to kill insects, bacteria, fungi, and other organisms in food products. Gamma rays pass through food without heating or cooking. Therefore, no heat-sensitive nutrients are lost in the process (Figure 33-12).

The cost of each of the preceding processes and techniques is different and must be taken into account by processing plants. Where energy costs are low, more sophisticated techniques may be used without incurring excessive costs. Researchers are always looking at ways to preserve food more economically in relation to energy costs.

INTERNET KEY WORDS:
food additives

Food Additives to Enhance Sales of Food Commodities

Processing of some foods may reduce their natural nutritional value. To compensate for that loss, vitamins and minerals are added to foods to restore their nutritional value. Bread, noodles, and rice have vitamins and minerals added to them before they are packaged. Vitamins A and D may be added to fluid milk before it leaves the processor.

A food additive is anything that is added to a food product during processing and before it goes into a package. In addition to food additives that restore nutritional

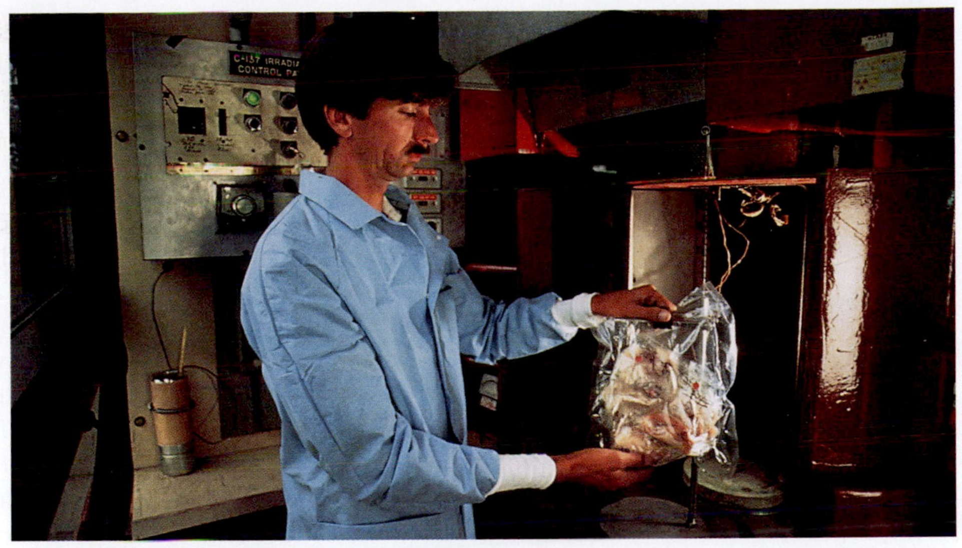

FIGURE 33-12 Irradiation is now an approved method for controlling *Salmonella* bacteria in chicken meat.

value, preservatives are also added to foods to extend their shelf life. Shelf life refers to the amount of time before spoilage begins. Food additives also enhance the color or appearance of foods. Other food additives reduce the cooking time of foods such as oatmeal. Sugar is probably one of the most widely used food additives. It is found on the labels of many cereals and beverages. Food labels identify the contents of food products. The order of ingredients on a food label indicates the proportions of each, in descending order (Figure 33-13). The government is continually testing and evaluating the positive and negative effects of food additives. Some food additives are known to cause harm, and they are banned by the government: Among them are calamus extracts and oils, calcium cyclamate, coumarin, and safrole.

Grape Juice:

Grape juice, grape juice from concentrate, ascorbic acid (vitamin C). No artificial flavors or colors added.

This label tells you:

- mostly grape juice and juice concentrate
- vitamin C added

Grape Juice Drink: (10% Grape Juice)

Water; high fructose corn syrup; sugar; grape juice concentrate; fumaric; citric, and maltic acids (provide tartness); vitamin C; natural flavor; artificial color.

This label tells you:

- mostly added water, corn syrup, and sugar
- some grape juice
- vitamin C added, plus other things

Powdered Grape Drink:

Sugar, citric acid (provides tartness), natural and artificial flavor, artificial color, vitamin C.

This label tells you:

- mostly sugar
- no juice at all
- vitamin C added, plus other things

FIGURE 33-13 Products may look the same, but read the labels for the real differences. Ingredients are listed in order from the most to the least amount found in the products.

FOOD PREPARATION TECHNIQUES

Some foods can be eaten as they come in the package. Crackers and raisins are examples of foods that do not require additional processing at home to ensure safe eating. However, cooking is often required for other foods, whereas heating may only improve flavor in others. Raw meat should be cooked before it is eaten to ensure safety. How food is cooked is determined by the type of food and the appliances available. Two basic methods are used to cook foods. The appliances used to accomplish these methods vary from household to household. **Dry-heat cooking** involves surrounding the food with dry air in the oven or under the broiler. This method is usually used for tender cuts of meat having little connective tissue and for vegetables having high-moisture content, such as potatoes. **Moist-heat cooking** involves surrounding the food with hot liquid or by steaming, braising, boiling, or stewing the food. The warm moisture breaks down the connective tissues. Moist-heat cooking is a popular method used for less-tender cuts of meat and for vegetables with low moisture content.

Appliances have been designed to accommodate different types of food as well as different time and energy demands. The goal of preparing good food in a short time period, without expending excessive energy, has resulted in the development of several appliances that supplement the traditional gas or electric range and oven. Pressure cookers concentrate moisture by sealing it in, thus reducing cooking time. Crockpots and oven bags allow slow, moist cooking without the operator continually being near the cooking process.

Conventional ovens are used to cook small amounts of foods by dry heat. This requires less energy than large, traditional ovens. The microwave oven uses electromagnetic waves to heat and cook food. It offers a more energy-efficient way to cook foods that require both dry and moist heat. Approximately 50 percent of the energy goes to the food in the microwave, whereas only 6 to 14 percent of the energy used by a conventional range actually goes to the food.

Convection ovens heat food with the forced movement of hot air. Ovens that combine convection and microwave functions are now on the market. Again, energy efficiency, as well as convenience for the food preparer, is a constant goal.

Dehydrators remove moisture from food; **smokers** preserve food by keeping smoke in contact with the food for prolonged periods. The drying or smoking of food for family use is not very popular in the United States. Refrigeration is still required for most foods after they are dried or smoked.

Courtesy of Price Chopper Supermarkets.

FIGURE 33-14 Baked goods from grains form the basic diet in many cultures.

FOOD PRODUCTS FROM CROPS

Food from plant sources helps meet body requirements for food in four of the five food groups and in essential oils. Fruits, vegetables, grains, protein, and oils all come from crops (Figure 33-14).

Fruits, Vegetables, and Nuts

Fruits, vegetables, and nuts are nearly ready to eat when they are harvested. This can be as simple as pick, wash, and eat for items such as leafy vegetables, berries, and fruits. The process could include picking, cutting, peeling, shelling, washing, trimming, cooking the food, or eating it raw. These products may be processed for storage for as short as a few hours to as long as many years. Food processing may be done at home for the family or commercially for the billions of people in the world market.

For some food products, the journey from field to table may involve many processes. The first stage may be to simply cool the product and hold it in a temperature- and humidity-controlled environment until the next stage, or it may be to process the food without first precooling. Where processing occurs immediately, the product typically passes through washing equipment designed to avoid injury to the product. Washing may involve flotation, rotary, water-jet, or other cleaning procedures. Skins and hulls may be removed by hot water bath, steam, flame, or cutters. The product may then be trimmed, halved, quartered, sectioned, sliced, diced, or crushed for drying, canning, freezing, or processing into ready-to-cook or precooked products.

Processed foods must be packaged and correctly labeled. All along the way, nutritionists, chemists, inspectors, public health officials, and others monitor the process and product. The law requires that food be safe for consumption and of the weight, quality, and grade that is specified on the label.

Cereal Grains

Cereal grains compose the major diet of most of the world's people. Rice, wheat, corn, barley, oats, and other grains are consumed as whole grains, cracked or rolled grains, flour, bran, and many other products. Grains are economical to process because they can be left on the plant until nearly dry enough to prevent spoilage. If stored in a cool, dry place, properly dried grains can remain edible for many years. However, much of the world's food grain is lost every year because of rotting in the field and in storage or being consumed or ruined by insects, birds, and rodents.

Processing grain for human consumption generally means separating or milling the grain into its basic components—hulls, bran, flour, and germ. These components are then used to make breakfast cereals, breads, pastries, pasta, and thousands of other products that fill the grocery shelves.

Oil Crops

Soybeans, cottonseed, peanuts, rape, palm nuts, coconuts, olives, and corn are examples of crops that are rich in vegetable oils. These are used for cooking, frying, baking, and other food processes, as well as for many food products such as dressings, coffee creamers, and shortening. These oils are also used in the manufacture of paints, lacquers, plastics, and many products of industry. The seeds or nuts and other oil-rich plant parts are crushed or ground and heated. The oil is then extracted by solvents and purified for food and industrial uses. The meal is dried and ground mostly for livestock feed (Figure 33-15).

Those who do not eat meat products can still meet the requirements for food from all five food groups. People with allergies to milk products can meet their nutritional needs for protein and other nutrients from soybeans and other plant sources.

FOOD PRODUCTS FROM ANIMALS

Meats, fish, poultry, and dairy products provide high-quality protein and are nutritious foods from animal sources (Figure 33-16).

Dairy Products

Milk is used and consumed in a variety of ways. Approximately 37 percent of all milk consumed in the United States is in fluid form. The remainder is used to

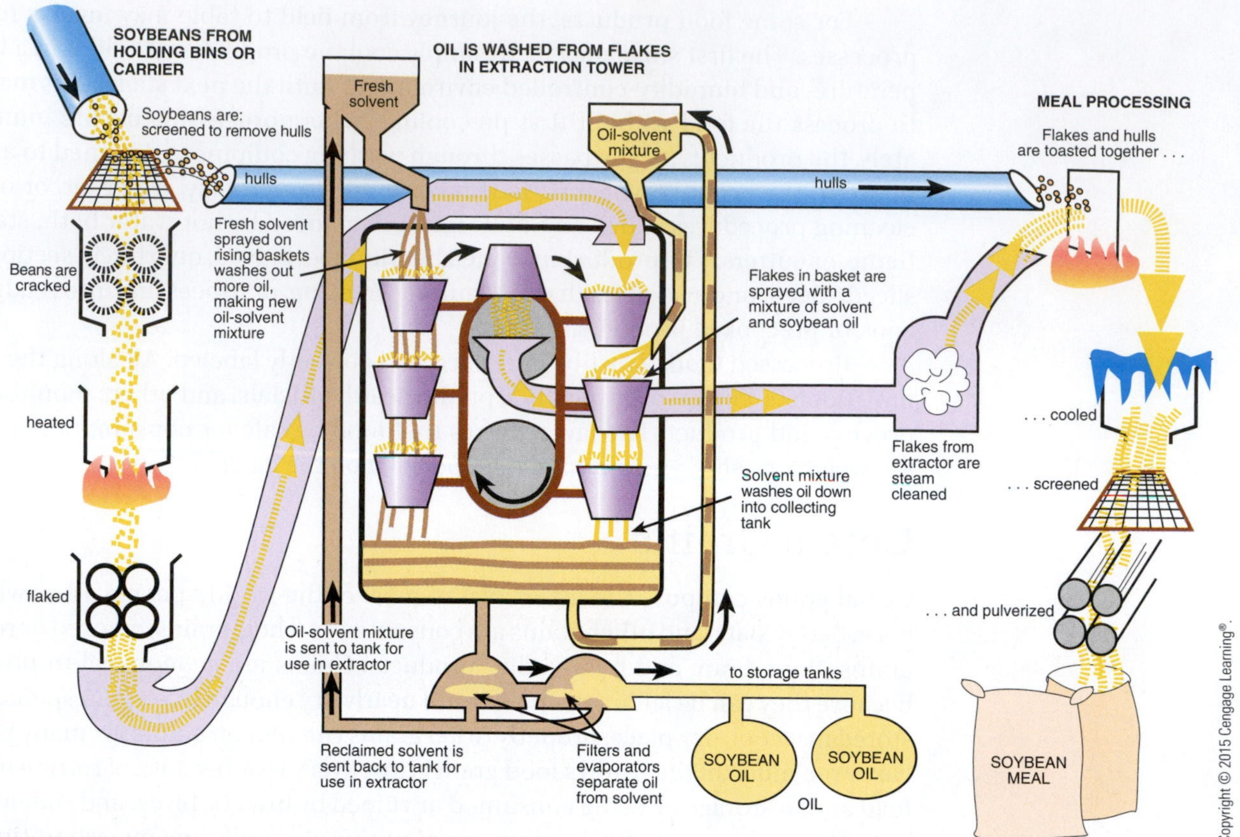

FIGURE 33-15 Processing of grains and oil-crop seeds releases the basic building blocks for many products.

FIGURE 33-16 Meats, fish, poultry, eggs, and dairy products are nutritious foods from animal sources.

FIGURE 33-17 Milk is pasteurized to kill bacteria, homogenized to keep the milk fat evenly suspended, and bottled for the convenience of the customer.

make many products, including cheese, yogurt, butter, frozen foods, dried whole milk, cottage cheese, evaporated milk, and condensed milk. The processing of milk in different ways results in these many different products (Figure 33-17).

Milk and Milk Products

Although recommendations for milk consumption are greatly influenced by calcium needs, milk contains some of all the essential food nutrients needed by the body. The major components of milk are water, fat, protein, sugar, and minerals. In addition, numerous other highly important dietary components are found in milk.

INTERNET KEY WORDS:
dairy products, nutrition

Water

Although cow's milk is a fluid containing about 88 percent water, it also contains about 12 percent total solids. This is comparable to the solid content of many other foods. Because milk is a food specifically prepared by nature for the nourishment of the very young, it provides the water necessary for life. The water also acts as a carrier for dissolved, suspended, and emulsified components.

Protein

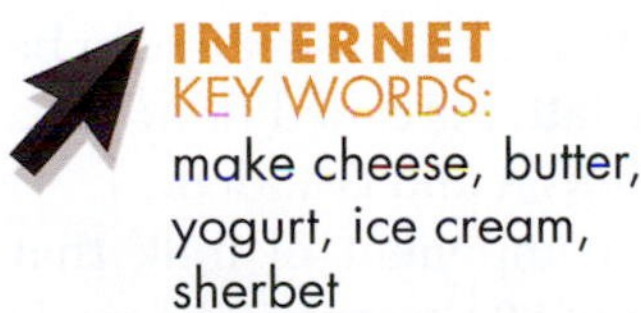

INTERNET KEY WORDS:
make cheese, butter, yogurt, ice cream, sherbet

Milk provides a substantial proportion of the total protein in our food supply. **Casein** is the most abundant protein in milk. It is found only in milk and makes up about 82 percent of the total milk protein. Casein exists in suspended form and is easily coagulated by the action of acids and enzymes. Milk proteins are high-quality proteins that contain all of the essential amino acids in proper balance for good nutrition.

Milk Fat

Fats are concentrated sources of energy. Many different fatty acids in milk give it the distinctive, pleasing flavor that complements many prepared foods. It also contains vitamins that are fat soluble—A, D, E, and K. Milk fat exists in a highly emulsified state, which facilitates its digestion.

Lactose

Lactose is milk's major carbohydrate and accounts for about half of the nonfat solids in milk. The relative sweetening power of lactose is about one-sixth of sucrose, the common table sugar.

Minerals

Milk contains seven minerals as major constituents and many more in minor or trace amounts. Calcium and phosphorus are essential in human nutrition for building bony structures and for certain metabolic processes. Milk is the chief source of food calcium in the diets of people in the United States. It has the added advantage of containing phosphorus in the same biological ratio to calcium as occurs in the growing skeleton. It is difficult to provide the recommended daily amounts of dietary calcium without using milk or milk products because calcium is poorly distributed among other foods.

Vitamins

All of the vitamins known to be required by humans are found in milk. Some are fat soluble and are associated with butterfat, whereas those that are water soluble are found in the nonfat portion of milk. Vitamin A and carotene are present in high concentration in milk fat. Carotene, from which vitamin A is formed in the body, gives milk fat its characteristic color. The vitamin D content of fresh milk is low. However, most commercially pasteurized milk is fortified with vitamin D to balance the product for best nutrition.

Milk is an abundant source of riboflavin (vitamin B_2) and an important source of niacin. Although the niacin content of milk is low, it is in a fully available form. Milk contains significant amounts of thiamine (vitamin B_1). Other vitamins of the vitamin B complex occurring in milk include pantothenic acid, pyridoxine (vitamin B_6), biotin, vitamin B_{12}, folic acid, and choline.

Processed Milk Products

The nutritional completeness of milk has led to its reputation as nature's most nearly perfect food. Where it falls short in minor ways, science and the food-processing industry have intervened to improve the options for the consumer.

Fluid milk sold in the United States is pasteurized to ensure safety from disease-causing organisms, and it is homogenized to keep the milk fat in suspension so it does not require stirring before each use. Whole milk is marketed with 3.5 percent milk fat. However, for those desiring less fat intake, milk can also be purchased with 2 percent, 1 percent, or no milk fat. The latter is called skim milk. Milk may also be purchased as fortified milk with vitamins A and D added.

The fat in milk is called butterfat. Cream is a component of milk that contains up to 40 percent butterfat. Butter, which is about 80 percent fat, is made from cream. Cream is recovered from whole milk by concentrating the fat portion of the milk. This is accomplished by passing milk through a cream separator. Whipping cream contains about 40 percent fat, table cream 18 to 20 percent, and "half-and-half" approximately 12 percent.

Ice cream, ice milk, and sherbet (also called sherbert) are also dairy products. They account for most of the frozen desserts in the United States.

Nonfat dried milk is used both as human food and animal feed. It is frequently used as an ingredient in dairy and other food products.

Courtesy of DeVere Burton.

FIGURE 33-18 Cheese is manufactured from milk by coagulating the milk solids to form curd. The curd is ground and pressed into larger blocks of cheese. Bacteria act on the cheese to give it the characteristic flavor that makes each variety unique.

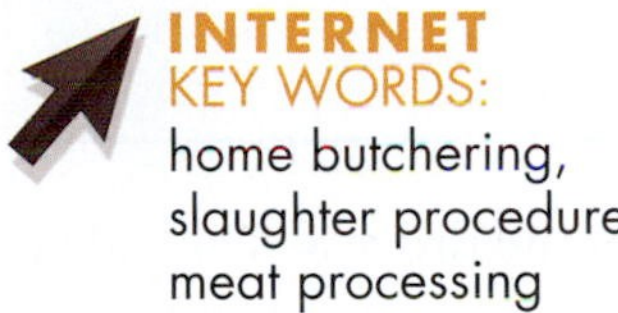

INTERNET KEY WORDS:
home butchering,
slaughter procedure
meat processing

Cheese is made by exposing milk to certain bacterial fermentations or by treating it with enzymes. Both methods are designed to coagulate some of the proteins found in milk. Cheese is available in many types and varieties (Figure 33-18). Cottage cheese is made from skimmed milk. Condensed and evaporated milks are canned milk products. Condensed milk and evaporated milk are both produced by removing large portions of water from the whole milk through a machine called a **vacuum pan**. Condensed milk is further treated by adding sugar. The sugar content makes condensed milk an important ingredient in the baking and ice cream industries. The many components of milk have resulted in the evolution of an extensive dairy industry. Consumers can now choose from a great array of dairy products (Figure 33-19).

Meat Products

Meat products, like milk, are enjoyed in a variety of forms. Animal species vary somewhat, but the procedures for slaughtering and processing each are similar. There are common by-products from the processing of animals, as well as the familiar meats that are important to U.S. eating habits.

Beef

Meats of all kinds come from slaughtered animals. Slaughter means to kill and process or dress animals for market purposes. The procedure is as follows for beef.

The first step is to render insensible by making the animal unable to sense pain. There are several methods of accomplishing this that comply with the Humane Slaughter Act of 1958. Packers that do not comply with provisions of this act are unable to sell meat to the federal government. The approved methods must be rapid and effective. Methods used in rendering insensible include a single blow or gunshot, electrical current, or use of carbon dioxide gas. In addition, rendering insensible may be accomplished by following the ritual requirements of religious faiths.

When the animal is insensible to pain, it is shackled, hoisted, and stuck to permit bleeding. **Shackles** are mechanical devices that confine the legs and prevent movement; hoist means to raise into position. To *stick* means to cut a major artery to permit the blood to drain from the body. A large artery is severed for efficient bleeding out. The head of the animal is removed during or after the bleeding-out process.

Courtesy of USDA.

FIGURE 33-19 Milk is a basic ingredient of many processed food items.

AGRI-PROFILE CAREER AREAS: FOOD SCIENTIST/FOOD TECHNICIAN/ NUTRITIONIST/DIETICIAN

Courtesy of USDA/ARS #K-48191.

Are these young visionaries contemplating careers in food science? Food scientists develop better ways to process, handle, package, prepare, and market food products.

Food science programs are available in high schools, colleges, technical institutes, and universities. Careers in food science may be in specialty areas, such as meats, fruits, vegetables, baked goods, dairy products, wine, or other beverages.

Food companies and consumers rely on food scientists and technicians to develop new products to meet the ever-changing needs of a busy world. Convenience products, such as instant- and freeze-dried coffee, dried and vacuum-packed fruits and meats, processed chicken tenders, boneless rolled meat, yogurt, ice cream, pasteurized and homogenized milk, shelf-safe milk, low-calorie and low-fat food, space-age food, unrefrigerated foods, and many other products are the handiworks of food science and technology.

Food scientists work for food companies, universities, research centers, food chain stores, dairies, radio and television stations, and government agencies. Some become authors and publish journal and magazine articles, cookbooks, and recipe books.

The removal of the **hide** or skin is the next step in the process. The hide is cut open at the median, or midline, of the belly of the animal, and hide pullers are used to remove the hide in one piece. The breast and rump bones are split at this time by sawing.

The term **viscera** refers to organs located in the body cavity of the animal. Organs that are removed include the heart, liver, and intestines. The kidneys are not removed at this time. Plants that are regulated by the USDA must have the carcass and viscera inspected to confirm good animal health. **Carcass** refers to the body meat of the animal—the part that is left after the offal has been removed. Offal consists of the non-meat material that is converted to by-products. It includes the blood, head, shanks, tail, viscera, hide, and loose fat.

The next step is to split the carcass by cutting through the center of the backbone and removing the tail. The **split carcass** (sides of the animal) is then washed with warm water under pressure.

A high-quality carcass should have a smooth appearance after it is cooled. To accomplish this, the hot carcass is wrapped with a large cloth called a **shroud**. Sides are cooled for a minimum of 24 hours before ribbing and further processing (Figure 33-20). The meat is kept at 34° F until it is sold and consumed.

Courtesy of USDA/ARS #K-4284-12.

FIGURE 33-20 After slaughter, the carcasses of meat animals are inspected by USDA inspectors and cooled before being cut into block or retail cuts.

Fresh beef is often processed and shipped within a few days. Small processors usually allow carcasses to hang in the cooler to allow them to age, or ripen. To **age**, or **ripen**, means to leave undisturbed for a period so that minor biological changes can take place while the beef cools. Fresh beef is not in its most tender state immediately after slaughter. While the beef is aging, evaporation (loss of moisture) and discoloration are kept to a minimum. A fairly thick covering of fat on the carcass helps the aging process. Three methods are used to age beef—traditional aging, fast aging, and vacuum packaging. Time, temperature, and technology help define these three methods.

Beef carcasses are generally disposed of in three ways: as block beef; fabricated, boxed beef; and processed meats. Traditionally, meat is shipped in exposed halves, quarters, or wholesale cuts to be cut into retail cuts in supermarkets. In this condition, it is referred to as **block beef**. It is ready for sale "over the block" or counter. This traditional method creates concern over sanitation, shrinkage, spoilage, and discoloration. Therefore, more packers are using the disassembly process. The **disassembly process** means that the carcass is divided into smaller cuts, vacuum sealed, boxed, moved into storage, and shipped to retailers. This process is also known as **fabrication and boxing**. Processed meats are made from scraps of meat that are not in suitable form for sale over the block. Such meats have the bones removed and are sold as boneless cuts. They can also be canned, made into sausage, dried, or smoked.

Sheep

Sheep are slaughtered in much the same way as beef. They are first rendered insensible and bled. Next, the front feet are removed, the pelt is removed, and the hind feet and head are removed. The opening of the carcass and removal of the viscera, called *evisceration*, are similar to the procedures outlined for beef animals. In view of the small size of a lamb, the forelegs are folded at the knees and are held in place by a skewer after evisceration. Washing and cooling procedures are similar to those used for beef.

Hogs

The procedure for slaughtering hogs is a little different from that used for cattle and sheep. Hogs are rendered insensible, shackled, hoisted, and bled. The carcasses are then plunged into hot water at 150° F for about 4 minutes. This process is required to loosen the hair and dry skin.

The hair of hogs is removed by mechanical scraping. A de-hairing machine can remove the hair from about 500 hogs per hour. After the hair is removed, the hog is returned to overhead racks and processing continues.

The hog is washed and singed before the removal of the head. Singe means to burn lightly to remove any remaining hair. Next, the carcass is opened and eviscerated before being split or halved with a cleaver or electric saw. The leaf fat is removed. Leaf fat consists of layers of fat inside the body cavity. Before the carcass is washed, the kidneys and facing hams are inspected. After washing, the carcass is sent to coolers at 34° F.

Unlike the fat from beef cattle, lard is considered a product along with the meat. Lard is the fat from hogs. It is used for a variety of cooking and baking products. Lard is often combined with other animal fats and with vegetable oils, such as cottonseed, soybean, peanut, and coconut. Such mixes are extensively used for baking, cooking, frying, and other food preparations and for commercial products.

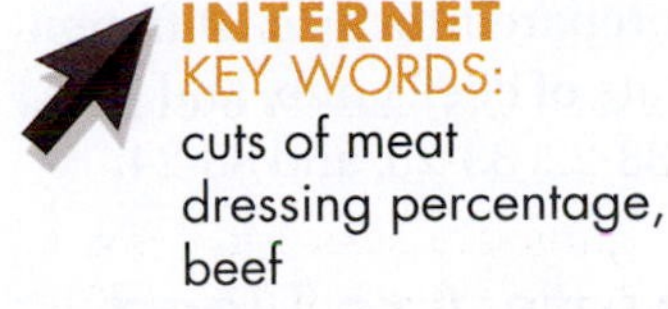

Poultry

The steps in poultry processing are similar in many ways to those required for other animals. The feathers that cover a bird, like the hair that covers the hog, must be removed before evisceration.

The process begins by securing the bird on a conveyor belt and bleeding out. Next, the bird is scalded before feather removal or picking. Singeing or lightly burning the skin is required to remove the fine hairs that cover a bird under its feathers. After the feathers and hair have been removed, the bird is washed and eviscerated and the giblets cleaned. **Giblets** are the heart, liver, and gizzard of a bird. The bird is then cooled to 40° F, usually with ice (Figure 33-21).

Courtesy of USDA.

FIGURE 33-21 Modern processing plants make it possible to process, inspect, and package the millions of broilers that are consumed fresh or frozen for later use.

Fish

After fish are caught or harvested, they, too, must be prepared for processing and consumption. The procedure depends on the type of fish. Evisceration is usually done after the scales and head are removed. Washing and cooling follow.

Fish, like other foods, are consumed in a variety of forms. Whether a tuna is to be consumed whole or processed to be eaten later determines whether the fish is left whole or is cut up. Some shellfish, such as crab and lobster, are kept alive until they are cooked. The heat of cooking kills them. The meat can then be removed and consumed or processed.

Kosher Slaughter

Kosher means right and proper. This type of animal processing is based on the religious ritual of the Jewish faith. It requires that animals be killed by a rabbi or a specially trained representative. The methods, as well as the time by which the meat of the animal must be sold, are based on ritual that relates to concerns for sanitation. Meat must be consumed quickly. Neither packers nor retailers are permitted to hold kosher meat for more than 216 hours (9 days). Washing is required every 72 hours.

Major Cuts of Meat

After an animal is slaughtered, the meat is further prepared for use. Different areas of the animal are useful for different purposes. Cuts of beef, lamb, and pork, with wholesale and retail terms, are shown in Figures 33-22, 33-23, and 33-24.

By-products—How Waste Products Are Used

Although most of the animal is consumed by humans, some parts are inedible. The **dressing percentage** is a term used to indicate the percentage or yield of hot carcass weight to the weight of the animal on foot. The offal is removed from the live animal to arrive at the dressing percentage. The formula is the following: hot

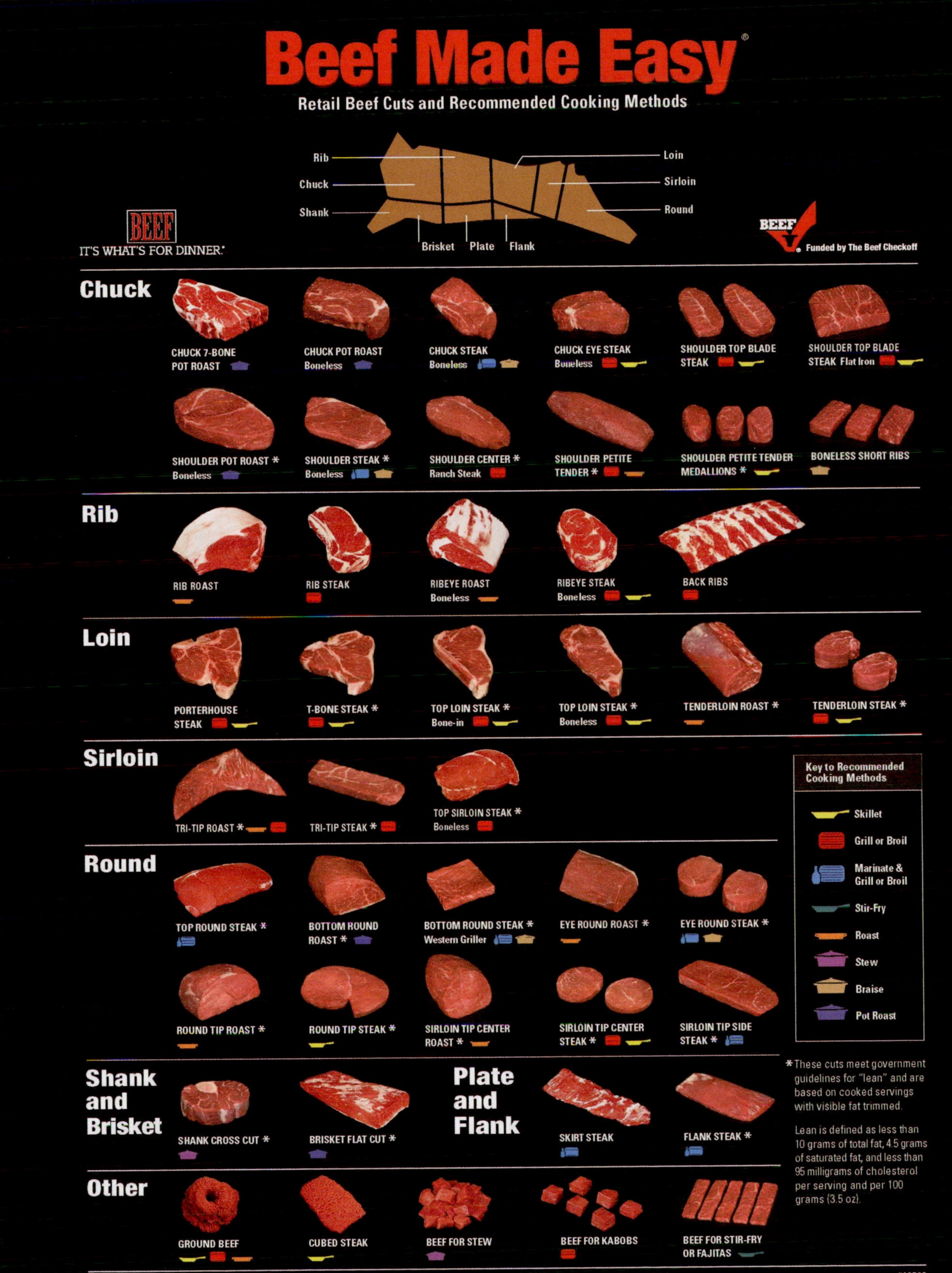

Chart courtesy of the Beef Checkoff Program.

FIGURE 33-22 Retail cuts of beef—where they come from and how to cook them.

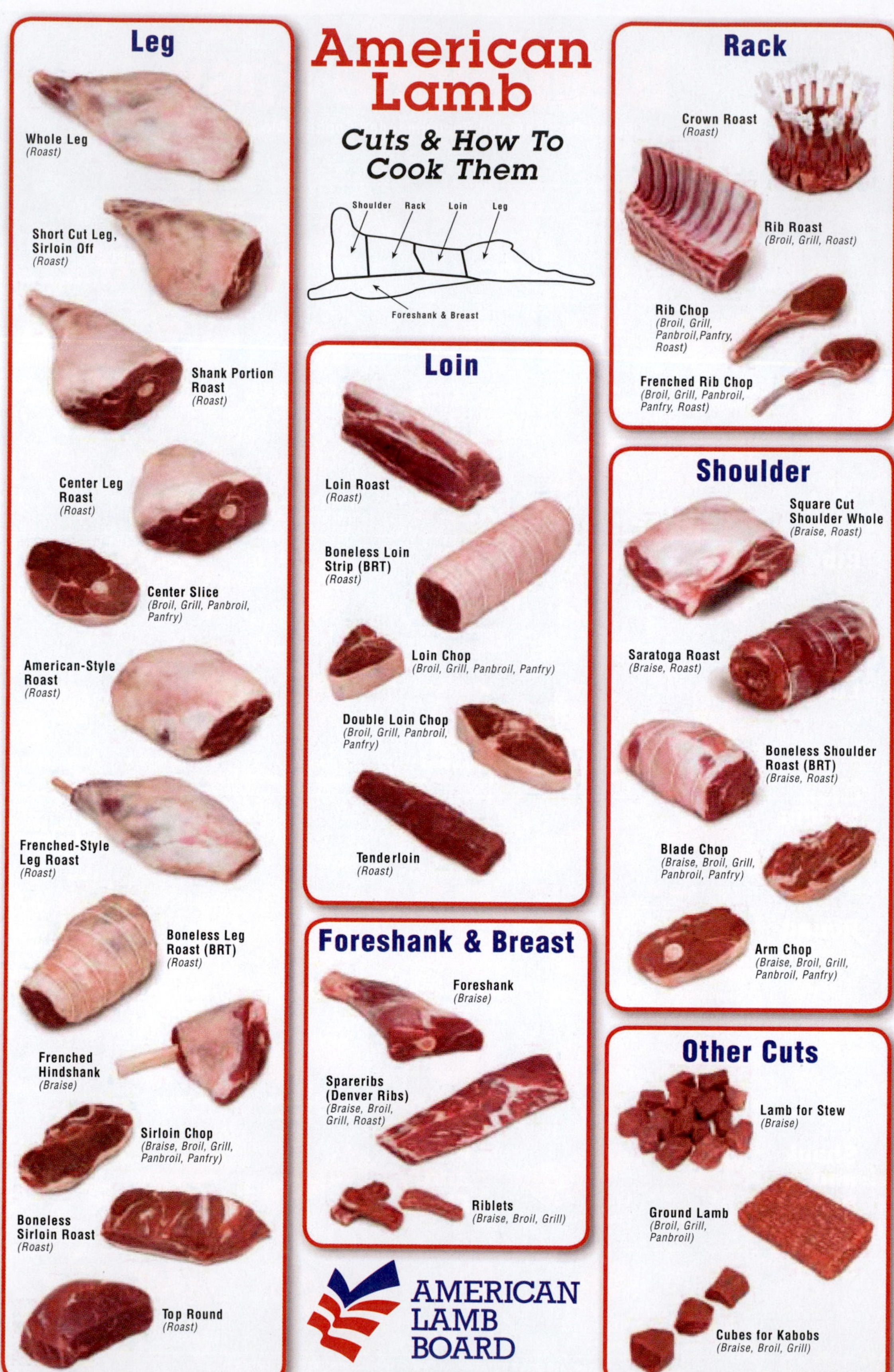

FIGURE 33-23 Retail cuts of lamb—where they come from and how to cook them.

Pork Basics

The Other White Meat®
Don't be blah.®

Shoulder Butt

Upper row (l-r):
Bone-in Blade Roast,
Boneless Blade Roast
Lower row (l-r):
Ground Pork (The Other Burger®),
Sausage, Blade Steak

Cooking Methods
Blade Roast/Boston butt –
roast, indirect heat on grill, braise, slow cooker
Blade Steak –
braise, broil, grill
Ground Pork –
broil, grill, roast (bake)

Picnic Shoulder

Upper row (l-r):
Smoked Picnic,
Arm Picnic Roast
Lower row:
Smoked Hocks

Cooking Methods
Smoked Picnic Roast –
roast, braise
Arm Picnic Roast –
roast, braise, slow cooker
Smoked Hocks –
braise, stew

Side

Top:
Spareribs
Bottom:
Slab Bacon, Sliced Bacon

Cooking Methods
Spareribs –
roast, indirect heat on grill, braise, slow cooker
Bacon –
broil, roast (bake), microwave

Leg

Upper row (l-r):
Bone-in Fresh Ham,
Smoked Ham
Lower row (l-r):
Leg Cutlets,
Fresh Boneless Ham Roast

Cooking Methods
Fresh Leg of Pork –
roast, indirect heat on grill, slow cooker
Smoked Ham –
roast, indirect heat on grill
Ham Steak –
broil, roast

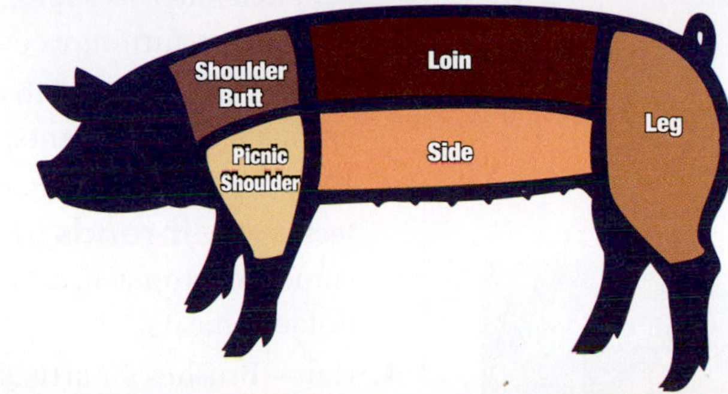

Roasts
No-fuss family dinner or holiday favorite

THE MANY SHAPES OF PORK
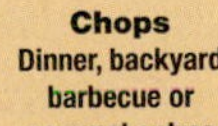
Cut Loose!

When shopping for pork, consider cutting traditional roasts into a variety of different shapes

Chops
Dinner, backyard barbecue or gourmet entree

Cubes
Great for kabobs, stew and chili
grill, stew, braise, broil

Strips
Super stir fry, fajitas and salads
grill, sauté, stir fry

Cutlets
Delicious breakfast chops and quick sandwiches
1/8 to 3/8 inch thick – sauté, grill

www.TheOtherWhiteMeat.com

#03341 04/2007

Courtesy of the National Pork Board.

FIGURE 33-24 Retail cuts of pork—where they come from and how to cook them.

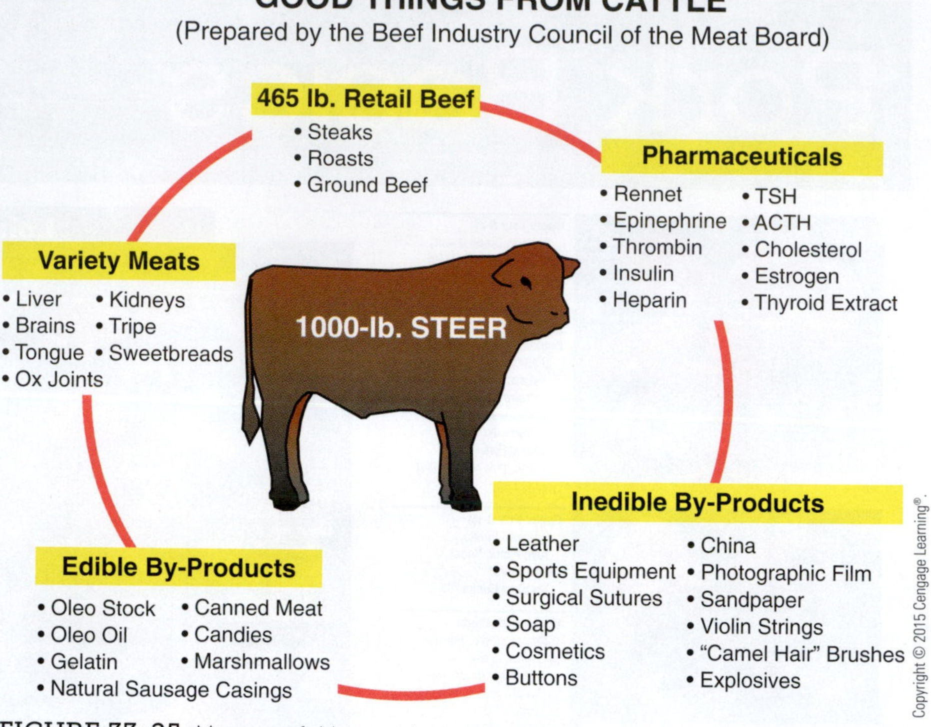

FIGURE 33-25 Many useful by-products come from animals.

carcass weight divided by the live weight times 100. The offal may be 40 percent of the live weight of the animal.

What happens to the offal accounts for many products that are used daily (Figure 33-25). By-products can generally be divided into 12 categories:

1. Hides—Leather from animal hides is used to make a variety of consumer products, such as shoes, harnesses, saddles, belting, clothing, sports equipment, furniture coverings, coats, hats, and gloves (Figure 33-26).
2. Fats—These are used to make products such as oleomargarine, soaps, animal feeds, lubricants, leather dressing, candles, and fertilizers.
3. Variety meats—The heart, liver, brains, kidneys, tongue, cheek meat, tail, feet, **sweetbreads** (thymus and pancreatic glands), and **tripe** (pickled rumen, or stomach, of cattle and sheep) are sold over the counter as variety or fancy meats.
4. Hair—Brushes for artists are made from the fine hairs on the inside of the ears of cattle. Other hair from cattle and hogs is used for toothbrushes, paintbrushes, mattresses, upholstery for residential and commercial furniture, and air filters.
5. Horns and hoofs—These items are used as a carving medium and are fashioned into decorative knife and umbrella handles, goblets, combs, and buttons.
6. Blood—Blood is used in making animal feeds, fertilizers, and shoe polish.
7. Meat scraps and muscle tissue—After separation from the fat, meat scraps and muscle tissue are most often made into meat-meal or tankage. **Tankage** is the dried animal residue used as fertilizer and feed for

© iStockphoto/susaro.

FIGURE 33-26 Leather is a by-product of the meat-processing industry. It is used for many purposes, including the manufacture of sporting goods, shoes, furniture, upholstery, belts, coats, saddles, and numerous other items.

animals other than cattle. The "mad cow disease" incident resulted in some recycled meat products being banned in cattle rations.

8. Bones—Bones are put to some of the same uses mentioned for horns and hoofs. In addition, bones are converted into stock feed, fertilizers, and glue.
9. Intestines and bladders—Sausage, lard, cheese, and snuff all use the intestines and bladders from cattle. Strings for musical instruments and tennis rackets are also made from these by-products.
10. Glands—The pharmaceutical industry relies heavily on animal glands for many medicinal drugs.
11. Collagen—**Collagen** is the chief constituent of the connective tissues. Glue and gelatin are made from collagen. These products, in various forms, are used in the furniture, photography, medical, and baking industries.
12. Contents of the stomach—Stomach contents of slaughtered animals are used primarily in the production of feed and fertilizer.

NEW FOOD PRODUCTS ON THE HORIZON

The foods that we eat and how they reach us are exciting areas of agriscience. New foods, and new versions of familiar foods, are arriving on the market daily. Research is being conducted on improving the nutritional values of foods and keeping our food costs at a minimum.

HOT TOPICS IN AGRISCIENCE

LIFE EXPECTANCY—AN INDICATOR OF FOOD SAFETY

A newborn baby in the United States has a life expectancy of 75.4 years for boys and 80.4 years for girls.

A newborn baby in the United States is entering the most dangerous period in life—infancy and early childhood. U.S. boys, on average, can expect to live to the age of 75.4 years, whereas U.S. girls can expect to live to the age of 80.4 years. These calculations are based on the records and charts of government and life insurance companies, which insure against death on the basis of their charts. Each year that you survive, these numbers will go up unless you take up habits (such as smoking) that are known to reduce survival rates.

The average life expectancy of an individual depends on adequate medical care, clean drinking water, sanitary facilities, and food of sufficient amount and quality. Countries in which these resources are unavailable usually experience high death rates, especially among infants and small children. Other factors such as wars also influence the average number of years a person may reasonably expect to live. Countries with the greatest life expectancy are those that are wealthy enough to feed and care for their people.

The average life expectancy at birth has been increasing in many countries of the world for quite a few years. The following examples of life expectancy at birth illustrate the importance of providing access to plenty of good-quality food, medical care, clean water, and sanitary facilities: Mozambique—39.2, Swaziland—39.6 years, years, India—64.7 years, United States and Western Europe—78.3 to 80.9 years, and Japan, Norway, Iceland, and Sweden—80.2 to 82.6 years.

The fruit industry uses an electronic, fruit-shaped beeswax sensor to log the bumps and bruises sustained by fruit during shipment. Improvements in handling equipment are resulting in better apples and are reducing loss for producers.

Eggs are used in the preparation of other foods as well as being eaten by themselves. Eggs were sometimes taken out of diets because of their high cholesterol content, which leads to heart problems. Low-carbohydrate diets have brought about a resurgence in consumption of eggs. Researchers have found ways to reduce the cholesterol level in eggs by changing the rations fed to laying hens. These eggs are now available on the market.

The conveniences that everyone enjoys are expected to improve because of continuing new developments in the food science sector of agriscience.

STUDENT ACTIVITIES

1. Write the Terms to Know and their meanings in your notebook.
2. Keep a diary of what you eat for a week. Note which foods represent each of the six food groups.
3. With your teacher, arrange a visit to a meat department of a local grocery store to observe meat processing.
4. Compare the end result when a food has been processed in a variety of ways—for example, fresh, canned, frozen, dehydrated, and freeze-dried potatoes.
5. Make a collage of items that exist because of animal by-products.
6. Using an outline of an animal, identify the major cuts of meat and where they come from on the animal.
7. Research the steps that must be taken to prepare kosher foods from food products other than meats. Write a brief summary outlining your findings.
8. Create a game that will help you remember the number of servings a person should eat daily from each food group. Use the *MyPlate* initiative as your model.

SELF-EVALUATION

A. MULTIPLE CHOICE

1. Which of the following is *not* a carbohydrate?
   a. sugar
   b. starch
   c. fiber
   d. meat
2. The number of minerals needed by the body is approximately
   a. 2.
   b. 12.
   c. 20.
   d. 200.
3. How much food from the protein foods group is recommended for teenagers?
   a. 2 to 3 oz.
   b. 3 to 4 oz.
   c. 5 to 6 oz.
   d. 7 to 8 oz.
4. The butterfat content of table cream is approximately
   a. 10 to 12 percent.
   b. 18 to 20 percent
   c. 30 to 32 percent.
   d. 38 to 40 percent.

5. To which milk product is sugar added during processing?
   a. evaporated milk
   b. condensed milk
   c. skimmed milk
   d. dried milk
6. Which mineral is added to most drinking water in the United States to assist in tooth development?
   a. zinc
   b. iron
   c. fluorine
   d. chlorine
7. Iron is important in the diet for the development of
   a. bones.
   b. vision.
   c. blood.
   d. hair.
8. Traditional aging of meat takes approximately
   a. 1 to 6 hours.
   b. 1 to 6 days.
   c. 1 to 6 weeks.
   d. 1 to 6 months.

B. MATCHING

________	1. Canning	a. Thymus and pancreatic glands
________	2. Shrouding	b. Internal organs
________	3. Dehydration	c. Added before packaging
________	4. Viscera	d. Stored in airtight containers
________	5. Sweetbreads	e. Reduced moisture content
________	6. Food additive	f. Wrap in cloth

C. COMPLETION

1. Vitamins __________, __________, __________, and __________ are fat soluble.
2. Minerals that assist in fluid regulation are __________, __________, and __________.
3. Meat from animals that have been slaughtered as prescribed by Jewish ritual is called __________.
4. The method of precooking food, removing moisture, and then freezing is known as __________.

SECTION 10

COMMUNICATIONS AND MANAGEMENT IN AGRISCIENCE

Looking at the broad picture, agriscience is the industry that feeds, clothes, shelters, and keeps the nation healthy. These benefits spill over to many other nations of the world. The United States shares the results and products of research in agriscience and donates or sells things such as food, fiber, crops, livestock, plant oils, lumber, medicines, breeding stock, germ plasm, seed, agricultural chemicals, machines, and an endless array of agriscience materials for producing food and fiber as well as the end products themselves. Our land-grant system of colleges and universities has provided a system of research in agriscience that is the envy of the world. In many countries of the world, the greatest prize in education is to be admitted to and graduate from a land-grant college or university in the United States.

The agriscience industry in the United States provides the nation with its playgrounds, outdoor fitness centers, and a source of renewal for a nation boxed in by modern skyscrapers, large industrial facilities, and clogged highways. These sources of renewal take the form of open spaces for softball, baseball, soccer, hockey, horseback riding, golf, hiking, hunting, biking, walking, fishing, bird-watching, recreational farming, and gardening. A significant number of people make their living as farmers, ranchers, greenhouse operators, aquaculturists, landscapers, nursery operators, zoo workers, and others in the outdoor segment of the food, fiber, and ornamental plant and exotic animal segments of the industry. Agriculture and its supporting industries employ about 20 percent of this nation's workers and provide the income for these workers to enjoy the fruits of our free-enterprise system. In contrast to many nations, where most of a household's income is consumed by the purchase of food, we spend less than 10 percent of our income on food and have the remaining income available for clothing, housing, transportation, recreation, health, appliances, communications, education, leisure pursuits, and gadgets. No other nation has the

Putting It All Together

abundance of opportunities that the United States offers under an agricultural industry that produces such affordable food and leaves so much income for other needs and preferences.

The driving force for this abundance is planning and management. In no other sector of the economy is the profit incentive of the free-enterprise system in greater evidence. In a system in which one can receive and use or accumulate the proceeds from work or innovation, the human spirit and initiative accelerates; the results are demonstrated in greater productivity. The adage that "where there's a will, there's a way" is proven correct every day as people apply their education, inventiveness, innovation, and energy in a workplace where a substantial amount of the proceeds come back to the individual. Whether one is the owner or a stockholder receiving the profits of the organization, a salesperson or manager in line for bonuses or commissions, a pieceworker or assembly-line worker whose income is tied directly to output, or a per-hour worker with the option to seek a better-paying position, the free enterprise system rewards productivity.

This section addresses the application of communication and management skills in agriscience. These skills transcend agriscience and culminate in the marketplace where the fruits of our labor are sold. The marketplace yields the cash to buy labor and other inputs for the business and yields profits for the inventors, entrepreneurs, and other risk takers.

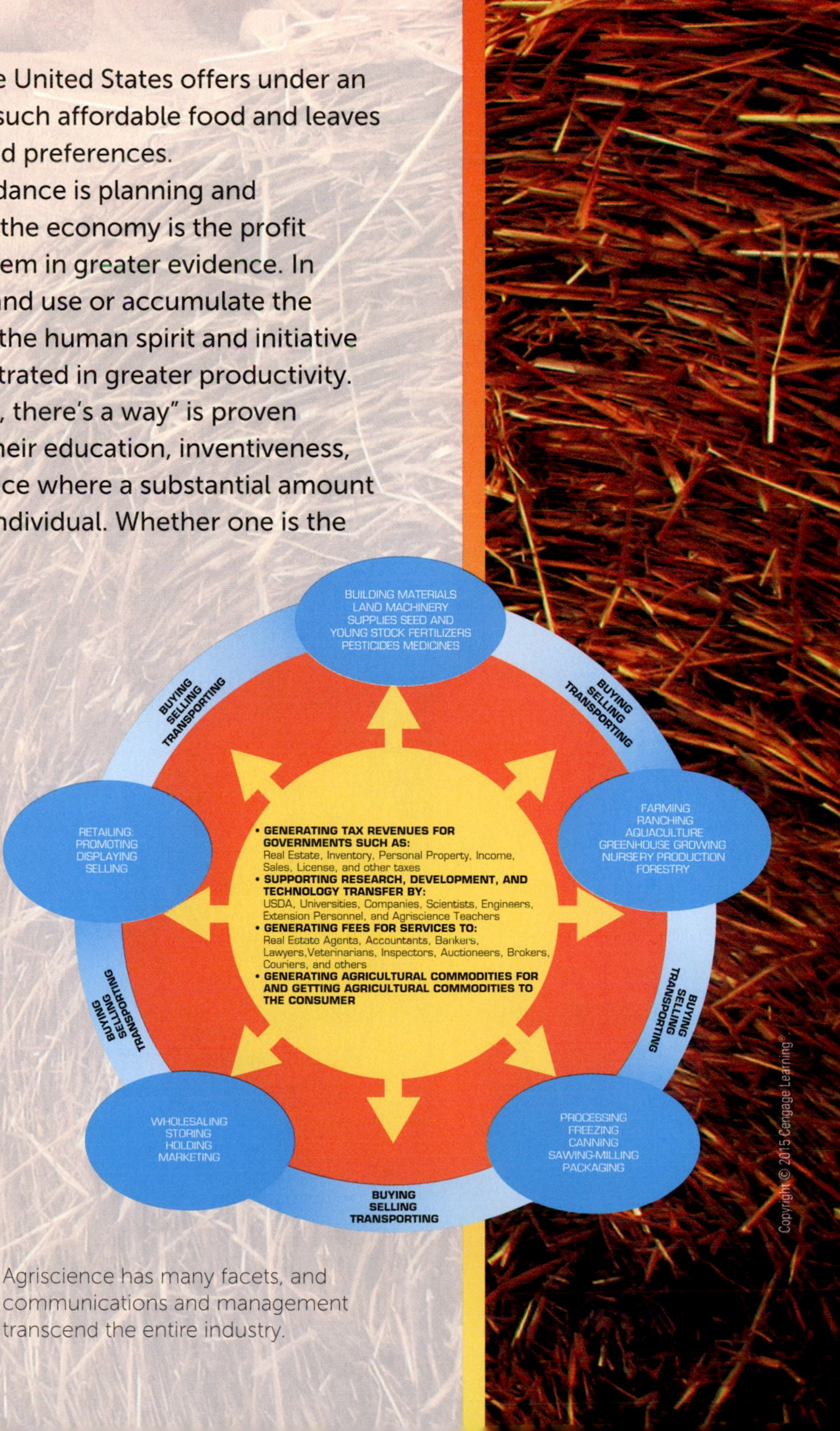

Agriscience has many facets, and communications and management transcend the entire industry.

UNIT 34

Marketing in Agriscience

OBJECTIVE

To determine the strategies and procedures for marketing agricultural commodities to maximize profits.

MATERIALS LIST

- newspaper and marketing reports
- Internet access

COMPETENCIES TO BE DEVELOPED

After studying this unit, you should be able to:

- describe the marketing strategies that maximize profits.
- describe various pricing strategies.
- distinguish between wholesale and retail marketing.
- describe some methods of marketing at farms, roadside stands, and farmers' markets.
- discuss advantages and disadvantages of terminal markets, auctions, and direct marketing.
- recognize fees, commissions, and other costs of marketing.
- list the procedures for handling livestock to minimize losses during marketing.
- understand the grades of popular agriscience commodities.
- recognize marketing trends and cycles.
- describe the use of futures in agriscience marketing.

SUGGESTED CLASS ACTIVITIES

1. Visit a livestock auction or a fruit and vegetable wholesale market. Observe how the price is established for the product that is offered for sale. Make arrangements ahead of time for the manager of the market to speak to the class about the services that are offered at the place of business. Ask about such business issues as insurance, bonding, financing, storage or yarding practices, and transportation. Have each class member prepare a written report that covers the basic points that were learned. Limit the report to one to two pages in length.
2. Obtain a videotape that is used in a video auction. Obtain permission to listen in to the phone call when a sale is actually

TERMS TO KNOW

supply
demand
consumer demographics
product advertising
institutional advertising
psychological pricing
penetration pricing
skimming
loss-leader pricing
prestige pricing
retail marketing
consumer
middlemen
wholesale marketing
terminal market
commission
auction markets
video merchandising
direct sales
cooperative
vertical integration
commodity exchange
futures market
opening a position
offsetting a position
out-of-the-market
profit

made. If the time of the sale is not convenient for the class, request permission to record the audio portion of the sale, and then play it as you observe the videotape. Discuss the advantages and disadvantages of using a video auction to market livestock.

3. Divide the class into small groups. Have each group design a new marketing campaign for a locally produced agricultural product. Make a commercial. Prepare a videotape or present it live to the class.

In our free-enterprise system, commodities, goods, and services are produced, modified, and improved. The expectation is to sell or market the commodity or service as value is added along the way to the final user or consumer. Many people receive financial benefits in the form of salaries, profits, or "in kind" benefits. Governments receive tax revenues from the various transactions. This unit explores the basics of marketing agriscience commodities to maximize profits.

A number of factors need to be considered when deciding how to market agriscience products. Some of these factors include the following:

- Demand for the product to be produced
- Supply of the product that is available
- Types and availability of markets in the area
- Competition from similar products
- Buying power of intended consumers
- Seasonal variations in demand

Supply and demand have long been the factors that determine whether the production and processing of a product can be profitable. **Supply** is the amount of a product available at a specific time and price. Some of the things that may determine supply are how many people are in the business of producing that product in the market area, how much of the product is coming into the market area from other areas, and the history of profitability for that product in the area.

Demand for a product is also determined by numerous factors. **Demand** is the amount of a product wanted at a specific time and price. It is often determined largely by price. The less expensive a product of a given quality is, the more it is in demand. However, other factors influence demand for a product. The amount of money available to consumers to buy the product is a factor that is often overlooked. Competition from similar products may reduce demand. Seasonal variations in demand also need to be considered. It is easier to sell ice cream in the summer than in the winter. Advertising, personal contacts, and product samples to create demand are also factors to be considered (Figure 34-1).

FOCUS ON CONSUMERS

Consumers ultimately determine marketing success. Commodities or services may be on the doorsteps of consumers, but if the consumer chooses not to buy and does not submit to persuasion or coercion, then there is no sale. In our democratic,

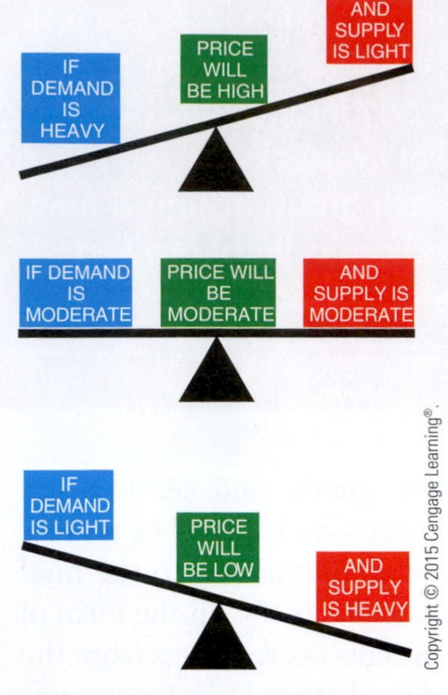

FIGURE 34-1 In a true free-market system, the relationship between supply and demand will determine price.

capitalistic system, we have the right to produce and market goods and services for personal profit within the state and federal legal frameworks. Similarly, the right of refusal to buy goods and services is a basic right of consumers. The marketing system involves the interplay between producers and consumers and all the people, processes, and jobs that come into play along the way.

Consumer Demographics

One of the first marketing functions is to decide who will be the target population. The target population consists of the people who will be given deliberate exposure to the product. Consider some of the following factors when determining target populations: Who are the consumers? How many consumers are there? What are their preferences? Where are they located? Categories of information about consumers or potential consumers are known as **consumer demographics**. Knowledge of consumer demographics is helpful in determining target populations and marketing strategies (Figure 34-2).

Some useful categories of consumer demographics for the marketing of agriscience products are population density, ethnic makeup of the population, family income, discretionary money held by individuals within the family, family size, eating preferences, who makes buying decisions in the household, occupation and work locations, tendencies to eat at home or away, clothing preferences, styles, and recreational preferences. People generally satisfy their needs for basic food, clothing, and shelter first and then move on to purchases that reflect group and individual tastes. Because the purchase of food in the United States requires less than 10 percent of the average family income, there is considerable room for choice in the purchase of additional goods and services.

Important Questions to Ask When Determining Who the Consumer Is

- Who are the consumers?
- How large is this consumer population?
- Where are they located?

FIGURE 34-2 Knowing who the customers are, as well as knowing their needs and preferences, is essential to developing an effective marketing plan.

ADVERTISING AND PROMOTION

Advertising is any form of nonpersonal presentation of a product or service. For maximum effectiveness, advertising must be coordinated with other marketing techniques, such as public relations, promotional programs, pricing, personal selling, product description and characteristics, sales promotions, and warranties. Effective advertising performs a variety of functions. These include creating an awareness of the product, educating the potential customer, motivating a customer to seek out the product, and reinforcing the value of purchases already made.

Product advertising focuses on the product itself. This might include the product's usefulness, durability, price, value, and the customer's need for the item (Figure 34-3). In contrast, **institutional advertising** is designed to create a favorable image of the firm or institution offering the products or services. Such advertising is likely to claim financial stability, longevity, reliability, and integrity of the company or institution and its employees.

Advertising efforts that seek to find new users for a given product, such as motor boats, are trying to create primary demand. Similarly, advertising intended to attract buyers to a specific brand of the product is attempting to create selective demand. The promotion of any one agricultural commodity from farms, ranches, ponds, nurseries, or greenhouses is expensive and difficult because there are so many small-volume producers. This is in contrast to other industries, such as automobiles, where there are relatively few manufacturers and nationwide advertising is practical and affordable.

INTERNET KEY WORDS:
beef checkoff
pork checkoff
lamb checkoff
milk checkoff
Got Milk? campaign

FIGURE 34-3 Advertising should always focus attention on the strengths of the product or service. Good advertising displays the product/service in attractive settings wherever concentrations of potential customers can be found.

To address the problem of many small-volume producers of individual agricultural commodities, the U.S. government has instituted a system that generates a pool of money from producers of a given commodity such as wheat, rice, cotton, tobacco, peanuts, beef, lamb, pork, or milk. Here, the seller contributes a small amount for each unit sold to an advertising pool to be used to purchase product advertising for the specific commodity. The program is referred to as the check-off program for each approved commodity. It is estimated that about 90 percent of all U.S. farmers contribute to some 300 federal and state generic promotion programs, covering approximately 80 commodities. The spending for research and promotion by these programs is reportedly near the half-billion-dollar mark annually. Media advertising for agricultural goods and services typically includes the use of newspapers, radio, television, outdoor billboards, signs, pamphlets, transportation advertising, direct mail, telephone book classified sections, breed journals, product catalogs, and magazines.

The promotion of commodities may be done in numerous ways. Product displays are common in stores and other retail outlets. Similarly, displays and sales of machinery, animals, food, flowers, ornaments, landscape designs, and other commodities and services are common at fairs, shows, open houses, trade shows, and professional meetings. Free samples may even be offered in stores, at sidewalk promotions, fairs, shows, public auctions, and wherever prospective customers may gather.

COMMODITY PRICING

Final determination of prices for agricultural commodities is not always under the control of the producer or representative of the producer. This is largely due to the perishable nature of unprocessed foods; the relationship between peak quality and price of fruits and vegetables; and the lack of control over production of crops,

SCIENCE PROFILE HOW MUCH WILL YOU PAY FOR IMPROVED HEALTH?

© wong sze yuen/Shutterstock.com.
A healthy family requires foods that furnish all of the required nutrients.

People are willing to pay more for health-promoting genetically modified (GM) food, according to a University of Purdue study. Jayson Lusk, an associate professor of agricultural economics at the University of Purdue, sent out a mail survey that asked consumers how much they would pay for golden rice. Golden rice is a GM food that has a gene from a daffodil in it. This gene triggers rice to produce a chemical that the human body can change into vitamin A. In his questionnaire, Lusk outlined the benefit of golden rice from a consumer's standpoint. Earlier studies indicated that consumers would pay more for foods that were not GM foods. In these studies, the benefits to farmers were outlined, but the potential benefits for consumers were not mentioned. The results of Lusk's study show that consumers will pay a greater price for golden rice if they understand the benefit. A valuable lesson can be learned from this study. As products like these become available to consumers, a shift in marketing must follow. Consumers need to be informed about the personal benefits they will get from the GM food. If they are not informed of the benefit, they will have no incentive to buy the food. Without proper marketing, GM foods tend to meet resistance. Conversely, if advertising is done correctly, foods such as golden rice can be a big success.

eggs, milk, and meat after the production cycle is started. Therefore, the producer may not have much control over the supply that is being collectively generated and may have to accept lower prices or loss of the product because of spoilage.

However, the growth of producer-owned marketing cooperatives has enabled producers to negotiate prices on a large-volume basis. Another benefit of cooperatives is that of owning processing plants so that fresh commodities can be processed and withheld, if necessary, until better prices are offered (Figure 34-4). Cooperatives, privately owned businesses, and individuals all use a variety of strategies when setting prices.

© iStockphoto/bigworld.
FIGURE 34-4 Modern marketing cooperatives have given small producers much more power to influence the prices they receive for their commodities.

Psychological pricing is a strategy designed to make the price seem lower or less significant than it is. For instance, the pricing of an item at \$1.99 seems to attract many buyers that may reject the same item priced at \$2.00. Needless to say, the profit from selling more items will easily offset the loss of one cent per item. Another technique is offering multiple items for a given price, such as three for a dollar. Here, the customer is lured into buying three items to save a small amount, when one item may be all the customer needs. Discount pricing and bulk selling are also used as techniques of psychological pricing. Sometimes customers are lured into buying more than they would normally buy because of the discount attraction, and then they also consume more than they would otherwise consume. Some consumers believe that larger packages are automatically priced at less cost per unit. Occasionally, sellers exploit this perception by pricing commodities in larger containers at greater costs per unit than in the standard package (Figure 34-5).

Penetration pricing is a strategy wherein price is set below that of competitors to entice customers to try an item. Although profits will be low temporarily, they can be regained if new customers continue to buy and the

FIGURE 34-5 Modern computers and scanning devices enable retailers to change prices quickly in accordance with promotional activities or changing market conditions.

price is gradually increased to new levels. **Skimming** is the opposite of the penetration pricing strategy. With this method, the price of a new product is set for unusually high profits at first, when affluent and willing customers are available. This enables the producer to recoup development expenses more quickly and gradually reduce the price as sales volume increases or other competing products enter the market. Skimming is also known as "sliding down the demand curve." **Loss-leader pricing** is a procedure wherein a popular commodity is offered for sale in a special sales promotion at prices less than cost—in other words, sold at a loss. The purpose is to attract customers into the business establishment with the hopes that they will buy other items priced at profitable levels. **Prestige pricing** is a procedure used to target buyers with special desires for quality, fashion, or image. The customer's notion or perception that a certain brand is better may be further reinforced by a higher price tag. Prestige pricing seems to work well in the garment and cosmetic industries, and it can be quite profitable. In general, in a nonregulated, free-enterprise system, there is a tendency to price commodities according to what the market will bear, rather than what it costs to produce the items. This pricing strategy is generally regarded as fair because our system rewards risk-taking with the potential for profit. If one must be prepared to shoulder losses when offering a commodity that does not reach sales expectations, then it only seems reasonable to permit that individual or company to benefit from high-profit opportunities.

MARKETING STRATEGIES FOR MAXIMIZING PROFITS

To make the most money from a production enterprise, successful marketing is a must. Strategies that can be used to market agricultural products most profitably include the following:

- Determine what types of markets are available to you.
- Determine the costs of various types of marketing.
- Determine transportation costs to market at each of the markets available to you, and sell where transportation costs are favorable.
- Determine the most profitable form in which to market your product (age, size, weight, degree of preparation).

- Advertise to create markets where none existed before.
- In seasonal markets, market your product at the peak of demand (Figure 34-6).

RETAIL MARKETING

Retail marketing is selling a product directly to **consumers** (people who use a product). This may take place on the farm, at roadside markets, or at the farmers' markets that have developed around many centers of population (Figure 34-7).

Retail marketing is generally used for ready-to-eat products. For example, ham, sausage, and fresh fruits and vegetables, rather than live pigs, are usually sold at a retail outlet. This creates many other problems that need to be addressed and that have an effect on the profitability of a production enterprise. When an animal product is processed for retail sales, various standards and state and federal regulations must be met. Also, changes in facilities are often necessary. For example, the sanitation standards for producing grade A milk for bottled milk sales require much more expensive facilities than those for producing milk for cheese production. Sanitary storage of processed products and the increased labor costs associated with retail customers must also be considered.

Retailing at the Farm

On-the-farm retail sales present special problems. If retail marketing is to take place on the farm where the product is produced, the farm must be well maintained and attractive to the customers buying the product. Facilities for parking are necessary. Animal and sales facilities must be clean, and the animals that are present must be well cared for and contented. An educational program

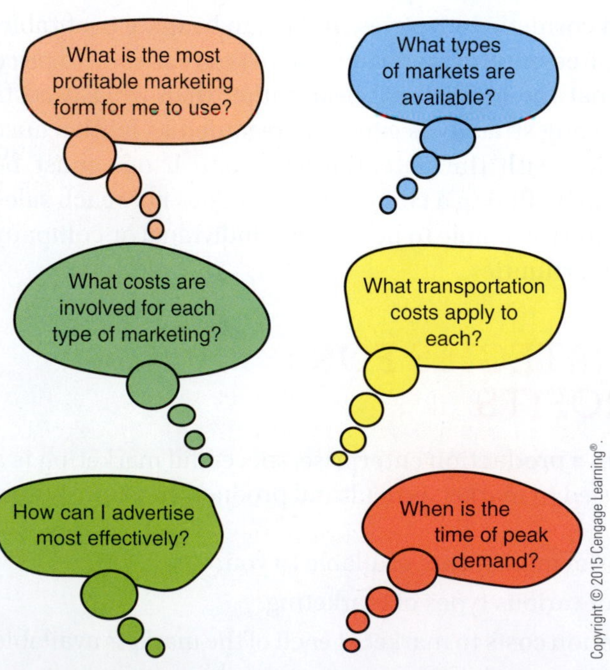

FIGURE 34-6 Good marketing strategies are developed by asking key questions about potential markets and the cost to access them.

FIGURE 34-7 Retail marketing by the producer adds extra dimensions to the business. It allows the producer to capture additional shares of the profits from the commodity.

AGRI-PROFILE CAREER AREAS: AUCTIONEER/CO-OP MANAGER/DEALER/ GROWER/PACKER/GRADER/MEAT CUTTER/ PRODUCE MANAGER/COMMUNICATIONS SPECIALIST

Courtesy of USDA/ARS #K-3998-5.

Agriscience researchers are developing new products that are rapidly solving environmental problems. Chemist Sevim Erhan evaluates biodegradable inks from soybeans. A new product requires attention to marketing strategies to convince potential consumers to try it.

The marketing of agricultural products occurs in all segments of our society on a continuing basis. The year-round displays of produce, dairy, meat, seafood, bread, frozen foods, canned goods, deli foods, and ready-to-eat foods in supermarkets are a testimony to the strength of the agriscience industry. The producers and breeders of vegetable and fruit crops, grains, flowers, ornamental shrubs, turfgrass, trees, pets, laboratory animals, livestock, horses, dairy animals, poultry, fish, wildlife, milk, eggs, wool, and fur all market the products of their businesses. This is only the beginning in the process of agricultural marketing. Whole systems of professional and scientific workers have evolved as a result of the broad array of agricultural production units. These include but are not limited to farm, ranch, pond, fish tank, feedlot, orchard, nursery, greenhouse and turf production units.

Marketing cooperatives, dealers, auction markets, auctioneers, livestock handlers, jobbers, brokers, clerks, truckers, railroad personnel, commodity market managers, futures traders, writers, and broadcasters all get in the marketing act. Government, as well as company inspectors, monitors products as they make the numerous transitions from production site to finishing site, pet shop, by-product, or dinner table. Teachers and extension workers provide instruction and professional-development programs for personnel.

Marketing careers may be launched through programs in technical schools, colleges, or universities in the plant and animal sciences, food science, agricultural economics, marketing, or communications. Studies in the biological sciences are appropriate for laboratory work associated with agricultural products and marketing.

to make consumers aware of good management techniques is often necessary so that they understand what is happening on the farm. Salespersons or the farm family themselves must be available to assist customers. Privacy is often difficult to achieve because retail customers often feel that when products are sold at the farm, it is acceptable to drop by at any time.

On the positive side, profits from farm sales are often greater because retail prices are higher. Also, middlemen are eliminated in the marketing procedure. **Middlemen** are people who handle an agricultural product between the farm and the consumer. Examples of middlemen include buyers, processors, truckers, and salespeople. Greater prices can often be charged when a superior product is produced. For farm families who like to meet new people, farm retail sales create excellent opportunities to do so. These types of sales provide opportunities for urban and suburban people to see the agricultural way of life firsthand.

Roadside Markets

Roadside marketing retains nearly all of the advantages of on-the-farm retail sales while eliminating some of the disadvantages. The sales unit is usually somewhat removed from the actual farm operation when roadside marketing is used.

By removing the customers from the production area, more efficient use of labor can take place. Less perfect care and maintenance of facilities can be tolerated.

On the less positive side, separate facilities for the retail sales unit must be maintained. More labor may be required because it may be inconvenient for the farmer to staff the roadside stand. The positive effects of the consumer seeing the actual farm in operation are diminished.

Courtesy of FFA.

FIGURE 34-8 Farmers' markets provide opportunities for producers to sell at retail without opening their production facilities to the general public.

Farmers' Markets

Farmers' markets have appeared in many large metropolitan areas to cater to the demands of urban consumers. They are normally operated 1 or 2 days per week and give urban and suburban consumers access to fresh agricultural products directly from the producers. Farmers' markets give the producers access to markets that would seldom be available otherwise. The producers also have the opportunity to educate the consumer concerning the value of good products (Figure 34-8).

Several costs and inconveniences are associated with marketing agricultural products at farmers' markets. Often, fees must be paid to cover the costs of operating farmers' markets. Competition may be greater, especially if several producers are selling the same type of product. Vehicles with heating, cooling, or both may be necessary to get products to the market as fresh as possible. Products may be subject to certain food regulations and packaging requirements to meet state and federal standards. Facilities to hold and display products must also be available at the market.

In summary, the decision to market agricultural products by retail methods requires much consideration. Although the returns are generally greater when products are marketed retail, the costs are also higher. Consideration must be given to the type of product, availability of markets, labor availability, and personal desires before a decision concerning an appropriate marketing method can be made.

WHOLESALE MARKETING

Wholesale marketing is the marketing of a product through a middleman who gets the product to the consumer. Most agricultural products in the United States are marketed in this way. Wholesale markets allow for marketing large volumes of products with comparatively little labor. Types of wholesale markets include terminal markets, auction markets, and direct sales.

Terminal Markets

A **terminal market** is sometimes a stockyard that functions as a place to hold animals until they are sold to another party. The terminal market never actually owns the animals. Animals that are delivered to the terminal market are consigned to a selling agent who does the actual selling of the animals. The terminal market charges the seller a fee for caring for the animals until they are sold. This is called a *yardage fee*. The selling agent also receives a fee, called a **commission**, for the service of selling the animals.

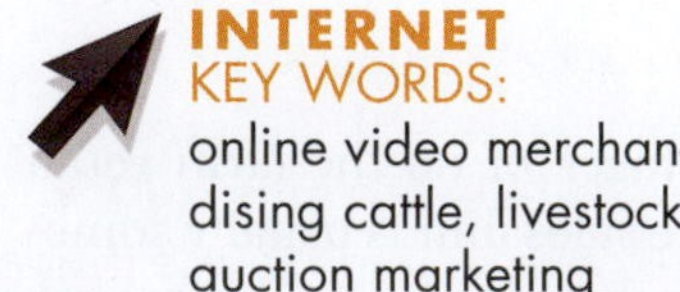

INTERNET KEY WORDS:
online video merchandising cattle, livestock auction marketing

The use of terminal markets to market animals has always been mostly confined to the Midwestern and western states. In recent years, the use of terminal markets to market animals has greatly decreased, and most livestock are currently marketed by other methods.

HOT TOPICS IN AGRISCIENCE VIDEO MERCHANDISING

Video merchandising is a marketing tool that allows buyers at different locations to bid in a competitive auction for a product that they have already viewed or that they are simultaneously viewing on video. The buyers base their bids on their own judgment from viewing online video and evaluating the description of the product by an expert judge of quality. Video merchandising has become especially popular as a way of selling livestock. Many feedlot operators in the Midwest Corn Belt now make use of video merchandising online or via telephone to purchase feeder calves and cattle from the western range states and other regions of the country.

An element of trust must be preserved among the buyer, the seller, and the on-site evaluator that the quality of the product is represented accurately. Many of the buyers will not see the product until a trucker delivers it to them several days later. Some buyers never see the product at all because they are acting as brokers by purchasing the product for one of their customers. They must be able to accurately represent the description and quality of the product to their clients. Online/video merchandising has found a market "niche" in the agricultural industry.

Auction Markets

At **auction markets**, animals are sold by public bidding on individual lots of animals. An auctioneer conducts the sales at an auction market. These markets are widespread and are convenient to most local communities. They have grown in popularity and now represent the most common means of marketing animals of importance to agriculture. Auction markets are usually most practical for the small livestock producer.

Video merchandising is a recent marketing innovation. **Video merchandising** is usually done by recording the livestock or other commodity on a digital video. The videos are loaded to the Internet or distributed to potential buyers together with a predetermined time when interested buyers can join a conference call or online site to bid on the consignment of merchandise. This type of marketing strategy has become popular in recent years as a good way to sell livestock and other agricultural commodities. It has the advantage of engaging potential buyers from a much larger geographic area than traditional livestock auctions are capable of doing.

As is true when selling in terminal markets, commissions are charged for selling animals through other markets. The amount of commission charged varies with the type and size of the animal. Because many auction markets are small, there may not be the competition that is present at terminal markets to buy small numbers of animals (Figure 34-9).

Direct Sales

Direct sales occur as the producer sells crops and animals directly to processors. This method of marketing has several distinct advantages. There are no commission fees to be paid to selling agents, and there are no yardage fees. Transportation costs are kept to a minimum because the buyer generally comes to the farm to make purchases. The animals look their best for the buyer because they have not been exposed to the stresses of hauling and contact with strange facilities and animals.

There are a few negative aspects to direct marketing of commodities. A producer must have fairly large numbers of animals to be marketed at one time

SCIENCE CONNECTION RAINBOW COTTONS, HIGH-TECH LEATHER, AND BIODEGRADABLE DETERGENTS

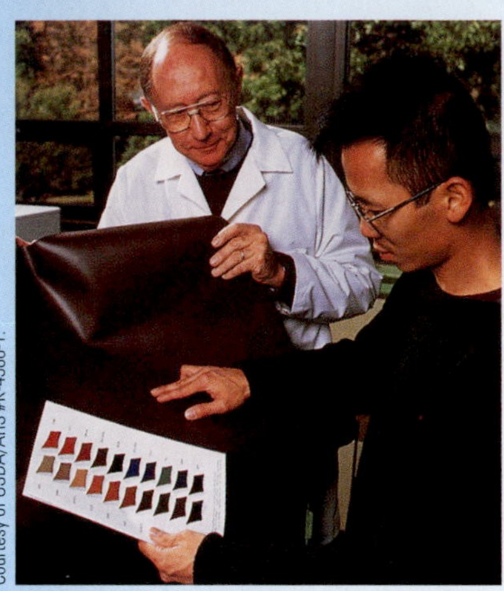
Courtesy of USDA/ARS #K-4388-1.

Frank Scholnick and James Chen of the U.S. Department of Agriculture examine the results of experimental treatments for leather.

Agriscience research is turning out new products. For example, new colors of no-wrinkle fabrics are made with cotton, leather is tanned by less-polluting methods, and a new detergent made from animal fats is biodegradable. Cotton has always been a preferred fabric, but its tendency to wrinkle has limited its use. In the past, cotton fabric was dyed before a no-wrinkle finish was applied because the chemical bond created before the refinishing process would repel dye if applied after the process. The cotton fabric must swell to accommodate the molecules of dye, but this cannot normally occur after the fabric has been heated and treated with a no-wrinkle finish. The fabric industry needs to be able to dye fabric just before clothing is manufactured to keep up with the rapid changes in clothing fashions. After all, who wants to get stuck with a warehouse full of fabric dyed in the hues and colors that are out of fashion?

before the purchaser comes to the farm or ranch to buy. The producer is generally at the mercy of the buyer concerning the price received for the animals sold, and the producer is often not paid for the animals until after they have been processed. Direct sales of animals, especially beef cattle, have increased significantly in recent years.

Courtesy of Bill Angell.

FIGURE 34-9 Auction markets are popular for selling livestock in many communities.

To solve the problem, scientists at the Agricultural Research Service (ARS) Textile Finishing Chemistry Research Unit in New Orleans have developed techniques that allow industry to apply a no-wrinkle finish to cotton fabric before dying. By adding a variety of quaternary ammonium salts to the no-wrinkle treatment solution and a positive charge to the fabric, they have developed a procedure that produces a no-wrinkle fabric that will accept dye. The procedure also broadens the choice of dyes that can be used, giving industry more options for supplies and a wider range of shades with deeper, more vibrant colors.

On another front, scientists at the ARS Eastern Regional Research Center in Philadelphia are looking for more environmentally friendly ways to decrease bacterial decay in hides and thus preserve leather. One of the most promising techniques is the use of electron beam irradiation. The technique promises to replace the salt and brine method of curing and extend the qualities of strength and elasticity for some new leather products. On yet another front, scientists have long sought laundry detergents that will not pollute groundwater when they are discharged with wastewater. Tallow is beef fat and a by-product of the beef industry. Agriscientists have developed a tallow-laced soap that is environmental friendly. The soap contains no phosphates; will not harm humans, domestic animals, or wildlife and will usually biodegrade in 24 hours. The product is as effective as phosphate detergents and is economical, too.

Other Wholesale Marketing Techniques

Approximately 83 percent of the milk marketed in the United States in 2009 was sold through farmer-owned, milk-marketing cooperatives. **Cooperatives** are businesses that are owned by groups of producers who join together to market a commodity. The cooperatives then either process the milk and sell it directly to consumers or sell it to other large processing plants. Marketing cooperatives have the ability to maintain product quality, arrange for transportation of products from farm to market, balance supply and demand of agricultural products, and plan advertising to increase sales.

Courtesy of USDA.

FIGURE 34-10 Vertical integration in the broiler industry takes the dressed bird directly to the retailer and eliminates the need for live-bird markets.

In the poultry market, almost 99 percent of the chickens produced for meat are grown under a system called **vertical integration** (several steps in the production, marketing, and processing of animals are joined together). The use of vertical integration in the production and marketing of agricultural products allows for extremely large systems of production that can be efficient. Only the number of animals that are expected to be in demand by consumers will be produced. Such production systems are completely market driven. There is less competition from other producers, and all phases of production can be controlled (Figure 34-10).

The oldest of all marketing strategies is bartering, or trading, agricultural commodities for other products or services. This system is used widely in developing countries of the world, and it is still used to a limited extent in rural communities across the United States. Another obvious method for marketing agricultural produce and livestock is the cash sale. It is still a common practice for people to buy beef or a hog for the freezer or to buy vegetables or fruit directly from the farm. Although this method would not be feasible for large farming operations, it does account for a significant number of sales, especially on small organic farms.

A number of methods are used to market agricultural commodities. The methods chosen by an individual producer are often a matter of what is available and what the producer prefers. Care should be taken to carefully choose the means of marketing. Intelligent marketing practices often make the difference between profit and loss in a competitive business

MARKETING FEES AND COMMISSIONS

Because there are numerous methods of marketing animals, the fees and commissions vary fairly widely. Livestock are usually sold by the head, and a fee is charged for each animal sold. This fee may vary according to the size of the animal and the area of the country. At many purebred livestock sales, up to 10 percent of the selling price goes for fees and commissions. If animals are kept at a terminal market, yardage fees are also charged for the feed and care of the animals. Charges for insurance may also be deducted from the seller's check.

For animals that are purchased on the farm for slaughter, a fee called *pencil shrink* is sometimes deducted from the selling price of the animals. Pencil shrink is the estimated amount of weight that an animal will lose when it is transported to market. This usually varies from 1 to 5 percent. Fees may also be charged for the sale of commodities based on a percentage of the gross amount received for the product.

MARKETING PROCEDURES

How produce is packaged and how animals are prepared for and transported to markets can often mean the difference between profit and loss. When shipping animals, certain procedures should be followed:

- Always handle live animals quietly and carefully. Animals that are calm and quiet lose less weight during transportation.
- Move animals when temperatures are moderate, if possible. Usually, this means at night during the summer.
- Do not overcrowd animals on trailers or in lots while they are waiting to go to market.
- Do not overfeed animals just before hauling them to market. Animals do not travel well on full stomachs. Do provide ample fresh, clean water.
- Avoid injuring or bruising animals when loading and unloading them. Dead or injured animals are worth little at the market.
- Sort animals according to sex and size before shipping them to market.
- Precondition animals for several days before marketing them.

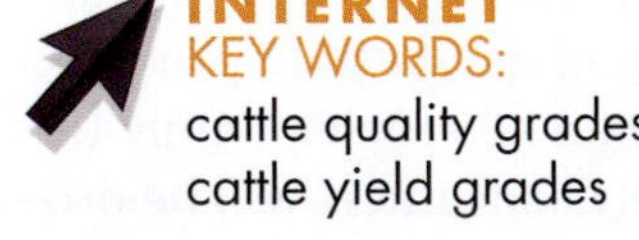

GRADES AND MARKET CLASSES OF ANIMALS

Cattle

Beef animals are classified as either calves or cattle. Calves are younger than 1 year, whereas cattle are older than 1 year. Calves are further divided into veal calves, feeder calves, and slaughter calves. Veal calves are younger than 3 months and are slaughtered for meat that is also called *veal.* They usually weigh less than 200 lbs.

Calves between 3 months and 1 year old that are marketed for meat are called slaughter calves. They have usually been fed at least some grain. Feeder calves are 6 months to 1 year old and are sold to people who feed them to market weight as slaughter cattle. The sex classes for feeder and slaughter calves are steers, heifers, and bulls. Steers are castrated male calves, heifers are young female calves, and bulls are unaltered male calves.

Cattle are divided into feeder cattle and slaughter cattle. Feeder cattle are further categorized into age groups of yearlings and 2-year-olds and older. Yearlings are between 1 and 2 years old, and 2-year-olds are 2 or more years old. These two classes of cattle can also be divided into six sex classes—steer, heifer, bull, bullock, cow, and stag. A cow is a female animal that has had a calf; a heifer is a female that has not had a calf. A steer is a male castrated before sexual maturity. A bull is a mature male; a bullock is a young intact male; and a stag is a male animal that was castrated after reaching sexual maturity.

Slaughter cattle are marketed for the purpose of being processed for meat. They are divided into the same age and sex classes as feeder cattle. They are also divided into quality and yield grades. Quality grades refer to the amount and distribution of finish (fat) on the animal. The quality grades for cattle are prime, choice, select, standard, commercial, utility, canner, and cutter (Figure 34-11). Yield grades are based on the amount of lean meat an animal will yield in relation to fat and bone (Figure 34-12). The yield grades are 1 through 5, with yield grade 1 producing the largest proportion of lean meat.

Swine

Two classes of swine are feeder pigs and slaughter hogs. In general, pigs are swine younger than 4 months old, whereas hogs are swine older than 4 months. Feeder pigs are sold to be fed to higher weights before being slaughtered. Slaughter hogs are sold for immediate slaughter.

The sex classes of swine are gilt, barrow, boar, sow, and stag. Gilts are young female swine. A barrow is a castrated male swine. Boars are unaltered males; stags are mature male swine that have been castrated after sexual maturity. Sows are mature female swine.

Swine are also graded according to quality, with the official USDA grades being U.S. No. 1 through U.S. No. 4. Animals with lean meat of an unacceptable quality are graded U.S. Utility. The highest grade of swine is U.S. No. 1 (Figures 34-13 and 34-14)

Sheep

Market sheep are classified according to age, use, sex, and weight. The age classes are lambs, yearlings, and sheep. Lambs are young sheep and can be divided into hothouse lambs, spring lambs, and lambs. Hothouse lambs are marketed below 3 months of age, usually for the Christmas or Easter holidays. Spring lambs are 3 to 7 months old, and lambs are 7 to 12 months old. Lambs can also be classified as feeder lambs or slaughter lambs, depending on whether they are to be fed to heavier weights or slaughtered immediately.

Yearlings are between 1 and 2 years of age, and sheep are older than 2 years. Sheep are divided into three sex classes: ram, ewe, and wether. Rams are unaltered male sheep, ewes are female sheep, and wethers are castrated male sheep.

Sheep are also graded according to yield and quality. The yield grades are 1 through 5, with 1 yielding the highest proportion of lean meat. Quality grades are prime, choice, good, utility, and cull (Figure 34-15).

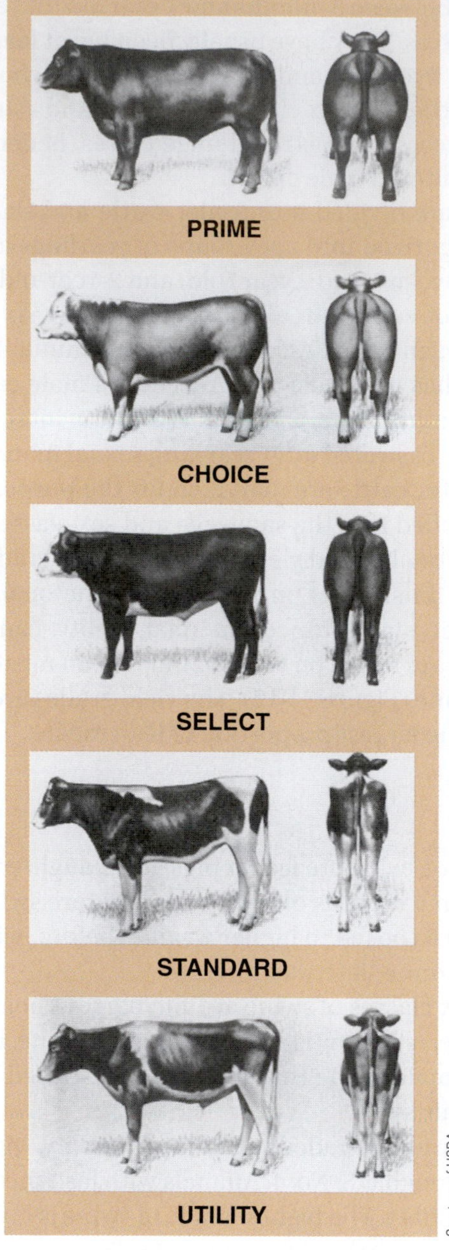

FIGURE 34-11 Quality grades of slaughter cattle are based on age or maturity and the amount and distribution of fat on the carcass.

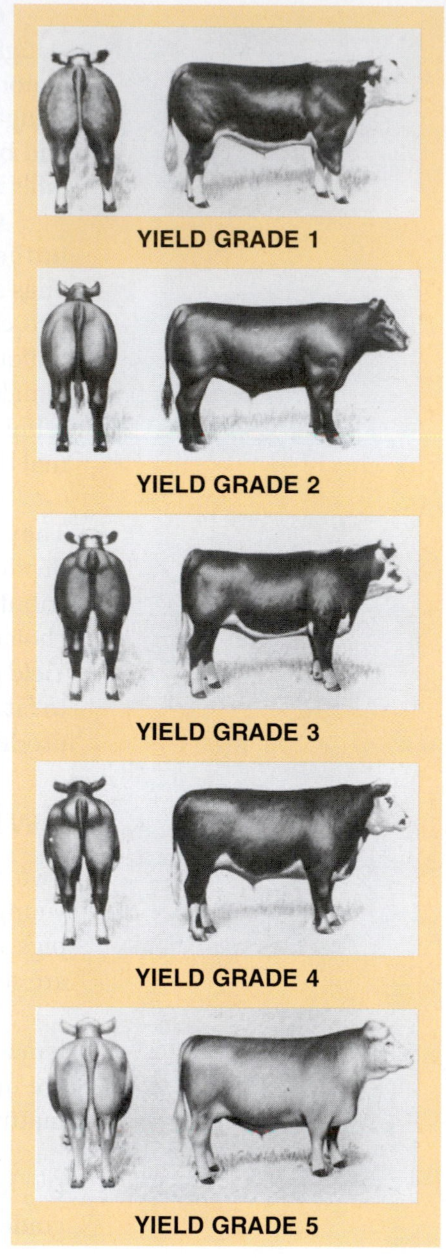

FIGURE 34-12 Yield grades of slaughter cattle are based on the yield of lean meat in proportion with the amount of fat and bone, with grade 1 being the highest and grade 5 being the lowest.

A review of useful terms regarding cattle, swine, and sheep is provided in Figure 34-16.

MARKETING TRENDS AND CYCLES

Agricultural commodity prices tend to increase and decrease on a fairly well-defined cycle. The cause of general price cycling is supply and demand or government intervention. However, cycles are sometimes interrupted, accelerated, or delayed by disasters or political events. For example, the discovery of "mad cow disease" interfered with both the beef and milk price cycles.

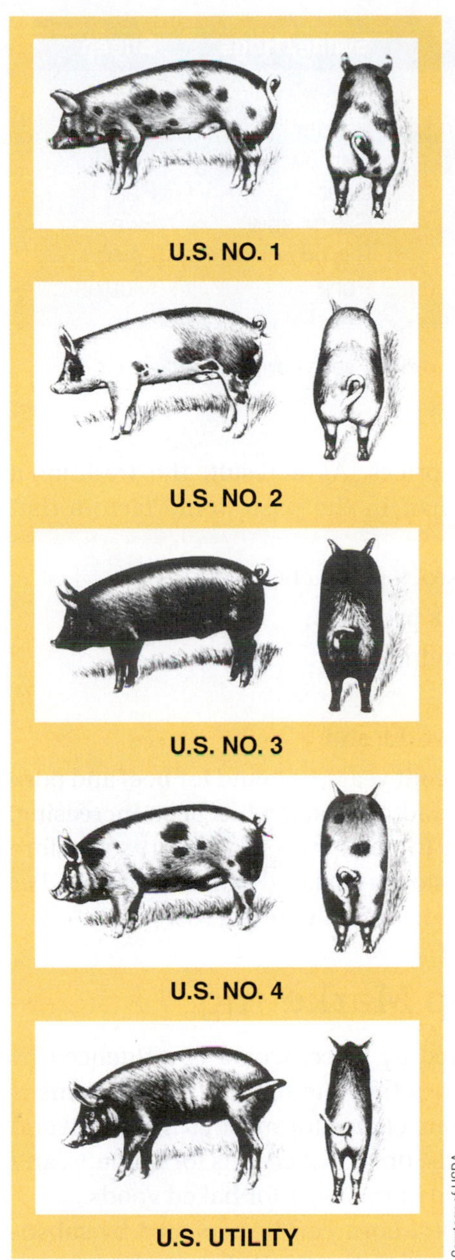

FIGURE 34-13 Quality grades of slaughter hogs.

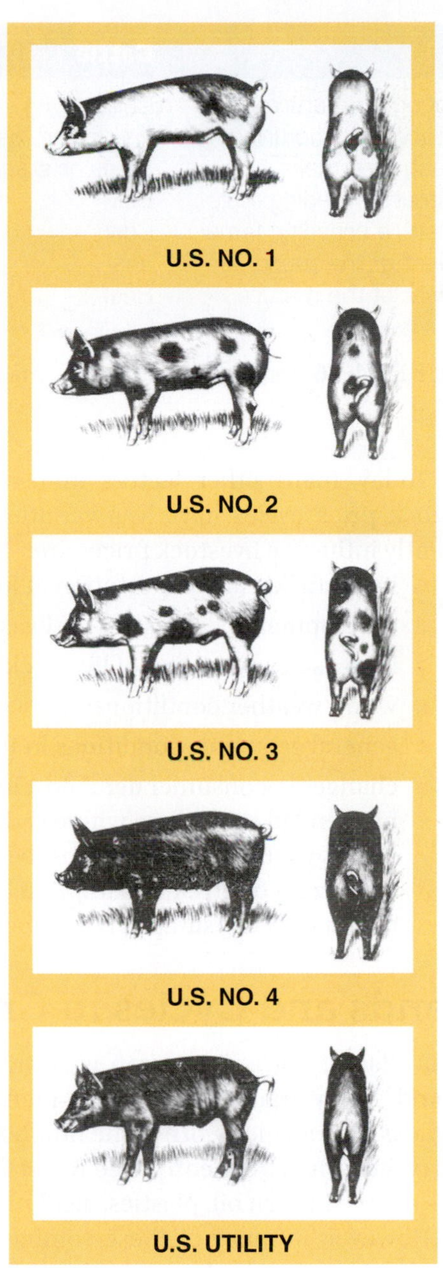

FIGURE 34-14 Quality grades of feeder pigs.

FIGURE 34-15 Quality grades of slaughter lambs.

Trends and Cycles in Animal Markets

When prices are high, producers tend to increase production of animals and animal products. At first, this action pushes the prices even higher because female animals that normally would have been marketed are retained in the breeding herd. When the results of the increased production reach the market and supply exceeds demand, the prices begin to decrease. When prices decrease to the point where production is not profitable, livestock producers reduce production by selling some of the breeding herd. Of course, this pushes prices even lower. With decreased supplies, the demand increases, and the cycle begins again.

	Cattle	Swine/Hogs	Sheep
Young, immature	Calf	Pig	Lamb
Young, but maturing	Feeder or yearling	Feeder	Yearling
Castrated male	Steer or stag	Barrow or stag	Wether
Mature breeding male	Bull	Boar	Ram
Mature breeding female	Cow	Gilt or sow	Ewe
Meat of the young	Veal	Pig or pork	Lamb
Meat of the mature	Beef	Pork	Mutton

FIGURE 34-16 Terms given to meat animals of various sexes and maturities.

Today, many other factors influence prices. As a result, the traditional livestock price cycles have less variation than in the past. Some factors that currently influence livestock prices are:

- importation and exportation of livestock products;
- development of new uses for livestock products;
- increased advertising of livestock products;
- world weather conditions;
- general economic conditions in the world; and
- changes in consumer demands. In recent years, demand for beef and pork has tended to decline, whereas demand for poultry has been increasing. Egg consumption per person began to decline in the mid-1980s before stabilizing briefly. Consumption has slowly declined again since 2005. Per capita milk consumption has declined gradually since 1985.

Trends and Cycles in Crop Marketing

The availability of a given crop is stimulated by price, which is influenced by demand. Many grain crops have so many uses that the market for the items is complex. For instance, corn is the number-one choice for most farm-animal and poultry feeds. It is frequently used in pet foods, breakfast cereals for humans, and as the source of corn oil, plastics, alcohol, and ingredients for baked goods.

However, in many of these commodities, corn can be replaced by substitutes. For instance, barley and other grains are substituted if the price of corn increases above a certain level. Similarly, when the price of crude oil increases to a certain level, the price of corn becomes competitive. Gasoline refiners will then use less petroleum and more grain alcohol in the fuel. Weather and other production factors influence the size of the world grain supply and influence prices in the United States and abroad. Trade policies greatly influence prices of a commodity because the flow of commodities into the world market greatly complicates the formulas that determine prices. The policy of permitting world supplies to flow into the United States has a similar affect on domestic prices.

A classic example of how U.S. government policy can affect prices, supplies, and farm conditions occurred in 2003. The U.S. government closed the border to shipments of cattle from Canada due to the first case of mad cow disease in North America, which occurred in Canada. Beef prices in the United States hit record highs in the following months. When a case of the disease was later found in Washington state, a temporary decline in consumer demand impacted the market again.

Courtesy of FFA.

FIGURE 34-17 Large trucks and modern highways move agricultural commodities quickly from the farm to the processor or market.

Many factors make agricultural marketing unpredictable: The use of rapid transportation on a modern system of interstate highways (Figure 34-17); air shipping of highly perishable foods; refrigerated shipping; food-processing procedures that lock in freshness; extensive, specialized, high-technology storage facilities; and an excellent marketing system all contribute to the availability of food and fiber commodities to nearly any part of the nation on a year-round basis. These factors help maintain markets and prices that are fairly stable and predictable.

GLOBAL MARKETING IN AGRICULTURE

Historically, agricultural producers have focused on producing commodities and managing their businesses in ways that hold the best promise for acceptable profits. Although producers always dream of favorable prices and fair profits, these do not always materialize. Telephones, radios, televisions, fax machines, computers, and other modern communication devices have permitted many changes in marketing during the last two decades. There have always been many middlemen in processing and marketing. The number and types of middlemen are increasing as the demand increases for products that are convenient and easy to use (Figure 34-18).

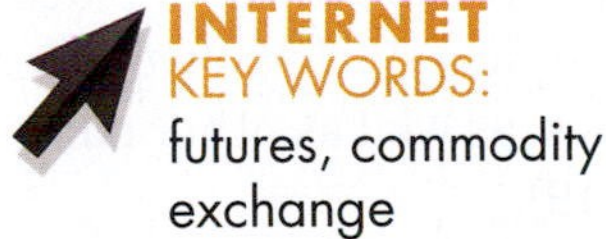

INTERNET KEY WORDS:
futures, commodity exchange

The Agricultural Commodity Futures Market

The use of futures contracts in the agricultural commodity marketing complex has become an important tool for both buyers and sellers of agricultural commodities. The Chicago Board of Trade, Chicago Mercantile Exchange, Winnipeg Commodity Exchange, and London International Financial Futures Exchange are all commodity exchanges. A **commodity exchange** is an organization licensed to manage the process of buying and selling commodities under specific laws using a system of licensed brokers. Commodity exchanges also manage futures markets. A **futures market** is a procedure conducted by commodity exchanges to provide networks and legal frameworks for sellers and buyers to work through brokers in making contracts called futures contracts, or simply futures.

Courtesy of DeVere Burton.

FIGURE 34-18 Despite all of the high-tech marketing strategies that are available, direct face-to-face marketing is still used to introduce selected products to potential buyers.

Buying and Selling Futures

Futures are defined as legally binding agreements, made on the trading floor of a futures exchange, to buy or sell something at a future date. Futures prices are determined by competitive bidding of brokers for prospective buyers and sellers from all over the world. Anyone with appropriate assets can work through a broker and buy futures or sell futures on the commodity market. The futures market floats up and down by the moment while the futures exchange is "open," or in session. Telephone and computer networks around the world permit agricultural managers, buyers, sellers, and brokers to communicate almost instantly to place their bids to buy or sell, and the commodity exchange handles the legal work to complete the transactions. Many agricultural commodities, as well as other commodities, are bought and sold in this manner (Figure 34-19).

INTERNET KEY WORDS:
agriculture, export, markets

Farmers, ranchers, cattle feeders, crop growers, and other agricultural owners and managers use the futures market as a tool to stabilize their businesses. The wise buying and selling of futures can protect them from losses caused by excessive price fluctuations. For instance, before deciding how much corn to grow, how much to spend on fertilizer and other inputs, or how much corn to keep for feeding cattle or hogs on the farm, the farmer should know the price per bushel for which the corn could be sold after it is grown. The other decisions are then easier to make.

By looking in the commodity-price sections of major business newspapers, such as the *Wall Street Journal*, the seller can determine the futures price for corn delivered in a given month. If the farmer regards that price as high enough to commit a certain number of bushels for sale, then he will inform his broker to "sell" that many bushels at the rate of the futures price for the month selected.

FIGURE 34-19 Commodity exchanges make possible the use of futures trading as a valuable marketing and management tool.

The broker will negotiate a contract that obligates the farmer to deliver that many bushels of corn on the specified date for the amount of the futures price. Now the farmer is guaranteed the price for the corn, and other decisions about the farming operation can be based on this selling price. If the price of corn is lower in December than the $6.30 futures contract that was made, the payment on delivery will still be $6.30 per bushel, and the farmer will have successfully protected the business against a price drop. However, if the price of corn is greater than $6.30 per bushel in December, the farmer might wonder if the purchase of a futures contract was a wise decision. Here is how a futures contract might be made for 5000 bushels of corn. Suppose the settlement price on the futures market for December is $6.30 per bushel. The value of the contract for 5000 bushels is 6.30 × 5000, or $31,500. The farmer is obligated to deliver the corn for $31,500 regardless of the market price of corn at the expiration of the contract.

Opening and Offsetting a Position

The purchase of a futures contract does not have to be a final transaction that plays through to the actual buying or selling of the commodity. In fact, in most cases, the goods are not delivered because the participant makes another contract that offsets the first one. The initial step in the futures market is called **opening a position**. At a later date, before the futures contract expires, the participant can take a second position to offset the first. Such action is known as **offsetting a position**. When the position is offset, the participant is said to be **out-of-the-market** and has no obligation to deliver or take delivery. To offset the position described previously, the farmer would need to buy 5000 bushels of corn for delivery on the same date as the first contract. This would offset the contract to sell 5000 bushels. If the farmer were fortunate enough to find a day when the futures price was less than $6.30 and buy futures for 5000 bushels for the specified date of delivery at that lower price, then the farmer could pocket the difference as income, or **profit**, from the business transaction.

EXPORT MARKETING

As transportation and communications improve, the world seems to get smaller. More people are familiar with more places and people around the world, so more business is being conducted between and among people and nations of different continents. Technology and computer languages are spreading rapidly, so different cultures are able to communicate in business, commerce, engineering, and other fields. Fiber-optics; lasers; satellites; computers; nationwide and worldwide telephone, television, and computer networks; international airlines and air routes; trade agreements; and improved education all contribute to the increase in international investment, business, trade, and commerce.

The United States now exports large amounts of its agricultural products. The livestock, meats, hides, dairy products, grain, lumber, fruits, vegetables, ornamentals, and processed plant and animal products that we export help to offset our trade deficit. It is interesting that U.S. imports of agricultural commodities have also increased in recent years, though not as rapidly as exports. Foreign trade activity has great implications for marketing products.

In recent years, there has been a substantial increase in travel abroad. State delegations, including governors and staff officials, legislators, educators, farmers, ranchers, horticulturists, and consultants of all kinds are traveling abroad. The purposes of such delegations typically are to develop goodwill; promote professional linkages, partnerships, and exchanges; and develop markets for commodities. When preparing for careers in agriscience or developing strategies for marketing agricultural products, it is important to consider the international and global opportunities.

The marketing of products is a complex operation if maximum profits are to be realized. There are many different markets for plants and animals and many ways of marketing them and their products. Planning for marketing should take place before entering into production, and it should be kept in mind during the entire production process.

STUDENT ACTIVITIES

1. Write the Terms to Know and their meanings in your notebook.
2. Obtain several copies of marketing reports and newspapers that have marketing reports. Compare the prices received for the various classes and grades of animals listed. Also compare the prices received from various markets.
3. Interview a livestock buyer. Ask how prices are determined and where animals are bought and sold. Also ask how to go about getting a job as a livestock buyer. Report your findings to the class.
4. Talk to a livestock grader about how to determine the various grades and classes of livestock. Try to grade some animals yourself.
5. Invite the operator of a retail market to class to talk about the issues involved in retail marketing of animals and animal products.
6. List the types of markets for agricultural commodities in your area.
7. Write a brief paper on the "best philosophy for pricing."
8. Develop an ad or other promotional plan for an agricultural commodity.

9. Build a display for retailing an agricultural commodity.
10. Research the procedures used to buy and sell in the futures market.
11. Do research on one or more major international trade agreements that influence the marketing of agriscience products.

SELF-EVALUATION

A. MULTIPLE CHOICE

1. Consumer demographics include
   a. populations, preferences, and product cost.
   b. family size, eating preferences, and discretionary money.
   c. customer location and tax rates.
   d. transportation costs.
2. The grading system based on the amount and distribution of finish on an animal is called
   a. prime.
   b. quality.
   c. yield.
   d. commercial.
3. The amount of a product that is available at a specific time and price is the
   a. supply.
   b. demand.
   c. commission.
   d. production.
4. The estimated amount of weight that an animal loses during transportation to market is
   a. yardage.
   b. stress.
   c. loss.
   d. pencil shrink.
5. Yardage is a fee paid to terminal markets for
   a. feed.
   b. insurance.
   c. selling.
   d. none of the above.
6. Lambs that are slaughtered when they are younger than 3 months old are called
   a. mutton.
   b. chevon.
   c. feeder lambs.
   d. hothouse lambs.
7. The least desirable yield grade for cattle is
   a. U.S. No. 1
   b. Utility.
   c. U.S. No. 4.
   d. 5.
8. The sex classes for sheep are
   a. stag, boar, and sow.
   b. lamb, sheep, and mutton.
   c. ram, ewe, and wether.
   d. none of the above.
9. Roadside stands are
   a. wholesale markets.
   b. terminal markets.
   c. retail markets.
   d. direct markets.
10. A male animal castrated after reaching sexual maturity is a
    a. wether.
    b. steer.
    c. stag.
    d. barrow.
11. An example of a wholesale market is a(n)
    a. terminal market.
    b. roadside stand.
    c. on-the-farm market.
    d. farmers' market.

12. A definite trend in agriscience marketing is
    a. cost per unit keeps going up.
    b. pricing is always based on cost.
    c. the number of commodities is decreasing.
    d. marketing is more global.

B. MATCHING

________	1. Prime	a. Castrated male pig
________	2. Ewe	b. Female sheep
________	3. U.S. No. 1	c. Quality grade of cattle
________	4. Barrow	d. Quality grade of swine
________	5. Veal	e. Calf meat
________	6. Futures purchase	f. Out-of-the-market
________	7. Offset position	g. Opening a position

C. COMPLETION

1. The lowest-quality grade of slaughter hogs is__________.
2. __________is the amount of a product desired at a specific place and time.
3. A professional livestock seller receives a __________ for selling animals for producers.
4. __________markets are those where animals are sold by competitive bidding.
5. Futures for agricultural commodities are purchased through a __________.
6. Rather than delivering goods sold on the futures market, most people take a(n) __________position.

UNIT 35

Agribusiness Planning

OBJECTIVE

To define management, determine management performance, determine how decisions are made, and describe economic principles that affect management.

MATERIALS LIST

- writing materials
- calculator
- income statement
- newspapers
- Internet access

COMPETENCIES TO BE DEVELOPED

After studying this unit, you should be able to:

- define management.
- describe the importance of management.
- describe kinds of agriscience management decisions.
- list eight steps in decision making.
- describe the economic principles of supply and demand, diminishing returns, comparative advantage, and resource substitutions.
- use capital and credit wisely in business management.

SUGGESTED CLASS ACTIVITIES

1. Invite the owner or manager of an agribusiness to make a presentation to the class on the subject of managing a successful business. Ask him or her to identify the key elements of successful agribusiness management. Also discuss the barriers to managing a successful business, and invite the guest to share examples of ways that he or she has overcome these barriers.
2. Develop a business plan for a school business such as producing flowers in a school greenhouse. Implement the plan by electing a board of directors, hiring a manager, and producing or obtaining a marketable product that can be sold to the public. Conduct the business enterprise for a specific period, and determine profits or losses for that time period. What steps were the most difficult to implement? Which elements of the business plan were most critical to the success of the business?

TERMS TO KNOW

agribusiness management
capital
credit
diminishing returns
comparative advantage
resource substitution
long-term loan
intermediate-term loan
capital investment
short-term loan
federal land bank
Production Credit Association (PCA)
Commodity Credit Corporation (CCC)
Farmers Home Administration (FHA)
Small Business Administration (SBA)
promissory note
discount loan
add-on-loan
amortized loan

3. Invite a banker or other agricultural lender to talk to the class about obtaining credit. Ask him or her to discuss small business loans, prequalifications for credit, interest rates, and repayment plans. The students should be prepared to ask appropriate questions and participate in the discussion.

AGRIBUSINESS MANAGEMENT

Organizing or restructuring an agribusiness is much like building a functional home. It starts with a plan or blueprint that is thoroughly examined for flaws. A prospective agribusiness owner should engage competent advisors to identify the strengths and weaknesses of the business plan in much the same way that building inspectors examine the plans for a house. A powerful advantage is gained when flaws are identified and corrected in a business plan before capital is invested.

Agribusiness management is the human element that carries out a plan to meet goals and objectives in an agricultural business or enterprise, generally referred to as an agribusiness. Management decides the types of business or production activities in which the business will engage, such as horticulture, aquaculture, and farm supplies. Agribusinesses include farming, ranching, nursery operations, landscaping, retail stores, service enterprises, marketing businesses, lending institutions, veterinary services, consulting activities, and a variety of others.

According to Dun and Bradstreet, 88 percent of all businesses that fail do so because of poor management. Poor management of an agribusiness is frequently related to having no set objectives or goals. Management is usually considered to be good when maximum profits are achieved from the available resources. Generally, good management has established goals and objectives to guide them in developing good action plans (Figure 35-1).

Influences on Agribusiness Management

Agribusiness management is influenced by the members who make up the boards of directors in corporations and cooperatives. Land, labor, and capital, being in limited supply, also influence management decisions. For instance, limited **capital**, which is money or property, may prevent the agribusiness manager from buying a new piece of equipment that would make the operation more efficient or protect the operation from excessive losses (Figure 35-2). Similarly, limited land will influence the agribusiness manager's selection of enterprises. For example, a cow–calf beef enterprise will need more land than a feedlot operation. Limited labor will influence the manager's selection of enterprises. For instance, if labor is limited, it will prevent the use of crop enterprises with high labor requirements, such as watermelons or tobacco. In contrast, grain production requires minimal labor.

Estimating a Manager's Performance

The performance or ability of the agribusiness manager can be estimated in several ways. Dollar income is the measurement most often used. According to a study conducted in Ohio, the managerial ability of an individual

FIGURE 35-1 Planning skills that are learned in an agriscience class can also be used to develop the goals and objectives for a business plan. Many businesses that fail did not have clearly defined goals and objectives.

FIGURE 35-2 Modern irrigation equipment permits profitable farming in areas that would otherwise be inefficient or too risky because of crop loss by drought.

- Income goal (Economic orientation)
- Willingness or tendency to make decisions (Decisiveness)
- Manager's ability to recognize alternatives and opportunities
- The extent of social activities or distractions from business goals

FIGURE 35-3 Factors for predicting the managerial performance of agribusiness managers.

can be determined by the manager's economic orientation, decisiveness, ability to identify alternatives, and extent of social activities or distractions from business goals (Figure 35-3).

Agribusiness is constantly changing, and it is expected to change even faster in the future. As a business becomes more complex, errors in management will be more costly. Therefore, the need for people who are well educated in agribusiness management is increasing.

CHARACTERISTICS OF DECISIONS IN AGRIBUSINESS

Although there are various types of agribusiness decisions, most decisions of managers are organizational or operational (Figure 35-4). Both types of decisions affect the success of the business. The following discussion covers four characteristics of a decision as well as steps in decision making.

Agribusiness Decisions

Organizational

- Should I rent more land, or should I borrow money and purchase land?
- Should the business be operated as a partnership or corporation?
- What lender offers the best source of borrowed capital?

Operational

- Should cattle be sold this week or next?
- Should I use high-magnesium lime or regular lime?
- When should I start planting corn? Soybeans? Small grain?
- When should I change the photoperiod for my poinsettias?

FIGURE 35-4 Kinds of agribusiness decisions.

AGRI-PROFILE CAREER AREAS: OWNER/MANAGER/CONSULTANT

© Richard Thornton/Shutterstock.com.

Agribusinesses are the economic backbones of many communities.

Agribusiness activities cut across the food and nonfood spectrums of agriculture. This includes the supply side, as well as the output side, of production. Career opportunities exist in every sector where goods are bought or sold.

Agribusinesses sell products such as feed, seed, pesticides, fertilizer, tools, equipment, plants, or animals. They also sell services such as animal care, crop spraying, or recreational fishing privileges. Careful planning is important before starting an agribusiness because many new businesses end in failure. Careful planning increases the chances of success.

Career opportunities in agribusiness planning include financial services such as banking, accounting services, management services, teaching, university extension service, marketing, market analysis, product specialization, product engineering, sales, ownership, and management. Agricultural economics, business management, accounting, finance, and personnel management are college programs that can lead to careers in agribusiness.

Importance

The importance of a decision may be determined by measuring the potential loss or gain that is likely to occur as a consequence of the decision. For example, the selection of an agribusiness enterprise or business venture is likely to be more important than the selection of a brand of a given commodity. The manager must spend a sufficient amount of time selecting a type of business because this decision cannot be changed easily after it is made. In contrast, if sales of a certain brand of wire or species of plant are not successful, a different brand or species can be added or substituted at little additional cost.

Frequency

The frequency, or how often a decision is made, can vary greatly. Determining the cost per day to rent a truck may be used as an example. The cost of renting the truck only for 1 day, for a one-time job, might not be an important decision. However, the cost of renting the truck every few weeks as an ongoing expense makes the decision more important.

INTERNET KEY WORDS: decision-making skills

Urgency

It is important that certain decisions be made immediately, while some decisions can be delayed. The urgency of a decision depends on the cost of waiting. Two examples of decisions with different degrees of urgency are hiring a new employee and buying a new file cabinet. If it is the busy time of year, an additional person may be needed immediately and should be hired. If you delay, you may lose sales. However, the delay in buying a file cabinet is not likely to affect the overall management and profit of the business.

Available Choices

Some situations offer several choices. If several choices are available, the manager should delete the least-likely courses of action and focus on the more promising ones. The manager should take time to gather important information regarding the options that are available. If no choices are available, there is no decision to be made.

Steps in Decision Making

Making decisions based on facts will increase your success as a manager. The agribusiness manager must determine the process to be used for making better decisions. The following eight steps should be helpful in developing a management process (Figure 35-5):

1. Start the management process with a situation or established goal.
2. Gather all available facts and information for accurate analysis of the problem.
3. Analyze the available resources. Human resources are labor, time, skills, and interest. Material resources are land, equipment, and capital. Reevaluate your goal and adjust it if appropriate.
4. Determine the possible ways of accomplishing the goal or solving the problem.
5. Make an informed decision concerning the direction(s) that will be taken.
6. Follow through with a plan of action.
7. Assume responsibility for implementing the plan.
8. Evaluate the results to determine whether the goals were accomplished. If not, could a better route have been selected? Should the goals be modified?

After going through these eight steps, the manager should determine whether the process worked well. If not, where were the weak links? What changes should be made to improve the process? These and other questions will enable the manager to fine-tune the process until it fits his or her style, the personalities of the staff, and the structure of the organization.

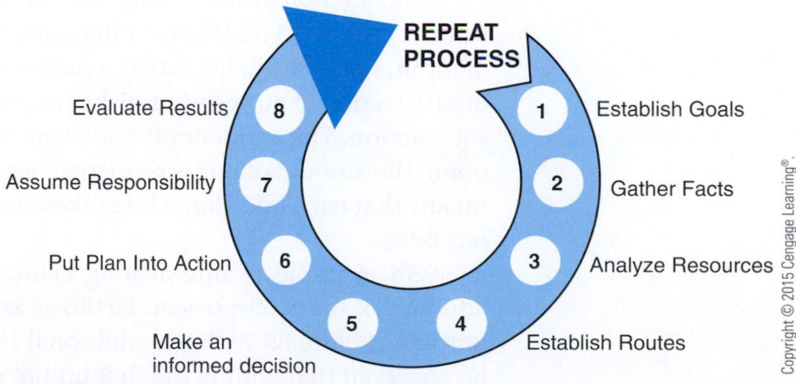

FIGURE 35-5 Eight steps in the decision-making process.

FUNDAMENTAL PRINCIPLES OF ECONOMICS

Price, Supply, and Demand

Price is the amount received for an item or service. The price is determined by three price-making factors: (1) the supply of the item, (2) the demand for the item, and (3) the general price level.No one factor can be used to explain all price changes.

The general price level of a commodity is influenced by the supply or availability of the item, the demand for it, the influence of wars, recessions, depressions, and many other factors. When supply and demand are in balance, a general price is established. An increase or decrease in either supply or demand is likely to influence the prevailing price.

The quantity of a product that is available to buyers at a given time is called supply. A producer of vegetables has control over some factors that influence the vegetable supply. However, one factor that the manager cannot control is the weather. If the rains do not come, the supply of vegetables will be reduced. The ability to buy items needed to produce the vegetables can also increase or reduce the supply. The ability to buy these items is also influenced by their prices. Cost and availability of credit may also influence supply. **Credit** is borrowed money. It permits the agribusiness manager to buy items needed for production or for business operations.

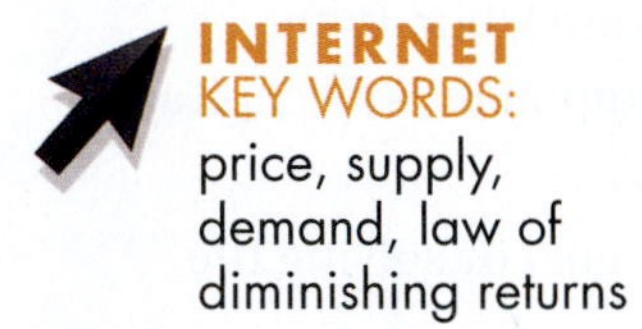

Demand may be defined as the quantity of a product that buyers will purchase at a specific price at a given time. The quantity that a buyer is willing to purchase depends on the quantity available and the price. Both the desire and the ability to purchase influence the extent of demand. The population growth may also influence demand. As the population of the country changes, the demand for various foods changes.

Diminishing Returns

The term **diminishing returns** is often used in economics. It refers to the amount of profit generated by additional inputs. Most of the time, the term is not completely understood. An understanding of the law of diminishing returns can be extremely helpful to the agribusiness manager in decision making.

The principle of diminishing returns has two parts: physical returns and economic returns. An explanation of both physical and economic returns follows.

Satisfaction from eating may be used to illustrate diminishing physical returns (Figure 35-6). When eating, each bite of food represents an additional unit of input. Each bite helps satisfy a part of the hunger. However, the added amount of satisfaction of hunger diminishes as you eat more. Each mouthful results in less satisfaction. This tendency is known as diminishing physical returns. At a certain point, the amount of hunger satisfied becomes negative with each bite taken. This means that each additional bite takes away from the satisfaction gained by previous bites.

An example of diminishing economic returns is demonstrated with the addition of units of nitrogen fertilizer as inputs and the effect on corn yields as outputs (Figure 35-7). Each additional input requires an additional cost. It may be observed that adding the first unit of nitrogen produced the greatest increase in yield. However, the rate of increase in yield diminished as more units of input were added. The decision that the manager needs to make is how many units of

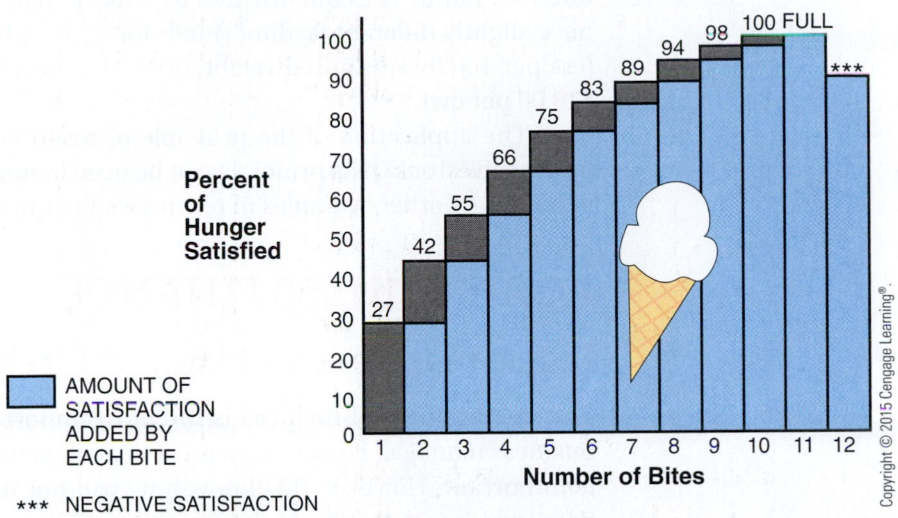

FIGURE 35-6 The law of diminishing returns.

Nitrogen Applied (lb.)	Yield (bu.)	Yield Increase (bu.)
0	67	—
60	103	+36
120	133	+30
180	154	+21
240	173	+19
300	169	−4

Effect of Nitrogen on Corn Yields*

*Hypothetical and not based on research

FIGURE 35-7 Effect of increased nitrogen inputs on corn yields.

nitrogen should be applied to obtain the greatest profit from growing the corn. If the number of inputs is too low or too high, the greatest potential income will not be achieved.

Comparative Advantage

The United States has nine major farming areas that have developed over a period of years. These areas have developed because of changes in demand or other factors. Within each area, different commodities are produced. Most operations in an area produce similar commodities and have similar systems of production. The reason for the similarity in operations is the comparative advantage found in following the programs that have evolved in that area.

Comparative advantage is the emphasis in a given area where the most returns can be achieved. Comparative advantage may be illustrated in several ways. An example has been the change in livestock production in New York State. At one time, most of the farms in New York raised sheep. Today, many of these farms are producing milk because of the high demand and favorable prices for milk in New York. It has become more advantageous to produce milk than to produce lambs and wool.

A similar situation exists for broiler production on farms. Fifty years ago, most farms raised enough chickens to supply their families with eggs and chicken for Sunday dinner. Few farms raise chickens anymore. The production of eggs and broilers has become so competitive that it is usually less expensive for farm families to buy eggs and broilers at the grocery store than it is to produce them in small numbers on their farm.

Resource Substitution

Resource substitution refers to the use of one resource or item to replace another, when the results are the same. It is often possible to substitute a less expensive item for a more expensive one.

For example, it may be possible to substitute barley for corn in making a cheaper dairy feed. This substitution can be done without affecting total milk production. Barley is frequently less expensive than corn. Because barley and corn have slightly different feeding values for dairy cattle, barley must sell for $9.00 or less per 100 lbs. (hundredweight, or cwt) to be a better buy than corn selling for $10.00 per cwt.

The application of the principle of resource substitution is not limited to feeding livestock. This principle can be used in most management situations. Can you think of other examples of resource substitutions?

AGRIBUSINESS FINANCE

Importance and Uses of Credit

The management of finances is the most important single function of the agribusiness manager. Possessing a great deal of technical knowledge and know-how is important. However, this know-how will not make one a successful agribusiness manager unless he or she is a competent money manager. Careful planning of finances is probably more important than planning other aspects of the agribusiness. Some good agribusinesses have failed because the manager was not a good money manager.

Credit is borrowed money. At one time, the manager who sought credit was considered a poor manager. Today, this is no longer true. The manager approaching a lending institution to negotiate for credit should not do so in an apologetic manner. The lending institution must sell its commodity—money—to stay in business. The use of credit is a privilege that must not be abused, however. Once labeled a poor credit risk, the reputation is difficult to overcome. Credit used wisely, however, is a valuable tool in business.

INTERNET KEY WORDS:
credit, short-term, intermediate-term, long-term

Classifications of Credit

Credit is classified according to its period of use. Loans may be classified as long-term, intermediate-term, or short-term (Figure 35-8).

Long-term loans are used to purchase land and buildings. The loan period ranges from 8 to 40 years. Interest rates for this type of loan are usually less than they are for other types. These loans are made by federal land banks, the Farmers Home Administration (FHA), insurance companies, local banks, and individuals.

Classifications of Credit		
Long-Term Loans	**Intermediate-Term Loans**	**Short-Term Loans**
Used to purchase land and buildings, and are for a period of 8 to 40 years	Extend for a period of 1 to 7 years for the purchase of breeding livestock, farm equipment, tractors,and similar items	Usually written for a period of 1 year or less to cover the cost of feeder livestock, feed, fertilizer, seed, fuel, and so forth

FIGURE 35-8 Classifications of agricultural credit.

Intermediate-term loans are for periods of 1 to 7 years. Capital investments are usually made with the money from this type of loan. **Capital investment** is money spent on commodities that are kept 6 months or longer. Examples are breeding stock, tractors, store equipment, and warehouse equipment. Production credit associations, the FHA, finance companies, and local banks are sources of this type of credit.

Short-term loans are for a period of 1 year or less. They are often referred to as production loans. Managers who borrow short-term money are usually required to pledge something of value for security. Managers and businesses with good reputations may be allowed to borrow without security. Money borrowed without security is referred to as a signature loan. The borrower signs a promissory note that the loan will be paid on or before a specified date. Production credit associations and local banks are sources of short-term loans. It is not uncommon for local agribusinesses to permit their reliable customers to make purchases on a short-term credit basis by simply signing a sales slip.

Types of Credit

Two types of credit are productive credit and consumptive credit (Figure 35-9). Productive credit is used to increase production or income. This is justifiable when the estimated increase in production will increase profits. This type of credit is used to purchase supplies, plants, flowers, livestock, land, equipment, storage facilities, seed, fertilizer, labor, and other materials.

© LivingCanvas/Shutterstock.com. © Elena Elisseeva/Shutterstock.com. © Krivosheev Vitaly/Shutterstock.com.

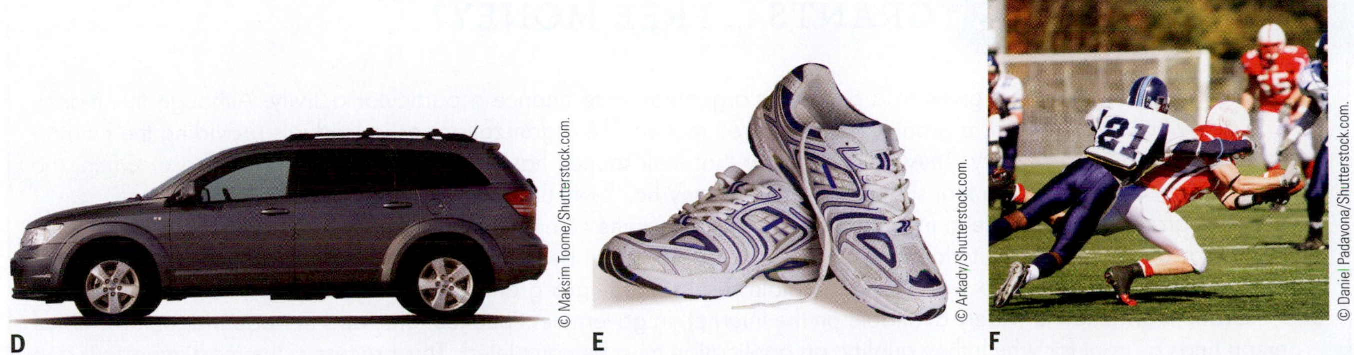

© Maksim Toome/Shutterstock.com. © Arkady/Shutterstock.com. © Daniel Padavona/Shutterstock.com.

FIGURE 35-9 Types of credit based on how the loans are used: Productive credit is used to pay for the cost of raising (A) livestock, (B) crops, and (C) equipment or to purchase livestock or products for resale. Consumptive credit is used to purchase goods or services such as (D) cars, (E) clothing, or (F) recreation that are used up with no expectation of earning a profit.

Consumptive credit is used to purchase consumable items used by the individual; it does not contribute to the business income. It is relatively easy for a family to abuse the use of consumptive credit. This type of credit can also limit the amount of productive credit available to a family business.

A common form of consumer credit that is widely used is credit card purchases. Most credit card purchases paid in 30 days or fewer are interest free. However, the cost of consumer credit is high if the balances are not paid off within the first billing cycle. Federal statutes require firms charging interest on credit card purchases to publish their interest rates.

INTERNET KEY WORDS:
federal land bank
farm credit system
production credit associations
Commodity Credit Corporations
Farmers Home Administration
Small Business Administration

Sources of Credit

Many sources of credit are available, and many differences exist in lending policies. Because of this, knowledge and understanding of credit and credit practices should be helpful in securing credit. Lenders differ in the interest rates they charge, the length of the loan period, and the purposes for which money is loaned.

When seeking a loan, remember that the interest rate is not the only important factor. Of substantial importance is the lender's willingness to extend a line of credit should an unexpected event occur. Another important factor is the lender's knowledge of the agribusiness.

Some agencies and institutions make only certain types of loans (Figure 35-10). Commercial banks are the most important source of credit. About 25 percent of the total agricultural debt is owed to banks. This percentage represents about 13.5 percent of the real-estate loans and 45 percent of other loans. Commercial banks lead in loan volume because they make all types of loans.

Individuals are the next most important source of agribusiness credit. About 36 percent of agricultural real estate mortgages are held by individuals. Interest rates, tax deferments and reductions, and payment options are some of the reasons that individuals are willing to finance real estate.

Retail merchants also supply non-real-estate credit. Their credit is usually in the form of an open account. Sometimes loans are made available by a business for purchasing equipment. The interest rates on such loans are usually higher from these sources than they are from commercial banks. The popularity of this type of credit is based on its convenience and availability.

AGRI-PROFILE GRANTS ... FREE MONEY?

A grant is money that is given to a person or organization to finance a particular activity. Although this money rarely has to be repaid, a grant is not really free money. The organizations or individuals providing the funding must see a success story. They need to know that their money has been used for a good purpose. Often, the grantor asks for documentation that shows that grant money has been used wisely.

Many grants are available to individuals in agriculture. Some common ones are educational, livestock promotion, value-added, and rural rehabilitation. These grants can be of great benefit to people whose goals match those of the grantor and who are willing to work hard to obtain results. Finding a grant for which you qualify will take time and effort. Grant information is widely available on the Internet, in government publications, and in trade magazines. Once a person finds a grant for which they qualify, an application must be completed. This process is the most important step in getting a grant. Many applications are usually accepted for each available grant. Educating yourself on grant writing will increase your chances of securing a grant. Many resources are available that provide valuable information on how to write successful grant proposals. Local colleges offer classes and workshops for grant writing.

Sources of Agricultural Credit and Typical Rates				
Source	Length of Loan	Annual Interest Rate*	Percent of Appraisal Loan Value	Purpose of Loan
Commercial Banks and Trust Companies	6–12 Months	8–13%	To 100%	Farm Production Items
	2–3 Years	8–12%	70–80%	Machinery, Equipment, and Livestock
	10–20 Years	9–11%	60–75%	Real Estate
Federal Land Banks	20–35 Years	8½% Variable	Up to 85%	Real Estate
Production Credit Association	1 Year	9% Variable	To 100%	Production Items
	3–7 Years	9%	Varies	Machinery, Equipment, and Livestock
Farmers Home Administration	40 Years	5%	To 100%	Real Estate
	7 Years	7–9%	To 100%	Machinery, Equipment, and Livestock
Insurance Companies	20–35 Years	Varies	To 75%	Real Estate
Individuals	15–30 Years	6–7%	To 75%	Real Estate
	2–5 Years	6–8%	To 80%	Machinery, Equipment, and Livestock
Equipment Manufacturers	1 Month–5 Years	17.5%	100%	Machinery and Equipment

*Annual interest rates fluctuate based on the current prime rate. These percentages are for example only and should not be considered current.

FIGURE 35-10 Sources of agribusiness credit.

Federal land banks were created by Congress in 1916 to provide long-term credit for agriculture. These banks are excellent sources of credit for the purchase of real estate because of the favorable interest rates.

Production Credit Associations (PCAs) were established by an Act of Congress in 1933. The purpose was to provide favorable, short-term credit for agriculture. These funds are secured from the federal intermediate credit banks, who in turn secure their money from private lenders. Life insurance companies have been excellent sources of credit for financing long-term real estate loans. Loans are made through brokers, correspondents, and company representatives. Insurance companies hold about 15 percent of agricultural real estate mortgages.

The **Commodity Credit Corporation (CCC)** is administered as an agency of the U.S. Department of Agriculture (USDA). The CCC is a government-owned corporation. Loans are made on eligible commodities, such as grain, cotton, peanuts, and tobacco. Farmers use the commodities as security for loans. Payment of loans is made by purchasing the loans or delivering the commodities to the CCC.

The **Farmers Home Administration (FHA)** was created by the government during the Depression. The FHA's original purpose was to assist tenant farmers in becoming landowners. The FHA provides financing to farmers who are unable to secure credit from any other sources. The advantages of borrowing money from the FHA are as follows:

- A large percentage of the total cost of the property can be borrowed.
- The repayment plan is based on the borrower's ability to repay.
- Supervision of the loan and assistance with planning are provided.

The **Small Business Administration (SBA)** also provides loans to agribusinesses. In 1976, Congress passed legislation that permits the SBA to make agricultural production loans. These loans have favorable interest rates.

Cost of Credit

Interest is the greatest expense in borrowing money. However, borrowers are also required to pay other fees. These fees include commissions, recording fees, title certification charges, insurance, and service charges. Many formulas are used for calculating interest (Figure 35-11).

A simple-interest loan refers to one in which the full amount of a loan is received by the borrower and is paid back with interest after a short period. The borrower generally signs a **promissory note** agreeing to the terms of the loan. If payments are made several times throughout the duration of the loan, the interest is paid to date and interest is charged on the remaining balance of the principal.

In the case of a **discount loan**, interest is subtracted from the principal at the time the loan is made. For example, on a $1000 loan, the borrower would receive only $900 of the $1000 if the interest were 10 percent. Because the full amount of the principal is not received, the resulting true interest rate is 11 percent.

The **add-on loan** method is used for calculating interest on consumer loans. The interest is charged for the entire amount of the principal for the entire length of time. The total principal plus total interest is divided into combined equal installments. This results in a very high rate of true interest. For example, a $1000 loan at 10 percent interest would require repayment of $1100 to be repaid in monthly installments of $91.67 ($1000 principal + $100 interest = $1100/12 = $91.67).

SCIENCE CONNECTION

WORKHORSES OF MODERN BUSINESS PLANNING

© NAN728/Shutterstock.com.

Computer programs and computer models help the business planner to test alternative practices without risking time and capital.

The microcomputer has become the workhorse of modern agribusiness planning as well as the workhorse of business operations. Capable of split-second computations and tremendous storage capacity, the microcomputer is found in nearly every business and generally serves as the nerve center for the business. After a small investment in software and a modest investment in accessories, the computer can be used to evaluate marketing alternatives, predict financial outcomes for proposed business ventures, and accumulate data on prospective customers. By using a good computer system, the manager and staff can handle complex mailing lists, individualize communications to large numbers of patrons, track sales and inventories, keep appropriate financial accounts, and develop presentations for the owners, board of directors, or patrons.

A small business may have just one computer with a high-quality printer and monitor. However, such businesses usually graduate to multiple computers. Color monitors, printers, and scanners are replacing

A Comparison of the Different Methods of Calculating Interest		
Example: A Loan of $600.00 for 12 Months at 10%. Repaid in 12 Equal Payments.		
	FORMULAS FOR CALCULATING INTEREST	INTEREST CALCULATED
SIMPLE INTEREST	Interest (I) = Principal X Rate X Time (I = P X R X T)	\$600.00 X .10 X 1 = \$60.00 = 10% True Interest
DISCOUNTED LOAN	Rate (R) = $\frac{\text{Interest (dollar cost)}}{\text{Principal X Time}}$ $\left(R = \frac{I}{P X T}\right)$	$\frac{60.00}{540.00}$ X 1 = .11, or 11% True Interest
ADD-ON LOAN	Rate (R) = $\frac{\text{2 X No. of Payments X Interest Charged in \$}}{\text{Beginning Principal X Years X (No. of Payments + 1)}}$	\$600.00 X .10 = \$60.00 660.00 = Total Payments $\frac{2 X 12 X 60}{600 X 1 X 13}$ = .184, or 18.4% True Interest

FIGURE 35-11 Formulas for calculating interest.

The **amortized loan** is generally used for the purchase of land, buildings, and other expensive items. Payments are made monthly and are computed so the interest owed plus the payment on the principal is equal throughout the repayment time. By this method, almost all of the payment at the beginning of the repayment period is for interest, whereas most of the payment near the end is for principal. In other words, the amount being paid on the principal increases proportionately as the amount due for interest decreases. A careful analysis of amortization schedules for a loan over various lengths of time will help the manager determine the cost of borrowing and will aid in making wise decisions on when, how much, at what rate, and for how long to borrow capital (Figure 35-12).

Seeking a Loan

Most agribusinesses must borrow money to expand their operations or increase income. The decision to incur debt is not easily reached. A careful examination of available alternatives must be made. and the best alternative should be chosen.

black-and-white systems because the price for computer hardware continues to decline even as memory capacity increases. The software for a business may be an integrated software package, typically with word-processing, database, spreadsheet, and communications software. Communications software permits the computer to use communication equipment to access outside networks and download data and programs from outside sources. This gives the business access to market quotations, continuous status reports on inventories, up-to-date marketing information, and databases on world business and commerce.

Data for tax computation, annual inventory, business reports, and other government obligations are readily available. Specialized computer programs can be purchased to calculate taxes and prepare tax returns, thereby cutting business costs for services and consultants.

Salespeople, buyers, company executives, and other on-the-road personnel use portable computers to gather and synthesize information from sales contacts and conferences and to phone the information to the home computer for staff to act on before the traveler returns to the office. With the arrival of voice-input capacity and computers that talk back, typical agribusinesses will continue to find new ways to use the computer, and its role as a business "workhorse" will undoubtedly expand.

A. 48 PAYMENTS OF $368.67 EACH

Pmt	Principal	Interest	Balance	Total Interest	Pmt	Principal	Interest	Balance	Total Interest
		12.000%	14,000.00				12.000%	14,000.00	
1	228.67	140.00	13,771.33	140.00	25	290.35	78.32	7,541.63	2,758.38
2	230.96	137.71	13,540.37	277.71	26	293.25	75.42	7,248.38	2,833.80
3	233.27	135.40	13,307.10	413.11	27	296.19	72.48	6,952.19	2,906.28
4	235.60	133.07	13,071.50	546.18	28	299.15	69.52	6,653.04	2,975.80
5	237.95	130.72	12,833.55	676.90	29	302.14	66.53	6,350.90	3,042.33
6	240.33	128.34	12,593.22	805.24	30	305.16	63.51	6,045.74	3,105.84
7	242.74	125.93	12,350.48	931.17	31	308.21	60.46	5,737.53	3,166.30
8	245.17	123.50	12,105.31	1,054.67	32	311.29	57.38	5,426.24	3,223.68
9	247.62	121.05	11,857.69	1,175.72	33	314.41	54.26	5,111.83	3,227.94
10	250.09	118.58	11,607.60	1,294.30	34	317.55	51.12	4,794.28	3,329.06
11	252.59	116.08	11,355.01	1,410.38	35	320.73	47.94	4,473.55	3,377.00
12	255.12	113.55	11,099.89	1,523.93	36	323.93	44.74	4,149.62	3,421.74
13	257.67	111.00	10,842.22	1,634.93	37	327.17	41.50	3,822.45	3,463.24
14	260.25	108.42	10,581.97	1,743.35	38	330.45	38.22	3,492.00	3,501.46
15	262.85	105.82	10,319.12	1,849.17	39	333.75	34.92	3,158.25	3,536.38
16	265.48	103.19	10,053.64	1,952.36	40	337.09	31.58	2,821.16	3,567.96
17	268.13	100.54	9,785.51	2,052.90	41	340.46	28.21	2,480.70	3,596.17
18	270.81	97.86	9,514.70	2,150.76	42	343.86	24.81	2,136.84	3,620.98
19	273.52	95.15	9,241.18	2,245.91	43	347.30	21.37	1,789.54	3,642.35
20	276.26	92.41	8,964.92	2,338.32	44	350.77	17.90	1,438.77	3,660.25
21	279.02	89.65	8,685.90	2,427.97	45	354.28	14.39	1,084.49	3,674.64
22	281.81	86.86	8,404.09	2,514.83	46	357.83	10.84	726.66	3,685.48
23	284.63	84.04	8,119.46	2,598.87	47	361.40	7.27	365.26	3,692.75
24	287.48	81.19	7,831.98	2,680.06	48	365.26	3.65	0.00	3,696.40

B. 24 PAYMENTS OF $659.03 EACH

Pmt	Principal	Interest	Balance	Total Interest	Pmt	Principal	Interest	Balance	Total Interest
		12.000%	14,000.00				12.000%	14,000.00	
1	519.03	140.00	13,480.97	140.00	13	584.86	74.17	6,832.53	1,399.92
2	524.22	134.81	12,956.75	274.81	14	590.70	68.33	6,241.83	1,468.25
3	529.46	129.57	12,427.29	404.38	15	596.61	62.42	5,645.22	1,530.67
4	534.76	124.27	11,892.53	528.65	16	602.58	56.45	5,042.64	1,587.12
5	540.10	118.93	11,352.43	647.58	17	608.60	50.43	4,434.04	1,637.55
6	545.51	113.52	10,806.92	761.10	18	614.69	44.34	3,819.35	1,681.89
7	550.96	108.07	10,225.96	869.17	19	620.84	38.19	3,198.51	1,720.08
8	556.47	102.56	9,699.49	971.73	20	627.04	31.99	2,571.47	1,752.07
9	562.04	96.99	9,137.45	1,068.72	21	633.32	25.71	1,938.15	1,777.78
10	567.66	91.37	8,569.79	1,160.09	22	639.65	19.38	1,298.50	1,797.16
11	573.33	85.70	7,996.46	1,245.79	23	646.04	12.99	652.46	1,810.15
12	579.07	79.96	7,417.39	1,325.75	24	652.46	6.52	0.00	1,816.67

FIGURE 35-12 Comparison of a loan for $14,000 at 12 percent interest paid over 4 years versus the same loan repaid in 2 years.

When seeking a loan, the agribusiness manager must be prepared to fully explain the benefits and risks to potential lenders. Many loan applications are not approved because the borrower does not present the details of the business. Other loan applications are not approved because the borrower does not effectively present the advantages of expanding. To be effective, the presentation must be organized, and the details of the operation should be put in writing. Lenders are concerned about the benefits and risks associated with providing a loan. Lenders are also concerned about a logical presentation of facts and the details concerning specific agreements in the loan contract.

Information that should be included in the presentation includes the following: (1) an agribusiness plan, (2) business records (income statements, expense records, net worth statement, and financial history), (3) terms of the loan, and (4) method of repayment.

A borrower should investigate several sources of credit before selecting a lender. The final selection should be one that best meets the needs of the borrower.

Selecting a Lender

Factors that should be considered in selecting a lender include the following:

- Lending institution representative's knowledge of agribusiness problems and practices
- Lending institution representative's experience in handling agricultural credit of a similar nature
- Reputation of the lending institution
- Loan policies (interest rate, repayment schedule, closing costs, penalty clause, optional prepayment clause, and policy regarding failure to meet payment because of circumstances beyond borrower's control)
- Date the loan would be advanced
- Possibility of increasing the loan
- Availability of credit for other purposes

Using Credit Wisely

There is a chance that the borrower will fail to meet the financial obligations of a loan. In this case, the borrower may lose part or all that is owned. Because of this, the borrower takes a much greater risk than the lender. The borrower, however, can minimize risk by applying certain rules:

- Production loans should be used to generate new income.
- The borrower should limit the amount borrowed on new or unfamiliar business ventures.
- The borrower should keep debt as low as possible while still maintaining efficiency.
- The borrower should keep abreast of markets and trends.
- A debt-to-net-worth ratio of 1:1 or less should be maintained.
- A proper debt-to-income relationship should be maintained. The income must be greater than the principal and interest payments.
- Dependability and terms of the loan should be considered when selecting a lender.

- The borrower should have a definite repayment plan and schedule supported by evidence of an adequate cash flow.
- The borrower should be businesslike, fair, and frank.
- The borrower should have adequate insurance to reduce the lender's risk. Property, liability, and crop and life insurance provide partial protection against risk.

Management is a vital part of the business and commerce associated with supplies, services, production, processing, distribution, and selling of plant, animal, and natural-resource commodities. Careful planning and the application of sound agribusiness principles and procedures are necessary for success in many agriscience careers.

HOT TOPICS IN AGRISCIENCE INTERNET SALES

© Angela Waye/Shutterstock.com.

Internet sales are gaining popularity because nearly any good or service is made available for delivery at competitive prices. Great effort has been made to assure that security is good for both the buyer and the seller.

Many businesses are taking advantage of the Internet to market and sell their products. This is a new way to connect a customer with a product, and nearly anything a person may want is available for sale on the Internet. The Internet is particularly adapted to the sale of unique products that may be hard to find, but it is also a source of products that are easily obtained at a store. For example, you can buy tires, groceries, insurance, and financing for your home on the Internet. Shopping is easy, and you can compare prices among competitors quickly.

A major effort has been made to protect the security of credit card information that is provided to companies with products for sale. The credit card is the key medium of exchange for Internet sales because it makes it possible to collect the payment instantly before an item is shipped or delivered. It also provides some protection to the buyer against fraudulent sales. Most major credit card companies will work with their cardholders to reverse a payment when there is evidence that a product was misrepresented. Internet sales are here to stay.

STUDENT ACTIVITIES

1. Write the Terms to Know and their meanings in your notebook.
2. Explain how the eight steps in decision making can be used in planning an FFA or other school money-raising activity.
3. Explain how the principal of resource substitution may be used in making an agricultural mechanics project.
4. Discuss with a banker the procedures used in applying for an agribusiness loan.
5. Arrange to have a banker discuss agribusiness capital and credit in a class presentation.
6. Arrange for an accountant to talk to the class about taxes to consider when planning an agribusiness.
7. Make an outline of this unit. Phrases and words that are bold should be included, together with a brief description of the key points of the text.
8. Search the Internet for grants for which you may qualify to help you set up or expand a business. Discuss with your teacher what steps you need to take to apply for a grant.

SELF-EVALUATION

A. MULTIPLE CHOICE

1. Most businesses fail because of
   a. death of the manager.
   b. lack of capital.
   c. labor problems.
   d. poor management.
2. Management is considered good if
   a. labor is adequate at all times.
   b. maximum profits are achieved.
   c. the business survives for 5 years.
   d. the business survives for 10 years.
3. The importance of a decision may be measured by the
   a. availability of the manager.
   b. inventory.
   c. potential for gain or loss.
   d. time of year.
4. The amount of profit generated by additional inputs is known as
   a. capital.
   b. diminishing returns.
   c. margin.
   d. profit.
5. Using a resource in the place of another is known as
   a. bailing out.
   b. integration.
   c. resource management.
   d. resource substitution.
6. Loans for 8 to 10 years are called
   a. amortized loans.
   b. intermediate-term loans.
   c. long-term loans.
   d. term loans.
7. The most important source of credit is
   a. commercial banks.
   b. individuals.
   c. insurance companies.
   d. land banks.
8. The formula for calculating simple interest is
   a. $I = P \times R \times T$.
   b. $I = P/R \times T$.
   c. $I = R/P \times T$.
   d. $I = P \times T$.
9. A debt-to-net-worth ratio should not exceed
   a. 5:1.
   b. 3:1.
   c. 2:1.
   d. 1:1.
10. The type of insurance that a borrower should have is
   a. liability.
   b. life.
   c. property.
   d. all of the above.

B. MATCHING

________	1. Capital	a. Money loaned
________	2. Price	b. Quantity desired
________	3. Supply	c. 1 year or less
________	4. Credit	d. Quantity available
________	5. Demand	e. 1 to 7 years
________	6. Short-term loan	f. Money property
________	7. Intermediate-term loan	g. Pay interest in the beginning
________	8. Generated income	h. Consumptive credit
________	9. Used for the individual	i. Amount received
________	10. Add-on loan	j. Productive credit

C. COMPLETION

1. List the eight steps in decision making, in their correct order.
2. Study Figure 35-12, and then answer the following questions:
   a. What is the total interest paid on the loan if repaid in 4 years (48 payments)? $__________
   b. What is the total interest paid on the loan if repaid in 2 years (24 payments)? $__________
   c. How much more does the loan cost if repaid in 4 years rather than 2 years? $__________
   d. What are your observations about stretching a loan out over more time?

UNIT 36

Entrepreneurship in Agriscience

OBJECTIVE

To define entrepreneurship and determine considerations for planning and operating an agribusiness.

MATERIALS LIST

- variety of publications with pictures
- materials to make a collage or bulletin board
- Internet access

COMPETENCIES TO BE DEVELOPED

After studying this unit, you should be able to:

- define and describe entrepreneurship.
- describe steps in planning a business venture.
- state five basic functions performed in the operation of a small business.
- select a product or service for a personal or group enterprise.
- determine the basic functions performed by small-business managers.
- analyze the outcome of a business venture.
- use small-business financial records.
- analyze the benefits of self-employment versus other types of employment.

SUGGESTED CLASS ACTIVITIES

1. Schedule a class period in a computer lab that has Internet access. Explore the Web for evidence of entrepreneur activities and have class members share the locations with each other. Print the best examples and compile the pages into a reference booklet.
2. Divide the class into groups of four to five students. Assign each group to brainstorm about a possible product or service that might be provided to a specific target population by an entrepreneur. Design a plan for establishing a business to market the new venture following the steps that are provided in this unit.
3. Invite a local successful entrepreneur to discuss with the class what steps he or she took to start his or her own business. The students should be prepared to ask appropriate questions and participate in the discussion.

TERMS TO KNOW

entrepreneur
entrepreneurship
buying function
selling function
promoting function
distribution function
financing function
short-term plan
long-term plan
actuating
balance sheet
asset
liability
net worth
profit and loss statement
cash flow statement
inventory report

Entrepreneurship in agriscience provides extensive career possibilities at all levels of effort. It provides opportunities for the inventor, risk taker, profit seeker, owner, manager, and employee. The products and services of entrepreneurs impact fresh food, processing plants, equipment sales, fishing, guide services, veterinarian work, forestry work, farming, ranching, teaching, research, mechanics, environmental efforts, law, insurance, real estate, and finance. Opportunities in agriscience exist worldwide (Figure 36-1).

© leedsn/Shutterstock.com.

FIGURE 36-1 An entrepreneur must learn to recognize and take advantage of business opportunities when they occur. Agribusiness opportunities are found throughout the world, and today's young entrepreneurs will find opportunities to do business in the international arena.

THE ENTREPRENEUR

The **entrepreneur** is the person who organizes a business or trade or improves an idea. The word is taken from the French word *entreprendre*, which means to "undertake." **Entrepreneurship** is the process of planning and organizing a small-business venture. It also involves managing people and resources to create, develop, and implement solutions to problems to meet the needs of people. An inventor is the person responsible for devising something new or for making an improvement to an existing idea or product.

The entrepreneur can be the inventor as well as the small-business manager. However, in many cases, they are different people, each with distinct talents. The entrepreneur functions as the liaison between the inventor and the manager or management team. The entrepreneur brings these two groups together for the purpose of getting the invention to the individuals it will serve.

It is the entrepreneur who visualizes the venture strategy and is willing to take the risk to get the venture off the ground. Inventors or business managers are not entrepreneurs unless they organize the venture. An understanding of the differences, as well as the similarities, among these roles is important to comprehending entrepreneurship as a career option.

ENTREPRENEURSHIP

Various types of business enterprises are part of the entrepreneurship system. The element of individual, partner, or corporate ownership means private control as opposed to government ownership control. The profits (or losses) from entrepreneurship go to the owners. However, owners hire managers and other employees; therefore, entrepreneurships provide jobs for everyone (Figures 36-2 and 36-3).

A sales project conducted by a school organization such as the FFA is a type of entrepreneurship. However, it differs from a typical small business venture in the following ways:

1. The sales project should be a valuable learning experience as well as a money-making activity.
2. The sales project is usually of a limited duration. It is completed within a short period. A small business typically is designed to operate indefinitely.
3. The sales project involves the voluntary participation of the class. The small business must hire and pay its employees.
4. The sales project usually involves little risk for financial loss to the individual. The small-business owner (entrepreneur) could lose his or her investment.
5. The sales project may need to be approved by the school administration, but it usually does not need to be licensed by the government, pay taxes, or file employee reports.

Photo courtesy of the National FFA Organization.

FIGURE 36-2 Many businesses start with high school occupational-experience programs in agriscience.

Courtesy of FFA.

FIGURE 36-3 Successful businesses generate job opportunities for managers and other employees.

© Kurhan/Shutterstock.com.

FIGURE 36-4 Group projects in school can help students develop valuable entrepreneurial skills.

With these distinctions in mind, you can plan, organize, and implement a sales project and gain insight into the world of entrepreneurship at the same time (Figure 36-4).

Operating Businesses

Five basic functions are performed in the operation of both a small business and a sales project. These functions represent the basic steps in moving a product or service from the supplier to the consumer. They are the buying function, selling function, promoting function, distribution function, and financing function.

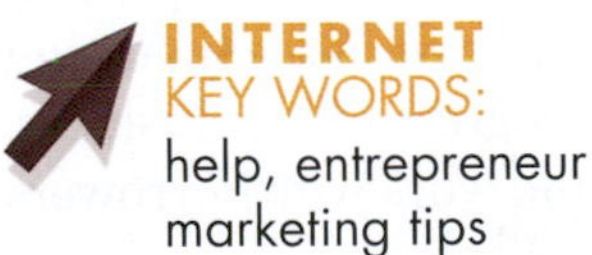

INTERNET KEY WORDS:
help, entrepreneur marketing tips

Buying Function

The **buying function** involves selecting a product or service to be marketed or sold for profit. The selection of a product or service is based on thorough marketing research, which determines consumer (customer) needs and wants. It also determines who may already be promoting the product or service.

Selling Function

The **selling function** includes studying the product or service to determine the reasons customers will want and need the product or service. This function also involves developing suitable customer approaches. Planning sales presentations, determining methods of overcoming objections, and planning for the close of a sale are all part of this function (Figure 36-5).

Promoting Function

The **promoting function** involves developing a plan to identify ways to make potential customers aware of the product or service to be offered. Examples of promotional activities include newspaper, TV, radio, and outdoor advertising.

Distribution Function

The **distribution function** involves physically organizing and delivering the selected product or service. For products, this includes receiving, storing, and distributing the merchandise. Many of the same activities are associated with services, particularly distributing to the customer (Figure 36-6).

HOT TOPICS IN AGRISCIENCE INTERNET MARKETING

© Kheng Guan Toh/Shutterstock.com.

The Internet has opened a new world to shopping from home. Nearly any product or service is at the fingertips of consumers, and millions of people shop for bargains nearly every day.

Internet marketing is a new strategy that has allowed entrepreneurs to market products directly to anyone who has access to a computer with an Internet connection. It allows a person to run a business from his or her own home, and it allows business organizations of all sizes to serve Internet shoppers. Internet marketing has opened up new advertising and promotion opportunities, and virtually anyone who has a product or service to sell can do so using the Internet.

Internet marketing has also introduced some new problems to the business world. Payment is made by credit card, but how do you keep credit card numbers secure from discovery by dishonest people? How do you protect Internet shoppers from being exploited by cybercrooks and others who develop lists of names by tracing the Internet sites that are visited? What privacy rights should be protected by Internet providers? What new laws are needed to protect consumers who use the Internet? These are only a few of many issues that are developing alongside this new marketing opportunity.

Financing Function

INTERNET KEY WORDS: why a business plan is needed, how to write a business plan

The **financing function** includes obtaining capital for the initial inventory, recording sales, maintaining inventory, computing profit or loss, and reporting the results of the venture. Accountability to lenders is always part of the financing function, and lenders should always be informed of the progress of the business. Informed lenders are more likely to finance new ventures and work with borrowers when problems arise.

Courtesy of FFA.

FIGURE 36-5 Effective selling and promotion are absolutely essential to the success of most businesses. How a product is displayed is important. The flowers at this greenhouse are eye-catching and visually appealing. This type of presentation is sure to attract customers.

© iStockphoto/Sean Locke.

FIGURE 36-6 Product distribution may be made to retail outlets or to other businesses, homes, farms, and many other destinations.

AGRI-PROFILE

CAREER AREAS: OWNER/MANAGER/ASSISTANT MANAGER/ MANAGEMENT TRAINEE

© iStockphoto/YinYang.

The entrepreneur is constantly seeking ways to provide better goods and services to the clientele.

Agribusiness management is seen by many as an ideal career. This career has the advantage of continuing in the work of a family business where one gains experience as he or she grows up. A person may also start a business and work in a locality where there may not be jobs in specialized areas.

Many high school and college students are well established in agribusinesses before they finish their educations. Some popular agribusinesses include lawn services, logging, lumber businesses, greenhouse or nursery operations, machinery repair, agricultural supplies, home and garden centers, florist shops, retail flower sales, livestock sales, farming, and ranching.

Preparation for careers in agribusiness management includes both formal education and on-the-job training. Effective programs should include classroom, laboratory, supervised agriscience experience, and leadership development. High school agribusiness and agriscience programs provide excellent training for business, but they do not take the place of higher education in business management. Advanced agribusiness programs may be taken at technical schools, colleges, or universities to obtain appropriate training in economics, finance, and management.

SELECTING A PRODUCT OR SERVICE

The first step in the process of establishing a new business or developing a fund-raising project is the selection of a product or service (buying function). This involves analyzing potential products/services to determine the level of demand (number of customers). The analysis of potential products or services should provide answers to these and other pertinent questions:

- Who are the potential purchasers of the proposed product or service? Customers can often be viewed in groups that have similar interests.
- On what basis does each group make decisions on the product or services? Typical responses to this question include price, quality, and continuing service after the sale.
- What are the competing products or services? How do they compare with the proposed product or service relative to price, quality, and service?
- What is the total estimated demand for the product or service?
- How much profit can be expected on the sale of this product or service? (Multiply the number of units that are expected to sell times the markup per unit and subtract any expenses incurred in the purchase, sale, and delivery of the product or service.)

ORGANIZATION AND MANAGEMENT

The role of a venture's organization and management is to make things happen so its goals can be achieved. To accomplish this, managers work mainly with data and people. Managing involves getting all the parts of the business—including personnel, marketing strategies, finances, and records—to function together to achieve the venture's goals.

No two managers have jobs that are the same. The jobs are shaped by the type of venture and the personality of the individual manager. However, managers perform many of the same functions. These functions include

- planning work,
- organizing people and resources for work,
- actuating work, and
- controlling and evaluating work.

Planning

When managers make plans, they set objectives or goals. They establish policy for the business, and they recommend strategies to achieve the goals. They develop procedures, methods, and/or programs to support policies and implement strategies leading to achievement of goals. Plans must be constantly reviewed and updated. No matter how thorough the plans, they do not guarantee success. In planning, managers must make short-term and long-term plans. **Short-term plans** are accomplished in several days or weeks. **Long-term plans** are accomplished over several months or years.

Organizing

After a plan is developed, the work must be organized. The manager must identify the people needed to carry out the plan and arrange for the necessary equipment and supplies. Staff development may also be required.

Actuating

Actuating simply means putting the plan into action. The manager needs to inform employees of the plan. All of the persons involved must understand their roles. Actuation also includes motivating people to work efficiently and effectively together. It also includes employee motivation to want to get the job done.

Controlling and Evaluating

Managers must carefully control implementation of plans after the work begins. The quality and quantity of all results must be evaluated. If the results are satisfactory, work can continue. If problems arise, changes must be made and alternate plans may need to be developed. Managers must be capable of making adjustments in personnel, equipment, policies, or procedures whenever necessary (Figure 36-7).

FIGURE 36-7 The manager must be familiar with all the functions and able to work with all the employees of the business.

HOT TOPICS IN AGRISCIENCE COOPERATION FOR THE GOOD OF ALL

© ruzanna/Shutterstock.com.

A farmers' cooperative pools the business efforts of many farmers, giving them the advantage of volume sales and purchases that none of the members could achieve alone.

A cooperative, or co-op, is a type of business organization that is created, owned, and controlled by its members. The function of a co-op is to ensure that the objectives of its members are met. There are three main types of co-ops: market, purchasing, and service based. A co-op that is organized for the purpose of marketing receives products from its members and then resells those products for the best possible price. Marketing co-ops are common in agriculture. One producer alone may not be able to produce enough crops or other agricultural commodities to obtain a large contract. But when many producers band together, they can ensure that they will have the ability to compete for large-order contracts.

Purchasing co-ops can make large-quantity purchases for members. By purchasing in bulk, the members can get the best possible prices for products they need. Items such as farm equipment, seed, fertilizers, dairy supplies, and much more can be purchased at reduced prices for members. Service co-ops provide low-cost service items to their members. For example, health insurance can be purchased at a much lower cost when a large number of people are becoming insured as opposed to a single family. This makes it possible to spread the cost of losses across many participants instead of just a few individuals. So, as the size of the insured pool increases, the cost per individual goes down. Co-ops are valuable tools to the farm and ranch families that elect to participate in them. Members have discovered that much more can be accomplished when working together than it is possible to do alone.

INTERNET KEY WORDS:
financial records needed, small business

In a very small venture, all management functions may be carried out by the same person. As a venture grows and its goals and objectives become more involved, additional managers may be required. When they are hired, organizational lines of responsibility must be developed and job descriptions written to identify the functions that are performed by each manager.

SMALL-BUSINESS FINANCIAL RECORDS

Financial records are invaluable tools to the entrepreneur and manager. Good financial management allows the manager to maintain control of the business venture and to increase profits or reduce losses.

Financial records can reveal which items are selling and which are not. These records can also show the extent of success of each person on the sales force. Financial records should indicate the amount of inventory on hand and how much has been sold. How much profit is made and the total value of the venture can be computed at any time using appropriate financial records.

The small-business owner or entrepreneur needs and uses detailed financial records. Financial records for items such as buildings, fixtures and equipment, credit, debt, and return on investments are examples of records maintained by small businesses. However, balance sheets, profit and loss statements, and inventory and sales reports may be useful for a group educational project as well as for a small business. Some typical financial records are discussed in the following sections (Figure 36-8).

FIGURE 36-8 Financial records are fundamental management tools.

Balance Sheet

Date: ______

ASSETS

Current: Cash ______
Merchandise ______
Accounts Receivable ______
Fixed: Land ______
Building ______
Machinery ______
Equipment ______
Other:
1. ______
2. ______
3. ______

TOTAL ASSETS: ______

LIABILITIES

Current: Notes Payable
Accounts Payable ______
Noncurrent:
Debts more than one-year maturity ______

TOTAL:

NET WORTH: ______

TOTAL LIABILITIES AND NET WORTH: ______

FIGURE 36-9 A balance sheet is a statement of a business owner's assets, liabilities, and net worth.

Balance Sheet

A **balance sheet** is like a photograph of the business at a point in time. It shows the assets, liabilities, and owner's investment on a particular date. The equation for the balance sheet follows: assets = net worth − liabilities. **Assets** include everything the venture owns, including cash on hand, equipment, and inventory. **Liabilities** include both current and long-term debts. **Net worth** is the owner's investment in the business, including profits as they occur. A sample balance sheet is shown in Figure 36-9.

SCIENCE CONNECTION — GOLDEN OPPORTUNITIES WAITING

It is widely known that research in agriscience plays a continuous role in keeping the food, fiber, and natural resources sectors in the United States productive and efficient. What is not so well known is the impact of agriscience research on the businesses and industries that the general public does not consider to be agribusinesses. In one tally, the USDA Agricultural Research Service documented nearly 75 companies that were manufacturing products or using processes from 53 new technologies developed by that agency. Some of these companies were formed to take advantage of new opportunities to meet apparent needs with emerging technologies. Others were established companies that were able to expand their product lines as a result of the research (Figure 36-10).

PRODUCT	COMPANY	LOCATION
No-calorie, high-fiber flour	Mt. Pulaski Products Canadian Harvest	Mt. Pulaski, IL Cambridge, MN
Plant virus test kit	Agdia, Inc.	Elkhart, IN
Super-Slurper (starch-derived absorbent material)	Super Absorbent Co. Grain Processing Co. Henkel Corporation	Lumberton, NC Muscatine, IA Kankakee, IL
Traps for stable flies	Tiger Farm Products	Hopedale, MA
Turkey hemorrhagic enteritis virus vaccine	Arko Laboratories Oxford Laboratories Willmar Poultry Co.	Jewell, IA Worthington, MN Willmar, MN
Dietary supplement	Monarch Nutri. Labs	Ogden, UT
Southwestern corn borer pheromone trap	Great Lakes IPM	Vestaburg, MI
Microbial insecticides	Reuter Labs	Haymarket, VA
Improved fire ant insecticide	Griffin Corp.	Valdosta, GA
Japanese beetle trap	Consep Membranes	Bend, OR
Direct marketing of fruits and vegetables	Three Rivers Produce	Southeastern, OK
Biochemical and physiological factors influencing turkey egg hatchability	Agrimatic Corp.	Paramount, CA
Improve flavor, texture, and juiciness of processed poultry meat	Continental Grain Co.	Pendergrass, GA
Discover and develop mycoparasites for biocontrol of selected soilborne plant pathogens	Agracetus, Inc.	Middleton, WI
Bee breeding and genetics for bee stock improvement	Weaver Apiaries, Inc.	Navasota, TX
Modification of starch to provide controlled delivery of agricultural chemicals	Illinois Cereal Mills, Inc.	Paris, IL
Cotton gin system design and evaluation to maximize product quality and minimize processing costs	Lummus Industries, Inc.	Columbus, GA
Systems to apply irrigation water efficiently, control intake, and reduce nitrate leaching	Elvert Edgar Oest and Co.	Fruita, CO
Biological control of weeds using exotic and endemic plant pathogens	Confederated Tribes, Coleville Reservation	Nespelem, WA
New engineering concepts for deciduous fruit production, harvesting, sorting	Agri-Tech	Woodstock, VA
Molecular biology of orbiviruses to diagnose and characterize bluetongue virus	Veterinary Diagnostic Technology, Inc.	Wheat Ridge, CO
Apparatus and method for rapid analyses of multiple samples	Lachat Chemicals Alpkem Corp.	Mequon, WI Clackamas, OR
Novel durable press finishing of textiles	Duro Finishing Corp.	Fall River, MI
Milk-like products from peanuts	Seabrook Blanching	Edenton, NC
Highly absorbent polymeric compositions derived from flour	Illinois Cereal Mills Industrial Services Venture Chemicals Polysorb	Paris, IL Bradenton, FL Lafayette, LA Smelterville, ID
Rope wick applicator	Rear's Mfg. Co. Brothers Equip. Co. BCP Manufacturing Co. Agman, Inc. Rodgers Sales Stanley Resser	Eugene, OR Friend, NV Winters, TX Riverside, MO Clarksdale, MI Weldon, IL
Starch-based semi-permeable films	Uni-Star Industries	Cuba, IL
Controlled bulk vegetable fermentation	Trumark, Inc.	Roseville, NJ
Reducing water content of emulsions, suspensions, and dispersions with highly absorbent starch-containing polymeric compositions	Super Absorbent Co. Grain Processing Co. Worne Biotech Promar, Inc.	Lumberton, NC Muscatine, IA Medford, NJ Milwaukie, OR
Removal of heavy metal ions from waste water	Tetrahedron, Inc.	Stanhope, NJ
In-ovo vaccination	Embrex, Inc.	Morrisville, NC
Super Slurper dewatering cartridge	Central Illinois Manufacturing Co.	Bement, IL
Turkey semen extender	Continental Plastic Corp.	Delavan, WI
Cotton and soybean seed	Delta and Pineland Seed Co.	Scott, MI
Recirculating wiper for agricultural chemicals	Apple Machine Co.	Fort Pierce, FL
Biopesticide for stone fruit	Fermenta	Painesville, OH
Marek's disease vaccine	Select Laboratories, Inc. Tri Bio Laboratories	Gainesville, GA State College, PA
Biodegradable plastic	Agri-Tech	Peoria, IL
Vaccination of chicken embryos for protection against avian coccidiosis	Embrex, Inc.	Morrisville, NC
Methods that affect molecular genetic transfer and transformation in crop plants	Pioneer Hi-Bred, Inc.	Johnston, IA
Physiology of agriculturally important spiroplasmas and mycoplasmas	Agdia, Inc.	Elkhart, IN
High-solids tomatoes and low-sugar potatoes using tissue culture	Northrup King Co.	Gilroy, CA
Construction of expression vectors from Marek's disease virus	Select Laboratories, Inc.	Gainesville, GA
Computerized control and management of center pivot irrigation systems	Valmont Industries	Valley, NV
Colonization factors expressed by *Campylobacter jejuni* in chicken	Southeastern Poultry and Egg Association	Decatur, GA
New technologies in cotton ginning	Moisture Systems Corp.	Hopkinton, MA
Integrated strategies for managing filth-breeding flies on dairy farms	Olson Products, Inc.	Medina, OH
Monoclonal antibodies for diagnosis of Marek's disease	Vineland Laboratories Intervet, Inc.	Vineland, NH Millsboro, DE
Development of avian expression vectors using serotypes 1, 2, and 3 of Marek's disease virus	Solvay Animal Health, Inc.	Charles City, IA
Techniques for automated nondestructive quality evaluation of horticultural crops	Process Equipment Co.	Tipp City, OH
Development of methods to protect packaged agricultural products from insect infestation	Trécé	Salinas, CA
Effects of residue decomposition in low-input sustainable agricultural systems	Crops Genetics Intl.	Hanover, MD
Retention of quality of lightly processed fruits and vegetables	Tree Top, Inc.	Selah, WA

Adapted from USDA/ARS.

FIGURE 36-10 Some new technologies developed by the Agricultural Research Service of USDA and at the companies that have marketed products using those technologies.

Profit and Loss Statement

The **profit and loss statement** projects costs and other expenses against sales and revenue over time. A profit and loss statement has five basic sections (Figure 36-11):

1. Total sales
2. Cost of goods sold
3. Gross profit
4. Expenses
5. Net profit

Cash Flow Statement

A **cash flow statement** describes the availability of funds for the purposes of running a business (Figure 36-12). When a positive cash flow exists, the income that is received by the business exceeds the expenses, and there is cash on hand. A negative cash flow indicates that the expenses that are generated by the business exceed the income at a given time. Nearly every business deals with

Profit and Loss Statement

For period ending ________________

TOTAL SALES	$75,000
COST OF GOODS SOLD	
Beginning Inventory	55,000
(Plus) Purchases	10,000
(Less) Ending Inventory	15,000
TOTAL COST OF GOODS SOLD	50,000
(Beginning Inventory plus Purchases minus Ending Inventory)	25,000
GROSS PROFIT	25,000
(Total Sales minus Total Cost of Goods Sold)	
EXPENSES	
Salaries	18,000
Payroll Taxes	2,500
Rent	4,500
Advertising	1,000
TOTAL EXPENSES	26,000
NET PROFIT (LOSS) (before taxes)	(1,000)
(Gross Profit minus Total Expenses)	

FIGURE 36-11 Sample profit and loss statement.

Inventory Report

Item No.	Description	Beginning Inventory	Items Sold	Ending Inventory	Total Cost Beg./Inv.	Total Cost End/Inv.	Total Sales	Total Profit
000	Tick Tack Tote	45	45	0	33.75	.00	45.00	11.25
001	Brown Tray	32	32	0	25.60	.00	48.00	22.40
002	Fish Tray	20	20	0	15.00	.00	30.00	15.00
003	Lion Tray	6	6	0	4.80	.00	9.00	4.20
004	Guitar Tray	8	8	0	6.40	.00	12.00	5.60
005	Asst. Animal Tray	34	34	0	25.50	.00	51.00	25.50
006	Elephant Tray	10	10	0	5.00	.00	15.00	10.00
007	Screwdriver Set	22	22	0	11.00	.00	33.00	22.00
008	Brush & Shoehorn	36	36	0	32.40	.00	81.00	48.60
009	Fully Auto. Umbrella	9	9	0	20.70	.00	31.50	10.80
010	Auto. Umbrella	8	8	0	16.00	.00	24.00	8.00
011	Dad's No. 1 Keychain	200	190	10	90.00	4.50	190.00	104.50
012	Grandpa Keychain	100	79	21	45.00	13.95	79.00	47.95
013	Dad's Pad	48	30	18	28.80	10.80	45.00	27.00
014	Mini Screwdriver Set	75	75	0	60.00	.00	150.00	90.00
015	Dad's Plaque	72	72	0	28.80	.00	108.00	79.20
016	Tic Tac Toe	36	36	0	21.60	.00	36.00	14.40
017	Dad's Pen	1	1	0	.80	.00	1.50	.70

FIGURE 36-12 Sample cash flow statement.

periods of negative cash flow, and it is at these times that an operating loan may be needed to tide the business over until income is received. The cash flow statement allows a business owner to estimate the need for credit and to establish when the credit will be needed. For example, a beef ranch operates with a negative cash flow for most of the year. Once the calves are ready for market, they are sold together with cows that are not being retained in the breeding herd. Most of the income is received within a few days or weeks. In contrast, a dairy farm has a fairly constant cash flow throughout the year from the sale of milk and excess animals. The cash flow statement is intended to help the business manager and the loan officer determine how much operating cash is needed, when it is needed, and when it will be repaid.

Inventory Report

The **inventory report** includes how many units of each product are on hand, how many were sold, each item's cost, total sales, and profit. The inventory report can be used to make decisions such as reducing the price of slow-selling items, reordering other items, and learning which items are yielding the best profits. The use of computers at checkout stands allows this report to be updated each time a sale is made. Some software is also capable of generating new orders for products as the inventory for a particular item is reduced to a predetermined level.

SELF-EMPLOYMENT VERSUS OTHER FORMS OF EMPLOYMENT

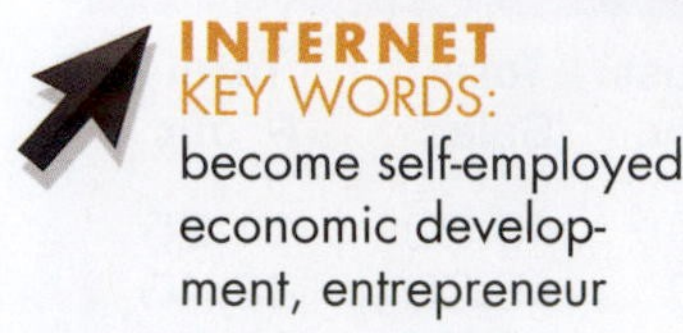

The decision to continue working for someone else or to open a business is a difficult one to make. A way to help make that decision is to look at the advantages and disadvantages of working for someone else and working for yourself.

Being an Employee

The advantages of being an employee, working for someone else, center on security. The salaried employee has no personal financial risk or responsibility to the company for which he or she works. Employees generally work regular hours. If they work additional hours, they may be paid overtime for those hours. In addition, employees are often guaranteed vacation time and fringe benefits such as life and health insurance. In some cases, they are offered retirement plans. An employee can count on a somewhat stable lifestyle with a fairly accurate idea of what the income will be from year to year. In addition, the employee may move up the career ladder within the company.

However, there are some disadvantages to working for someone else. The company does not have any financial responsibility to the employee should there be a recession resulting in a cutback in personnel. Some companies have a predetermined salary scale, so the employee could remain at a particular salary level until the right combination of years or experience is met. This is also true of promotions that could be based on years accumulated rather than on merit. The work pattern could become fairly routine. If no positions open within the company, the employee must wait until a position opens or look for a position with another company. Management controls these decisions, not the employee.

Being Self-employed

One of the major reasons generally cited for opening a business is that it gives the owner control over his or her destiny. The owner has the opportunity to set personal goals and recruit a team to help carry out those goals. Successfully meeting those goals results in a sense of achievement and, it is hoped, financial reward.

Owning the business and being responsible for the decisions related to the business give the entrepreneur a sense of independence (Figure 36-13). The owner's ability to make money is not restricted to a particular level.

The disadvantages of owning a business are not as obvious as the advantages. For instance, the necessary capital outlay may jeopardize family savings or even the family home. The number of hours required to run the business will mean a definite commitment from the owner as well as the family members involved. Should the company have difficulty, the responsibility for both the management and the financial problems rests with the owner.

Contributions of Small Businesses

Entrepreneurs have been credited with being the cornerstone of the U.S. enterprise system. Furthermore, many see entrepreneurs as the self-renewing agents of the economic environment in the United States. In recent years, numerous socialistic countries of the world, such as China, Russia, and Eastern European countries, have encouraged entrepreneurship after forbidding capitalism for decades.

© mangostock/Shutterstock.com.

FIGURE 36-13 Ownership of a small business encourages innovation and gives the owner a sense of independence.

© iStockphoto/YinYang.

FIGURE 36-14 The small businesses that dot the landscape of the United States are highly productive and responsive to the needs of their customers.

The role of entrepreneurs in the United States is highlighted by the following:

- Most businesses in the United States (95 percent) are classified as small by the Small Business Administration.
- New businesses are formed at a rapid rate (about 600,000 per year).
- Small businesses generate almost half (48 percent) of the U.S. gross national product (GNP).
- Small businesses employ about half of all U.S. workers.
- Small businesses created approximately two-thirds of the new jobs in our economy in the last two decades.
- Small businesses produce 13 times as many patents (products, services, techniques) as large businesses (Figure 36-14).

The U.S. economy fluctuates up and down in response to many influences. However, as the economy continues toward an emphasis on services, the role and importance of small business and entrepreneurship is expected to increase. This is likely to occur because small businesses are especially dominant in the service sector of the economy.

New businesses can be the center of innovation because they are not generally tied to existing ways of doing things. They have a sense of energy, urgency, and vitality that comes from the entrepreneurial spirit.

The contribution of small businesses to the U.S. private-enterprise system is important, and it is likely to remain so in the future.

STUDENT ACTIVITIES

1. Write the Terms to Know and their meanings in your notebook.
2. Develop a collage on the bulletin board illustrating entrepreneurship opportunities in agriscience.
3. Make a table showing the advantages and disadvantages of being (1) an employee and (2) an entrepreneur. The following format is suggested:

Being an Employee	Being Self-employed
Advantages	Advantages
1.	1.
2.	2.
etc.	etc.
Disadvantages	Disadvantages
1.	1.
2.	2.
etc.	etc.

4. Participate in a group sales project in the FFA, some other school group, 4-H, or some other community group. Encourage the group to operate the project to help all participants obtain useful business skills.
5. Organize a cooperative business within your class or FFA for buying or marketing a product of interest to the group.
6. Become familiar with the record book and record-keeping system used by members of your local FFA chapter or 4-H club for keeping records on individual projects.
7. Become an entrepreneur—own and operate your own business venture.
8. Get a summer or after-school job with a company similar to one you may be considering for your own future career. While working for that company, learn as much about running a business as you can.
9. Create an individual business plan using the information you have learned from this unit. Discuss with your parents and trusted advisors what steps you will need to take to make your business plan a reality.

SELF-EVALUATION

A. MULTIPLE CHOICE

1. The selection of a product or service to be sold for profit is called the
   a. buying function.
   b. distribution function.
   c. promoting function.
   d. selling function.
2. Determining reasons that customers may wish to buy a product or service is called the
   a. buying function.
   b. distribution function.
   c. promoting function.
   d. selling function.
3. Development of a plan to identify ways to make potential customers aware of a product or service is called the
   a. buying function.
   b. distribution function.
   c. promoting function.
   d. selling function.

4. When analyzing competing products or services, which is not a factor?
   a. price
   b. quality
   c. service
   d. supply
5. Which is not a function of managers?
   a. actuating
   b. controlling and evaluating
   c. organizing people and resources
   d. setting policy
6. A disadvantage of being an employee is
   a. overtime pay.
   b. regular hours.
   c. little or no control over future job.
   d. fringe benefits.
7. A major advantage of owning a business is
   a. control over your destiny.
   b. financial responsibility.
   c. the relationship of business income and family finances.
   d. a shorter work week.

B. MATCHING

__________	1. Entrepreneur	a. Business venture
__________	2. Entrepreneurship	b. Organizer; risk taker
__________	3. Balance sheet	c. Expenses versus revenues
__________	4. Profit and loss statement	d. One who devises
__________	5. Inventor	e. Photograph of a business

APPENDIX A
Developing a Personal Budget

One of the most important parts of becoming an adult is creating a budget to handle your personal finances. A budget is a plan of action that includes projections of income and expenses for all or part of business or personal expenses. The process in this appendix is designed to help you understand a personal budget. Many of the same principles can be used when developing your supervised agricultural experience (SAE).

There are several reasons for developing and using a budget. Among them are the following:

- Helps you plan for the lifespan of assets
- Is an excellent device for organizing
- Is useful to obtain credit
- Allows experimenting with different outcomes
- Identifies costs and income
- Helps refine and organize
- Is a good management tool for anyone with limited experience

LIMITATIONS OF BUDGETS

People do not budget for several reasons. Being aware of these limitations will help you avoid the mistake of not carefully budgeting. Following is a list of some of these reasons:

- Budgeting takes time.
- It is necessary to search for accurate information.
- It is sometimes difficult to predict future actions.
- Managers tend to underestimate costs and overestimate income.

THE BUDGET PROCESS

Budgets do not need to be complicated. The process can be as simple as determining, listing, and summarizing all of your anticipated expenses and all of your anticipated income. Obviously, the goal should be for your income to exceed your expenses. If this is not the case, you need to change the level of either your income or your expenses.

A simple worksheet follows that may help you organize a budget. List your estimated costs for the items that apply to you or your business.

MONTHLY EXPENSES

Home:	Rent/Own	______
	Maintenance	______
Utilities:	Phone	______
	Electricity	______
	Cable	______
	Gas	______
Insurance:	Home	______
	Vehicle	______
	Life	______
	Health	______
	Disability	______
Vehicle:	Payment	______
	Gas	______
	Oil/Tires/Maintenance	______
Food:		______
Clothing:		______
Savings:		______
Entertainment/Recreation:		______
Emergencies:		______
Cleaning/Toiletry/Beauty:		______
Gifts/Contributions:		______
Medical Care:		______
Installment Payments:	Credit Cards	______
	Bank Notes	______
	Department Stores	______

MONTHLY INCOME

Salary:	______
Odd jobs:	______
Gifts:	______
Sales:	______
Other Income:	______
Total Income:	______
BALANCE: *(income minus expenses)*:	______

Portions of Appendix A were adapted from the Georgia Agricultural Education Curriculum Guide, Atlanta, GA.

APPENDIX B
Plan Supervised Agricultural Experience

This appendix includes plans for building birdhouses and nesting boxes. Both of these units can be constructed using various types of wood. Because the houses of the structures are constantly exposed to the elements, it is essential that wood used in the construction be carefully selected. Use your problem-solving skills to determine which materials are best suited for the various parts. Comparisons of different types of woods are included to aid in your decisions. Safety is a constant concern when constructing projects or conducting experimentation. Rules governing safety are included in this appendix.

GUIDELINES FOR SAFETY IN THE LABORATORY

The following rules must be observed to ensure your safety and the safety of others in the agricultural education classroom and laboratory.

1. Your concern for safety should begin even before the first laboratory or shop activity. Always read and think about safety before starting an activity.
2. Perform laboratory work only when your teacher is present. Unauthorized or unsupervised laboratory experimenting is not allowed.
3. Know the location and use of all safety equipment in your laboratory. These should include the eye wash station, first-aid kit, fire extinguisher, and safety shower (if available).
4. Wear safety glasses or goggles at all times in the wood and metal laboratory. Wear protective clothing along with ear-and-eye protection accessories at all times as directed by the instructor. Tie back loose hair and secure loose clothing.
5. Check chemical labels twice to make sure you have the correct substance. Pay attention to the hazard classification shown on the label.
6. Avoid unnecessary movement and talk in the laboratory.
7. Gum, food, or drinks should not be brought into the laboratory.
8. Never sniff chemicals, and do not place your nose near the opening of a chemical container.
9. Any laboratory accident, however small, should be reported immediately to your teacher.
10. In case of a chemical spill on your skin or clothing, rinse the affected area with plenty of water. If the eyes are affected, water-washing must begin immediately and continue for 10 to 15 minutes or until professional assistance is obtained.
11. When discarding used chemicals or hazardous materials, carefully follow the instructions provided.
12. Know where the Material Safety Data Sheet forms are located and learn how to use them.
13. No horseplay, running, and so on are allowed at any time in the laboratory/shop or classroom.
14. Do not use equipment unless you have been instructed on its use and safety precautions and have been given permission to do so.
15. Keep all tools stored in their correct locations while not in use.
16. Make sure all safety devices, guards, and other items of protective equipment are in place and working at all times.

17. Make sure compressed gas cylinders are kept chained to a wall or on a secure rack with caps tightly on.
18. Before making adjustments on any piece of equipment, make sure it is turned off and unplugged.
19. Keep your work area clean and clear of hazards.
20. If water has been spilled on the floor, clean it up immediately. If a hazardous chemical has been spilled, notify your teacher but do not attempt to clean it up until you are told to do so and advised of the proper procedures.
21. Do not put hands and fingers closer than 2 inches to a blade while using saws, drills, and other similar equipment.
22. Know where all tools, chemicals, and equipment are located in the shop.
23. Always use the proper tool for the job.

Portions of Appendix B were adapted from the Georgia Agricultural Education Curriculum Guide, Atlanta, GA.

24. Do not stand closer than 6 feet to power equipment being operated by someone else.
25. Recognize the color coding for shops, and observe all regulations regarding color coding:

    Safety RED = *DANGER*
    Safety ORANGE = *WARNING*
    Safety YELLOW = *CAUTION*
    Safety BLUE = *INFORMATION*
    Safety GREEN = *SAFETY*
    Safety WHITE = *TRAFFIC MARKINGS*
26. Lift heavy objects safely using proper lifting techniques.
27. Never throw anything in the classroom or laboratory.
28. Take responsibility for the safety of yourself and others. Follow tool and equipment safety manuals. Obey rules and guidelines as well as the instructions of your teacher. Use good judgment for safety's sake. If in doubt, ask!

CONSTRUCTION PROJECT: FLICKER/ WOODPECKER HOUSES

Plan for Flicker/Woodpecker Houses

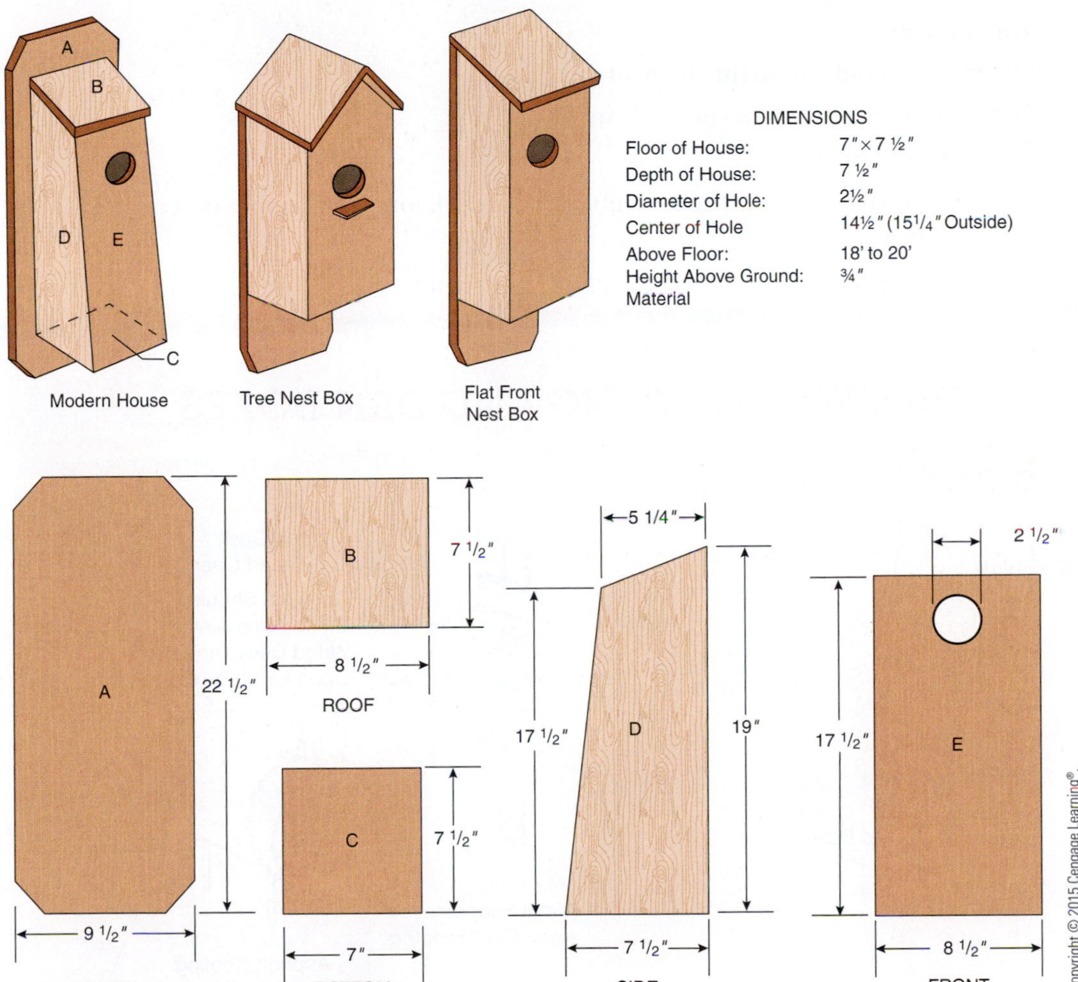

BILL OF MATERIALS (FOR MODERN HOUSE)			
MATERIALS NEEDED	**QUANTITY**	**DIMENSIONS**	**DESCRIPTION OR USE**
Lumber (rough and rustic)	1	3/4" × 8−1/2" × 17−1/2"	Front (A)
"	2	3/4" × 7−1/2" × 19"	Sides (B)
"	1	3/4" × 7" × 7−1/2"	Bottom (C)
"	1	3/4" × 7−1/2" × 8−1/2"	Roof (D)
"	1	3/4" × 9−1/2" × 22−1/2"	Back (E)
Box nails or finishing nails		4d or 6d	Attach

The flicker will nest readily in a well-placed birdhouse of the proper dimensions. Roughen the interior of the nest box to assist the young woodpeckers to reach the entrance hole. Cover the bottom with sawdust so that the mother bird can shape the nest for eggs and young birds. Material may be one-half-inch instead of three-fourths-inch thick, in which case the width of the front and roof will be 8 inches.

Construction Procedure for Flicker/Woodpecker Houses

1. Select pine, redwood, or other lumber that will withstand weather without paint. Rough lumber is generally preferred and will provide a rustic appearance. Planed lumber should be used if the house is to be painted for decorative purposes, but this is not recommended for most species except Martin houses.
2. Cut all parts from a 3/4"×9−1/2"×8' board.
3. Cut the sides to the proper angle.
4. Bevel the top edge of the front to conform to the angle of the sides.
5. Bevel the upper edge of the roof to conform to the back and sides.
6. Nail the front, sides, and bottom together.
7. Nail the back to the sides and bottom. (Center the assembly on the back for a good appearance. Leave 1 inch at the bottom and 2.5 inches at the top.)
8. Attach the roof with two hinges or nail it lightly so the roof is easily removed for periodic cleaning of the nest area.

CONSTRUCTION PROJECT: NESTING AND DEN BOXES

Plan for Nesting and Den Boxes

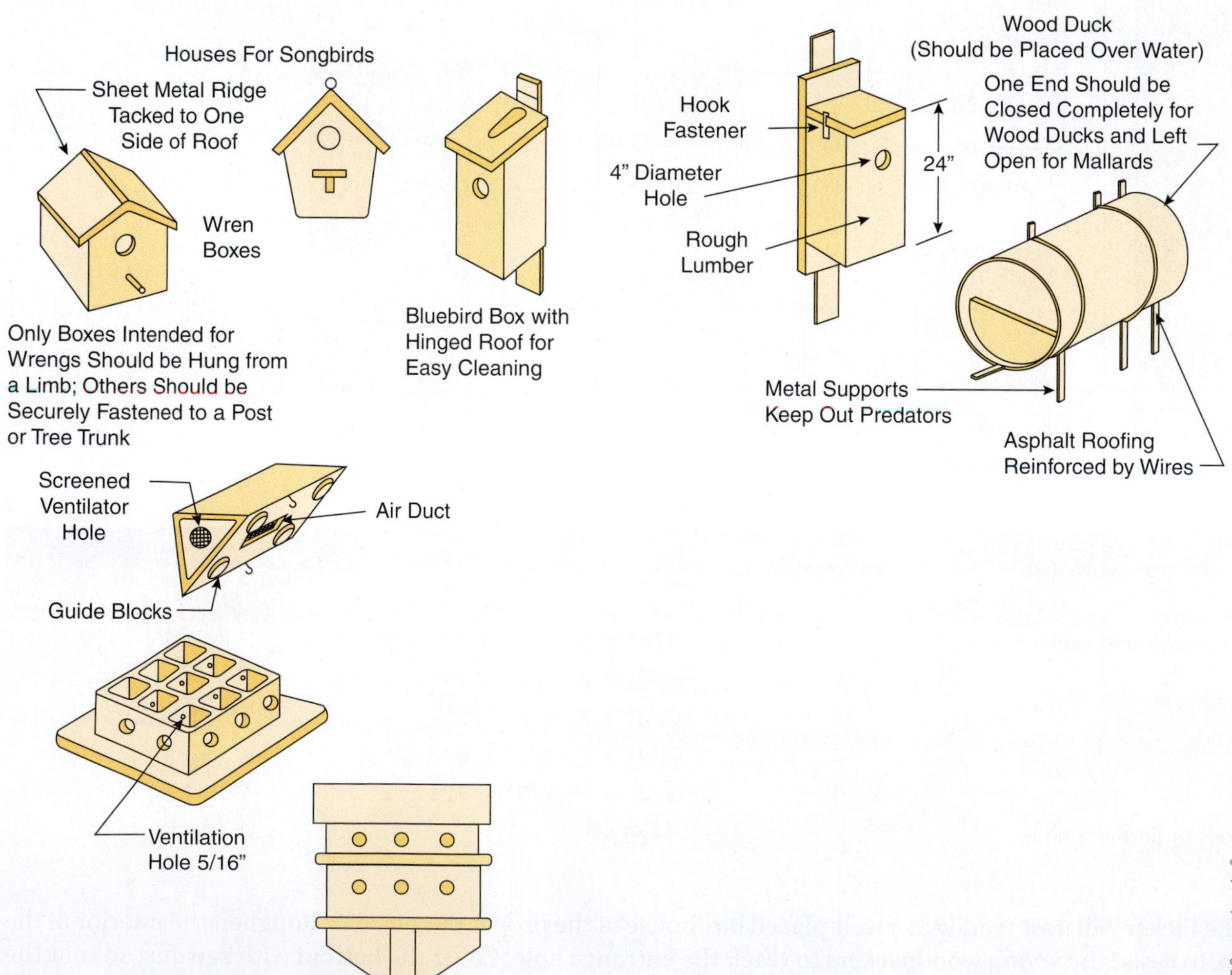

Bill of Materials

Bills of materials are not provided with these nesting and den boxes because of the variety of types pictured. Nesting and den boxes should be planned from the information provided. An appropriate bill of materials may then be developed for the plan.

Construction Procedure for Nesting and Den Boxes

Construction procedures will vary with the nesting or den box being constructed. However, the procedure outlined for the flicker/woodpecker houses may be helpful.

CHARACTERISTICS OF COMMON WOODS			
SPECIES	HARDNESS	KNOWN FOR	SOME MAJOR USES
Birch	Hard	Surface veneer for panels	Cabinets and doors
Cedar, red	Medium	Pleasant odor	Furniture, chests, and birdhouses
Cherry	Hard	Red grain	Fine furniture
Cypress	Medium	Rot resistance	Structural material in wet places, birdhouses
Fir and Hemlock	Soft	Light, straight, strong	Construction framing, siding, sheathing
Locust, black	Hard	Rot resistance	Fence posts, birdhouses
Maple	Hard	Light grain	Floors, bowling alleys, durable furniture
Mahogany	Medium	Reddish color	Fine furniture
Oak	Hard	Toughness, strength	Floors, barrels, wagon bodies, feeders, farm buildings
Pine, yellow	Medium	Wear resistance, tough	Floors, stairs, trim
Pine, white	Soft	Easy to work, straight	Shelving, siding, trim
Redwood	Soft	Excellent rot resistance	Yard posts, fences, birdhouses
Walnut, black	Hard	Brown grain	Fine furniture
Willow, black	Soft	Brown grain, easy to work, walnut look	Furniture

TYPES OF NAILS

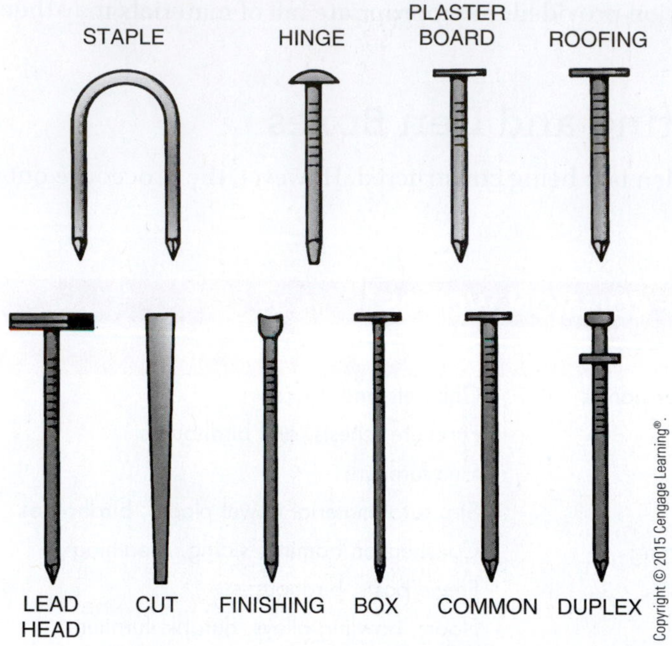

LENGTH OF NAILS

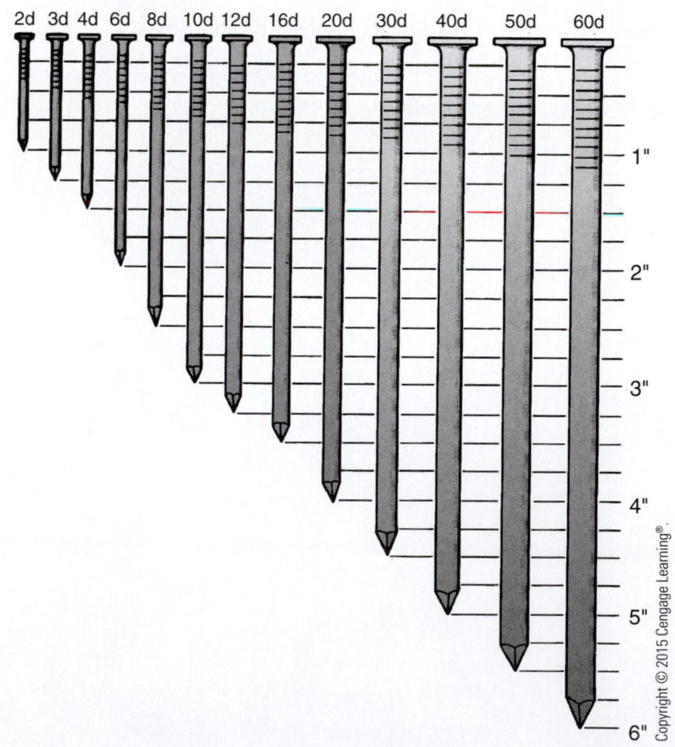

GAUGES AND TYPES OF WOOD SCREWS

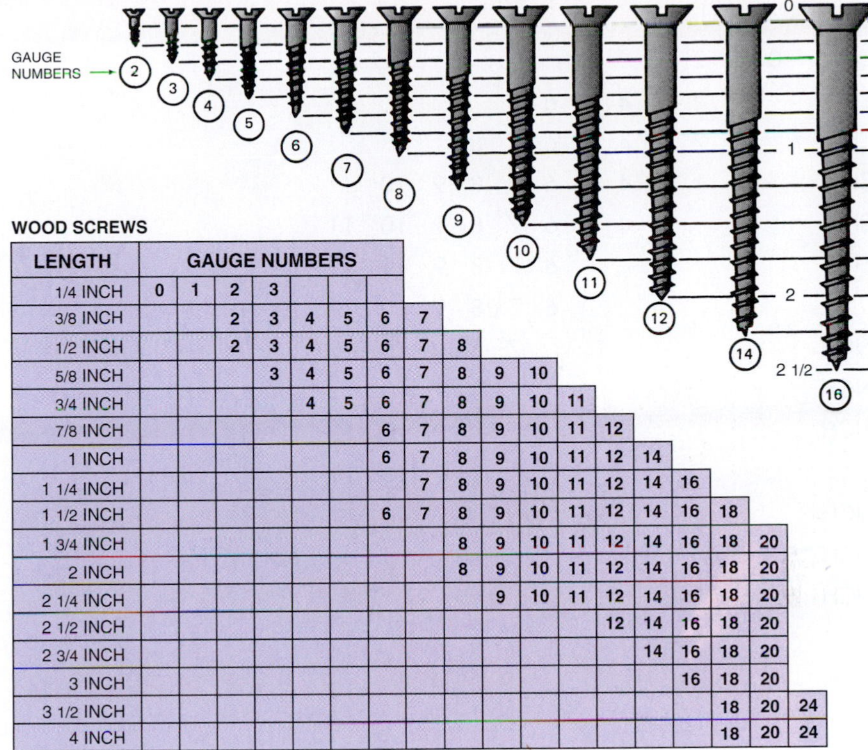

LENGTH	GAUGE NUMBERS																	
1/4 INCH	0	1	2	3														
3/8 INCH			2	3	4	5	6	7										
1/2 INCH			2	3	4	5	6	7	8									
5/8 INCH				3	4	5	6	7	8	9	10							
3/4 INCH					4	5	6	7	8	9	10	11						
7/8 INCH							6	7	8	9	10	11	12					
1 INCH							6	7	8	9	10	11	12	14				
1 1/4 INCH								7	8	9	10	11	12	14	16			
1 1/2 INCH							6	7	8	9	10	11	12	14	16	18		
1 3/4 INCH									8	9	10	11	12	14	16	18	20	
2 INCH									8	9	10	11	12	14	16	18	20	
2 1/4 INCH										9	10	11	12	14	16	18	20	
2 1/2 INCH													12	14	16	18	20	
2 3/4 INCH														14	16	18	20	
3 INCH															16	18	20	
3 1/2 INCH																18	20	24
4 INCH																18	20	24

When You Buy Screws Specify (1) Length, (2) Gauge Number, (3) Type of Head–Flat, Round, or Oval, (4) Material–Steel, Brass, Bronze, etc., (5) Finish–Bright, Steel Blued, Cadmium, Nickel, or Chromium Plated.

WOOD SCREWS

LENGTH	GAUGE NUMBERS
1/4 INCH	0 1 2 3
3/8 INCH	2 3 4 5 6 7
1/2 INCH	2 3 4 5 6 7 8
5/8 INCH	3 4 5 6 7 8 9 10
3/4 INCH	4 5 6 7 8 9 10 11
7/8 INCH	6 7 8 9 10 11 12
1 INCH	6 7 8 9 10 11 12 14
1 1/4 INCH	7 8 9 10 11 12 14 16
1 1/2 INCH	6 7 8 9 10 11 12 14 16 18
1 3/4 INCH	8 9 10 11 12 14 16 18 20
2 INCH	8 9 10 11 12 14 16 18 20
2 1/4 INCH	9 10 11 12 14 16 18 20
2 1/2 INCH	12 14 16 18 20
2 3/4 INCH	14 16 18 20
3 INCH	16 18 20
3 1/2 INCH	18 20 24
4 INCH	18 20 24

When You Buy Screws Specify (1) Length, (2) Gauge Number, (3) Type of Head–Flat, Round, or Oval, (4) Material–Steel, Brass, Bronze, etc., (5) Finish–Bright, Steel Blued, Cadmium, Nickel, or Chromium Plated.

REFERENCES

Bolstad, P. (April–May 1993). Alaskan Adventure: FFA prepared this member to live and work in our northernmost state. *FFA New Horizons*. Retrieved from https://archives.iupui.edu/bitstream/handle/2450/5453/FFANewHorizons_41_4_AprMay1993.pdf?sequence=10

Bryan, M. D. and Drea, J. (March 12, 1992). Targeting Euonymus Scale for Biocontrol. *Agricultural Research, 3*.

Bureau of Labor Statistics, U.S. Department of Labor, *Occupational Outlook Handbook, 2012–13 Edition*, Agricultural Workers. Retrieved from http://www.bls.gov/ooh/farming-fishing-and-forestry/agricultural-workers.htm

Food and Agriculture Organization of the United Nations. (2004). *The State of Food and Agriculture, 2003–2004*. Retrieved from ftp://ftp.fao.org/DOCREP/fao/006/y5160e/y5160e00.pdf

Futch, S. H., Derrick, K. S., and Brlansky, R. H. (February 2005. Reviewed March 2008 and March 2011). Field Identification of Citrus Blight. [Document HS995]. Horticultural Sciences Department, Florida Cooperative Extension Service, Institute of Food and Agricultural Sciences, University of Florida. Retrieved from http://edis.ifas.ufl.edu/pdffiles/HS/HS24100.pdf

Hershey, D. (February 1, 1994). Solution Culture Hydroponics: History & Inexpensive Equipment. *The American Biology Teacher 56* (2). Retrieved from http://www.jstor.org/discover/10.2307/4449764?uid=3739832&uid=2&uid=4&uid=3739256&sid=21101429448203

Hoagland, D. R. and Arnon, D. I. (1950). The Water-Culture Method for Growing Plants Without Soil. California Agricultural Experiment Station Circular 347.

Intergovernmental Panel on Climate Change. (1996). *Climate Change 1995: The Science of Climate Change. Contribution of the Working Group I to the Second Assessment Report of the Intergovernmental Panel on Climate Change*. Houghton, J. T., Meira Filhon, L. G., Callander, B. A., Harris, N., Kattenberg, A., and Maskell, K. (eds.). Cambridge, United Kingdom and New York, NY, USA: Cambridge University Press. Retrieved from http://www.ipcc.ch/publications_and_data/publications_and_data_reports.shtml

Intergovernmental Panel on Climate Change (2007). *Climate Change 2007: The Physical Science Basis. Contribution of Working Group I to the Fourth Assessment Report of the Intergovernmental Panel on Climate Change*. Solomon, S., D. Qin, M. Manning, Z. Chen, M. Marquis, K. B. Averyt, M. Tignor, and H. L. Miller (eds.). Cambridge, United Kingdom and New York, NY, USA: Cambridge University Press. Retrieved from http://www.ipcc.ch/publications_and_data/ar4/wg1/en/contents.html

Moore, P. (2003, June 18). Nature vs. Politics. *The Wall Street Journal*. Retrieved from http://www.cccinc-7candlesticks.org/CCCIncfamily3forest_management.html

Natural Resources Defense Council. (March 2011). *Why We Need Bees: Nature's tiny workers put food on our tables*. Bee Facts [Data Sheet]. Retrieved from http://www.nrdc.org/wildlife/animals/files/bees.pdf

Nielsen, E. G. and Lee, L. K. (October 1987). *The Magnitude and Costs of Groundwater Contamination from Agricultural Chemicals: A national perspective*. Resources and Technology Division, Econonriic Research Service, U.S. Department of Agriculture. Agricultural Economic Report No. 576. Retrieved from http://naldc.nal.usda.gov/download/CAT88907300/PDF

The Secretary's Commission on Achieving Necessary Skills. (June 1991). *What Work Requires of Schools, A SCANS Report for America 2000*. U.S. Department of Labor. Retrieved from http://wdr.doleta.gov/SCANS/whatwork/whatwork.pdf

Schieder, S. (1990). Prudent planning for a warmer plant. *New Scientist*, 128 (1743).

Scientists' Statement, Global Climatic Disruption. (June 18, 1997). Open Letter. The Woods Hole Research Center. Retrieved from http://www.whrc.org/resources/essays/pdf/1997_climate_stmt.pdf#search="statement on global climatic disruption"

Spread of Africanized Honey Bees in the United States. (July 23, 2012). [Interactive Map]. *National Atlas of the United States*. Retrieved from http://nationalatlas.gov/mld/afrbeep.html

Spencer, R. (1998, August 14). Measuring the temperature of Earth from space: even with needed corrections, data still don't show the expected signature of global warming. *Space Science News* [a NASA publication]. Retrieved from http://spacescience.spaceref.com/newhome/headlines/notebook/essd13aug98_1.htm

Tooker, J. (May 2009). Entomological Notes, Alfalfa Weevil. [Insect Fact Sheet]. Penn State College of Agricultural Sciences, Cooperative Extension, Department of Entomology. Retrieved from http://ento.psu.edu/extension/factsheets/pdf/Alfalfa%20Weevil.pdf

United Nations Framework Convention on Climate Change (December 12, 2003). *Press Release*: Milan conference concludes as ministers call for urgent and coordinated action on climate *change*. Retrieved from http://unfccc.int/press/press_releases_advisories/items/3476.php

University of Maryland Baltimore County Department of Marine Biotechnology (UMBC). (n.d.). Aquaculture and Fisheries Biotechnology. Retrieved from http://www.umbc.edu/marinebiotech/fish_tech_more.html

U.S. Environmental Protection Agency. (January 2012). The EPA and Food Security. Retrieved from http://www.epa.gov/pesticides/factsheets/securty.htm

USDA, ARS. (Draft-Revised April 2004). The Commercial Storage of Fruits, Vegetables, and Florist and Nursery Stocks. Gross, K. C., Wang, C. Y., and Saltveit, M. (Eds.). *Agriculture Handbook 66*. Retrieved from http://www.ba.ars.usda.gov/hb66/contents.html

Wilcox, E. and Giuliano, W. M. (April 2006). *Red Imported Fire Ants and Their Impacts on Wildlife*. Document WEC 207. Department of Wildlife Ecology and Conservation, Florida Cooperative Extension Service, Institute of Food and Agricultural Sciences (IFAS), University of Florida. Retrieved from http://edis.ifas.ufl.edu/pdffiles/UW/UW24200.pdf

World Health Organization. (2012). Malaria [Fact Sheet]. Retrieved from http://www.who.int/mediacentre/factsheets/fs094/en/index.html

GLOSSARY

A

A horizon — layer near the soil surface consisting of mineral and organic matter.

horizonte A — capa cerca de la superficie que consta de materia mineral y orgánica.

abiotic — nonliving disease.

abiótica — enfermedad causada por factores no vivientes.

abortion — loss of a fetus before it is viable.

aborto — pérdida de un feto antes de que sea viable.

accent — distinctive feature or quality.

acento — rasgo o calidad distintiva.

accent color — attention-getting color.

color acento — color llamativo.

acid — pH of less than 7.0.

ácido — pH menor de 7.0.

acidity — sourness or the tendency toward a low pH, less than 7.0.

ácidez — ácido o la tendencia a tener un pH bajo menor de 7.0.

active ingredient — a component that achieves one or more purposes of the mixture.

ingrediente activo — componente que alcanza uno o más de los objetivos de la mezcla.

actuating — putting a plan into action.

accionar — poner un plan en acción.

acute toxicity — a measurement of the immediate effects of a single exposure to a chemical.

toxidad aguda — una medida de los efectos inmediatos de una sola exposición a una sustancia química.

adaptation — a process that occurs as heritable traits favoring the survival of an organism are passed from one generation to the next.

adaptación — un proceso que ocurre a medida que las características que favorecen la supervivencia de un organismo se pasan de una generación a la siguiente.

add-on-loan — the method used for calculating interest on consumer loans. The loan is repaid in installments of equal payments. Interest is added on at the beginning of the payments.

préstamo adicional — el método que se usa para calcular el interés sobre los préstamos al consumidor. El préstamo se paga a plazos en cuotas iguales. El interés se agrega al comienzo de los pagos.

adenine (A) — a base in genes designated by the letter *A*.

adenino — una base en los genes designado con la letra «A».

adjourn — a motion used to close a meeting.

suspender — una moción que se usa para levantar una junta.

adventitious root — root other than the primary root or a branch of a primary root.

raíz adventicia — una raíz diferente de la raíz fundamental o una rama de la raíz fundamental.

aeration — the mixing of air into water or soil to improve the oxygen supply of plants and other organisms.

airear — mezclar el aire con agua o tierra para mejorar la provisión de oxígeno de las platas y otros organismos.

aeroponics — the growing of plants in which their roots hang in the air and are misted regularly with a nutrient solution.

cultivo aeropónico — el cultivo de plantas donde las raíces de las plantas cuelgan en el aire y son rociadas con regularidad con una solución con nutrientes.

agar — a plant nutrient medium in which plant tissues are placed during the tissue-culture process.

agar — medio de nutrición de plantas donde se colocan los tejidos de las plantas durante el proceso de cultivo de tejidos.

age (ripen) — to leave undisturbed for a period.

madurar — dejar sin tocar por un período.

agregado — unidad de tierra que incluye principalmente partículas de tierra arcillosa, légamo y arena adheridos por una sustancia tipo gel formada por materia orgánica.

agribusiness — commercial firms that have developed with or stem from agriculture.

agroindustria — las firmas comerciales que se han desarrollado con o se derivan de la agricultura.

agribusiness management — the human element that carries out a plan to meet goals and objectives in an agriscience business.

administración de agroindustria — el elemento humano que lleva a cabo un programa para cumplir con las metas y los objetivos de un negocio de agriciencia.

agricultural — business, employment, or trade in agriculture, agribusiness, or renewable natural resources.

agrícola — negocio, empleo, o industria de la agricultura, la agroindustria, o los recursos naturales renovables.

agricultural economics — management of agricultural resources, including farms and agribusinesses.

economía agrícola — administración de recursos agrícolas, incluyendo granjas y agroindustrias.

agricultural education — teaching and program management in agriculture.

educación agrícola — enseñanza y administración de programa en la agricultura.

agricultural engineering — application of engineering principles in agricultural settings.

ingeniería agrícola — aplicación de los principios de ingeniería dentro del ambiente agrícola.

agricultural mechanics — design, operation, maintenance, service, selling, and use of power units, machinery, equipment, structures, and utilities in agriscience.

mecánica agrícola — diseño, funcionamiento, mantenimiento, servicio, venta y uso de aparatos mecánicos, maquinaria, equipos, estructuras y servicios públicos en la agriciencia.

agricultural processing, products, and distribution — industry that hauls, grades, processes, packages, and markets commodities from production sources.

procesamiento, productos, y distribución agrícolas — industria que transporta, clasifica, empaca y comercializa productos básicos de las fuentes de producción.

agricultural supplies and services — businesses that sell supplies and agencies that provide services for people in agriscience.

provisiones y servicios agrícolas — los negocios que venden provisiones y las agencias que prestan servicios para las personas en la industria de la agrociencia.

agriculture — activities concerned with the production of plants and animals, and the related supplies, services, mechanics, products, processing, and marketing.

agricultura — actividades que tratan de la producción de plantas y de animales, y las provisiones, los servicios, maquinaria, productos, procesamiento y comercialización relacionados.

agriscience — the application of scientific principles and new technologies to agriculture.

agriciencia — la aplicación de los principios científicos y nuevas tecnologías a la agricultura.

agriscience literacy — education in or understanding about agriscience.

educación agrocientífica — el estudio o la comprensión de la agrociencia.

agriscience professions — professional jobs dealing with agriscience situations.

profesiones de agrociencia — puestos profesionales que se ocupan de los asuntos de agrociencia.

agriscience research project — an original agricultural research project consisting of identifying a problem, a review of scientific literature, conducting a research project, and reporting the results.

proyecto de investigación de agrociencia — un proyecto de investigación agrícola original que consiste en identificar un problema, analizar literatura científica, llevar a cabo un proyecto de investigación y reportar los resultados.

agronomy — science and economics of managing land and field crops.

agronomía — ciencia y economía de la administración de tierras y cultivos del campo.

air — colorless, odorless, and tasteless mixture of gases.

aire — mezcla de gases incolora, inodora e insípida.

air layering — plant propagation by girdling a plant stem, wrapping with sphagnum peat, and protecting with plastic.

acodo al aire — propagación de las mediante la realización de una incisión anular en el tallo, la envoltura del tallo con turba de esfagno y su protección con plástico.

alkaline — pH of more than 7.0.

alcalino — que tiene valor pH mayor a 7.0.

alkalinity — sweetness, or the tendency toward a high pH greater than 7.0.

alcalinidad — dulzura o la tendencia a un alto nivel de pH superior a 7.0.

alluvial deposit — soil transported by streams.

depósito aluvial — depósito de tierras transportadas por arroyos.

amend — a type of motion used to add to, subtract from, or strike out words in a main motion.

enmendar — un tipo de moción que se usa para agregar, substraer, o quitar palabras de una moción principal.

ammonia/nitrite/nitrate — the chemical components generated during biological breakdown of animal wastes.

amoníaco/nitrito/nitrato — los componentes químicos que se producen durante la descomposición biológica de los desechos animales.

amortized loan — loan repaid in equal installments of principal and interest.

préstamo amoritzado — préstamo pagado en cuotas iguales de capital principal e interés.

amphibian(s) — organisms that complete part of their life cycle in water and part on land.

anfibios — organismos que cumplen parte del ciclo de la vida en el agua y otra parte en la tierra.

anatomy — the structure and arrangement of the various parts of the body of an animal or plant.

anatomía — la estructura y composición de las diversas partes del cuerpo de un animal o una planta.

angiosperm — a plant with its seeds enclosed in a pod or seed case.

angiosperma — planta cuyas semillas están envueltas por una vaina o una cápsula.

Angora — wool from wool-producing rabbits or goats.

Angora — lana de los conejos o cabras que dan la lana.

animal science — animal growth, care, and management.

ciencia de animales — cría, cuido, y administración de animales.

animal science technology — use of modern principles and practices in animal growth and management.

tecnología de las ciencias de animales — uso de principios y métodos modernos en la cría y la administración de animales.

anion — ion that is negatively charged.

anión — ion con carga negativa.

annual — a plant with a life cycle that is completed in one growing season.

anual — una planta cuyo ciclo de vida que se completa en una temporada de crecimiento.

annual ring — ring in a cross section of a tree root, trunk, or limb representing 1 year's growth.

anillo de desarrollo anual — anillo formado en un corte transversal de raíz, tronco, o rama de un árbol, que representa el crecimiento de un año.

annual weed — a weed that completes its life cycle within 1 year.

mala hierba anual — una mala hierba que completa su ciclo de vida en un año.

anther — portion of the male part that contains the pollen.

antera — porción de la parte macho que contiene el polen.

antibiotic — substance used to help prevent or control infections and diseases of animals.

antibiótico — sustancia que se usa para evitar o controlar infecciones y enfermedades de los animales.

apiary — area where beehives are kept.

apiario — sitio en donde setienen las colmenas.

apiculture — beekeeping.

apicultura — arte de criar abejas.

aquaculturist — trained professional involved in the production of aquatic plants and animals.

acuiculturista — profesional entrenado, dedicado a la producción de plantas y animales acuáticos.

aquaculture — raising of finfish, shellfish, and other aquatic animals under controlled conditions. Also, the management of the aquatic environment for production of plants and animals.

acuicultura — la crianza de peces, mariscos, crustáceos, y otros animales acuáticos en circunstancias controladas. Adicionalmente, la administración del ambiente acuático para la producción de plantas y animales.

aquifer — water-bearing rock formation.

acuífero — formación rocosa que contiene agua.

arachnid (arthropod) — a living creature such as a spider or mite that is distinguished from insects by having eight legs.

arácnido — criatura viva tal como las arañas o los ácaros que se distingue de los insectos por tener ocho patas.

arboriculture — care and management of trees for ornamental purposes.

arboricultura — cuidado y administración de árboles para fines ornamentales.

arid — an area deficient in rainfall; dry.

árido — un área que carece de precipitación; seca.

asbestos — heat- and friction-resistant material.

asbesto — materia que es resistente al calor y fricción.

aseptic – sterile or free of microorganisms.
aséptica — estéril o libre de microorganismos.
asexual reproduction — propagation using a part or parts of one parent plant.
reproducción asexual — reproducción que utiliza una o más partes de una sola planta madre.
assets — anything the business owns.
activos — total de lo que posee una empresa.
asymmetrical — not equal on both sides of center.
asimétrico — que ambos lados no son iguales relativos al centro.
auction market — market where products are sold by public bidding.
mercado de subasta — mercado donde se venden los bienes mediante subastas públicas.
axil — the upper angle between the leaf or flower stem and the stalk of the plant.
axila — el ángulo superior entre el tallo de la hoja o la flor y el tallo de la planta.

B

B horizon — soil below the A horizon or topsoil and generally referred to as subsoil.
horizonte B — la tierra debajo del horizonte A o de la capa superficial del suelo, generalmente conocido con el nombre de subsuelo.
bacteria — one-celled, microscopic organisms.
bacteria — microorganismos unicelulares.
balance — state of quality and calm between items.
equilibrio — estado de bienestar y tranquilidad entre los ítems.
balance sheet — a statement of the assets and liabilities of a business on a specific date.
hoja de balance — un extracto de cuenta de los bienes y deudas de un negocio en una fecha específica.
balled and burlapped (B & B) — plants that have been dug in the field, wrapped in burlap, and laced with a heavy twine.
empacado y harpillerado — plantas que han sido arrancadas del campo, envueltas en arpillera, y atadas con cordel grueso.
balling gun — a device used to place a pill in an animal's throat.
pistola de «balling» — aparato que se usa para meter una pastilla en la garganta de un animal.
bare-root — plant that is dug for transplanting with little soil remaining on the roots.
planta de raíces expuestas — plantas que son desenterradas para el transplante con sólo un poco de tierra que queda en las raíces.
bases — genetic material that connects strands of DNA.
bases — material genético que conecta fibras de ADN.
basic color — background color.
color básico — color de fondo.
bedrock — the area below horizon C consisting of large soil particles.
roca de fondo — el área debajo del horizonte C que consiste en partículas de tierra más grandes.
BelRus — superior baking potato bred to grow well in the Northeast.
«BelRus» — patata de hornear de calidad superior producida para cultivarse bien en el noreste de los EE.UU.
biennial — a plant that takes two growing seasons or 2 years from seed to complete its life cycle.
bianual — una planta que toma dos temporadas de crecimiento o dos años desde semilla para completar su ciclo de vida.
biennial weed — a weed that will live for 2 years.
mala hierba bienal — una mala hierba que vive por dos años.
binomial — having two names.
binomio — que tiene dos nombres.
bio — life or living.
bio — vida o viviente.
biochemistry — chemistry as it applies to living matter.
bioquímica — química que se aplica a la materia viviente.
biological control — pest control that uses natural control agents.
control biológico — control de los insectos o animales nocivos que utiliza agentes de control natural.
biology — basic science of the plant and animal kingdoms.
biología — la ciencia básica de los reinos vegetal y animal.
biotechnology — use of cells or components of cells to produce products or processes.
biotecnología — uso de células o componentes de células para producir productos o procesos.
biotic disease — disease caused by living organisms.
enfermedad biótica — enfermedad causada por organismos vivientes.
blade — the upper portion of the grass leaf.
brizna — la parte superior de la hoja de pasto.
blanching — the brief scalding of food before freezing.
escaldar — el sumergir los alimentos por un tiempo breve en agua hirviendo antes de congelarlos.

blemish — in horses, any abnormality that does not affect the use of the horse.
defecto — en los caballos, cualquier abnormalidad que no afecte el uso del caballo.

block beef — meat sold over the counter to consumers.
carne descuartizada — carne de res que se vende al por menor a los consumidores.

boar — male animal of the swine family.
verraco — el macho de la familia de los porcinos.

board foot — a unit of measurement for lumber that equals 1 × 12 × 12 inches.
«board foot» (b.ft.) — unidad de medida de los maderos que equivale a 1 × 11 × 12 pulgadas.

border planting — a planting that is used to separate some part of the landscape from another. It might also be used as a fence or a windbreak.
cantero de plantas — una sección de plantas que se usa para separar alguna parte del paisaje de la otra. Se puede usar como cerca o también protección contra el viento.

bovine somatotropin (BST) — hormone that stimulates increased milk production in cows.
somatotropina bovina — una hormona que estimula (or, if you must separate in syllables, estimula) un aumento de secreción láctea en las vacas.

brackish water — waters influenced by tide and river flow with intermediate salinity of 3 to 22 percent.
agua salobre — aguas afectadas por la marea y el flujo de río, que tienen salinidad del 3 al 22 por ciento.

bran — skin or covering of a wheat kernel.
salvado — cascarilla del grano de trigo.

broiler — young chicken grown for meat.
pollo tomatero — gallina jóven que se cría para comer.

buck — male animal of the goat, deer, or rabbit family.
cabrón — macho de la cabra.
conejo — macho de la coneja. Gamo – ciervo macho.

bud grafting — the union of a small piece of plant tissue containing a bud and a plant rootstock.
injerto de brote — la unión de un pequeño pedazo de tejido vegetal que contiene un brote y una rama de una planta enraízada.

buffer — a substance in a solution that tends to stabilize the pH.
estabilizador — substancia en una solución que tiende a estabilizar el pH.

bulb — short underground stem surrounded by many overlapping, fleshy leaves.
bulbo — tallo corto, subterráneo envuelto por muchas hojas carnosas y traslapadas.

bull — male animal of the cattle family.
toro — macho del ganado vacuno.

business meeting — a gathering of people working together to make decisions.
sesión (de negocio) — asamblea de personas que trabajan juntas para tomar decisiones.

buying function — selection of a product or service to be marketed or sold for a profit.
función adquisitiva — selección de un producto o servicio para que se comercialice o venda por una ganancia.

C

C horizon — soil below the B horizon; it is important for storing and releasing water to the upper layers of the soil.
horizonte C — la tierra debajo del horizonte B; es importante para guardar y soltar agua a las capas superiores de la tierra.

calf — young member of the cattle family.
becerro — un jóven del ganado vacuno.

calyx — group of sepals of a flower.
cáliz — grupo de los sépalos de la flor.

cambium — growth layer in a tree root, trunk, or limb.
cambium — capa de crecimiento de raíz, tronco, o rama del árbol.

cane cutting(s) — stems are cane-like and cuttings are cut into sections that have one or two eyes, or nodes.
esqueje a la caña — los tallos son parecidos a la caña y se cortan en secciones que tienen uno o dos nudos.

cane — mature wood in grapevines and some other fruits that has produced fruit and lost the leaves. The next year's new shoots and fruit grow from the cane.
caña — madera madura en viñedos u otras frutas que han producido y perdido sus hojas. Los retoños del próximo año crecen de las cañas.

canning — storing food in airtight containers.
enlatado — guardar los comestibles en recipientes herméticos.

cannula — blunt needle.
cánula — aguja sin punta afilada.

canopy — the top of the plant that has the framework and leaves.
copa o bóveda de ramas — la parte superior de la planta que tiene la estructura y las hojas.

capability class — soil classification indicating the most intensive, but safe, land use.
clase de capacidad — clasificación de suelos indicando el uso más intensivo pero seguro de la tierra.

capillary water — water held by soil particles and available for plant use.

agua capilar — agua sujetada por partículas de tierra y disponible para el uso de las plantas.

capital — money or property.

capital — dinero o propiedad.

capital investment — money spent on commodities that are kept 6 months or longer.

inversión de capital — dinero gastado en productos básicos que se guarda por seis meses o más.

carbohydrate — starches and sugars that provide energy in the diet.

carbohidrato — almidones y azúcares que proporcionan energía en la dieta.

carbon monoxide — colorless, odorless, and highly poisonous gas; carbon dioxide and water combine to make plant food and release oxygen.

monóxido de carbono — gas incoloro, inodoro, y sumamente tóxico; el bióxido de carbono y agua se combinan para hacer alimento vegetal y soltar oxígeno.

carcass — body of meat after the animal has been eviscerated.

res abierta en canal — el cuerpo de carne después de haber sido destripado el animal.

carcinogen — a chemical capable of producing a tumor.

carcinógeno — una sustancia química capaz de producir un tumor.

career — a person's occupation or profession.

carrera — la profesión u ocupación de una persona.

career exploration — learning about occupations and jobs as part of the process of choosing one's life work.

exploración de carrera — aprender sobre ocupaciones y trabajos como parte del proceso de elegir el trabajo de la vida de un individuo.

carrying capacity — the number of animals that a pasture will provide feed for.

capacidad de carga — la cantidad de animales que pueden alimentarse en un pastizal.

casein — predominant protein in milk.

caseína — proteina prevalente de la leche.

cash crop — a crop grown for cash sale.

cultivo comercial — una cosecha cultivada para vender por pago al contado.

cash flow statement — a financial document that describes the availability of operating funds for a business at different points in time.

estado de dinero efectivo en caja — documento financiero que describe los fondos disponibles para negocios en diferentes puntos del tiempo.

cation — ion that is positively charged.

catión — ion de carga positiva.

cell — a unit of protoplasmic material with a nucleus and cell walls.

célula — unidad de materia protoplásmica con núcleo y paredes.

cell membrane — a thin tissue surrounding the contents of a cell through which nutrients pass in and waste materials pass out.

membrana celular — un tejido delgado que rodea el contenido de una célula, a través del cual entran los nutrientes y salen los materiales de desecho.

cellulose — woody fiber parts that make up plant cell walls.

celulosa — partes leñosas de fibra que constituyen las paredes de célula vegetal.

central nervous system — the brain and the spinal cord.

sistema nervioso central — el cerebro y la médula espinal.

cereal crop — grasses grown for their edible seeds.

cultivo de cereal — hierbas que son cultivadas por sus semillas comestibles.

cheese — milk that is exposed to bacterial fermentation.

queso — leche que se expone a la fermentación bacterial.

chemical control — the use of pesticides for pest control.

control químico — el uso de pesticidas para control de los insectos y plagas nocivos.

chemistry — science dealing with the characteristics of elements or simple substances.

química — la ciencia que trata de las características de los elementos o sustancias sencillas.

chevon — meat from goats.

carne de chivo — carne de ganado caprino.

chick — newborn chicken or pheasant.

polluelo — gallina o faisán recién nacido.

chlorofluorocarbon (CFC) — any of a group of compounds consisting of chlorine, fluorine, carbon, and hydrogen used as aerosol propellants and refrigeration gas.

cloroflurocarbonos — cualquiera de un grupo de compuestos que se consisten en cloro, flúor, carbono, e hidrógeno usados como propulsores de aerosol y de gas de refrigeración.

chlorophyll — green pigment in leaves.

clorofila — pigmento verde de las hojas.

chloroplast — membrane-bound body inside a cell containing chlorophyll pigment; necessary for photosynthesis.

cloroplasto — dentro de la célula, cuerpo envuelto por una membrana que contiene la clorofila; necesario para la fotosíntesis.

chlorosis — yellowing of the leaf.

clorosis — amarilleo de la hoja.

chromosome — the rod-like carrier for genes.

cromosoma — cuerpo en forma de bastoncillo que lleva los genes.

chronic toxicity — a measurement of the effect of a chemical over a long period and under lower exposure doses.

toxidad crónica — medición del efecto de una sustancia química durante un período largo de tiempo y a dosis de exposición más bajas.

circulatory system — the system that provides food and oxygen to the cells of the body and filters waste materials from the body.

aparato circulatorio — el sistema que proporciona los alimentos y oxígeno a las células del cuerpo, y filtra los residuos del cuerpo.

citizenship — functioning and interacting in a society in a positive way.

ciudadanía — funcionamiento e interacción en una sociedad en una forma positiva.

clay — smallest of soil particles; less than 0.002 mm.

arcilla — las partículas más pequeñas del suelo; menos de 0, (This is proper Spanish mathematical notation. If client prefers English standards, then leave the period.) 002 mm.

clear cut — removal of all marketable trees from an area.

corte raso — el cortar y eliminar todos los árboles comerciales de un área.

climate — the weather conditions of a specific region.

clima — las condiciones atmosféricas de una región específica.

climatic conditions — temperature, temperature range, and precipitation.

condiciones climáticas — temperatura, variedad de temperatura y precipitación.

clod — a lump or mass of soil.

terrón — masa de tierra.

clone — exact duplicate.

clon — una duplicación exacta.

cockerel — young male chicken or pheasant.

gallo jóven — gallo jóven.

faisán jóven — faisán jóven.

Note: Spanish does not have an equivalent to "cockerel"—a cockerel is referred to as either a "young male chicken" or "young male pheasant."

cold frame — a bottomless wood box with a sloping glass in which early-season plants are protected from freezing temperatures.

bastidor frío — caja de madera sin fondo con un vidrio inclinado en el cual se protegen de las temperaturas congeladas a las plantas de temporada temprana.

collagen — chief component of connective tissue.

colágeno — componente principal del tejido conjuntivo.

collar — light green or white banded area on the outside of a leaf blade.

collar — banda verde pálida o blanca en la parte exterior de la hoja.

colluvial deposit — soil deposited by gravity.

depósito coluvial — suelo depositado por la gravedad.

color breed — breed of horses based on color.

cría selectiva basada en el color — cría selectiva basada en el color.

Note: The translation for «color breed» is the same as that of the English definition provided.

colostrum — first milk produced by mammals; high in antibodies.

calostro — primera leche producida por mamíferos; tiene niveles elevados de anticuerpos.

colt — young male horse or pony.

potro — caballo o pony macho jóven.

combine — machine that is used to cut and thresh seed crops such as grain.

segadora trilladora — máquina que siega y trilla el grano en el campo.

commensalism — one type of wildlife living in, on, or with another without either harming or helping it.

comensalismo — una clase de fauna que vive en, encima de, o con, otra clase sin dañarla ni tampoco ayudarla.

commission — fee for selling a product.

comisión — honorario por la venta de un producto.

Commodity Credit Corporation (CCC) — institution that lends money for production of farm commodities.

corporación de crédito para productos básicos — institución que presta dinero para la producción de mercancías agrícolas.

commodity exchange — organization licensed to manage the buying and selling of commodities.

bolsa de comercio — organización autorizada para administrar la compra y venta de productos básicos.

comparative advantage — the ability to produce a commodity with greater returns than those of competitor because of a favorable condition such as climate.

ventaja comparativa — la habilidad de producir un producto básico con mayor ganancia que la del competidor debido a condiciones favorables tales como el clima.

competition — two types of wildlife eating the same source of food.

competición — dos clases de fauna que consumen la misma fuente de alimentación.

compost — mixture of partially decayed organic matter.

composte — mezcla de material orgánico parcialmente descompuesto.

compound — a chemical substance that is composed of more than one element.

compuesto — una sustancia química compuesta por más de un elemento.

compound leaf — two or more leaves arising from the same part of the stem.

hoja digitada — dos o más hojas que brotan del mismo tallo.

concentrate — feed high in total digestible nutrients and low in fiber.

concentrado — alimento con alto nivel de nutrientes digeribles totales y bajo en fibra.

condominium — apartment building or unit in which the apartments are individually owned.

condominio — edificio de apartamentos o unidad en la cual los apartamentos tiene propietarios individuales.

conifer — evergreen tree that has needle-like leaves and produces cones.

conífera — árboles de hoja perene con forma de aguja que produce piñas.

conservation tillage — techniques of soil preparation, planting, and cultivation that disturb the soil the least and leave the maximum amount of plant residue on the surface.

labranza de conservación — técnicas de preparación del terreno, siembra y cultivo que menos remueven la tierra y dejan la cantidad máxima de residuos vegetales en la superficie.

consumer demographics — categories of information about preferences of consumers or potential consumers.

datos demográficos del consumidor — catego-rías de información sobre preferencias de los consumidores o clientes potenciales.

consumer — person who uses a product.

consumidor — persona que usa un producto.

contact herbicide — a herbicide that will not move or translocate within the plant.

herbicida de contacto — un herbicida que no se mueve ni se desplaza dentro de la planta.

contagious — diseases that can be spread by contact.

contagiosas — enfermedades que se pueden transmitir por contacto.

containerized plant — plant that is grown in a pot or other type of container and shipped in the container.

cultivadas en recipiente — plantas que crecen en macetas u otros tipos de recipientes y son transportadas en el recipiente.

contaminate — to add material that will change the purity or usefulness of a substance.

contaminar — añadir material que cambiará la pureza o utilidad de una sustancia.

continuous-flow system — the nutrient solution flows constantly over the plant roots.

sistemas de flujo continuo — la solución nutritiva fluye continuamente sobre las raíces de las plantas.

contour — level line around a hill in which all points along the line are the same elevation.

contorno — linea de nivel alrededor de una colina en la cual todos los puntos a lo largo de la línea tienen la misma elevación.

contour farming — operations such as plowing, discing, planting, cultivating, and harvesting across the slope and on the level.

cultivo del contorno — operaciones tales como el arado, el enterramiento, sembrar, cultivar, y la recolección de la pendiente y el nivel.

convection oven — an oven that heats food by circulating hot air.

horno de convección — un horno que calienta la comida mediante la circulación de aire caliente.

conventional tillage — land is plowed, turning over all crop residues.

labranza clásica — se ara la tierra, volteando todos los residuos de la cosecha.

Cooperative Extension System — an educational agency of the U.S. Department of Agriculture and an arm of land-grant state universities.

System Cooperativo de Extensión — una entidad educativa del Departamento de Agricultura de los Estados Unidos, y un ramo de las universidades estatales de concesión de terrenos.

cooperative — group of producers who join together to market a commodity and/or to purchase supplies.

cooperativa — grupo de productores que se reunen para comercializar un producto básico y/o para comprar abastecimientos.

corm — short, flattened underground stem surrounded by scaly leaves.

bulbo — tallo corto aplanado, subterréaneo envuelto por hojas escamosas.

corn picker — machine that removes ears of corn from stalks.

recolectora de maíz — máquina que recoge las espigas de maíz de los tallos.

cotton gin — machine that removes cotton seed from cotton fiber.

desmotadora — máquina que saca la semilla a la fibra del algodón.

courage — willingness to proceed under difficult conditions.

valor — la voluntad de seguir adelante a pesar de las circunstancias difíciles.

cow — female of the cattle family that has given birth.

vaca — hembra de la familia vacuna que ha parido.

credit — money borrowed.

crédito — dinero prestado.

crop rotation — planting of different crops in a given field every year or every several years.

rotación de cultivos — siembra de cultivos diferentes en un campo determinado cada año o cada dos o tres años.

crop science — use of modern principles in growing and managing crops.

ciencia de cultivos — el uso de principios modernos en el cultivo y la administración de los cultivos.

crown — an unelongated stem of major meristematic tissue of turfgrass.

corona — un tallo no alargado de tejido meristemático (or, if you must separate it in syllables: me-ristemático) mayor del césped.

crustacean — aquatic organism with an exoskeleton that molts during growth.

crustáceo — organismo acuático con un dermatoesqueleto que se muda a medida que el organizmo va creciendo.

cultivar — a group of plants with a particular species that has been cultivated and is distinguished by one or more characteristics; through sexual or asexual propagation, it will keep these characteristics.

variedad obtenida por selección — un grupo de plantas de una especie particular que ha sido cultivada y que se distingue por una o más características (or, if you must separate in syllables: ca-racterísticas) las características se mantendrán con la propagación sexual o asexual.

cultivation — the act of preparing and working soil.

cultivo — la acción de preparar y labrar la tierra.

cultural control — pest control that adapts farming practices to better control pests.

control del cultivo — control de las plagas o los insectos nocivos que adapta los métodos agrícolas para controlar mejor las plagas y los insectos nocivos.

cuticle — topmost layer of the leaf; waxy protective covering of the leaf.

cutícula — la superficie de la hoja; capa protectora cerosa de la hoja.

cutting — vegetative part removed from the parent plant and managed so it will regenerate itself.

tala — la parte vegetativa que se quita de la planta progenitora para que se regenere.

cytosine (C) — a base in genes designated by the letter *C*.

citosino — una base en los genes denominada con la letra «C».

D

deciduous — plants that lose their leaves every year.

caduco — plantas cuyas hojas se caen cada año.

decomposer — an organism that is capable of breaking down dead plant and animal matter into soil components.

descompositor — un organismo que es capaz de descomponer la materia muerta de la planta y del animal y convertirla en componentes del suelo.

deficiency disease — condition resulting from improper levels or balances of nutrients.

enfermedade por carencia — enfermedad resultante de niveles bajos o desequilibrios de los nutrientes.

dehydration — removing moisture with heat.

deshidratación — quitar el agua usando calor.

dehydrator — a device for drying food.

deshidratador — un aparato para secar alimentos.

dehydrofrozen product — removing moisture after partial cooking, and then freezing.

producto deshidrocongelaado — quitar la humedad después de cocinar parcialmente y luego congelar.

demand — the amount of a product wanted at a specific time and price.

demanda — cantidad de producto pedido en un tiempo específico y por un precio específico.

dicotyledon (dicot) — plant with two seed leaves.

dicotiledónea — plantas con dos hojas de semilla.

digestive system — system that provides food for the body of the animal and for all of its systems.

aparato digestivo — sistema que proporciona los alimentos al cuerpo del animal y para todos sus sistemas.

diminishing returns — point at which each additional unit of input decreases the output or returns.

rendimientos decrecientes — el punto en donde cada unidad adicional de insumo disminuye el rendimiento o los ingresos.

direct sale — the selling of crops or animals directly to a processor by the producer.

venta directa — a la venta de animales o cultivos directamente a los procesadores por el productor.

disassembly process — dividing the carcass into smaller cuts.

carnear — tomar la res abierta en canal y descuartizarla.

discount loan — a loan in which interest is subtracted from the principal at the time the loan is made.

préstamo con descuento — un empréstito en que se substrae el interés del principal en el momento en que se realiza el préstamo.

disease triangle — the term applied to the relationship of the host, pathogen, and the environment in disease development.

triángulo de enfermedad — aplícase a la relación entre el portador, el agente patógeno y el medio ambiente en el desarrollo de la enfermedad.

disinfectant — material that destroys infective agents such as bacteria and viruses.

desinfectante — material que mata a los agentes infecciosos como bacterias y virus.

distribution function — physical organization and delivery of product or service.

función de distribución — organización física y entrega de un producto o servicio.

distributor — person or business storing food for transport to regional markets.

distribuidor — persona o empresa que almacena alimentos para transportarlos a mercados regionales.

DNA (deoxyribonucleic acid) — coded genetic material in a cell.

ácido desoxirribonucléico (ADN) — la materia genética codificada de una célula.

dock — remove or shorten tails of certain animals.

cortar la cola — quitar o acortar las colas de ciertos animales.

doe — female of the goat, deer, or rabbit family.

cabra — hembra del cabrón.

coneja — hembra del conejo.

domestic — a tame animal for use by humans.

domestico — a domesticar animales para su uso por los seres humanos.

dominant — gene that expresses itself to the exclusion of other genes.

dominante — el gen que se manifiesta a exclusión de otros genes.

donkey — member of the horse family with long ears and a short erect mane.

burro — miembro de la familia de los caballos con orejas grandes y crines erectas.

dormant — resting stage, no active growth.

latente — estado de descanso; no hay crecimiento activo.

double-eye cutting — used when the plant has leaves that are opposite.

esqueje de yema doble — empleado cuando la planta tiene hojas simétricas.

draft — animals used for work.

tiro — animales que se usan para cargar.

draft horse — type of horse bred for work.

caballo de tiro — tipo de caballo criado para trabajar.

drake — male duck.

pato — el macho de los patos.

drench — a process of administering drugs orally to animals.

administración oral — el proceso de administrar medicamentos oralmente a los animales.

dressing percentage — the proportion of the live weight of an animal to the weight of the carcass before cooling.

porcentaje de rendimiento del canal — la proporción del peso vivo del animal en relación al peso de la canal antes de refrigerarse.

drip irrigation — the use of small tubes to deliver irrigation water to the roots of crop plants in a uniform manner.

irrigación de goteo — el uso de pequeños tubos que llevan agua de irrigación de manera uniforme a las raíces de plantas de cultivo.

drip line — the edge of the tree where the branches stop.

línea de goteo — el límite del árbol donde terminan las ramas.

drone — male bee.

zángano — macho de la abeja.

drupe — a stone fruit.

drupa — fruta que contiene un carozo.

dry-heat cooking — surrounding food with dry air while cooking.

cocinar al calor seco — rodear la comida con aire seco para cocinarla.

dry matter — material left after all water is removed from a feed material.

materia seca — materia que queda después de quitar toda el agua del material de alimento.

duckling — young duck.

patito — cría del pato.

dwarf — tree that has rootstock that limits growth to 10 feet or fewer.

enano — árbol que tiene un rizoma que limita el crecimiento a 10 pies o menos.

E

economic threshold level — the level of pest damage to justify the cost of a control measure.

«precio de umbral» — el nivel de daño causado por plagas o insectos nocivos que justifica el costo de una medida de control.

edible — fit to eat or consume by mouth.

comestible — apropiado para comerse o tomar por boca.

element — a uniform substance that cannot be further decomposed by ordinary means.

elemento — una sustancia uniforme que no puede descomponerse por medios ordinarios.

embryo— a fertilized egg.

embrion - óvulo fecundado.

endocrine or hormone system — a group of ductless glands that release hormones into the body.

sistema endocrino u hormonal — conjunto de glándulas de secreción interna las cuales secretan hormonas en el cuerpo.

endoplasmic reticulum — a cell structure that stores proteins and facilitates their movement to other parts of the cell as needed.

retículo endoplásmico — una estructura celular que almacena las proteínas y facilita su circulación a otras partes de la célula como sea necesario.

endosperm — interior of a wheat kernel that will become wheat flour.

endosperma — parte interior de un grano de trigo que llegará a ser harina de trigo.

enterprise — commercial business to generate profits.

empresa — establecimiento comercial que produce ganancias.

enthusiasm — energy to do a job and inspiration to encourage others.

entusiasmo — energía para hacer una obra y inspiración para animar a otras personas.

entomology — science of insect life.

entomología — ciencia de la vida de los insectos.

entomophagous — insects that feed on other insects.

entomófagos — insectos que se alimentan con otros insectos.

entrepreneur — person organizing a business, trade, or entertainment.

empresario — persona que organiza un negocio, comercio, o espectáculo.

entrepreneurship — process of planning and organizing a business.

capacidad empresarial — el proceso de planear y organizar un negocio.

entrepreneurship supervised agricultural experience — enterprises that are supervised by teachers and conducted by students as owners or managers of businesses based on agriscience.

experiencia agrocientífica supervisada por empresarios — empresas supervisadas por maestros y conducidas por estudiantes como dueños o gerentes de negocios basados en la agrociencia.

environment — All the conditions, circumstances, and influences surrounding and affecting an organism.

medio ambiente — todas las circunstancias, condiciones e influencias que rodean y afectan al organismo.

epidermis — surface layer on the lower and upper sides of the leaf.

epidermis — capa de la cara superior e inferior de la hoja.

equine — having characteristics of a horse.

equino — que tiene características de un caballo.

equitation — the art of riding on horseback.

equitación — el arte de montar caballo.

eradicant fungicide — a fungicide applied after disease infection has occurred.

fungicida erradicador — un fungicida aplicado después de que ha ocurrido la infección.

eradication — complete control or removal of a pest from a given area.

erradicación — control completo o extirpación de una plaga o insecto dañino de un área determinada.

erosion — wearing away.

erosión — desgaste o deterioro.

estrogen — hormone that regulates the heat period.

estrógeno — hormona que regula el período en que el animal está en celo.

estrus — heat period or time when female animal is receptive to the male animal.

celo — período o tiempo fecundo cuando la hembra está receptiva al macho.

estuary — ecological system including bays, streams, and tidal areas influenced by brackish water.

estuario — sistema ecológico que incluye bahías, arroyos, y tierras bajas del litoral que son afectados por agua salobre.

evergreen — plants that do not lose their leaves on a yearly basis.

planta perenne — plantas a que no se deshojan cada año.

ewe — female animal of the sheep family.

oveja — hembra del carnero.

exoskeleton — the external body wall of an insect.

dermatoesqueleto — caparazón o pared exterior del cuerpo de un insecto.

exploratory supervised agricultural experience — a program in which the teacher conducts activities that allow students to become involved in learning a variety of subjects about agriscience and careers related to agriculture.

experiencia agrocientífica exploratoria supervisada — un programa en el cual el maestro conduce actividades que permiten a los estudiantes participar en el aprendizaje de una variedad de materias sobre agrociencia y carreras relacionadas a la agricultura.

extemporaneous speaking — a form of public speaking wherein some preparation is made, but the speech is not written or memorized.

discurso improvisado — una forma de discurso público en donde se hace una cierta preparación, pero el discurso no se pone por escrito ni se memoriza.

extravaginal growth — turfgrass growth in which growth originates from an axillary bud on the crown.

crecimiento extravaginal — crecimiento de césped que brota de un botón axilar en la corona.

F

fabrication and boxing — vacuum sealing of meat in boxes before shipment.

fabricación y empaque — envasar al vacío la carne en cajas antes de transportarla.

famine — widespread starvation.

hambruna — hambre general.

Farmers Home Administration (FHA) — government agency that assists farmers to become landowners.

Departamento de Hogares de Agricultores — organismo de gobierno que ayuda a los granjeros a volverse propietarios de tierra.

farrier — a person who shoes horses.

herrador de caballos — persona que clava y ajusta las herraduras a un caballo.

farrow — giving birth in the swine family.

parir a cerdos — parir a cerdos.

Note: In Spanish, the word *parir* means "to give birth in the nonhuman mammal." Thus, no other word is used to differentiate the birth process of a particular species of animal.

fat — nutrients that have 2.25 times as much energy as carbohydrates.

grasa — alimentos nutritivos que tienen 2.25 veces más energía que los carbohidratos.

federal land bank — lending institution that provides long-term credit for agriculture.

banco federal de tierras — institución prestamista que provee crédito de largo plazo para la agricultura.

feed additive — a nonnutritive substance added to feed to improve growth, increase feed efficiency, or maintain health.

aditivos para alimentos — una sustancia no nutritiva añadida al alimento para mejorar el crecimiento, aumentar la eficacia del alimento, o para mantener la salud.

feedstuff — any edible material used for animal feed.

forraje — cualquiera material comestible que sirve para el alimento.

fermentation — a chemical change that results in gas release.

fermentación — un cambio químico que resulta en desprendimiento de gases.

fertilizer — material that supplies nutrients for plants.

abono — material que proporciona nutrientes para las las plantas.

fertilizer grade — percentages of primary nutrients in fertilizer.

grado del abono — proporción de nutrientes en el abono.

fetus — embryo from the time of attachment to the uterine wall until birth.

feto — embrión desde el momento en el que se pega a la pared uterina hasta el nacimiento.

FFA — a national intracurricular organization for students enrolled in agriscience programs in their schools.

AFF — una organización dentro del programa para estudiantes matriculados en los programas de agriciencia.

fibrous root — one of the two major root systems, consisting of many fine, hair-like roots.

raíz fasfciculada — uno de dos sistemas principales de raíz, que consiste de muchas raíces finas.

field crop — a class of crop that includes grains, oil crops, and specialty crops.

cultivo de campo — una clase de cultivo que incluye granos, cultivos de aceite y cultivos de especialización.

filament — structure that supports the anther.

filamento — estructura que sostiene la antera.

filly — young female horse or pony.

potra — hembra jóven del caballo o pony.

financing function — obtaining capital for a business.

función de financiar — obtener capital para un negocio.

fine-textured (clay) soil — soil that usually forms very hard lumps or clods when dry; plastic when wet.

terreno de textura fina (arcilla) — usualmente forma terrones duros cuando seca; es plástico cuando está mojado.

floriculture — production and distribution of cut flowers, potted plants, greenery, and flowering herbaceous plants.

floricultura — producción y distribución de flores cortadas, de plantas en maceta, ramas y hojas verdes, y plantas herbáceas florecientes.

flower — reproductive part of the plant.

flor — parte reproductora de la planta.

flowering bud — a terminal bud on a plant that produces flowers.

brote de flor — un botón terminal en una planta que produce flores.

fluke — very small, flat worm that is a parasite.

trematodos — gusanos de cuerpo plano muy pequeños que son parásitos.

foal — newborn horse or pony. Also, to give birth in horses.

potro — caballo o pony recién nacido.

food additive — anything added to food before packaging.

suplemento aditivo — cualquiera cosa añadida al alimento antes de envasarlo.

food chain — interdependence of plants or animals on each other for food.

cadena trófica — dependencia recíproca de plantas o animales para alimentarse.

food industry — production, processing, storage, preparation, and distribution of food.

industrias alimenticias — producción, procesado, almacenaje, preparación, y distribución de alimentos.

foot-candle — amount of light found 1 foot from a standard candle or candela.

candela por pie cuadrado — cantidad de luz que se encuentra de un pie de una candela o vela estándar.

forage — crop plants grown for their vegetative growth and fed to animals.

forraje — plantas cultivadas por su vegetación y dadas de comer a los animales.

forest — large group of trees and shrubs.

bosque — gran conjunto de árboles y arbustos.

forest land — land at least 10 percent stocked by forest trees.

tierra forestal — tierra poblada con por lo menos 10 por ciento de árboles forestales.

forestry — industry that grows, manages, and harvests trees for lumber, poles, posts, panels, paper, and many other commodities.

silvicultura — industria que cultiva, administra, y cosecha los árboles para maderos, varas, palos, tableros, papel y muchos otros productos.

formal — equal in size and number.

formal — igual en cantidad y tamaño.

formulation — the physical properties of the pesticide and its inert ingredients.

formulación — las propiedades físicas del pesticida y sus ingredientes inertes.

4-H — network of youth clubs directed by Cooperative Extension System personnel to enhance personal development and provide skill development in many areas.

4-H — una red de asociaciones dirigida por el personal del Servicio Cooperativo de Extensión para fomentar el desarrollo personal y enseñar técnicas en muchas áreas.

free water — water that drains out of soil after it has been wetted.

agua libre — agua que se drena del suelo después de mojar el suelo.

freeze-drying — removing moisture with cold.

deshidratar por congelación — quitar la humedad usando frío.

fresh water — water that flows from the land to oceans and contains little or no salt.

agua dulce — agua que corre desde la tierra hasta los mares y que tiene poca o ninguna sal.

fructose — simple fruit sugar.

fructosa — azúcar simple de frutas.

fruit — mature ovary; seed.

fruto — óvulo maduro, semilla.

fry — small, newly hatched fish.
alevines — peces recién salidos del huevo.
fungal endophyte — microscopic plant growing within a plant.
endofito fungoso — planta microscópica que crece dentro de una planta.
fungi — members of a major group of lower plant life that lack chlorophyll and obtain nourishment from either live or decaying organic matter.
hongos — miembros de un gran grupo de plantas inferiores que no producen clorofila y obtienen nutrición de organismos vivos o materias orgánicas en descomposición.
fungicide — a material used to destroy fungi or protect plants against their attack.
fungicida — una sustancia que se usa para destruir hongos o proteger a las plantas de su ataque.
furrow — groove made in the soil.
surco — hendidura hecha en la tierra.
futures market — legal frameworks for sellers and buyers to buy and sell futures contracts.
mercado a futuro — estructura legal para vendedores y compradores para comprar y vender contratos a futuro.

G

gait — way of moving.
paso — manera de moverse.
galactose — simple milk sugar.
galactosa — azúcar simple de leche.
gamete — a reproductive cell.
gameto — una célula reproductiva.
gander — male goose.
ganso — macho de la gansa.
gavel — a small wooden hammer-like tool that is used by the presiding officer to direct a meeting.
martillo — un pequeño martillo de madera usado por un funcionario que dirige una reunión.
gelding — castrated male animal of the horse family.
caballo castrado — caballo al cual se le han extirpado los testículos.
Note: Because there is no term for "gelding" in Spanish other than "castrated horse," the Spanish definition reads, "castrated horse—horse whose organs necessary for procreation have been removed."
gene — a unit of hereditary material located on a chromosome.
gene — unidad de materia hereditaria ubicada en un cromosoma.
gene mapping — finding and recording the locations of genes in a cell.
trazar un mapa de genes — localización y registro de las posiciones de genes en una célula.
generation — the collective offspring of common parents.
generación — la descendencia colectiva de padres comunes.
gene splicing — the process of removing and inserting genes in cells.
empalme de genes — el proceso de extirpar y meter genes en las células.
genetic control — pest control using resistant varieties of crops.
control genético — control de plagas o insectos dañinos usando variedades resistentes de cultivos.
genetic engineering — movement of genes from one cell to another.
ingeniería genética — traslado de genes de una célula a otra.
genetics — the biology of heredity.
genética — la biología de la herencia.
genotype — the genetic makeup of an individual or group of organisms.
genotipo — el material genético de un individuo o grupo de organismos.
genus — (plural, *genera*) a closely related and definable group of animals or plants comprising one or more species.
género — un grupo definible y estrechamente relacionado de plantas o animales que consta de úna o más especies.
germ — new wheat plant inside the kernel.
germen — progenie de trigo dentro del grano.
germinate — a seed sprouting or starting to grow.
germinar — una semilla que brota o empieza a crecer.
gestation — length of pregnancy.
gestación — tiempo que dura el embarazo.
giblets — heart, liver, and gizzard of poultry.
menudillos — corazón, higadillo, y molleja de las aves.
gills — organ of aquatic animal that absorbs oxygen from the water.
agalla — órgano de animal acuático que absorbe el oxígeno del agua.
gilt — female of the swine family that has not given birth.
cerda jóven nulípara — hembra de la familia porcina que no ha parido.
ginning — the process of removing the seeds from cotton.
desmotar — el proceso de sacar la semilla al algodón.

Girl Scouts and Boy Scouts — youth organizations that provide opportunities for leadership development and skill development.

Niñas Exploradoras y Niños Exploradores — asociaciones que dan oportunidades de desarrollo de capacidades de liderazgo y capacidades generales.

glacial deposit — soil deposited by ice.

depósitos glaciales — tierras depositadas por hielo.

glucose — simple sugar and the building blocks for other nutrients.

glucosa — azúcar simple y molécula fundamental de otros nutrientes.

goose — female of the goose family.

gansa — hembra del ganso.

gosling — young goose.

ansarino — ganso jóven.

grade — quality standard.

grado — estándar de calidad.

grader — person who inspects food for freshness, size, and quality.

clasificador — persona que inspecciona los alimentos para evaluar su frescura, tamaño, y calidad.

gradient — a measurable change in an amount over time or distance.

índice — un cambio mensurable de una cantidad en el tiempo o la distancia.

grafting — joining two plant parts together so that they will grow as one.

injertar — juntar dos partes de plantas para que crezcan como una sola planta.

graft union — the location where a scion and rootstock meet when parts of two plants are grafted together.

unión de injerto — sitio donde una vara y un tallo de dos plantas diferentes se juntan.

grass waterway — strip of grass growing in the low area of a field where water can gather and cause erosion.

estepa de gramíneas — franja de hierba que crece en la parte baja del campo donde el agua puede acumularse y causar erosión.

gravitational water — water that drains out of soil after it has been wetted.

agua de gravitación — agua que se drena del suelo después de mojarlo.

greenhouse effect — a buildup of heat at the Earth's surface caused by energy from sunlight becoming trapped in the atmosphere.

efecto de invernadero — concentración de calor en la superficie de la tierra debido a la energía de la luz del sol que queda encerrada en la atmósfera.

green manure (crop) — crop grown to be plowed under to provide organic matter to the soil.

plantas para abono verde — plantas cultivadas para ser enterradas con el objetivo de proporcionar materia orgánica al suelo.

Green Revolution — process in which many countries became self-sufficient in food production.

revolución verde — proceso en el que muchos países logran a ser autosuficientes en la producción de alimentos.

group planting — trees or shrubs planted together so as to point out some special feature or to provide privacy or a small garden area.

siembra por agrupación — árboles o arbustos plantados juntos para señalar alguna característica especial o para proveer intimidad o un jardín pequeño.

guanine (G) — a base in genes designated by the letter *G*.

guanino (G) — una base en los genes denominada con la letra «G».

guard cells — cells that surround the stoma.

células guardias — células que rodean el estoma.

H

habitat — area or type of environment in which an organism or biological population normally lives.

hábitat — área o tipo de medio ambiente en que normalmente vive un organismo o una población biológica.

hand — unit of measurement for horses equal to 4 inches.

palmo — unidad de medir a los caballos que es igual a cuatro pulgadas.

cubrición a mano — el sistema de acoplamiento en donde el macho y la hembra son separados excepto durante la época de celo.

hardness — wood's resistance to compression.

dureza de la madera — la resistencia de la madera a la compresión.

hardwood — wood from deciduous trees.

madera dura — madera de los árboles caducos.

hardy — the ability of a plant to survive and grow in a given environment.

resistente — la capacidad de una planta de sobrevivir y crecer en un ambiente determinado.

harvester — person responsible for taking products from plants in the field.

cosechador — persona responsable de recoger el fruto del campo.

harvesting — taking a product from the plant where it was grown or produced.

cosechar — recoger un producto de la planta donde fue cultivado o producido.

hay — forages that have been cut and dried to a low level of moisture, used for animal feed.

heno — forrajes que han sido cortados y secados a un nivel bajo de humedad, usado para alimentar animales.

haylage — silage made from forages dried to 40 to 55 percent moisture.

hierba presecada y ensilada — ensilaje hecho de forrajes secados a 40 ó 55 por ciento de humedad.

heartwood — inactive core of a tree trunk or limb.

duramen — núcleo inactivo de un tronco o rama de árbol.

heel cutting — a shield-shaped cut made about halfway through the wood around the leaf and the axial bud.

escudete — un corte en forma de escudo hecho a medio camino a través de la madera alrededor de la hoja y el botón axilar.

heel in — place plant roots in a trench that is deep enough to cover the roots with moist soil to protect them until they can be permanently planted.

heifer — female animal of the cattle family that has not given birth.

vaquilla — hembra de ganado vacuno que no ha parido.

heritability — the capacity to be passed down from parent to offspring.

capacidad hereditaria — la capacidad de ser transmitido del progenitor a la progenie.

hen — adult female chicken, duck, turkey, and pheasant.

gallina — hembra adulta del gallo.

Note: Spanish does not have a specific word for turkey or pheasant female.

herb — plant kept for aroma, medicine, or seasoning.

hierba — planta guardada para fines aromáticos, medicinales o como condimento.

herbaceous — a plant that has a stem that does not turn woody; it is more or less soft and succulent, and often lacks winter hardiness.

herbáceo — una planta que tiene un tallo que no se vuelve leñoso; es más o menos blando y carnoso, tampoco es resistente al invierno.

herbicide — a substance for killing weeds.

herbicida — una sustancia empleada para matar las malas hierbas.

heredity — transmission of characteristics from parent to offspring.

herencia — la transmisión de rasgos de los progenitores a la progenie.

heterozygous — pairs of genes that are different.

hetercigótico — pares de genes que son diferentes.

hide — skin of an animal.

cuero — piel de los animales.

high technology — use of electronics and ultra-modern equipment to perform tasks and control machinery and processes.

alta tecnología — el uso de electrónica y equipo ultra-moderno para desempeñar funciones y controlar maquinaria y procesos.

hinny — cross between a stallion and a jennet.

burdégano — un cruce entre un semental y una burra.

home gardening — production of vegetables for use by a single family.

jardinería en el hogar — producción de vegetales para el uso de una sola familia.

homozygous — pairs of genes that are alike.

homocigótico — pares de genes que son semejantes.

horizon — layer.

horizonte — capa.

hormone — chemical that regulates activities of the body.

hormona — sustancia química que regula las actividades del cuerpo.

horse — member of the horse family 14.2 or more hands tall.

caballo — miembro de la familia de los caballos que mide 14.2 ó más palmos de alto.

horticulture — the science of producing, processing, and marketing fruits, vegetables, and ornamental plants.

horticultura — la ciencia de producir, procesar, y comercializar frutas, verduras, y plantas ornamentales.

host animal — a species of animal in or on which diseases or parasites can live.

animal huésped — una especie de animal en o sobre el cual viven parásitos o enfermedades.

hotbed — a cold frame that has a source of artificial heat.

superficie caliente — un bastidor frío que tiene un surtidor de calor artificial.

hybrid — plant or animal offspring from crossing two different species or varieties.

híbrido — progenie vegetal o animal que procede del cruzar dos distintas especies o razas.

hybrid vigor — offspring of greater strength and potential resulting when two different breeds or varieties are crossed.
vigor híbrido — progenie de más fuerza y potencial que resulta de cruzar dos distintas razas o variedades.

hydrocarbon — organic compound containing hydrogen and carbon.
hidrocarburo — compuesto orgánico que contienen hidrógeno y carbono.

hydrocooling — removal of heat by immersing in cold water.
enfriamiento con agua — quitar del calor por inmersión en agua fría.

hydroponics — the practice of growing plants without soil.
cultivo hidropónico — la costumbre de cultivar plantas sin tierra.

hygroscopic water — water that is held too tightly by soil particles for plant roots to absorb.
agua higroscópica — agua sujetada con demasiado rigor por las partículas de tierra para que las raíces de la planta la absorban.

hyphae — the thread-like vegetative structure of fungi.
hifas — la estructura vegetal filiforme de los hongos y mohos.

I

imbibition — the absorption of water into the cell causing it to swell.
embebición — la absorción de agua en la célula, lo cual hace que se hinche.

immune— not affected by.
inmune — que no es afectado.

implant — a substance placed under the skin and slowly absorbed to improve growth of animals.
implantar — una sustancia comercializads bajo la piel y lentamente absorbida para mejorar el crecimiento de los animales.

improvement activity — project that improves beauty, convenience, safety, value, or efficiency learned outside the regularly scheduled classroom or laboratory classes.
actividad de mejoramiento — proyecto que mejora la belleza, conveniencia, seguridad, valor, o eficacia y que se aprende fuera de las clases de escuela o laboratorios programados normalmente.

improvement by selection — picking the best parents for the next generation.
mejoramiento por selección — seleccionar a los mejores padres para la generación siguiente.

inbreeding — mating of animals that are related.
procrear en consanguinidad — apareamiento de animales emparentados.

incomplete dominance — neither gene expresses itself to the exclusion of the other.
dominancia incompleta — ninguno de los genes se manifiesta a exclusión al otro.

induction — a specific set of conditions must occur to cause flower development.
inducción — deben darse circunstancias específicas para causar el desarrollo de la flor.

inflorescence — the flowers and ultimately the seed area of the plant.
florescencia — son las flores y en última instancia el área de las semillas de la planta.

informal — in landscaping, planting areas that contain elements that are not equal in number, size, or texture.
informal — cuando se refiere al paisajismo, se refiere a las áreas que contienen elementos que no son iguales en cantidad, tamaño, ni textura.

infusion — the process for treating udder problems through the teat canal.
infusión — el proceso usado para tratar problemas de la ubre por el canal de la teta.

inner bark — contains the cambium from which the tree grows and the bark is renewed; transports food within a tree.
corteza interna — contiene el cambium del que crece el árbol y la corteza se renueva; transporta alimentos dentro del árbol.

inorganic compound — a compound that does not contain carbon.
compuesto inorgánico — un compuesto que no contiene carbono.

insecticide — a material used to kill insects or protect against their attack.
insecticida — un material que se usa para matar insectos o proteger contra su ataque.

insemination — the placement of semen in the female reproductive tract.
inseminación — la introducción de semen en las vías genitales de la hembra.

instar — the state of the insect during the period between molts.
estados larvarios — la fase del insecto durante el período entre las mudas.

institutional advertising — advertising designed to create a favorable image of the firm or institution.
publicidad institucional — publicidad creada para establecer un concepto favorable de la empresa o institución.

integrated pest management (IPM) — pest-control program based on multiple control practices.
control integrado de plagas y animales nocivos — programa de control de los insectos y plagas nocivos basado en múltiples métodos de control.
integrity — capable of personally upholding a high moral standard.
integridad — capaz de sostener personalmente un alto nivel moral.
intermediate-term loan — loan with payment periods that range from 1 to 7 years.
préstamo de plazo intermedio — préstamo que tienen plazos de pago que oscilan entre uno y siete años.
internode — area between two nodes.
entrenudos — área entre dos nudos.
internship — a career experience that places a student in a business for a specific period to learn technical skills through actual work experiences.
internado — una experiencia que coloca al estudiante (or, if you must divide in syllables: estudiante) en un negocio por un plazo específico de tiempo para aprender habilidades técnicas a través de experiencias reales en el trabajo.
intradermal — between layers of skin.
intradérmico — entre las capas de la piel.
intramuscular — in a muscle.
intramuscular — en los músculos.
intraperitoneal — in the abdominal cavity.
intraperitoneal — en la cavidad abdominal.
intraruminal — in the rumen.
intraruminal — en el rumen.
intravaginal growth — a type of growth in which shoots develop within the lower leaf sheath at a crown's axillary bud.
crecimiento intravaginal — un tipo de renuevo que desarrolla brotes en el interior de la vaina de la hoja inferior, en la corona del botón axilar.
intravenous — in a vein.
intravenoso — en la vena.
inventory report — units received and sold, unit cost, total sales, and profit.
inventario — unidades recibidas y vendidas, coste por unidad, ventas totales y ganancia.
involuntary muscle — muscle that operates in the body without control by the will of the animal.
músculo involuntario — músculo que funciona en el cuerpo sin control de la voluntad del animal.
ion — atom or a group of atoms that has an electrical charge.
iones — átomo o grupo de átomos que tienen una carga eléctrica.
irradiation — treating food with gamma rays.
irradiación — tratamiento de comestibles con rayos gamma.
irrigation — addition of water to plants to supplement that provided by rain or snow.
irrigación — agregado de agua a los cultivos para suplementar la proporcionada por la lluvia o la nieve.

J

jack — male donkey or mule.
burro — macho de la burra o mula.
jennet — female donkey or mule.
burra — hembra del burro o mula.
jungle fowl — wild ancestor of the chicken.
ave de la selva — antepasado salvaje de la gallina.

K

Katahdin — popular potato variety of the 1930s.
«Katahdin» — variedad popular de patata de la década de 1930.
kid — young goat.
chivo — cabrío jóven.
knowledge — familiarity, awareness, and understanding.
conocimiento — familiaridad, conciencia, y entendimiento.
kosher — prepared in accordance with Jewish dietary laws.
«Kosher» — preparado según las leyes dietéticas judaicas.

L

lactation period — period when mammals are producing milk.
período de lactancia — período en que los mamíferos producen leche.
lactose — compound milk sugar.
lactosa — compuesto de azúcar de leche.
lacustrine deposit — soil deposited by lakes.
depósito lacustre — los suelo depositado por los lagos.
lamb — member of the sheep family younger than 1 year; also, meat from young sheep.
cordero — miembro de la familia de las ovejas que tiene menos de un año de edad; también la carne de la oveja jóven.

larvae — mobile organisms that become fixed and grow into nonmobile adults.

larvas — organismos movibles que se vuelven fijos y se desarrollan y convierten en adultos no móviles.

laser — an intense, narrow beam of light that is used to measure nutrient deficiencies in plants from an orbiting satellite.

laser — un haz de luz intenso, estrecho que se utiliza para medir deficiencias en nutrientes en plantas desde un satélite en órbita.

lay on the table — a motion used to stop discussion on a motion until the next meeting. The way to table a motion is to say, "I move to table the motion."

colocar sobre la mesa — una moción usada para detener el debate sobre una moción hasta la próxima reunión. La manera de presentaesta moción es decir, "Propongo que se coloque el asunto sobre la mesa"

Note: This is a literal translation. No such concept exists in Spanish.

layer — chicken developed for the purpose of laying eggs.

gallina ponedora — una gallina criada para el objetivo de poner huevos.

LC_{50} — lethal concentration of a pesticide in the air required to kill 50 percent of the test population.

CM_{50} — concentración mortal de un pesticida en el aire necesaria para matar el 50 por ciento de una población prueba.

LD_{50} — lethal dose of a pesticide required to kill 50 percent of a test population.

DM_{50} — dosis mortal de un pesticida necesaria para matar el 50 por ciento de una población prueba.

leadership — the capacity or ability to lead.

dirección — la capacidad o habilidad de guiar.

leaf blade — the wide portion of a leaf in which photosynthesis occurs.

limbo — la porción de una hoja donde ocurre la fotosíntesis.

leaf cutting — a cutting made from a leaf without a petiole.

corte de hoja — un corte hecho de una hoja sin pecíolo.

leaf petiole cutting — a plant consisting of a leaf and petiole on which a rooting compound is applied and the cutting is placed in a soil medium to develop into a new plant.

esqueje de pecícolo de hojas — una planta que consiste de una hoja y un pecícolo en la cual se aplica un compuesto para enrraizar, y el esqueje se coloca en un medio de tierra parar desarrollar (if it must be divided in syllables: desarro- llar) una nueva planta.

leaf section cutting — a section of a plant leaf containing a vein that is placed in a plant growth medium to generate a new plant.

esqueje de sección de hoja — una sección de una hoja de la planta que contiene un nervio, que se coloca en un medio de crecimiento para generar una nueva planta.

legume — plant in which certain nitrogen-fixing bacteria use nitrogen gas from the air and convert it to nitrates that the plant can use as food.

legumbre — planta en la cual ciertas bacterias fijadora del nitrógeno utilizan el gas de nitrógeno del aire y lo convierten a nitratos que la planta puede usar para alimentarse.

lenticels — pores in the stem that allow the passage of gases in and out of the plant.

lenticelas — poros en el tallo o el pedúnculo que permiten que los gases entren y salgan de la planta.

liabilities — current and long-term debts.

pasivos — deudas actuales y a largo plazo.

light horse — type of horse developed for riding.

caballo de silla — tipo de caballo criado para montar.

light intensity — brightness of light.

intensidad de la luz — luminosidad de la luz.

ligule — a membranous or hairy structure located on the inside of the leaf.

lígula — está ubicada en la vaina foliar interior y es una estructura membranosa y pilosa.

lime — material that reduces the acid content of soil and supplies nutrients such as calcium and magnesium to improve plant growth.

cal — material que reduce el contenido de ácido del suelo y proporciona minerales tales como calcio y magnesio para mejorar el crecimiento de las plantas.

linen — fabric produced from fibers in flax plants.

lino — tejido producido de las fibras de las plantas lináceas.

linseed oil — oil produced from flax seed.

aceite de linaza — aceite producido de la semilla de lino.

lipid — fat droplets inside a cell.

lípidos — gotitas de grasa dentro de una célula.

loam — a granular soil containing a good balance of sand, silt, and clay.

suelo arcilloso — un suelo granular que contiene una mezcla balanceada de arena, légamo y arcilla.

loess deposit — soil deposited by wind.

depósito loésico — tierra depositada por el viento.

long-term loans — loans with payment periods that range from 8 to 40 years.

préstamos a largo plazo — préstamos cuyos plazos de pago oscilan entre 8 a 40 años.

long-term plans — plans accomplished over months or years.

planes a largo plazo — planes realizados por meses o años.

loss-leader pricing — commodity offered for sale at prices less than the cost level.

precio de lanzamiento — producto básico que se ofrece a la venta a un precio menor al nivel de costo.

loyalty — reliable support for an individual, group, or cause.

lealtad — apoyo seguro para un individuo, un grupo o una causa.

lumber — boards cut from trees.

maderos — trozos cortados de los árboles.

M

macronutrient — element used in relatively large quantities.

macronutriente — elemento usado en cantidades relativamente grandes.

main motion — a basic motion used to present a proposal for the first time. The way to state it is to get recognized, and then say, "I move that ..."

moción principal — la moción básica para presentar una propuesta por primera vez. La manera de declararla es ser reconocido y luego decir, «Yo propongo que ...

malting — process of preparing grain for the production of beer and alcohol.

malteado — proceso de preparar el grano para la producción de cerveza y alcohol.

maltose — compound milk sugar.

maltosa — azúcar de la leche compuesto.

mammal — milk-producing animal.

mamífero — animal que produce la leche.

mange — crusty skin condition caused by mites.

sarna — enfermedad de la piel que es causada por los acáridos.

mare — adult female horse or pony.

yegua — hembra adulta del caballo o poney.

margin — edge of the leaf.

reborde — el borde de la hoja.

market gardening — production of a wide variety of vegetables that are generally sold in local roadside markets.

jardinería comercial — producción de una amplia variedad de vegetales que generalmente se venden en mercados locales ubicados al costado de los caminos.

maturity — the state or quality of being fully grown.

madurez — estado o cualidad de estar completamente desarrollado.

mechanical pest control — pest control that affects the pest's environment or the pest itself.

control mecánico de plagas o insectos nocivos — control de plagas o insectos nocivos que afecta el ambiente de la plaga o insecto nocivo o la plaga o insecto en sí.

medium — surrounding environment in which something functions and thrives.

medio ambiente — los alrededores en los cuales funciona y crece alguna cosa.

meiosis — cell division that results in the formation of gametes.

meiosis — división celular que resulta en la formación de los gametos.

meristematic tissue — plant tissue responsible for plant growth.

tejido meristemático — tejido vegetal que causa el crecimiento de la planta.

mesophyll — tissue of the leaf where photosynthesis occurs.

mesofilo — tejido de la hoja en donde ocurre la fotosíntesis.

metamorphosis — the change in growth stages of an insect.

metamorfosis — el cambio en las etapas de crecimiento del insecto.

microbe — living organism that requires the aid of a microscope to be seen.

microbio — organismo viviente que requiere la ayuda del microscopio para verse.

micronutrient — element used in very small quantities.

micronutriente — elemento usado en cantidades muy pequeñas.

microorganism — tiny plant or animal that may contribute to food spoilage.

microorganismo — microbio animal o vegetal que puede contribuir a la putrefacción de los comestibles.

middlemen — people who handle an agricultural product between the farm and the consumer.

intermediarios — personas que manejan un producto agrícola entre la granja y el consumidor.

milking machine — machine that milks cows and goats.

máquina de ordeñar — aparato o máquina que extrae la leche de la ubre de las vacas y cabras.

mineral — element essential for normal body functions.
mineral — elemento esencial para las funciones normales de cuerpo.
minimum tillage — soil is worked only enough so that seed will germinate.
labranza mínima — la tierra es labrada justo lo suficiente para que germinen las semillas.
minutes — the official written record of a business meeting.
actas — el registro oficial escrito de una reunión de negocios.
mitochondrion — plays a role in converting animal food to usable energy.
mitocondria — juega un papel importante en la conversión de los alimentos de origen animal energía aprovechable.
mitosis — simple cell division for growth.
mitosis — división celular simple que resulta en el crecimiento.
mohair — hair from Angora goats used to make a shiny, heavy, woolly fabric.
mohair — pelo de la cabra de Angora usado para hacer un tejido lanoso, grueso y con brillo.
moist-heat cooking — surrounding the food with liquid while cooking.
cocinar con calor húmedo — rodear la comida con líquido mientras se cocina.
moldboard plow — plow with a curved bottom that will turn prairie soils.
arado de vertedera — arado con orejera capaz de arar el suelo de llanura.
molt — softening and cracking of the exoskeleton so that crustaceans may escape and grow a larger exoskeleton.
mudar — proceso de ablandamiento y agrietamiento del dermatoesquéleto para que los crustáceos puedan escaparse y desarrollar un dermatoesquéleto más grande.
monoclonal antibody — natural substance in blood that fights diseases and infections.
anticuerpo monoclonal — substancia natural en la sangre que lucha contra enfermedades e infecciones.
monocotyledon (monocot) — plant with one seed leaf.
monocotiledóneo — planta con solamente una hoja de semilla.
monogastric — animal with a single-compartment stomach.
monogástrico — animal con estómago de un solo compartimento.
mosaic — a virus disease of plants in which a leaf shows a symptom of light and dark-green mottling of the foliage.
mosaico — una enfermedad viral de las plantas en donde la hoja manifiesta un síntoma de vetas manchas de verde claro y oscuro en el follaje.
motion — a proposal for group action that is presented in a meeting to be acted on by an organization.
moción — una propuesta para acción de un grupo que es presentada en una junta para ser puesta en acción por la organización.
mulch — material placed on soil to break the fall of raindrops (preventing erosion), prevent weeds from growing, or improve the appearance of the area.
pajote — material puesto encima del suelo para cortar la caída de las gotas de lluvia (para prevenir la erosión), para prevenir el crecimiento de malas hierbas, o para mejorar el aspecto del área.
mule — cross between a jack and a mare.
mula — una cruce entre el burro macho y la yegua.
muscular system — the lean meat of the animal.
sistema muscular — la carne sin grasa del animal.
mutation — change in genes.
mutación — cambios en los genes.
mutton — meat from mature sheep.
cordero — carne de cordero mayor.
mutualism — two types of wildlife living together for the mutual benefit of both.
mutualismo — dos clases de fauna que viven juntas para el beneficio mutuo de las dos.
mycelium — a collection of fungal hyphae.
micelio — un grupo de hifa fungosa.
MyPlate — MyPlate illustrates the five food groups that are the building blocks for a healthy diet using a familiar image—a place setting for a meal. USDA's Choose My Plate campaign features selected messages to focus consumers on key behaviors such as: balancing calories by reducing portion sizes; increasing consumption of healthy foods, such as fruits, vegetables, and whole grains; and identifying foods to reduce, including high sodium foods and sugary drinks.
Mi plato — Mi plato utiliza una imagen familiar para mostrar los cinco grupos de alimentos que son la base fundamental de una dieta saludable: el entorno de una comida. La campaña "*Choose My Plate*" [Elige Mi plato] del Departamento de Agricultura de Estados Unidos presenta mensajes seleccionados para que los consumidores se enfoquen en comportamientos clave, tal como equilibrar las calorías mediante la reducción del tamaño de las porciones; aumentar el consumo de alimentos saludables, como frutas, verduras y granos integrales e identificar los alimentos que se deben reducir, incluyendo los alimentos ricos en sodio y las bebidas azucaradas.

N

nematode — a type of tiny roundworm that causes damage to specific kinds of plants.

nematodo — un pequeño tipo de ascáride que causa daños a un tipo específico de plantas.

net worth — the value of an owner's assets minus liabilities (debts).

valor neto — el valor de los activos de un propietario menos los pasivos.

neutral — neither acidic nor alkaline.

neutro — ni ácido ni alcalino.

nitrate — a form of nitrogen used by plants.

nitrato — un tipo de nitrógeno usado por plantas.

nitrogen fixation — conversion of nitrogen gas to nitrate by bacteria.

fijación del nitrógeno — la conversión de gas de nitrógeno en nitrato realizada por bacterias.

nitrous oxide — a compound containing nitrogen and oxygen; they make up about 5 percent of the pollutants in automobile exhaust.

óxido nitroso — un compuesto que contienen nitrógeno y oxígeno; constituyen aproximadamente el 5 por ciento de los agentes contaminantes en el gas de escape de los automóviles.

node — portion of the stem that is swollen or slightly enlarged that gives rise to buds.

nudo — la parte del tallo o pedúnculo que está hinchada o levemente agrandada que produce los botones.

nomenclature — a systematic method of naming plants or animals.

nomenclatura — un método sistemático de dar nombres a las plantas o los animales.

noncontagious — diseases that cannot be spread to other animals.

no contagiosas — enfermedades que se transmiten a otros animales.

no-till — seed is planted directly into the residue of the previous crop, without exposing the soil surface.

sin labranza — dícese de cuando se siembra la semilla directamente en el residuo de la cosecha previa, sin exponer la superficie del suelo.

noxious weed — plant that is highly damaging to an environment and that is controlled under authority of state law.

hierba tóxica — planta que es sumamente dañina al medio ambiente que es controlada bajo la autoridad de la ley estatal.

nuclear membrane — a protective membrane that forms a barrier surrounding the cell nucleus.

membrana nuclear — una membrana protectora que forma una barrera en torno al núcleo de la célula.

nucleic acid — original name for deoxyribonucleic acid.

ácido nucleico — nombre original por ácido desoxirribonucléico.

nucleus — a cell structure that contains pairs of chromosomes on which genes are located at specific locations.

núcleo — una estructura celular que contiene pares de cromosomas en cuales de los genes están ubicados en lugares específicos.

nurse crop — a crop used to protect another crop until it can get established.

cultivo protector — un cultivo que se usa para proteger a otro cultivo hasta que éste pueda establecerse.

nursery — a place where young trees, shrubs, and other plants are grown.

vivero — un lugar donde se crían árboles, arbustos, y otras plantas jóvenes.

nutrient — substance necessary for the functioning of an organism.

nutriente — sustancia necesaria para el funcionamiento del organismo.

nutrition — the process whereby all body parts receive materials needed for function, growth, and renewal.

nutrición — el proceso en que todas las partes del cuerpo reciben materiales necesarios para función, crecimiento y renovación.

O

O horizon — the soil layer on the surface that is composed of organic matter and a small amount of mineral matter.

horizonte O — la capa de tierra que está en la superficie compuesta de substancias orgánicas y una pequeña cantidad de substancias minerales.

offsetting a position — taking a second position on the futures market to offset a first.

compensar una posición — tomar una posición secundaria en el mercado a futuro para compensar una primera.

oilseed crop — crop produced for the oil content of their seeds.

cultivo de semilla oleaginosa — cultivo cultivado por el contenido de aceite de sus semillas.

olericulturist — one who studies the cultivation of vegetables.

oleicultor — uno que estudia el cultivo de verduras.

olericulture — the cultivation of vegetables.

oleicultura — el cultivo de verduras.

online video merchandising — the use of online video recordings of livestock that are offered for sale at distant locations, thus enabling a potential buyer to bid for the livestock online or via telephone.

vídeo en línea merchandising — el uso de vídeo en línea las grabaciones del ganado que se ponen a la venta en lugar lejanos, permitiendo así que un comprador potencial para ofertar por el ganado en línea o por teléfono.

on-the-job training — experience obtained while working in an actual job setting.

entrenamiento práctico — experiencia obtenida mientras se trabaja en un empleo verdadero.

opening a position — initial step in the futures market.

una posición inicial — paso inicial en el mercado a futuro.

order of business — the items and sequence of activities conducted at a meeting.

orden del día — los ítems y la secuencia de actividades conducidos en una reunión.

organic compound — a compound that contains carbon.

compuesto orgánico — un compuesto que contiene carbono.

ornamental — a plant grown for its appearance and beauty.

ornamentales — planta que se cultiva por su apariencia y belleza.

osmosis — the flow of a fluid through a semipermeable membrane separating two solutions, which permits the passage of the solvent but not the dissolved substance. The liquid will flow from a weaker to a stronger solution, thus tending to equalize concentrations.

osmosis — el flujo de un líquido a través de una membrana semipermeable que separa dos soluciones, que permite el paso del solvente pero no de la sustancia disuelta. El líquido fluirá de una solución más débil a una más fuerte, así tendiendo a igualar concentraciones.

out-of-the-market — when a futures position is offset by a second position.

fuera del mercado — cuando una posición a futuro es compensada por una posición segunda.

ova — female reproductive cells or eggs from the ovary of an animal.

óvulo — células reproductivas femeninas o huevos del ovario de un animal.

ovary — female organ that produces eggs or female sex cells; also, that portion of the flower that contains the ovules or seeds.

ovario — órgano de la hembra que produce los óvulos o células sexuales femeninas; además aquella parte de la flor que contiene los óvulos o semillas.

overripe — beyond maturity.

demasiado maduro — más allá de madurez.

overseeding — seeding a second crop into one that is already growing.

sobresembrar — sembrar un cultivo segundo encima de uno que ya está creciendo.

ovulation — process of releasing mature eggs from the ovary.

ovulación — el proceso de liberar óvulos maduros del ovario.

ovule — unfertilized seed.

óvulo (vegetal) — semilla ya no fecundada.

oxidative deterioration — decay resulting from exposure to air.

deterioro oxidativo — deterioro resultante de la exposición al aire.

ozone — compound that exists in limited quantities about 15 miles above the Earth's surface.

ozono — compuesto que existe en cantidades limitadas aproximadamente quince millas encima de la superficie de la tierra.

P

packer — person or firm responsible for preparing commodities for shipment.

empaquetador — persona o firma responsable de preparar productos básicos para envío.

parasite — organism that lives in or on another organism with no benefit to the host.

parásito — organismo que vive en o encima de otro organismo sin beneficio al huésped.

parasitism — one type of wildlife living and feeding on another without killing it.

parasitismo — una clase de fauna que vive en y se alimenta de otra sin matarla.

parent material — the horizon of unconsolidated material from which soil develops.

materiales madres del suelo — el horizonte de material no consolidado del que se forma el suelo.

parliamentary procedure — a system of guidelines or rules for making group decisions in business meetings.

procedimiento parlamentario — un sistema de principios o normas para tomar decisiones de grupo en reuniones de negocio.

particulate — a small particle that is suspended in the air.
las partículas — una pequeña partícula que queda suspendida en el aire.

parturition — the birthing process.
alumbramiento — el proceso de dar a luz.

pasture — forages that are harvested by the livestock itself.
pasto — forrajes que se cosechan por el ganado mismo.

pathogen — organism that produces disease.
patógeno — organismos que producen las enfermedades.

peat moss — a type of organic matter made from sphagnum moss.
turba — un tipo de materia orgánica formada de esfagno.

penetration pricing — a strategy in which price is set less than that of competitors.
precio de penetración — una estrategia donde se fija el precio más bajo que el de los competidores.

percolation — movement of water through the soil.
filtración — paso del agua a través de la tierra.

perennial — a plant that lives from year to year.
perene — una planta que vive de año a año.

perennial weed — a weed that lives for more than 2 years.
mala hierba perenne — una mala hierba que vive más de dos años.

perfect flower — flower containing all of the parts: stamen, pistil, petals, and sepals.
flor perfecta — flor que consta de todas las partes: estambre, pistilo, pétalos y sépalos.

peripheral nervous system — system that controls the functions of the body tissues, including the organs.
sistema nervioso periférico — sistema que controla las funciones de los tejidos de cuerpo, incluyendo los órganos.

perlite — natural volcanic glass material having water-holding capabilities.
perlita — material vítreo volcánico natural que tiene capacidades de contener el agua.

pesticide — chemical used to control pests.
pesticida — sustancia química empleada para combatir plagas e insectos nocivos.

pesticide resistance — the ability of a pest to tolerate a lethal level of a pesticide.
resistencia al pesticida — la capacidad de plagas o insectos nocivos para tolerar un nivel mortal de pesticida.

pest resurgence — ability of a pest population to recover.
resurgimiento de plaga o insecto nocivo — la capacidad de recuperarse de una población de plagas o insectos nocivos.

petal — brightly colored, sometimes fragrant portion of the flower.
pétalo — parte de la flor de color vivo y a veces fragrante.

petiole — the slender leaf stock that supports the blade, attaching it to the stem of a plant.
pecíolo — el esbelto rabo de la hoja que sostiene la brizna, uniéndola al tallo de la planta.

pH — measurement of acidity or alkalinity from 1 to 14.
valor pH — medida de la acidez o alcalinidad desde 1 a 14.

phenotype — physical appearance of an individual.
fenotipo — características externas, o aspecto de un individuo.

pheromone — a chemical secreted by an organism to cause a specific reaction by another organism of the same species.
feromona — una sustancia química secretada por un organismo para causar una reacción específica por otro organismo de la misma especie.

phloem — cells forming conductive tissues that carry manufactured food to areas of the plant where it is stored or used.
floema — las células que forman los tejidos conductores que transportan productos alimenticios elaborados en zonas de la planta en la que se almacena o utilizado.

photodecomposition — chemical breakdown caused by exposure to light.
fotodescomposición — la descomposición química causada por la exposición a la luz.

photosynthesis — process in which chlorophyll in green plants enables those plants to use light to manufacture sugar from carbon dioxide and water.
fotosíntesis — proceso en donde la clorofila de las plantas verdes las capacita para usar la luz para fabricar azúcar del dióxido de carbono y agua.

phototropism — a process by which a plant leaf is capable of adjusting its angle of exposure to the sun.
fototropismo — un proceso por el cual la hoja de una planta es capaz de ajustar su ángulo de exposición al sol.

physiology — study of the functions and vital processes of living creatures and their organs.
fisiología — el estudio de las funciones y los procesos vitales de criaturas vivientes y sus órganos.

pistil — female part of the flower consisting of stigma, style, ovary, and ovule.

pistilo — parte femenina de las flores que consta de estigma, estilo, ovario, y óvulo.

placement supervised agricultural experience — career experiences that place students with employers who are conducting agricultural business such as farming ranching, greenhouses, and others.

colocación supervisada de experiencia agrocientífica — experiencias de carrera que colocan a estudiantes con empleadores que conducen negocios agriculturales tales como agricultura, ganadería, viveros y otros.

plan — to think through.

planear — pensar muy bien en cómo se va a hacer algo.

plant hardiness zone map — a map developed by the U.S. Department of Agriculture, dividing the country into zones based on average winter temperatures.

mapa de zonas de resistencia de las plantas — mapa desarrollado por el Departamento de Agricultura de los EE.UU., que divide al país en zonas basándose en las temperaturas medias de invierno.

plant nutrition — provision of elements to plants.

nutrición vegetal — el suministro de elementos a las plantas.

plantscaping — the design and arrangement of plants and structures in an indoor area.

paisajismo de interiores — el diseño y el arreglo de plantas y de estructuras en un área de interior.

plugging — establishment of turf by using small pieces of existing turfgrass.

cubrir de trasplantes de tepe — establecer el césped usando pequeños pedazos de césped ya existente.

plywood — construction material made of thin layers of wood glued together.

madera contrapechada — material de construcción hecho de hojas delgadas de madera adheridas.

point of order — a procedure used to object to some item in or about a meeting that is not being done properly. The procedure to use is to say, "point of order." The presiding officer should then recognize the member by saying, "State your point."

cuestión de orden — un procedimiento que sirve para oponerse a algún ítem de o sobre la reunión que no está haciéndose adecuadamente. El procedimiento es decir, «¡Cuestión de orden!» Entonces el funcionario que preside reconoce al miembro y dice «Exprese su cuestión.»

pollen — small male sperm or grains that are necessary for fertilization in the flower.

polen — esperma pequeño o polvillo macho que es necesario para la fecundación de la flor.

pollination — transfer of pollen from anther to stigma.

polinización — transporte del polen del antero hasta el estigma.

polluted — containing harmful chemicals or organisms.

contaminado — que contiene sustancias químicas u organismos dañinos.

pome — fleshy fruits with embedded core and seeds.

pomo — frutos carnosos con carozo y semillas.

pomologist — a fruit grower or scientist.

pomólogo — científico o cultivador de frutos.

pony — member of the horse family less than 14.2 hands tall.

pony — miembro de la familia de los caballos que mide menos de 14.2 palmos de alto.

porcine somatotropin (PST) — hormone that increases meat production in swine.

somatotropino porcino — una hormona que aumenta la producción de carne en los cerdos.

pore — spaces between soil particles through which plant roots penetrate, and in which air, water, and nutrients are stored.

poro — espacio entre partículas de tierra en los que penetran las raíces y donde se almacenan el aire, el agua y los nutrientes.

pork — meat from swine.

puerco — carne del cerdo.

postemergence herbicide — a herbicide applied after the weed or crop is present.

herbicida postemergente — un herbicida aplicado después de que se presenta la mala hierba o el cultivo.

potable — drinkable—that is, free of harmful chemicals and organisms.

potable — que puede beberse, o sea que está libre de organismos y sustancias químicas dañinas.

poult — young turkey.

pavipollo — pavo jóven.

ppm — parts per million.

partes por millón — partes por millón.

ppt — parts per thousand.

partes por mil — partes por mil.

Note: In Spanish, abbreviations are not used for "PPM" and "PPT" because they would both be "PPM," translated verbatim.

precipitate — a chemical action that results in the dropout of solids in a solution.

precipitado — una reacción química que resulta en el depósito de sedimento en una solución.

pre-cooling — rapid removal of heat before storage or shipment.
prerefrigeración — eliminación rápida del calor antes de almacenamiento o transporte.
predation — a way of life where one type of wildlife eats another type.
depredación — una manera de vivir en que un tipo de fauna se alimenta de otro tipo.
predator — an animal that feeds on a smaller or weaker animal.
predador — un animal que se alimenta con otro animal más pequeño o más débil.
preemergence herbicide — a herbicide applied before weed or crop germination.
herbicida pre-emergente — un herbicida aplicado antes de la germinación de malas hierbas o cultivos.
presiding officer — a president, vice president, or chairperson who is designated to lead a business meeting.
funcionario — el presidente, vicepresidente, o presidente interino designado para dirigir una sesión de negocio.
prestige pricing — pricing to buyers with special desires for quality, fashion, or image.
precio de prestigio — valoración para los compradores que tienen deseos especiales de calidad, moda o imagen.
prey — animal eaten by another animal.
presa — animal comido por otro animal.
primary nutrients — in agriculture, nitrogen, phosphorus, and potash.
nutrientes principales — en la agricultura, nitrógeno, fósforo y potasa.
processing — turning raw agricultural products into consumable food.
procesado — convertir los productos agrícolas en bruto en alimentos comestibles.
processor — person or business cleaning, separating, handling, and preparing a product before it is sold to a distributor.
procesador — persona o empresa que limpia, separa, maneja, y prepara un producto antes de que se venda al distribuidor.
producer — person who grows a crop.
productor — persona que cultiva el cultivo.
product advertising — advertising that focuses on the product itself.
publicidad de producto — publicidad que se concentra en el producto mismo.
production agriculture — farming and ranching.
agricultura de producción — cultivo y ganadería.
Production Credit Association(s) (PCA) — lending institution that provides short-term credit for agriculture.
Asociación(es) de Crédito para Productores — institución de préstamo que provee crédito a corto plazo para la agricultura.
production or productive enterprise — project conducted for wages or profit.
producción o empresa productiva — proyecto conducido por salarios o lucro.
profession — occupation requiring an education, especially law, medicine, teaching, or the ministry.
profesión — vocación que exige una enseñanza superior, especialmente el derecho, la medicina, la enseñanza, o el ministerio.
profit — proceeds from a business transaction.
beneficio — ganancias de una transacción comercial.
profit and loss statement — projection of costs and expenses against sales and revenue over time.
estado de ganancias y pérdidas — proyección de los costos y gastos comparados con las ventas e ingresos por un tiempo.
progeny — offspring.
progenie — descendecia.
progesterone — hormone that prevents estrus during pregnancy and causes development of the mammary system.
progesterona — hormona que impide el estro durante el embarazo y causa el desarrollo del sistema mamario.
project — a series of activities related to a single objective or enterprise.
proyecto — una serie de actividades relativas a un sólo objetivo o empresa.
promissory note — note signed by borrower agreeing to the terms of the loan.
pagaré — obligación escrita firmada por el prestatario en la cual acepta las condiciones y los plazos del préstamo.
promoting function — plan to make potential customers aware of the product or service.
función de promocionar — plan para hacer que los clientes potenciales se den cuenta del producto o servicio.
propagation — process of increasing the numbers of a species or perpetuating a species.
propagación — el proceso de aumentar la población de una especie o de perpetuar una especie.
protectant fungicide — a fungicide applied before disease infection.

fungicida preventivo — un fungicida aplicado antes de la infección.

protein — nutrient made up of amino acids and essential for maintenance, growth, and reproduction.

proteina — nutriente que consta de aminoácidos y que es esencial para mantener la vida, el crecimiento, y la reproducción.

protozoa — microscopic, one-celled animals that are parasites of animals.

protozarios — microorganismos unicelulares que son parásitos de otros animales.

pruning — the removal of dead, broken, unwanted, diseased, or insect-infested wood.

podar — eliminar la madera muerta, trozada, no deseada, enferma o plagada por los insectos.

psychological pricing — a strategy designed to make a price seem lower or less significant.

valoración psicológica — una estrategia creada para hacer que un precio parezca más bajo o menos significativo.

pullet — young female chicken.

polla — hembra jóven del pollo.

pulpwood — wood used for making fiber for paper and other products.

madera para pasta de papel — madera usada para hacer fibras para papel y otros productos.

Q

quality grade — grade based on amount and distribution of finish on an animal.

nivel de calidad — nivel basado en la cantidad y distribución del acabado de engorde de un animal.

quarantine — isolation of pest-infested material.

cuarentena — aislamiento de material plagado.

queen — only fertile, egg-laying female bee in the hive.

reina — la única abeja hembra del panal que es fecunda y pone huevos.

R

radioactive material — material that is emitting radiation.

material radioactivo — material que emite la radiación.

radon — colorless, radioactive gas formed by disintegration of radium.

radón — gas radioactivo incoloro, formado por la desintegración del rádium.

ram — male member of the sheep family.

carnero — macho de la oveja.

ration — the amount of feed fed in one day.

ración — la porción de alimento que se da de comer en un día.

real-world experience — an activity conducted in the daily routine of our society.

experiencia de la vida diaria — una actividad realizada dentro de la rutina diaria de nuestra sociedad.

reaper — machine that cuts grain.

segador — máquina que corta cereales.

recessive — a gene that remains hidden and expresses itself only in the absence of a dominant gene.

recesivo — un gen que queda oculto y se expresa sólo en ausencia de un gen dominante.

recombinant DNA technology — gene splicing.

tecnología de recombinación del ADN — empalme de genes.

recuperative potential — the ability of a plant to recover after being damaged.

potencial de recuperación — la capacidad de la planta para reestablecerse después de ser dañada.

refer — a motion used to refer some other motion to a committee or person for finding more information and/or taking action on the motion on behalf of the members. The way to state a referral is to say, "I move to refer this motion to . . ."

referir — una moción usada para referir otra moción a una comisión para averiguar más información y/o para tomar acción sobre la moción en nombre de los socios. La manera de declarar proponer una referencia es decir: decir, «Yo propongo referir esta moción a . . .s»

renewable natural resources — resources provided by nature that can replace themselves.

recursos naturales renovables — recursos suministrados por la naturaleza que pueden recuperarse solos.

reproduction — the generation of a new plant or animal.

reproducción — la creación de una planta o animal nuevo.

residual soil — parent material formed in place.

suelo residual — materiales madres del suelo formados en el lugar.

resource substitution — the use of a resource or item for another when the results are the same.

sustitución de recursos — el uso de un recurso o cuerpo en vez de otro cuando los resultados son semejantes.

respiration — a process in which energy and carbon dioxide are released due to digestion or the breakdown of plant tissues during periods of darkness.

respiración — un proceso que libera energía y bióxido de carbono gracias a la digestión o descomposición de tejidos vegetales durante períodos de oscuridad.

respiratory system — system that provides oxygen to the blood of the animal.

sistema respiratorio — sistema que proporciona el oxígeno a la sangre del animal.

resumé — a summary of a person's education, technical skills, and career experiences.

currículum vitae — un resumen de la educación de una persona, sus habilidades técnicas y experiencias de carrera.

retailer — person or store that sells directly to the consumer.

minorista — persona o tienda que vende directamente al consumidor.

retail marketing — the selling of a product directly to consumers.

comercialización minorista — la venta de un producto directamente al consumidor.

retortable pouch — package with foil between two layers of plastic.

embalaje de retortas — paquete que tiene una hoja de aluminio entre dos hojas de plástico.

rhizome — horizontal, underground stems.

rizoma — tallo subterráneo, horizontal.

ribosome — a cell structure that is responsible for synthesizing proteins.

ribosoma — una estructura celular que es responsable de la síntesis de las proteínas.

rooster — adult male chicken or pheasant.

gallo — adulto macho del pollo.

faisán — adulto macho de los faisanes.

rootbound — roots restricted by a container.

raíces restringidas por el recipiente — raíces restringidas por un recipiente.

Note: The translation of "rootbound" is the same as that of the English definition provided.

root cap — the outermost part of a root that protects the tender tip of a growing root as it penetrates the soil.

capucha de raíz — la parte más externa de una raíz que protege la tierna punta de una raíz en crecimiento mientras penetra la tierra.

root crop — crop grown for the thick, fleshy storage root that it produces.

cosecha de vegetales de raíces — cultivo realizado para obtener el tubérculo grueso carnoso que produce.

root cutting — section of a root cut used for propagation purposes.

estaca de raíz — cuando se corte parte de la raíz y se usa para fines de propagación.

root hairs — small microscopic roots that arise from the cells located on the surface of a plant root.

peluza de raíz — pequeñas raíces microscópicas que surgen de las células localizadas en la superficie de la ríz de una planta.

rooting hormone — chemical used to stimulate root formation on a cutting.

hormona radicular — sustancia química usada para estimular la formación de raíces en un esqueje.

root pruning — systematic cutting of the roots by hand or machine to encourage the roots to develop within the root ball range.

poda de raíces — el podar sistematicamente las raíces a mano o a máquina para alentar a que las raíces se desarrollen dentro del área de la raíz.

rootstock — root system and stem of a plant on which another plant is grafted.

portainjerto — sistema radicular y tallo de una planta sobre la cual se injerta otra planta.

roughage — grass, hay, or silage and other feeds high in fiber and low in TDN.

forraje — hierba, heno, o ensilaje y otros alimentos con alto contenido de fibra y bajo contenido de NDT.

roundworm — slender worm that is tapered on both ends.

ascáride — lombriz delgado que tiene aspecto afilado en sus dos extremos.

rumen — one of the compartments of the stomach in cattle, deer, and sheep that is responsible for the breakdown of cellulose in the feed that is consumed.

rumen — uno de los compartimentos del estómago de ganado, venados, y ovejas responsable de descomponer la celulosa que contiene el alimento consumido.

ruminant — animal that has a stomach with four digestive compartments.

rumiante — animal que tiene un estómago con cuatro compartimentos.

russet — a baking potato variety best adapted for growth in sandy vocanic soils.

russet — una variedad de la patata para el horno adaptada para el crecimiento en suelos volcánicos arenosos.

S

salmonid — any of the family of soft-finned fish such as salmon or trout that have the last vertebrae upturned.

salmónicos — cualquiera de los integrantes de la familia de peces de aletas blandas, tales como el salmón y la trucha, que tienen la última vértebra doblada hacia arriba.

saltwater marsh — the ecological system of plants and animals influenced by tidal waters with salinity of 15 to 34 parts per million.

salina — sistemas ecológicos de plantas y animales influidos por mareas con salinidades de 15 a 34 partes por mil.

sand — largest soil particles; 1 to 0.05 mm.

arena — las partículas de tierra más grandes; de 1 a 0.05 mm.

sapwood — transports water and minerals upward in tree roots, trunks, and stems.

sámago — traslada agua y minerales hacia arriba en las raíces, los troncos y tallos de árbol.

scale — size of items.

escala — tamaño de un cuerpo.

scarify — to soak, keep moist, or mechanically scrape a seed coat to aid germination.

escarificar — empapar, mantener húmedo, o raspar a máquina la capa de una semilla para alentar la germinación.

scientific method — a procedure for investigating problems of a scientific nature.

método cientifico — un procedimiento para investigar problemas de naturaleza científica.

scion — the top of the stem and leaves of a plant that is to be propagated.

codo o varilla de injerto — la punta del tallo y hojas de una planta que se va a propagar.

scurvy — vitamin C deficiency disease.

escorbuto — una enfermedad causada por la deficiencia de vitamina C.

seasonal — pertaining to a particular season of the year.

estacional — que pertenece a una estación determinada del año.

secondary host — a plant or animal that carries a disease or parasite during part of the life cycle.

huésped secundario — una planta o un animal que porta una enfermedad o parásito durante parte del ciclo de la vida.

secondary nutrient — nutrients required by plants in moderate amounts; they include calcium, magnesium, and sulfur.

nutrientes secundarios — los nutrientes requeridos por las plantas en cantidades moderadas; incluyen el calcio, el magnesio y azufre.

seed blend — combination of different cultivars of the same species.

mezcla de semilla — combinación de cultivos diferentes de la misma especie.

seed culm — stem that supports the inflorescence of the plant.

caña de semilla — tallo que permite la florescencia de la planta.

seed legume crop — crop that is nitrogen-fixing and produces edible seeds.

legumbre — cultivo que es fijador del nitrógeno y produce semillas comestibles.

seedling — young plant grown from seed.

plántula — planta joven cultivada de semilla.

seed mixture — combination of two or more species.

mixtura de semilla — combinación de dos o más especies.

selective breeding — mating adults who have characteristics desired in the offspring.

crianza selectiva — selección de padres para apareamiento que tienen los rasgos que se desean para la descendencia.

selflessness — placing the desires and welfare of others above oneself.

abnegación — colocar los deseos y el bienestar de los demás por encima de los de uno mismo.

selling function — market research, sales plans, and sales closures.

función de venta — las investigaciones del mercado, las ventas, los planes y cierres de venta.

semen — the sperm cells and accompanying fluids.

semen — las células de los espermatazoides y los líquidos que los acompañan.

semiarid — an area partially deficient in rainfall; dry.

semiárido — un área que parcialmente carece de precipitación, seca.

semidwarf — tree that has rootstock that limits growth to 15 feet or fewer.

semienano — árbol que tiene un rizoma que limita su crecimiento a 15 pies o menos.

seminal root — root formed at the time of seed germination to anchor the seed in soil.

raíz seminal — raíz formada en el momento de la germinación de la semilla para fijar la semilla en el suelo.

semipermeable membrane — membrane that permits a solution to move through it.

membrana semipermeable — membrana que permite que la penetre una solución.

sepal — small, green, leaf-like structure found at the base of the flower.
sépalo — cuerpo verde pequeño y de aspecto de hoja que se encuentran en la base de la flor.
sequence — related or continuous series.
sucesión — series continuas o relacionadas.
sewage system — receives and treats human waste.
sistema de cloacas — recibe y trata los desechos humanos.
sex-linked — genes carried on chromosomes that determine sex.
(herencia) ligada al sexo — genes que se transmiten por los cromosomas que determinan el sexo.
sexual reproduction — union of an egg and sperm to produce a seed or fertilized egg.
reproducción sexual — unión de un óvulo y un esperma para producir una semilla u óvulo fecundado.
shackles — mechanical devices that restrict movement.
grilletes — dispositivos mecánicos que restringen el movimiento.
sheath — lower portion of the turfgrass leaf.
vaina foliar — parte a la base de la hoja de césped.
sheet erosion — removal of soil from broad areas of the land.
erosión laminar — eliminación del suelo de áreas amplias del terreno.
shellfish — any aquatic animal having a shell or shell-like exoskeleton.
mariscos — cualquier animal acuático que tiene cáscara o dermatoesqueleto parecido a un cascarón.
short-term loan — loan with a payment period of 1 year or fewer.
préstamo a corto plazo — préstamo con plazos de pago de un año o menos.
short-term plan — plan accomplished in days or weeks.
plan a corto plazo — plan realizado dentro de días o semanas.
shrinkage — changes in dimensions of wood as it reacts to changes in humidity and temperature.
contracción — disminución de volumen de la madera cuando reacciona a los cambios de humedad y temperatura.
shrouded — wrapped tightly with a cloth.
amortajado — envuelto ajustadamente con un paño.
shrub — a woody perennial plant that normally produces many stems or shoots from the base and does not reach more than 15 feet in height.
arbusto — plantas perennes leñosas que normalmente producen muchas ramas o brotes desde la base, y no alcanzan a más que 15 pies de alto.
signal word — required words on the label that denotes the relative toxicity of the product.
palabra señalizadora — palabras que se deben incluir en la etiqueta para indicar la toxicidad relativa del producto.
silage — feed resulting from the storage and fermentation of green crops in the absence of air.
ensilaje — alimento formado por la colocación en silos y fermentación de cultivos verdes en ausencia de aire.
silo — airtight storage facility for silage.
silo — almacenamiento estanco para el ensilaje.
silt — intermediate soil particles; 0.05 to 0.002 mm.
légamo — partículas de tierra intermedias; 0.05 a 0.002 mm.
silviculture — the scientific management of forests.
silvicultura — la administración científica de los bosques.
simple layering — stem is bent to the ground, held in place, and covered with soil.
acodo simple — el tallo es doblado hasta tocar la tierra, sujetado y cubierto con tierra.
simple leaf — a single leaf arising from a plant stem.
hoja simple — una hoja singular que nace de un tallo de planta.
single-eye cutting — cutting made with the node of a plant with alternate leaves.
injertos con un solo nudo — injerto realizado con la yema de una planta que tiene hojas alternadas.
skeletal system — bones joined together by cartilage and ligaments.
sistema esqueletal — huesos unidos por cartílago y ligamentos.
skimming — setting the price of a new product for unusually high profits at first when affluent and willing customers are available.
fijación de precios de nivel elevado — fijar el precio de un producto nuevo para que se saquen ganancias extraordinariamente altas al principio, cuando hay clientes adinerados y dispuestos.
Small Business Administration (SBA) — institution that provides loans to agribusinesses.
Departamento de Pequeñas Empresas — hace empréstitos a las agroindustrias.
smoker — device used to add smoke flavor and taste to food.
secador al humo — dispositivo que se usa para añadir el sabor ahumado a los comestibles.

sodding — removal of a portion of the soil and turfgrass plant for vegetative establishment purposes.

cubrir con césped — excavación de una parte del suelo y césped para fines de establecimiento vegetal.

softwood — wood from conifers.

madera blanda — la madera de las coníferas.

soil — top layer of the Earth's surface suitable for the growth of plant life.

tierra, suelo — capa de la superficie de la tierra que adaptada al crecimiento de vida vegetal.

soil amendment — additive that improves the soil.

enmienda — además de; rectificación en de los suelos.

soil profile — a cross-sectional view of soil.

pérfil del suelo — una vista del suelo por corte transversal.

soil science — study of the properties and management of soil to grow plants.

ciencia de estudio de los suelos — el estudio de las propiedades y administración de los suelos para cultivar plantas.

soil structure — the way soil granules bind together in clumps; includes the pore arrangement in the aggregates.

estructura de los suelos — en la manera grã¡nulos del suelo enlazar juntos en bosquetes de poro; incluye el arreglo de los agregados.

sow — a mature female pig that has given birth.

cerda — un cerdo hembra adulto que ha dado a luz.

spawn — egg-laying process of fish.

poner huevos — en los peces, el proceso de poner huevos.

species — the basic unit in the classification system whose members have similar structure, common ancestors, and maintain their characteristics; subgroup of genus.

especie — la unidad básica del sistema de clasificación cuyos miembros tienen estructura semejante, antepasados comunes, y mantienen sus características; subgrupo de género.

specimen plant — plant that is used as a single plant to highlight it or some other special feature of the landscape.

planta maestra — planta que se usa como planta singular para mostrarla o para mostrar otro rasgo especial del paisaje.

sperm — male reproductive units.

esperma — unidades reproductivas del macho.

sphagnum — pale and ashy mosses used to condition soil.

esfagno — musgo pálido y ceniciento usado para acondicionar el suelo.

split carcass — halves of an animal after it is killed.

res abierta en canal — las mitades del animal después de la matanza.

split-vein cutting — cutting made by slitting a large leaf on the veins before placing in a rooting medium.

esqueje de nervio partido — un injerto realizado cortando una hoja grande por sus nervios antes de colocarla en un medio para echar raíces.

sprigging — planting a section of a rhizome or stolon (sprig) in the soil.

propagación por estacas de raíz — plantar una parte de un rizoma o estolón (ramito) en el suelo.

sprinkler — device that sprays water on crops; its effect is much like rain.

aspersor — dispositivos que rocían agua al aire para regar los cultivos; tiene casi el mismo efecto que la lluvia.

sprinkler irrigation — the use of pipes to deliver water under pressure to a sprinkler head for the purpose of watering plants that are remote from a water source.

irrigación por aspersión — el uso de cañerías para entregar el agua bajo presión a una cabeza de aspersión con el fin de regar plantas que están lejos de una fuente de agua.

squab — young pigeon that is butchered for meat before it is old enough to fly.

pichón — aves jóvenes que son matadas por su carne antes de ser suficientemente maduras para volar.

stake — driving a wooden or metal rod into the ground near the plant and tying the plant to the rod.

estaca — clavar una estaca o varilla de metal en el suelo cerca de la planta y amarrar la planta a la varilla o estaca.

stallion — adult male horse or pony.

semental — adulto macho del caballo o pony.

stamen — male part of the flower that contains the pollen, anther, and filament.

estambre — órgano sexual masculino de la flor que consiste en el polen, antero, y filamento.

starch — major energy source in livestock feeds.

almidón — fuente principal de energía en los alimentos.

starter solution — diluted mixture of single or complete fertilizers used when plants are transplanted.

abono de material inicial — mixtura diluida de abonos individuales o completos que se usan cuando se trasplantan las plantas.

steer — castrated male member of the cattle family.
bueyezuelo — macho castrado de la familia vacuna.

stem section cutting — a section of plant stem containing a node that is placed in a plant medium to generate a new plant.
corte de sección del tallo — una sección del tallo de la planta que contiene un nudo que es colocado en un medio de cultivo para generar una nueva planta.

stem tip cutting — cutting taken from the end of a stem or branch, normally including the terminal bud.
esqueje de punta de tallo — un esqueje sacado del extremo de un tallo o ramo, normalmente incluye el botón terminal.

stigma — part of the pistil that receives the pollen.
estigma — parte del pistilo que recibe el polen.

stimulant crop — crop that stimulates the senses of users.
cosecha estimulante — cosecha que estimula los sentidos de los usuarios.

stolon — a stem that grows aboveground.
estolón — un tallo que crece encima de la tierra.

stolonizing — type of vegetative establishment in which sprigs are broadcast onto the soil surface.
sembrar estolones al voleo — tipo de establecimiento vegetal en que los ramitos son sembrados al voleo en la superficie del suelo.

stoma — small openings, usually on the lower side of the leaf, that control movement of gases.
estomas — aberturas pequeñas, usualmente en la vaina exterior de la hoja, que controlan el movimiento (if you must divide in syllables: movi-miento) de gases.

strip cropping — alternating strips of row crops with strips of close-growing crops.
cultivo en fajas — fajas alternas de cultivos en hilera con fajas de cultivos que crecen muy juntos.

style — enlarged terminal part of the pistil.
estilo — parte terminal agrandada del pistilo.

subcutaneous — under the skin.
subcutáneo — debajo de la piel.

subsoil — a soil layer that corresponds to the B horizon that is composed almost entirely of mineral with generally large chunky soil structure.
subsuelo — una capa del suelo que corresponde al horizonte B que está compuesto casi completamente por minerales y grandes trozos de tierra.

succulent — in horticulture, thick, fleshy leaves or stems that store moisture.
planta suculenta — en la horticultura: hojas gruesas y carnosas que almacenan agua.

sucrose — compound cane sugar.
sucrosa — azúcar de caña compuesto.

sulfur — pale-yellow element occurring widely in nature.
azufre — elemento amarillo pálido muy común en la naturaleza.

supervised agricultural experience — SAE; all supervised agriscience experiences.
experiencia agrícola supervisado —EAS; todas experiencias agrícolas supervisadas.

supply — the quantity of a product that is available to buyers at a given time.
oferta — la cantidad de un producto que está disponible a los consumidores en un momento dado.

surface irrigation — water flows over the soil surface to the crop.
riego de superficie — el agua fluye sobre la superficie del suelo al cultivo.

sweetbreads — thymus and pancreatic glands of animals.
mollejas — glándulas de timo y páncreas de los animales.

symbol — a warning illustration such as a skull and crossbones located on a chemical container that warns of chemical toxicity to humans and animals.
símbolo — una ilustración de advertencia como una calavera y huesos localizados en un recipiente de químicos que advierte de toxicidad química para animales y humanos.

symmetrical — in landscaping, a planting that is equal in number, size, and texture on both sides.
simétrico — en el paisajismo, colocación de plantas igual cantidad, tamaño y textura a ambos lados de un eje central.

syringing — a light application of water to a turfgrass.
aplicación con mango de riego — una aplicación ligera de agua al césped.

systemic herbicide — a herbicide that is absorbed by the roots of a plant and translocated throughout the plant.
herbicida sistemático — un herbicida que es absorbido por las raíces de la planta y transportado por la planta.

T

tack — equipment or gear for horses.
arreo — guarniciones o aparejo para caballos.

tact — skill of encouraging others in positive ways.
tacto — característica o facilidad de animar a los demás de maneras positivas.

TAN — measurement of total ammonial nitrogen.
NAT — medida de nitrógeno amoniacal total.

tankage — dried animal residue after slaughter.
fertilizante orgánico — los desechos secados del animal después de la matanza.

taproot — large main root of the system; usually has little or no branch roots.
raíz eje — gran raíz primaria del sistema radicular; usualmente tiene pocas o ninguna raíz secundaria.

targeted pest — identified pest that, if introduced, poses a major economic threat.
plaga o insecto nocivo objetivo — plaga o insecto nocivo que, si se introduce, representa una amenaza económica mayor.

taxonomy — systematic classification of plants and animals.
taxonomía — clasificación sistemática de plantas y animales.

T-budding — a grafting method in which a bud is placed on a rootstock using a vertical cut in the stem with a horizontal cut across the top forming a slit into which the bud is placed.
retoños en T — un método de injerto en el cual un codo es unido a un tallo usando un corte vertical en el tallo con un corte horizontal a lo largo de la parte superior formando una ranura dentro de la cual se coloca el injerto.

TDN — total digestible nutrients; the measure of digestibility of feed.
NDT — nutrientes digeribles totales; medición de la digestibilidad del alimento.

technology — application of science to an industrial or commercial objective; also, the equipment and expertise to cultivate, harvest, store, process, and transport crops for consumption.
tecnología — la aplicación de ciencia a un objetivo industrial o comercial; también, el equipo y pericia para cultivar, cosechar, almacenar, procesar, y transportar las cosechas para su consumo.

terminal — growing at the end of a branch or stem.
terminal — dícese de algo que está creciendo en el extremo de una rama o un tallo.

terminal bud — bud at the end of a twig or branch.
botón terminal — yema que está en el extremo de una rama o ramita.

terminal market — a stockyard that acts as a place to hold animals until they are sold to another party.
mercado terminal — un corral que sirve para guardar a los animales hasta que se vendan a otro individuo.

terrace — soil or wall structure built across the slope to capture water and move it safely to areas where it will not cause erosion.
bancal — estructura de tierra o pared construida a lo largo del declive para coger el agua y transportarla seguramente a las áreas en donde no causará erosión.

terrestrial — land organism.
terrícola — organismo que vive en la tierra.

testosterone — male sex hormone.
testosterona — hormona sexual masculina.

tetraethyl lead — colorless, poisonous, and oily liquid.
plomo tetraetílico — líquido incoloro, tóxico, y aceitoso.

texture — visual or surface quality.
textura — cualidad visual o de la superficie de algo.

thatch — building of organic matter on the soil and around turfgrass plants.
fertilización del suelo — la fomentación de materia orgánica en el suelo y alrededor de las plantas de césped.

thermal requirement — classification of plants according to growing season required.
requisito termal — clasificación de las plantas según la temporada de cultivo necesaria.

thymine (T) — a base in genes designated by the letter *T*.
timino — una base en los genes nominada por la letra «T».

tidewater — water that flows up the mouth of a river as the ocean tide rises and comes in.
agua de marea — agua que fluye a la boca de un río mientras la marea sube y entra.

tiller — a young plant shoot that grows from the crown of a plant.
retoño — brote joven de una planta que emerge del botón axilar de la corona de una planta.

timberland — forest land capable of producing more than 20 cf3 of industrial wood per year.
bosque productor de madera — tierra forestal capaz de producir más de 20 pies cúbicos de maderos industriales por año.

tip layering — tip of a shoot placed in media and covered.
acodo de punta — la punta un brote que se mete en un medio de cultivo y se tapa.

tissue culture — plant reproduction using small, actively growing plant parts under sterile conditions and medium.
cultivo de tejidos — reproducción vegetal usando partes pequeñas y de crecimiento activo de la planta bajo condiciones y medios estériles.

tofu — food made by boiling and crushing soybeans and letting them coagulate into curds.
tofu — comestible elaborado hirviendo y triturando la soja y dejando que se coagule para formar requesón.

tom — male turkey.
pavo — macho de la pava.

topsoil — desirable proportions of plant nutrients, chemicals, and living organisms located near the surface that support good plant growth.
capa superficial del suelo — apropiadas de nutrientes vegetales, sustancias químicas y organismos vivos ubicados cerca de la superficie que permiten un buen crecimiento vegetal.

townhouse — one of a row of houses connected by common side walls.
vivienda condominio — una de una hilera de casas conectadas por paredes laterales comunes.

toxicity — a measurement of how poisonous a chemical is.
toxicidad — una medida de la toxidad de una sustancia química.

toxin(s) — a poisonous substance, causing injury to animals or plants.
toxina — una sustancia capaz de producir efectos tóxicos, resultando en daño a animales o plantas.

tractor — source of power for belt-driven machines, as well as for pulling.
locomotora de tracción — fuente de potencia para máquinas propulsadas por correa, y también para tirar.

transpiration — process by which a plant loses water vapor.
transpiración — proceso por el cual la planta pierde u vapor de agua.

transplant — plant grown from seed in a special container.
trasplante — planta cultivada de semilla en un recipiente especial.

tree — a woody plant that produces a main trunk and has a more or less distinct and elevated head (a height of 15 feet or more).
árbol — planta leñosa que produce un tronco principal y que tiene una copa más o menos diferenciada y elevada (de 15 pies o más de alto).

tripe — the pickled rumen of cattle and sheep.
callos — el rumen adobado de ganado y corderos.

truck cropping — large-scale production of a few selected vegetable crops that are shipped to wholesale markets.
producción agrícola a granel — producción de gran escala de pocos vegetales seleccionados de cultivos que son enviados a mercados de abasto.

trucker — person transporting commodities from farm to consumer.
camionero — persona que transporta productos básicos de la finca al consumidor.

tuber — specialized food-storage stem that grows underground.
tubérculos radiculares — raíz especializada de almacenamiento de alimentación que crece subterráneamente.

turbidity — measure of suspended solids.
turbiedad — medida de sólidos suspendidos.

turfgrass — grass that is mowed frequently for a short and even appearance.
césped — hierbas que se siegan frecuentemente para obtener un aspecto corto y plano.

turgor — swollen or still condition as a result of being filled with liquid.
turgencia — condición hinchada o rígida resultante de llenarse de líquido.

U

underripe — as applied to vegetation, fruits or seeds, any that is not mature.
insuficientemente madura — según aplica a la vegetación, cualquiera que no esté madura.

unsoundness — in horses, any abnormality that affects the use of the horse.
defectuoso — en los caballos, cualquier abnormalidad que afecta el uso del caballo.

urinary system — system that removes waste materials from the blood.
sistema urinario — conjunto de órganos que elimina los desechos de la sangre.

V

vacuum pan — a device used to remove water from milk.
tacho de vacío — un dispositivo usado para quitar el agua de la leche.

variegated — a leaf having streaks, marks, or patches of color.
jaspeado — una hoja que tiene rayas, marcas, o manchas de color.

veal — young calves or the meat from young calves.
ternera — terneros jóvenes o el carne de ellos.

vector — a living organism that transmits a disease.
vector — un organismo viviente que transmite una enfermedad.

vegetable — any herbaceous plant whose fruit, seeds, roots, tubers, bulbs, stems, leaves, or flower parts are used as food.
vegetal — cualquier planta herbácea cuyos frutos, semillas, raíces, bulbos, tallos, hojas, o partes de flor, son usados como comestibles.

vegetative bud — a terminal plant bud that produces stem and leaf growth.

botón vegetativo — el botón de una planta terminal que produce crecimiento de tallo y hoja.

vegetative reproduction — reproduction in plants other than by seed.

propagación vegetativa — reproducción de las plantas mediante un método distinto al de la.

veneer — thin sheet of wood used in paneling and furniture.

chapa — hoja delgada de madera que se usa en revestimiento de madera y en mueblería.

vermiculite — mineral matter used for starting plant seeds and cuttings.

vermiculita — materia mineral usada para empezar semillas e injertos de plantas.

vertebrate — an animal with a backbone.

vertebrados — animales que tienen columnas vertebrales.

vertical integration — occurs when several steps in the production, marketing, and processing of animals are joined together.

integración vertical — ocurre cuando varios pasos en la producción, mercadeo, y procesamiento de animales son unidos.

vice — bad habit in horses.

vicio — mala costumbre de un caballo.

virus — pathogenic entity consisting of nucleic acid and a protein sheath.

virus — seres patógenos que consisten de ácido nucléico y una vaina de proteína.

viscera — internal organs of an animal, including heart, liver, and intestines.

víscera — órganos internos del animal incluyendo el corazón, higado e intestinos.

vitamin — complex chemical essential for normal body functions.

vitamina — sustancias químicas esenciales para funciones normales del cuerpo.

voluntary muscle — muscle that can be controlled by animals to do things such as walk and eat food.

músculo voluntario — músculo que pueden ser controlados por los animales para hacer cosas como caminar y comer alimentos.

W

warp — the tendency of wood to bend permanently because of moisture change.

alabearse — la tendencia de la madera de torcerse permanentemente debido al cambio de la humedad.

water — clear, colorless, tasteless, and nearly odorless liquid.

agua — líquido claro, incoloro, insípido, y casi inodoro.

water cycle — movement of water to surface, to atmosphere, to surface.

ciclo del agua — movimiento de agua desde la superficie, hasta la atmósfera, y regreso a la superficie.

waterfowl — ducks and geese.

aves acuáticas — patos y gansos.

water resources — all aspects of water conservation and management.

recursos de agua — todos los aspectos de la conservación y administración del agua.

water table — level below which soil is saturated or filled with water.

nivel hidrostático — nivel debajo del cual el suelo está saturado de agua.

watershed — a large land area from which water flows or from which it is absorbed from rain or melting snow, and from which water drains as it emerges to the surface through springs.

cuenca — una área grande de tierra de donde fluye el agua o donde es absorvida de la lluvia o nieve derretida, y de donde el agua se drena a medida que emerge a la superficie por manantiales.

wetland — a lowland area often associated with ponds or creeks that is saturated with freshwater.

tierras mojadas — una región de tierra baja que muchas veces es asociada con charcos o arroyos, que está saturada con agua dulce.

wholesale marketing — the marketing of a product through a middleman.

comercialización al por mayor — el mercadeo de un producto por un intermediario.

wholesaler — person who sells to the retailer.

mayorista — persona que vende al minorista.

wildlife — animals that are adapted to live in a natural environment without the help of humans.

fauna — animales que se han adaptado a vivir en un ambiente natural sin la ayuda de los seres humanos.

woodlot — small, privately owned forest.

bosque privado — bosque pequeño que tiene propietario individual.

wool — modified hair obtained from sheep and some other animals. It is a fiber with good insulating qualities that is used to make cloth.

lana — pelo modificado obtenido de borregos y otros animales. Es una fibra con muy buenas cualidades de aislamiento que se usa para hacer telas.

worker — female bee that does the work in the hive.
obrera — abeja hembra que hace el trabajo de la colmena.

X

X-Gal — compound that causes marked bacteria to turn blue.
«X-Gal» — una compuesto que causa que se vuelva azúl la bacteria marcada.

xylem — vessels of the vascular bundle that transport the water and nutrients within plants from roots to leaves.
xilema — los buques de los haces vasculares que transportan el agua y nutrientes en las plantas de raíces en las hojas.

Y

yield grade — grade based on the amount of lean meat in relation to the amount of fat and bone in cattle and sheep.
proporción carne/grasa — niveles basados en la cantidad de carne magra en proporción con la cantidad de grasa y hueso en el ganado vacuno y las ovejas.

Z

zygote — fertilized egg.
cigoto — óvulo fecundado.

INDEX

Page numbers in italics refer to figures.

B

I

Q

R